Student Solutions Manual & Study Guide

to accompany

COLLEGE ALGEBRA & TRIGONOMETRY

THIRD EDITION

Kolman Levitan Shapiro

by

Cheryl Roberts

Northern Virginia Community College

SAUNDERS COLLEGE PUBLISHING

Harcourt Brace College Publishers

Fort Worth • Philadelphia • Boston • New York • Chicago • Orlando

San Francisco • Atlanta • Dallas • London • Toronto • Austin • San Antonio

Printed in the United States of America.

Roberts: Student Solutions Manual & Study Guide to accompany COLLEGE ALGEBRA AND TRIGONOMETRY, 3/e, by Kolman

ISBN 0-03-052827-5

345 017 987654321

PREFACE

This manual contains solutions to every odd exercise in **College Algebra and Trigonometry 3/e**. It also includes solutions to the progress checks, review tests, and cumulative review exercises. Additional practice exercises and tests are located at the end of every chapter. The Writing Exercises will help further your understanding of this material.

I would like to thank Mary Jeanne King for the typing of this manual and Emily Cameron for proofreading.

Great care has been taken to avoid errors. However, if you do find an error or have a suggestions for improving the presentation of a problem, please send your comments to: Mathematics Editor, Saunders College Publishing, The Public Ledger Building, 620 Chestnut Street, Suite 560, Philadelphia, PA 19106.

Cheryl Roberts
Northern Virginia Community College
Manassas Campus

TABLE OF CONTENTS

(Additional Practice Exercises and Practice Tests are located at the end of each chapter.)

CHAPTER 1 SECTION 1

SECTION 1.1 PROGRESS CHECK

1a). (Page 3)

$V = \{a, e, i, o, u\}$

1b).

No; since k doesn't appear in any of the words in the sentence.

1c).

Yes; u appears in the word *particular*.

1d),

$\{a, e, i, o\}, \{a, e, i, u\}, \{a, e, o, u\}$

$\{a, i, o, u\}, \{e, i, o, u\}$

Form sets containing 4 elements by leaving out one of the original elements each time.

2a).(Page 8)

$$\frac{3}{5}+\frac{1}{4}=\frac{3}{5}\cdot\frac{(4)}{(4)}+\frac{1}{4}\cdot\frac{(5)}{(5)}=\frac{12}{20}+\frac{5}{20}=\frac{17}{20}$$

2b).

$$\frac{5}{2}\cdot\frac{4}{15}=\frac{5}{2}\cdot\frac{2\cdot 2}{3\cdot 5}=\frac{1}{1}\cdot\frac{2}{3}=\frac{2}{3}$$

2c).

$$\frac{2}{3}+\frac{3}{7}=\frac{2}{3}\cdot\frac{(7)}{(7)}+\frac{3}{7}\cdot\frac{(3)}{(3)}=\frac{14}{21}+\frac{9}{21}=\frac{23}{21}$$

EXERCISE SET 1.1

1.
$\{3, 4, 5, 6, 7\}$

3.
$\{-9\}$

5.
$\{1, 2\}$

7.
$\{1, 3, 7\}$

9.
False; $-14 \notin \{1, 2, 3, \ldots\}$.

11.
False; $\frac{\pi}{3}$ is irrational.

13.
True.

15.
True.

17.
True.

19.
False; $\pi + (-\pi) = 0$
and 0 is a rational number.

21.
False; $\pi \cdot \frac{1}{\pi} = 1$ and 1 is a rational number.

23.
commutative (addition)

25.
distributive

27.
associative (addition)

29.
closure (multiplication)

31.
commutative (multiplication)

33.
associative, commutative (multiplication)

35.
multiplicative inverse

37.
$4 - 3 \neq 3 - 4$
since $1 \neq -1$

39.
$$2(3+4) \neq 2(3)+4$$
$$2(7) \neq 6+4$$
$$14 \neq 10$$

41.
symmetric

43.
transitive or substitution

45.
Prove:

If $-a=-b$, then $ac=bc$.

Proof:

Statement	Reason
$-ac$ is a real number	Closure property for multiplication
$-ac=-ac$	Reflexive property
$-ac=-bc$	Substitution property with $-a=-b$
$(-1)(-ac)=(-1)(-bc)$	Multiplication property of Equality
$ac=bc$	Rule of signs and multiplication

47.
Prove:

If $a-c=b-c$, then $a=b$.

Proof:

Statement	Reason
$(a-c),(b-c)$ are real numbers	Closure property
$(a-c)+c=(b-c)+c$	Additive property of Equality
$a+(-c+c)=b+(-c+c)$	Associative law
$a+0=b+0$	Additive inverse
$a=b$	Additive identity

49.
Prove:

The real number 0 does not have a reciprocal.

Proof:

Statement	Reason
$1=0\cdot\frac{1}{0}$	Multiplicative inverse
$=0\cdot b$	Substitution
$=0$	Multiplication by 0.
$1\neq 0$	Contradiction

Therefore original assumption must be false, thus $\frac{1}{0}$ is not the reciprocal of 0.

51a).

$\frac{1}{3},\frac{2}{3}$ since $\frac{1}{3}+\frac{2}{3}=1$

$\frac{2}{5},\frac{3}{5}$ since $\frac{2}{5}+\frac{3}{5}=1$

$-\frac{1}{4},-\frac{7}{4}$ since $-\frac{1}{4}+\left(-\frac{7}{4}\right)=-2$

51b).

$\pi,-\pi$ since $\pi+(-\pi)=0$

$\sqrt{2},-\sqrt{2}$ since $\sqrt{2}+(-\sqrt{2})=0$

$\sqrt{5},-\sqrt{5}$ since $\sqrt{5}+(-\sqrt{5})=0$

53a).
$$(-8)+(13)=5$$

53b).
$$(-8)+(-13)=-21$$

53c).
$$8-(-13)=8+13=21$$

53d).
$$(-5)(3)-(-12)=-15+12=-3$$

53e).
$$\left(\frac{8}{9}+3\right)+\left(-\frac{5}{9}\right)=\left(\frac{8}{9}+\frac{3}{1}\cdot\frac{(9)}{(9)}\right)+\left(-\frac{5}{9}\right)$$
$$=\left(\frac{8}{9}+\frac{27}{9}\right)+\left(-\frac{5}{9}\right)$$
$$=\frac{35}{9}+\left(-\frac{5}{9}\right)$$
$$=\frac{30}{9}$$
$$=\frac{10}{3}$$

53f).
$$\frac{-5}{\frac{3}{2}}=\frac{-5}{\frac{3}{2}}\cdot\frac{(2)}{(2)}=\frac{-10}{3}$$

53g).
$$\frac{\frac{5}{8}}{\frac{1}{2}}=\frac{\frac{5}{8}}{\frac{1}{2}}\cdot\frac{(8)}{(8)}=\frac{5}{4}$$

53h).
$$\frac{\frac{-2}{3}}{\frac{-4}{3}}=\frac{\frac{-2}{3}}{\frac{-4}{3}}\cdot\frac{(3)}{(3)}=\frac{-2}{-4}=\frac{1}{2}$$

53i).
$$\left(\frac{3}{4}\right)\left(\frac{21}{37}\right)+\left(\frac{3}{4}\right)\left(\frac{16}{37}\right)=\frac{63}{148}+\frac{48}{148}=\frac{111}{148}$$

53j).
$$\frac{\frac{1}{3}-\left(-\frac{1}{4}\right)}{\frac{7}{8}-\frac{3}{16}}=\frac{\frac{1}{3}+\frac{1}{4}}{\frac{7}{8}-\frac{3}{16}}=\frac{\frac{1}{3}\cdot\frac{(4)}{(4)}+\frac{1}{4}\cdot\frac{(3)}{(3)}}{\frac{7}{8}\cdot\frac{(2)}{(2)}-\frac{3}{16}}$$
$$=\frac{\frac{4}{12}+\frac{3}{12}}{\frac{14}{16}-\frac{3}{16}}=\frac{\frac{7}{12}}{\frac{11}{16}}\cdot\frac{(48)}{(48)}=\frac{28}{33}$$

53k).
$$\frac{\left(\frac{3}{5}\right)\left(\frac{1}{7}\right)}{\frac{1}{2}+\frac{1}{3}}=\frac{\frac{3}{35}}{\frac{1}{2}\cdot\frac{(3)}{(3)}+\frac{1}{3}\cdot\frac{(2)}{(2)}}=\frac{\frac{3}{35}}{\frac{3}{6}+\frac{2}{6}}=\frac{\frac{3}{35}}{\frac{5}{6}}$$
$$=\frac{3}{35}\cdot\frac{6}{5}=\frac{18}{175}$$

53l).

$$\frac{2}{5}\left(\frac{3}{2}\cdot\frac{4}{7}\right)=\frac{2}{5}\left(\frac{3}{1}\cdot\frac{2}{7}\right)$$

$$=\frac{2}{5}\left(\frac{6}{7}\right)=\frac{12}{35}$$

55a).

$$\frac{1}{4}=0.25$$

55b).

$$-\frac{3}{5}=-0.6$$

55c).

$$\frac{10}{13}=0.7692308$$

55d).

$$\frac{2}{7}=0.2857143$$

57.

$$\frac{\text{inches}}{\text{miles}}:\ \frac{1}{10}=\frac{3.5}{x}$$

$$1x=(10)(3.5)$$

$$x=35$$

35 miles

59.

length of first piece : x

length of second piece : $10-x$

$$\frac{x}{10-x}=\frac{2}{3}$$

$$3x=2(10-x)$$

$$3x=20-2x$$

$$5x=20$$

$$x=4$$

$$10-x=10-4=6$$

4 feet and 6 feet

61.

1 lb. 3 oz. = 19 oz.

Compare $\frac{\text{price}}{\text{pound}}$: $\frac{49}{19}$ compared to $\frac{35}{13}$

Changing to decimals: 2.58 compared to 2.69

\$2.58 per lb. is the better buy, thus

1 lb. 3 oz. for 49 cents is the better value.

63.

18% is deducted from 100% pay leaving $100\%-18\%=82\%$ for net pay.

Thus 210 is 82% of her gross earnings.

$$210=0.82x$$

$$\frac{210}{0.82}=x$$

$$x=256.10$$

Her pay before the deduction is \$256.10

65.

$$\begin{array}{r} S\ E\ N\ D \\ +\ M\ O\ R\ E \\ \hline M\ O\ N\ E\ Y \end{array}$$

When adding two digits, the most that can be carried is 1. Thus M=1.

$$\begin{array}{r} S\ E\ N\ D \\ +\ 1\ O\ R\ E \\ \hline 1\ O\ N\ E\ Y \end{array}$$

Since we need to carry, S must equal 9 or 8 if there is a previous carry.

Try 8.

$$\begin{array}{r} 8\ E\ N\ D \\ +\ 1\ 0\ R\ E \\ \hline 1\ 0\ N\ E\ Y \end{array}$$

8+1+carry=0 with a carry i.e. O=0, but if O=0 the 0+E wouldn't produce a carry thus S=9.

$$\begin{array}{r} 9\ E\ N\ D \\ +\ 1\ 0\ R\ E \\ \hline 1\ 0\ N\ E\ Y \end{array}$$

For E+0=N≠E we must have a carry. Also, 1+E=N thus N is one more than E. Out of the remaining unused digits {2, 3 ,4 ,5 ,6 ,7 ,8}, only 5, 6, 7, 8 are possibilities.

65. (Continued)

Try E=5, N=6 (Consider the remaining two pairs to see where they go wrong.)

$$\begin{array}{r} 9\ {}^{1}5\ 6\ D \\ +\ 1\ 0\ R\ 5 \\ \hline 1\ 0\ 6\ 5\ Y \end{array}$$

6+R=15

R=9 (used already)

Must have a carry, so R=8

$$\begin{array}{r} 9\ {}^{1}5\ {}^{1}6\ D \\ +\ 1\ 0\ 8\ 5 \\ \hline 1\ 0\ 6\ 5\ Y \end{array}$$

D+5=Y must produce a carry so out of the remaining unused digits {2, 3, 4, 7} only D=7 works.

$$\begin{array}{r} 9\ {}^{1}5\ {}^{1}6\ 7 \\ +\ 1\ 0\ 8\ 5 \\ \hline 1\ 0\ 6\ 5\ Y \end{array}$$

7+5=Y

Y=2

$$\begin{array}{r} 9\ {}^{1}5\ {}^{1}6\ 7 \\ +\ 1\ 0\ 8\ 5 \\ \hline 1\ 0\ 6\ 5\ 2 \end{array}$$

$10,652 is the amount of money the student requested.

CHAPTER 1 SECTION 2

SECTION 1.2 PROGRESS CHECK

1a). (Page 13)

Algebraic Expression	Equivalent Statement	Geometric Statement
$-1<3$	$-1-3=-4$ is negative	-1 lies to the left of 3

1b).

Algebraic Expression	Equivalent Statement	Geometric Statement
$2\le 2$	$2-2=0$	2 coincides with 2

1c).

Algebraic Expression	Equivalent Statement	Geometric Statement
$-2.7<-1.2$	$-2.7-(-1.2)=-1.5$ is negative	-2.7 lies to the left of -1.2

1d).

Algebraic Expression	Equivalent Statement	Geometric Statement
$-4<-2<0$	$-2-(-4)=2$ is positive and $-2-0=-2$ is negative	-2 lies to the right of -4 and to the left of 0

1e).

Algebraic Expression	Equivalent Statement	Geometric Statement
$-\frac{7}{2}<\frac{7}{2}<7$	$\frac{7}{2}-\left(-\frac{7}{2}\right)=7$ is positive and $\frac{7}{2}-7=-\frac{7}{2}$ is negative	$\frac{7}{2}$ lies to the right of $-\frac{7}{2}$ and to the left of 7

2a).(Page 16)

$\overline{PR}=|-6-6|=|-12|=12$

2b).

$\overline{QP}=|4-(-6)|=|4+6|=|10|=10$

2c).

$\overline{PQ}=\overline{QP}=10$

EXERCISE SET 1.2

1.

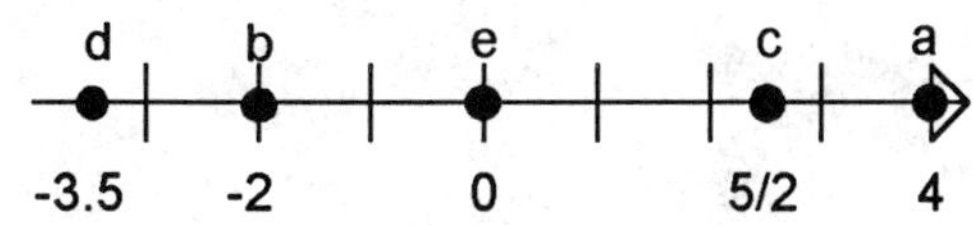

3.

A:1, B:2.5, C:−2, D:4, E:−3.5, O:0

5.
4 is to the left of 6

7.
-2 is to the left of $\frac{3}{4}$

9.
-5 is to the left of $-\frac{2}{3}$

11.

0 1 2 3 4 5 6 7 8

13.

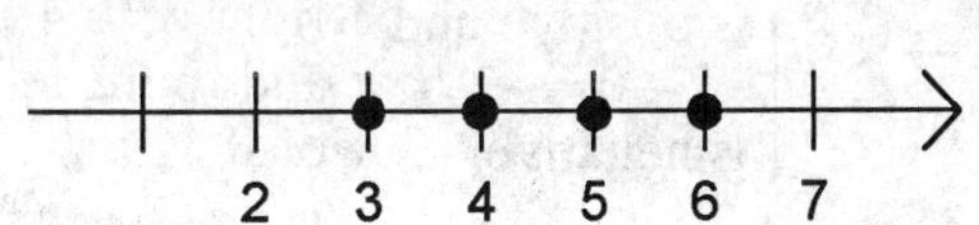

15.
$10 > 9.99$

17.
$a \geq 0$

19.
$x > 0$

21.
$\frac{1}{4} < a < \frac{1}{2}$

23.
$b \geq 5$

25.
Multiplication by a negative number reverses the inequality sign.

27.
Multiplication by a negative number reverses the inequality sign.

29.
Multiplication by a positive number preserves the inequality sign.

31.
$|2| = 2$

33.
$|1.5| = 1.5$

35.
$-|2| = -(2) = -2$

37.
$$|2-3|=|-1|=1$$

39.
$$|2-(-2)|=|2+2|=|4|=4$$

41.
$$\frac{|14-8|}{|-3|}=\frac{|6|}{3}=\frac{6}{3}=2$$

43.
$$\frac{|3|-|2|}{|3|+|2|}=\frac{3-2}{3+2}=\frac{1}{5}$$

45.
$$\overline{AB}=|2-5|=|-3|=3$$

47.
$$\overline{AB}=|-3-(-1)|=|-3+1|=|-2|=2$$

49.

$$\overline{AB} = \left|-\frac{4}{5}-\frac{4}{5}\right| = \left|-\frac{8}{5}\right| = \frac{8}{5}$$

51.

Consider $|x+y| = |x|+|y|$

case i: If $x \geq 0,\ y \geq 0$ then $x+y \geq 0$.

Thus $|x+y| = x+y$

$|x| = x$

$|y| = y$

$|x+y| = |x|+|y| = x+y$

For example: $x = 2,\ y = 3$

$|2+3| = 5 = |2|+|3|$

case ii: If $x < 0,\ y < 0$ then $x+y < 0$

Thus $|x+y| = -(x+y) = -x-y$.

$|x| = -x$

$|y| = -y$

$|x+y| = |x|+|y| = -x-y$

For example: $x = -2,\ y = -3$

$|-2+(-3)| = |-5| = 5$

$|-2|+|-3| = |2|+|3| = 2+3 = 5$

If the values of x and y share the same sign then $|x+y| = |x|+|y|$.

case iii: Consider $|x+y| < |x|+|y|$

Suppose $x = 2,\ y = -3$

Then $|x+y| = |2+(-3)| = |-1| = 1$

$|x| = |2| = 2$

$|y| = |-3| = 3$

$|x+y| = 1 < |x|+|y| = 2+3 = 5$

If the values of x and y have opposite signs then $|x+y| < |x|+|y|$.

53.

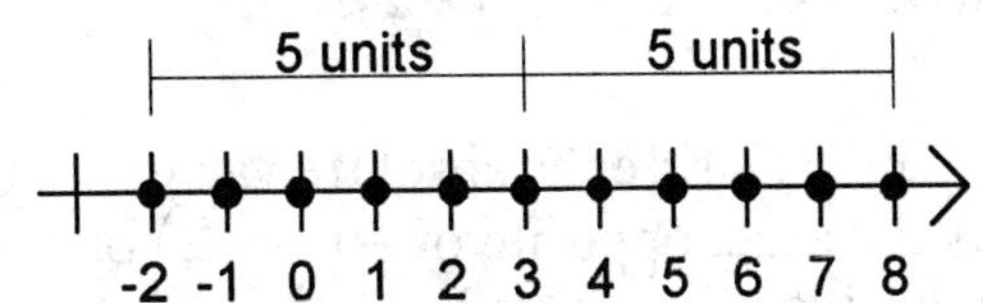

$\{-2, -1, 0, 1, 2, 3, 4, 5, 6, 7, 8\}$

55a).

$$-1 < 5$$
$$-1+2 < 5+2$$
$$1 < 7$$

55b).

$$-1 < 5$$
$$-1-5 < 5-5$$
$$-6 < 0$$

55c).

$$-1 < 5$$
$$(-1)(2) < (5)(2)$$
$$-2 < 10$$

55d).

$$-1 < 5$$
$$(-1)(-5) > (5)(-5)$$
$$5 > -25$$

55e).

$$-1 < 5$$
$$\frac{-1}{-1} > \frac{5}{-1}$$
$$1 > -5$$

55f).

$$-1 < 5$$
$$-\frac{1}{2} < \frac{5}{2}$$

55g).

$$-1 < 5$$
$$(-1)^2 < (5)^2$$
$$1 < 25$$

In Exercises 57 - 61, recall that the midpoint of AB is $\frac{a+b}{2}$.

57.

$$\frac{2+5}{2} = \frac{7}{2}$$

59.

$$\frac{-3+(-1)}{2} = \frac{-4}{2} = -2$$

61.

$$\frac{-\frac{4}{5}+\frac{4}{5}}{2} = \frac{0}{2} = 0$$

63a).

When does $|3-x| = 3-x$?

Note $(3-x)$ inside the absolute value bars is the exact same as $(3-x)$ on the right side of =. This happens if the expression inside the absolute value bars is greater than or equal to 0.

i.e. $3-x \geq 0$

$3 \geq x$

or $x \leq 3$

63b).

When does $|5x-2| = -(5x-2)$?

Note $(5x-2)$ inside the absolute value bars is the exact opposite of $-(5x-2)$ on the right side of =. This happens if the expression inside the absolute value bars is less than or equal to 0.

i.e. $5x-2 \leq 0$

$5x \leq 2$

$x \leq \frac{2}{5}$

CHAPTER 1 SECTION 3

SECTION 1.3 PROGRESS CHECK

1a).(Page 20)

$$x^5 \cdot x^2 = x^{5+2} = x^7$$

1b).

$$\begin{aligned}(2x^6)(-2x^4) &= (2)(-2)(x^6)(x^4)\\ &= -4x^{6+4}\\ &= -4x^{10}\end{aligned}$$

2a).(Page 23)

$$\begin{aligned}&(x^2+2)(x^2-3x+1)\\ &= x^2(x^2-3x+1)+2(x^2-3x+1)\\ &= x^4-3x^3+x^2+2x^2-6x+2\\ &= x^4-3x^3+3x^2-6x+2\end{aligned}$$

2b).

$$\begin{aligned}&(x^2-2xy+y)(2x+y)\\ &= x^2(2x+y)-2xy(2x+y)+y(2x+y)\\ &= 2x^3+x^2y-4x^2y-2xy^2+2xy+y^2\\ &= 2x^3-3x^2y+2xy-2xy^2+y^2\end{aligned}$$

3a).(Page 24)

$$\begin{aligned}&(2x^2-xy+y^2)(3x+y)\\ &= 2x^2(3x+y)-xy(3x+y)+y^2(3x+y)\\ &= 6x^3+2x^2y-3x^2y-xy^2+3xy^2+y^3\\ &= 6x^3-x^2y+2xy^2+y^3\end{aligned}$$

3b)

$$\begin{aligned}&(2x-3)(3x-2)\\ &= (2x)(3x)+(2x)(-2)+(-3)(3x)+(-3)(-2)\\ &= 6x^2-4x-9x+6\\ &= 6x^2-13x+6\end{aligned}$$

EXERCISE SET 1.3

1.

$$r+2s+t = 2+2(-3)+4 = 2-6+4 = 0$$

3.

$$\frac{rst}{r+s+t} = \frac{(2)(-3)(4)}{2+(-3)+4} = \frac{-24}{3} = -8$$

5.

$$\frac{r+s}{rt} = \frac{2+(-3)}{(2)(4)} = -\frac{1}{8}$$

7.

$$\frac{2}{3}r+5 = \frac{2}{3}(12)+5 = 8+5 = 13$$

9a).

$$\begin{aligned}\text{Amount } A &= P + Prt\\ &= 2000 + (2000)(0.08)(1)\\ &= 2000 + 160\\ &= 2160\end{aligned}$$

\$2160

9b).

$$\begin{aligned}\text{Amount } A &= P + Prt\\ &= 2000 + (2000)(0.08)\left(\frac{1}{2}\right)\\ &= 2000 + 80\\ &= 2080\end{aligned}$$

\$2080

9c).

$$\begin{aligned}\text{Amount } A &= P + Prt\\ &= 2000 + (2000)(0.08)\left(\frac{8}{12}\right);\ \left(8\text{months} = \frac{8}{12}\text{yr}\right)\\ &= 2000 + 106.67\\ &= 2106.67\end{aligned}$$

\$2106.67

11.

$$\begin{aligned}&0.02r + 0.314st + 2.25t\\ &= 0.02(2.5) + 0.314(3.4)(2.81) + 2.25(2.81)\\ &= 9.37\end{aligned}$$

13.

$$\begin{aligned}|x| - |x|\cdot|y| &= |-3| - |-3|\cdot|4|\\ &= 3 - (3)(4)\\ &= 3 - 12\\ &= -9\end{aligned}$$

15.

$$\frac{|a-2b|}{2a} = \frac{|1-2(2)|}{2(1)} = \frac{|1-4|}{2} = \frac{|-3|}{2} = \frac{3}{2}$$

17.

$$\begin{aligned}\frac{-|a-2b|}{|a+b|} &= \frac{-|-2-2(-1)|}{|-2+(-1)|} = \frac{-|-2+2|}{|-3|}\\ &= \frac{-|0|}{3} = \frac{0}{3} = 0\end{aligned}$$

19.

$$b^5 \cdot b^2 = b^{5+2} = b^7$$

21.

$$\begin{aligned}(4y^3)(-5y^6) &= (4)(-5)(y^3)(y^6)\\ &= -20y^{3+6} = -20y^9\end{aligned}$$

23.

$$\left(\frac{3}{2}x^3\right)(-2x) = \left(\frac{3}{2}\right)(-2)(x^3)(x)$$
$$= -3x^{3+1} = -3x^4$$

25.

$$1^3 = 1$$
$$10^8 = 100{,}000{,}000$$
$$2^5 = 32$$
$$7^1 = 7$$

27.

(c) since $\frac{2}{3}$ is used as an exponent

(d) since -4 is used as an exponent

29.

2; degree = 3

31.

$\frac{3}{5}$; degree = 4

33.

$$\underbrace{}_{\substack{\text{degree}\\\text{of terms}}}\ \underbrace{3x^2y}_{3} - \underbrace{4x^2}_{2} - \underbrace{2y}_{1} + \underbrace{4}_{0}$$

degree of polynomial: 3

35.

$$\underbrace{}_{\substack{\text{degree}\\\text{of terms}}}\ \underbrace{2xy^3}_{4} - \underbrace{y^3}_{3} + \underbrace{3x^2}_{2} - \underbrace{2}_{0}$$

degree of polynomial: 4

37.

$$3x^2y^2 + 2xy - x + 2y + 7$$
$$= 3(2)^2(-1)^2 + 2(2)(-1) - 2 + 2(-1) + 7$$
$$= 12 - 4 - 2 - 2 + 7$$
$$= 11$$

39.

$$2.1x^3 + 3.3x^2 - 4.1x - 7.2$$
$$= 2.1(4.1)^3 + 3.3(4.1)^2 - 4.1(4.1) - 7.2$$
$$= 176.20$$

41.

$$\frac{1}{2}bh$$

43.

$$\underbrace{76.50x}_{\substack{\text{cost of}\\\text{G.E. stock}\\\text{purchase}}} + \underbrace{61y}_{\substack{\text{cost of}\\\text{Exxon Stock}\\\text{purchase}}} + \underbrace{45z}_{\substack{\text{cost of}\\\text{AT\&T stock}\\\text{purchase}}} = \text{total cost of stock purchase}$$

45.

$$(2x^2 + 3x + 8) - (5 - 2x + 2x^2)$$
$$= 2x^2 + 3x + 8 - 5 + 2x - 2x^2$$
$$= 5x + 3$$

47.

$$\left(2s^2t^3 - st^2 + st - s + t\right) - \left(3s^2t^2 - 2s^2t - 4st^2 - t + 3\right)$$
$$= 2s^2t^3 - st^2 + st - s + t - 3s^2t^2 + 2s^2t + 4st^2 + t - 3$$
$$= 2s^2t^3 - 3s^2t^2 + 2s^2t + 3st^2 + st - s + 2t - 3$$

49.

$$a^2bc + ab^2c + 2ab^3 - 3a^2bc - 4ab^3 + 3$$
$$= -2a^2bc + ab^2c - 2ab^3 + 3$$

51.

$$(2 - x)\left(2x^3 + x - 2\right)$$
$$= 2\left(2x^3 + x - 2\right) - x\left(2x^3 + x - 2\right)$$
$$= 4x^3 + 2x - 4 - 2x^4 - x^2 + 2x$$
$$= -2x^4 + 4x^3 - x^2 + 4x - 4$$

53.

$$(-3s + 2)\left(-2s^2 - s + 3\right)$$
$$= -3s\left(-2s^2 - s + 3\right) + 2\left(-2s^2 - s + 3\right)$$
$$= 6s^3 + 3s^2 - 9s - 4s^2 - 2s + 6$$
$$= 6s^3 - s^2 - 11s + 6$$

55.

$$\left(2y^2 + y\right)\left(-2y^3 + y - 3\right)$$
$$= 2y^2\left(-2y^3 + y - 3\right) + y\left(-2y^3 + y - 3\right)$$
$$= -4y^5 + 2y^3 - 6y^2 - 2y^4 + y^2 - 3y$$
$$= -4y^5 - 2y^4 + 2y^3 - 5y^2 - 3y$$

57.

$$\left(a^2 - 4a + 3\right)\left(4a^3 + 2a + 5\right)$$
$$= a^2\left(4a^3 + 2a + 5\right) - 4a\left(4a^3 + 2a + 5\right)$$
$$+3\left(4a^3 + 2a + 5\right)$$
$$= 4a^5 + 2a^3 + 5a^2 - 16a^4 - 8a^2 - 20a$$
$$+12a^3 + 6a + 15$$
$$= 4a^5 - 16a^4 + 14a^3 - 3a^2 - 14a + 15$$

59.

$$\left(-3a + ab + b^2\right)\left(3b^2 + 2b + 2\right)$$
$$= -3a\left(3b^2 + 2b + 2\right) + ab\left(3b^2 + 2b + 2\right)$$
$$+b^2\left(3b^2 + 2b + 2\right)$$
$$= -9ab^2 - 6ab - 6a + 3ab^3 + 2ab^2 + 2ab$$
$$+3b^4 + 2b^3 + 2b^2$$
$$= 3b^4 + 3ab^3 + 2b^3 - 7ab^2 + 2b^2 - 4ab - 6a$$

61.

$$2(3x - 2)(3 - x)$$
$$= 2\left(9x - 3x^2 - 6 + 2x\right)$$
$$= 2\left(-3x^2 + 11x - 6\right)$$
$$= -6x^2 + 22x - 12$$

63.

pays for IBM stock : $-100x$
sells AT&T stock : $+\ 45y$
sells TRW stock : $+\ 50z$

$-100x+45y+50z$

65.

$(x-1)(x+3)=x^2+2x-3$

67.

$(2x+1)(2x+3)=4x^2+8x+3$

69.

$(3x-2)(x-1)=3x^2-5x+2$

71.

$(x+y)^2=x^2+2xy+y^2$

73.

$(3x-1)^2=9x^2-6x+1$

75.

$(2x+1)(2x-1)=4x^2-1$

77.

$(x^2+y^2)^2=x^4+2x^2y^2+y^4$

79a).

$3^{10}+3^{10}+3^{10}=3(3^{10})=3^{1+10}=3^{11}$

79b).

$2^n+2^n+2^n+2^n=4(2^n)=2^2(2^n)=2^{2+n}$

81a).

$$\left(\frac{2}{x}-1\right)\left(\frac{2}{x}+1\right)$$
$$=\frac{4}{x^2}-1$$

81b).

$$\left(\frac{wx}{y}-z\right)^2$$
$$=\left(\frac{wx}{y}-z\right)\left(\frac{wx}{y}-z\right)$$
$$=\frac{w^2x^2}{y^2}-\frac{2wxz}{y}+z^2$$

81c).

$$(x+y+z)(x+y-z)$$
$$=[(x+y)+z][(x+y)-z]$$
$$=(x+y)^2-z^2$$
$$=x^2+2xy+y^2-z^2$$

83.

Eric : $\dfrac{1\text{ mi}}{4.23\text{ min}}=0.236\text{ mi/min}$

Benjamin : $\dfrac{4.23\text{ mi}}{60\text{ min}}=0.07\text{ mi/min}$

Eric is the faster runner.

85a).

$A(B+17)$

$=8(32+17)$

$=392$

85b).

$5B-A^2$

$=5(32)-(8)^2$

$=96$

85c).

$A^A=8^8=16{,}777{,}216$

85d).

16^A-3AB

$=16^8-3(8)(32)$

$=4{,}294{,}966{,}528$

CHAPTER 1 SECTION 4

SECTION 1.4 PROGRESS CHECK

1a). (Page 28)

$$4x^2 - x = x \cdot 4x - x \cdot 1 = x(4x-1)$$

1b).

$$3x^4 - 9x^2 = 3x^2 \cdot x^2 - 3x^2 \cdot 3 = 3x^2(x^2-3)$$

1c).

$$3m(2x-3y) - n(2x-3y) = (2x-3y)(3m-n)$$

2a). (Page 29)

$$\begin{aligned} &2m^3n + m^2 + 2mn^2 + n \\ &= (2m^3n + m^2) + (2mn^2 + n) \\ &= m^2(2mn+1) + n(2mn+1) \\ &= (2mn+1)(m^2+n) \end{aligned}$$

2b).

$$\begin{aligned} &2a^2 - 4ab^2 - ab + 2b^3 \\ &= (2a^2 - 4ab^2) - (ab - 2b^3) \\ &= 2a(a-2b^2) - b(a-2b^2) \\ &= (a-2b^2)(2a-b) \end{aligned}$$

3a).(Page 31)

$$\begin{aligned} 3x^2 - 16x + 21 &= (3x- \quad)(x- \quad) \\ &= (3x-7)(x-3) \end{aligned}$$

$[\text{Try } 1\cdot 21,\ 21\cdot 1,\ 3\cdot 7,\ 7\cdot 3]$

3b).

$$2x^2 + 3x - 9 = \begin{Bmatrix} (2x+ \quad)(x- \quad) \\ \text{OR} \\ (2x- \quad)(x+ \quad) \end{Bmatrix}$$

$[\text{Try } 1\cdot 9,\ 9\cdot 1,\ 3\cdot 3]$

$$2x^2 + 3x - 9 = (2x-3)(x+3)$$

4a). (Page 32)

$$x^2 - 49 = (x)^2 - (7)^2 = (x+7)(x-7)$$

4b).

$$16x^2 - 9 = (4x)^2 - (3)^2 = (4x+3)(4x-3)$$

4c).

$$25x^2 - y^2 = (5x)^2 - (y)^2 = (5x+y)(5x-y)$$

5a).(Page 34)

$$x^3+5x^2-6x = x(x^2+5x-6)$$
$$= x(x+6)(x-1)$$

5b).

$$2x^3-2x^2y-4xy^2 = 2x(x^2-xy-2y^2)$$
$$= 2x(x+y)(x-2y)$$

5c).

$$-3x(x+1)+(x+1)(2x^2+1)$$
$$=(x+1)(-3x+2x^2+1)$$
$$=(x+1)(2x^2-3x+1)$$

To complete the factoring we must try to factor $2x^2-3x+1$ further.

Step 1. $(2x \quad)(2x \quad)$

Step 2. $a \cdot c = (2)(1) = 2$

Step 3. Two integers whose product is 2 and whose sum is -3 are -1 and -2.

$(2x-1)(2x-2)$

Step 4. Reducing $2x-2$ to $x-1$ by discarding the common factor 2, we have

$2x^2-3x+1=(2x-1)(x-1)$

Thus $-3x(x+1)+(x+1)(2x^2+1)$

$=(x+1)(2x-1)(x-1)$

EXERCISE SET 1.4

1.

$$5x-15=5(x-3)$$

3.

$$-2x-8y=-2(x+4y)$$

5.

$$5bc+25b=5b(c+5)$$

7.

$$-3y^2-4y^5=-y^2(3+4y^3)$$

9.

$$3x^2+6x^2y-9x^2z=3x^2(1+2y-3z)$$

11.

$$x^2+4x+3=(x+3)(x+1)$$

13.

$$y^2-8y+15=(y-3)(y-5)$$

15.

$$a^2-7ab+12b^2=(a-3b)(a-4b)$$

17.

$$y^2 - \frac{1}{9} = (y)^2 - \left(\frac{1}{3}\right)^2$$
$$= \left(y + \frac{1}{3}\right)\left(y - \frac{1}{3}\right)$$

19.

$$9 - x^2 = (3)^2 - (x)^2$$
$$= (3+x)(3-x)$$

21.

$$x^2 - 5x - 14 = (x-7)(x+2)$$

23.

$$\frac{1}{16} - y^2 = \left(\frac{1}{4}\right)^2 - (y)^2$$
$$= \left(\frac{1}{4} + y\right)\left(\frac{1}{4} - y\right)$$

25.

$$x^2 - 6x + 9 = x^2 - 2\cdot 3\cdot x + (3)^2$$
$$= (x-3)^2$$

27.

$$x^2 - 12x + 20 = (x-10)(x-2)$$

29.

$$x^2 + 11x + 24 = (x+8)(x+3)$$

31.

$$2x^2 - 3x - 2: \quad (2x \quad\ )(2x \quad\ )$$
$$a\cdot c = 2\cdot(-2) = -4$$
$$-4\cdot 1 = -4 \qquad -4+1=-3$$
$$(2x-4)(2x+1)$$

Discard common factor 2 out of $2x-4$

$$2x^2 - 3x - 2 = (x-2)(2x+1)$$

33.

$3a^2-11a+6$: $(3a\quad)(3a\quad)$

$a\cdot c=3\cdot 6=18$

$-9\cdot(-2)=18\quad -9+(-2)=-11$

$(3a-9)(3a-2)$

Discard common factor 3 out of $3a-9$

$3a^2-11a+6=(a-3)(3a-2)$

35.

$6x^2+13x+6$: $(6x\quad)(6x\quad)$

$a\cdot c=6\cdot 6=36$

$9\cdot 4=36\qquad 9+4=13$

$(6x+9)(6x+4)$

Discard common factor 3 out of $6x+9$ and discard common factor 2 out of $6x+4$

$6x^2+13x+6=(2x+3)(3x+2)$

37.

$8m^2-6m-9$: $(8m\quad)(8m\quad)$

$a\cdot c=8\cdot(-9)=-72$

$-12\cdot 6=-72\quad -12+6=-6$

$(8m-12)(8m+6)$

Discard common factor 4 out of $8m-12$ and discard common factor 2 out of $8m+6$

$8m^2-6m-9=(2m-3)(4m+3)$

39.

$10x^2-13x-3$: $(10x\quad)(10x\quad)$

$a\cdot c=10\cdot(-3)=-30$

$-15\cdot 2=-30\qquad -15+2=-13$

$(10x-15)(10x+2)$

Discard common factor 5 out of $10x-15$ and discard common factor 2 out of $10x+2$

$10x^2-13x-3=(2x-3)(5x+1)$

41.

$6a^2-5ab-6b^2$: $(6a\quad b)(6a\quad b)$

$a\cdot c=6\cdot(-6)=-36$

$-9\cdot 4=-36\qquad -9+4=-5$

$(6a-9b)(6a+4b)$

Discard common factor 3 out of $6a-9b$ and discard common factor 2 out of $6a+4b$

$6a^2-5ab-6b^2=(2a-3b)(3a+2b)$

43.

$10r^2s^2+9rst+2t^2$:

$(10rs\quad t)(10rs\quad t)$

$a\cdot c=10\cdot 2=20$

$5\cdot 4=20\qquad 5+4=9$

$(10rs+5t)(10rs+4t)$

Discard common factor 5 out of $10rs+5t$ and discard common factor 2 out of $10rs+4t$

$10r^2s^2+9rst+2t^2=(2rs+t)(5rs+2t)$

45.

$$16-9x^2y^2=(4)^2-(3xy)^2$$
$$=(4+3xy)(4-3xy)$$

47.

$8n^2-18n-5$: $(8n \quad)(8n \quad)$

$a\cdot c=8\cdot(-5)=-40$

$(-20)\cdot 2=-40 \qquad (-20)+2=18$

$(8n-20)(8n+2)$

Discard common factor 4 out of $8n-20$ and discard common factor 2 out of $8n+2$

$$8n^2-18n-5=(2n-5)(4n+1)$$

49.

$$2x^2-2x-12=2(x^2-x-6)$$
$$=2(x-3)(x+2)$$

51.

$30x^2-35x+10=5(6x^2-7x+2)$:

$5(6x \quad)(6x \quad)$

$a\cdot c=6\cdot 2=12$

$(-4)\cdot(-3)=12 \qquad -4+(-3)=-7$

$5(6x-4)(6x-3)$

Discard common factor 2 out of $6x-4$ and discard common factor 3 out of $6x-3$

$$30x^2-35x+10=5(3x-2)(2x-1)$$

53.

$18x^2m+33xm+9m=3m(6x^2+11x+3)$:

$3m(6x \quad)(6x \quad)$

$a\cdot c=6\cdot 3=18$

$9\cdot 2=18 \qquad 9+2=11$

$3m(6x+9)(6x+2)$

Discard common factor 3 out of $6x+9$ and discard common factor 2 out of $6x+2$

$$18x^2m+33xm+9m=3m(2x+3)(3x+1)$$

55.

$12x^2-22x^3-20x^4=2x^2(6-11x-10x^2)$:

$2x^2(\quad -10x)(\quad -10x)$

$a\cdot c=-10\cdot 6=-60$

$-15\cdot 4=-60 \qquad -15+4=-11$

$2x^2(-15-10x)(4-10x)$

Discard common factor -5 out of $-15-10x$ and discard common factor 2 out of $4-10x$

$$12x^2-22x^3-20x^4=2x^2(3+2x)(2-5x)$$

57.

$$\begin{aligned} x^4 - y^4 &= \left(x^2\right)^2 - \left(y^2\right)^2 \\ &= \left(x^2 + y^2\right)\left(x^2 - y^2\right) \\ &= \left(x^2 + y^2\right)(x + y)(x - y) \end{aligned}$$

59.

$$b^4 + 2b^2 - 8 = \left(b^2 + 4\right)\left(b^2 - 2\right)$$

61.

$$\begin{aligned} x^3 + 27y^3 &= (x)^3 + (3y)^3 \\ &= (x + 3y)\left[(x)^2 - (x)(3y) + (3y)^2\right] \\ &= (x + 3y)\left(x^2 - 3xy + 9y^2\right) \end{aligned}$$

63.

$$\begin{aligned} 27x^3 - y^3 &= (3x)^3 - (y)^3 \\ &= (3x - y)\left[(3x)^2 + (3x)(y) + (y)^2\right] \\ &= (3x - y)\left(9x^2 + 3xy + y^2\right) \end{aligned}$$

65.

$$\begin{aligned} a^3 + 8 &= (a)^3 - (2)^3 \\ &= (a + 2)\left[(a)^2 - (a)(2) + (2)^2\right] \\ &= (a + 2)\left(a^2 - 2a + 4\right) \end{aligned}$$

67.

$$\begin{aligned} \frac{1}{8}m^3 - 8n^3 &= \left(\frac{1}{2}m\right)^3 - (2n)^3 \\ &= \left(\frac{1}{2}m - 2n\right)\left[\left(\frac{1}{2}m\right)^2 + \left(\frac{1}{2}m\right)(2n) + (2n)^2\right] \\ &= \left(\frac{1}{2}m - 2n\right)\left(\frac{1}{4}m^2 + mn + 4n^2\right) \end{aligned}$$

69.

$$\begin{aligned} (x + y)^3 - 8 &= (x + y)^3 - (2)^3 \\ &= \left[(x + y) - 2\right]\left[(x + y)^2 + (x + y)(2) + (2)^2\right] \\ &= (x + y - 2)\left(x^2 + 2xy + y^2 + 2x + 2y + 4\right) \end{aligned}$$

71.

$$\begin{aligned} 8x^6 - 125y^6 &= \left(2x^2\right)^3 - \left(5y^2\right)^3 \\ &= \left(2x^2 - 5y^2\right)\left[\left(2x^2\right)^2 + \left(2x^2\right)\left(5y^2\right) + \left(5y^2\right)^2\right] \\ &= \left(2x^2 - 5y^2\right)\left(4x^4 + 10x^2y^2 + 25y^4\right) \end{aligned}$$

73.

$$\begin{aligned} 4(x + 1)(y + 2) - 8(y + 2) &= 4(y + 2)\left[(x + 1) - 2\right] \\ &= 4(y + 2)(x - 1) \end{aligned}$$

75.

$$3(x+2)^2(x-1)-4(x+2)^2(2x+7)=(x+2)^2[3(x-1)-4(2x+7)]$$
$$=(x+2)^2(3x-3-8x-28)$$
$$=(x+2)^2(-5x-31)$$
$$=-(x+2)^2(5x+31)$$

77.

1st odd integer:

$$2x+1$$

2nd consecutive odd integer:

$$(2x+1)+2=2x+3$$

$$(2x+1)^2-(2x+3)^2$$
$$=4x^2+4x+1-(4x^2+12x+9)$$
$$=4x^2+4x+1-4x^2-12x-9$$
$$=-8x-8$$
$$=8(-x-1)$$

Thus the difference is divisib le by 8.

79.

$1+n(n+1)(n+2)(n+3)$ Try $n=2$

$$1+(2)(2+1)(2+2)(2+3)=1+2(3)(4)(5)$$
$$=1+120$$
$$=121=11^2$$

In fact any natural numb er, n, will work. See exercise 80.

81a).

$$(x+h)^3-x^3$$
$$=[(x+h)-x][(x+h)^2+(x+h)(x)+(x)^2]$$
$$=h(x^2+2xh+h^2+x^2+xh+x^2)$$
$$=h(3x^2+3xh+h^2)$$

81b).

$$2^n+2^{n+1}+2^{n+2}=2^n(1+2+2^2)$$
$$=2^n(7)$$

81c).

$$16-81x^{12}=(4)^2-(9x^6)^2$$
$$=(4+9x^6)(4-9x^6)$$
$$=(4+9x^6)[(2)^2-(3x^3)^2]$$
$$=(4+9x^6)(2+3x^3)(2-3x^3)$$

81d).

$$z^2-x^2+2xy-y^2=z^2-(x^2-2xy+y^2)$$
$$=z^2-(x-y)(x-y)$$
$$=z^2-(x-y)^2$$
$$=[z+(x-y)][z-(x-y)]$$
$$=(z+x-y)(z-x+y)$$

83a).

$$C\left[(R+1)^2-r^2\right]$$
$$=C\{[(R+1)+r][(R+1)-r]\}$$
$$=C(R+1+r)(R+1-r)$$

83b).

$$pa^2+(1-p)b^2-\left[pa+(1-p)b\right]^2$$

Let $t=1-p$ and substitute.

We now have

$$pa^2+tb^2-[pa+tb]^2$$
$$=pa^2+tb^2-(pa+tb)(pa+tb)$$
$$=pa^2+tb^2-(p^2a^2+2tbpa+t^2b^2)$$
$$=pa^2+tb^2-p^2a^2-2tbpa-t^2b^2$$

Rearranging, we have

$$=pa^2-p^2a^2+tb^2-t^2b^2-2tbpa$$
$$=pa^2(1-p)+tb^2(1-t)-2tbpa$$

Since $t=1-p$, $1-t=p$

Substituting, we have

$$pa^2t+tb^2p-2tbpa$$
$$=pt(a^2+b^2-2ab)$$
$$=pt(a^2-2ab+b^2)$$
$$=pt(a-b)(a-b)$$
$$=pt(a-b)^2$$

Substituting $(1-p)$ for t, we have

$$=p(1-p)(a-b)^2$$

83c).

$$X^2-3LX+2L^2=(X-L)(X-2L)$$

83d).

$$(R_1+R_2)^2-2r(R_1+R_2)$$
$$=(R_1+R_2)[(R_1+R_2)-2r]$$
$$=(R_1+R_2)(R_1+R_2-2r)$$

83e).

$$-16t^2+64t+336=-16(t^2-4t-21)$$
$$=-16(t-7)(t+3)$$

CHAPTER 1 SECTION 5

SECTION 1.5 PROGRESS CHECK

1a). (Page 38)

$$\frac{4-x^2}{x^2-x-6}=\frac{(2-x)(2+x)}{(x-3)(x+2)}=\frac{2-x}{x-3},\ x\neq -2$$

1b).

$$\frac{8-2x}{y}\div\frac{x^2-16}{y}=$$

$$=\frac{\frac{2(4-x)}{y}}{\frac{(x+4)(x-4)}{y}}\cdot\frac{(y)}{(y)}=\frac{2(4-x)}{(x+4)(x-4)}$$

$$=\frac{-2(x-4)}{(x+4)(x-4)}=\frac{-2}{x+4}=-\frac{2}{x+4},$$

$y\neq 0,\ x\neq 4$

2.(Page 40)

$$\frac{2a}{(3a^2+12a+12)b}\qquad\frac{-7b}{a(4b^2-8b+4)}\qquad\frac{3}{ab^3+2b^3}$$

$$\frac{2a}{3(a^2+4a+4)b}\qquad\frac{-7b}{4a(b^2-2b+1)}\qquad\frac{3}{b^3(a+2)}$$

$$\frac{2a}{3(a+2)^2b}\qquad\frac{-7b}{4a(b-1)^2}\qquad\frac{3}{b^3(a+2)}$$

LCD: $3\cdot 4ab^3(a+2)^2(b-1)^2=12ab^3(a+2)^2(b-1)^2$

3a).(Page 41)

$$\frac{x-8}{x^2-4}+\frac{3}{x^2-2x}$$
$$=\frac{(x-8)}{(x+2)(x-2)}+\frac{3}{x(x-2)} \qquad LCD\text{: } x(x+2)(x-2)$$
$$=\frac{x-8}{(x+2)(x-2)}\cdot\frac{(x)}{(x)}+\frac{3}{x(x-2)}\cdot\frac{(x+2)}{(x+2)}$$
$$=\frac{x^2-8x}{x(x+2)(x-2)}+\frac{3x+6}{x(x+2)(x-2)}$$
$$=\frac{x^2-8x+3x+6}{x(x+2)(x-2)}$$
$$=\frac{x^2-5x+6}{x(x+2)(x-2)}$$
$$=\frac{(x-3)(x-2)}{x(x+2)(x-2)}$$
$$=\frac{x-3}{x(x+2)},\ x\neq 2$$

3b).

$$\frac{4r-3}{9r^3}-\frac{2r+1}{4r^2}+\frac{2}{3r} \qquad LCD\text{: } 36r^3$$
$$=\frac{(4r-3)}{9r^3}\cdot\frac{(4)}{(4)}-\frac{(2r+1)}{4r^2}\cdot\frac{(9r)}{(9r)}+\frac{2}{3r}\cdot\frac{(12r^2)}{(12r^2)}$$
$$=\frac{16r-12}{36r^3}-\frac{(18r^2+9r)}{36r^3}+\frac{24r^2}{36r^3}$$
$$=\frac{16r-12-(18r^2+9r)+24r^2}{36r^3}$$
$$=\frac{16r-12-18r^2-9r+24r^2}{36r^3}$$
$$=\frac{6r^2+7r-12}{36r^3}$$

4a).(Page 42)

$$\frac{2+\frac{1}{x}}{1-\frac{2}{x}}=\frac{2+\frac{1}{x}}{1-\frac{2}{x}}\cdot\frac{(x)}{(x)}$$
$$=\frac{2x+1}{x-2},\ x\neq 0$$

4b).

$$\frac{\frac{a}{b}+\frac{b}{a}}{\frac{1}{a}-\frac{1}{b}}=\frac{\frac{a}{b}+\frac{b}{a}}{\frac{1}{a}-\frac{1}{b}}\cdot\frac{(ab)}{(ab)}$$

$$=\frac{a^2+b^2}{b-a}$$

OR

$$=\frac{a^2+b^2}{-(a-b)}$$

$$=-\frac{a^2+b^2}{a-b},\ a\neq 0,\ b\neq 0$$

EXERCISE SET 1.5

1.

$$\frac{x+4}{x^2-16}=\frac{x+4}{(x+4)(x-4)}$$

$$=\frac{1}{x-4},\quad x\neq -4$$

3.

$$\frac{x^2-8x+16}{x-4}=\frac{(x-4)(x-4)}{x-4}$$

$$=x-4,\quad x\neq 4$$

5.

$$\frac{6x^2-x-1}{2x^2+3x-2}=\frac{(2x-1)(3x+1)}{(2x-1)(x+2)}$$

$$=\frac{3x+1}{x+2},\quad x\neq\frac{1}{2}$$

7.

$$\frac{2}{3x-6}\div\frac{3}{2x-4}$$

$$=\frac{\frac{2}{3(x-2)}}{\frac{3}{2(x-2)}}\cdot\frac{(6)(x-2)}{(6)(x-2)}$$

$$=\frac{4}{9},\quad x\neq 2$$

9.

$$\frac{25-a^2}{b+3}\cdot\frac{2b^2+6b}{a-5}$$
$$=\frac{(5+a)(5-a)}{b+3}\cdot\frac{2b(b+3)}{a-5}$$
$$=\frac{(5+a)(-1)(a-5)(2b)}{(a-5)}$$
$$=-2b(5+a),\quad a\neq 5,\quad b\neq -3$$

11.

$$\frac{x+2}{3y}\div\frac{x^2-2x-8}{15y^2}$$
$$=\frac{\frac{x+2}{3y}}{\frac{(x-4)(x+2)}{15y^2}}\cdot\frac{15y^2}{15y^2}$$
$$=\frac{5y(x+2)}{(x-4)(x+2)}$$
$$=\frac{5y}{x-4}$$

13.

$$\frac{6x^2-x-2}{2x^2-5x+3}\cdot\frac{2x^2-7x+6}{3x^2+x-2}$$
$$=\frac{(3x-2)(2x+1)}{(2x-3)(x-1)}\cdot\frac{(2x-3)(x-2)}{(3x-2)(x+1)}$$
$$=\frac{(2x+1)(x-2)}{(x-1)(x+1)}$$

15.

$$\left(x^2-4\right)\cdot\frac{2x+3}{x^2+2x-8}$$
$$=\frac{(x+2)(x-2)}{1}\cdot\frac{(2x+3)}{(x+4)(x-2)}$$
$$=\frac{(x+2)(2x+3)}{(x+4)}$$

17.

$$\left(x^2-2x-15\right)\div\frac{x^2-7x+10}{x^2+1}$$
$$=\frac{\frac{(x-5)(x+3)}{1}}{\frac{(x-5)(x-2)}{x^2+1}}\cdot\frac{(x^2+1)}{(x^2+1)}$$
$$=\frac{(x-5)(x+3)(x^2+1)}{(x-5)(x-2)}$$
$$=\frac{(x+3)(x^2+1)}{x-2}$$

19.

$$\frac{x^2-4}{x^2+2x-3}\cdot\frac{x^2+3x-4}{x^2-7x+10}\cdot\frac{x+3}{x^2+3x+2}$$
$$=\frac{(x+2)(x-2)}{(x+3)(x-1)}\cdot\frac{(x+4)(x-1)}{(x-5)(x-2)}\cdot\frac{x+3}{(x+2)(x+1)}$$
$$=\frac{x+4}{(x-5)(x+1)}$$

21.

$$\frac{4}{x},\ \frac{x-2}{y}$$

LCD: xy

23.

$$\frac{5-a}{a},\ \frac{7}{2a}$$

LCD: $2a$

25.

$$\frac{2b}{b-1}, \frac{3}{(b-1)^2}$$

LCD: $(b-1)^2$

27.

$$\frac{4x}{x-2}, \frac{5}{x^2+x-6}$$

$$\frac{4x}{x-2}, \frac{5}{(x+3)(x-2)}$$

LCD: $(x+3)(x-2)$

29.

$$\frac{3}{x+1}, \frac{2}{x}, \frac{x}{x-1}$$

LCD: $x(x+1)(x-1)$

31.

$$\frac{8}{a-2}+\frac{4}{2-a}$$

$$=\frac{8}{a-2}+\frac{4}{(-1)(a-2)}$$

$$=\frac{8}{a-2}-\frac{4}{a-2}$$

$$=\frac{8-4}{a-2}$$

$$=\frac{4}{a-2}$$

33.

$$\frac{x-1}{3}+2$$

$$=\frac{x-1}{3}+\frac{2}{1}$$

$$=\frac{x-1}{3}+\frac{2}{1}\cdot\frac{(3)}{(3)}$$

$$=\frac{x-1}{3}+\frac{6}{3}$$

$$=\frac{x-1+6}{3}$$

$$=\frac{x+5}{3}$$

35.

$$\frac{1}{a+2}+\frac{3}{a-2}$$

$$=\frac{1}{(a+2)}\cdot\frac{(a-2)}{(a-2)}+\frac{3}{(a-2)}\cdot\frac{(a+2)}{(a+2)}$$

$$=\frac{a-2}{(a+2)(a-2)}+\frac{3a+6}{(a+2)(a-2)}$$

$$=\frac{a-2+3a+6}{(a+2)(a-2)}$$

$$=\frac{4a+4}{(a+2)(a-2)}$$

OR $\dfrac{4(a+1)}{(a+2)(a-2)}$

37.

$$\frac{4}{3x}-\frac{5}{xy}$$
$$=\frac{4}{3x}\cdot\frac{(y)}{(y)}-\frac{5}{xy}\cdot\frac{(3)}{(3)}$$
$$=\frac{4y}{3xy}-\frac{15}{3xy}$$
$$=\frac{4y-15}{3xy}$$

39.

$$\frac{5}{2x+6}-\frac{x}{x+3}$$
$$=\frac{5}{2(x+3)}-\frac{x}{x+3}$$
$$=\frac{5}{2(x+3)}-\frac{x}{x+3}\cdot\frac{(2)}{(2)}$$
$$=\frac{5}{2(x+3)}-\frac{2x}{2(x+3)}$$
$$=\frac{5-2x}{2(x+3)}$$

41.

$$\frac{5x}{2x^2-18}+\frac{4}{3x-9}$$
$$=\frac{5x}{2(x+3)(x-3)}+\frac{4}{3(x-3)}$$
$$=\frac{5x}{2(x+3)(x-3)}\cdot\frac{(3)}{(3)}+\frac{4}{3(x-3)}\cdot\frac{2(x+3)}{2(x+3)}$$
$$=\frac{15x}{6(x+3)(x-3)}+\frac{8x+24}{6(x+3)(x-3)}$$
$$=\frac{15x+8x+24}{6(x+3)(x-3)}$$
$$=\frac{23x+24}{6(x+3)(x-3)}$$

43.

$$\frac{1}{x-1}+\frac{2x-1}{(x-2)(x+1)}$$
$$=\frac{1}{x-1}\cdot\frac{(x-2)(x+1)}{(x-2)(x+1)}+\frac{(2x-1)}{(x-2)(x+1)}\cdot\frac{(x-1)}{(x-1)}$$
$$=\frac{x^2-x-2}{(x-1)(x-2)(x+1)}+\frac{2x^2-3x+1}{(x-1)(x-2)(x+1)}$$
$$=\frac{x^2-x-2+2x^2-3x+1}{(x-1)(x-2)(x+1)}$$
$$=\frac{3x^2-4x-1}{(x-1)(x-2)(x+1)}$$

45.

$$\frac{2x}{x^2+x-2}+\frac{3}{(x+2)}$$
$$=\frac{2x}{(x+2)(x-1)}+\frac{3}{(x+2)}$$
$$=\frac{2x}{(x+2)(x-1)}+\frac{3}{(x+2)}\cdot\frac{(x-1)}{(x-1)}$$
$$=\frac{2x}{(x+2)(x-1)}+\frac{3x-3}{(x+2)(x-1)}$$
$$=\frac{2x+3x-3}{(x+2)(x-1)}$$
$$=\frac{5x-3}{(x+2)(x-1)}$$

47.

$$\frac{2x-1}{x^2+5x+6}-\frac{x-2}{x^2+4x+3}=\frac{2x-1}{(x+2)(x+3)}-\frac{x-2}{(x+3)(x+1)}$$
$$=\frac{2x-1}{(x+2)(x+3)}\cdot\frac{(x+1)}{(x+1)}-\frac{x-2}{(x+3)(x+1)}\cdot\frac{(x+2)}{(x+2)}$$
$$=\frac{2x^2+x-1}{(x+2)(x+3)(x+1)}-\frac{x^2-4}{(x+2)(x+3)(x+1)}$$
$$=\frac{2x^2+x-1-(x^2-4)}{(x+2)(x+3)(x+1)}$$
$$=\frac{2x^2+x-1-x^2+4}{(x+2)(x+3)(x+1)}$$
$$=\frac{x^2+x+3}{(x+2)(x+3)(x+1)}$$

49.

$$\frac{2x}{x^2-1}+\frac{x+1}{x^2+3x-4}=\frac{2x}{(x+1)(x-1)}+\frac{x+1}{(x+4)(x-1)}$$
$$=\frac{2x}{(x+1)(x-1)}\cdot\frac{(x+4)}{(x+4)}+\frac{(x+1)}{(x+4)(x-1)}\cdot\frac{(x+1)}{(x+1)}$$
$$=\frac{2x^2+8x}{(x+1)(x-1)(x+4)}+\frac{x^2+2x+1}{(x+1)(x-1)(x+4)}$$
$$=\frac{2x^2+8x+x^2+2x+1}{(x+1)(x-1)(x+4)}$$
$$=\frac{3x^2+10x+1}{(x+1)(x-1)(x+4)}$$

51.

$$\frac{1+\frac{2}{x}}{1-\frac{3}{x}}=\frac{1+\frac{2}{x}}{1-\frac{3}{x}}\cdot\frac{(x)}{(x)}$$
$$=\frac{x+2}{x-3}$$

53.

$$\frac{x+1}{1-\frac{1}{x}}=\frac{x+1}{1-\frac{1}{x}}\cdot\frac{(x)}{(x)}$$
$$=\frac{x^2+x}{x-1}$$

55.

$$\frac{x^2-16}{\frac{1}{4}-\frac{1}{x}}=\frac{(x+4)(x-4)}{\frac{1}{4}-\frac{1}{x}}\cdot\frac{(4x)}{(4x)}$$
$$=\frac{4x(x+4)(x-4)}{x-4}$$
$$=4x(x+4)$$

57.

$$2-\frac{1}{1+\frac{1}{a}}=2-\frac{1}{1+\frac{1}{a}}\cdot\frac{(a)}{(a)}$$
$$=2-\frac{a}{a+1}$$
$$=\frac{2}{1}\cdot\frac{(a+1)}{(a+1)}-\frac{a}{(a+1)}$$
$$=\frac{2a+2}{a+1}-\frac{a}{a+1}$$
$$=\frac{2a+2-a}{a+1}$$
$$=\frac{a+2}{a+1}$$

59.

$$\frac{\frac{a}{b}-\frac{b}{a}}{\frac{1}{a}+\frac{1}{b}}=\frac{\frac{a}{b}-\frac{b}{a}}{\frac{1}{a}+\frac{1}{b}}\cdot\frac{(ab)}{(ab)}$$
$$=\frac{a^2-b^2}{b+a}$$
$$=\frac{(a+b)(a-b)}{a+b}$$
$$=a-b$$

61.

$$3-\frac{2}{1-\frac{1}{1+x}}=3-\left[\frac{2}{\left(1-\frac{1}{1+x}\right)}\cdot\frac{(1+x)}{(1+x)}\right]$$
$$=3-\frac{2+2x}{(1+x)-1}$$
$$=3-\frac{2+2x}{x}$$
$$=\frac{3}{1}\cdot\frac{(x)}{(x)}-\frac{2+2x}{x}$$
$$=\frac{3x-(2+2x)}{x}$$
$$=\frac{3x-2-2x}{x}$$
$$=\frac{x-2}{x}$$

63.

$$\frac{y-\frac{1}{1-\frac{1}{y}}}{y+\frac{1}{1+\frac{1}{y}}}=\frac{y-\left[\frac{1}{1-\frac{1}{y}}\cdot\frac{(y)}{(y)}\right]}{y+\left[\frac{1}{1+\frac{1}{y}}\cdot\frac{(y)}{(y)}\right]}$$

$$=\frac{y-\frac{y}{y-1}}{y+\frac{y}{y+1}}$$

$$=\left[\frac{y-\frac{y}{y-1}}{y+\frac{y}{y+1}}\right]\cdot\frac{(y+1)(y-1)}{(y+1)(y-1)}$$

$$=\frac{y(y+1)(y-1)-y(y+1)}{y(y+1)(y-1)+y(y-1)}$$

$$=\frac{y(y+1)[(y-1)-1]}{y(y-1)[(y+1)+1]}$$

$$=\frac{(y+1)(y-2)}{(y-1)(y+2)}$$

65.

$$1-\frac{1}{1+\frac{1}{1-\frac{1}{1+x}}}=1-\frac{1}{1+\left[\frac{1}{1-\frac{1}{1+x}}\cdot\frac{(1+x)}{(1+x)}\right]}$$

$$=1-\frac{1}{1+\frac{1+x}{(1+x)-1}}$$

$$=1-\frac{1}{1+\frac{1+x}{x}}$$

$$=1-\left[\frac{1}{1+\frac{1+x}{x}}\cdot\frac{(x)}{(x)}\right]$$

$$=1-\frac{x}{x+1+x}$$

$$=\frac{1}{1}\cdot\frac{(2x+1)}{(2x+1)}-\frac{x}{2x+1}$$

$$=\frac{2x+1-x}{2x+1}$$

$$=\frac{x+1}{2x+1}$$

67a).

$$\frac{1}{R_1}+\frac{1}{R_2}+\frac{1}{R_3}+\frac{1}{R_4}$$

$$=\frac{1}{R_1}\cdot\frac{(R_2R_3R_4)}{(R_2R_3R_4)}+\frac{1}{R_2}\cdot\frac{(R_1R_3R_4)}{(R_1R_3R_4)}+\frac{1}{R_3}\cdot\frac{(R_1R_2R_4)}{(R_1R_2R_4)}$$

$$+\frac{1}{R_4}\cdot\frac{(R_1R_2R_3)}{(R_1R_2R_3)}$$

$$=\frac{R_2R_3R_4}{R_1R_2R_3R_4}+\frac{R_1R_3R_4}{R_1R_2R_3R_4}+\frac{R_1R_2R_4}{R_1R_2R_3R_4}+\frac{R_1R_2R_3}{R_1R_2R_3R_4}$$

$$=\frac{R_2R_3R_4+R_1R_3R_4+R_1R_2R_4+R_1R_2R_3}{R_1R_2R_3R_4}$$

67b).

$$\frac{3}{c-d}+\frac{4d}{(c-d)^2}-\frac{5d^2}{(c-d)^3}$$

$$=\frac{3}{c-d}\cdot\frac{(c-d)^2}{(c-d)^2}+\frac{4d}{(c-d)^2}\cdot\frac{(c-d)}{(c-d)}-\frac{5d^2}{(c-d)^3}$$

$$=\frac{3(c^2-2cd+d^2)}{(c-d)^3}+\frac{4cd-4d^2}{(c-d)^3}-\frac{5d^2}{(c-d)^3}$$

$$=\frac{3c^2-6cd+3d^2+4cd-4d^2-5d^2}{(c-d)^3}$$

$$=\frac{3c^2-2cd-6d^2}{(c-d)^3}$$

67c).

$$\frac{6}{k+3}+k-2$$
$$=\frac{6}{k+3}+\frac{k}{1}\cdot\frac{(k+3)}{(k+3)}-\frac{2}{1}\cdot\frac{(k+3)}{(k+3)}$$
$$=\frac{6}{k+3}+\frac{k^2+3k}{k+3}-\frac{2k+6}{k+3}$$
$$=\frac{6+k^2+3k-(2k+6)}{k+3}$$
$$=\frac{6+k^2+3k-2k-6}{k+3}$$
$$=\frac{k^2+k}{k+3}$$
OR $\dfrac{k(k+1)}{k+3}$

69a).

$$\frac{\frac{1}{b}}{\frac{1}{a}+\frac{1}{b}}=\frac{1}{\frac{1}{a}}$$

$\frac{1}{b}$ has been incorrectly cancelled

$$\frac{\frac{1}{\text{b}}}{\frac{1}{a}+\frac{1}{b}}=\frac{\frac{1}{b}}{\frac{1}{a}+\frac{1}{b}}\cdot\frac{(ab)}{(ab)}$$
$$=\frac{a}{b+a}$$

69b).

$$\frac{\frac{1}{b}}{\frac{1}{a}+\frac{1}{b}}=\left(\frac{1}{b}\right)\left(\frac{a}{1}+\frac{b}{1}\right)$$

Incorrect reciprocal of denominator has been used

$$\frac{\frac{1}{\text{b}}}{\frac{1}{a}+\frac{1}{b}}=\frac{\frac{1}{b}}{\frac{1}{a}+\frac{1}{b}}\cdot\frac{(ab)}{(ab)}$$
$$=\frac{a}{b+a}$$

69c).

$$\frac{a^2(3b+4a)}{a^2+b^2}=\frac{3b+4a}{b^2}$$

a^2 has been incorrectly cancelled

The left side cannot be simplified further.

69d).

$$\frac{1-x}{1+x}=-1$$

1 has been incorrectly cancelled

The left side cannot be simplified further.

69e).

$$\left(x^2-y^2\right)^2=x^4-y^4$$

Both terms on the left side have been squared, but to square a binomial FOIL must be used.

$$\left(x^2-y^2\right)^2=\left(x^2-y^2\right)\left(x^2-y^2\right)$$
$$=x^4-2x^2y^2+y^4$$

69f).

$$\frac{a+b}{a}=a$$

b has been incorrectly cancelled

The left side cannot be simplified further.

CHAPTER 1 SECTION 6

SECTION 1.6 PROGRESS CHECK

1a). (Page 45)

$$\left(x^3\right)^4 = x^{3\cdot 4} = x^{12}$$

1b).

$$x^4\left(x^2\right)^3 = x^4 \cdot x^6 = x^{4+6} = x^{10}$$

1c).

$$\frac{a^{14}}{a^8} = a^{14-8} = a^6$$

1d).

$$\frac{-2(x+1)^n}{(x+1)^{2n}} = \frac{-2}{(x+1)^{2n-n}} = \frac{-2}{(x+1)^n} = -\frac{2}{(x+1)^n}$$

1e).

$$\left(3ab^2\right)^3 = 3^3a^3\left(b^2\right)^3 = 27a^3b^6$$

1f).

$$\left(\frac{-ab^2}{c^3}\right)^3 = \frac{(-a)^3\left(b^2\right)^3}{\left(c^3\right)^3} = \frac{-a^3b^6}{c^9} = -\frac{a^3b^6}{c^9}$$

2a). (Page 47)

$$x^{-2}y^{-3} = \frac{1}{x^2}\cdot\frac{1}{y^3} = \frac{1}{x^2y^3}$$

2b).

$$\frac{-3x^4y^{-2}}{9x^{-8}y^6} = \frac{-3}{9}\cdot x^{4-(-8)}\cdot y^{-2-6}$$
$$= -\frac{1}{3}x^{12}y^{-8}$$
$$= -\frac{x^{12}}{3y^8}$$

2c).

$$\left(\frac{x^{-3}}{x^{-4}}\right)^{-1} = \frac{\left(x^{-3}\right)^{-1}}{\left(x^{-4}\right)^{-1}} = \frac{x^3}{x^4} = \frac{1}{x}$$

3a). (Page. 49)

$$0.02834952 = 2.8\times 10^{-2}$$

3b).

$$0.02834952 = 2.83\times 10^{-2}$$

3c).

$$0.02834952 = 2.8350\times 10^{-2}$$

EXERCISE SET 1.6

1.
$$x^2 \cdot x^4 = x^{2+4} = x^6$$

3.
$$\frac{b^6}{b^2} = b^{6-2} = b^4$$

5.
$$(2x)^4 = 2^4 x^4 = 16x^4$$

7.
$$\left(-\frac{1}{2}\right)^4\left(-\frac{1}{2}\right)^3 = \left(-\frac{1}{2}\right)^7$$
$$= \frac{(-1)^7}{2^7}$$
$$= -\frac{1}{128}$$

9.
$$\left(y^4\right)^{2n} = y^{8n}$$

11.
$$-\left(\frac{x}{y}\right)^3 = -\frac{x^3}{y^3}$$

13.
$$\left(x^3\right)^5 \cdot x^4 = x^{15} \cdot x^4 = x^{19}$$

15.
$$\left(-2x^2\right)^5 = (-2)^5\left(x^2\right)^5$$
$$= -32x^{10}$$

17.
$$x^{3n} \cdot x^n = x^{4n}$$

19.
$$\frac{x^n}{x^{n+2}} = \frac{1}{x^{(n+2)-n}} = \frac{1}{x^2}$$

21.
$$\left(-5x^3\right)\left(-6x^5\right) = 30x^8$$

23.
$$\frac{\left(r^2\right)^4}{\left(r^4\right)^2} = \frac{r^8}{r^8} = 1$$

25.

$$\left(\frac{3}{2}x^2y^3\right)^n = \left(\frac{3}{2}\right)^n\left(x^2\right)^n\left(y^3\right)^n$$
$$= \left(\frac{3}{2}\right)^n x^{2n}y^{3n}$$

27.

$$(2x+1)^3(2x+1)^7 = (2x+1)^{10}$$

29.

$$\left(-2a^2b^3\right)^{2n} = (-2)^{2n}\left(a^2\right)^{2n}\left(b^3\right)^{2n}$$
$$= (-1)^{2n}(2)^{2n}a^{4n}b^{6n}$$
$$= 2^{2n}a^{4n}b^{6n}$$

31.

$$2^0 + 3^{-1} = 1 + \frac{1}{3} = \frac{4}{3}$$

33.

$$\frac{3}{\left(2x^2+1\right)^0} = \frac{3}{1} = 3$$

35.

$$\frac{1}{3^{-4}} = 3^4 = 81$$

37.

$$(-x)^3 = (-1)^3x^3 = -x^3$$

39.

$$\frac{1}{y^{-6}} = y^6$$

41.

$$5^{-3}5^5 = 5^2 = 25$$

43.

$$\left(3^2\right)^{-3} = 3^{-6} = \frac{1}{3^6} = \frac{1}{729}$$

45.

$$\left(x^{-3}\right)^{-3} = x^9$$

47.

$$\frac{2^2}{2^{-3}} = 2^2\cdot 2^3 = 2^5 = 32$$

49.

$$\frac{2x^4y^{-2}}{x^2y^{-3}} = 2x^{4-2}y^{-2-(-3)}$$
$$= 2x^2y$$

51.

$$\left(3a^{-2}b^{-3}\right)^{-2} = (3)^{-2}\left(a^{-2}\right)^{-2}\left(b^{-3}\right)^{-2}$$
$$= \frac{1}{3^2}a^4b^6$$
$$= \frac{a^4b^6}{9}$$

53.

$$\left(-\frac{1}{2}x^3y^{-4}\right)^{-3} = \left(-\frac{1}{2}\right)^{-3}\left(x^3\right)^{-3}\left(y^{-4}\right)^{-3}$$
$$= (-2)^3 x^{-9} y^{12}$$
$$= \frac{-8y^{12}}{x^9}$$

55.

$$\frac{3a^5b^{-2}}{9a^{-4}b^2} = \frac{a^5 \cdot a^4}{3b^2 \cdot b^2} = \frac{a^9}{3b^4}$$

57.

$$\left(\frac{2a^2b^{-4}}{a^{-3}c^{-3}}\right)^2 = \frac{2^2\left(a^2\right)^2\left(b^{-4}\right)^2}{\left(a^{-3}\right)^2\left(c^{-3}\right)^2}$$
$$= \frac{4a^4b^{-8}}{a^{-6}c^{-6}}$$
$$= \frac{4a^4a^6c^6}{b^8}$$
$$= \frac{4a^{10}c^6}{b^8}$$

59.

$$\left(a - 2b^2\right)^{-1} = \frac{1}{a - 2b^2}$$

61.

$$\frac{(a+b)^{-1}}{(a-b)^{-2}} = \frac{(a-b)^2}{a+b}$$

63.

$$\frac{a^{-1}+b^{-1}}{a^{-1}-b^{-1}} = \frac{\frac{1}{a}+\frac{1}{b}}{\frac{1}{a}-\frac{1}{b}}$$
$$= \frac{\frac{1}{a}+\frac{1}{b}}{\frac{1}{a}-\frac{1}{b}} \cdot \frac{(ab)}{(ab)}$$
$$= \frac{b+a}{b-a}$$

65.

$$\left(\frac{a}{b}\right)^{-n} = \frac{1}{\left(\frac{a}{b}\right)^n} = \frac{1}{\frac{a^n}{b^n}} = \frac{1}{\frac{a^n}{b^n}} \cdot \frac{\left(b^n\right)}{\left(b^n\right)} = \frac{b^n}{a^n} = \left(\frac{b}{a}\right)^n$$

In Exercises 67 - 69, [x^2] represents squaring key and [x^y] represents raise to exponent key. Find corresponding keys on your calculator, their appearance may vary.

67.

[−3.67] [x^2] [x^y] [−1] = 0.074

69.

[( ] [4.46] [x^2] [÷] [4.46] [x^y] [−1] [ )] [x^y] [−1] = 0.0113

71.

$0.0091 = 9.1 \times 10^{-3}$

73.

$23. = 2.3 \times 10^1$

75.

$0.8 \times 10^{-3} = 8.0 \times 10^{-1} \times 10^{-3} = 8.0 \times 10^{-4}$

77.

$8.93 \times 10^{-4} = 0008.93 \times 10^{-4} = 0.000893$

79.

$145 \times 10^3 = 145.000 \times 10^3 = 145{,}000$

81.

$1253 \times 10^{-6} = 001253. \times 10^{-6} = 0.001253$

83.

$$\begin{aligned} V &= \frac{4}{3}\pi r^3 \\ &= \frac{4}{3}(3.14)(0.09)^3 \\ &= 0.0030521 \\ &= 3.0521 \times 10^{-3} \text{ cu in} \end{aligned}$$

85.

$$\frac{\text{people}}{\text{sq miles}} : \frac{2{,}600{,}000}{240} = \frac{2.6 \times 10^6}{2.4 \times 10^2}$$

$$= 1.083 \times 10^{6-2}$$

$$= 1.083 \times 10^4$$

persons per sq mile

87a).

$$\frac{2^{n+3}+2^n+2^n}{4\left(2^{n+3}-2^{n+1}\right)}=\frac{2^n\left(2^3+1+1\right)}{4\left(2^{n+1}\right)\left(2^2-1\right)}$$

$$=\frac{2^n(10)}{4\left(2^{n+1}\right)(3)}$$

$$=\frac{10}{4(2)(3)}$$

$$=\frac{5}{12}$$

87b).

$$\frac{a\left(1-r^3\right)}{1-r}=\frac{a(1-r)\left(1+r+r^2\right)}{1-r}=a\left(1+r+r^2\right)$$

87c).

$$\frac{9^{6m}}{3^{2m}}=\frac{\left(3^2\right)^{6m}}{3^{2m}}=\frac{3^{12m}}{3^{2m}}=3^{12m-2m}=3^{10m}$$

87d).

$$\frac{8^4+8^4+8^4+8^4}{4^4}=\frac{4\left(8^4\right)}{4^4}=\frac{8^4}{4^3}=\frac{(4\cdot 2)^4}{4^3}$$

$$=\frac{4^4\cdot 2^4}{4^3}=4\cdot 2^4=64$$

87e).

$$\frac{\left(6\times 10^{-2}\right)\left(2\times 10^{-3}\right)}{3\times 10^8}=\frac{(6)(2)}{3}\times\frac{\left(10^{-2}\right)\left(10^{-3}\right)}{\left(10^8\right)}$$

$$=4\times\frac{10^{-5}}{10^8}$$

$$=4\times 10^{-13}$$

CHAPTER 1 SECTION 7

SECTION 1.7 PROGRESS CHECK

1a). (Page 54)

$$27^{\frac{4}{3}}=\left(27^{\frac{1}{3}}\right)^4=3^4=81$$

1b).

$$\left(a^{\frac{1}{2}}b^{-2}\right)^{-2}=\left(a^{\frac{1}{2}}\right)^{-2}\left(b^{-2}\right)^{-2}=a^{-1}b^4=\frac{b^4}{a}$$

1c).

$$\left(\frac{x^{\frac{1}{3}}y^{\frac{2}{3}}}{z^{\frac{5}{6}}}\right)^{12}=\frac{\left(x^{\frac{1}{3}}\right)^{12}\left(y^{\frac{2}{3}}\right)^{12}}{\left(z^{\frac{5}{6}}\right)^{12}}$$

$$=\frac{x^{\frac{1}{3}\cdot 12}\,y^{\frac{2}{3}\cdot 12}}{z^{\frac{5}{6}\cdot 12}}$$

$$=\frac{x^4y^8}{z^{10}}$$

2a). (Page 55)

$$\sqrt[4]{2rs^3}=\left(2rs^3\right)^{\frac{1}{4}}=\left(2r\right)^{\frac{1}{4}}\left(s^3\right)^{\frac{1}{4}}=\left(2r\right)^{\frac{1}{4}}s^{\frac{3}{4}}$$

2b).

$$(x+y)^{\frac{5}{2}}=\sqrt{(x+y)^5}$$

2c).

$$y^{-\frac{5}{4}}=\frac{1}{y^{\frac{5}{4}}}=\frac{1}{\sqrt[4]{y^5}}$$

2d).

$$\frac{1}{\sqrt[4]{m^5}}=\frac{1}{m^{\frac{5}{4}}}=m^{-\frac{5}{4}}$$

3a). (Page 57)

$$\frac{-9xy^3}{\sqrt{3xy}}=\frac{-9xy^3}{\sqrt{3xy}}\cdot\frac{\sqrt{3xy}}{\sqrt{3xy}}$$

$$=\frac{-9xy^3\sqrt{3xy}}{3xy}$$

$$=-3y^2\sqrt{3xy}$$

3b).

$$\frac{-6}{\sqrt{2}+\sqrt{6}}=\frac{-6}{\sqrt{2}+\sqrt{6}}\cdot\frac{\sqrt{2}-\sqrt{6}}{\sqrt{2}-\sqrt{6}}$$
$$=\frac{-6\left(\sqrt{2}-\sqrt{6}\right)}{2-6}$$
$$=\frac{-6\left(\sqrt{2}-\sqrt{6}\right)}{-4}$$
$$=\frac{-3\left(\sqrt{2}-\sqrt{6}\right)}{-2}\text{ or }\frac{3}{2}\left(\sqrt{2}-\sqrt{6}\right)$$

3c).

$$\frac{4}{\sqrt{x}-\sqrt{y}}=\frac{4}{\sqrt{x}-\sqrt{y}}\cdot\frac{\sqrt{x}+\sqrt{y}}{\sqrt{x}+\sqrt{y}}$$
$$=\frac{4\left(\sqrt{x}+\sqrt{y}\right)}{x-y}$$

4a). (Page 59)

$$\sqrt{75}=\sqrt{25\cdot 3}=\sqrt{25}\sqrt{3}=5\sqrt{3}$$

4b).

$$\sqrt{\frac{18x^6}{y}}=\frac{\sqrt{18x^6}}{\sqrt{y}}=\frac{\sqrt{(9x^6)(2)}}{\sqrt{y}}$$
$$=\frac{\sqrt{9x^6}\sqrt{2}}{\sqrt{y}}=\frac{3|x|^3\sqrt{2}}{\sqrt{y}}$$
$$=\frac{3|x|^3\sqrt{2}}{\sqrt{y}}\cdot\frac{\sqrt{y}}{\sqrt{y}}=\frac{3|x|^3\sqrt{2y}}{y}$$

4c).

$$\sqrt[3]{ab^4c^7}=\sqrt[3]{a\cdot b^3\cdot b\cdot c^6\cdot c}$$
$$=\sqrt[3]{b^3c^6}\sqrt[3]{abc}$$
$$=bc^2\sqrt[3]{abc}$$

4d).

$$\frac{-2xy^3}{\sqrt[4]{32x^3y^5}}=\frac{-2xy^3}{\sqrt[4]{16\cdot 2x^3y^4\cdot y}}=\frac{-2xy^3}{\sqrt[4]{16}\sqrt[4]{y^4}\sqrt[4]{2x^3y}}$$
$$=\frac{-2xy^3}{2y\sqrt[4]{2x^3y}}=\frac{-xy^2}{\sqrt[4]{2x^3y}}\cdot\frac{\sqrt[4]{2^3xy^3}}{\sqrt[4]{2^3xy^3}}$$
$$=\frac{-xy^2\sqrt[4]{8xy^3}}{2xy}=-\frac{y\sqrt[4]{8xy^3}}{2}$$

EXERCISE SET 1.7

1.

$$16^{\frac{3}{4}}=\left(16^{\frac{1}{4}}\right)^3=2^3=8$$

3.

$$(-64)^{-\frac{2}{3}}=\frac{1}{(-64)^{\frac{2}{3}}}=\frac{1}{\left[(-64)^{\frac{1}{3}}\right]^2}\frac{1}{(-4)^2}=\frac{1}{16}$$

5.

$$\frac{2x^{\frac{1}{3}}}{x^{-\frac{3}{4}}}=2x^{\frac{1}{3}-\left(-\frac{3}{4}\right)}=2x^{\frac{4}{12}+\frac{9}{12}}=2x^{\frac{13}{12}}$$

7.

$$\left(\frac{x^{\frac{3}{2}}}{x^{\frac{2}{3}}}\right)^{\frac{1}{6}}=\frac{\left(x^{\frac{3}{2}}\right)^{\frac{1}{6}}}{\left(x^{\frac{2}{3}}\right)^{\frac{1}{6}}}=\frac{x^{\frac{1}{4}}}{x^{\frac{1}{9}}}=x^{\frac{1}{4}-\frac{1}{9}}=x^{\frac{9}{36}-\frac{4}{36}}=x^{\frac{5}{36}}$$

9.

$$\left(x^{\frac{1}{3}}y^{2}\right)^{6}=\left(x^{\frac{1}{3}}\right)^{6}\left(y^{2}\right)^{6}=x^{2}y^{12}$$

11.

$$\left(\frac{x^{15}}{y^{10}}\right)^{\frac{3}{5}}=\frac{\left(x^{15}\right)^{\frac{3}{5}}}{\left(y^{10}\right)^{\frac{3}{5}}}=\frac{x^{9}}{y^{6}}$$

13.

$$\left(\frac{1}{4}\right)^{\frac{2}{5}}=\sqrt[5]{\left(\frac{1}{4}\right)^{2}}=\sqrt[5]{\frac{1}{16}}$$

15.

$$a^{\frac{3}{4}}=\sqrt[4]{a^{3}}$$

17.

$$\left(12x^{3}y^{-2}\right)^{\frac{2}{3}}=\sqrt[3]{\left(12x^{3}y^{-2}\right)^{2}}$$
$$=\sqrt[3]{144x^{6}y^{-4}}$$
$$=\sqrt[3]{\frac{144x^{6}}{y^{4}}}$$

19.

$$\sqrt[4]{8^{3}}=8^{\frac{3}{4}}$$

21.

$$\frac{1}{\sqrt[5]{(-8)^{2}}}=\frac{1}{(-8)^{\frac{2}{5}}}=(-8)^{-\frac{2}{5}}$$

23.

$$\frac{1}{\sqrt[4]{\frac{4}{9}a^{3}}}=\frac{1}{\left(\frac{4}{9}a^{3}\right)^{\frac{1}{4}}}=\left(\frac{4}{9}a^{3}\right)^{-\frac{1}{4}}$$

25.

$$\sqrt{\frac{4}{9}}=\frac{\sqrt{4}}{\sqrt{9}}=\frac{2}{3}$$

27.

$\sqrt[4]{-81}$ Not Real

29.

$$\sqrt{(-5)^2} = |-5| = 5$$

31.

$$\sqrt{\left(\frac{5}{4}\right)^2} = \frac{5}{4}$$

33.

$$(14.43)^{\frac{3}{2}} = \left(\sqrt{14.43}\right)^3 = 54.82$$

(Calculator)

35.

$$\sqrt{x^2+y^2} \neq x+y$$

Let $x = 2, \quad y = 3$:

$$\sqrt{(2)^2+(3)^2} \neq 2+3$$

$$\sqrt{4+9} \neq 5$$

$$\sqrt{13} \neq 5$$

37.

$$\sqrt{48} = \sqrt{16 \cdot 3} = \sqrt{16}\sqrt{3} = 4\sqrt{3}$$

39.

$$\sqrt[3]{54} = \sqrt[3]{27 \cdot 2} = \sqrt[3]{27}\sqrt[3]{2} = 3\sqrt[3]{2}$$

41.

$$\sqrt[3]{y^7} = \sqrt[3]{y^6 \cdot y} = \sqrt[3]{y^6}\sqrt[3]{y}$$

$$= \sqrt[3]{\left(y^2\right)^3}\sqrt[3]{y} = y^2\sqrt[3]{y}$$

43.

$$\sqrt[4]{96x^{10}} = \sqrt[4]{16 \cdot 6 \cdot x^8 \cdot x^2} = \sqrt[4]{16}\sqrt[4]{x^8}\sqrt[4]{6}\sqrt[4]{x^2}$$

$$= 2\sqrt[4]{\left(x^2\right)^4}\sqrt[4]{6}x^{\frac{2}{4}} = 2x^2\sqrt[4]{6}x^{\frac{1}{2}}$$

$$= 2x^2\sqrt[4]{6}\sqrt{x}$$

45.

$$\sqrt{x^5y^3} = \sqrt{x^4 \cdot x \cdot y^2 \cdot y} = \sqrt{x^4}\sqrt{y^2}\sqrt{xy}$$

$$= \sqrt{\left(x^2\right)^2} \cdot y\sqrt{xy} = x^2y\sqrt{xy}$$

47.

$$\sqrt[4]{16x^8y^5} = \sqrt[4]{16x^8y^4 \cdot y}$$

$$= \sqrt[4]{16}\sqrt[4]{x^8}\sqrt[4]{y^4}\sqrt[4]{y}$$

$$= 2\sqrt[4]{\left(x^2\right)^4} \cdot y\sqrt[4]{y}$$

$$= 2x^2y\sqrt[4]{y}$$

49.

$$\sqrt{\frac{1}{5}} = \frac{\sqrt{1}}{\sqrt{5}} = \frac{1}{\sqrt{5}} \cdot \frac{\sqrt{5}}{\sqrt{5}} = \frac{\sqrt{5}}{\sqrt{(5)^2}} = \frac{\sqrt{5}}{5}$$

51.

$$\frac{1}{\sqrt{3y}} = \frac{1}{\sqrt{3y}} \cdot \frac{\sqrt{3y}}{\sqrt{3y}}$$

$$= \frac{\sqrt{3y}}{\sqrt{(3y)^2}} = \frac{\sqrt{3y}}{3y}$$

53.

$$\frac{4x^2}{\sqrt{2x}} = \frac{4x^2}{\sqrt{2x}} \cdot \frac{\sqrt{2x}}{\sqrt{2x}}$$
$$= \frac{4x^2\sqrt{2x}}{\sqrt{(2x)^2}} = \frac{4x^2\sqrt{2x}}{2x} = 2x\sqrt{2x}$$

55.

$$\sqrt[3]{x^2y^7} = \sqrt[3]{x^2 \cdot y^6 \cdot y} = \sqrt[3]{y^6}\sqrt[3]{x^2y}$$
$$= \sqrt[3]{(y^2)^3}\sqrt[3]{x^2y} = y^2\sqrt[3]{x^2y}$$

57.

$$2\sqrt{3} + 5\sqrt{3} = (2+5)\sqrt{3} = 7\sqrt{3}$$

59.

$$3\sqrt{x} + 4\sqrt{x} = (3+4)\sqrt{x} = 7\sqrt{x}$$

61.

$$2\sqrt{27} + \sqrt{12} - \sqrt{48}$$
$$= 2\sqrt{9\cdot 3} + \sqrt{4\cdot 3} - \sqrt{16\cdot 3}$$
$$= 6\sqrt{3} + 2\sqrt{3} - 4\sqrt{3}$$
$$= (6+2-4)\sqrt{3}$$
$$= 4\sqrt{3}$$

63.

$$\sqrt[3]{40} + \sqrt{45} - \sqrt[3]{135} + 2\sqrt{80}$$
$$= \sqrt[3]{8\cdot 5} + \sqrt{9\cdot 5} - \sqrt[3]{27\cdot 5} + 2\sqrt{16\cdot 5}$$
$$= 2\sqrt[3]{5} + 3\sqrt{5} - 3\sqrt[3]{5} + 8\sqrt{5}$$
$$= (2-3)\sqrt[3]{5} + (3+8)\sqrt{5}$$
$$= -\sqrt[3]{5} + 11\sqrt{5} \quad \text{or} \quad 11\sqrt{5} - \sqrt[3]{5}$$

65.

$$2\sqrt{5} - (3\sqrt{5} + 4\sqrt{5})$$
$$= 2\sqrt{5} - 3\sqrt{5} - 4\sqrt{5}$$
$$= (2-3-4)\sqrt{5}$$
$$= -5\sqrt{5}$$

67.

$$\sqrt{3}(\sqrt{3} + 4)$$
$$= \sqrt{3^2} + 4\sqrt{3}$$
$$= 3 + 4\sqrt{3}$$

69.

$$3\sqrt[3]{x^2y}\sqrt[3]{xy^2} = 3\sqrt[3]{x^3y^3} = 3xy$$

71.

$$(\sqrt{2} - \sqrt{3})^2 = (\sqrt{2} - \sqrt{3})(\sqrt{2} - \sqrt{3})$$
$$= 2 - \sqrt{6} - \sqrt{6} + 3$$
$$= 5 - 2\sqrt{6}$$

73.

$$\left(\sqrt{3x}+\sqrt{2y}\right)\left(\sqrt{3x}-2\sqrt{2y}\right)$$
$$=3x-2\sqrt{6xy}+\sqrt{6xy}-2(2y)$$
$$=3x-\sqrt{6xy}-4y$$
$$\text{or}\quad 3x-4y-\sqrt{6xy}$$

75.

$$\frac{3}{\sqrt{2}+3}=\frac{3}{\sqrt{2}+3}\cdot\frac{\sqrt{2}-3}{\sqrt{2}-3}$$
$$=\frac{3\left(\sqrt{2}-3\right)}{2-9}=\frac{3}{-7}\left(\sqrt{2}-3\right)$$
$$=-\frac{3}{7}\left(\sqrt{2}-3\right)\text{ or }\frac{3}{7}\left(3-\sqrt{2}\right)$$

77.

$$\frac{-2}{\sqrt{3}-4}=\frac{-2}{\sqrt{3}-4}\cdot\frac{\sqrt{3}+4}{\sqrt{3}+4}$$
$$=\frac{-2\left(\sqrt{3}+4\right)}{3-16}=\frac{-2}{-13}\left(\sqrt{3}+4\right)$$
$$=\frac{2}{13}\left(\sqrt{3}+4\right)$$

79.

$$\frac{-3}{3\sqrt{a}+1}=\frac{-3}{3\sqrt{a}+1}\cdot\frac{3\sqrt{a}-1}{3\sqrt{a}-1}=\frac{-3\left(3\sqrt{a}-1\right)}{9a-1}$$

81.

$$\frac{-3}{5+\sqrt{5y}}=\frac{-3}{5+\sqrt{5y}}\cdot\frac{5-\sqrt{5y}}{5-\sqrt{5y}}$$
$$=\frac{-3\left(5-\sqrt{5y}\right)}{25-5y}=-\frac{3\left(5-\sqrt{5y}\right)}{5(5-y)}$$

83.

$$\frac{\sqrt{2}+1}{\sqrt{2}-1}=\frac{\sqrt{2}+1}{\sqrt{2}-1}\cdot\frac{\sqrt{2}+1}{\sqrt{2}+1}$$
$$=\frac{2+2\sqrt{2}+1}{2-1}$$
$$=\frac{3+2\sqrt{2}}{1}=3+2\sqrt{2}$$

85.

$$\frac{\sqrt{6}+\sqrt{2}}{\sqrt{3}-\sqrt{2}}=\frac{\sqrt{6}+\sqrt{2}}{\sqrt{3}-\sqrt{2}}\cdot\frac{\sqrt{3}+\sqrt{2}}{\sqrt{3}+\sqrt{2}}$$
$$=\frac{\sqrt{18}+\sqrt{12}+\sqrt{6}+2}{3-2}$$
$$=\frac{3\sqrt{2}+2\sqrt{3}+\sqrt{6}+2}{1}$$
$$=3\sqrt{2}+2\sqrt{3}+\sqrt{6}+2$$

87.

$$\sqrt{12}-\sqrt{10}=\frac{2\sqrt{3}-\sqrt{10}}{1}$$
$$=\frac{2\sqrt{3}-\sqrt{10}}{1}\cdot\frac{2\sqrt{3}+\sqrt{10}}{2\sqrt{3}+\sqrt{10}}$$
$$=\frac{12-10}{2\sqrt{3}+\sqrt{10}}=\frac{2}{2\sqrt{3}+\sqrt{10}}$$

89.

$$\frac{\sqrt{x}-4}{16-x}=\frac{\sqrt{x}-4}{16-x}\cdot\frac{\sqrt{x}+4}{\sqrt{x}+4}$$
$$=\frac{x-16}{(16-x)(\sqrt{x}+4)}$$
$$=\frac{-1(16-x)}{(16-x)(\sqrt{x}+4)}$$
$$=\frac{-1}{\sqrt{x}+4}$$

91.

$$\sqrt{x}+\sqrt{y}\neq\sqrt{x+y}$$

Let $x=4,\ y=9$:

$$\sqrt{4}+\sqrt{9}\neq\sqrt{4+9}$$
$$2+3\neq\sqrt{13}$$
$$5\neq\sqrt{13}$$

93.

The property $\sqrt[n]{ab}=\sqrt[n]{a}\sqrt[n]{b}$ doesn't apply if n is even and either a or b is negative.

$$\sqrt{(-2)(-2)}\neq\sqrt{-2}\sqrt{-2}$$

95a).

$$\sqrt{x\sqrt{x\sqrt{x}}}=\sqrt{x\sqrt{x\cdot x^{\frac{1}{2}}}}$$
$$=\sqrt{x\sqrt{x^{\frac{3}{2}}}}$$
$$=\sqrt{x\cdot\left(x^{\frac{3}{2}}\right)^{\frac{1}{2}}}$$
$$=\sqrt{x\cdot x^{\frac{3}{4}}}$$
$$=\sqrt{x^{\frac{7}{4}}}$$
$$=\left(x^{\frac{7}{4}}\right)^{\frac{1}{2}}$$
$$=x^{\frac{7}{8}}\text{ or }\sqrt[8]{x^7}$$

95b).

$$\left(x^{\frac{1}{2}}-x^{-\frac{1}{2}}\right)^2=\left(x^{\frac{1}{2}}-x^{-\frac{1}{2}}\right)\left(x^{\frac{1}{2}}-x^{-\frac{1}{2}}\right)$$
$$=x-2x^{\frac{1}{2}}x^{-\frac{1}{2}}+x^{-1}$$
$$=x-2x^0+x^{-1}$$
$$=x-2+\frac{1}{x}$$
$$=\frac{x^2}{x}-\frac{2x}{x}+\frac{1}{x}$$
$$=\frac{x^2-2x+1}{x}$$
$$=\frac{(x-1)^2}{x}$$

95c).

$$\sqrt{1+x^2}-\frac{\sqrt{1+x^2}}{2}=\left(1-\frac{1}{2}\right)\sqrt{1+x^2}$$
$$=\frac{1}{2}\sqrt{1+x^2}\text{ or }\frac{\sqrt{1+x^2}}{2}$$

95d).

$$\sqrt[5]{\frac{3^4+3^4+3^4}{5^4+5^4+5^4+5^4+5^4}}=\sqrt[5]{\frac{3(3^4)}{5(5^4)}}$$
$$=\frac{\sqrt[5]{3^5}}{\sqrt[5]{5^5}}=\frac{3}{5}$$

95e).

$$\frac{5(1+x^2)^{\frac{1}{2}}-5x^2(1+x^2)^{-\frac{1}{2}}}{1+x^2}$$
$$=\frac{5(1+x^2)^{-\frac{1}{2}}\left[(1+x^2)-x^2\right]}{1+x^2}$$
$$=\frac{5(1)}{(1+x^2)^{\frac{3}{2}}}=\frac{5}{(1+x^2)^{\frac{3}{2}}}$$

97.

$$\frac{1}{2\pi}\sqrt{\frac{Lc_1c_2}{c_1+c_2}}=\frac{1}{2\pi}\frac{\sqrt{Lc_1c_2}}{\sqrt{c_1+c_2}}$$
$$=\frac{1}{2\pi}\frac{\sqrt{Lc_1c_2}}{\sqrt{c_1+c_2}}\cdot\frac{\sqrt{c_1+c_2}}{\sqrt{c_1+c_2}}$$
$$=\frac{1}{2\pi}\frac{\sqrt{Lc_1c_2(c_1+c_2)}}{c_1+c_2}$$
$$=\frac{\sqrt{Lc_1c_2(c_1+c_2)}}{2\pi(c_1+c_2)}$$

CHAPTER 1 SECTION 8

SECTION 1.8 PROGRESS CHECK

1a). (Page 65)

$$(-9+3i)+(6-2i)=(-9+6)+(3-2)i$$
$$=-3+i$$

1b).

$$7i-(3+9i)=7i-3-9i$$
$$=-3+(7-9)i$$
$$=-3-2i$$

2a). (Page 66)

$$(-3-i)(4-2i)=-3(4-2i)-i(4-2i)$$
$$=-12+6i-4i+2i^2$$
$$=-12+2i+2(-1)$$
$$=-14+2i$$

2b).

$$(-4-2i)(2-3i)=-4(2-3i)-2i(2-3i)$$
$$=-8+12i-4i+6i^2$$
$$=-8+8i+6(-1)$$
$$=-14+8i$$

3a). (Page 67)

$$\frac{4-2i}{5+2i}=\frac{4-2i}{5+2i}\cdot\frac{5-2i}{5-2i}$$
$$=\frac{20-8i-10i+4i^2}{5^2+2^2}$$
$$=\frac{20-18i+4(-1)}{25+4}$$
$$=\frac{16-18i}{29}$$
$$=\frac{16}{29}-\frac{18}{29}i$$

3b).

$$\frac{1}{2-3i}=\frac{1}{2-3i}\cdot\frac{2+3i}{2+3i}$$
$$=\frac{2+3i}{2^2+3^2}$$
$$=\frac{2+3i}{4+9}$$
$$=\frac{2+3i}{13}$$
$$=\frac{2}{13}+\frac{3}{13}i$$

3c).

$$\frac{-3i}{3+5i}=\frac{-3i}{3+5i}\cdot\frac{3-5i}{3-5i}$$
$$=\frac{-9i+15i^2}{3^2+5^2}$$
$$=\frac{-9i+15(-1)}{9+25}$$
$$=\frac{-9i-15}{34}$$
$$=-\frac{15}{34}-\frac{9}{34}i$$

EXERCISE SET 1.8

1.
$$i^{60} = \left(i^4\right)^{15} = (1)^{15} = 1$$

3.
$$i^{83} = i^{80} \cdot i^3 = \left(i^4\right)^{20} \cdot i^3 = (1)^{20} \cdot i^3 = i^3 = -i$$

5.
$$-i^{33} = -i^{32} \cdot i = -\left(i^4\right)^8 \cdot i = -(1)^8 \cdot i = -i$$

7.
$$i^{-84} = \frac{1}{i^{84}} = \frac{1}{\left(i^4\right)^{21}} = \frac{1}{(1)^{21}} = 1$$

9.
$$-i^{-25} = \frac{-1}{i^{25}} = \frac{-1}{i^{24} \cdot i} = \frac{-1}{\left(i^4\right)^6 \cdot i} = \frac{-1}{(1)^6 \cdot i}$$
$$= \frac{-1}{i} = \frac{-1}{i} \cdot \frac{-i}{-i} = \frac{i}{-i^2} = \frac{i}{-(-1)} = \frac{i}{1} = i$$

11.
$$-\frac{3}{4} = -\frac{3}{4} + 0i$$

13.
$$\sqrt{-25} = i\sqrt{25} = 5i = 0 + 5i$$

15.
$$-\sqrt{-36} = -i\sqrt{36} = -6i = 0 - 6i$$

17.
$$3 - \sqrt{-49} = 3 - i\sqrt{49} = 3 - 7i$$

19.
$$0.3 - \sqrt{-98} = 0.3 - i\sqrt{98} = 0.3 - 7\sqrt{2}i$$

21.
$$-2 - \sqrt{-16} = -2 - i\sqrt{16} = -2 - 4i$$

23.
$$(3x-1) + (y+5)i = 1 - 3i$$
$$3x - 1 = 1 \qquad y + 5 = -3$$
$$3x = 2 \qquad y = -8$$
$$x = \frac{2}{3}$$

25.
$$\begin{aligned}(2y+1)-(2x-1)i&=-8+3i\\ 2y+1=-8 \qquad & -(2x-1)=3\\ 2y=-9 \qquad & -2x+1=3\\ y=-\frac{9}{2} \qquad & -2x=2\\ & x=-1\end{aligned}$$

27.
$$\begin{aligned}2i+(3-i)&=3+(2-1)i\\ &=3+i\end{aligned}$$

29.
$$\begin{aligned}2+3i+(3-2i)&=(2+3)+(3-2)i\\ &=5+i\end{aligned}$$

31.
$$\begin{aligned}-3-5i-(2-i)&=-3-5i-2+i\\ &=(-3-2)+(-5+1)i\\ &=-5-4i\end{aligned}$$

33.
$$\begin{aligned}-2i(3+i)&=-6i-2i^2\\ &=-6i-2(-1)\\ &=2-6i\end{aligned}$$

35.
$$\begin{aligned}i\left(-\frac{1}{2}+i\right)&=-\frac{1}{2}i+i^2\\ &=-\frac{1}{2}i+(-1)\\ &=-1-\frac{1}{2}i\end{aligned}$$

37.
$$\begin{aligned}(2-i)(2+i)&=2^2+1^2\\ &=4+1\\ &=5\\ &=5+0i\end{aligned}$$

39.
$$\begin{aligned}(-2-2i)(-4-3i)&=8+6i+8i+6i^2\\ &=8+14i+6(-1)\\ &=2+14i\end{aligned}$$

41.
$$\begin{aligned}(3-2i)(2-i)&=6-3i-4i+2i^2\\ &=6-7i+2(-1)\\ &=4-7i\end{aligned}$$

43.
$$\begin{aligned}(2-i)(2+i)&=2^2+1^2\\ &=4+1\\ &=5\end{aligned}$$

45.
$$\begin{aligned}(3+4i)(3-4i)&=3^2+4^2\\ &=9+16\\ &=25\end{aligned}$$

47.
$$\begin{aligned}(-4-2i)(-4+2i)&=(-4)^2+2^2\\ &=16+4\\ &=20\end{aligned}$$

49.

$$\frac{2+5i}{1-3i}=\frac{2+5i}{1-3i}\cdot\frac{1+3i}{1+3i}$$
$$=\frac{2+6i+5i+15i^2}{1^2+3^2}$$
$$=\frac{2+11i+15(-1)}{1+9}$$
$$=\frac{-13+11i}{10}$$
$$=-\frac{13}{10}+\frac{11}{10}i$$

51.

$$\frac{3-4i}{3+4i}=\frac{3-4i}{3+4i}\cdot\frac{3-4i}{3-4i}$$
$$=\frac{9-12i-12i+16i^2}{3^2+4^2}$$
$$=\frac{9-24i+16(-1)}{9+16}$$
$$=\frac{-7-24i}{25}$$
$$=-\frac{7}{25}-\frac{24}{25}i$$

53.

$$\frac{3-2i}{2-i}=\frac{3-2i}{2-i}\cdot\frac{2+i}{2+i}$$
$$=\frac{6+3i-4i-2i^2}{2^2+1^2}$$
$$=\frac{6-i-2(-1)}{4+1}$$
$$=\frac{8-i}{5}$$
$$=\frac{8}{5}-\frac{1}{5}i$$

55.

$$\frac{2+5i}{3i}=\frac{2+5i}{3i}\cdot\frac{-3i}{-3i}$$
$$=\frac{-6i-15i^2}{-9i^2}$$
$$=\frac{-6i-15(-1)}{-9(-1)}$$
$$=\frac{-6i+15}{9}$$
$$=\frac{15}{9}-\frac{6}{9}i$$
$$=\frac{5}{3}-\frac{2}{3}i$$

57.

$$\frac{4i}{2+i}=\frac{4i}{2+i}\cdot\frac{2-i}{2-i}$$
$$=\frac{8i-4i^2}{2^2+1^2}$$
$$=\frac{8i-4(-1)}{4+1}$$
$$=\frac{4+8i}{5}$$
$$=\frac{4}{5}+\frac{8}{5}i$$

59.

$$\frac{1}{4+3i}=\frac{1}{4+3i}\cdot\frac{4-3i}{4-3i}$$
$$=\frac{4-3i}{4^2+3^2}$$
$$=\frac{4-3i}{16+9}$$
$$=\frac{4-3i}{25}$$
$$=\frac{4}{25}-\frac{3}{25}i$$

61.

$$\frac{1}{1-\frac{1}{3}i}=\frac{1}{1-\frac{1}{3}i}\cdot\frac{1+\frac{1}{3}i}{1+\frac{1}{3}i}$$
$$=\frac{1+\frac{1}{3}i}{1^2+\left(\frac{1}{3}\right)^2}$$
$$=\frac{1+\frac{1}{3}i}{1+\frac{1}{9}}$$
$$=\frac{1+\frac{1}{3}i}{\frac{10}{9}}$$
$$=\frac{1}{\frac{10}{9}}+\frac{\frac{1}{3}}{\frac{10}{9}}i$$
$$=\frac{9}{10}+\frac{3}{10}i$$

63.

$$\frac{1}{-5i}=\frac{1}{-5i}\cdot\frac{5i}{5i}$$
$$=\frac{5i}{5^2}$$
$$=\frac{1}{5}i$$
$$=0+\frac{1}{5}i$$

65.

$$\begin{aligned}x^2-2x+5&=(1+2i)^2-2(1+2i)+5\\&=(1+2i)(1+2i)-2-4i+5\\&=1+4i+4i^2+3-4i\\&=4+4(-1)\\&=4-4\\&=0\end{aligned}$$

67.

$$\begin{aligned}x^2-2x+5&=(1-i)^2-2(1-i)+5\\&=(1-i)(1-i)-2+2i+5\\&=1-2i+i^2+3+2i\\&=4+(-1)\\&=3\end{aligned}$$

69.

$$\begin{aligned}(a+bi)+(c+di)&=(a+c)+(b+d)i\\&=(c+a)+(d+b)i\\&=(c+di)+(a+bi)\end{aligned}$$

71.

$$(a+bi)+(0+0i)=(a+0)+(b+0)i=a+bi$$
$$(0+0i)+(a+bi)=(0+a)+(0+b)i=a+bi$$

$$(a+bi)(1+0i)=a+0i+bi+0i^2=a+bi$$
$$(1+0i)(a+bi)=a+bi+0i+0i^2=a+bi$$

73.

$$
\begin{aligned}
&(a+bi)[(c+di)+(e+fi)] \\
&=(a+bi)[(c+e)+(d+f)i] \\
&=a(c+e)+a(d+f)i+b(c+e)i+b(d+f)i^2 \\
&=ac+ae+adi+afi+bci+bei+(bd+bf)(-1) \\
&=(ac+ae-bd-bf)+(ad+af+bc+be)i
\end{aligned}
$$

$$
\begin{aligned}
&(a+bi)(c+di)+(a+bi)(e+fi) \\
&=ac+adi+bci+bdi^2+ae+afi+bei+bfi^2 \\
&=(ac+ae)+(ad+bc+af+be)i+(bd+bf)(-1) \\
&=(ac+ae-bd-bf)+(ad+bc+af+be)i
\end{aligned}
$$

Thus

$$(a+bi)[(c+di)+(e+fi)]=(a+bi)(c+di)+(a+bi)(e+fi)$$

75.

$$
\begin{aligned}
2y-10 &\geq 0 \\
2y &\geq 10 \\
y &\geq 5
\end{aligned}
$$

CHAPTER 1 REVIEW EXERCISES

1. $\{1,2,3,4\}$
(The natural numbers start with 1.)

2. $\{-3,-2,-1\}$

3. $\{2\}$

4. True
(Irrational numbers are a subset of the real numbers.)

5. False, $-35 < 0$

6. False, -14 is an integer

7. False, 0 is rational

8. additive inverse

9. distributive

10. commutative (addition)

11. multiplicative identity

12.

13.

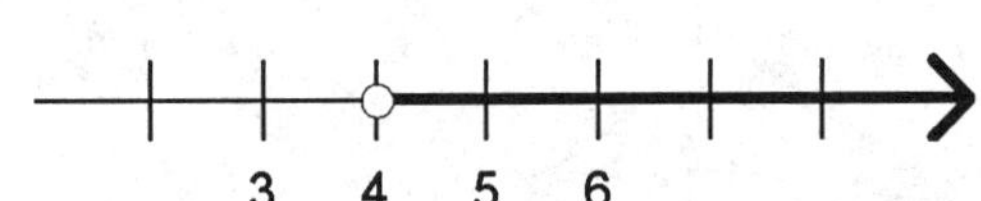

14.

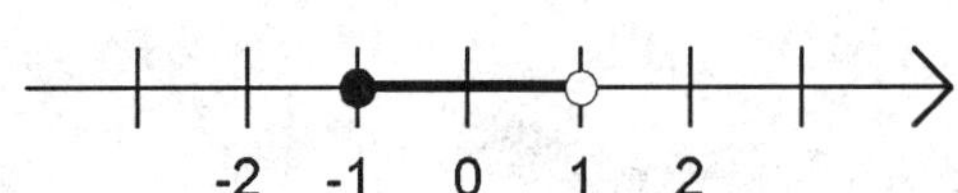

15.

$$|-3|-|1-5|$$
$$=3-|-4|$$
$$=3-4$$
$$=-1$$

16.

$$\overline{PQ}=\left|\frac{9}{2}-6\right|$$
$$=\left|\frac{9}{2}-\frac{12}{2}\right|$$
$$=\left|-\frac{3}{2}\right|$$
$$=\frac{3}{2}$$

17.

$$3.25x+0.15y$$
$$=3.25(12)+0.15(80)$$
$$=39+12$$
$$=51$$

\$51

18.

(c); since $-\frac{1}{2}$ is used as an exponent

19.

-0.5; degree 7

20.

-7; degree 5

21.

$$\left(3a^2b^2-a^2b+2b-a\right)-\left(2a^2b^2+2a^2b-2b-a\right)$$
$$=3a^2b^2-a^2b+2b-a-2a^2b^2-2a^2b+2b+a$$
$$=a^2b^2-3a^2b+4b$$

22.

$$x(2x-1)(x+2)$$
$$=x\left(2x^2+4x-x-2\right)$$
$$=x\left(2x^2+3x-2\right)$$
$$=2x^3+3x^2-2x$$

23.

$$3x(2x+1)^2$$
$$=3x(2x+1)(2x+1)$$
$$=3x\left(4x^2+4x+1\right)$$
$$=12x^3+12x^2+3x$$

24.

$$2x^2-2=2\left(x^2-1\right)$$
$$=2(x+1)(x-1)$$

25.

$$x^2-25y^2=x^2-(5y)^2$$
$$=(x+5y)(x-5y)$$

26.

$$2a^2+3ab+6a+9b$$
$$=(2a^2+3ab)+(6a+9b)$$
$$=a(2a+3b)+3(2a+3b)$$
$$=(2a+3b)(a+3)$$

27.

$4x^2+19x-5:(4x\quad)(4x\quad)$

$a\cdot c=4\cdot(-5)=-20$

$-1\cdot 20=-20\quad -1+20=19$

$(4x-1)(4x+20)$

Discard common factor 4

out of $4x+20$

$4x^2+19x-5=(4x-1)(x+5)$

28.

$$x^8-1=(x^4)^2-1^2$$
$$=(x^4+1)(x^4-1)$$
$$=(x^4+1)\left[(x^2)^2-1\right]$$
$$=(x^4+1)(x^2+1)(x^2-1)$$
$$=(x^4+1)(x^2+1)(x+1)(x-1)$$

29.

$$27r^6+8s^6$$
$$=(3r^2)^3+(2s^2)^3$$
$$=(3r^2+2s^2)\left[(3r^2)^2-(3r^2)(2s^2)+(2s^2)^2\right]$$
$$=(3r^2+2s^2)(9r^4-6r^2s^2+4s^4)$$

30.

$$\frac{14(y-1)}{3(x^2-y^2)}\cdot\frac{9(x+y)}{-7xy^2}$$
$$=\frac{-2(y-1)\cdot 3(x+y)}{(x+y)(x-y)(xy^2)}$$
$$=-\frac{6(y-1)}{xy^2(x-y)}$$

31.

$$\frac{4-x^2}{2y^2} \div \frac{x-2}{3y}$$

$$= \frac{\frac{(2-x)(2+x)}{2y^2}}{\frac{x-2}{3y}} \cdot \frac{6y^2}{6y^2}$$

$$= \frac{3(2-x)(2+x)}{2y(x-2)}$$

$$= \frac{-3(x-2)(2+x)}{2y(x-2)}$$

$$= -\frac{3(2+x)}{2y}$$

32.

$$\frac{x^2-2x-3}{2x^2-x} \div \frac{x^2-4x+3}{3x^3-3x^2}$$

$$= \frac{\frac{(x-3)(x+1)}{x(2x-1)}}{\frac{(x-3)(x-1)}{3x^2(x-1)}} \cdot \frac{3x^2(2x-1)(x-1)}{3x^2(2x-1)(x-1)}$$

$$= \frac{3x(x-3)(x+1)(x-1)}{(x-3)(x-1)(2x-1)}$$

$$= \frac{3x(x+1)}{2x-1}$$

33.

$$\frac{a+b}{a+2b} \cdot \frac{a^2-4b^2}{a^2-b^2}$$

$$= \frac{a+b}{a+2b} \cdot \frac{(a+2b)(a-2b)}{(a+b)(a-b)}$$

$$= \frac{a-2b}{a-b}$$

34.

$$-\frac{1}{2x^2}, \frac{2}{x^2-4}, \frac{3}{x-2}$$

$$-\frac{1}{2x^2}, \frac{2}{(x+2)(x-2)}, \frac{3}{x-2}$$

LCD: $2x^2(x+2)(x-2)$

35.

$$\frac{4}{x}, \frac{5}{x^2-x}, \frac{-3}{(x-1)^2}$$

$$\frac{4}{x}, \frac{5}{x(x-1)}, -\frac{3}{(x-1)^2}$$

LCD: $x(x-1)^2$

36.

$$\frac{2}{(x-1)y}, -\frac{4}{y^2}, \frac{x+2}{5(x-1)^2}$$

LCD: $5y^2(x-1)^2$

37.

$$\frac{y-1}{x^2(y+1)}, \quad \frac{x-2}{2xy-2x}, \quad \frac{3x}{4y^2+8y+4}$$

$$\frac{y-1}{x^2(y+1)}, \quad \frac{x-2}{2x(y-1)}, \quad \frac{3x}{4(y+1)^2}$$

LCD: $4x^2(y-1)(y+1)^2$

38.

$$2+\frac{4}{a^2-4}=2+\frac{4}{(a+2)(a-2)}$$

$$=\frac{2}{1}\cdot\frac{(a+2)(a-2)}{(a+2)(a-2)}+\frac{4}{(a+2)(a-2)}$$

$$=\frac{2(a^2-4)+4}{(a+2)(a-2)}$$

$$=\frac{2a^2-8+4}{(a+2)(a-2)}$$

$$=\frac{2a^2-4}{(a+2)(a-2)} \text{ or } \frac{2(a^2-2)}{(a+2)(a-2)}$$

39.

$$\frac{3}{x^2-16}-\frac{2}{x-4}=\frac{3}{(x+4)(x-4)}-\frac{2}{x-4}$$

$$=\frac{3}{(x+4)(x-4)}-\frac{2}{(x-4)}\cdot\frac{(x+4)}{(x+4)}$$

$$=\frac{3}{(x+4)(x-4)}-\frac{2x+8}{(x+4)(x-4)}$$

$$=\frac{3-(2x+8)}{(x+4)(x-4)}$$

$$=\frac{3-2x-8}{(x+4)(x-4)}$$

$$=\frac{-2x-5}{(x+4)(x-4)}$$

40.

$$\frac{\frac{3}{x+2}-\frac{2}{x-1}}{x-1}=\frac{\frac{3}{x+2}-\frac{2}{x-1}}{x-1}\cdot\frac{(x+2)(x-1)}{(x+2)(x-1)}$$

$$=\frac{3(x-1)-2(x+2)}{(x+2)(x-1)^2}$$

$$=\frac{3x-3-2x-4}{(x+2)(x-1)^2}$$

$$=\frac{x-7}{(x+2)(x-1)^2}$$

41.

$$x^2+\frac{\frac{1}{x}+1}{x-\frac{1}{x}}=x^2+\left[\frac{\frac{1}{x}+1}{x-\frac{1}{x}}\cdot\frac{x}{x}\right]$$
$$=x^2+\frac{1+x}{x^2-1}$$
$$=x^2+\frac{1+x}{(x+1)(x-1)}$$
$$=x^2+\frac{1}{x-1}$$
$$=\frac{x^2}{1}\cdot\frac{(x-1)}{(x-1)}+\frac{1}{x-1}$$
$$=\frac{x^3-x^2}{x-1}+\frac{1}{x-1}$$
$$=\frac{x^3-x^2+1}{x-1}$$

42.

$$\left(2a^2b^{-3}\right)^{-3}=2^{-3}\left(a^2\right)^{-3}\left(b^{-3}\right)^{-3}$$
$$=\frac{1}{2^3}a^{-6}b^9$$
$$=\frac{b^9}{8a^6}$$

43.

$$2\left(a^2-1\right)^0=2(1)=2$$

44.

$$\left(\frac{x^3}{y^{-6}}\right)^{-\frac{4}{3}}=\frac{\left(x^3\right)^{-\frac{4}{3}}}{\left(y^{-6}\right)^{-\frac{4}{3}}}$$
$$=\frac{x^{-4}}{y^8}$$
$$=\frac{1}{x^4y^8}$$

45.

$$\frac{x^{3+n}}{x^n}=x^{(3+n)-n}=x^3$$

46.

$$\sqrt{80}=\sqrt{16\cdot 5}=\sqrt{16}\sqrt{5}=4\sqrt{5}$$

47.

$$\frac{2}{\sqrt{12}}=\frac{2}{\sqrt{4\cdot 3}}=\frac{2}{\sqrt{4}\sqrt{3}}=\frac{2}{2\sqrt{3}}$$
$$=\frac{1}{\sqrt{3}}=\frac{1}{\sqrt{3}}\cdot\frac{\sqrt{3}}{\sqrt{3}}=\frac{\sqrt{3}}{3}$$

48.

$$\sqrt{x^7y^5}=\sqrt{\left(x^3\right)^2\cdot x\cdot\left(y^2\right)^2\cdot y}$$
$$=\sqrt{\left(x^3\right)^2}\sqrt{\left(y^2\right)^2}\sqrt{xy}$$
$$=x^3y^2\sqrt{xy}$$

49.

$$\sqrt[4]{32x^8y^6} = \sqrt[4]{16\cdot 2\left(x^2\right)^4 y^4 \cdot y^2}$$
$$= \sqrt[4]{16}\ \sqrt[4]{2}\ \sqrt[4]{\left(x^2\right)^4}\ \sqrt[4]{y^4}\ \sqrt[4]{y^2}$$
$$= 2\ \sqrt[4]{2}x^2y\cdot\sqrt[4]{y^2}$$
$$= 2x^2y\ \sqrt[4]{2y^2}$$

50.

$$\frac{\sqrt{x}}{\sqrt{x}+\sqrt{y}} = \frac{\sqrt{x}}{\sqrt{x}+\sqrt{y}}\cdot\frac{\sqrt{x}-\sqrt{y}}{\sqrt{x}-\sqrt{y}}$$
$$= \frac{\sqrt{x}\left(\sqrt{x}-\sqrt{y}\right)}{x-y}$$
$$= \frac{x-\sqrt{xy}}{x-y}$$

51.

$$\frac{\left(5.1\times 10^7\right)\left(3.45\times 10^{-2}\right)}{7.1\times 10^4}$$
$$= \frac{(5.1)(3.45)}{7.1}\times\frac{\left(10^7\right)\left(10^{-2}\right)}{10^4}$$
$$= 2.48\times 10$$

52.

$$\frac{\sqrt{x}-\sqrt{y}}{x-y} = \frac{\sqrt{x}-\sqrt{y}}{x-y}\cdot\frac{\sqrt{x}+\sqrt{y}}{\sqrt{x}+\sqrt{y}}$$
$$= \frac{x-y}{(x-y)\left(\sqrt{x}+\sqrt{y}\right)}$$
$$= \frac{1}{\sqrt{x}+\sqrt{y}}$$

53.

$$\sqrt[4]{x^2y^2} + 2\sqrt[4]{x^2y^2} = (1+2)\left(x^2y^2\right)^{\frac{1}{4}}$$
$$= 3\left(x^2\right)^{\frac{1}{4}}\left(y^2\right)^{\frac{1}{4}}$$
$$= 3|x|^{\frac{1}{2}}|y|^{\frac{1}{2}}$$
$$= 3\sqrt{|x||y|}$$
$$= 3\sqrt{|xy|}$$

54.

$$\left(\sqrt{3}+\sqrt{5}\right)^2 = \left(\sqrt{3}+\sqrt{5}\right)\left(\sqrt{3}+\sqrt{5}\right)$$
$$= 3+\sqrt{15}+\sqrt{15}+5$$
$$= 8+2\sqrt{15}$$

In Exercise 55, the following keystrokes will be utilized:

[ABS] : absolute value

[$\sqrt{\ }$] : square root

[x^y] : raise x to the yth power

[$\sqrt[x]{\ }$] : xth root

(Note: The order of keystrokes is based on the use of a graphics calculator.)

55a).

[12] [÷] [5] [−] [3] [÷] [7] = 1.97

55b).

[ABS] [(] [(] [−4] [)] [x^y] [3] [−] [5] [x^y] [6] [)] = 15,689

55c)

[$\sqrt{\ }$] [8] = 2.83

55d).

[π] [x^y] [8] = 9488.53

55e).

[5] [$\sqrt[x]{\ }$] [−27] = −1.93

55f).

[ABS] [(] [2] [+] [$\sqrt{\ }$] [3] [)] [÷] [−6] = −0.62

55g).

[3] [$\sqrt[x]{\ }$] [4] [+] [$\sqrt{\ }$] [(] [1] [÷] [8] [)] = 1.94

55h).

[10] [$\sqrt[x]{\ }$] [0.5] = 0.93

55i).

[(] [2] [÷] [3] [)] [x^y] [4] = 0.20

55j).

[9] [x^y] [(] [5] [÷] [8] [)] = 3.95

56.

$$(x-2)+(2y-1)i=-4+7i$$

$$x-2=-4 \qquad 2y-1=7$$

$$x=-2 \qquad 2y=8$$

$$y=4$$

57.

$$i^{47}=i^{44}\cdot i^{3}=\left(i^{4}\right)^{11}\cdot i^{3}=(1)^{11}\cdot i^{3}=i^{3}=-i$$

58.

$$2+(6-i)=(2+6)-i=8-i$$

59.

$$\begin{aligned}(2+i)^2&=(2+i)(2+i)\\&=4+2i+2i+i^2\\&=4+4i+(-1)\\&=3+4i\end{aligned}$$

60.

$$\begin{aligned}(4-3i)(2+3i)&=8+12i-6i-9i^2\\&=8+6i-9(-1)\\&=17+6i\end{aligned}$$

61.

$$\begin{aligned}\frac{4-3i}{2+3i}&=\frac{4-3i}{2+3i}\cdot\frac{2-3i}{2-3i}\\&=\frac{8-12i-6i+9i^2}{2^2+3^2}\\&=\frac{8-18i+9(-1)}{4+9}\\&=\frac{-1-18i}{13}\\&=-\frac{1}{13}-\frac{18}{13}i\end{aligned}$$

62a).

$$\begin{aligned}&\frac{1}{a}+\frac{1}{b}+\frac{1}{c}\\&=\frac{1}{a}\cdot\frac{(bc)}{(bc)}+\frac{1}{b}\cdot\frac{(ac)}{(ac)}+\frac{1}{c}\cdot\frac{(ab)}{(ab)}\\&=\frac{bc}{abc}+\frac{ac}{abc}+\frac{ab}{abc}\\&=\frac{bc+ac+ab}{abc}\end{aligned}$$

62b).

$$\begin{aligned}\frac{\frac{1}{a}+\frac{1}{b}}{\frac{1}{c}+\frac{1}{d}}&=\frac{\frac{1}{a}+\frac{1}{b}}{\frac{1}{c}+\frac{1}{d}}\cdot\frac{(abcd)}{(abcd)}\\&=\frac{bcd+acd}{abd+abc}\\&=\frac{cd(b+a)}{ab(d+c)}\end{aligned}$$

63.

number of pounds to lose: $200-180=20$

number of pounds needed to lose per week: $\frac{20}{8}=\frac{5}{2}$ (8 weeks until reunion)

decrease in calories/ day to lose 1 lb: $2400-1900=500$

$$\frac{\text{lbs / week}}{\text{decrease in cal / day}}:\frac{1}{500}=\frac{\frac{5}{2}}{x}$$

$$(1)(x)=(500)\left(\frac{5}{2}\right)$$

$$x=1250$$

He must decrease his caloric intake per day. by 1250 calories, thus his caloric intake per day should be $2400-1250=1150$ calories

64a).

$\{p,v,s,t\}$

committees of 3:

$\{p,v,s\},\{p,v,t\},\{p,s,t\},\{v,s,t\}$

4 ways to form a committee of 3

64b).

president has 2 votes, each of the others has 1 vote

{president, vice - president}

{president, secretary}

{president, treasurer}

{vice - president, secretary, treasurer}

65.

$$\frac{6000 \text{ sec}}{60} = 100 \text{ min}$$

$$\frac{100 \text{ min}}{60} = \frac{5}{3} \text{ hr}$$

1 child eats 1 hotdog in $\frac{1}{10}$ hr,

so 1 child eats 10 hotdogs in 1 hr

rate per child: hotdogs / hr

$$\frac{\text{hotdogs}}{\text{hr}} : \frac{10}{1} = \frac{x}{\frac{5}{3}}$$

$$(10)\left(\frac{5}{3}\right) = (x)(1)$$

$$x = \frac{50}{3} \text{ hotdogs}$$

In $\frac{5}{3}$ hr (6000 sec): $\frac{\text{children}}{\text{hotdogs}} : \frac{1}{\frac{50}{3}} = \frac{y}{100}$

$$(100)(1) = \left(\frac{50}{3}\right)(y)$$

$$100\left(\frac{3}{50}\right) = y$$

$$y = 6$$

6 children can eat 100 hotdogs in 6000 seconds

66.

cost: \$80

25% of $80 = (0.25)(80) = \$20.00$

lowest selling price:

$\$80 + \$20.00 = \$100.00$

67.

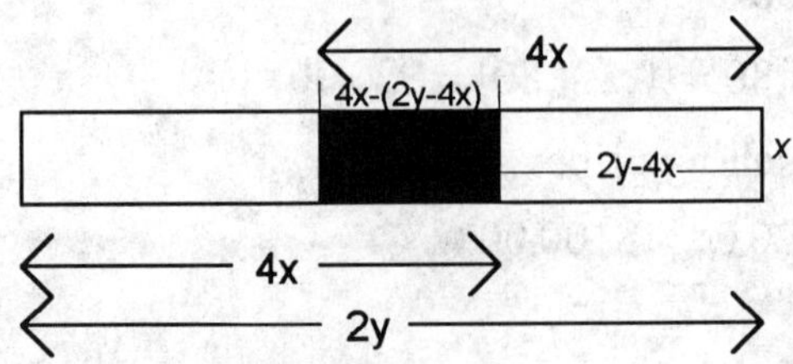

$$\begin{aligned}\text{Area} &= \text{width} \cdot \text{length} \\ &= x \cdot [4x - (2y - 4x)] \\ &= x[4x - 2y + 4x] \\ &= x(8x - 2y) \\ &= 8x^2 - 2xy\end{aligned}$$

68.

width: $4 - (s + s) = 4 - 2s$

length: $5 - (s + s) = 5 - 2s$

height: s

$$\begin{aligned}\text{volume} &= \text{width} \cdot \text{length} \cdot \text{height} \\ &= (4 - 2s)(5 - 2s)s\end{aligned}$$

$4 - 2s > 0$ and $5 - 2s > 0$ and $s > 0$

$-2s > -4$ and $-2s > -5$ and $s > 0$

$s < 2$ and $s < \frac{5}{2}$ and $s > 0$

Thus $0 < s < 2$

69a).

$$\begin{aligned}(x - y)(x^2 + xy + y^2) &= x(x^2 + xy + y^2) - y(x^2 + xy + y^2) \\ &= x^3 + x^2y + xy^2 - x^2y - xy^2 - y^3 \\ &= x^3 - y^3\end{aligned}$$

69b).

$$(x-y)(x^3+x^2y+xy^2+y^3)$$
$$=x(x^3+x^2y+xy^2+y^3)-y(x^3+x^2y+xy^2+y^3)$$
$$=x^4+x^3y+x^2y^2+xy^3-x^3y-x^2y^2-xy^3-y^4$$
$$=x^4-y^4$$

69c).

$$(x-y)(x^4+x^3y+x^2y^2+xy^3+y^4)$$
$$=x(x^4+x^3y+x^2y^2+xy^3+y^4)$$
$$-y(x^4+x^3y+x^2y^2+xy^3+y^4)$$
$$=x^5+x^4y+x^3y^2+x^2y^3+xy^4$$
$$-x^4y-x^3y^2-x^2y^3-xy^4-y^5$$
$$=x^5-y^5$$

70.

$$x^n-y^n=$$
$$(x-y)(x^{n-1}+x^{n-2}y+x^{n-3}y^2+\cdots+x^2y^{n-3}+xy^{n-2}+y^{n-1})$$

Note: exponents of x are decreasing by 1 in each term and exponents of y are increasing by 1 in each term

71.

$$13\times 17=(20-7)(20-3)$$
$$=(20)(20)-(20)(3)-(20)(7)+21$$
$$=400-60-140+21$$
$$=221$$

72.

$$ac+ad-bc-bd=(ac+ad)+(-bc-bd)$$
$$=a(c+d)-b(c+d)$$
$$=(c+d)(a-b)$$

$$ac+ad-bc-bd=ac-bc+ad-bd$$
$$=(ac-bc)+(ad-bd)$$
$$=c(a-b)+d(a-b)$$
$$=(a-b)(c+d)$$

73.

When adding three digits the maximum carry is 2.

$$\begin{array}{rccccc} & F & O & R & T & Y \\ & & & T & E & N \\ + & & & T & E & N \\ \hline & S & I & X & T & Y \end{array}$$

$Y+N+N=Y$

so the last digit of the sum of $N+N$ must be 0.

Thus $N=0$ or $N=5$.

If

$N=$	0	5
	$2E+T=T$	$1+2E+T=T$ (1 from carry)
	$2E=0$	
	$E=0$ ⊗	$\underbrace{1+2E}_{\text{odd}}=\underbrace{0}_{\text{even}}$ ⊗
	or	
	$E=5$	

Thus $N=0$ and $E=5$

$$\begin{array}{rccccc} & F & O & R & T & Y \\ & & & T & 5 & 0 \\ + & & & T & 5 & 0 \\ \hline & S & I & X & T & Y \end{array}$$

unused digits: $\{1,2,3,4,6,7,8,9\}$

73. (Continued)

$O \neq I$ so must have carry of 1 or 2 which means $I=O+1$ or $I=O+2$ and this sum must cause a carry. Possibilities then for O are

O	with carry	last digit of sum
8	2	0 ⊗
9	1	0 ⊗
9	2	$1=I$ ✓

Unused digits: $\{2,3,4,6,7,8\}$

$$\begin{array}{rccccc} & {}^{1}F & {}^{2}9 & {}^{1}R & T & Y \\ & & & T & 5 & 0 \\ + & & & T & 5 & 0 \\ \hline & S & 1 & X & T & Y \end{array}$$

$F \neq S$ thus must have a carry and F and S are consecutive such that they do not themselves create a carry.

Possibilities

F	S	
2	3	3 will be used
3	4	

Unused digits: $\{2,4,6,7,8\}$

$1+R+T+T$ must create a carry of 2

73. (Continued)

T	R	$1+R+T+T$	Resulting last digit
8	7	$1+7+8+8$	4
8	6	$1+6+8+8$	3 ⊗
8	4	$1+4+8+8$	1 ⊗

$$\begin{array}{rccccc} & {}^1F & {}^29 & {}^17 & 8 & Y \\ & & & 8 & 5 & 0 \\ + & & & 8 & 5 & 0 \\ \hline & S & 1 & X=4 & 8 & Y \end{array}$$

Unused digits: $\{\underline{\underline{2}},6\}$

73. (Continued)

$F=2,\ S=3$

$$\begin{array}{rccccc} & 2 & 9 & 7 & 8 & Y \\ & & & 8 & 5 & 0 \\ + & & & 8 & 5 & 0 \\ \hline & 3 & 1 & 4 & 8 & Y \end{array}$$

$Y=6$

$$\begin{array}{rccccc} & {}^12 & {}^29 & {}^17 & 8 & 6 \\ & & & 8 & 5 & 0 \\ + & & & 8 & 5 & 0 \\ \hline & 3 & 1 & 4 & 8 & 6 \end{array}$$

$F=2,\ O=9,\ R=7,\ T=8,\ Y=6,$
$E=5,\ N=0,\ S=3,\ I=1,\ X=4$

74.

$28=1+2+4+7+14$

$2^{p-1}(2^p-1)$ for $p=2$:

$$\begin{aligned} 2^{2-1}(2^2-1) &= 2(4-1) \\ &= 2(3) \\ &= 6 \end{aligned}$$

when $p=3$, $2^3-1=7$ is prime:

$$\begin{aligned} 2^{3-1}(2^3-1) &= 2^2(8-1) \\ &= 4(7) \\ &= 28 \text{ (2nd perfect number)} \end{aligned}$$

74. (Continued).

when $p=5$, $2^5-1=31$ is prime:

$$\begin{aligned} 2^{5-1}(2^5-1) &= 2^4(32-1) \\ &= 16(31) \\ &= 496 \end{aligned}$$

when $p=7$, $2^7-1=127$ is prime:

$$\begin{aligned} 2^{7-1}(2^7-1) &= 2^6(128-1) \\ &= 64(127) \\ &= 8128 \end{aligned}$$

when $p=13$, $2^{13}-1=8191$ is prime:

$$\begin{aligned} 2^{13-1}(2^{13}-1) &= 2^{12}(8192-1) \\ &= 4096(8191) \\ &= 33{,}550{,}336 \end{aligned}$$

75a).

$$\begin{aligned} \text{time} &= \frac{\text{distance}}{\text{rate}} \\ &= \frac{1\times 10^{26}}{3\times 10^{8}} \\ &= \frac{1}{3}\times 10^{26-8} \\ &= \frac{1}{3}\times 10^{18} \\ &= 0.33\times 10^{18} \\ &= 3.3\times 10^{17}\ \text{ seconds} \end{aligned}$$

75b).

$$(365\text{ days})\left(24\,\frac{\text{hrs}}{\text{day}}\right) = 8.76\times 10^{3}$$

$$(8.76\times 10^{3}\,\text{hrs})\left(60\,\frac{\text{min}}{\text{hr}}\right) = 5.256\times 10^{5}$$

$$(5.256\times 10^{5}\,\text{min})\left(60\,\frac{\text{sec}}{\text{min}}\right) = 3.1536\times 10^{7}\,\text{seconds}$$

75c).

$$\frac{3.3\times 10^{17}\,\text{sec}}{3.1536\times 10^{7}\left(\dfrac{\text{sec}}{\text{yr}}\right)} = 1.0464\times 10^{10}\,\text{yr}$$

76.

$$\begin{aligned} \sqrt{x+\sqrt{x+\sqrt{x}}} &= \sqrt{x+\sqrt{x+x^{\frac{1}{2}}}} \\ &= \sqrt{x+\left(x+x^{\frac{1}{2}}\right)^{\frac{1}{2}}} \\ &= \left[x+\left(x+x^{\frac{1}{2}}\right)^{\frac{1}{2}}\right]^{\frac{1}{2}} \end{aligned}$$

77.

$$\left(\sqrt{5-\sqrt{24}}\right)^2 = 5-\sqrt{24} = 5-2\sqrt{6}$$

Equals

$$\begin{aligned}\left(\sqrt{2}-\sqrt{3}\right)^2 &= \left(\sqrt{2}-\sqrt{3}\right)\left(\sqrt{2}-\sqrt{3}\right)\\ &= 2-\sqrt{6}-\sqrt{6}+3\\ &= 5-2\sqrt{6}\end{aligned}$$

78.

$$T = 1+\frac{1}{T}$$

$$\frac{\sqrt{5}+1}{2} = 1+\frac{1}{\frac{\sqrt{5}+1}{2}} \quad ?$$

$$\begin{aligned}1+\frac{1}{\frac{\sqrt{5}+1}{2}} &= 1+\left[\frac{1}{\frac{\sqrt{5}+1}{2}}\cdot\frac{(2)}{(2)}\right]\\ &= 1+\frac{2}{\sqrt{5}+1}\\ &= 1+\left[\frac{2}{\sqrt{5}+1}\cdot\frac{\sqrt{5}-1}{\sqrt{5}-1}\right]\\ &= 1+\frac{2\left(\sqrt{5}-1\right)}{5-1}\\ &= 1+\frac{2\left(\sqrt{5}-1\right)}{4}\\ &= \frac{4}{4}+\frac{2\sqrt{5}-2}{4}\\ &= \frac{4+2\sqrt{5}-2}{4}\\ &= \frac{2+2\sqrt{5}}{4}\\ &= \frac{2\left(1+\sqrt{5}\right)}{4}\\ &= \frac{1+\sqrt{5}}{2}\end{aligned}$$

79a).

$$\frac{\sqrt{x+h+1}-\sqrt{x-1}}{h}$$

$$=\frac{\sqrt{x+h+1}-\sqrt{x-1}}{h}\cdot\frac{\sqrt{x+h+1}+\sqrt{x-1}}{\sqrt{x+h+1}+\sqrt{x-1}}$$

$$=\frac{x+h+1-(x-1)}{h\left(\sqrt{x+h+1}+\sqrt{x-1}\right)}$$

$$=\frac{x+h+1-x+1}{h\left(\sqrt{x+h+1}+\sqrt{x-1}\right)}$$

$$=\frac{h+2}{h\left(\sqrt{x+h+1}+\sqrt{x-1}\right)}$$

79b).

$$\frac{\sqrt{3+x}-\sqrt{3}}{x}=\frac{\sqrt{3+x}-\sqrt{3}}{x}\cdot\frac{\sqrt{3+x}+\sqrt{3}}{\sqrt{3+x}+\sqrt{3}}$$

$$=\frac{3+x-3}{x\left(\sqrt{3+x}+\sqrt{3}\right)}$$

$$=\frac{x}{x\left(\sqrt{3+x}+\sqrt{3}\right)}$$

$$=\frac{1}{\sqrt{3+x}+\sqrt{3}}$$

80.

$$V = IZ$$
$$= (2-3i)(6+2i)$$
$$= 12+4i-18i-6i^2$$
$$= 12-14i-6(-1)$$
$$= 18-14i$$

$(18-14i)$ volts

CHAPTER 1 REVIEW TEST

1.
$\{2,4,6,8,10,12\}$

2.
$\{3\}$

3.
F; -1.36 is rational

4.
F; π is irrational

5.
Commutative (multiplication)

6.
Multiplicative inverse

7.

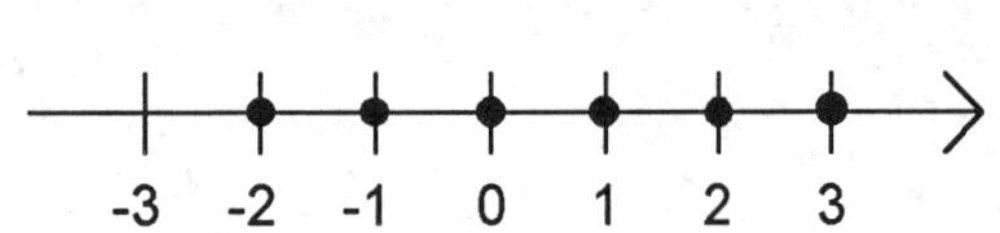

8.

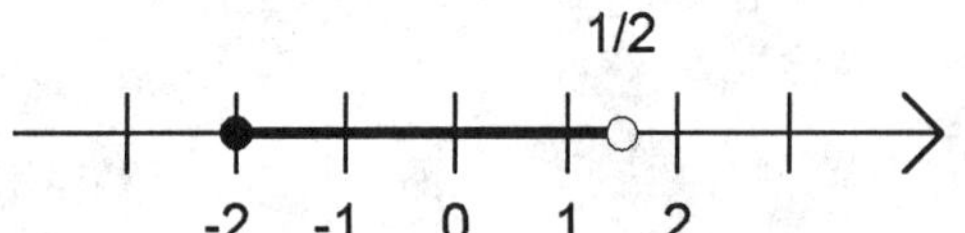

9.
$$\begin{aligned}|2-3|-|4-2| &= |-1|-|2| \\ &= 1-2 \\ &= -1\end{aligned}$$

10.
$$\begin{aligned}\overline{AB} &= |-6-(-4)| \\ &= |-6+4| \\ &= |-2| \\ &= 2\end{aligned}$$

11.
$$\begin{aligned}3x^2 - xy &= 3(5)^2 - (5)(10) \\ &= 3(25) - 50 \\ &= 75 - 50 \\ &= 25\end{aligned}$$

12.
$$\begin{aligned}\frac{-|y-2x|}{|xy|} &= \frac{-|-1-2(3)|}{|3(-1)|} \\ &= \frac{-|-1-6|}{|-3|} \\ &= \frac{-|-7|}{3} \\ &= -\frac{7}{3}\end{aligned}$$

13.
b; since -4 is used as an exponent

14.
-2.2; degree 5

15.
14; degree 6

16.
$$\begin{aligned}&3xy+2x+3y+2-(1-y-x-xy)\\&=3xy+2x+3y+2-1+y+x+xy\\&=4xy+3x+4y+1\end{aligned}$$

17.
$$\begin{aligned}&(a+2)(3a^2-a+5)\\&=a(3a^2-a+5)+2(3a^2-a+5)\\&=3a^3-a^2+5a+6a^2-2a+10\\&=3a^3+5a^2+3a+10\end{aligned}$$

18.
$$\begin{aligned}&8a^3b^5-12a^5b^2+16a^2b\\&=4a^2b(2ab^4-3a^3b+4)\end{aligned}$$

19.
$$\begin{aligned}4-9x^2&=2^2-(3x)^2\\&=(2+3x)(2-3x)\end{aligned}$$

20.
$$\begin{aligned}\frac{m^4}{3n^2}\div\left(\frac{m^2}{9n}\cdot\frac{n}{2m^3}\right)&=\frac{m^4}{3n^2}\div\frac{1}{18m}\\&=\frac{\frac{m^4}{3n^2}}{\frac{1}{18m}}\cdot\frac{(18mn^2)}{(18mn^2)}\\&=\frac{6m^5}{n^2}\end{aligned}$$

21.
$$\begin{aligned}\frac{16-x^2}{x^2-3x-4}\cdot\frac{x-1}{x+4}&=\frac{(4+x)(4-x)(x-1)}{(x-4)(x+1)(x+4)}\\&=\frac{-1(x-4)(x-1)}{(x-4)(x+1)}\\&=\frac{-(x-1)}{x+1}\\&\text{or }\frac{1-x}{x+1}\end{aligned}$$

22.
$$-\frac{1}{2x^2},\ \frac{2}{4x^2-4},\ \frac{3}{x-2}$$
$$-\frac{1}{2x^2},\ \frac{2}{4(x+1)(x-1)},\ \frac{3}{x-2}$$
LCD: $4x^2(x+1)(x-1)(x-2)$

23.

$$\frac{2x}{x^2-9}+\frac{5}{3x+9}=\frac{2x}{(x+3)(x-3)}+\frac{5}{3(x+3)}$$
$$=\frac{2x}{(x+3)(x-3)}\cdot\frac{(3)}{(3)}+\frac{5}{3(x+3)}\cdot\frac{(x-3)}{(x-3)}$$
$$=\frac{6x}{3(x+3)(x-3)}+\frac{5x-15}{3(x+3)(x-3)}$$
$$=\frac{6x+5x-15}{3(x+3)(x-3)}$$
$$=\frac{11x-15}{3(x+3)(x-3)}$$

24.

$$\frac{2-\frac{4}{x+1}}{x-1}=\frac{2-\frac{4}{x+1}}{x-1}\cdot\frac{(x+1)}{(x+1)}$$
$$=\frac{2(x+1)-4}{(x-1)(x+1)}$$
$$=\frac{2x+2-4}{(x-1)(x+1)}$$
$$=\frac{2x-2}{(x-1)(x+1)}$$
$$=\frac{2(x-1)}{(x-1)(x+1)}$$
$$=\frac{2}{x+1}$$

25.

$$\left(\frac{x^{\frac{7}{2}}}{x^{\frac{2}{3}}}\right)^{-6}=\frac{\left(x^{\frac{7}{2}}\right)^{-6}}{\left(x^{\frac{2}{3}}\right)^{-6}}$$
$$=\frac{x^{-21}}{x^{-4}}=x^{-21-(-4)}=x^{-17}=\frac{1}{x^{17}}$$

26.

$$\frac{y^{2n}}{y^{n-1}}=y^{2n-(n-1)}=y^{2n-n+1}=y^{n+1}$$

27.

$$\frac{-1}{(x-1)^0}=\frac{-1}{1}=-1$$

28.

$$\left(2a^2b^{-1}\right)^2 = 2^2\left(a^2\right)^2\left(b^{-1}\right)^2 = 4a^4b^{-2} = \frac{4a^4}{b^2}$$

29.

$$\begin{aligned} 3\sqrt[3]{24} - 2\sqrt[3]{81} &= 3\sqrt[3]{8\cdot 3} - 2\sqrt[3]{27\cdot 3} \\ &= 3\sqrt[3]{8}\sqrt[3]{3} - 2\sqrt[3]{27}\sqrt[3]{3} \\ &= 6\sqrt[3]{3} - 6\sqrt[3]{3} \\ &= (6-6)\sqrt[3]{3} \\ &= 0\left(\sqrt[3]{3}\right) \\ &= 0 \end{aligned}$$

30.

$$\begin{aligned} \left(\sqrt{7}-5\right)^2 &= \left(\sqrt{7}-5\right)\left(\sqrt{7}-5\right) \\ &= 7 - 5\sqrt{7} - 5\sqrt{7} + 25 \\ &= 32 - 10\sqrt{7} \end{aligned}$$

31.

$$\begin{aligned} \frac{1}{2}\sqrt{\frac{xy}{4}} - \sqrt{9xy} &= \frac{1}{2}\cdot\frac{\sqrt{xy}}{\sqrt{4}} - \sqrt{9}\sqrt{xy} \\ &= \frac{1}{2}\cdot\frac{\sqrt{xy}}{2} - 3\sqrt{xy} \\ &= \left(\frac{1}{4} - 3\right)\sqrt{xy} \\ &= -\frac{11}{4}\sqrt{xy} \end{aligned}$$

32.

$$\begin{aligned} 2 - x &\ge 0 \\ -x &\ge -2 \\ x &\le 2 \end{aligned}$$

33.

$$\begin{aligned} &(2-i)+(-3+i) \\ &= \left[2+(-3)\right]+(-i+i) \\ &= -1+0i \end{aligned}$$

34.

$$\begin{aligned} (5+2i)(2-3i) &= 10 - 15i + 4i - 6i^2 \\ &= 10 - 11i - 6(-1) \\ &= 16 - 11i \end{aligned}$$

35.

$$\begin{aligned} \frac{5+2i}{2-i} &= \frac{5+2i}{2-i}\cdot\frac{2+i}{2+i} \\ &= \frac{10+5i+4i+2i^2}{2^2+1^2} \\ &= \frac{10+9i+2(-1)}{4+1} \\ &= \frac{8+9i}{5} \\ &= \frac{8}{5}+\frac{9}{5}i \end{aligned}$$

CHAPTER 1 ADDITIONAL PRACTICE EXERCISES

In Exercises 1-4, perform the indicated operations.

1.
$$(7a^3b^2 - 6a^2b^3 + 4a - b) - (6a^3b^3 + 5a^2b^3 - 3a - 6b)$$

2.
$$5x(3x-1)^2$$

3.
$$\frac{5(a-3)^2}{25(a^2-9)} \cdot \frac{4(a+3)}{a^2-a-6}$$

4.
$$3+\frac{y}{2+\frac{1}{1+y}}$$

In Exercises 5-7, factor each expression.

5.
$$16a^2 - 121b^2$$

6.
$$3x^2 + 4xy - 6x - 8y$$

7.
$$-2x^3 - 250$$

In Exercises 8-10, simplify and express the answers using only positive exponents. All variables represent positive numbers.

8.
$$(5x^3y^{-2})^{-3}$$

9.
$$\sqrt[3]{56a^4b^7c}$$

10.
$$\frac{a}{\sqrt{a}-b}$$

CHAPTER 1 PRACTICE TEST

1. Find $\overline{PQ}$ if the coordinates of P and Q are -23 and 12, respectively

2. Find the value of $|16-25|-|1-7|$.

3. Evaluate $\frac{-|3a-2b|}{5|a|}$ when $a=-4$ and $b=7$.

4. Find the LCD of
$\frac{a}{a^2-2a-15}, \frac{-7}{(a+3)^2}, \frac{4}{a+1}$

In Exercises 5-13, perform the indicated operations.

5. $(-7x+5y-11xy)-(8y-4x+13xy)$

6. $(c-3)(c^2+6c-2)$

7. $\frac{x^5}{4y^3} \div \left(\frac{x^6}{2y} \cdot \frac{6y^4}{5x}\right)$

8. $\frac{x+2}{x^2+2x-3} \cdot \frac{x^2-2x-15}{x^2-4}$

9. $\frac{5}{x+1}+\frac{3}{x-1}$

10. $\frac{a}{a^2-5a+6}-\frac{2}{a^2-4}$

11. $1-\frac{7}{x^2-9}$

12. $\dfrac{\frac{6}{x-3}-\frac{1}{x-1}}{x-3}$

13. $5-\dfrac{2}{3-\frac{1}{1+x}}$

In Exercises 14-17, factor completely.

14.
$16x^3 - 54$

15.
$2ac - 3ad + 2bc - 3bd$

16.
$8x^2 + 37xy - 15y^2$

17.
$9(x-2)(y+3) - 6(y+3)$

18.
Simplify $\left(64x^{-3}y^{6}\right)^{-\frac{2}{3}}$

19.
Rationalize the denominator: $\dfrac{\sqrt{3}-2}{\sqrt{5}+\sqrt{6}}$

20.
Divide: $\dfrac{2+3i}{7-i}$

CHAPTER 2 SECTION 1

SECTION 2.1 PROGRESS CHECK

1a). (Page 80)

$$-\frac{2}{3}(x-5)=\frac{3}{2}(x+1)$$
$$\left[-\frac{2}{3}(x-5)\right](6)=\left[\frac{3}{2}(x+1)\right](6)$$
$$-4(x-5)=9(x+1)$$
$$-4x+20=9x+9$$
$$11=13x$$
$$\frac{11}{13}=x$$

1b).

$$\frac{1}{3}x+2-3\left(\frac{x}{2}+4\right)=2\left(\frac{x}{4}-1\right)$$
$$\left[\frac{1}{3}x+2-3\left(\frac{x}{2}+4\right)\right](12)=\left[2\left(\frac{x}{4}-1\right)\right](12)$$
$$4x+24-3(6x+48)=2(3x-12)$$
$$4x+24-18x-144=6x-24$$
$$-14x-120=6x-24$$
$$-20x=96$$
$$x=-\frac{24}{5}$$

2a) (Page 82)

$$\frac{3}{x}-1=\frac{1}{2}-\frac{6}{x}$$
$$\left(\frac{3}{x}-1\right)(2x)=\left(\frac{1}{2}-\frac{6}{x}\right)(2x)$$
$$6-2x=x-12$$
$$-3x=-18$$
$$x=6$$

2b).

$$-\frac{2x}{x+1}=1+\frac{2}{x+1}$$
$$\left(-\frac{2x}{x+1}\right)(x+1)=\left(1+\frac{2}{x+1}\right)(x+1)$$
$$-2x=x+1+2$$
$$-3x=3$$
$$x=-1$$

Checking, $x=-1$ causes a denominator of zero, thus the equation has no solution.

EXERCISE SET 2.1

1.

$$\begin{aligned} 2(-5)+3 &= -7 \\ -10+3 &= -7 \\ -7 &= -7 \end{aligned}$$

T

3.

$$\begin{aligned} k\left(\frac{6}{4-k}\right)+6 &= 4\left(\frac{6}{4-k}\right) \\ \left[k\left(\frac{6}{4-k}\right)+6\right](4-k) &= \left[4\left(\frac{6}{4-k}\right)\right](4-k) \\ 6k+6(4-k) &= 24 \\ 6k+24-6k &= 24 \\ 24 &= 24 \end{aligned}$$

T

5.

$$\begin{aligned} 3x+5 &= -1 \\ 3x &= -6 \\ x &= -2 \end{aligned}$$

7.

$$\begin{aligned} 2 &= 3x+4 \\ -2 &= 3x \\ -\frac{2}{3} &= x \end{aligned}$$

9.

$$\begin{aligned} \frac{3}{2}t-2 &= 7 \\ \frac{3}{2}t &= 9 \\ t &= 6 \end{aligned}$$

11.

$$\begin{aligned} 0 &= -\frac{1}{2}a-\frac{2}{3} \\ \frac{1}{2}a &= -\frac{2}{3} \\ a &= -\frac{4}{3} \end{aligned}$$

13.

$$\begin{aligned} -5x+8 &= 3x-4 \\ -8x &= -12 \\ x &= \frac{3}{2} \end{aligned}$$

15.

$$\begin{aligned} -2x+6 &= -5x-4 \\ 3x &= -10 \\ x &= -\frac{10}{3} \end{aligned}$$

17.

$$\begin{aligned} 2(3b+1) &= 3b-4 \\ 6b+2 &= 3b-4 \\ 3b &= -6 \\ b &= -2 \end{aligned}$$

19.

$$\begin{aligned} 4(x-1) &= 2(x+3) \\ 4x-4 &= 2x+6 \\ 2x &= 10 \\ x &= 5 \end{aligned}$$

21.

$$\begin{aligned} 2(x+4)-1 &= 0 \\ 2x+8-1 &= 0 \\ 2x &= -7 \\ x &= -\frac{7}{2} \end{aligned}$$

23.

$$\begin{aligned} -4(2x+1)-(x-2) &= -11 \\ -8x-4-x+2 &= -11 \\ -9x-2 &= -11 \\ -9x &= -9 \\ x &= 1 \end{aligned}$$

25.

$$\begin{aligned} kx+8 &= 5x \\ 8 &= 5x-kx \\ 8 &= x(5-k) \\ \frac{8}{5-k} &= x, \ k \neq 5 \end{aligned}$$

27.

$$\begin{aligned} 2-k+5(x-1) &= 3 \\ 2-k+5x-5 &= 3 \\ 5x &= 6+k \\ x &= \frac{6+k}{5} \end{aligned}$$

29.

$$\begin{aligned} \frac{x}{2} &= \frac{5}{3} \\ \left(\frac{x}{2}\right)(6) &= \left(\frac{5}{3}\right)(6) \\ 3x &= 10 \\ x &= \frac{10}{3} \end{aligned}$$

31.

$$\begin{aligned} \frac{2}{x}+1 &= \frac{3}{x} \\ \left(\frac{2}{x}+1\right)(x) &= \left(\frac{3}{x}\right)(x) \\ 2+x &= 3 \\ x &= 1 \end{aligned}$$

33.

$$\frac{2y-3}{y+3}=\frac{5}{7}$$

$$\left(\frac{2y-3}{y+3}\right)[7(y+3)]=\left(\frac{5}{7}\right)[7(y+3)]$$

$$7(2y-3)=5(y+3)$$

$$14y-21=5y+15$$

$$9y=36$$

$$y=4$$

35.

$$\frac{1}{x-2}+\frac{1}{2}=\frac{2}{x-2}$$

$$\left(\frac{1}{x-2}+\frac{1}{2}\right)[(2)(x-2)]=\left(\frac{2}{x-2}\right)[2(x-2)]$$

$$2+x-2=4$$

$$x=4$$

37.

$$\frac{2}{x-2}+\frac{2}{x^2-4}=\frac{3}{x+2}$$

$$\frac{2}{x-2}+\frac{2}{(x+2)(x-2)}=\frac{3}{x+2}$$

$$\left[\frac{2}{x-2}+\frac{2}{(x+2)(x-2)}\right][(x+2)(x-2)]=\left(\frac{3}{x+2}\right)[(x+2)(x-2)]$$

$$2(x+2)+2=3(x-2)$$

$$2x+4+2=3x-6$$

$$12=x$$

39.

$$\frac{x}{x-1}-1=\frac{3}{x+1}$$

$$\left[\frac{x}{x-1}-1\right][(x-1)(x+1)]=\left(\frac{3}{x+1}\right)[(x-1)(x+1)]$$

$$x(x+1)-(x-1)(x+1)=3(x-1)$$

$$x^2+x-(x^2-1)=3x-3$$

$$x^2+x-x^2+1=3x-3$$

$$x+1=3x-3$$

$$4=2x$$

$$2=x$$

41.

$$\frac{4}{b}-\frac{1}{b+3}=\frac{3b+2}{b^2+2b-3}$$

$$\frac{4}{b}-\frac{1}{b+3}=\frac{3b+2}{(b+3)(b-1)}$$

$$\left[\frac{4}{b}-\frac{1}{b+3}\right][b(b+3)(b-1)]=\left[\frac{3b+2}{(b+3)(b-1)}\right][b(b+3)(b-1)]$$

$$4(b^2+2b-3)-(b^2-b)=3b^2+2b$$

$$4b^2+8b-12-b^2+b=3b^2+2b$$

$$3b^2+9b-12=3b^2+2b$$

$$-12=-7b$$

$$\frac{12}{7}=b$$

43.

$$\frac{3r+1}{r+3}+2=\frac{5r-2}{r+3}$$

$$\left[\frac{3r+1}{r+3}+2\right](r+3)=\left(\frac{5r-2}{r+3}\right)(r+3)$$

$$3r+1+2r+6=5r-2$$

$$5r+7=5r-2$$

$$7=-2$$

No solution.

45.

$$x^2+x-2=(x+2)(x-1)$$

$$(x+2)(x-1)=(x+2)(x-1)$$

I

47.

$$2x+1=3x-1$$

$$2=x$$

C

49.

$2x-3=5$ $\quad 2x=8 \quad x=4 \quad$ T

$2x=8 \quad x=4$

$x=4$

51.

$x(x-1)=5x \quad x-1=5 \quad x=6 \quad$ F

$x^2-x=5x \quad$ Linear

Quadratic

53.

$$3(x^2+2x+1)=-6$$
$$x^2+2x+1=-2$$

$$x^2+2x+1=-2$$

$$(x+1)^2=-2$$
$$x^2+2x+1=-2$$

T

55a).

$$N=0.22\overline{2}$$
$$10N=2.22\overline{2}$$
$$-\ N=0.22\overline{2}$$
$$9N=2$$
$$N=\frac{2}{9}$$

55b).

$$N=\ \ 0.123\overline{123}$$
$$1000N=123.123\overline{123}$$
$$-\ \ N=\ \ 0.123\overline{123}$$
$$999N=123$$
$$N=\frac{123}{999}=\frac{41}{333}$$

55c).

$$N=1.35\overline{35}$$
$$100N=135.35\overline{35}$$
$$-\ \ N=\ \ 1.35\overline{35}$$
$$99N=134$$
$$N=\frac{134}{99}$$

55d).

$$N=0.9\overline{9}$$
$$10N=9.9\overline{9}$$
$$-\ N=0.9\overline{9}$$
$$9N=9$$
$$N=1$$

57a).

$$\frac{w-c}{w-d}=\frac{c^2}{d^2}$$
$$d^2(w-c)=c^2(w-d)$$
$$wd^2-cd^2=wc^2-c^2d$$
$$wd^2-wc^2=cd^2-c^2d$$
$$w(d^2-c^2)=cd(d-c)$$
$$w=\frac{cd(d-c)}{(d-c)(d+c)}$$
$$w=\frac{cd}{d+c}$$

57b).

$$a^2=\frac{a+c}{x}+c^2$$
$$a^2-c^2=\frac{a+c}{x}$$
$$(a^2-c^2)(x)=\left(\frac{a+c}{x}\right)(x)$$
$$(a^2-c^2)(x)=a+c$$
$$x=\frac{a+c}{(a+c)(a-c)}$$
$$x=\frac{1}{a-c}$$

57c).

$$(a-y)(y+b)-c(y+c)=(c-y)(y+c)+ab$$
$$ay+ab-y^2-by-cy-c^2=yc+c^2-y^2-cy+ab$$
$$(ay-y^2-by-cy)+ab-c^2=c^2-y^2+ab$$
$$ay-by-cy=2c^2$$
$$y(a-b-c)=2c^2$$
$$y=\frac{2c^2}{a-b-c}$$

59.

$$T=1+\cfrac{1}{1+\cfrac{1}{1+\cfrac{1}{T}}}$$

$$T=1+\cfrac{1}{1+\cfrac{1}{1+\cfrac{1}{T}}\cdot\cfrac{T}{T}}$$

$$T=1+\cfrac{1}{1+\cfrac{T}{T+1}}$$

$$T=1+\cfrac{1}{1+\cfrac{T}{T+1}}\cdot\frac{T+1}{T+1}$$

$$T=1+\frac{T+1}{T+1+T}$$

$$T=1+\frac{T+1}{2T+1}$$

$$T=\frac{2T+1+T+1}{2T+1}$$

$$T=\frac{3T+2}{2T+1}$$

59. (continued)

Substituting: $\frac{1+\sqrt{5}}{2}=\frac{3\frac{(1+\sqrt{5})}{2}+2}{2\left(\frac{1+\sqrt{5}}{2}\right)+1}$?

$$=\left[\frac{3\left(\frac{1+\sqrt{5}}{2}\right)+2}{2\left(\frac{1+\sqrt{5}}{2}\right)+1}\right]\left(\frac{2}{2}\right) \quad ?$$

$$=\frac{3+3\sqrt{5}+4}{2+2\sqrt{5}+2} \quad ?$$

$$=\frac{7+3\sqrt{5}}{4+2\sqrt{5}} \quad ?$$

$$=\frac{7+3\sqrt{5}}{4+2\sqrt{5}}\cdot\frac{4-2\sqrt{5}}{4-2\sqrt{5}} \quad ?$$

$$=\frac{28-2\sqrt{5}-30}{16-20} \quad ?$$

$$=\frac{-2-2\sqrt{5}}{-4} \quad ?$$

$$\frac{1+\sqrt{5}}{2}=\frac{1+\sqrt{5}}{2}$$

CHAPTER 2 SECTION 2

EXERCISE SET 2.2

1.

Number of red chips : n

Number of blue chips : $3+2n$

3.

$6n-5=26$

5.

	current age	age 4 years ago
Fred	x	$x-4$
John	$x+12$	$(x+12)-4$

$$(x+12)-4=2(x-4)$$
$$x+8=2x-8$$
$$16=x$$
$$x+12=28$$

Fred : 16 years old

John : 28 years old

7.

1st number : x

2nd number : $x+1$

3rd number : $x+2$

$$(x)+(x+1)+(x+2)=21$$
$$3x+3=21$$
$$3x=18$$
$$x=6$$
$$x+1=7$$
$$x+2=8$$

The three consecutive numbers are 6, 7 and 8.

9.

Sunday's temperature : x

$$\frac{90+82+x}{3}=80$$
$$172+x=240$$
$$x=68$$

The temperature must be 68° F.

11.

Length of 1st piece : x

Length of 2nd piece : $x+4$

$$(x)+(x+4)=12$$
$$2x+4=12$$
$$2x=8$$
$$x=4$$
$$x+4=8$$

The lengths are 4 m and 8 m.

13.

Number of nickels : n

Number of dimes : $5+2n$

$$\begin{aligned} 5n+10(5+2n) &= 300 \\ 5n+50+20n &= 300 \\ 25n &= 250 \\ n &= 10 \\ 5+2n &= 25 \end{aligned}$$

There are 10 nickels and 25 dimes.

15.

	cost per ticket	·	# of tickets	=	revenue
adult	3		x		$3x$
child	1.50		$700-x$		$1.50(700-x)$

$$\begin{aligned} 3x+1.50(700-x) &= 1650 \\ 3x+1050-1.50x &= 1650 \\ 1.5x &= 600 \\ x &= 400 \\ 700-x &= 300 \end{aligned}$$

There are 400 adult and 300 children's tickets sold.

17.

Number of \$6 tickets : x

Number of \$5 tickets : $2x$

Number of \$3 tickets : $1+3x$

$$\begin{aligned} 6x+5(2x)+3(1+3x) &= 503 \\ 6x+10x+3+9x &= 503 \\ 25x &= 500 \\ x &= 20 \\ 2x &= 40 \\ 1+3x &= 61 \end{aligned}$$

There are 20 \$6 tickets,
40 \$5 tickets
and 61 \$3 tickets.

19.

Three - speed bicycle sales : x

Ten - speed bicycle sales : $16000-x$

$$\begin{aligned} 0.11(x)+0.22(16000-x) &= 0.19(16000) \\ 100[0.11x+0.22(16000-x)] &= 100[0.19(16000)] \\ 11x+22(16000-x) &= 19(16000) \\ 11x+352000-22x &= 304000 \\ -11x &= -48000 \\ x &= 4363.64 \\ 16000-x &= 11636.36 \end{aligned}$$

The three - speed bicycle sales are \$4,363.64 and the ten - speed bicycle sales are \$11,636.36.

21.

Amount returned : x

$$0.08(12000)+0.08(12000-x)(2)=1760$$
$$960+1920-0.16x=1760$$
$$-0.16x=-1120$$
$$x=7000$$

\$7,000 was returned after the first year.

23.

	rate	·	time	=	distance
1st truck	50		$x+2$		$50(x+2)$
2nd truck	55		x		$55x$

$$50(x+2)=55x$$
$$50x+100=55x$$
$$100=5x$$
$$20=x$$

It takes 20 hours for the second truck to overtake the first truck.

25.

	rate	·	time	=	distance
1st prof	x		6.5		$6.5x$
2nd prof	$x+4$		6.5		$6.5(x+4)$

$$6.5x+6.5(x+4)=676$$
$$6.5x+6.5x+26=676$$
$$13x=650$$
$$x=50$$
$$x+4=54$$

Their speeds are 50 mph and 54 mph.

27.

	time	·	rate	=	distance
express train	4		$2x$		$4(2x)$
local train	4		x		$4x$

$$4(2x)+4x=480$$
$$8x+4x=480$$
$$12x=480$$
$$x=40$$
$$2x=80$$

The express train travels 80 kph and the local train travels 40 kph.

29.

Ounces of Ceylon tea : x

Ounces of Formosa tea: $8-x$

$$1.50x+2.00(8-x)=1.85(8)$$
$$1.50x+16-2x=14.8$$
$$-0.5x=-1.2$$
$$x=2.4$$
$$8-x=5.6$$

The mixture should contain 2.4 oz of Ceylon tea and 5.6 oz of Formosa tea.

31.

27 gallons of water + 9 gallons of acetic acid = 36 gallons of solution that is 25% acetic acid.

$$\overbrace{\boxed{36\text{ gal}}}^{25\%}-\overbrace{\boxed{x}}^{0\%}=\overbrace{\boxed{36-x}}^{40\%}$$

$$0.25(36)-0(x)=0.40(36-x)$$
$$9=14.4-0.40x$$
$$-5.4=-0.40x$$
$$13.5=x$$

13.5 gal of water must be evaporated.

33.

1st number : x

2nd number : $3x$

$$\frac{1}{x}-\frac{1}{3x}=8$$
$$3x\left[\frac{1}{x}-\frac{1}{3x}\right]=3x(8)$$
$$3-1=24x$$
$$2=24x$$
$$\frac{1}{12}=x$$
$$\frac{1}{4}=3x$$

$$\frac{1}{3x}-\frac{1}{x}=8$$
$$3x\left[\frac{1}{3x}-\frac{1}{x}\right]=3x(8)$$
$$1-3=24x$$
$$-2=24x$$
$$-\frac{1}{12}=x$$
$$-\frac{1}{4}=3x$$

The numbers are $\frac{1}{12}$ and $\frac{1}{4}$ or $-\frac{1}{12}$ and $-\frac{1}{4}$.

35.

	rate	·	time	=	work done
Computer A	$\frac{1}{4}$		x		$\frac{1}{4}x$
Computer B	$\frac{1}{6}$		x		$\frac{1}{6}x$

$$\frac{1}{4}x+\frac{1}{6}x=1$$
$$12\left[\frac{1}{4}x+\frac{1}{6}x\right]=12(1)$$
$$3x+2x=12$$
$$5x=12$$
$$x=\frac{12}{5}$$

It takes $\frac{12}{5}=2.4$ hours to complete the work.

37.

Time for senior editor alone : x

Time for junior editor alone : $2x$

	rate	·	time	=	work done
Jr. Editor	$\frac{1}{2x}$		3		$\frac{3}{2x}$
Sr. Editor	$\frac{1}{x}$		3		$\frac{3}{x}$

$$\frac{3}{2x}+\frac{3}{x}=1$$
$$2x\left[\frac{3}{2x}+\frac{3}{x}\right]=2x(1)$$
$$3+6=2x$$
$$9=2x$$
$$\frac{9}{2}=x$$
$$9=2x$$

It would take the senior editor $\frac{9}{2}=4\frac{1}{2}$ days and the junior editor 9 days.

39.

Time for Press B alone : x

	rate	·	time	=	workdone
Press A	$\frac{1}{8}$		6		$\frac{3}{4}$
Press B	$\frac{1}{x}$		2		$\frac{2}{x}$

$$\frac{3}{4}+\frac{2}{x}=1$$

$$4x\left[\frac{3}{4}+\frac{2}{x}\right]=4x(1)$$

$$3x+8=4x$$

$$8=x$$

It would take Press B 8 hours to do the job alone.

41.

Wind's speed : 20 mph

Airplane's speed in still air : x

	distance	÷	rate	=	time
against wind	300		$x-20$		$\frac{300}{x-20}$
with wind	400		$x+20$		$\frac{400}{x+20}$

$$\frac{300}{x-20}=\frac{400}{x+20}$$

$$300(x+20)=400(x-20)$$

$$300x+6000=400x-8000$$

$$-100x=-14000$$

$$x=140$$

The airplane's speed in still air is 140 mph.

43.

$$C=2\pi r$$

$$\frac{1}{2\pi}(C)=\frac{1}{2\pi}(2\pi r)$$

$$\frac{C}{2\pi}=r$$

45.

$$F=\frac{9}{5}C+32$$

$$F-32=\frac{9}{5}C$$

$$\frac{5}{9}(F-32)=C$$

47.

$$A=\frac{1}{2}h(b+b')$$
$$\frac{2}{h}(\text{A})=\frac{2}{h}\left[\frac{1}{2}h(b+b')\right]$$
$$\frac{2A}{h}=b+b'$$
$$\frac{2A}{h}-b'=b$$

49.

$$\frac{1}{f}=\frac{1}{f_1}+\frac{1}{f_2}$$
$$\frac{1}{f}-\frac{1}{f_1}=\frac{1}{f_2}$$
$$\frac{f_1-f}{ff_1}=\frac{1}{f_2}$$

Take reciprocal of both sides.

$$\frac{ff_1}{f_1-f}=f_2$$

51.

$$S=\frac{a-rL}{L-r}$$
$$S(L-r)=a-rL$$
$$SL-Sr=a-rL$$
$$SL+rL=a+Sr$$
$$L(S+r)=a+Sr$$
$$L=\frac{a+Sr}{S+r}$$

53a).
$2(r+s)$

53b).
$0.05|r-s|$

53c).
$2s-5$

53d).
$\frac{r}{s}$

53e).
r^2+s^2

53f).
$\frac{r+s}{2}$

53g).
$6r-4s$

55.

1st consecutive, even number : x

2nd consecutive, even number: $x+2$

3rd consecutive, even number: $x+4$

$$\begin{aligned} 2x+3(x+2) &= 4(x+4) \\ 2x+3x+6 &= 4x+16 \\ 5x+6 &= 4x+16 \\ x &= 10 \\ x+2 &= 12 \\ x+4 &= 14 \end{aligned}$$

The numbers are 10, 12 and 14.

57.

$$\overbrace{\boxed{10-x}}^{30\%\text{ antifreeze}} + \overbrace{\boxed{x}}^{100\%\text{ antifreeze}} = \overbrace{\boxed{10}}^{2(30\%)\text{ antifreeze}}$$

$$\begin{aligned} 0.30(10-x)+1(x) &= 0.60(10) \\ 3-0.30x+x &= 6 \\ 0.7x &= 3 \\ x &= 4.29 \end{aligned}$$

4.29 qts should be drained and replaced.

59.

	r	t	d
bicyclist A	20	t	$20t$
bicyclist B	30	t	$30t$

$$20t + 30t = 100$$
$$50t = 100$$
$$t = 2$$

Bicyclists A and B will meet in 2 hours. The bird is in the air flying at 40 miles per hour for 2 hours. Hence the bird travels $(40)(2) = 80$ miles.

61.

	distance ÷	rate =	time
Stephan	$\frac{1}{3}d$	20	$\frac{\frac{1}{3}d}{20} = \frac{d}{60}^*$
	$\frac{2}{3}d$	35	$\frac{\frac{2}{3}d}{35} = \frac{2d}{105}^*$
Enrique	d	30	$\frac{d}{30}$

*Stephan's total time $= \frac{d}{60} + \frac{2d}{105} = \frac{d}{28}$

$\frac{d}{30} < \frac{d}{28}$, hence Enrique finished before Stephen.

63.

$$Wf=\left(\frac{W}{k}-1\right)\left(\frac{1}{k}\right)$$

$$Wf=\frac{W}{k^2}-\frac{1}{k}$$

$$\frac{1}{k}=\frac{W}{k^2}-Wf$$

$$\frac{1}{k}=W\left(\frac{1}{k^2}-f\right)$$

$$\frac{1}{k}=W\left(\frac{1-fk^2}{k^2}\right)$$

$$\frac{k^2}{k\left(1-fk^2\right)}=W$$

$$\frac{k}{1-fk^2}=W$$

65.

$$\frac{E}{c}=\frac{R+r}{r}$$

$$Er=cR+cr$$

$$Er-cr=cR$$

$$r(E-c)=cR$$

$$r=\frac{cR}{E-c}$$

CHAPTER 2 SECTION 3

SECTION 2.3 PROGRESS CHECK

1a). (Page 98)

$$4x^2 - x = 0$$
$$x(4x-1) = 0$$
$$x = 0 \text{ or } 4x = 0$$
$$x = \frac{1}{4}$$

1b).

$$3x^2 - 11x - 4 = 0$$
$$(3x+1)(x-4) = 0$$
$$3x+1 = 0 \text{ or } x-4 = 0$$
$$x = -\frac{1}{3} \text{ or } x = 4$$

2a). (Page 101)

$$5x^2 + 13 = 0$$
$$x^2 = -\frac{13}{5}$$
$$x = \pm\sqrt{-\frac{13}{5}}$$
$$x = \pm\frac{\sqrt{13}}{\sqrt{5}}i$$
$$x = \pm\frac{\sqrt{65}}{5}i$$

2b).

$$(2x-7)^2 - 5 = 0$$
$$(2x-7)^2 = 5$$
$$2x - 7 = \pm\sqrt{5}$$
$$2x = 7 \pm \sqrt{5}$$
$$x = \frac{7 \pm \sqrt{5}}{2}$$

3a). (Page 103)

$$x^2 - 3x + 2 = 0$$
$$x^2 - 3x \quad = -2$$
$$\left[\frac{1}{2}(-3)\right]^2 = \frac{9}{4}$$
$$x^2 - 3x + \frac{9}{4} = -2 + \frac{9}{4}$$
$$\left(x - \frac{3}{2}\right)^2 = \frac{1}{4}$$
$$x - \frac{3}{2} = \pm\frac{1}{2}$$
$$x = \frac{3}{2} \pm \frac{1}{2}$$
$$x = \frac{3+1}{2} = 2 \text{ or } x = \frac{3-1}{2} = 1$$

3b).

$$3x^2 - 4x + 2 = 0$$
$$x^2 - \frac{4}{3}x \quad = -\frac{2}{3}$$
$$\left[\frac{1}{2}\left(-\frac{4}{3}\right)\right]^2 \qquad x^2 - \frac{4}{3}x + \frac{4}{9} = -\frac{2}{3} + \frac{4}{9}$$
$$\left(x - \frac{2}{3}\right)^2 = -\frac{2}{9}$$
$$x - \frac{2}{3} = \pm\frac{\sqrt{2}}{3}i$$
$$x = \frac{2}{3} \pm \frac{\sqrt{2}}{3}i$$
$$x = \frac{2 \pm \sqrt{2}i}{3}$$

4a). (Page 105)

$$x^2 - 8x = -10$$
$$x^2 - 8x + 10 = 0$$

$$a = 1 \quad b = -8 \quad c = 10$$

$$x = \frac{-b \pm \sqrt{b^2 - 4ac}}{2a}$$
$$= \frac{-(-8) \pm \sqrt{(-8)^2 - 4(1)(10)}}{2(1)}$$
$$= \frac{8 \pm \sqrt{24}}{2}$$
$$= \frac{8 \pm 2\sqrt{6}}{2}$$
$$= 4 \pm \sqrt{6}$$

4b).

$$4x^2 - 2x + 1 = 0$$

$$a = 4 \quad b = -2 \quad c = 1$$

$$x = \frac{-b \pm \sqrt{b^2 - 4ac}}{2a}$$
$$= -\frac{(-2) \pm \sqrt{(-2)^2 - 4(4)(1)}}{2(4)}$$
$$= \frac{2 \pm \sqrt{-12}}{8}$$
$$= \frac{2 \pm 2\sqrt{3}i}{8}$$
$$= \frac{1 \pm \sqrt{3}i}{4}$$

5a). (Page 107)

$$4x^2 - 20x + 25 = 0$$
$$b^2 - 4ac = (-20)^2 - 4(4)(25) = 0$$

one real, double root

5b).

$$5x^2 - 6x = -2$$
$$5x^2 - 6x + 2 = 0$$
$$b^2 - 4ac = (-6)^2 - 4(5)(2) = -4 < 0$$

two complex roots

5c).

$$10x^2 = x + 2$$
$$10x^2 - x - 2 = 0$$
$$b^2 - 4ac = (-1)^2 - 4(10)(-2) = 81 > 0$$

(perfect square)

two real, rational roots

5d).

$$x^2 + x - 1 = 0$$
$$b^2 - 4ac = (1)^2 - 4(1)(-1) = 5 > 0$$

(not a perfect square)

two real, irrational roots

6. (Page 108)

$$x-\sqrt{1-x}=-5$$
$$x+5=\sqrt{1-x}$$
$$(x+5)^2=1-x$$
$$x^2+10x+25=1-x$$
$$x^2+11x+24=0$$
$$(x+8)(x+3)=0$$
$$x=-8 \quad x=-3$$

Check: $x=-8$

$$-8-\sqrt{1-(-8)}=-5 \quad ?$$
$$-8-\sqrt{9}=-5 \quad ?$$
$$-8-3=-5 \quad ?$$
$$-11\neq-5$$

Check: $x=-3$

$$-3-\sqrt{1-(-3)}=-5 \quad ?$$
$$-3-\sqrt{4}=-5 \quad ?$$
$$-3-2=-5 \quad ?$$
$$-5=-5$$

Solution: $x=-3$

7. (Page 109)

$$\sqrt{5x-1}-\sqrt{x+2}=1$$
$$\sqrt{5x-1}=1+\sqrt{x+2}$$
$$5x-1=\left(1+\sqrt{x+2}\right)^2$$
$$5x-1=1+2\sqrt{x+2}+x+2$$
$$5x-1=x+3+2\sqrt{x+2}$$
$$4x-4=2\sqrt{x+2}$$
$$2x-2=\sqrt{x+2}$$
$$(2x-2)^2=x+2$$
$$4x^2-8x+4=x+2$$
$$4x^2-9x+2=0$$
$$(x-2)(4x-1)=0$$
$$x=2 \qquad x=\frac{1}{4}$$

Check: $x=2$

$$\sqrt{5(2)-1}-\sqrt{2+2}=1 \quad ?$$
$$\sqrt{9}-\sqrt{4}=1 \quad ?$$
$$3-2=1 \quad ?$$
$$1=1$$

Check : $x=\frac{1}{4}$

$$\sqrt{5\left(\frac{1}{4}\right)-1}-\sqrt{\frac{1}{4}+2}=1 \quad ?$$
$$\sqrt{\frac{1}{4}}-\sqrt{\frac{9}{4}}=1 \quad ?$$
$$\frac{1}{2}-\frac{3}{2}=1 \quad ?$$
$$-1\neq1$$

Solution : $x=2$

8a). (Page 110)

$$3x^4 - 10x^2 - 8 = 0$$
let $u = x^2$
$$3u^2 - 10u - 8 = 0$$
$$(3u+2)(u-4) = 0$$

$$u = -\frac{2}{3} \qquad u = 4$$
$$x^2 = -\frac{2}{3} \qquad x^2 = 4$$
$$x = \pm\sqrt{\frac{2}{3}}i \qquad x = \pm 2$$
$$x = \pm\frac{\sqrt{6}}{3}i$$

8b).

$$4x^{\frac{2}{3}} + 7x^{\frac{1}{3}} - 2 = 0$$
Let $u = x^{\frac{1}{3}}$
$$4u^2 + 7u - 2 = 0$$
$$(u+2)(4u-1) = 0$$
$$u = -2 \qquad u = \frac{1}{4}$$
$$x^{\frac{1}{3}} = -2 \qquad x^{\frac{1}{3}} = \frac{1}{4}$$
$$x = -8 \qquad x = \frac{1}{64}$$

8c).

$$\frac{2}{x^2} + \frac{1}{x} - 10 = 0$$
let $u = \frac{1}{x}$
$$2u^2 + u - 10 = 0$$
$$(2u+5)(u-2) = 0$$
$$u = -\frac{5}{2} \qquad u = 2$$
$$\frac{1}{x} = -\frac{5}{2} \qquad \frac{1}{\mathrm{x}} = 2$$
$$x = -\frac{2}{5} \qquad x = \frac{1}{2}$$

8d).

$$\left(1+\frac{2}{x}\right)^2 - 8\left(1+\frac{2}{x}\right) + 15 = 0$$
let $u = 1 + \frac{2}{x}$
$$u^2 - 8u + 15 = 0$$
$$(u-5)(u-3) = 0$$
$$u = 5 \qquad u = 3$$
$$1 + \frac{2}{x} = 5 \qquad 1 + \frac{2}{x} = 3$$
$$x + 2 = 5x \qquad x + 2 = 3x$$
$$2 = 4x \qquad 2 = 2x$$
$$\frac{1}{2} = x \qquad 1 = x$$

EXERCISE SET 2.3

1.

$$x^2 - 3x + 2 = 0$$
$$(x-2)(x-1) = 0$$
$$x - 2 = 0 \text{ or } x - 1 = 0$$
$$x = 2 \text{ or } x = 1$$

3.

$$x^2 + x - 2 = 0$$
$$(x+2)(x-1) = 0$$
$$x + 2 = 0 \text{ or } x - 1 = 0$$
$$x = -2 \text{ or } x = 1$$

5.

$$x^2 + 6x = -8$$
$$x^2 + 6x + 8 = 0$$
$$(x+4)(x+2) = 0$$
$$x + 4 = 0 \text{ or } x + 2 = 0$$
$$x = -4 \text{ or } x = -2$$

7.

$$y^2 - 4y = 0$$
$$y(y-4) = 0$$
$$y = 0 \text{ or } y - 4 = 0$$
$$y = 0 \text{ or } y = 4$$

9.

$$2x^2 - 5x = -2$$
$$2x^2 - 5x + 2 = 0$$
$$(2x-1)(x-2) = 0$$
$$2x - 1 = 0 \text{ or } x - 2 = 0$$
$$x = \frac{1}{2} \text{ or } x = 2$$

11.

$$t^2 - 4 = 0$$
$$(t+2)(t-2) = 0$$
$$t + 2 = 0 \text{ or } t - 2 = 0$$
$$t = -2 \text{ or } t = 2$$

13.

$$6x^2 - 5x + 1 = 0$$
$$(2x-1)(3x-1) = 0$$
$$2x - 1 = 0 \text{ or } 3x - 1 = 0$$
$$x = \frac{1}{2} \text{ or } x = \frac{1}{3}$$

15.

$$3x^2 - 27 = 0$$
$$3x^2 = 27$$
$$x^2 = 9$$
$$x = \pm 3$$

17.

$$\begin{aligned} 5y^2 - 25 &= 0 \\ 5y^2 &= 25 \\ y^2 &= 5 \\ y &= \pm\sqrt{5} \end{aligned}$$

19.

$$\begin{aligned} (2r+5)^2 &= 8 \\ 2r+5 &= \pm\sqrt{8} \\ 2r &= -5 \pm 2\sqrt{2} \\ r &= \frac{-5 \pm 2\sqrt{2}}{2} \end{aligned}$$

21.

$$\begin{aligned} (3x-5)^2 - 8 &= 0 \\ (3x-5)^2 &= 8 \\ 3x-5 &= \pm\sqrt{8} \\ 3x &= 5 \pm 2\sqrt{2} \\ x &= \frac{5 \pm 2\sqrt{2}}{3} \end{aligned}$$

23.

$$\begin{aligned} 9x^2 + 64 &= 0 \\ 9x^2 &= -64 \\ x^2 &= -\frac{64}{9} \\ x &= \pm\sqrt{-\frac{64}{9}} \\ x &= \pm\frac{8}{3}i \end{aligned}$$

25.

$$\begin{aligned} x^2 - 2x &= 8 \\ \left[\frac{1}{2}(-2)\right]^2 &= 1 \\ x^2 - 2x + 1 &= 8 + 1 \\ (x-1)^2 &= 9 \\ x - 1 &= \pm 3 \\ x &= 1 \pm 3 \\ x &= 4 \text{ or } x = -2 \end{aligned}$$

27.

$$\begin{aligned} 2r^2 - 7r &= 4 \\ r^2 - \frac{7}{2}r &= 2 \\ \left[\frac{1}{2}\left(-\frac{7}{2}\right)\right]^2 &= \frac{49}{16} \\ r^2 - \frac{7}{2}r + \frac{49}{16} &= 2 + \frac{49}{16} \\ \left(r - \frac{7}{4}\right)^2 &= \frac{81}{16} \\ r - \frac{7}{4} &= \pm\frac{9}{4} \\ r &= \frac{7}{4} \pm \frac{9}{4} \\ r &= 4 \text{ or } r = -\frac{1}{2} \end{aligned}$$

29.

$$
\begin{aligned}
3x^2+8x &= 3\\
x^2+\frac{8}{3}x &= 1\\
\left[\frac{1}{2}\left(\frac{8}{3}\right)\right]^2 &= \frac{16}{9}\\
x^2+\frac{8}{3}x+\frac{16}{9} &= 1+\frac{16}{9}\\
\left(x+\frac{4}{3}\right)^2 &= \frac{25}{9}\\
x+\frac{4}{3} &= \pm\frac{5}{3}\\
x &= -\frac{4}{3}\pm\frac{5}{3}\\
x &= \frac{1}{3} \text{ or } x=-3
\end{aligned}
$$

31.

$$
\begin{aligned}
2y^2+2y &= -1\\
y^2+y &= -\frac{1}{2}\\
\left[\frac{1}{2}(1)\right]^2 &= \frac{1}{4}\\
y^2+y+\frac{1}{4} &= -\frac{1}{2}+\frac{1}{4}\\
\left(y+\frac{1}{2}\right)^2 &= -\frac{1}{4}\\
y+\frac{1}{2} &= \pm\sqrt{-\frac{1}{4}}\\
y &= -\frac{1}{2}\pm\frac{1}{2}i\\
y &= \frac{-1\pm i}{2}
\end{aligned}
$$

33.

$$
\begin{aligned}
4x^2-x &= 3\\
x^2-\frac{1}{4}x &= \frac{3}{4}\\
\left[\frac{1}{2}\left(-\frac{1}{4}\right)\right]^2 &= \frac{1}{64}\\
x^2-\frac{1}{4}x+\frac{1}{64} &= \frac{3}{4}+\frac{1}{64}\\
\left(x-\frac{1}{8}\right)^2 &= \frac{49}{64}\\
x-\frac{1}{8} &= \pm\frac{7}{8}\\
x &= \frac{1}{8}\pm\frac{7}{8}\\
x &= 1 \text{ or } x=-\frac{3}{4}
\end{aligned}
$$

35.

$$
\begin{aligned}
3x^2+2x &= -1\\
x^2+\frac{2}{3}x &= -\frac{1}{3}\\
\left[\frac{1}{2}\left(\frac{2}{3}\right)\right]^2 &= \frac{1}{9}\\
x^2+\frac{2}{3}x+\frac{1}{9} &= -\frac{1}{3}+\frac{1}{9}\\
\left(x+\frac{1}{3}\right)^2 &= -\frac{2}{9}\\
x+\frac{1}{3} &= \pm\sqrt{-\frac{2}{9}}\\
x &= -\frac{1}{3}\pm\frac{\sqrt{2}i}{3}\\
x &= \frac{-1\pm\sqrt{2}i}{3}
\end{aligned}
$$

37.

$$2x^2+3x=0$$

$$a=2 \quad b=3 \quad c=0$$

$$x=\frac{-b\pm\sqrt{b^2-4ac}}{2a}$$

$$=\frac{-3\pm\sqrt{3^2-4(2)(0)}}{2(2)}$$

$$=\frac{-3\pm\sqrt{9}}{4}$$

$$=\frac{-3\pm 3}{4}$$

$$x=0 \text{ or } x=-\frac{3}{2}$$

39.

$$5x^2-4x+3=0$$

$$a=5 \quad b=-4 \quad c=3$$

$$x=\frac{-b\pm\sqrt{b^2-4ac}}{2a}$$

$$=\frac{-(-4)\pm\sqrt{(-4)^2-4(5)(3)}}{2(5)}$$

$$=\frac{4\pm\sqrt{-44}}{10}$$

$$=\frac{4\pm 2\sqrt{11}i}{10}$$

$$=\frac{2\pm\sqrt{11}i}{5}$$

41.

$$5y^2-4y+5=0$$

$$a=5 \quad b=-4 \quad c=5$$

$$y=\frac{-b\pm\sqrt{b^2-4ac}}{2a}$$

$$=\frac{-(-4)\pm\sqrt{(-4)^2-4(5)(5)}}{2(5)}$$

$$=\frac{4\pm\sqrt{-84}}{10}$$

$$=\frac{4\pm 2\sqrt{21}i}{10}$$

$$=\frac{2\pm\sqrt{21}i}{5}$$

43.

$$3x^2+x-2=0$$

$$a=3 \quad b=1 \quad c=-2$$

$$x=\frac{-b\pm\sqrt{b^2-4ac}}{2a}$$

$$=\frac{-1\pm\sqrt{1^2-4(3)(2)}}{2(3)}$$

$$=\frac{-1\pm\sqrt{25}}{6}$$

$$=\frac{-1\pm 5}{6}$$

$$x=\frac{2}{3} \text{ or } x=-1$$

45.

$3y^2 - 4 = 0$

$a = 3 \quad b = 0 \quad c = -4$

$$\begin{aligned} y &= \frac{-b \pm \sqrt{b^2 - 4ac}}{2a} \\ &= \frac{-0 \pm \sqrt{0^2 - 4(3)(-4)}}{2(3)} \\ &= \frac{\pm\sqrt{48}}{6} \\ &= \frac{\pm 4\sqrt{3}}{6} \\ &= \frac{\pm 2\sqrt{3}}{3} \end{aligned}$$

47.

$4u^2 + 3u = 0$

$a = 4 \quad b = 3 \quad c = 0$

$$\begin{aligned} u &= \frac{-b \pm \sqrt{b^2 - 4ac}}{2a} \\ &= \frac{-3 \pm \sqrt{3^2 - 4(4)(0)}}{2(4)} \\ &= \frac{-3 \pm \sqrt{9}}{8} \\ &= \frac{-3 \pm 3}{8} \end{aligned}$$

$u = 0$ or $u = -\dfrac{3}{4}$

49.

$2x^2 + 2x - 5 = 0$

$a = 2 \quad b = 2 \quad c = -5$

$$\begin{aligned} x &= \frac{-b \pm \sqrt{b^2 - 4ac}}{2a} \\ &= \frac{-2 \pm \sqrt{2^2 - 4(2)(-5)}}{2(2)} \\ &= \frac{-2 \pm \sqrt{44}}{4} \\ &= \frac{-2 \pm 2\sqrt{11}}{4} \\ &= \frac{-1 \pm \sqrt{11}}{2} \end{aligned}$$

51.

$3x^2 + 4x - 4 = 0$

$(3x - 2)(x + 2) = 0$

$3x - 2 = 0$ or $x + 2 = 0$

$x = \dfrac{2}{3}$ or $x = -2$

53.

$$2x^2+5x+4=0$$

$$a=2 \quad b=5 \quad c=4$$

$$\begin{aligned} x &= \frac{-b\pm\sqrt{b^2-4ac}}{2a} \\ &= \frac{-5\pm\sqrt{5^2-4(2)(4)}}{2(2)} \\ &= \frac{-5\pm\sqrt{-7}}{4} \\ &= \frac{-5\pm\sqrt{7}i}{4} \end{aligned}$$

55.

$$\begin{aligned} 4u^2-1&=0 \\ 4u^2&=1 \\ u^2&=\frac{1}{4} \\ u&=\pm\sqrt{\frac{1}{4}} \\ u&=\pm\frac{1}{2} \end{aligned}$$

57.

$$4x^3+2x^2+3x=0$$

$$x\left(4x^2+2x+3\right)=0$$

$$x=0 \text{ or } 4x^2+2x+3=0$$

$$a=4 \quad b=2 \quad c=3$$

$$\begin{aligned} x &= \frac{-b\pm\sqrt{b^2-4ac}}{2a} \\ &= \frac{-2\pm\sqrt{2^2-4(4)(3)}}{2(4)} \\ &= \frac{-2\pm\sqrt{-44}}{8} \\ &= \frac{-2\pm2\sqrt{11}i}{8} \\ &= \frac{-1\pm\sqrt{11}i}{4} \end{aligned}$$

$$x=0 \text{ or } x=\frac{-1\pm\sqrt{11}i}{4}$$

59.

$$\begin{aligned} a^2+b^2&=c^2 \\ b^2&=c^2-a^2 \\ b&=\pm\sqrt{c^2-a^2} \end{aligned}$$

61.

$$V=\frac{1}{3}\pi r^2 h$$

$$r^2=\frac{3V}{\pi h}$$

$$r=\pm\sqrt{\frac{3V}{\pi h}}$$

$$r=\frac{\pm\sqrt{3V\pi h}}{\pi h}$$

63.

$$s=\frac{1}{2}gt^2+vt$$

$$0=\frac{1}{2}gt^2+vt-s$$

$$0=gt^2+2vt-2s$$

$$a=g \quad b=2v \quad c=-2s$$

$$t=\frac{-b\pm\sqrt{b^2-4ac}}{2a}$$

$$=\frac{-2v\pm\sqrt{(2v)^2-4(g)(-2s)}}{2g}$$

$$=\frac{-2v\pm\sqrt{4v^2+8gs}}{2g}$$

$$=\frac{-2v\pm\sqrt{4(v^2+2gs)}}{2g}$$

$$=\frac{-2v\pm 2\sqrt{v^2+2gs}}{2g}$$

$$=\frac{-v\pm\sqrt{v^2+2gs}}{g}$$

65.

$$x^2-2x+3=0$$

$$a=1 \quad b=-2 \quad c=3$$

$$b^2-4ac=(-2)^2-4(1)(3)=-8<0$$

two complex roots

67.

$$4x^2-12x+9=0$$

$$a=4 \quad b=-12 \quad c=9$$

$$b^2-4ac=(-12)^2-4(4)(9)=0$$

one double, real root

69.

$$-3x^2+2x+5=0$$

$$a=-3 \quad b=2 \quad c=5$$

$$b^2-4ac=2^2-4(-3)(5)=64>0$$

(perfect square)

two rational, real roots

71.

$$3x^2+2x=0$$

$$a=3 \quad b=2 \quad c=0$$

$$b^2-4ac=2^2-4(3)(0)=4>0$$

(perfect square)

two rational, real roots

73.

$2r^2 = r - 4$

$2r^2 - r + 4 = 0$

$a = 2 \quad b = -1 \quad c = 4$

$b^2 - 4ac = (-1)^2 - 4(2)(4) = -31 < 0$

two complex roots

75.

$3x^2 + 6 = 0$

$a = 3 \quad b = 0 \quad c = 6$

$b^2 - 4ac = 0^2 - 4(3)(6) = -72 < 0$

two complex roots

77.

$6r = 3r^2 + 1$

$3r^2 - 6r + 1 = 0$

$a = 3 \quad b = -6 \quad c = 1$

$b^2 - 4ac = (-6)^2 - 4(3)(1) = 24 > 0$

(not a perfect square)

two irrational, real roots

79.

$12x = 9x^2 + 4$

$9x^2 - 12x + 4 = 0$

$a = 9 \quad b = -12 \quad c = 4$

$b^2 - 4ac = (-12)^2 - 4(9)(4) = 0$

one double, real root

In Exercises 81-83, $b^2 - 4ac = 0$ in order to have a double root.

81.

$kx^2 - 4x + 1 = 0$

$a = k \quad b = -4 \quad c = 1$

$b^2 - 4ac = 0$

$(-4)^2 - 4(k)(1) = 0$

$16 - 4k = 0$

$k = 4$

83.

$x^2 - kx - 2k = 0$

$a = 1 \quad b = -k \quad c = -2k$

$b^2 - 4ac = 0$

$(-k)^2 - 4(1)(-2k) = 0$

$k^2 + 8k = 0$

$k(k+8) = 0$

$k = 0$ or $k + 8 = 0$

$k = 0$ or $k = -8$

85.

$$x+\sqrt{x+5}=7$$
$$\sqrt{x+5}=7-x$$
$$x+5=(7-x)^2$$
$$x+5=49-14x+x^2$$
$$0=x^2-15x+44$$
$$0=(x-11)(x-4)$$
$$x=11 \qquad x=4$$

Check : $x=11$ $\qquad$ $x=4$

$$11+\sqrt{11+5}=7 \ ? \qquad 4+\sqrt{4+5}=7 \ ?$$
$$11+\sqrt{16}=7 \ ? \qquad 4+\sqrt{9}=7 \ ?$$
$$11+4=7 \ ? \qquad 4+3=7 \ ?$$
$$15\neq 7 \qquad 7=7$$

Solution : $x=4$

87.

$$2x+\sqrt{x+1}=8$$
$$\sqrt{x+1}=8-2x$$
$$x+1=(8-2x)^2$$
$$x+1=64-32x+4x^2$$
$$0=4x^2-33x+63$$
$$0=(4x-21)(x-3)$$
$$x=\frac{21}{4} \qquad x=3$$

Check : $x=\frac{21}{4}$ $\qquad$ $x=3$

$$2\left(\frac{21}{4}\right)+\sqrt{\frac{21}{4}+1}=8 \ ? \qquad 2(3)+\sqrt{3+1}=8 \ ?$$
$$\frac{21}{2}+\sqrt{\frac{25}{4}}=8 \ ? \qquad 6+\sqrt{4}=8 \ ?$$
$$\frac{21}{2}+\frac{5}{2}=8 \ ? \qquad 6+2=8 \ ?$$
$$13\neq 8 \qquad 8=8$$

Solution : $x=3$

89.

$$\sqrt{3x+4}-\sqrt{2x+1}=1$$
$$\sqrt{3x+4}=1+\sqrt{2x+1}$$
$$3x+4=1+2\sqrt{2x+1}+2x+1$$
$$x+2=2\sqrt{2x+1}$$
$$x^2+4x+4=4(2x+1)$$
$$x^2-4x=0$$
$$x(x-4)=0$$
$$x=0 \qquad x=4$$

Check : $x=0$ $\qquad x=4$

$$\sqrt{3(0)+4}-\sqrt{2(0)+1}=1\ ? \qquad \sqrt{3(4)+4}-\sqrt{2(4)+1}=1\ ?$$
$$2-1=1\ ? \qquad 4-3=1\ ?$$
$$1=1 \qquad 1=1$$

Solutions : $x=0$ or $x=4$

91.

$$\sqrt{2x-1}+\sqrt{x-4}=4$$
$$\sqrt{2x-1}=4-\sqrt{x-4}$$
$$2x-1=16-8\sqrt{x-4}+x-4$$
$$x-13=-8\sqrt{x-4}$$
$$x^2-26x+169=64(x-4)$$
$$x^2-26x+169=64x-256$$
$$x^2-90x+425=0$$
$$(x-85)(x-5)=0$$
$$x=85 \qquad x=5$$

Check : $x=85$ $\qquad x=5$

$$\sqrt{2(85)-1}+\sqrt{85-4}=4\ ? \qquad \sqrt{2(5)-1}+\sqrt{5-4}=4\ ?$$
$$\sqrt{169}+\sqrt{81}=4\ ? \qquad \sqrt{9}+\sqrt{1}=4\ ?$$
$$13+9=4\ ? \qquad 3+1=4\ ?$$
$$22\neq 4 \qquad 4=4$$

Solution : $x=5$

93.

$$3x^4+5x^2-2=0$$

Let $u=x^2$

$$3u^2+5u-2=0$$
$$(3u-1)(u+2)=0$$

$$u=\frac{1}{3} \text{ or } u=-2$$
$$x^2=\frac{1}{3} \text{ or } x^2=-2$$
$$x=\pm\sqrt{\frac{1}{3}} \text{ or } x=\pm\sqrt{-2}$$
$$x=\pm\frac{\sqrt{3}}{3} \text{ or } x=\pm\sqrt{2}i$$

95.

$$\frac{6}{x^2}+\frac{1}{x}-2=0$$

Let $u=\frac{1}{x}$

$$6u^2+u-2=0$$
$$(3u+2)(2u-1)=0$$

$$u=-\frac{2}{3} \text{ or } u=\frac{1}{2}$$
$$\frac{1}{x}=-\frac{2}{3} \text{ or } \frac{1}{x}=\frac{1}{2}$$
$$x=-\frac{3}{2} \text{ or } x=2$$

97.

$$2x^{\frac{2}{5}}+5x^{\frac{1}{5}}+2=0$$

Let $u=x^{\frac{1}{5}}$

$$2u^2+5u+2=0$$
$$(2u+1)(u+2)=0$$

$$u=-\frac{1}{2} \text{ or } u=-2$$
$$x^{\frac{1}{5}}=-\frac{1}{2} \text{ or } x^{\frac{1}{5}}=-2$$
$$x=-\frac{1}{32} \text{ or } x=-32$$

99.

$$2\left(\frac{1}{x}+1\right)^2-3\left(\frac{1}{x}+1\right)-20=0$$

Let $u=\frac{1}{x}+1$

$$2u^2-3u-20=0$$
$$(2u+5)(u-4)=0$$

$$u=-\frac{5}{2} \qquad u=4$$
$$\frac{1}{x}+1=-\frac{5}{2} \qquad \frac{1}{x}+1=4$$
$$\frac{1}{x}=-\frac{7}{2} \qquad \frac{1}{x}=3$$
$$x=-\frac{2}{7} \qquad x=\frac{1}{3}$$

101a).

Prove : if r_1 and r_2 are the roots of

$ax^2+bx+c=0$ then $r_1r_2=\frac{c}{a}$.

The roots of $ax^2+bx+c=0$ are

$\frac{-b\pm\sqrt{b^2-4ac}}{2a}$.

Let $r_1=\frac{-b+\sqrt{b^2-4ac}}{2a}$

and $r_2=\frac{-b-\sqrt{b^2-4ac}}{2a}$.

$$r_1r_2=\left(\frac{-b+\sqrt{b^2-4ac}}{2a}\right)\left(\frac{-b-\sqrt{b^2-4ac}}{2a}\right)$$

$$=\frac{b^2-(b^2-4ac)}{4a^2}$$

$$=\frac{4ac}{4a^2}$$

$$=\frac{c}{a}$$

101b).

Prove : If r_1 and r_2 are the roots of

$ax^2+bx+c=0$ then $r_1+r_2=-\frac{b}{a}$

Using (a) $r_1=\frac{-b+\sqrt{b^2-4ac}}{2a}$ and

$r_2=\frac{-b-\sqrt{b^2-4ac}}{2a}$.

Then $r_1+r_2=\frac{-b+\sqrt{b^2-4ac}}{2a}+\frac{-b-\sqrt{b^2-4ac}}{2a}$

$$=\frac{-2b}{2a}$$

$$=-\frac{b}{a}$$

103.

$kx^2+3x+5=0$; sum of the roots is 6

$$r_1+r_2=6$$

$$r_1+r_2=-\frac{b}{a}=-\frac{3}{k}$$

$$6=-\frac{3}{k}$$

$$k=-\frac{1}{2}$$

105.

$3x^2-10x+2k=0$; product of the roots is -4

$$r_1r_2=-4$$

$$r_1r_2=\frac{c}{a}=\frac{2k}{3}$$

$$-4=\frac{2k}{3}$$

$$-6=k$$

107.

$2x^2 - kx + 9 = 0$; one root is double the other

$$r_1 = 2r_2$$

$$r_1 r_2 = 2r_2 r_2 = \frac{c}{a} = \frac{9}{2}$$

$$2r_2^{\,2} = \frac{9}{2}$$

$$r_2^{\,2} = \frac{9}{4}$$

$$r_2 = \pm\frac{3}{2}$$

$$r_1 = \pm 3$$

$$r_1 = 3,\ r_2 = \frac{3}{2} \qquad r_1 = -3, \qquad r_2 = -\frac{3}{2}$$

$$(x-3)(2x-3) = 0 \qquad (x+3)(2x+3) = 0$$

$$2x^2 - 9x + 9 = 0 \qquad 2x^2 + 9x + 9 = 0$$

$$k = \pm 9$$

109.

$6x^2 - 13x + k = 0$; one root is the reciprocal of the other

$$r_1 = \frac{1}{r_2}$$

$$r_1 + r_2 = \frac{1}{r_2} + r_2 = -\frac{b}{a} = \frac{13}{6}$$

$$\frac{1 + r_2^{\,2}}{r_2} = \frac{13}{6}$$

$$6 + 6r_2^{\,2} = 13r_2$$

$$6r_2^{\,2} - 13r_2 + 6 = 0$$

$$(3r_2 - 2)(2r_2 - 3) = 0$$

$$r_2 = \frac{2}{3} \qquad r_2 = \frac{3}{2}$$

$$(3x-2)(2x-3) = 0$$

$$6x^2 - 13x + 6 = 0$$

$$k = 6$$

111.

False. $-10 \le 6$ but $(-10)^2 \le 36$

113.

False. Suppose $x = \frac{1}{2}$, then $\frac{1}{x} = 2$

$$\frac{1}{2} \ge 2$$

115a).

$$55 = \frac{n(n+1)}{2}$$

$$110 = n^2 + n$$

$$0 = n^2 + n - 110$$

$$0 = (n+11)(n-10)$$

$$n = -11 \text{ or } n = 10$$

Since $n > 0$, $n = 10$.

115b).

$$36 = \frac{n(n+1)}{2}$$

$$72 = n^2 + n$$

$$0 = n^2 + n - 72$$

$$0 = (n+9)(n-8)$$

$$n = -9 \text{ or } n = 8$$

Since $n > 0$, $n = 8$

115c).

$$2+3+4+...+n=\frac{n(n+1)}{2}-1$$
$$=\frac{n^2+n-2}{2}$$

117a).

$$121=n^2$$
$$n=11$$

117b).

$$1+3+5+7+9+11+13+15+17+19+21$$

CHAPTER 2 SECTION 4

EXERCISE SET 2.4

1.

Time for A alone : $x-3$

Time for B alone : x

	rate	·	time	=	work done
Computer A	$\frac{1}{x-3}$		2		$\frac{2}{x-3}$
Computer B	$\frac{1}{x}$		2		$\frac{2}{x}$

$$\frac{2}{x-3}+\frac{2}{x}=1$$

$$x(x-3)\left[\frac{2}{x-3}+\frac{2}{x}\right]=x(x-3)(1)$$

$$2x+2x-6=x^2-3x$$

$$0=x^2-7x+6$$

$$0=(x-6)(x-1)$$

$$x=6 \quad \text{or} \quad \cancel{x=1}$$

$$x-3=3 \qquad \cancel{x-3=-2}$$

It takes Computer A 3 hours and Computer B 6 hours.

3.

Time for roofer alone : $x-6$

Time for assistant alone : x

	rate	·	time	=	work done
roofer	$\frac{1}{x-6}$		4		$\frac{4}{x-6}$
assistant	$\frac{1}{x}$		4		$\frac{4}{x}$

$$\frac{4}{x-6}+\frac{4}{x}=1$$

$$x(x-6)\left[\frac{4}{x-6}+\frac{4}{x}\right]=x(x-6)(1)$$

$$4x+4x-24=x^2-6x$$

$$0=x^2-14x+24$$

$$0=(x-12)(x-2)$$

$$x=12 \quad \text{or} \quad \cancel{x=2}$$

$$x-6=6 \qquad \cancel{x-6=-4}$$

The roofer takes 6 hours and the assistant takes 12 hours.

5.

Width : x

Length : $2x+4$

$$x(2x+4)=48$$
$$2x^2+4x-48=0$$
$$x^2+2x-24=0$$
$$(x+6)(x-4)=0$$
$$\cancel{x=-6} \quad \text{or} \quad x=4$$
$$2x+4=12$$

The width is 4 feet and the length is 12 feet.

7.

Original width : x

Original length : $\dfrac{48}{x}$

Increase each dimension by 4

New width : $x+4$

New length : $\dfrac{48}{x}+4$

$$(x+4)\left(\frac{48}{x}+4\right)=120$$
$$(x+4)\left(\frac{48}{x}+4\right)(x)=120(x)$$
$$(x+4)(48+4x)=120x$$
$$(x+4)(12+x)(4)=120x$$
$$(x+4)(12+x)=30x$$
$$x^2+16x+48=30x$$
$$x^2-14x+48=0$$
$$(x-6)(x-8)=0$$

$$x=6 \quad \text{or} \quad x=8$$
$$\frac{48}{x}=8 \qquad \frac{48}{x}=6$$

The dimensions are 6 cm by 8 cm.

9.

Width of strip : x

$$(60-2x)(80-2x)=\frac{1}{2}(60)(80)$$
$$2(30-x)2(40-x)=2400$$
$$1200-70x+x^2=600$$
$$x^2-70x+600=0$$
$$(x-60)(x-10)=0$$
$$x=60 \quad \text{or} \quad x=10$$

If $x=60$ then $60-2x=-60$ which isn't possible. The width of the strip is 10 feet.

11.

Number : x

$$x+\frac{1}{x}=\frac{26}{5}$$
$$5x^2+5=26x$$
$$5x^2-26x+5=0$$
$$(5x-1)(x-5)=0$$

$$x=\frac{1}{5} \quad \text{or} \quad x=5$$
$$\frac{1}{x}=5 \qquad \frac{1}{x}=\frac{1}{5}$$

The number is 5 or $\frac{1}{5}$.

13.

Smaller number : $x-4$

Larger number : x

$$(x-4)^2+x^2=58$$
$$x^2-8x+16+x^2=58$$
$$2x^2-8x-42=0$$
$$x^2-4x-21=0$$
$$(x-7)(x+3)=0$$

$$x=7 \quad \text{or} \quad x=-3$$
$$x-4=3 \qquad x-4=-7$$

The numbers are 3 and 7 or -3 and -7.

15.

1st number : x

2nd consecutive even number : $x+2$

$$\frac{1}{x}+\frac{1}{x+2}=\frac{7}{24}$$

$$24x(x+2)\left[\frac{1}{x}+\frac{1}{x+2}\right]=24x(x+2)\left(\frac{7}{24}\right)$$
$$24x+48+24x=7x^2+14x$$
$$0=7x^2-34x-48$$
$$0=(7x+8)(x-6)$$
$$x=-\frac{8}{7} \text{ (crossed out)} \quad \text{or} \quad x=6$$
$$x+2=8$$

The numbers are 6 and 8.

17.

Number of shares bought : x

Cost per share : $\dfrac{1200}{x}$

$$\left(\frac{1200}{x}+2\right)(x-30)=1200$$

$$x\left[\left(\frac{1200}{x}+2\right)(x-30)\right]=x(1200)$$

$$(1200+2x)(x-30)=1200x$$

$$2(600+x)(x-30)=1200x$$

$$x^2+570x-18000=600x$$

$$x^2-30x-18000=0$$

$$(x-150)(x+120)=0$$

$x=150$ or ~~$x=-120$~~

The investor bought 150 shares.

19.

Number of days worked: x

Rate per day: $\dfrac{192}{x}$

$$\left(\frac{192}{x}+8\right)(x-2)=192$$

$$x\left[\left(\frac{192}{x}+8\right)(x-2)\right]=x(192)$$

$$(192+8x)(x-2)=192x$$

$$8(24+x)(x-2)=192x$$

$$x^2+22x-48=24x$$

$$x^2-2x-48=0$$

$$(x-8)(x+6)=0$$

$x=8$ or ~~$x=-6$~~

He worked 8 days.

21.

1st square's side length : $\frac{x}{4}$

2nd square's side length : $\frac{48-x}{4}$

$$\left(\frac{x}{4}\right)^2+\left(\frac{48-x}{4}\right)^2=80$$
$$\frac{x^2}{16}+\frac{2304-96x+x^2}{16}=80$$
$$x^2+x^2-96x+2304=1280$$
$$2x^2-96x+1024=0$$
$$x^2-48x+512=0$$
$$(x-16)(x-32)=0$$

$$x=16 \quad \text{or} \quad x=32$$
$$48-x=32 \qquad 48-x=16$$

The wire should be cut 16 cm from the end.

23.

$$v_0=68$$
$$d=630$$

$$d=16t^2+v_0t$$
$$630=16t^2+68t$$
$$0=16t^2+68t-630$$
$$0=8t^2+34t-315$$
$$0=(2t-9)(4t+35)$$
$$t=\frac{9}{2} \quad \text{or} \quad \cancel{t=\frac{35}{4}}$$

It will take $\frac{9}{2}=4.5$ seconds.

25.

$$P = 7000$$
$$S = 9100$$
$$t = 2$$

$$S = P(1+r)^2$$
$$9100 = 7000(1+r)^2$$
$$1.3 = (1+r)^2$$
$$\pm 1.14 = 1+r$$
$$r = -1 \pm 1.14$$
$$r = 0.14 \quad \text{or} \quad \cancel{r = -2.14}$$

The rate is 14%.

27.

$$t_e = 80.8$$
$$t_s = 11.4$$
$$11.4 = 80.8\sqrt{1-\frac{v^2}{c^2}} \quad \text{Solve for v.}$$
$$0.14 = \sqrt{1-\frac{v^2}{c^2}}$$
$$0.02 = 1-\frac{v^2}{c^2}$$
$$\frac{v^2}{c^2} = 0.98$$
$$v^2 = 0.98c^2$$
$$v = 0.99c$$

29a).

$$\theta = \theta_0 + \omega_0 t + \frac{1}{2}\alpha_0 t^2$$
$$11 = 1 + 3t + \frac{1}{2}(2)t^2$$
$$0 = t^2 + 3t - 10$$
$$0 = (t+5)(t-2)$$
$$\cancel{t = -5} \quad \text{or} \quad t = 2$$

The time is 2 seconds.

29b).

$$\theta = \theta_0 + \omega_0 t + \frac{1}{2}\alpha_0 t^2$$
$$0 = \frac{1}{2}\alpha_0 t^2 + \omega_0 t + (\theta_0 - \theta)$$
$$a = \frac{1}{2}\alpha_0 \quad b = \omega_0 \quad c = \theta_0 - \theta$$
$$t = \frac{-\omega_0 \pm \sqrt{\omega_0^{\,2} - 4\left(\frac{1}{2}\alpha_0\right)(\theta_0 - \theta)}}{2\left(\frac{1}{2}\alpha_0\right)}$$
$$t = \frac{-\omega_0 \pm \sqrt{\omega_0^{\,2} - 2\alpha_0(\theta_0 - \theta)}}{\alpha_0}$$

31.

$$z = \frac{x - np}{\sqrt{np(1-p)}}$$

$$-2.48 = \frac{1-16p}{\sqrt{16p(1-p)}}$$

$$-2.48\sqrt{16p(1-p)} = 1 - 16p$$

$$6.15(16p)(1-p) = (1-16p)^2$$

$$98.4p(1-p) = 1 - 32p + 256p^2$$

$$98.4p - 98.4p^2 = 1 - 32p + 256p^2$$

$$0 = 354.4p^2 - 130.4p + 1$$

$$p = \frac{-(-130.4) \pm \sqrt{(-130.4)^2 - 4(354.4)(1)}}{2(354.4)}$$

$$p = 0.008 \quad \text{or} \quad p = 0.36$$

33.

$$A = 10\pi$$

$$\pi r^2 = 10\pi$$

$$r^2 = 10$$

$$r = \sqrt{10}$$

The radius is $\sqrt{10}$.

35a).

$$\frac{L+W}{L}=\frac{L}{W}$$
$$\frac{L+5}{L}=\frac{L}{5}$$
$$5(L+5)=L^2$$
$$5L+25=L^2$$
$$0=L^2-5L-25$$
$$L=\frac{-(-5)\pm\sqrt{(-5)^2-4(1)(-25)}}{2(1)}$$
$$=\frac{5\pm\sqrt{125}}{2}$$
$$=\frac{5\pm5\sqrt{5}}{2}$$

35b).

$$\frac{L+W}{L}=\frac{L}{W}$$
$$W(L+W)=L^2$$
$$WL+W^2=L^2$$
$$W^2+WL-L^2=0$$

$a=1 \quad b=L \quad c=-L^2$

$$W=\frac{-b\pm\sqrt{b^2-4ac}}{2a}$$
$$W=\frac{-L\pm\sqrt{L^2+4(1)(-L^2)}}{2(1)}$$
$$W=\frac{-L\pm\sqrt{L^2+4L^2}}{2}$$
$$W=\frac{-L\pm\sqrt{5L^2}}{2}$$
$$W=\frac{-L\pm L\sqrt{5}}{2}$$

CHAPTER 2 SECTION 5

SECTION 2.5 PROGRESS CHECK

1. (Page 118)

$$3x-2 \ge 5x+4$$
$$-2x-2 \ge 4$$
$$-2x \ge 6$$
$$x \le -3$$

2a). (Page 121)

$$\frac{3x-1}{4}+1 > 2+\frac{x}{3}$$
$$12\left[\frac{3x-1}{4}+1\right] > 12\left[2+\frac{x}{3}\right]$$
$$3(3x-1)+12 > 24+4x$$
$$9x-3+12 > 24+4x$$
$$9x+9 > 24+4x$$
$$5x+9 > 24$$
$$5x > 15$$
$$x > 3$$
$$(3,\infty)$$

2b).

$$\frac{2x-3}{2} \ge x+\frac{2}{5}$$
$$10\left(\frac{2x-3}{2}\right) \ge 10\left(x+\frac{2}{5}\right)$$
$$5(2x-3) \ge 10x+4$$
$$10x-15 \ge 10x+4$$
$$-15 \ge 4$$

No solution

3. (Page 122)

Unlimited local calls : 20
Flexible rate : $8+0.06x$
Where x represents the number of calls.

$$20 < 8+0.06x$$
$$12 < 0.06x$$
$$200 < x$$

Unlimited service costs less when the number of message units is greater than 200.

4. (Page 122)

$$-5 < 2-3x < -1$$
$$-7 < -3x < -3$$
$$\frac{7}{3} > x > 1$$
$$\left(1, \frac{7}{3}\right)$$

5a). (Page 125)

$$2x^2 \ge 5x+3$$
$$2x^2 = 5x+3$$
$$2x^2 - 5x - 3 = 0$$
$$(2x+1)(x-3) = 0$$
$$x = -\frac{1}{2} \quad x = 3$$

Interval, Critical Value	Test Point	Subsitution	Verification
$x < -\frac{1}{2}$	$x = -1$	$2(-1)^2 \ge 5(-1)+3$	True
$x = -\frac{1}{2}$	$x = -\frac{1}{2}$	$2\left(-\frac{1}{2}\right)^2 \ge 5\left(-\frac{1}{2}\right)+3$	True
$-\frac{1}{2} < x < 3$	$x = 0$	$2(0)^2 \ge 5(0)+3$	False
$x = 3$	$x = 3$	$2(3)^2 \ge 5(3)+3$	True
$x > 3$	$x = 4$	$2(4)^2 \ge 5(4)+3$	True

$$\left\{x \middle| x \le -\frac{1}{2} \text{ or } x \ge 3\right\}$$

5b).

$$\frac{2x-3}{1-2x} \geq 0 \qquad \text{Not defined when } x = \frac{1}{2}$$

$$\frac{2x-3}{1-2x} = 0$$

$$2x-3 = 0$$

$$x = \frac{3}{2}$$

Interval, Critical Value	Test Point	Substitution	Verification
$x < \frac{1}{2}$	$x = 0$	$\frac{2(0)-3}{1-2(0)} \geq 0$	False
$x = \frac{1}{2}$	$x = \frac{1}{2}$	$\frac{2\left(\frac{1}{2}\right)-3}{1-2\left(\frac{1}{2}\right)} \geq 0$	False
$\frac{1}{2} < x < \frac{3}{2}$	$x = 1$	$\frac{2(1)-3}{1-2(1)} \geq 0$	True
$x = \frac{3}{2}$	$x = \frac{3}{2}$	$\frac{2\left(\frac{3}{2}\right)-3}{1-2\left(\frac{3}{2}\right)} \geq 0$	True
$x > \frac{3}{2}$	$x = 2$	$\frac{2(2)-3}{1-2(2)} \geq 0$	False

$$\left\{x \middle| \frac{1}{2} < x \leq \frac{3}{2}\right\}$$

6. (Page 127)

$$(2y-9)(6-y)(y+5)\geq 0$$
$$(2y-9)(6-y)(y+5)=0$$
$$y=\frac{9}{2}\quad y=6\quad y=-5$$

Interval, Critical Value	Test Point	Substitution	Verification
$y<-5$	$y=-6$	$[2(-6)-9][6-(-6)][-6+5]\geq 0$	True
$y=-5$	$y=-5$	$[2(-5)-9][6-(-5)][-5+5]\geq 0$	True
$-5<y<\frac{9}{2}$	$y=0$	$[2(0)-9][6-0][0+5]\geq 0$	False
$y=\frac{9}{2}$	$y=\frac{9}{2}$	$\left[2\left(\frac{9}{2}\right)-9\right]\left[6-\frac{9}{2}\right]\left[\frac{9}{2}+5\right]\geq 0$	True
$\frac{9}{2}<y<6$	$y=5$	$[2(5)-9][6-5][5+5]\geq 0$	True
$y=6$	$y=6$	$[2(6)-9][6-6][6+5]\geq 0$	True
$y>6$	$y=7$	$[2(7)-9][6-7][7+5]\geq 0$	False

$$\left\{y \middle| y\leq -5 \text{ or } \frac{9}{2}\leq y\leq 6\right\}$$

or

$$[-\infty,-5],\left[\frac{9}{2},6\right]$$

EXERCISE SET 2.5

1. $[-5,1)$

3. $(9,\infty)$

5. $[-12,-3]$

7. $(3,7)$

9. $(-6,-4]$

11. $5\leq x\leq 8$

13.
$x > 3$

15.
$x \le 5$

17.
$x \ge 0$

19.
$x + 4 < 8$

$x < 4$

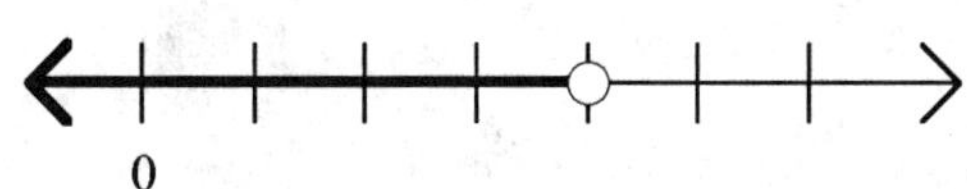

21.
$x + 3 < -3$

$x < -6$

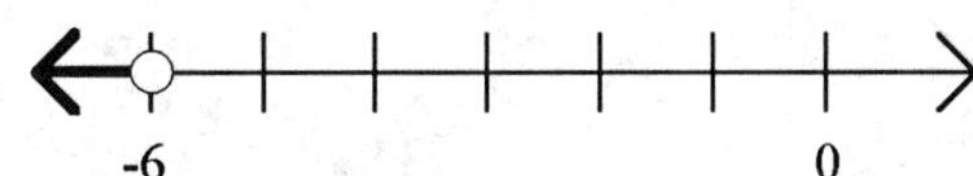

23.
$x - 3 \ge 2$

$x \ge 5$

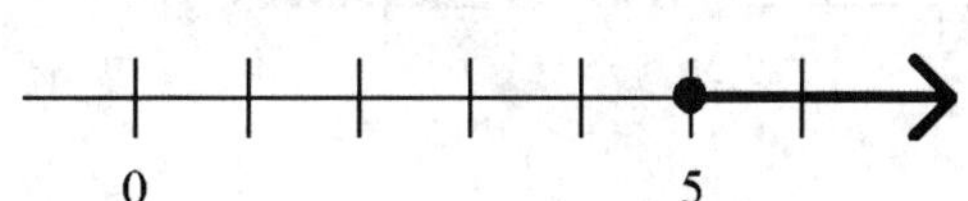

25.
$2 < a + 3$

$-1 < a$

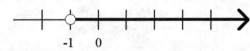

27.
$2y < -1$

$y < -\frac{1}{2}$

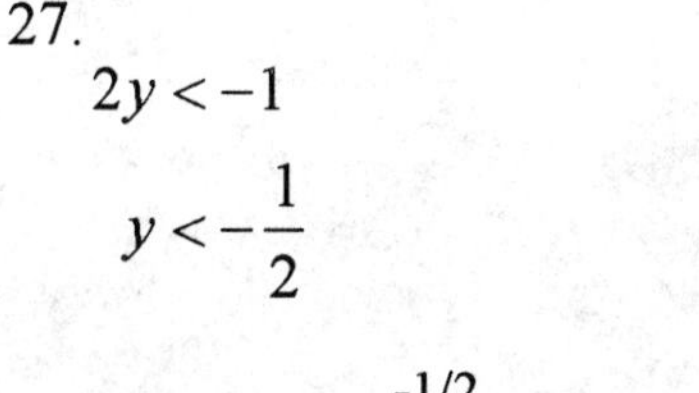

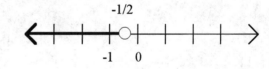

29.

$2x \geq 0$

$x \geq 0$

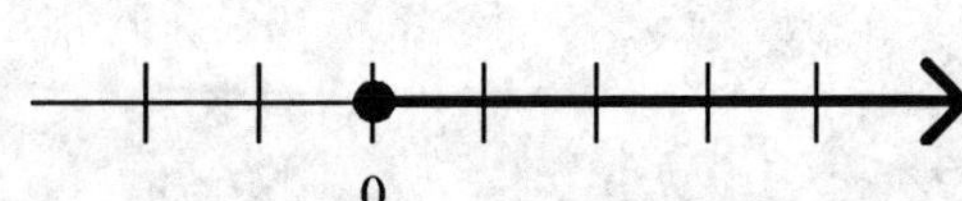

31.

$2r + 5 < 9$

$2r < 4$

$r < 2$

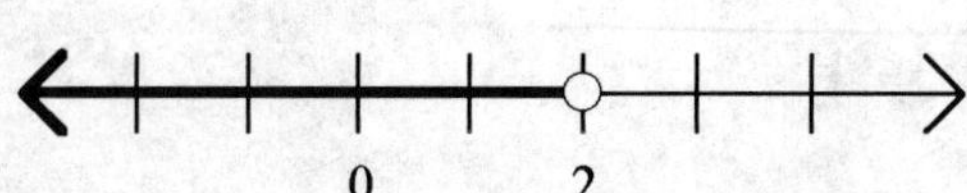

33.

$3x - 1 \geq 2$

$3x \geq 3$

$x \geq 1$

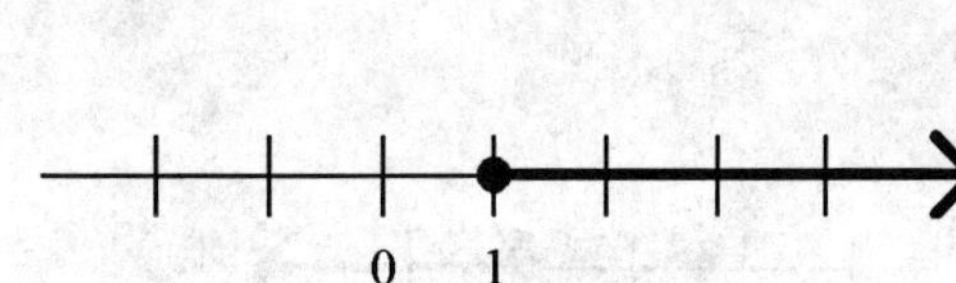

35.

$\frac{4}{5 - 3x} < 0$

Since the numerator $4 > 0$, the denominator must be less than zero for the quotient to be negative.

$5 - 3x < 0$

$-3x < -5$

$x > \frac{5}{3}$

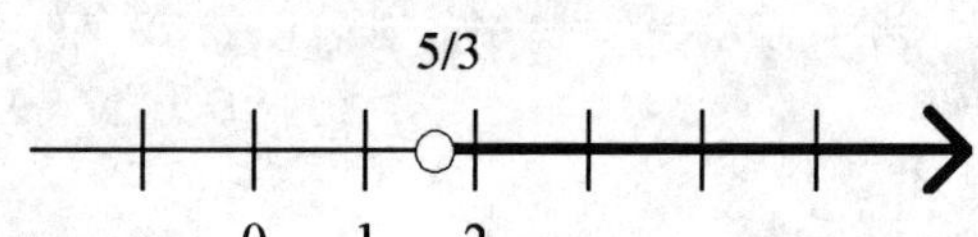

37.

$4x + 3 \leq 11$

$4x \leq 8$

$x \leq 2$

$(-\infty, 2]$

39.

$\frac{3}{2}x + 1 \geq 4$

$3x + 2 \geq 8$

$3x \geq 6$

$x \geq 2$

$[2, \infty)$

41.

$$\begin{aligned} 4(2x+1) &< 16 \\ 2x+1 &< 4 \\ 2x &< 3 \\ x &< \frac{3}{2} \end{aligned}$$

$$\left(-\infty, \frac{3}{2}\right)$$

43.

$$\begin{aligned} 2(x-3) &< 3(x+2) \\ 2x-6 &< 3x+6 \\ -x &< 12 \\ x &> -12 \end{aligned}$$

$$(-12, \infty)$$

45.

$$\begin{aligned} 3(2a-1) &> 4(2a-3) \\ 6a-3 &> 8a-12 \\ -2a &> -9 \\ a &< \frac{9}{2} \end{aligned}$$

$$\left(-\infty, \frac{9}{2}\right)$$

47.

$$\begin{aligned} \frac{2}{3}(x+1)+\frac{5}{6} &\geq \frac{1}{2}(2x-1)+4 \\ 12\left[\frac{2}{3}(x+1)+\frac{5}{6}\right] &\geq 12\left[\frac{1}{2}(2x+1)+4\right] \\ 8(x+1)+10 &\geq 6(2x-1)+48 \\ 8x+8+10 &\geq 12x-6+48 \\ 8x+18 &\geq 12x+42 \\ -4x &\geq 24 \\ x &\leq -6 \end{aligned}$$

$$(-\infty, -6]$$

49.

$$\begin{aligned} \frac{x-1}{3}+\frac{1}{5} &< \frac{x+2}{5}-\frac{1}{3} \\ 15\left[\frac{x-1}{3}+\frac{1}{5}\right] &< 15\left[\frac{x+2}{5}-\frac{1}{3}\right] \\ 5(x-1)+3 &< 3(x+2)-5 \\ 5x-5+3 &< 3x+6-5 \\ 5x-2 &< 3x+1 \\ 2x &< 3 \\ x &< \frac{3}{2} \end{aligned}$$

$$\left(-\infty, \frac{3}{2}\right)$$

51.

$$\begin{aligned} 3(x+1)+6 &\geq 2(2x-1)+4 \\ 3x+3+6 &\geq 4x-2+4 \\ 3x+9 &\geq 4x+2 \\ -x &\geq -7 \\ x &\leq 7 \end{aligned}$$

$$(-\infty, 7]$$

53.

$$-2 < 4x \le 5$$

$$-\frac{1}{2} < x \le \frac{5}{4}$$

$$\left(-\frac{1}{2}, \frac{5}{4}\right]$$

55.

$$-4 \le 2x + 2 \le -2$$

$$-6 \le 2x \le -4$$

$$-3 \le x \le -2$$

$$[-3, -2]$$

57.

$$3 \le 1 - 2x < 7$$

$$2 \le -2x < 6$$

$$-1 \ge x > -3$$

$$(-3, -1]$$

59.

$$-8 < 2 - 5x \le 7$$

$$-10 < -5x \le 5$$

$$2 > x \ge -1$$

$$[-1, 2)$$

61.

Third exam score : x

$$\frac{42 + 70 + x}{3} = 70$$

$$112 + x = 210$$

$$x = 98$$

The minimum score is 98.

63.

Number of appliances : x

$$30 + 25x > 130$$

$$25x > 100$$

$$x > 4$$

He must sell more than 4 appliances.

65.

Number of copies sold : x

$$25x > 38000 + 12x$$

$$13x > 38000$$

$$x > 2923.08$$

At least 2924 copies must be sold.

67.

$$2(15) + 2L \le 70$$

$$30 + 2L \le 70$$

$$2L \le 40$$

$$L \le 20 \text{ meters}$$

69.

$$x^2 + 3x - 4 \le 0$$

$$(x+4)(x-1) = 0$$

$$x = -4 \quad x = 1$$

Interval, Critical Value	Test Point	Substitution	Verification
$x < -4$	$x = -5$	$(-5)^2 + 3(-5) - 4 \le 0$	False
$x = -4$	$x = -4$	$(-4)^2 + 3(-4) - 4 \le 0$	True
$-4 < x < 1$	$x = 0$	$(0)^2 + 3(0) - 4 \le 0$	True
$x = 1$	$x = 1$	$(1)^2 + 3(1) - 4 \le 0$	True
$x > 1$	$x = 2$	$(2)^2 + 3(2) - 4 \le 0$	False

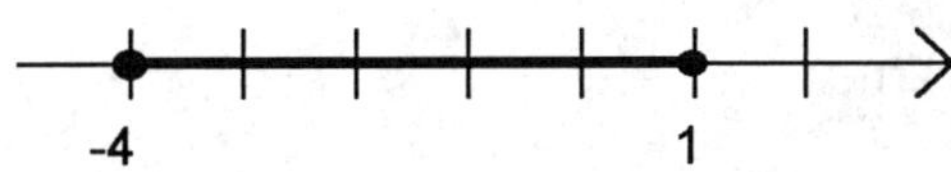

71.

$$3x^2 - 4x - 4 \geq 0$$
$$(3x+2)(x-2) = 0$$

$$x = -\frac{2}{3} \quad x = 2$$

Interval, Critical Value	Test Point	Substitution	Verification
$x < -\frac{2}{3}$	$x = -1$	$3(-1)^2 - 4(-1) - 4 \geq 0$	True
$x = -\frac{2}{3}$	$x = -\frac{2}{3}$	$3\left(-\frac{2}{3}\right)^2 - 4\left(-\frac{2}{3}\right) - 4 \geq 0$	True
$-\frac{2}{3} < x < 2$	$x = 0$	$3(0)^2 - 4(0) - 4 \geq 0$	False
$x = 2$	$x = 2$	$3(2)^2 - 4(2) - 4 \geq 0$	True
$x > 2$	$x = 3$	$3(3)^2 - 4(3) - 4 \geq 0$	True

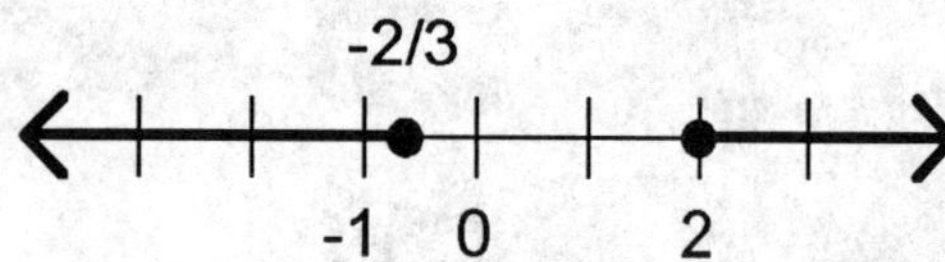

73.

$r^2 + 4r \geq 0$

$r(r+4) = 0$

$r = 0 \quad r = -4$

Interval, Critical Value	Test Point	Substitution	Verification
$r < -4$	$r = -5$	$(-5)^2 + 4(-5) \geq 0$	True
$r = -4$	$r = -4$	$(-4)^2 + 4(-4) \geq 0$	True
$-4 < r < 0$	$r = -1$	$(-1)^2 + 4(-1) \geq 0$	False
$r = 0$	$r = 0$	$(0)^2 + 4(0) \geq 0$	True
$r > 0$	$r = 1$	$(1)^2 + 4(1) \geq 0$	True

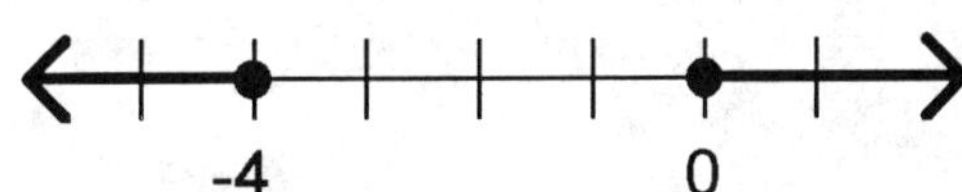

75.

$\frac{x-6}{x+4} \geq 0$ Not defined when $x+4=0$

$x=-4$

$\frac{x-6}{x+4}=0$

$x-6=0$

$x=6$

Interval, Critical Value	Test Point	Substitution	Verification
$x<-4$	$x=-5$	$\frac{-5-6}{-5+4} \geq 0$	True
$x=-4$	$x=-4$	$\frac{-4-6}{-4+4} \geq 0$	False
$-4<x<6$	$x=0$	$\frac{0-6}{0+4} \geq 0$	False
$x=6$	$x=6$	$\frac{6-6}{6+4} \geq 0$	True
$x>6$	$x=7$	$\frac{7-6}{7+4} \geq 0$	True

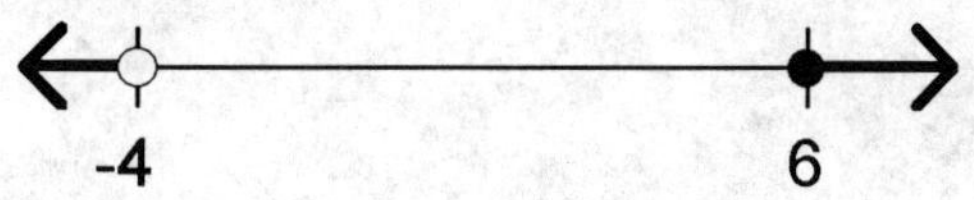

77.

$$\frac{x-1}{2x-3} \geq 0 \quad \text{Not defined when } 2x-3=0$$

$$x = \frac{3}{2}$$

$$\frac{x-1}{2x-3} = 0$$

$$x-1=0$$

$$x=1$$

Interval,

Critical Value	Test Point	Substitution	Verification
$x<1$	$x=0$	$\frac{0-1}{2(0)-3} \geq 0$	True
$x=1$	$x=1$	$\frac{1-1}{2(1)-3} \geq 0$	True
$1<x<\frac{3}{2}$	$x=\frac{5}{4}$	$\frac{\frac{5}{4}-1}{2\left(\frac{5}{4}\right)-3} \geq 0$	False
$x=\frac{3}{2}$	$x=\frac{3}{2}$	$\frac{\frac{3}{2}-1}{2\left(\frac{3}{2}\right)-3} \geq 0$	False
$x>\frac{3}{2}$	$x=2$	$\frac{2-1}{2(2)-3} \geq 0$	True

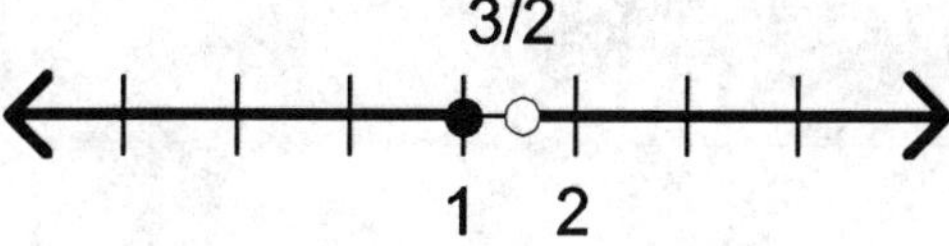

79.

$$\frac{4x+5}{x^2} \le 0 \quad \text{Not defined when } x = 0$$

$$\frac{4x+5}{x^2} = 0$$

$$4x+5 = 0$$

$$x = -\frac{5}{4}$$

Interval, Critical Value	Test Point	Substitution	Verification
$x < -\frac{5}{4}$	$x = -2$	$\frac{4(-2)+5}{(-2)^2} \le 0$	True
$x = -\frac{5}{4}$	$x = -\frac{5}{4}$	$\frac{4\left(-\frac{5}{4}\right)+5}{\left(-\frac{5}{4}\right)^2} \le 0$	True
$-\frac{5}{4} < x < 0$	$x = -1$	$\frac{4(-1)+5}{(-1)^2} \le 0$	False
$x = 0$	$x = 0$	$\frac{4(0)+5}{(0)^2} \le 0$	False
$x > 0$	$x = 1$	$\frac{4(1)+5}{(1)^2} \le 0$	False

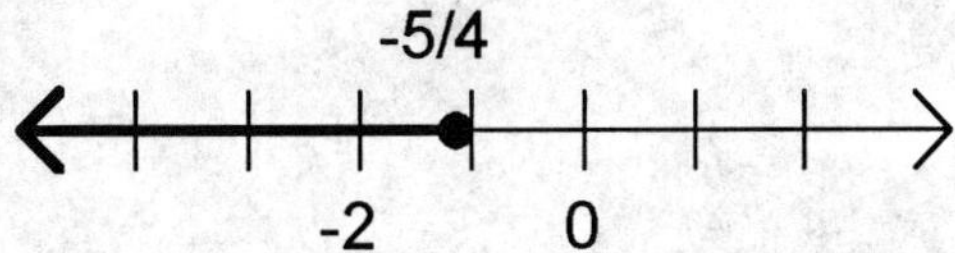

81.

$$(x-4)(2x+5)(2-x)\le 0$$
$$(x-4)(2x+5)(2-x)=0$$

$$x=4 \quad x=-\frac{5}{2} \quad x=2$$

Interval, Critical Value	Test Point	Substitution	Verification
$x<-\frac{5}{2}$	$x=-3$	$(-3-4)[2(-3)+5][2-(-3)]\le 0$	False
$x=-\frac{5}{2}$	$x=-\frac{5}{2}$	$\left(-\frac{5}{2}-4\right)\left[2\left(-\frac{5}{2}\right)+5\right]\left[2-\left(-\frac{5}{2}\right)\le 0\right]$	True
$-\frac{5}{2}<x<2$	$x=0$	$(0-4)[2(0)+5](2-0)\le 0$	True
$x=2$	$x=2$	$(2-4)[2(2)+5](2-2)\le 0$	True
$2<x<4$	$x=3$	$(3-4)[2(3)+5](2-3)\le 0$	False
$x=4$	$x=4$	$(4-4)[2(4)+5](2-4)\le 0$	True
$x>4$	$x=5$	$(5-4)[2(5)+5](2-5)\le 0$	True

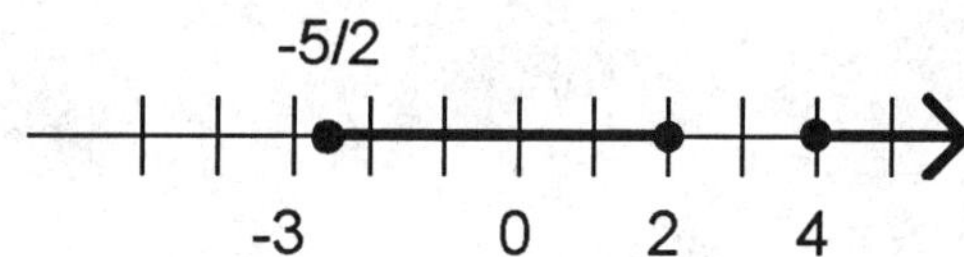

83.

$$x^2 - 3x - 10 \ge 0$$
$$(x-5)(x+2) = 0$$

$x = 5 \quad x = -2$

Interval,

Critical Value	Test Point	Substitution	Verification
$x < -2$	$x = -3$	$(-3)^2 - 3(-3) - 10 \ge 0$	True
$x = -2$	$x = -2$	$(-2)^2 - 3(-2) - 10 \ge 0$	True
$-2 < x < 5$	$x = 0$	$(0)^2 - 3(0) - 10 \ge 0$	False
$x = 5$	$x = 5$	$(5)^2 - 3(5) - 10 \ge 0$	True
$x > 5$	$x = 6$	$(6)^2 - 3(6) - 10 \ge 0$	True

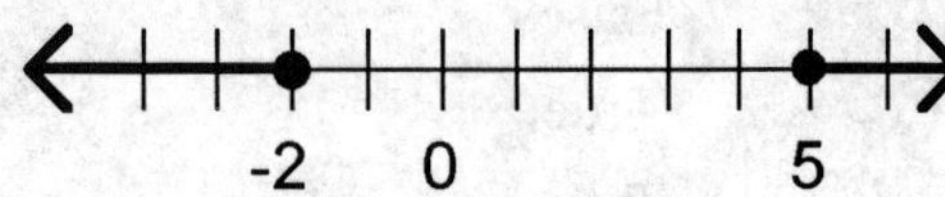

85.

$$3x^2 - 4x - 4 \le 0$$

$$(3x+2)(x-2) = 0$$

$$x = -\frac{2}{3} \quad x = 2$$

Interval, Critical Value	Test Point	Substitution	Verification
$x < -\frac{2}{3}$	$x = -1$	$3(-1)^2 - 4(-1) - 4 \le 0$	False
$x = -\frac{2}{3}$	$x = -\frac{2}{3}$	$3\left(-\frac{2}{3}\right)^2 - 4\left(-\frac{2}{3}\right) - 4 \le 0$	True
$-\frac{2}{3} < x < 2$	$x = 0$	$3(0)^2 - 4(0) - 4 \le 0$	True
$x = 2$	$x = 2$	$3(2)^2 - 4(2) - 4 \le 0$	True
$x > 2$	$x = 3$	$3(3)^2 - 4(3) - 4 \le 0$	False

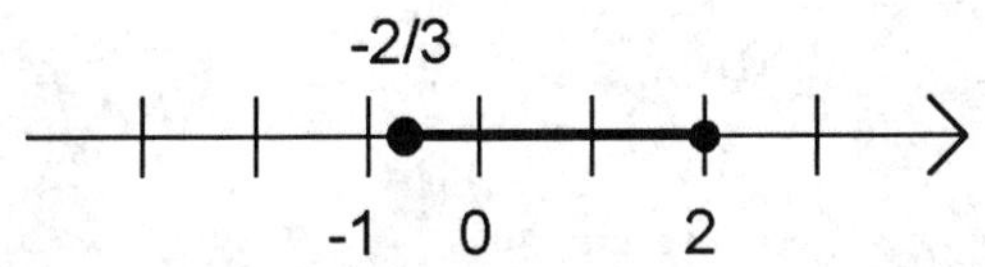

87.

$\frac{3x+2}{2x-3} \ge 0$ Not defined when $2x-3=0$

$$x = \frac{3}{2}$$

$$\frac{3x+2}{2x-3} = 0$$
$$3x+2=0$$
$$x = -\frac{2}{3}$$

Interval, Critical Value	Test Point	Substitution	Verification
$x < -\frac{2}{3}$	$x=-1$	$\frac{3(-1)+2}{2(-1)-3} \ge 0$	True
$x = -\frac{2}{3}$	$x = -\frac{2}{3}$	$\frac{3\left(-\frac{2}{3}\right)+2}{2\left(-\frac{2}{3}\right)-3} \ge 0$	True
$-\frac{2}{3} < x < \frac{3}{2}$	$x=0$	$\frac{3(0)+2}{2(0)-3} \ge 0$	False
$x = \frac{3}{2}$	$x = \frac{3}{2}$	$\frac{3\left(\frac{3}{2}+2\right)}{2\left(\frac{3}{2}\right)-3} \ge 0$	False
$x > \frac{3}{2}$	$x=2$	$\frac{3(2)+2}{2(2)-3} \ge 0$	True

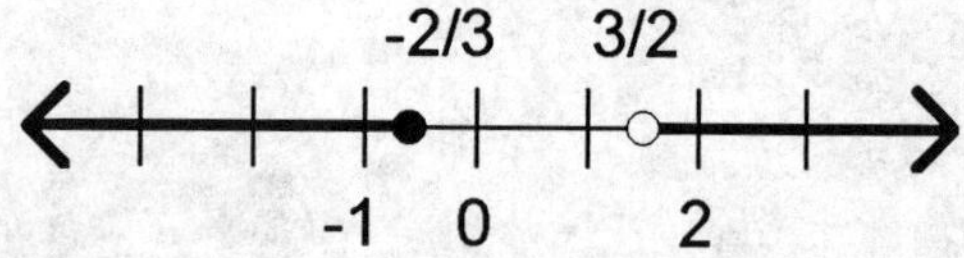

89.

$$\frac{2x-1}{x+2} \le 0 \qquad \text{Not defined when } x+2=0$$

$$x=-2$$

$$\frac{2x-1}{x+2}=0$$

$$2x-1=0$$

$$x=\frac{1}{2}$$

Interval, Critical Value	Test Point	Substitution	Verification
$x<-2$	$x=-3$	$\frac{2(-3)-1}{-3+2} \le 0$	False
$x=-2$	$x=-2$	$\frac{2(-2)-1}{-2+2} \le 0$	False
$-2<x<\frac{1}{2}$	$x=0$	$\frac{2(0)-1}{0+2} \le 0$	True
$x=\frac{1}{2}$	$x=\frac{1}{2}$	$\frac{2\left(\frac{1}{2}\right)-1}{\frac{1}{2}+2} \le 0$	True
$x>\frac{1}{2}$	$x=1$	$\frac{2(1)-1}{1+2} \le 0$	False

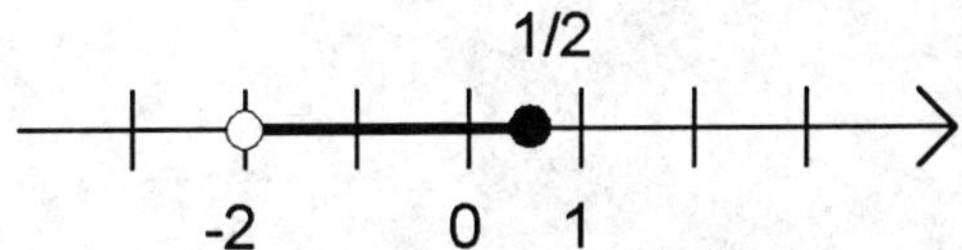

91.

$$2x^2 + 5x + 2 \le 0$$

$$(2x+1)(x+2) = 0$$

$$x = -\frac{1}{2} \quad x = -2$$

Interval, Critical Value	Test Point	Substitution	Verification
$x < -2$	$x = -3$	$2(-3)^2 + 5(-3) + 2 \le 0$	False
$x = -2$	$x = -2$	$2(-2)^2 + 5(-2) + 2 \le 0$	True
$-2 < x < -\frac{1}{2}$	$x = -1$	$2(-1)^2 + 5(-1) + 2 \le 0$	True
$x = -\frac{1}{2}$	$x = -\frac{1}{2}$	$2\left(-\frac{1}{2}\right)^2 + 5\left(-\frac{1}{2}\right) + 2 \le 0$	True
$x > -\frac{1}{2}$	$x = 0$	$2(0)^2 + 5(0) + 2 \le 0$	False

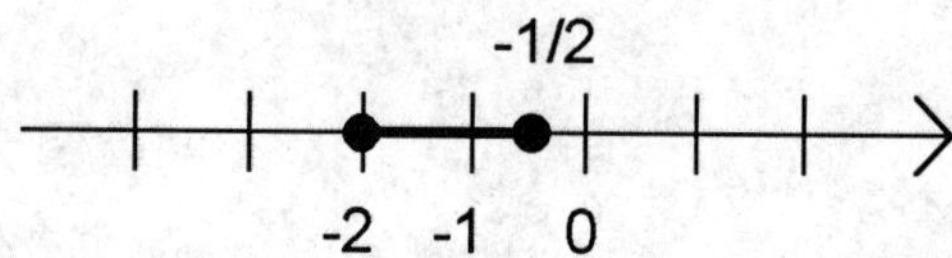

93.

$(2x+5)(3x-2)(x+1)<0$

$(2x+5)(3x-2)(x+1)=0$

$x=-\frac{5}{2}\quad x=\frac{2}{3}\quad x=-1$

Interval, Critical Value	Test Point	Substitution	Verification
$x<-\frac{5}{2}$	$x=-3$	$[2(-3)+5][3(-3)-2][-3+1]<0$	True
$x=-\frac{5}{2}$	$x=-\frac{5}{2}$	$\left[2\left(-\frac{5}{2}\right)+5\right]\left[3\left(-\frac{5}{2}\right)-2\right]\left[-\frac{5}{2}+1\right]<0$	False
$-\frac{5}{2}<x<-1$	$x=-2$	$[2(-2)+5][3(-2)-2][-2+1]<0$	False
$x=-1$	$x=-1$	$[2(-1)+5][3(-1)-2][-1+1]<0$	False
$-1<x<\frac{2}{3}$	$x=0$	$[2(0)+5][3(0)-2][0+1]<0$	True
$x=\frac{2}{3}$	$x=\frac{2}{3}$	$\left[2\left(\frac{2}{3}\right)+5\right]\left[3\left(\frac{2}{3}\right)-2\right]\left[\frac{2}{3}+1\right]<0$	False
$x>\frac{2}{3}$	$x=1$	$[2(1)+5][3(1)-2][1+1]<0$	False

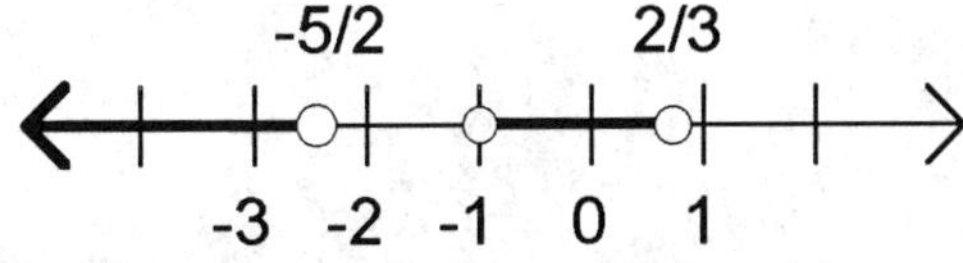

95.

$$(1-2x)(2x+1)(x-3)\le 0$$
$$(1-2x)(2x+1)(x-3)=0$$
$$x=\frac{1}{2}\quad x=-\frac{1}{2}\quad x=3$$

Interval,

Critical Value	Test Point	Substitution	Verification
$x<-\frac{1}{2}$	$x=-1$	$[1-2(-1)][2(-1)+1][-1-3]\le 0$	False
$x=-\frac{1}{2}$	$x=-\frac{1}{2}$	$\left[1-2\left(-\frac{1}{2}\right)\right]\left[2\left(-\frac{1}{2}\right)+1\right]\left[-\frac{1}{2}-3\right]\le 0$	True
$-\frac{1}{2}<x<\frac{1}{2}$	$x=0$	$[1-2(0)][2(0)+1][0-3]\le 0$	True
$x=\frac{1}{2}$	$x=\frac{1}{2}$	$\left[1-2\left(\frac{1}{2}\right)\right]\left[2\left(\frac{1}{2}\right)+1\right]\left[\frac{1}{2}-3\right]\le 0$	True
$\frac{1}{2}<x<3$	$x=1$	$[1-2(1)][2(1)+1][1-3]\le 0$	False
$x=3$	$x=3$	$[1-2(3)][2(3)+1][3-3]\le 0$	True
$x>3$	$x=4$	$[1-2(4)][2(4)+1][4-3]\le 0$	True

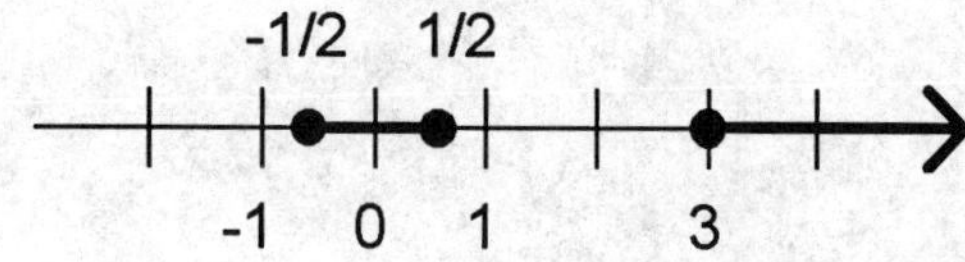

97.

$\sqrt{(2x+1)(x-3)}$

$(2x+1)(x-3)\geq 0$

$(2x+1)(x-3)=0$

$x=-\frac{1}{2}\quad x=3$

Interval,

Critical Value	Test Point	Substitution	Verification
$x<-\frac{1}{2}$	$x=-1$	$[2(-1)+1][-1-3]\geq 0$	True
$x=-\frac{1}{2}$	$x=-\frac{1}{2}$	$\left[2\left(-\frac{1}{2}\right)+1\right]\left[-\frac{1}{2}-3\right]\geq 0$	True
$-\frac{1}{2}<x<3$	$x=0$	$[2(0)+1][0-3]\geq 0$	False
$x=3$	$x=3$	$[2(3)+1][3-3]\geq 0$	True
$x>3$	$x=4$	$[2(4)+1][4-3]\geq 0$	True

$\left\{x \middle| x\leq -\frac{1}{2} \text{ or } x\geq 3\right\}$

99.

$\sqrt{2x^2+3x+1}$

$$2x^2+3x+1\geq 0$$
$$(2x+1)(x+1)=0$$

$$x=-\frac{1}{2} \qquad x=-1$$

Interval, Critical Value	Test Point	Substitution	Verification
$x<-1$	$x=-2$	$2(-2)^2+3(-2)+1\geq 0$	True
$x=-1$	$x=-1$	$2(-1)^2+3(-1)+1\geq 0$	True
$-1<x<-\frac{1}{2}$	$x=-\frac{3}{4}$	$2\left(-\frac{3}{4}\right)^2+3\left(-\frac{3}{4}\right)+1\geq 0$	False
$x=-\frac{1}{2}$	$x=-\frac{1}{2}$	$2\left(-\frac{1}{2}\right)^2+3\left(-\frac{1}{2}\right)+1\geq 0$	True
$x>-\frac{1}{2}$	$x=0$	$2(0)^2+3(0)+1\geq 0$	True

$$\left\{x \middle| x\leq -1 \text{ or } x\geq -\frac{1}{2}\right\}$$

101.

$40t - 16t^2 \geq 16 \qquad t \geq 0$

$0 \geq 16t^2 - 40t + 16$

$16t^2 - 40t + 16 = 0$

$2t^2 - 5t + 2 = 0$

$(2t-1)(t-2) = 0$

$t = \frac{1}{2} \qquad t = 2$

Interval, Critical Value	Test Point	Substitution	Verification
$t=0$	$t=0$	$40(0)-16(0)^2 \geq 16$	False
$0<t<\frac{1}{2}$	$t=\frac{1}{4}$	$40\left(\frac{1}{4}\right)-16\left(\frac{1}{4}\right)^2 \geq 16$	False
$t=\frac{1}{2}$	$t=\frac{1}{2}$	$40\left(\frac{1}{2}\right)-16\left(\frac{1}{2}\right)^2 \geq 16$	True
$\frac{1}{2}<t<2$	$t=1$	$40(1)-16(1)^2 \geq 16$	True
$t=2$	$t=2$	$40(2)-16(2)^2 \geq 16$	True
$t>2$	$t=3$	$40(3)-16(3)^2 \geq 16$	False

$\left\{t \middle| \frac{1}{2} \leq t \leq 2\right\}$

103.

Length of base : x

Length of side : $10 + \frac{1}{2}x$

$x + 2\left(10 + \frac{1}{2}x\right) \leq 100$

$x + 20 + x \leq 100$

$2x \leq 80$

$x \leq 40$

The maximum length of the base is 40 cm.

105.

Length of small bolt : x

Length of large bolt : $x + 8$

$10(x+8) \leq 200$

$x + 8 \leq 20$

$x \leq 12$

The largest length is 12 feet.

107.

$$P \le 1200 \qquad I \ge 0$$
$$I^2R \le 1200$$
$$I^2(48) \le 1200$$
$$I^2 \le 25$$
$$0 \le I \le 5$$

The maximum current is 5 amperes.

109.

$$50 < P < 75 \qquad V \ge 0$$
$$50 < \frac{150}{V} < 75$$
$$50 < \frac{150}{V} \text{ and } \frac{150}{V} < 75$$
$$V < 3 \quad \text{and } 2 < V$$

The volume is between 2 in^3 and 3 in^3.

111.

Final exam score : x

80% of 600 $= 480$

$$x + 310 \ge 480$$
$$x \ge 170$$

The lowest possible score is 170.

113.

$$-10 \le C \le 20$$
$$F = \frac{9}{5}C + 32$$
$$F - 32 = \frac{9}{5}C$$
$$\frac{5}{9}(F - 32) = C$$

$$-10 \le \frac{5}{9}(F - 32) \le 20$$
$$-18 \le F - 32 \le 36$$
$$14 \le F \le 68$$
$$14^\circ \le F \le 68^\circ$$

115.

$$67 - 1 < C < 67 + 1$$
$$66 < C < 68$$
$$66 < \frac{100W}{10} < 68$$
$$66 < 10W < 68$$
$$6.6 < W < 6.8$$

The smallest width is 6.6 inches and the largest width is 6.8 inches.

CHAPTER 2 SECTION 6

SECTION 2.6 PROGRESS CHECK

1a). (Page 131)

$|x+8|=9$

$x+8=9 \qquad x+8=-9$

$x=1 \qquad x=-17$

1b.

$|3x-4|=7$

$3x-4=7 \qquad 3x-4=-7$

$3x=11 \qquad 3x=-3$

$x=\frac{11}{3} \qquad x=-1$

2a. (Page 132)

$|x|<3$

$|x|=3$

$x=3 \qquad x=-3$

Interval, Critical Value	Test Point	Substitution	Verification
$x<-3$	$x=-4$	$\|-4\|<3$	False
$x=-3$	$x=-3$	$\|-3\|<3$	False
$-3<x<3$	$x=0$	$\|0\|<3$	True
$x=3$	$x=3$	$\|3\|<3$	False
$x>3$	$x=4$	$\|4\|<3$	False

$(-3, 3)$

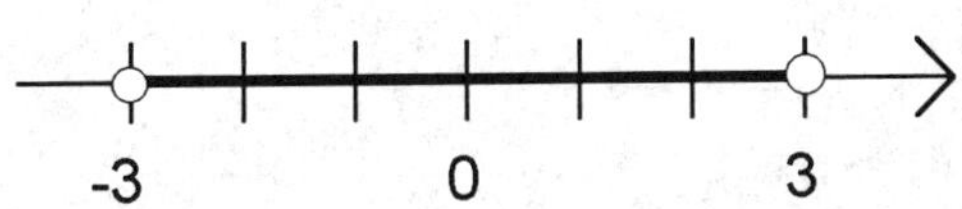

2b).

$|3x-1| \le 8$

$|3x-1| = 8$

$3x-1=8 \qquad 3x-1=-8$

$3x = 9 \qquad 3x=-7$

$x = 3 \qquad x = -\frac{7}{3}$

Interval,

Critical Value	Test Point	Substitution	Verification
$x < -\frac{7}{3}$	$x=-3$	$\|3(-3)-1\| \le 8$	False
$x = -\frac{7}{3}$	$x = -\frac{7}{3}$	$\left\|3\left(-\frac{7}{3}\right)-1\right\| \le 8$	True
$-\frac{7}{3} < x < 3$	$x=0$	$\|3(0)-1\| \le 8$	True
$x=3$	$x=3$	$\|3(3)-1\| \le 8$	True
$x>3$	$x=4$	$\|3(4)-1\| \le 8$	False

$\left[-\frac{7}{3}, 3\right]$

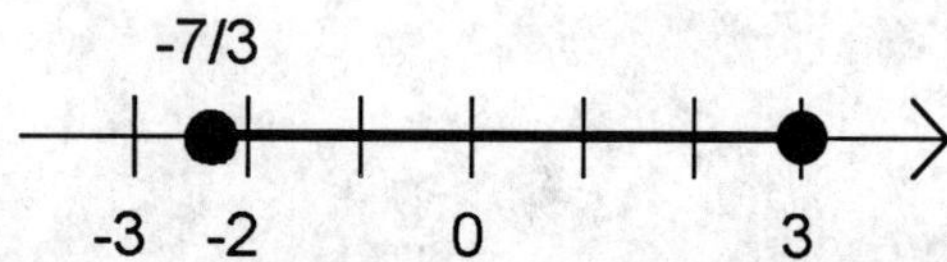

2c).

$|x| < -2$

No solution since absolute value is always nonnegative.

3a). (Page 134)

$|5x-6|>9$

$|5x-6|=9$

$5x-6=9 \qquad 5x-6=-9$

$5x=15 \qquad 5x=-3$

$x=3 \qquad x=-\frac{3}{5}$

Interval,

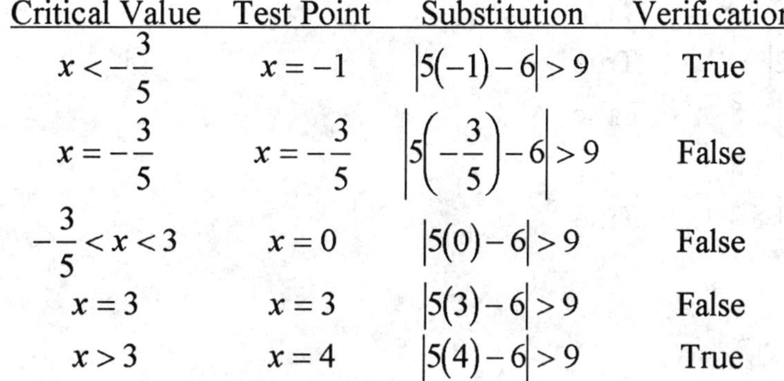

Critical Value	Test Point	Substitution	Verification
$x<-\frac{3}{5}$	$x=-1$	$\lvert 5(-1)-6\rvert>9$	True
$x=-\frac{3}{5}$	$x=-\frac{3}{5}$	$\left\lvert 5\left(-\frac{3}{5}\right)-6\right\rvert>9$	False
$-\frac{3}{5}<x<3$	$x=0$	$\lvert 5(0)-6\rvert>9$	False
$x=3$	$x=3$	$\lvert 5(3)-6\rvert>9$	False
$x>3$	$x=4$	$\lvert 5(4)-6\rvert>9$	True

$\left(-\infty,-\frac{3}{5}\right), \quad (3,\infty)$

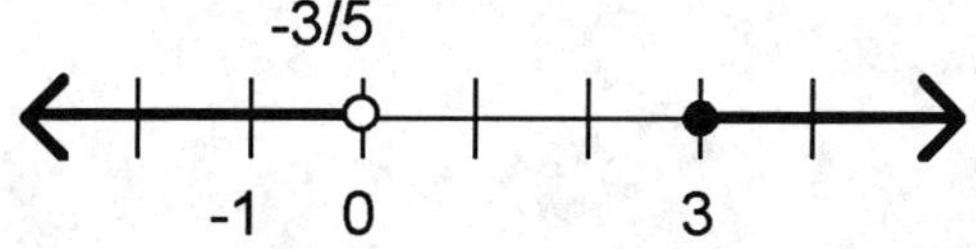

3b).

$$|2x-2| \geq 8$$

$$|2x-2| = 8$$

$2x-2=8 \qquad 2x-2=-8$

$2x=10 \qquad 2x=-6$

$x=5 \qquad x=-3$

Interval, Critical Value	Test Point	Substitution	Verification
$x<-3$	$x=-4$	$\|2(-4)-2\| \geq 8$	True
$x=-3$	$x=-3$	$\|2(-3)-2\| \geq 8$	True
$-3<x<5$	$x=0$	$\|2(0)-2\| \geq 8$	False
$x=5$	$x=5$	$\|2(5)-2\| \geq 8$	True
$x>5$	$x=6$	$\|2(6)-2\| \geq 8$	True

$(-\infty, -3], \quad [5, \infty)$

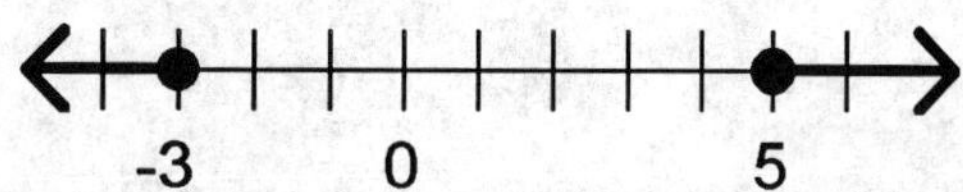

EXERCISE SET 2.6

1.

$$|x+2| = 3$$

$x+2=3 \qquad x+2=-3$

$x=1 \qquad x=-5$

3.

$$|2x-4| = 2$$

$2x-4=2 \qquad 2x-4=-2$

$2x=6 \qquad 2x=2$

$x=3 \qquad x=1$

5.

$|-3x+1|=5$

$-3x+1=5 \qquad -3x+1=-5$

$-3x=4 \qquad -3x=-6$

$x=-\frac{4}{3} \qquad x=2$

7.

$3|-4x-3|=27$

$|-4x-3|=9$

$-4x-3=9 \qquad -4x-3=-9$

$-4x=12 \qquad -4x=-6$

$x=-3 \qquad x=\frac{3}{2}$

9.

$\frac{1}{|s-1|}=\frac{1}{3}$

$3=|s-1|$

$s-1=3 \qquad s-1=-3$

$s=4 \qquad s=-2$

11.

$|x+1|>3$

$|x+1|=3$

$x+1=3 \qquad x+1=-3$

$x=2 \qquad x=-4$

Interval, Critical Value	Test Point	Substitution	Verification
$x<-4$	$x=-5$	$\|-5+1\|>3$	True
$x=-4$	$x=-4$	$\|-4+1\|>3$	False
$-4<x<2$	$x=0$	$\|0+1\|>3$	False
$x=2$	$x=2$	$\|2+1\|>3$	False
$x>2$	$x=3$	$\|3+1\|>3$	True

$x<-4$ or $x>2$

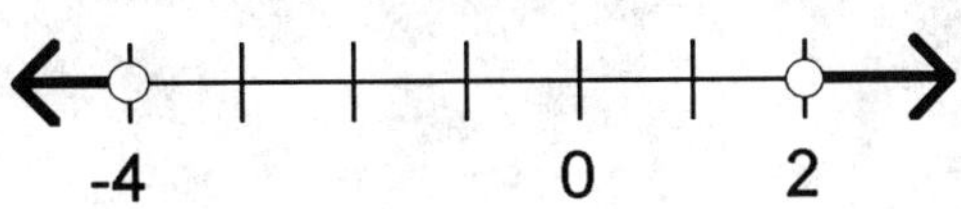

13.

$|4x-1|>3$

$|4x-1|=3$

$4x-1=3 \qquad 4x-1=-3$

$4x=4 \qquad 4x=-2$

$x=1 \qquad x=-\frac{1}{2}$

Interval,

Critical Value	Test Point	Substitution	Verification
$x<-\frac{1}{2}$	$x=-1$	$\lvert 4(-1)-1\rvert>3$	True
$x=-\frac{1}{2}$	$x=-\frac{1}{2}$	$\left\lvert 4\left(-\frac{1}{2}\right)-1\right\rvert>3$	False
$-\frac{1}{2}<x<1$	$x=0$	$\lvert 4(0)-1\rvert>3$	False
$x=1$	$x=1$	$\lvert 4(1)-1\rvert>3$	False
$x>1$	$x=2$	$\lvert 4(2)-1\rvert>3$	True

$x<-\frac{1}{2}$ or $x>1$

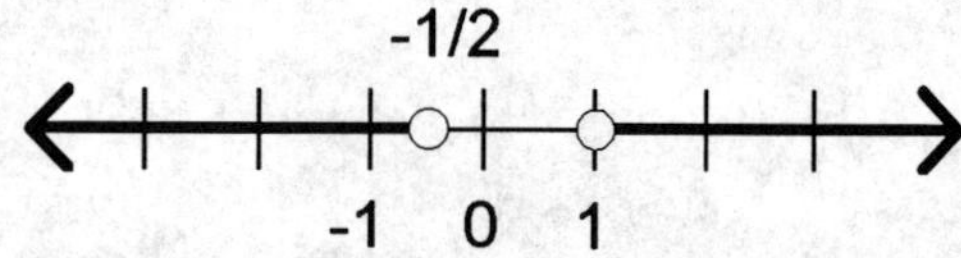

15.

$$\left|\frac{1}{3}-x\right|<\frac{2}{3}$$

$$\left|\frac{1}{3}-x\right|=\frac{2}{3}$$

$$\frac{1}{3}-x=\frac{2}{3} \qquad \frac{1}{3}-x=-\frac{2}{3}$$

$$-x=\frac{1}{3} \qquad -x=-1$$

$$x=-\frac{1}{3} \qquad x=1$$

Interval, Critical Value	Test Point	Substitution	Verification
$x<-\frac{1}{3}$	$x=-1$	$\left\lvert\frac{1}{3}-(-1)\right\rvert<\frac{2}{3}$	False
$x=-\frac{1}{3}$	$x=-\frac{1}{3}$	$\left\lvert\frac{1}{3}-\left(-\frac{1}{3}\right)\right\rvert<\frac{2}{3}$	False
$-\frac{1}{3}<x<1$	$x=0$	$\left\lvert\frac{1}{3}-0\right\rvert<\frac{2}{3}$	True
$x=1$	$x=1$	$\left\lvert\frac{1}{3}-1\right\rvert<\frac{2}{3}$	False
$x>1$	$x=2$	$\left\lvert\frac{1}{3}-2\right\rvert<\frac{2}{3}$	False

$$-\frac{1}{3}<x<1$$

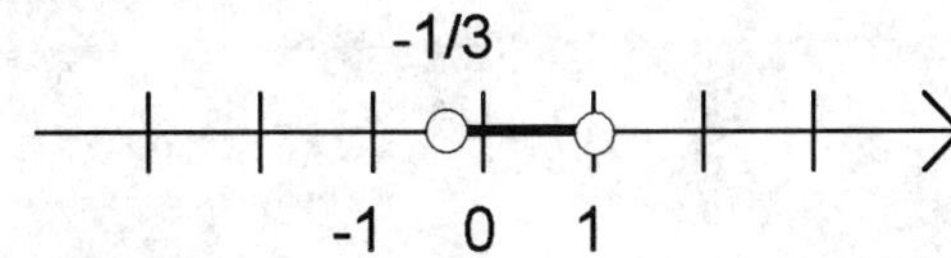

17.

$|x-3| \geq 4$

$|x-3| = 4$

$x-3=4 \qquad x-3=-4$

$x=7 \qquad x=-1$

Interval,

Critical Value	Test Point	Substitution	Verification
$x<-1$	$x=-2$	$\|-2-3\| \geq 4$	True
$x=-1$	$x=-1$	$\|-1-3\| \geq 4$	True
$-1<x<7$	$x=0$	$\|0-3\| \geq 4$	False
$x=7$	$x=7$	$\|7-3\| \geq 4$	True
$x>7$	$x=8$	$\|8-3\| \geq 4$	True

$(-\infty, -1], [7, \infty)$

19.

$$\frac{|2x-1|}{4}<2$$

$$|2x-1|<8$$

$$|2x-1|=8$$

$$2x-1=8 \qquad 2x-1=-8$$

$$2x=9 \qquad 2x=-7$$

$$x=\frac{9}{2} \qquad x=-\frac{7}{2}$$

Interval,

Critical Value	Test Point	Substitution	Verification
$x<-\frac{7}{2}$	$x=-4$	$\lvert 2(-4)-1\rvert<8$	False
$x=-\frac{7}{2}$	$x=-\frac{7}{2}$	$\left\lvert 2\left(-\frac{7}{2}\right)-1\right\rvert<8$	False
$-\frac{7}{2}<x<\frac{9}{2}$	$x=0$	$\lvert 2(0)-1\rvert<8$	True
$x=\frac{9}{2}$	$x=\frac{9}{2}$	$\left\lvert 2\left(\frac{9}{2}\right)-1\right\rvert<8$	False
$x>\frac{9}{2}$	$x=5$	$\lvert 2(5)-1\rvert<8$	False

$$\left(-\frac{7}{2},\frac{9}{2}\right)$$

21.

$$\frac{|2x+1|}{3}<0$$

$$|2x+1|<0$$

$|2x+1|$ is always greater than or equal to 0, hence there is no solution.

23.

$$\left|\frac{5-x}{3}\right|>4$$

$$\left|\frac{5-x}{3}\right|=4$$

$$\frac{5-x}{3}=4 \qquad \frac{5-x}{3}=-4$$

$$5-x=12 \qquad 5-x=-12$$

$$-7=x \qquad 17=x$$

Interval, Critical Value	Test Point	Substitution	Verification
$x<-7$	$x=-8$	$\left\|\frac{5-(-8)}{3}\right\|>4$	True
$x=-7$	$x=-7$	$\left\|\frac{5-(-7)}{3}\right\|>4$	False
$-7<x<17$	$x=0$	$\left\|\frac{5-0}{3}\right\|>4$	False
$x=17$	$x=17$	$\left\|\frac{5-17}{3}\right\|>4$	False
$x>17$	$x=18$	$\left\|\frac{5-18}{3}\right\|>4$	True

$(-\infty,-7),(17,\infty)$

25.

$$|2x+1|-3=-2$$

$$|2x+1|=1$$

$$2x+1=1 \qquad 2x+1=-1$$

$$2x=0 \qquad 2x=-2$$

$$x=0 \qquad x=-1$$

27.

$$2|3-x|+3=5$$

$$2|3-x|=2$$

$$|3-x|=1$$

$$3-x=1 \qquad 3-x=-1$$

$$-x=-2 \qquad -x=-4$$

$$x=2 \qquad x=4$$

29.

Prove : $\left|\frac{x}{y}\right| = \frac{|x|}{|y|}$, $y \neq 0$

Case i : If $x \geq 0,\ y > 0,$ then $\frac{x}{y} \geq 0.$

$$\text{Thus } \left|\frac{x}{y}\right| = \frac{x}{y} = \frac{|x|}{|y|}$$

Case ii : If $x \geq 0,\ y < 0,$ then $\frac{x}{y} \leq 0.$

$$\text{Thus } \left|\frac{x}{y}\right| = -\frac{x}{y} = \frac{x}{-y} = \frac{|x|}{|y|}.$$

Case iii : If $x < 0,\ y > 0,$ then $\frac{x}{y} < 0.$

$$\text{Thus } \left|\frac{x}{y}\right| = -\frac{x}{y} = \frac{-x}{y} = \frac{|x|}{|y|}$$

Case iv : If $x < 0,\ y < 0,$ then $\frac{x}{y} > 0.$

$$\text{Thus } \left|\frac{x}{y}\right| = \frac{x}{y} = \frac{-x}{-y} = \frac{|x|}{|y|}.$$

$$\text{Hence } \left|\frac{x}{y}\right| = \frac{|x|}{|y|},\ y \neq 0.$$

31.

$|x - 100| \leq 2$ $\quad\quad x \geq 0$

$|x - 100| = 2$

$x - 100 = 2 \quad x - 100 = -2$

$x = 102 \quad\quad x = 98$

Interval,

Critical Value	Test Point	Substitution	Verification
$x < 98$	$x = 0$	$\|0 - 100\| \leq 2$	False
$x = 98$	$x = 98$	$\|98 - 100\| \leq 2$	True
$98 < x < 102$	$x = 100$	$\|100 - 100\| \leq 2$	True
$x = 102$	$x = 102$	$\|102 - 100\| \leq 2$	True
$x > 102$	$x = 103$	$\|103 - 100\| \leq 2$	False

$98 \leq x \leq 102$

33.

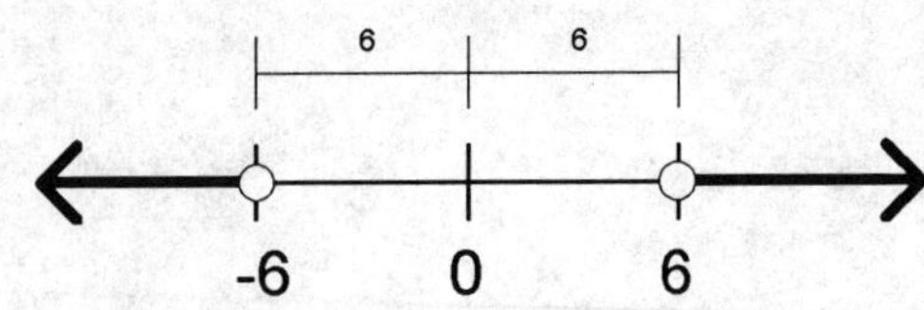

$|x-0|>6$

$|x|>6$

35.

$x \geq 5$ or $x \leq 1$

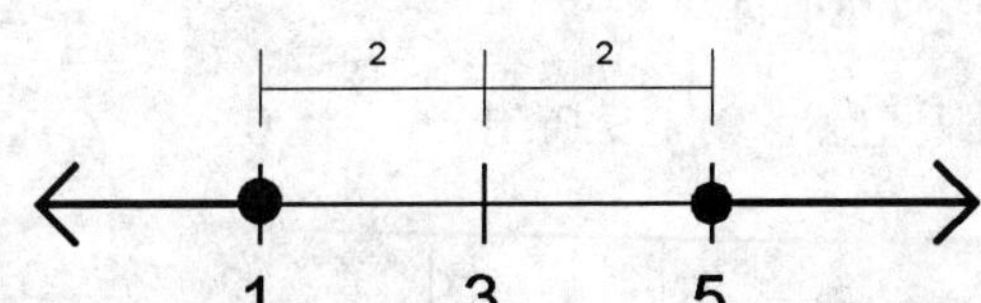

$|x-3| \geq 2$

37.

$|x-4|=10$

$x-4=10 \qquad x-4=-10$

$x=14 \qquad x=-6$

39.

The distance between x and 4 is 3 times the distance between $2x$ and 6.

$|x-4|=3|2x-6|$

$x-4=3(2x-6) \qquad -(x-4)=3(2x-6)$

$x-4=6x-18 \qquad -x+4=6x-18$

$-5x=-14 \qquad -7x=-22$

$x=\frac{14}{5} \qquad x=\frac{22}{7}$

41.

$|x-\mu|<k\sigma$

$-k\sigma<x-\mu<k\sigma$

$\mu-k\sigma<x<\mu+k\sigma$

CHAPTER 2 REVIEW EXERCISES

1.

$$3x - 5 = 3$$
$$3x = 8$$
$$x = \frac{8}{3}$$

2.

$$2(2x-3) - 3(x+1) = -9$$
$$4x - 6 - 3x - 3 = -9$$
$$x - 9 = -9$$
$$x = 0$$

3.

$$\frac{2-x}{3-x} = 4$$
$$2 - x = 4(3-x)$$
$$2 - x = 12 - 4x$$
$$3x = 10$$
$$x = \frac{10}{3}$$

4.

$$k - 2x = 4kx$$
$$k = 4kx + 2x$$
$$k = x(4k+2)$$
$$\frac{k}{4k+2} = x$$
$$\text{or } x = \frac{k}{2(2k+1)}$$

5.

Length : x

Width : $2x - 4$

$$2(x) + 2(2x-4) = 12$$
$$2x + 4x - 8 = 12$$
$$6x = 20$$
$$x = \frac{10}{3}$$
$$2x - 4 = \frac{8}{3}$$

The width is $\frac{8}{3}$ cm and the length is $\frac{10}{3}$ cm.

6.

Number of quarters : x

Number of dimes : $2x + 4$

$$25x + 10(2x+4) = 265$$
$$25x + 20x + 40 = 265$$
$$45x = 225$$
$$x = 5$$
$$2x + 4 = 14$$

There are 5 quarters and 14 dimes.

7.

	rate	·	time	=	distance
Going	150		t		$150t$
Returning	100		$4-t$		$100(4-t)$

$$150t = 100(4-t)$$
$$150t = 400-100t$$
$$250t = 400$$
$$t = \frac{8}{5}$$
$$150t = 240$$

The village is 240 miles from home base.

8.

Time for B alone : x

	rate	·	time	=	work done
A	$\frac{1}{3}$		2		$\frac{2}{3}$
B	$\frac{1}{x}$		2		$\frac{2}{x}$

$$\frac{2}{3}+\frac{2}{x}=1$$
$$3x\left[\frac{2}{3}+\frac{2}{x}\right]=3x(1)$$
$$2x+6=3x$$
$$6=x$$

It would take B 6 hours to do the job working alone.

9.

False : it is true only for two values of x, $\pm\sqrt{3}$.

10.

False : $3(3)-1=8\neq 10$

11.

$$x^2-x-20=0$$
$$(x-5)(x+4)=0$$

$x-5=0$ or $x+4=0$

$x=5$ or $x=-4$

12.

$$6x^2-11x+4=0$$
$$(2x-1)(3x-4)=0$$

$2x-1=0$ or $3x-4=0$

$2x=1$ or $3x=4$

$x=\frac{1}{2}$ or $x=\frac{4}{3}$

13.

$$x^2 - 2x + 6 = 0$$
$$x^2 - 2x = -6$$
$$\left[\frac{1}{2}(-2)\right]^2 = 1$$
$$x^2 - 2x + 1 = -6 + 1$$
$$(x-1)^2 = -5$$
$$x - 1 = \pm\sqrt{-5}$$
$$x = 1 \pm i\sqrt{5}$$

14.

$$2x^2 - 4x + 3 = 0$$
$$a = 2 \quad b = -4 \quad c = 3$$
$$x = \frac{-b \pm \sqrt{b^2 - 4ac}}{2a}$$
$$= \frac{-(-4) \pm \sqrt{(-4)^2 - 4(2)(3)}}{2(2)}$$
$$= \frac{4 \pm \sqrt{-8}}{4}$$
$$= \frac{4 \pm 2i\sqrt{2}}{4}$$
$$= \frac{2 \pm i\sqrt{2}}{2}$$

15.

$$3x^2 + 2x - 1 = 0$$
$$a = 3 \quad b = 2 \quad c = -1$$
$$x = \frac{-b \pm \sqrt{b^2 - 4ac}}{2a}$$
$$= \frac{-2 \pm \sqrt{2^2 - 4(3)(-1)}}{2(3)}$$
$$= \frac{-2 \pm \sqrt{16}}{6}$$
$$x = \frac{-2+4}{6} \quad \text{or} \quad x = \frac{-2-4}{6}$$
$$x = \frac{1}{3} \quad \text{or} \quad x = -1$$

16.

$$49x^2 - 9 = 0$$
$$49x^2 = 9$$
$$x^2 = \frac{9}{49}$$
$$x = \pm\frac{3}{7}$$

17.

$$kx^2 - 3\pi = 0$$
$$kx^2 = 3\pi$$
$$x^2 = \frac{3\pi}{k}$$
$$x = \pm\sqrt{\frac{3\pi}{k}}$$
$$x = \pm\frac{\sqrt{3\pi k}}{k}$$

18.

$$x^2 + x = 12$$
$$x^2 + x - 12 = 0$$
$$(x+4)(x-3) = 0$$
$$x + 4 = 0 \qquad x - 3 = 0$$
$$x = -4 \qquad x = 3$$

19.

$$3r^2 = 2r + 5$$
$$3r^2 - 2r - 5 = 0$$
$$b^2 - 4ac = (-2)^2 - 4(3)(-5) = 64 > 0 \text{(perfect square)}$$

two rational, real roots.

20.

$$4x^2 + 20x + 25 = 0$$
$$b^2 - 4ac = (20)^2 - 4(4)(25) = 0$$

one real, double root.

21.

$$6y^2 - 2y = -7$$
$$6y^2 - 2y + 7 = 0$$
$$b^2 - 4ac = (-2)^2 - 4(6)(7) = -164 < 0$$

two complex roots

22.

$$\sqrt{x} + 2 = x$$

Let $u = \sqrt{x}$

$$u + 2 = u^2$$
$$0 = u^2 - u - 2$$
$$0 = (u-2)(u+1)$$
$$u = 2 \qquad u = -1$$
$$\sqrt{x} = 2 \qquad \sqrt{x} = -1$$
$$x = 4 \qquad \text{no solution}$$
$$x = 4$$

23.

$$\sqrt{x+3}+\sqrt{2x-3}=6$$
$$\sqrt{x+3}=6-\sqrt{2x-3}$$
$$x+3=36-12\sqrt{2x-3}+2x-3$$
$$x+3=33+2x-12\sqrt{2x-3}$$
$$-x-30=-12\sqrt{2x-3}$$
$$x+30=12\sqrt{2x-3}$$

$$x^2+60x+900=144(2x-3)$$
$$x^2+60x+900=288x-432$$

$$x^2-228x+1332=0$$
$$(x-6)(x-222)=0$$
$$x=6 \qquad x=222$$

Check : $x=6$ $\qquad$ $x=222$

$$\sqrt{6+3}+\sqrt{2(6)-3}=6 \quad ? \qquad \sqrt{222+3}+\sqrt{2(222)-3}=6 \quad ?$$
$$\sqrt{9}+\sqrt{9}=6 \quad ? \qquad \sqrt{225}+\sqrt{441}=6 \quad ?$$
$$3+3=6 \quad ? \qquad 15+21=6 \quad ?$$
$$6=6 \qquad 36\neq 6$$

Solution : $x=6$

24.

$$x^4 - 4x^2 + 3 = 0$$

Let $u = x^2$

$$u^2 - 4u + 3 = 0$$
$$(u-3)(u-1) = 0$$

$u = 3$ $\quad$ $u = 1$

$x^2 = 3$ $\quad$ $x^2 = 1$

$x = \pm\sqrt{3}$ $\quad$ $x = \pm 1$

25.

$$\left(1-\frac{2}{x}\right)^2 - 8\left(1-\frac{2}{x}\right) + 15 = 0$$

Let $u = 1 - \dfrac{2}{x}$

$$u^2 - 8u + 15 = 0$$
$$(u-5)(u-3) = 0$$

$u = 5$ $\quad$ $u = 3$

$1 - \dfrac{2}{x} = 5$ $\quad$ $1 - \dfrac{2}{x} = 3$

$-\dfrac{2}{x} = 4$ $\quad$ $-\dfrac{2}{x} = 2$

$-\dfrac{1}{2} = x$ $\quad$ $-1 = x$

26.

Number of people attending : x

Cost per person : $\dfrac{420}{x}$

$$(x+10)\left(\frac{420}{x} - 1\right) = 420$$
$$(x+10)\left(\frac{420}{x} - 1\right)(x) = 420(x)$$
$$(x+10)(420 - x) = 420x$$
$$-x^2 + 410x + 4200 = 420x$$
$$0 = x^2 + 10x - 4200$$
$$0 = (x+70)(x-60)$$
$$x = -70 \text{ or } x = 60$$

60 people attended the meeting.

27.

$$3 \le 2x + 1$$
$$2 \le 2x$$
$$1 \le x$$

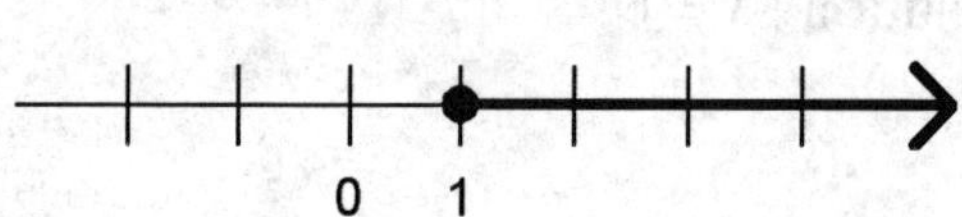

28.

$$-4 < -2x + 1 \le 10$$
$$-5 < -2x \le 9$$
$$\frac{5}{2} > x \ge -\frac{9}{2}$$
$$-\frac{9}{2} \le x < \frac{5}{2}$$

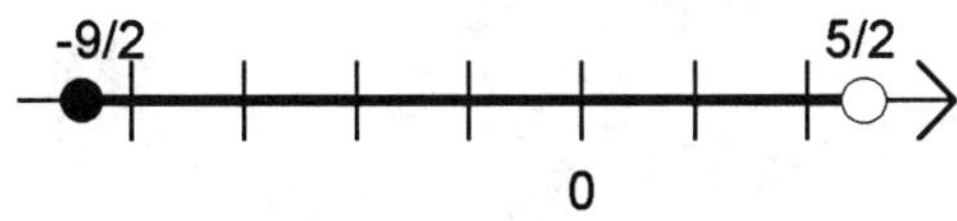

29.

$$2(a+5) > 3a+2$$
$$2a+10 > 3a+2$$
$$8 > a$$

$(-\infty, 8)$

30.

$$\frac{-1}{2x-5} \le 0 \qquad \text{Undefined when } 2x-5=0$$
$$x = \frac{5}{2}$$

Interval,

Critical Value	Test Point	Substitution	Verification
$x < \frac{5}{2}$	$x = 0$	$\frac{-1}{2(0)-5} \le 0$	False
$x = \frac{5}{2}$	$x = \frac{5}{2}$	$\frac{-1}{2\left(\frac{5}{2}\right)-5} \le 0$	False
$x > \frac{5}{2}$	$x = 3$	$\frac{-1}{2(3)-5} \le 0$	True

$\left(\frac{5}{2}, \infty\right)$

31.

$$\frac{2x}{3}+\frac{1}{2}\ge\frac{x}{2}-1$$

$$6\left(\frac{2x}{3}+\frac{1}{2}\right)\ge 6\left(\frac{x}{2}-1\right)$$

$$4x+3\ge 3x-6$$

$$x\ge -9$$

$[-9,\ \infty)$

32.

$$|3x+2|=7$$

$$3x+2=7 \qquad 3x+2=-7$$

$$3x=5 \qquad 3x=-9$$

$$x=\frac{5}{3} \qquad x=-3$$

33.

$$|4x-1|=5$$

$$4x-1=5 \qquad 4x-1=-5$$

$$4x=6 \qquad 4x=-4$$

$$x=\frac{3}{2} \qquad x=-1$$

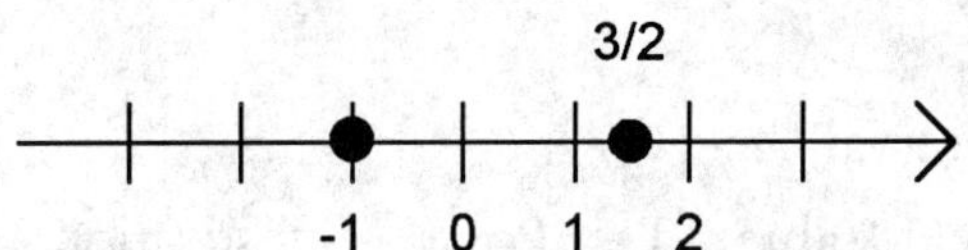

34.

$|2x+1| > 7$

$|2x+1| = 7$

$2x+1=7 \qquad 2x+1=-7$

$2x=6 \qquad 2x=-8$

$x=3 \qquad x=-4$

Interval, Critical Value	Test Point	Substitution	Verification
$x<-4$	$x=-5$	$\lvert 2(-5)+1\rvert > 7$	True
$x=-4$	$x=-4$	$\lvert 2(-4)+1\rvert > 7$	False
$-4<x<3$	$x=0$	$\lvert 2(0)+1\rvert > 7$	False
$x=3$	$x=3$	$\lvert 2(3)+1\rvert > 7$	False
$x>3$	$x=4$	$\lvert 2(4)+1\rvert > 7$	True

$x<-4$ or $x>3$

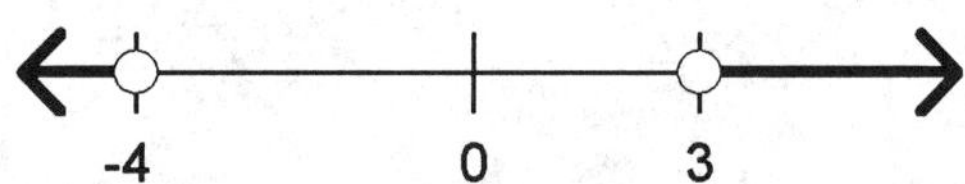

35.

$$|2-5x|<1$$
$$|2-5x|=1$$

$$2-5x=1 \qquad 2-5x=-1$$
$$-5x=-1 \qquad -5x=-3$$
$$x=\frac{1}{5} \qquad x=\frac{3}{5}$$

Interval,

Critical Value	Test Point	Substitution	Verification
$x<\frac{1}{5}$	$x=0$	$\lvert 2-5(0)\rvert<1$	False
$x=\frac{1}{5}$	$x=\frac{1}{5}$	$\left\lvert 2-5\left(\frac{1}{5}\right)\right\rvert<1$	False
$\frac{1}{5}<x<\frac{3}{5}$	$x=\frac{2}{5}$	$\left\lvert 2-5\left(\frac{2}{5}\right)\right\rvert<1$	True
$x=\frac{3}{5}$	$x=\frac{3}{5}$	$\left\lvert 2-5\left(\frac{3}{5}\right)\right\rvert<1$	False
$x>\frac{3}{5}$	$x=1$	$\lvert 2-5(1)\rvert<1$	False

$$\left(\frac{1}{5},\frac{3}{5}\right)$$

36.

$$|3x-2| \geq 6$$

$$|3x-2| = 6$$

$$3x-2=6 \qquad 3x-2=-6$$

$$3x=8 \qquad 3x=-4$$

$$x=\frac{8}{3} \qquad x=-\frac{4}{3}$$

Interval,

Critical Value	Test Point	Substitution	Verification
$x<-\frac{4}{3}$	$x=-2$	$\|3(-2)-2\| \geq 6$	True
$x=-\frac{4}{3}$	$x=-\frac{4}{3}$	$\left\|3\left(-\frac{4}{3}\right)-2\right\| \geq 6$	True
$-\frac{4}{3}<x<\frac{8}{3}$	$x=0$	$\|3(-0)-2\| \geq 6$	False
$x=\frac{8}{3}$	$x=\frac{8}{3}$	$\left\|3\left(\frac{8}{3}\right)-2\right\| \geq 6$	True
$x>\frac{8}{3}$	$x=3$	$\|3(3)-2\| \geq 6$	True

$$\left(-\infty,-\frac{4}{3}\right],\left[\frac{8}{3},\infty\right)$$

37.

$\sqrt{2x^2 - x - 6}$

$2x^2 - x - 6 \geq 0$

$(2x+3)(x-2) = 0$

$x = -\frac{3}{2} \qquad x = 2$

Interval, Critical Value	Test Point	Substitution	Verification
$x < -\frac{3}{2}$	$x = -2$	$2(-2)^2 - (-2) - 6 \geq 0$	True
$x = -\frac{3}{2}$	$x = -\frac{3}{2}$	$2\left(-\frac{3}{2}\right)^2 - \left(-\frac{3}{2}\right) - 6 \geq 0$	True
$-\frac{3}{2} < x < 2$	$x = 0$	$2(0)^2 - 0 - 6 \geq 0$	False
$x = 2$	$x = 2$	$2(2)^2 - 2 - 6 \geq 0$	True
$x > 2$	$x = 3$	$2(3)^2 - 3 - 6 \geq 0$	True

$x \leq -\frac{3}{2}$ or $x \geq 2$

38.

$x^2 + 4x - 5 \leq 0$

$(x+5)(x-1) = 0$

$x = -5 \qquad x = 1$

Interval, Critical Value	Test Point	Substitution	Verification
$x < -5$	$x = -6$	$(-6)^2 + 4(-6) - 5 \leq 0$	False
$x = -5$	$x = -5$	$(-5)^2 + 4(-5) - 5 \leq 0$	True
$-5 < x < 1$	$x = 0$	$(0)^2 + 4(0) - 5 \leq 0$	True
$x = 1$	$x = 1$	$(1)^2 + 4(1) - 5 \leq 0$	True
$x > 1$	$x = 2$	$(2)^2 + 4(2) - 5 \leq 0$	False

$[-5, 1]$

39.

$\frac{2x+1}{x+5} \geq 0$ Undefined when $x+5=0$

$x=-5$

$$\frac{2x+1}{x+5}=0$$
$$2x+1=0$$
$$x=-\frac{1}{2}$$

Interval,

Critical Value	Test Point	Substitution	Verification
$x<-5$	$x=-6$	$\frac{2(-6)+1}{-6+5} \geq 0$	True
$x=-5$	$x=-5$	$\frac{2(-5)+1}{-5+5} \geq 0$	False
$-5<x<-\frac{1}{2}$	$x=-1$	$\frac{2(-1)+1}{-1+5} \geq 0$	False
$x=-\frac{1}{2}$	$x=-\frac{1}{2}$	$\frac{2\left(-\frac{1}{2}\right)+1}{-\frac{1}{2}+5} \geq 0$	True
$x>-\frac{1}{2}$	$x=0$	$\frac{2(0)+1}{0+5} \geq 0$	True

$(-\infty,-5),\left[-\frac{1}{2},\infty\right)$

40.

$(3-x)(2x+3)(x+2)<0$

$(3-x)(2x+3)(x+2)=0$

$x=3 \qquad x=-\frac{3}{2} \qquad x=-2$

Interval,

Critical Value	Test Point	Substitution	Verification
$x<-2$	$x=-3$	$[3-(-3)][2(-3)+3][-3+2]<0$	False
$x=-2$	$x=-2$	$[3-(-2)][2(-2)+3][-2+2]<0$	False
$-2<x<-\frac{3}{2}$	$x=-\frac{7}{4}$	$\left[3-\left(-\frac{7}{4}\right)\right]\left[2\left(-\frac{7}{4}\right)+3\right]\left[-\frac{7}{4}+2\right]<0$	True
$x=-\frac{3}{2}$	$x=-\frac{3}{2}$	$\left[3-\left(-\frac{3}{2}\right)\right]\left[2\left(-\frac{3}{2}\right)+3\right]\left[-\frac{3}{2}+2\right]<0$	False
$-\frac{3}{2}<x<3$	$x=0$	$[3-0][2(0)+3][0+2]<0$	False
$x=3$	$x=3$	$[3-3][2(3)+3][3+2]<0$	False
$x>3$	$x=4$	$[3-4][2(4)+3][4+2]<0$	True

$\left(-2, -\frac{3}{2}\right), (3, \infty)$

41.

(10 rows)(15 seats to a row) = 150 seats

double seating : 2(150) = 300 seats

number of chairs added per row : x

$(10+x)(15+x)=300$

$150+25x+x^2=300$

$x^2+25x-150=0$

$(x+30)(x-5)=0$

$x=-30 \qquad x=5$

5 chairs need to be added to each row.

42.

Side of square : x

perimeter of square : $P=4x$

circumferance of circle : $P=2\pi r$

$4x=2\pi r$

$x=\frac{\pi r}{2}$

area of square : $x^2=\left(\frac{\pi r}{2}\right)^2=\frac{\pi^2 r^2}{4}$

area of circle : πr^2

Comparing areas : $\pi r^2 > \frac{1}{4}\pi \cdot \pi r^2$

$3.14r^2 > 2.47r^2$

The area of the circle is greater.

43.

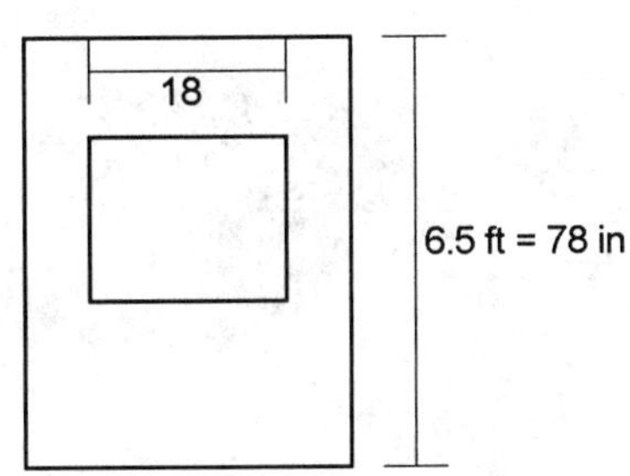

Distance above window : x
Distance below window : $2x-6$
Height of window : $78-[x+(2x-6)]=84-3x$

Area $=18(84-3x)\geq 324$

$$1512-54x\geq 324$$
$$-54x\geq -1188$$
$$x\leq 22$$

The maximum distance is 22 inches.

44.

$$\left|\frac{\mu-\bar{x}}{\frac{\sigma}{\sqrt{n}}}\right|<1.96$$

$$-1.96<\frac{\mu-\bar{x}}{\frac{\sigma}{\sqrt{n}}}<1.96$$

$$\frac{-1.96\sigma}{\sqrt{n}}<\mu-\bar{x}<\frac{1.96\sigma}{\sqrt{n}}$$

$$\bar{x}-\frac{1.96\sigma}{\sqrt{n}}<\mu<\bar{x}+\frac{1.96\sigma}{\sqrt{n}}$$

45a).

$$E=\frac{1}{2}mv^2$$
$$5000=\frac{1}{2}(10000)v^2$$
$$5000=5000v^2$$
$$1=v^2$$
$$v=\pm 1$$

The speed is 1 m/sec.

45b).

$$E=\frac{1}{2}mv^2$$
$$5000=\frac{1}{2}\left[\frac{1}{2}(10000)\right]v^2$$
$$5000=2500v^2$$
$$2=v^2$$
$$v=\pm\sqrt{2}$$

The speed is $\sqrt{2}$ m/sec.

46.

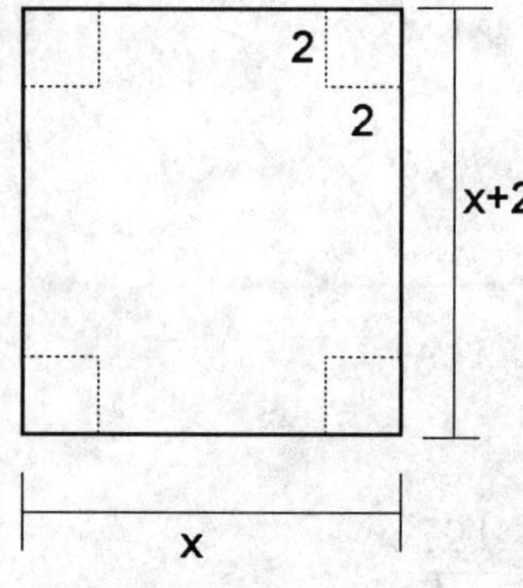

$$V = (\text{width})(\text{length})(\text{height})$$
$$96 = (x-4)[(x+2)-4](2)$$
$$96 = (x-4)(x-2)(2)$$
$$48 = x^2 - 6x + 8$$
$$0 = x^2 - 6x - 40$$
$$0 = (x-10)(x+4)$$
$$x = 10 \quad \text{or} \quad x = -4$$
$$x - 4 = 6$$
$$(x+2) - 4 = 8$$

The dimensions are
6 inches by 8 inches by 2 inches.

CHAPTER 2 REVIEW TEST

1.

$$5-4y=2$$
$$-4y=-3$$
$$y=\frac{3}{4}$$

2.

$$\frac{2+5y}{3y-1}=6$$
$$2+5y=6(3y-1)$$
$$2+5y=18y-6$$
$$-13y=-8$$
$$y=\frac{8}{13}$$

3.

Length of base : x

Length of one side : $x-2$

Length of other side : $3+\frac{1}{2}x$

$$(x)+(x-2)+\left(3+\frac{1}{2}x\right)=15$$
$$\frac{5}{2}x+1=15$$
$$\frac{5}{2}x=14$$
$$x=\frac{28}{5}$$
$$x-2=\frac{18}{5}$$
$$3+\frac{1}{2}x=\frac{29}{5}$$

The sides have lengths

$\frac{28}{5}$ m, $\frac{18}{5}$ m, and $\frac{29}{5}$ m.

4.

Amount at 6.5% : x

Amount at 7.5% : $x+200$

Amount at 9% : $2x+300$

$$0.065x+0.075(x+200)+0.09(2x+300)=1962$$
$$0.065x+0.075x+15+0.18x+27=1962$$
$$0.32x=1920$$
$$x=6{,}000$$
$$x+200=6{,}200$$
$$2x+300=12{,}300$$

\$6,000 was invested at 6.5%, \$6,200 was invested at 7.5% and \$12,300 was invested at 9%.

5.

$$(2x-1)^2 = 4x^2 - 4x + 1$$
$$4x^2 - 4x + 1 = 4x^2 - 4x + 1$$

True, it is an identity.

6.

$$x^2 - 5x = 14$$
$$x^2 - 5x - 14 = 0$$
$$(x-7)(x+2) = 0$$
$$x - 7 = 0 \quad \text{or} \quad x + 2 = 0$$
$$x = 7 \quad \text{or} \quad x = -2$$

7.

$$5x^2 - x + 4 = 0$$
$$x^2 - \frac{1}{5}x = -\frac{4}{5}$$
$$\left[\frac{1}{2}\left(-\frac{1}{5}\right)\right]^2 = \frac{1}{100}$$
$$x^2 - \frac{1}{5}x + \frac{1}{100} = -\frac{4}{5} + \frac{1}{100}$$
$$\left(x - \frac{1}{10}\right)^2 = \frac{-79}{100}$$
$$x - \frac{1}{10} = \pm\sqrt{\frac{-79}{100}}$$
$$x = \frac{1}{10} \pm \frac{\sqrt{79}i}{10}$$
$$x = \frac{1 \pm i\sqrt{79}}{10}$$

8.

$$12x^2 + 5x - 3 = 0$$
$$a = 12 \quad b = 5 \quad c = -3$$
$$x = \frac{-b \pm \sqrt{\mathrm{b}^2 - 4ac}}{2a}$$
$$= \frac{-5 \pm \sqrt{5^2 - 4(12)(-3)}}{2(12)}$$
$$= \frac{-5 \pm \sqrt{169}}{24}$$
$$x = \frac{-5+13}{24} \qquad x = \frac{-5-13}{24}$$
$$x = \frac{1}{3} \qquad x = -\frac{3}{4}$$

9.

$$(2x-5)^2+9=0$$
$$(2x-5)^2=-9$$
$$2x-5=\pm\sqrt{-9}$$
$$2x=5\pm 3i$$
$$x=\frac{5\pm 3i}{2}$$

10.

$$2+\frac{1}{x}-\frac{3}{x^2}=0$$

Let $u=\frac{1}{x}$

$$2+u-3u^2=0$$
$$(2+3u)(1-u)=0$$

$u=-\frac{2}{3}$ $\quad$ $u=1$

$\frac{1}{x}=-\frac{2}{3}$ $\quad$ $\frac{1}{x}=1$

$x=-\frac{3}{2}$ $\quad$ $x=1$

11.

$6x^2+x-2=0$

$b^2-4ac=(1)^2-4(6)(-2)=49>0$ (Perfect square)

two rational, real roots

12.

$3x^2-2x=-6$

$3x^2-2x+6=0$

$b^2-4ac=(-2)^2-4(3)(6)=-68<0$

two complex roots

13.

$$x-\sqrt{4-3x}=-8$$
$$x+8=\sqrt{4-3x}$$
$$x^2+16x+64=4-3x$$
$$x^2+19x+60=0$$
$$(x+4)(x+15)=0$$

$x=-4 \qquad x=-15$

Check : $x=-4$ $\qquad$ $x=-15$

$$-4-\sqrt{4-3(-4)}=-8 \ ? \qquad -15-\sqrt{4-3(-15)}=-8 \ ?$$
$$-4-\sqrt{16}=-8 \ ? \qquad -15-\sqrt{49}=-8 \ ?$$
$$-4-4=-8 \ ? \qquad -15-7=-8 \ ?$$
$$-8=-8 \qquad -22\neq-8$$

Solution : $x=-4$

14.

$$3x^4+5x^2-2=0$$

Let $u=x^2$

$$3u^2+5u-2=0$$
$$(3u-1)(u+2)=0$$

$$u=\frac{1}{3} \qquad u=-2$$
$$x^2=\frac{1}{3} \qquad x^2=-2$$
$$x=\pm\sqrt{\frac{1}{3}} \qquad x=\pm\sqrt{-2}$$
$$x=\pm\frac{\sqrt{3}}{3} \qquad x=\pm i\sqrt{2}$$

15.

Width : x

Length : $\frac{96}{x}$

$$(x+2)\left(\frac{96}{x}+2\right)=140$$
$$(x+2)\left(\frac{96}{x}+2\right)(x)=140(x)$$
$$(x+2)(96+2x)=140x$$
$$(x+2)(48+x)(2)=140x$$
$$x^2+50x+96=70x$$
$$x^2-20x+96=0$$
$$(x-8)(x-12)=0$$

$$x=8 \qquad x=12$$
$$\frac{96}{x}=12 \qquad \frac{96}{x}=8$$

The dimensions are 8 m by 12 m.

16.

$$-1 \le 2x + 3 < 5$$
$$-4 \le 2x < 2$$
$$-2 \le x < 1$$

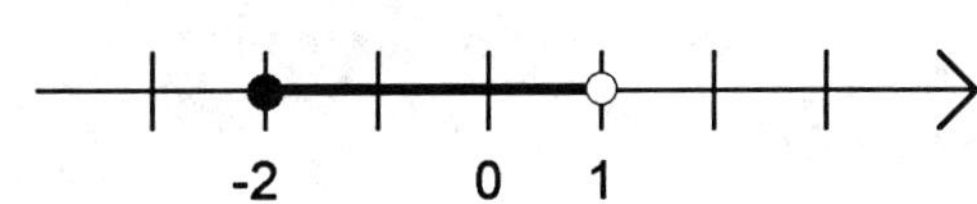

17.

$$3(2a-1) - 4(a+2) \le 4$$
$$6a - 3 - 4a - 8 \le 4$$
$$2a - 11 \le 4$$
$$2a \le 15$$
$$a \le \frac{15}{2}$$

$$\left(-\infty, \frac{15}{2}\right]$$

18.

$$-2 \le 2 - x \le 6$$
$$-4 \le -x \le 4$$
$$4 \ge x \ge -4$$

$$[-4, 4]$$

19.

$$|4x - 1| = 9$$

$$4x - 1 = 9 \qquad 4x - 1 = -9$$
$$4x = 10 \qquad 4x = -8$$
$$x = \frac{5}{2} \qquad x = -2$$

20.

$|2x-1| \le 5$

$|2x-1| = 5$

$2x-1=5 \qquad 2x-1=-5$

$2x=6 \qquad 2x=-4$

$x=3 \qquad x=-2$

Interval,

Critical Value	Test Point	Substitution	Verification
$x<-2$	$x=-3$	$\|2(-3)-1\| \le 5$	False
$x=-2$	$x=-2$	$\|2(-2)-1\| \le 5$	True
$-2<x<3$	$x=0$	$\|2(0)-1\| \le 5$	True
$x=3$	$x=3$	$\|2(3)-1\| \le 5$	True
$x>3$	$x=4$	$\|2(4)-1\| \le 5$	False

$-2 \le x \le 3$

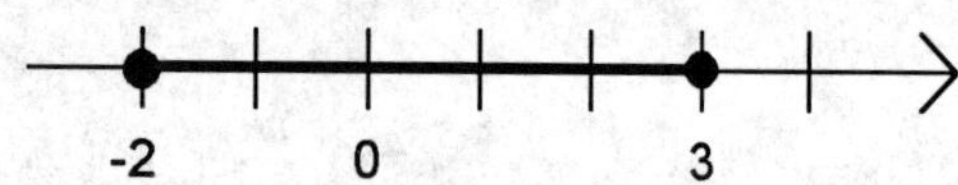

21.

$|1-3x|>5$

$|1-3x|=5$

$1-3x=5 \qquad 1-3x=-5$

$-3x=4 \qquad -3x=-6$

$x=-\dfrac{4}{3} \qquad x=2$

Interval, Critical Value	Test Point	Substitution	Verification
$x<-\frac{4}{3}$	$x=-2$	$\lvert 1-3(-2)\rvert>5$	True
$x=-\frac{4}{3}$	$x=-\frac{4}{3}$	$\left\lvert 1-3\left(-\frac{4}{3}\right)\right\rvert>5$	False
$-\frac{4}{3}<x<2$	$x=0$	$\lvert 1-3(0)\rvert>5$	False
$x=2$	$x=2$	$\lvert 1-3(2)\rvert>5$	False
$x>2$	$x=3$	$\lvert 1-3(3)\rvert>5$	True

$\left(-\infty,-\dfrac{4}{3}\right),(2,\infty)$

22.

$\sqrt{3x^2 - 4x + 1}$

$$3x^2 - 4x + 1 \geq 0$$

$$(3x - 1)(x - 1) = 0$$

$x = \frac{1}{3}$ $\quad\quad$ $x = 1$

Interval, Critical Value	Test Point	Substitution	Verification
$x < \frac{1}{3}$	$x = 0$	$3(0)^2 - 4(0) + 1 \geq 0$	True
$x = \frac{1}{3}$	$x = \frac{1}{3}$	$3\left(\frac{1}{3}\right)^2 - 4\left(\frac{1}{3}\right) + 1 \geq 0$	True
$\frac{1}{3} < x < 1$	$x = \frac{2}{3}$	$3\left(\frac{2}{3}\right)^2 - 4\left(\frac{2}{3}\right) + 1 \geq 0$	False
$x = 1$	$x = 1$	$3(1)^2 - 4(1) + 1 \geq 0$	True
$x > 1$	$x = 2$	$3(2)^2 - 4(2) + 1 \geq 0$	True

$x \leq \frac{1}{3}$ or $x \geq 1$

23.

$$-2x^2+3x-1\le 0$$

$$(-2x+1)(x-1)=0$$

$$x=\frac{1}{2} \qquad x=1$$

Interval,

Critical Value	Test Point	Substitution	Verification
$x<\frac{1}{2}$	$x=0$	$-2(0)^2+3(0)-1\le 0$	True
$x=\frac{1}{2}$	$x=\frac{1}{2}$	$-2\left(\frac{1}{2}\right)^2+3\left(\frac{1}{2}\right)-1\le 0$	True
$\frac{1}{2}<x<1$	$x=\frac{3}{4}$	$-2\left(\frac{3}{4}\right)^2+3\left(\frac{3}{4}\right)-1\le 0$	False
$x=1$	$x=1$	$-2(1)^2+3(1)-1\le 0$	True
$x>1$	$x=2$	$-2(2)^2+3(2)-1\le 0$	True

$$\left(-\infty,\ \frac{1}{2}\right],\ [1,\ \infty)$$

24.

$$(x-1)(2-3x)(x+2)\le 0$$

$$(x-1)(2-3x)(x+2)=0$$

$$x=1 \qquad x=\frac{2}{3} \qquad x=-2$$

Interval, Critical Value	Test Point	Substitution	Verification
$x<-2$	$x=-3$	$[-3-1][2-3(-3)][-3+2]\le 0$	False
$x=-2$	$x=-2$	$[-2-1][2-3(-2)][-2+2]\le 0$	True
$-2<x<\frac{2}{3}$	$x=0$	$[0-1][2-3(0)][0+2]\le 0$	True
$x=\frac{2}{3}$	$x=\frac{2}{3}$	$\left[\frac{2}{3}-1\right]\left[2-3\left(\frac{2}{3}\right)\right]\left[\frac{2}{3}+2\right]\le 0$	True
$\frac{2}{3}<x<1$	$x=\frac{3}{4}$	$\left[\frac{3}{4}-1\right]\left[2-3\left(\frac{3}{4}\right)\right]\left[\frac{3}{4}+2\right]\le 0$	False
$x=1$	$x=1$	$[1-1][2-3(1)][1+2]\le 0$	True
$x>1$	$x=2$	$[2-1][2-3(2)][2+2]\le 0$	True

$$\left[-2, \frac{2}{3}\right], [1, \infty)$$

25.

$$\frac{2x-5}{x+1} > -\frac{1}{3} \qquad \text{Not defined when } x = -1$$

$$\frac{2x-5}{x+1} = -\frac{1}{3}$$

$$3(2x-5) = -(x+1)$$

$$6x-15 = -x-1$$

$$7x = 14$$

$$x = 2$$

Interval, Critical Value	Test Point	Substitution	Verification
$x<-1$	$x=-2$	$\frac{2(-2)-5}{-2+1} > -\frac{1}{3}$	True
$x=-1$	$x=-1$	$\frac{2(-1)-5}{-1+1} > -\frac{1}{3}$	False
$-1<x<2$	$x=0$	$\frac{2(0)-5}{0+1} > -\frac{1}{3}$	False
$x=2$	$x=2$	$\frac{2(2)-5}{2+1} > -\frac{1}{3}$	False
$x>2$	$x=3$	$\frac{2(3)-5}{3+1} > -\frac{1}{3}$	True

$(-\infty, -1), (2, \infty)$

CHAPTER 2 ADDITIONAL PRACTICE EXERCISES

In Exercises 1-6, solve and check.

1. $3(x+8)-5=2x+17$

2. $\frac{4}{x}-1=\frac{5}{x}$

3. $2x^2-3x=8$

4. $1=\sqrt{x+7}-x$

5. $9|x-7|=36$

6. $\frac{3}{|2x+1|}=9$

In Exercises 7-9, solve the inequality and write the solution set using interval notation.

7. $4(3x+2)<5(6x-1)+8$

8. $x^2+3x-54\leq 0$

9. $\left|\frac{1}{2}+x\right|\geq\frac{3}{4}$

10. The larger of two numbers is 5 more than three times the smaller number. If their sum is 29, find the numbers.

CHAPTER 2 PRACTICE TEST

In Exercises 1-8, solve and check.

1.
$$7+2(6-x)=15+3(5-x)$$

2.
$$\frac{6}{x+1}-5=\frac{1}{x}$$

3.
$$20x^2+31x=7$$

4.
$$2x^2-6x+18=0$$

5.
$$\sqrt{x+3}-1=x$$

6.
$$|x+6|-4=2$$

7.
$$\frac{5}{|6-x|}=-1$$

8.
$$\left(1+\frac{3}{x}\right)^2-4\left(1+\frac{3}{x}\right)-21=0$$

9.
$V=\pi r^2 h$ Solve for h.

10.
Without solving, determine the nature of the roots of $5x^2-6x=10$.

11.
Solve $|3x-1|\le 7$ and graph the solution.

12.
Solve $\left|\frac{x+2}{3}\right|>4$ and write the solution in interval notation.

13.
Find the values of x for which $\sqrt{3x^2+5x-2}$ has real values.

In Exercises 14-16, solve the inequality and express the solution set in interval notation.

14.
$5(3x+6)-9 \geq 14x+4$

15.
$\frac{x+4}{x-4} \leq 1$

16.
$(x-5)(3-x)(2x+1)>0$

17.
The length of a rectangle is 5 more than twice its width. If the perimeter is 58 cm, find the dimensions.

18.
The sum of three consecutive even integers is 72. Find the numbers.

19.
Sally and Kim working together can complete a typing job in 2 hours. Sally working alone can do the job in 3 hours less than Kim working alone. How long does it take Kim to do the job alone?

20.
The sum of the reciprocals of two consecutive numbers is $\frac{25}{156}$. Find the numbers.

CHAPTER 3 SECTION 1

SECTION 3.1 PROGRESS CHECK

1. (Page 145)

$$\overline{PQ} = \sqrt{[-3-4]^2 + [2-(-2)]^2}$$
$$= \sqrt{(-7)^2 + (4)^2}$$
$$= \sqrt{65}$$

2. (Page 147)

$$\text{Midpoint } = \left(\frac{-2+1}{2}, \frac{-3+2}{2}\right) = \left(-\frac{1}{2}, -\frac{1}{2}\right)$$

3. (Page 148)

$$P(0, 1) \text{ , } Q(x, y) \text{ , Midpoint } = (1, 2)$$
$$(1, 2) = \left(\frac{0+x}{2}, \frac{1+y}{2}\right)$$
$$1 = \frac{x}{2} \qquad 2 = \frac{1+y}{2}$$
$$2 = x \qquad 4 = 1+y$$
$$3 = y$$
$$Q(2, 3)$$

4a). (Page 154)

$$x^2 - (-y)^2 = 1 \qquad (-x)^2 - y^2 = 1 \qquad (-x)^2 - (-y)^2 = 1$$
$$x^2 - y^2 = 1 \qquad x^2 - y^2 = 1 \qquad x^2 - y^2 = 1$$

yields original yields original yields original

Symmetry with respect to x - axis, y - axis, and the origin.

4b).

$$x+(-y)=10 \qquad (-x)+y=10 \qquad (-x)+(-y)=10$$
$$x-y=10 \qquad -x+y=10 \qquad -x-y=10$$

doesn't yield original   doesn't yield original   doesn't yield original

Not symmetric with respect to either axis or the origin.

4c).

$$(-y)=x+\frac{1}{x} \qquad y=(-x)+\frac{1}{(-x)} \qquad (-y)=(-x)+\frac{1}{(-x)}$$
$$-y=-x-\frac{1}{x}$$
$$-y=x+\frac{1}{x} \qquad y=-x-\frac{1}{x} \qquad y=x+\frac{1}{x}$$

doesn't yield original   doesn't yield original   yields original

Symmetric with respect to the origin.

EXERCISE SET 3.1

1.

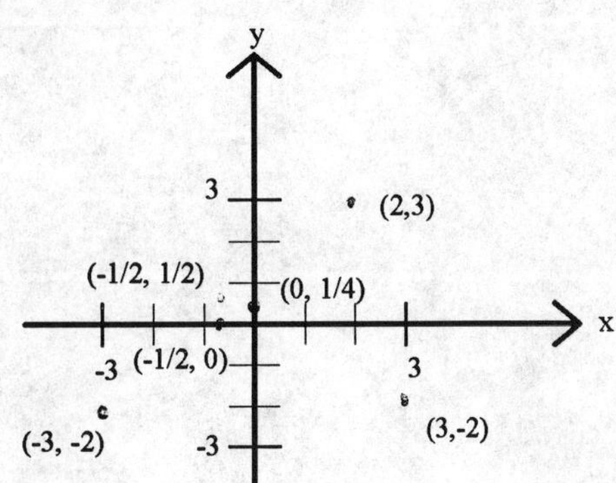

3.

$$
\begin{aligned}
d &= \sqrt{(5-2)^2+(4-1)^2} \\
&= \sqrt{3^2+3^2} \\
&= \sqrt{18} \\
&= 3\sqrt{2}
\end{aligned}
$$

The midpoint is

$$\left(\frac{5+2}{2}, \frac{4+1}{2}\right) = \left(\frac{7}{2}, \frac{5}{2}\right)$$

5.

$$
\begin{aligned}
d &= \sqrt{[-1-(-5)]^2+[-5-(-1)]^2} \\
&= \sqrt{4^2+(-4)^2} \\
&= \sqrt{32} \\
&= 4\sqrt{2}
\end{aligned}
$$

The midpoint is

$$\left(\frac{-1-5}{2}, \frac{-5-1}{2}\right) = (-3, -3)$$

7.

$$
\begin{aligned}
d &= \sqrt{\left[\frac{2}{3}-(-2)\right]^2+\left[\frac{3}{2}-(-4)\right]^2} \\
&= \sqrt{\left(\frac{8}{3}\right)^2+\left(\frac{11}{2}\right)^2} \\
&= \sqrt{\frac{64}{9}+\frac{121}{4}} \\
&= \sqrt{\frac{1345}{36}} \\
&= \frac{\sqrt{1345}}{6}
\end{aligned}
$$

The midpoint is

$$\left(\frac{\frac{2}{3}-2}{2}, \frac{\frac{3}{2}-4}{2}\right) = \left(-\frac{2}{3}, -\frac{5}{4}\right)$$

9.

$$\overline{AB} = \sqrt{[6-(-1)]^2 + [2-4]^2}$$
$$= \sqrt{7^2 + (-2)^2}$$
$$= \sqrt{53}$$

$$\overline{AC} = \sqrt{[6-0]^2 + [2-(-2)]^2}$$
$$= \sqrt{6^2 + 4^2}$$
$$= \sqrt{52}$$
$$= 2\sqrt{13}$$

$$\overline{BC} = \sqrt{[-1-0]^2 + [4-(-2)]^2}$$
$$= \sqrt{(-1)^2 + 6^2}$$
$$= \sqrt{37}$$

The shortest side length is $\overline{BC} = \sqrt{37}$.

11.

$$\overline{RS} = \sqrt{\left[-1-\left(-\frac{3}{2}\right)\right]^2 + \left[\frac{1}{2}-1\right]^2}$$
$$= \sqrt{\left(\frac{1}{2}\right)^2 + \left(-\frac{1}{2}\right)^2}$$
$$= \sqrt{\frac{2}{4}}$$
$$= \frac{\sqrt{2}}{2}$$

$$\overline{RT} = \sqrt{[-1-2]^2 + \left[\frac{1}{2}-(-1)\right]^2}$$
$$= \sqrt{(-3)^2 + \left(\frac{3}{2}\right)^2}$$
$$= \sqrt{\frac{45}{4}}$$
$$= \frac{3\sqrt{5}}{2}$$

$$\overline{ST} = \sqrt{\left[-\frac{3}{2}-2\right]^2 + [1-(-1)]^2}$$
$$= \sqrt{\left(-\frac{7}{2}\right)^2 + 2^2}$$
$$= \sqrt{\frac{65}{4}}$$
$$= \frac{\sqrt{65}}{2}$$

The shortest side length is $\overline{RS} = \frac{\sqrt{2}}{2}$.

In Exercises 13-15, the square of the longest side must equal the sum of the squares of the other two sides if ΔABC is a right triangle.

13.

Label the points : $A(1,-2),\ B(5,2),\ C(2,1)$

$$\overline{AB}=\sqrt{(1-5)^2+(-2-2)^2}=\sqrt{(-4)^2+(-4)^2}$$
$$=\sqrt{32}=4\sqrt{2}$$

$$\overline{AC}=\sqrt{(1-2)^2+(-2-1)^2}=\sqrt{(-1)^2+(-3)^2}$$
$$=\sqrt{10}$$

$$\overline{BC}=\sqrt{(5-2)^2+(2-1)^2}=\sqrt{3^2+1^2}=\sqrt{10}$$

$$\left(\overline{AB}\right)^2 \quad = \quad \left(\overline{AC}\right)^2+\left(\overline{BC}\right)^2 \quad ?$$
$$\left(4\sqrt{2}\right)^2 \quad = \quad \left(\sqrt{10}\right)^2+\left(\sqrt{10}\right)^2 \quad ?$$
$$32 \quad \neq \quad 20$$

ΔABC is not a right triangle.

15.

Label the points : $A(-4,1),\ B(1,4),\ C(4,-1)$

$$\overline{AB}=\sqrt{[-4-1]^2+[1-4]^2}=\sqrt{(-5)^2+(-3)^2}=\sqrt{34}$$

$$\overline{AC}=\sqrt{[-4-4]^2+[1-(-1)]^2}=\sqrt{(-8)^2+2^2}=2\sqrt{17}$$

$$\overline{BC}=\sqrt{[1-4]^2+[4-(-1)]^2}=\sqrt{(-3)^2+5^2}=\sqrt{34}$$

$$\begin{aligned}(\overline{AC})^2 &= (\overline{AB})^2+(\overline{BC})^2 \quad ?\\ (2\sqrt{17})^2 &= (\sqrt{34})^2+(\sqrt{34})^2 \quad ?\\ 68 &= 68\end{aligned}$$

$\Delta\ ABC$ is a right triangle.

In Exercises 17 - 19, the longest length must equal the sum of the other two lengths if $A,\ B,\ C$ are collinear.

17.

Label the points : $A(-1,2),\ B(1,1),\ C(5,-1)$

$$\overline{AB}=\sqrt{[-1-1]^2+[2-1]^2}=\sqrt{(-2)^2+1^2}=\sqrt{5}$$

$$\overline{AC}=\sqrt{[-1-5]^2+[2-(-1)]^2}=\sqrt{(-6)^2+3^2}=\sqrt{45}=3\sqrt{5}$$

$$\overline{BC}=\sqrt{[1-5]^2+[1-(-1)]^2}=\sqrt{(-4)^2+2^2}=\sqrt{20}=2\sqrt{5}$$

$$\begin{aligned}\overline{AC} &= \overline{AB}+\overline{BC} \quad ?\\ 3\sqrt{5} &= \sqrt{5}+2\sqrt{5} \quad ?\\ 3\sqrt{5} &= 3\sqrt{5}\end{aligned}$$

A, B, C are collinear.

19.

Label the points : $A(-1,2),\ B(1,5),\ C\left(-2,\frac{1}{2}\right)$

$$\overline{AB}=\sqrt{[-1-1]^2+[2-5]^2}=\sqrt{(-2)^2+(-3)^2}=\sqrt{13}$$

$$\overline{AC}=\sqrt{[-1-(-2)]^2+\left[2-\frac{1}{2}\right]^2}=\sqrt{1^2+\left(\frac{3}{2}\right)^2}=\sqrt{\frac{13}{4}}=\frac{\sqrt{13}}{2}$$

$$\overline{BC}=\sqrt{[1-(-2)]^2+\left[5-\frac{1}{2}\right]^2}=\sqrt{3^2+\left(\frac{9}{2}\right)^2}=\sqrt{\frac{117}{4}}=\frac{3\sqrt{13}}{2}$$

$$\overline{BC} = \overline{AB}+\overline{AC} \ ?$$

$$\frac{3\sqrt{13}}{2} = \sqrt{13}+\frac{\sqrt{13}}{2} \ ?$$

$$\frac{3\sqrt{13}}{2} = \frac{3\sqrt{13}}{2}$$

A, B, C are collinear.

21.

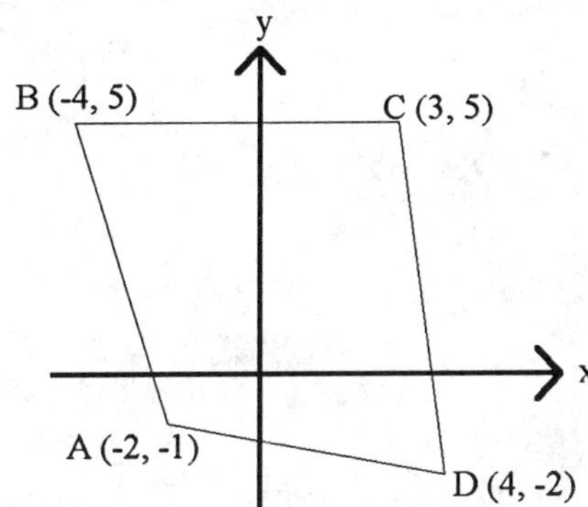

$$\overline{AB}=\sqrt{[-2-(-4)]^2+[-1-5]^2}=\sqrt{2^2+(-6)^2}=\sqrt{40}=2\sqrt{10}$$

$$\overline{BC}=\sqrt{[-4-3]^2+[5-5]^2}=\sqrt{(-7)^2}=\sqrt{49}=7$$

$$\overline{CD}=\sqrt{[3-4]^2+[5-(-2)]^2}=\sqrt{(-1)^2+7^2}=\sqrt{50}=5\sqrt{2}$$

$$\overline{DA}=\sqrt{[4-(-2)]^2+[-2-(-1)]^2}=\sqrt{6^2+(-1)^2}=\sqrt{37}$$

$$\text{Perimeter} = \overline{AB}+\overline{BC}+\overline{CD}+\overline{DA}$$
$$= 2\sqrt{10}+7+5\sqrt{2}+\sqrt{37}$$

23.

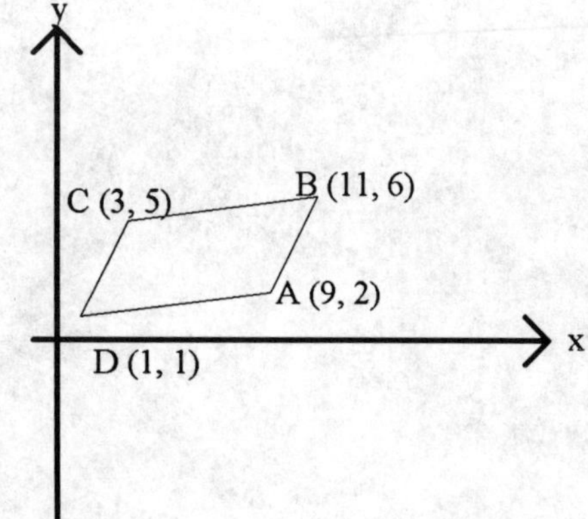

$$\overline{AB} = \sqrt{(9-11)^2 + (2-6)^2} = \sqrt{(-2)^2 + (-4)^2} = \sqrt{20} = 2\sqrt{5}$$

$$\overline{CD} = \sqrt{(3-1)^2 + (5-1)^2} = \sqrt{2^2 + 4^2} = \sqrt{20} = 2\sqrt{5}$$

$$\overline{BC} = \sqrt{(11-3)^2 + (6-5)^2} = \sqrt{8^2 + 1^2} = \sqrt{65}$$

$$\overline{AD} = \sqrt{(9-1)^2 + (2-1)^2} = \sqrt{8^2 + 1^2} = \sqrt{65}$$

$\overline{AB} = \overline{CD}$ and $\overline{BC} = \overline{AD}$

A, B, C, D are vertices of a parallelogram.

25.

$A(1, 7)\ B(4, 3)\ C(x, 5)$

$$\overline{AB} = \sqrt{(1-4)^2 + (7-3)^2} = \sqrt{(-3)^2 + 4^2} = \sqrt{25} = 5$$

$$\overline{BC} = \sqrt{(4-x)^2 + (3-5)^2} = \sqrt{(4-x)^2 + (-2)^2} = \sqrt{(4-x)^2 + 4}$$

$$\overline{AC} = \sqrt{(1-x)^2 + (7-5)^2} = \sqrt{(1-x)^2 + 2^2} = \sqrt{(1-x)^2 + 4}$$

$$\left(\overline{AB}\right)^2 = \left(\overline{AC}\right)^2 + \left(\overline{BC}\right)^2$$

$$5^2 = \left(\sqrt{(1-x)^2 + 4}\right)^2 + \left(\sqrt{(4-x)^2 + 4}\right)^2$$

$$25 = (1-x)^2 + 4 + (4-x)^2 + 4$$

$$25 = 1 - 2x + x^2 + 8 + 16 - 8x + x^2$$

$$25 = 2x^2 - 10x + 25$$

$$0 = 2x^2 - 10x$$

$$0 = 2x(x-5)$$

$2x = 0 \qquad x - 5 = 0$

$x = 0 \qquad x = 5$

27.

$x-\text{intercept}: \text{Let } y=0$

$0=2x+4$

$-4=2x$

$-2=x$

$y-\text{intercept}: \text{Let } x=0$

$y=2(0)+4$

$y=4$

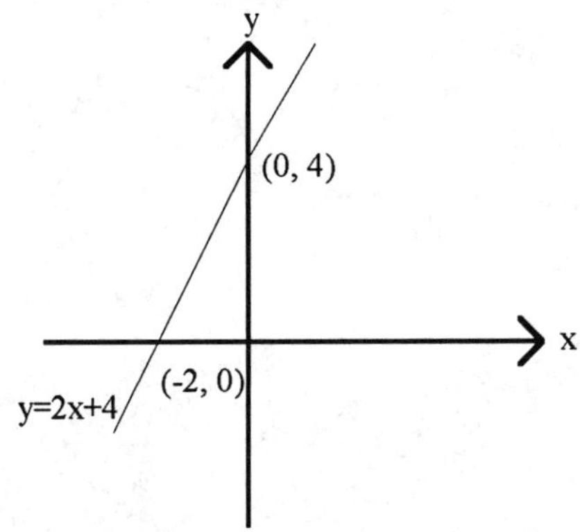

29.

$x-\text{intercept}: \text{Let } y=0$

$0=\sqrt{x}$

$0=x$

$y-\text{intercept}: \text{Let } x=0$

$y=\sqrt{0}$

$y=0$

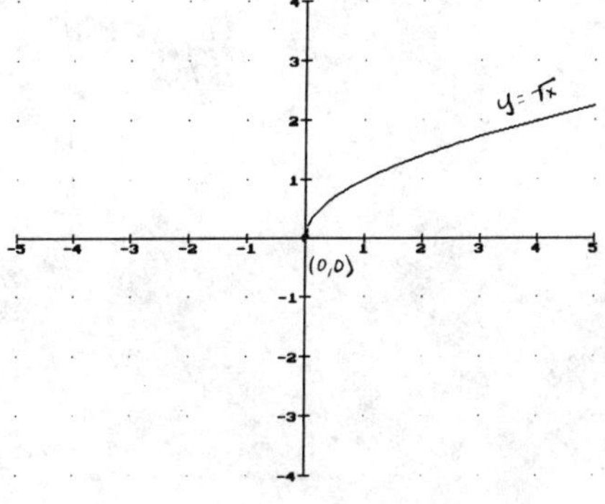

31.

x – intercept : Let $y = 0$

$0 = |x+3|$

$x = -3$

y – intercept : Let $x = 0$

$y = |0+3|$

$y = 3$

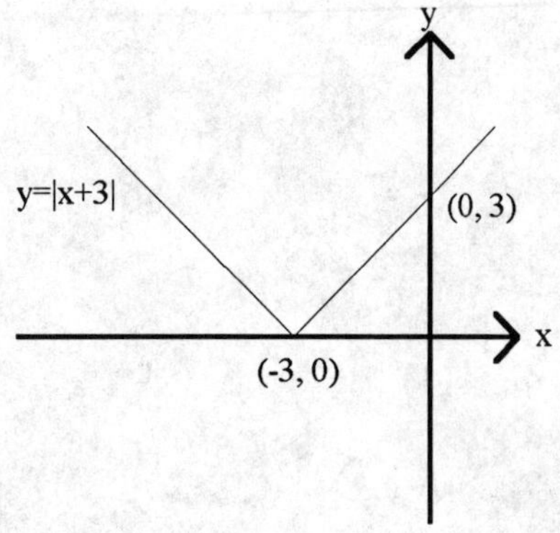

33.

x – intercept : Let $y = 0$

$0 = 3 - x^2$

$x^2 = 3$

$x = \pm\sqrt{3}$

y – intercept : Let $x = 0$

$y = 3 - 0^2$

$y = 3$

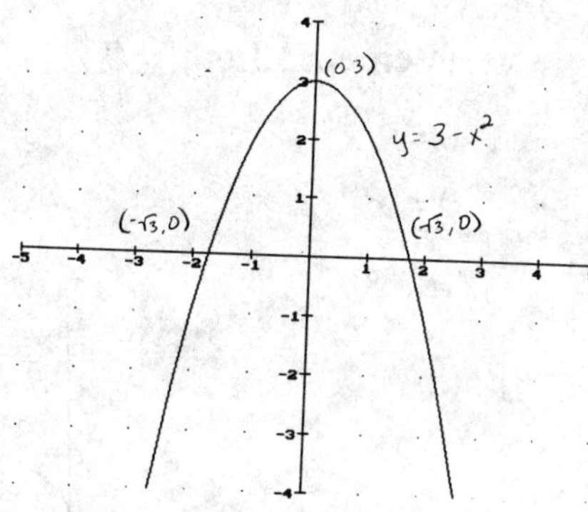

The curve is symmetric with respect to the y - axis since replacing x with $-x$ in the original equation doesn't alter the equation.

$$y = 3 - (-x)^2 = 3 - x^2$$

35.

$x-$intercept : Let $y=0$

$0=x^3+1$

$0=(x+1)(x^2-x+1)$

$x=-1$ only real solution

$y-$intercept : Let $x=0$

$y=0^3+1$

$y=1$

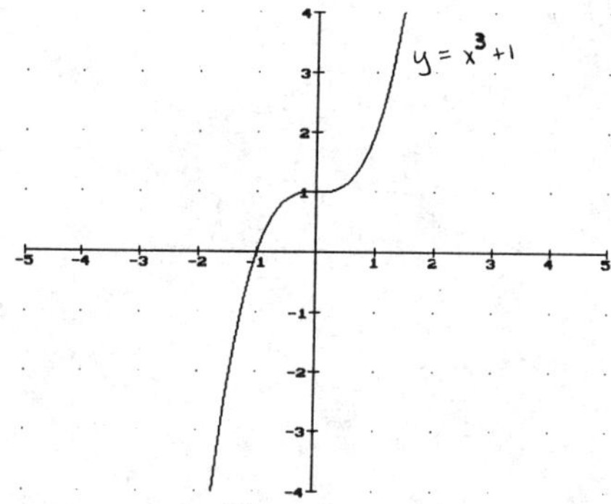

37.

$x-$intercept : Let $y = 0$

$x = 0^2 - 1$

$x = -1$

$y-$intercept : Let $x = 0$

$0 = y^2 - 1$

$1 = y^2$

$y = \pm 1$

The curve is symmetric with respect to the x - axis since replacing y with $-y$ in the original equation doesn't alter the equation.

$$x = (-y)^2 - 1 = y^2 - 1$$

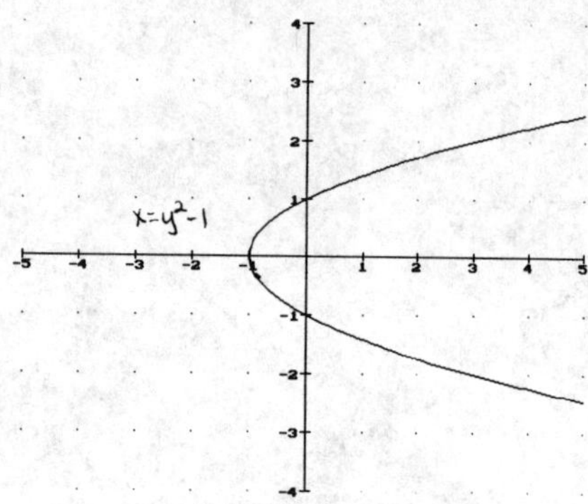

In Exercises 39-43, the RANGE values given show important features of the functions. Other RANGE values are also acceptable. The display on your calculator may vary from that shown.

39.

XSCL = 5

YSCL = 5

GRAPH $Y = 7 - X$

$x-$ intercept : 7

$y-$ intercept : 7

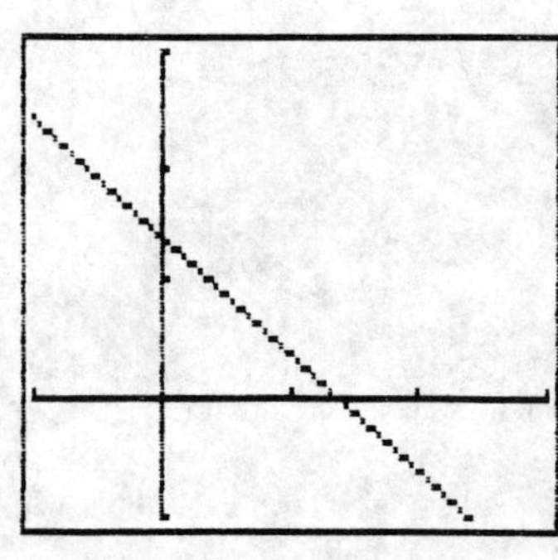

41.

XSCL = 1

YSCL = 1

GRAPH $Y = ABS\ X - X^2$

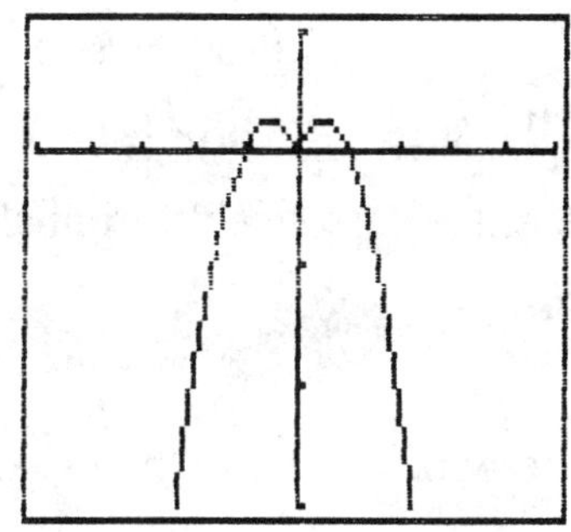

$x-$ intercepts : $-1, 0, 1$

$y-$ intercept : 0

43.

XSCL = 1

YSCL = 1

GRAPH $Y = X^3 - X^2 - X$

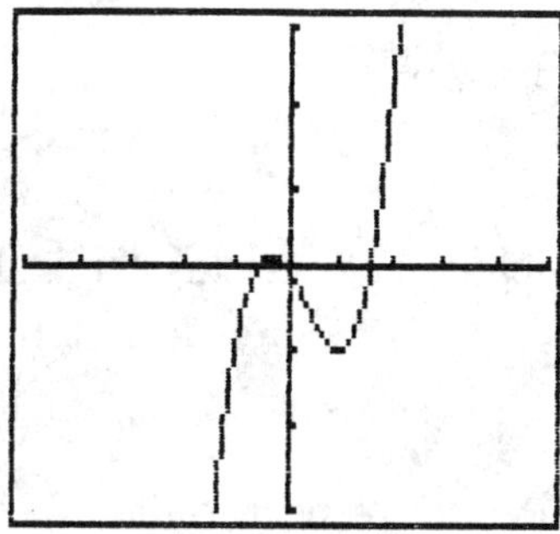

$x-$ intercepts : $-0.6, 0, 1.6$

$y-$ intercept : 0

45.

$3(-x)+2y=5$	$3x+2(-y)=5$	$3(-x)+2(-y)=5$
$-3x+2y=5$	$3x-2y=5$	$-3x-2y=5$
doesn't yield original	doesn't yield original	doesn't yield original

Symmetry : none of these

47.

$y^2=(-x)-4$	$(-y)^2=x-4$	$(-y)^2=(-x)-4$
$y^2=-x-4$	$y^2=x-4$	$y^2=-x-4$
doesn't yield original	yields original	doesn't yield original

Symmetric with respect to the x-axis

49.

$y^2 = 1 + (-x)^3$	$(-y)^2 = 1 + x^3$	$(-y)^2 = 1 + (-x)^3$
$y^2 = 1 - x^3$	$y^2 = 1 + x^3$	$y^2 = 1 - x^3$
doesn't yield original	yields original	doesn't yield original

Symmetric with respect to the x-axis

51.

$y^2 = (-x-2)^2$	$(-y)^2 = (x-2)^2$	$(-y)^2 = (-x-2)^2$
$y^2 = [-1(x+2)]^2$	$y^2 = (x-2)^2$	$y^2 = [-1(x+2)]^2$
$y^2 = (x+2)^2$		$y^2 = (x+2)^2$
doesn't yield original	yields original	doesn't yield original

Symmetric with respect to the x-axis

53.

$y^2(-x) + 2(-x)^2 = 4(-x)^2 y$	$(-y)^2 x + 2x^2 = 4x^2(-y)$	$(-y)^2(-x) + 2(-x)^2 = 4(-x)^2(-y)$
$-xy^2 + 2x^2 = 4x^2 y$	$y^2 x + 2x^2 = -4x^2 y$	$-xy^2 + 2x^2 = -4x^2 y$
doesn't yield original	doesn't yield original	doesn't yield original

Symmetry: none of these

55.

$y = \dfrac{(-x)^2 + 4}{(-x)^2 - 4}$	$(-y) = \dfrac{x^2 + 4}{x^2 - 4}$	$(-y) = \dfrac{(-x)^2 + 4}{(-x)^2 - 4}$
$y = \dfrac{x^2 + 4}{x^2 - 4}$	doesn't yield original	$-y = \dfrac{x^2 + 4}{x^2 - 4}$
yields original		doesn't yield original

Symmetric with respect to the y-axis

57.

$$y^2 = \frac{(-x)^2+1}{(-x)^2-1} \qquad (-y)^2 = \frac{x^2+1}{x^2-1} \qquad (-y)^2 = \frac{(-x)^2+1}{(-x)^2-1}$$

$$y^2 = \frac{x^2+1}{x^2-1} \qquad y^2 = \frac{x^2+1}{x^2-1} \qquad y^2 = \frac{x^2+1}{x^2-1}$$

yields original | yields original | yields original

Symmetric with respect to the y-axis, x-axis, origin

59.

$$(-x)y = 4 \qquad x(-y) = 4 \qquad (-x)(-y) = 4$$

$$-xy = 4 \qquad -xy = 4 \qquad xy = 4$$

doesn't yield original | doesn't yield original | yields original

Symmetric with respect to the origin.

61.

Label the points : $A(0,0)$, $B(4,3)$, $C(3,4)$

$$\overline{AB} = \sqrt{(0-4)^2+(0-3)^2} = \sqrt{(-4)^2+(-3)^2} = \sqrt{25} = 5$$
$$\overline{AC} = \sqrt{(0-3)^2+(0-4)^2} = \sqrt{(-3)^2+(-4)^2} = \sqrt{25} = 5$$
$$\overline{BC} = \sqrt{(4-3)^2+(3-4)^2} = \sqrt{1^2+(-1)^2} = \sqrt{2}$$

$\Delta\ ABC$ is an isosceles triangle since $\overline{AB} = \overline{AC} \neq \overline{BC}$

63.

$$\left(\frac{-3+5}{2}, \frac{6+(-2)}{2}\right) = (1, 2)$$

65.

$$(2, 7) = \left(\frac{x+6}{2}, \frac{10+y}{2}\right)$$

$$2 = \frac{x+6}{2} \qquad 7 = \frac{10+y}{2}$$

$$4 = x+6 \qquad 14 = 10+y$$

$$-2 = x \qquad 4 = y$$

67.

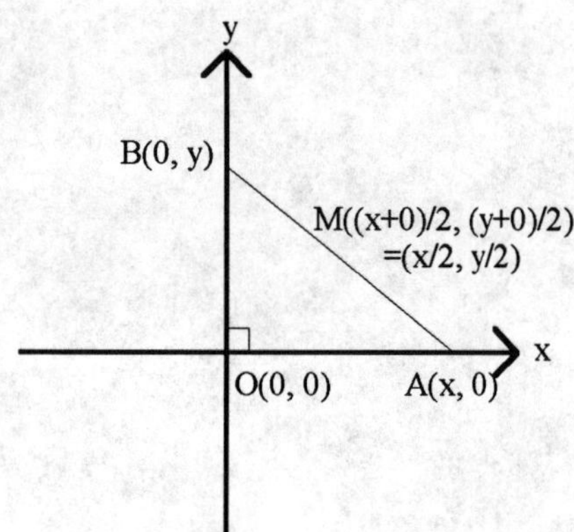

$d(A, M) = d(M, B)$ since M is the midpoint of AB

$$d(A, M) = \sqrt{\left(x - \frac{x}{2}\right)^2 + \left(0 - \frac{y}{2}\right)^2} = \sqrt{\left(\frac{x}{2}\right)^2 + \left(-\frac{y}{2}\right)^2} = \sqrt{\frac{x^2}{4} + \frac{y^2}{4}} = \frac{\sqrt{x^2 + y^2}}{2}$$

$$d(M, O) = \sqrt{\left(\frac{x}{2} - 0\right)^2 + \left(\frac{y}{2} - 0\right)^2} = \sqrt{\left(\frac{x}{2}\right)^2 + \left(\frac{y}{2}\right)^2} = \sqrt{\frac{x^2}{4} + \frac{y^2}{4}} = \frac{\sqrt{x^2 + y^2}}{2}$$

$$d(A, M) = d(M, B) = d(M, O)$$

69.

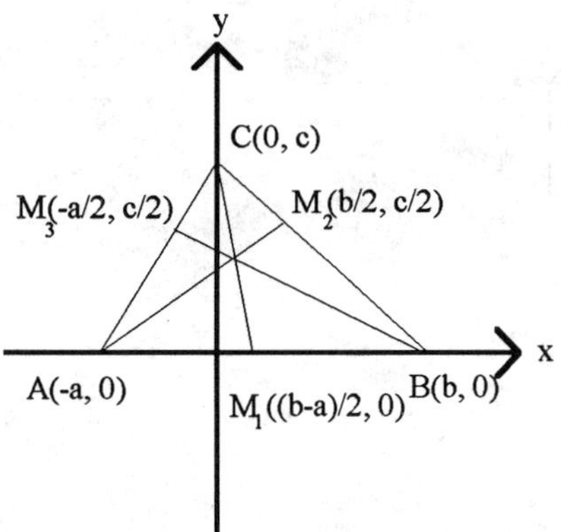

$$\overline{AB} = \sqrt{(-a-b)^2 + (0-0)^2} = |-a-b| = a+b$$

$$\overline{BC} = \sqrt{(b-0)^2 + (0-c)^2} = \sqrt{b^2+c^2}$$

$$\overline{AC} = \sqrt{(-a-0)^2 + (0-c)^2} = \sqrt{a^2+c^2}$$

Medians

$$\overline{AM_2} = \sqrt{\left(\frac{b}{2} - (-a)\right)^2 + \left(\frac{c}{2} - 0\right)^2} = \sqrt{\left(\frac{b}{2} + a\right)^2 + \frac{c^2}{4}}$$

$$\overline{BM_3} = \sqrt{\left(b - \left(-\frac{a}{2}\right)\right)^2 + \left(0 - \frac{c}{2}\right)^2} = \sqrt{\left(b + \frac{a}{2}\right)^2 + \frac{c^2}{4}}$$

$$\overline{CM_1} = \sqrt{\left(\frac{b-a}{2} - 0\right)^2 + (0-c)^2} = \sqrt{\frac{(b-a)^2}{4} + c^2}$$

69.(Continued)

$$\left(\overline{AM_2}\right)^2+\left(\overline{BM_3}\right)^2+\left(\overline{CM_1}\right)^2$$

$$=\left(\sqrt{\left(\frac{b}{2}+a\right)^2+\frac{c^2}{4}}\right)^2+\left(\sqrt{\left(b+\frac{a}{2}\right)^2+\frac{c^2}{4}}\right)^2+\left(\sqrt{\frac{(b-a)^2}{4}+c^2}\right)^2$$

$$=\left(\frac{b}{2}+a\right)^2+\frac{c^2}{4}+\left(b+\frac{a}{2}\right)^2+\frac{c^2}{4}+\frac{(b-a)^2}{4}+c^2$$

$$=\frac{b^2}{4}+ab+a^2+\frac{3c^2}{2}+b^2+ab+\frac{a^2}{4}+\frac{b^2-2ab+a^2}{4}$$

$$=\frac{3b^2}{2}+\frac{3ab}{2}+\frac{3a^2}{2}+\frac{3c^2}{2}=\frac{3}{2}\left[b^2+ab+a^2+c^2\right]$$

$$\frac{3}{4}\left[\left(\overline{AB}\right)^2+\left(\overline{BC}\right)^2+\left(\overline{AC}\right)^2\right]=$$

$$\frac{3}{4}\left[(a+b)^2+\left(\sqrt{b^2+c^2}\right)^2+\left(\sqrt{a^2+c^2}\right)^2\right]=$$

$$\frac{3}{4}\left[a^2+2ab+b^2+b^2+c^2+a^2+c^2\right]=\frac{3}{4}\left[2a^2+2ab+2b^2+2c^2\right]=\frac{3}{2}\left[a^2+ab+b^2+c^2\right]$$

Thus $\left(\overline{AM_2}\right)^2+\left(\overline{BM_3}\right)^2+\left(\overline{CM_1}\right)^2=\frac{3}{4}\left[\left(\overline{AB}\right)^2+\left(\overline{BC}\right)^2+\left(\overline{AC}\right)^2\right]$

71.

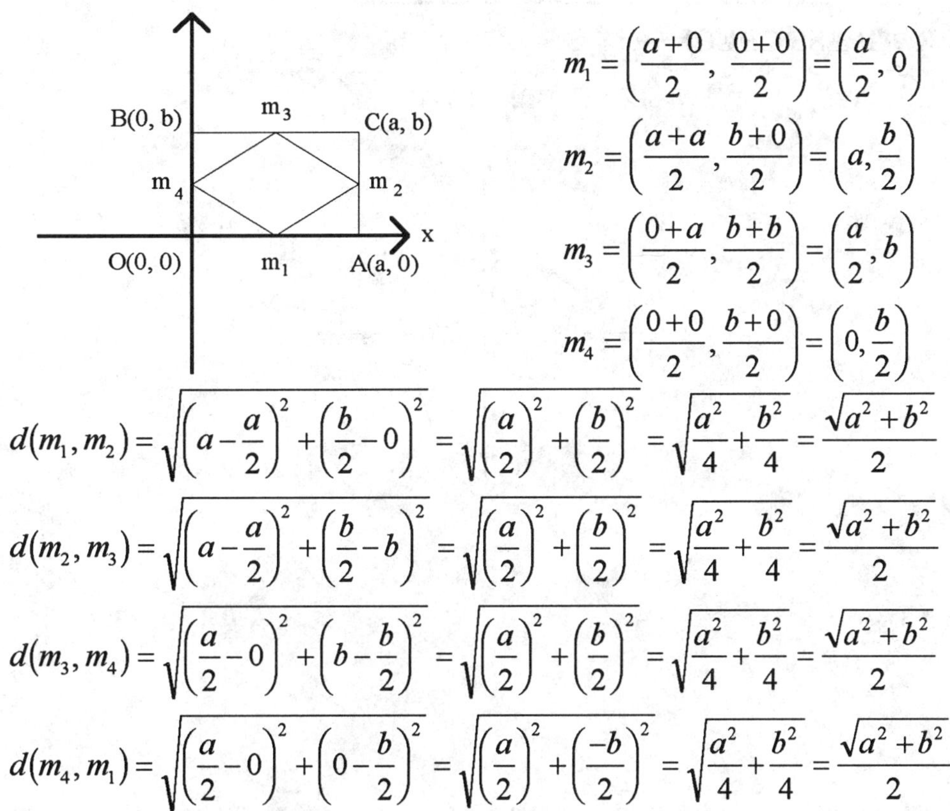

$$m_1 = \left(\frac{a+0}{2}, \frac{0+0}{2}\right) = \left(\frac{a}{2}, 0\right)$$

$$m_2 = \left(\frac{a+a}{2}, \frac{b+0}{2}\right) = \left(a, \frac{b}{2}\right)$$

$$m_3 = \left(\frac{0+a}{2}, \frac{b+b}{2}\right) = \left(\frac{a}{2}, b\right)$$

$$m_4 = \left(\frac{0+0}{2}, \frac{b+0}{2}\right) = \left(0, \frac{b}{2}\right)$$

$$d(m_1, m_2) = \sqrt{\left(a-\frac{a}{2}\right)^2 + \left(\frac{b}{2}-0\right)^2} = \sqrt{\left(\frac{a}{2}\right)^2 + \left(\frac{b}{2}\right)^2} = \sqrt{\frac{a^2}{4}+\frac{b^2}{4}} = \frac{\sqrt{a^2+b^2}}{2}$$

$$d(m_2, m_3) = \sqrt{\left(a-\frac{a}{2}\right)^2 + \left(\frac{b}{2}-b\right)^2} = \sqrt{\left(\frac{a}{2}\right)^2 + \left(\frac{b}{2}\right)^2} = \sqrt{\frac{a^2}{4}+\frac{b^2}{4}} = \frac{\sqrt{a^2+b^2}}{2}$$

$$d(m_3, m_4) = \sqrt{\left(\frac{a}{2}-0\right)^2 + \left(b-\frac{b}{2}\right)^2} = \sqrt{\left(\frac{a}{2}\right)^2 + \left(\frac{b}{2}\right)^2} = \sqrt{\frac{a^2}{4}+\frac{b^2}{4}} = \frac{\sqrt{a^2+b^2}}{2}$$

$$d(m_4, m_1) = \sqrt{\left(\frac{a}{2}-0\right)^2 + \left(0-\frac{b}{2}\right)^2} = \sqrt{\left(\frac{a}{2}\right)^2 + \left(\frac{-b}{2}\right)^2} = \sqrt{\frac{a^2}{4}+\frac{b^2}{4}} = \frac{\sqrt{a^2+b^2}}{2}$$

m_1, m_2, m_3, m_4 are vertices of a rhombus
since $d(m_1, m_2) = d(m_2, m_3) = d(m_3, m_4) = d(m_4, m_1)$ 4 equal sides

CHAPTER 3 SECTION 2

SECTION 3.2 PROGRESS CHECK

1. (Page 159)

x	y
-3	5
-2	0
-1	-3
0	-4
1	-3
2	0
3	5

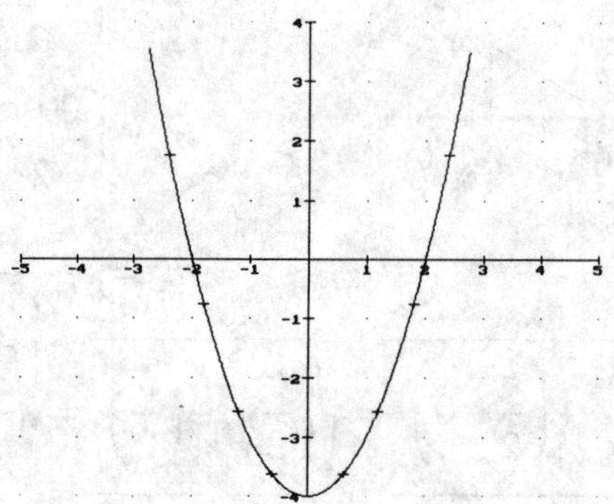

Domain : $\{x \mid -3 \le x \le 3\}$

Range : $\{y \mid -4 \le y \le 5\}$

2a). (Page 160)

$$f(-2) = (-2)^3 + 3(-2) - 4$$
$$= -8 - 6 - 4$$
$$= -18$$

2b).

$$f(t-1) = (t-1)^2 + 1$$
$$= t^2 - 2t + 1 + 1$$
$$= t^2 - 2t + 2$$

EXERCISE SET 3.2

1.

Domain : $\{x \mid -\infty < x < \infty\}$

Range : $\{y \mid -\infty < y < \infty\}$

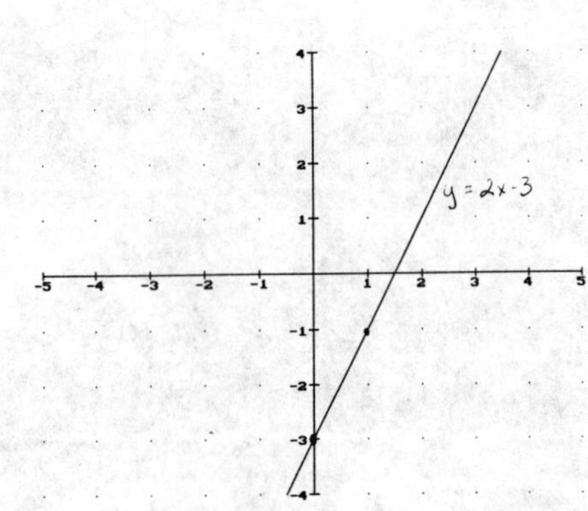

3.

Domain : $\{x \mid -\infty < x < \infty\}$

Range : $\{y \mid -\infty < y < \infty\}$

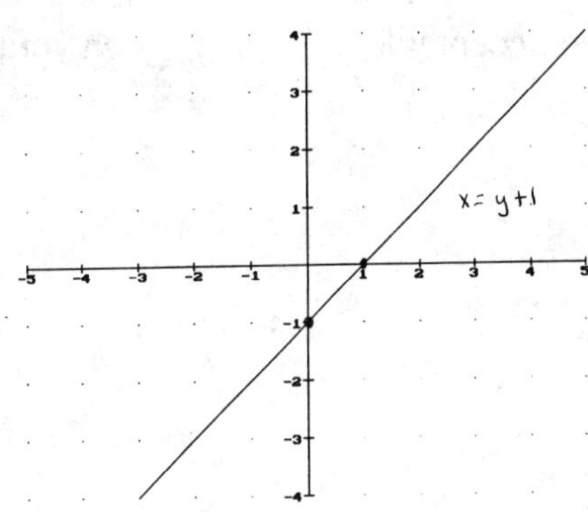

5.

Domain : $\{x \mid x \geq 1\}$

Range : $\{y \mid y \geq 0\}$

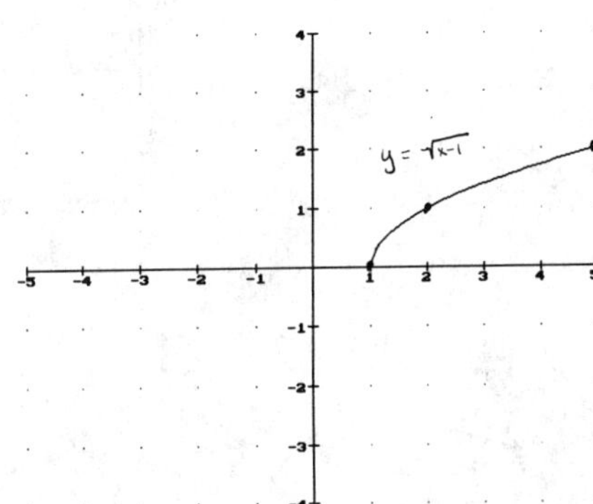

In Exercises 7-11, the RANGE values given show important features of the functions. Other RANGE values are also acceptable. The display on your calculator may vary from that shown.

7.

$f(x) = \sqrt{2x-3}$

domain : $2x - 3 \geq 0$

$2x \geq 3$

$x \geq \frac{3}{2}$

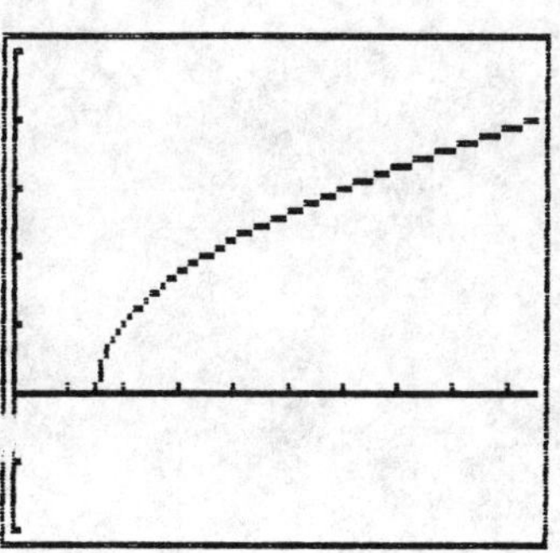

XMIN = 0

XMAX = 9.5

XSCL = 1

YMIN = −2

YMAX = 5

YSCL = 1

GRAPH $Y = \sqrt{\ }(2X-3)$

9.

$f(x) = \frac{1}{\sqrt{x-2}}$

domain : $x - 2 > 0$

$x > 2$

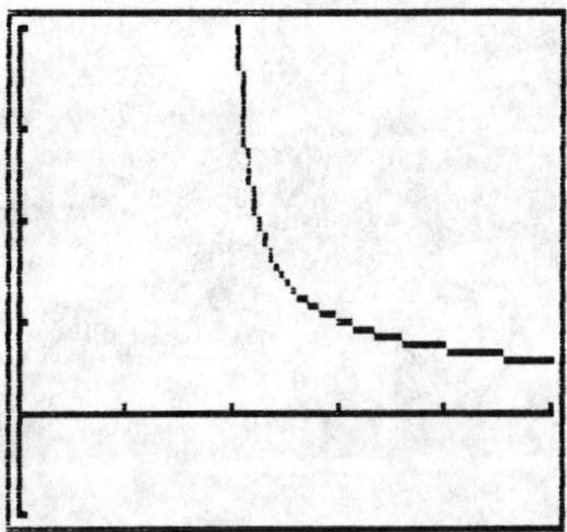

XMIN = 0

XMAX = 5

XSCL = 1

YMIN = −1

YMAX = 4

YSCL = 1

GRAPH $Y = 1 \div \left(\sqrt{\ }(X-2)\right)$

11.

$$f(x)=\frac{\sqrt{x-1}}{x-2}$$

domain : $x-1\geq 0$ $\quad x-2\neq 0$

$x\geq 1$ $\quad x\neq 2$

XMIN = 0
XMAX = 5
XSCL = 1
YMIN = −5
YMAX = 5
YSCL = 1

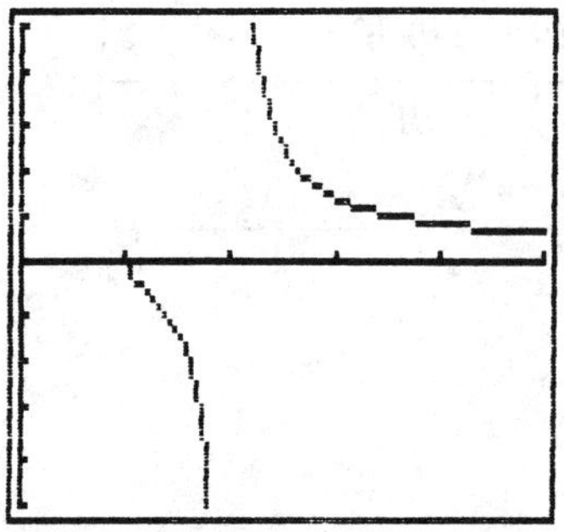

GRAPH $Y=\sqrt{\ }(X-1)\div(X-2)$

13.
Find the value of x for which $f(x)=2$

$$2=2x-5$$
$$7=2x$$
$$x=\frac{7}{2}$$

15.
Find the value of x for which $f(x)=2$

$$2=\frac{1}{x-1}$$
$$2(x-1)=1$$
$$2x-2=1$$
$$2x=3$$
$$x=\frac{3}{2}$$

17.

$$f(0)=2(0)^2+5$$
$$=0+5$$
$$=5$$

19.

$$f(a)=2(a)^2+5$$
$$=2a^2+5$$

21.

$$3f(x)=3(2x^2+5)$$
$$=6x^2+15$$

23.

$$f(a+h)=2(a+h)^2+5 \qquad f(a)=2a^2+5$$
$$=2(a^2+2ah+h^2)+5$$
$$=2a^2+4ah+2h^2+5$$

$$\frac{f(a+h)-f(a)}{h}=\frac{2a^2+4ah+2h^2+5-(2a^2+5)}{h}$$
$$=\frac{2a^2+4ah+2h^2+5-2a^2-5}{h}$$
$$=\frac{4ah+2h^2}{h}$$
$$=4a+2h$$

25.

$$g\left(\frac{1}{x}\right)=\left(\frac{1}{x}\right)^2+2\left(\frac{1}{x}\right)$$
$$=\frac{1}{x^2}+\frac{2}{x}$$

27.

$$g(-x)=(-x)^2+2(-x)$$
$$=x^2-2x$$

29.

$$g(a+h)=(a+h)^2+2(a+h) \qquad g(a)=a^2+2a$$
$$=a^2+2ah+h^2+2a+2h$$
$$\frac{g(a+h)-g(a)}{h}=\frac{a^2+2ah+h^2+2a+2h-(a^2+2a)}{h}$$
$$=\frac{a^2+2ah+h^2+2a+2h-a^2-2a}{h}$$
$$=\frac{2ah+h^2+2h}{h}$$
$$=2a+h+2$$

31.

$$\frac{1}{F(x)} = \frac{1}{\frac{x^2+1}{3x-1}}$$
$$= 1 \cdot \frac{3x-1}{x^2+1}$$
$$= \frac{3x-1}{x^2+1}$$

33.

$$2F(2x) = 2\left[\frac{(2x)^2+1}{3(2x)-1}\right]$$
$$= 2\left(\frac{4x^2+1}{6x-1}\right)$$
$$= \frac{8x^2+2}{6x-1}$$
$$\text{or } \frac{2(4x^2+1)}{6x-1}$$

35.

$$r(-8.27) = \frac{-8.27-2}{(-8.27)^2+2(-8.27)-3}$$
$$= -0.210$$

37.

$$r(2a) = \frac{2a-2}{(2a)^2+2(2a)-3}$$
$$= \frac{2a-2}{4a^2+4a-3}$$
$$\text{or } \frac{2(a-1)}{4a^2+4a-3}$$

39.

$$r(a+1) = \frac{a+1-2}{(a+1)^2+2(a+1)-3}$$
$$= \frac{a-1}{a^2+2a+1+2a+2-3}$$
$$= \frac{a-1}{a^2+4a}$$
$$\text{or } \frac{a-1}{a(a+4)}$$

41.

$$I = prt$$
$$I = (x)(0.07)(4)$$
$$I = 0.28x$$
$$I(x) = 0.28x$$

43.

$$C = \pi d$$
$$\frac{C}{\pi} = d$$
$$d = \frac{C}{\pi}$$
$$d(C) = \frac{C}{\pi}$$

45.

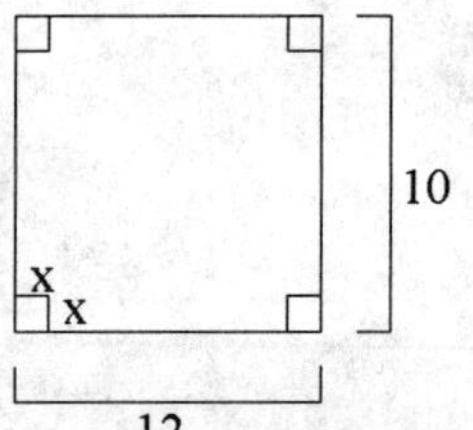

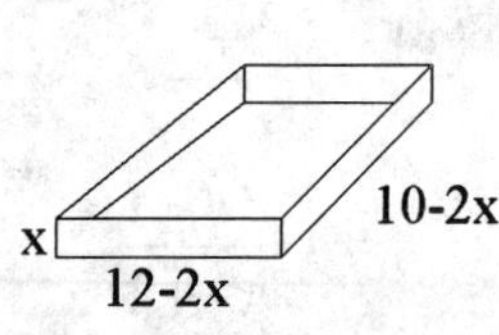

depth : x

$$V = lwh$$
$$V = (12-2x)(10-2x)(x)$$
$$= (12-2x)(10x-2x^2)$$
$$= 120x - 24x^2 - 20x^2 + 4x^3$$

$$V = 4x^3 - 44x^2 + 120x$$

47.

$x =$ number of \$5 decreases

Cost per player after x \$5 decreases : $300 - 5x$

Number of players sold after x \$5 decreases : $400 + 10x$

Gross Sales : S $\quad$ $S =$ (cost per player)(# of players)

$$= (300-5x)(400+10x)$$
$$= 120000 + 1000x - 50x^2$$

47a).

$$S = 120000 + 1000x - 50x^2$$

47b).

$$x = 10$$
$$S = 120000 + 1000(10) - 50(10)^2$$
$$= 125000$$

The gross sales would be \$125,000.

49a).
W : number of grams of H_2O_2

$H = 0.06W$

49b).
$O = 0.94W$

49c).
$W = 250g$

$H = 0.06(250) = 15$
$O = 0.94(250) = 235$

$15g$ of hydrogen; $235g$ of oxygen

CHAPTER 3 SECTION 3

SECTION 3.3 PROGRESS CHECK

1. (Page 170)

$|x-1|$ will shift the graph of $y=|x|$ one unit to the right.

$-|x-1|$: the minus in front of $|x-1|$ reflects the graph about the $x-$axis.

$-|x-1|+3$: adding 3, shifts the graph up 3 units.

2. (Page 173)

If $x < \frac{1}{2}$ there is not enough money to purchase any stamps,
so $N = 0$

If $\frac{1}{2} \le x < 1$, there is enough money to purchase one stamp only,
so $N = 1$

If $x = 1$, there is enough money to purchase two stamps,
so $N = 2$

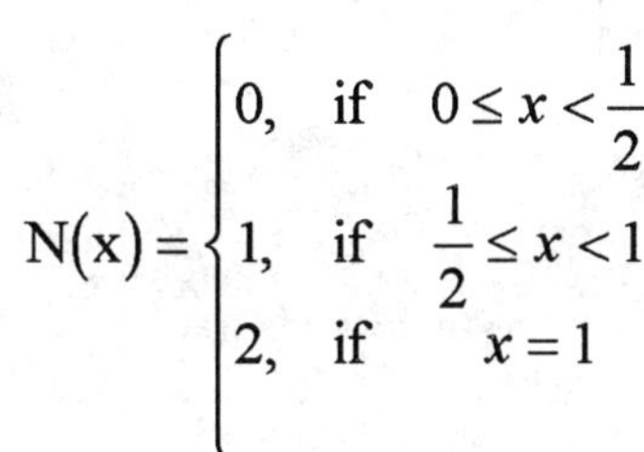

$$N(x) = \begin{cases} 0, & \text{if } 0 \le x < \frac{1}{2} \\ 1, & \text{if } \frac{1}{2} \le x < 1 \\ 2, & \text{if } x = 1 \end{cases}$$

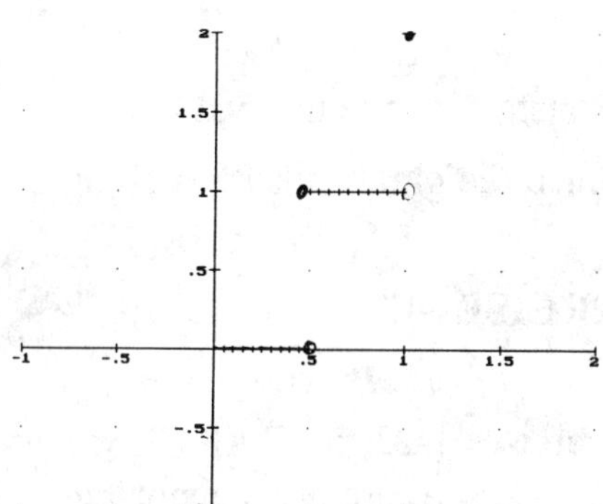

3. (Page 175)

$$f(x) = \begin{cases} 2x+1 & \text{if} \quad x < -1 \\ 0 & \text{if} \quad -1 \le x \le 3 \\ -2x+1 & \text{if} \quad x > 3 \end{cases}$$

Increasing on the interval $(-\infty, -1)$
Since the graph is rising

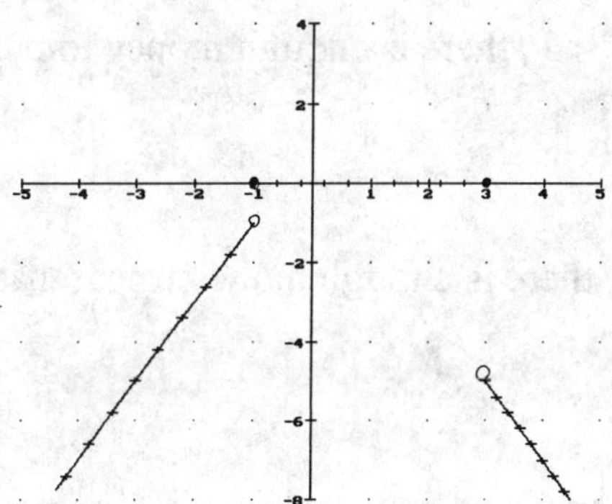

Decreasing on the interval $(3, \infty)$
Since the graph os falli ng

Constant on the interval $[-1, 3]$
Since the graph is neither rising nor falling.

EXERCISE SET 3.3

In Exercises 1-23, the RANGE values given show important features of the functions. Other RANGE values are also acceptable. The display on your calculator may vary from that shown.

1.
TENS viewing rectangle
XSCL = 1
YSCL = 1

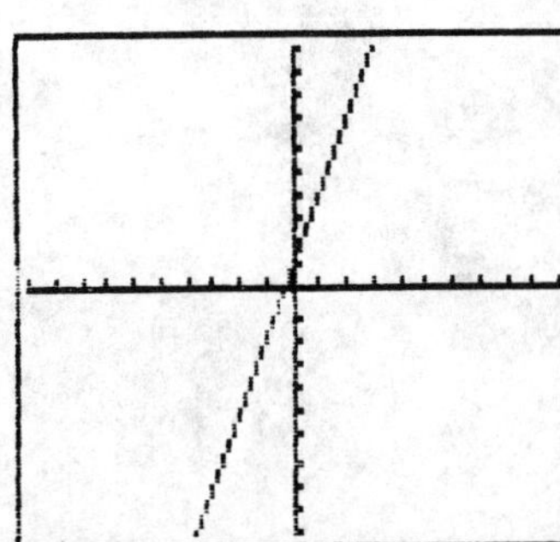

GRAPH $Y = 3X + 1$
increasing : $(-\infty, \infty)$

3.

XMIN = −5

XMAX = 5

XSCL = 1

YMIN = −2

YMAX = 10

YSCL = 1

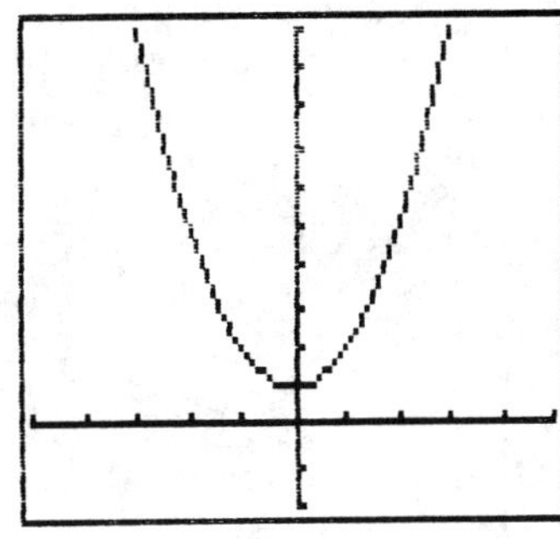

GRAPH $Y = X^2 + 1$

increasing : $[0, \infty])$

decreasing : $(-\infty, 0]$

5.

XMIN = −5

XMAX = 5

XSCL = 1

YMIN = −10

YMAX = 10

YSCL = 1

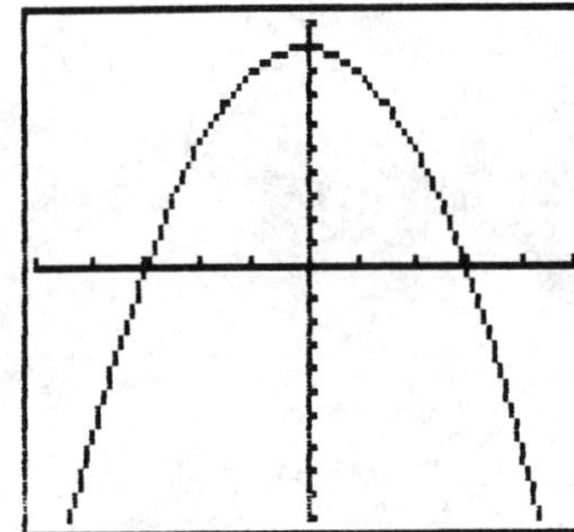

GRAPH $Y = 9 - X^2$

increasing : $(-\infty, 0]$

decreasing : $[0, \infty)$

7.

XMIN = −6
XMAX = 6
XSCL = 1
YMIN = −4
YMAX = 4
YSCL = 1

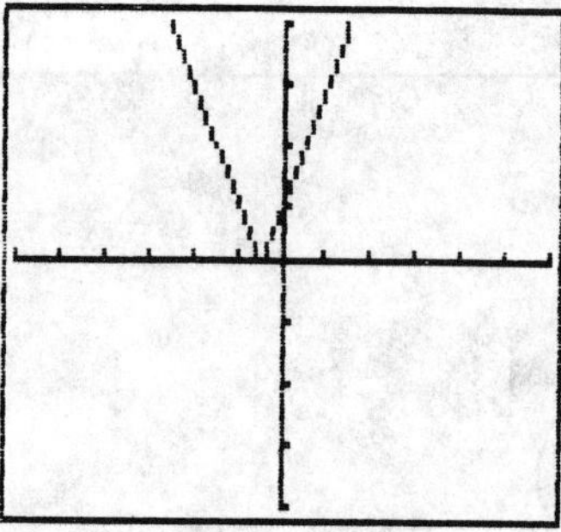

GRAPH $Y = ABS(2X+1)$

increasing : $\left[-\frac{1}{2}, \infty\right)$

decreasing : $\left(-\infty, -\frac{1}{2}\right]$

9.

XMIN = −6
XMAX = 6
XSCL = 1
YMIN = −4
YMAX = 4
YSCL = 1

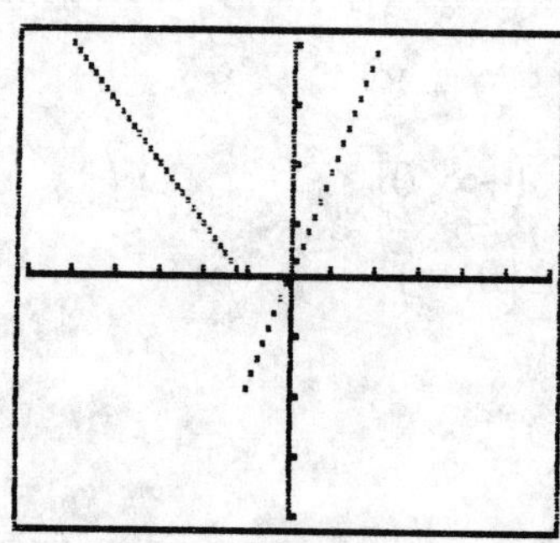

GRAPH $Y = \begin{cases} 2X & , \quad X > -1 \\ -X-1 & , \quad X \leq -1 \end{cases}$

by entering $Y = (2X)(X > 1) + (-X-1)(X \leq -1)$

increasing : $(-1, \infty)$

decreasing : $(-\infty, -1]$

11.

XMIN = −4

XMAX = 8

XSCL = 1

YMIN = −4

YMAX = 4

YSCL = 1

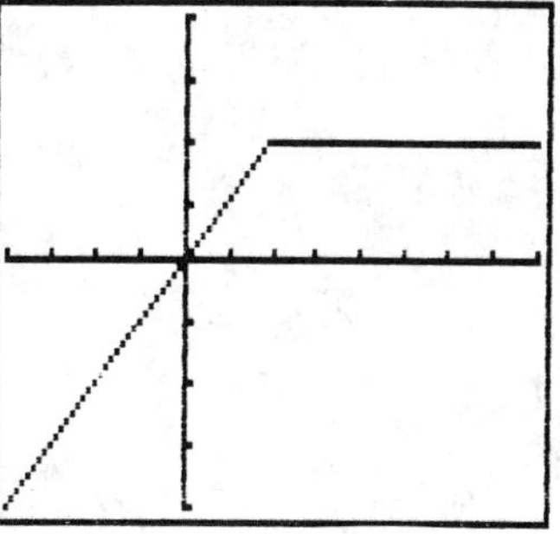

GRAPH $Y = \begin{cases} X & , \quad X < 2 \\ 2 & , \quad X \geq 2 \end{cases}$

by entering $Y = X(X < 2) + 2(X \geq 2)$

increasing : $(-\infty, 2]$

constant: $[2, \infty)$

13.

XMIN = −5

XMAX = 5

XSCL = 1

YMIN = −12

YMAX = 1

YSCL = 1

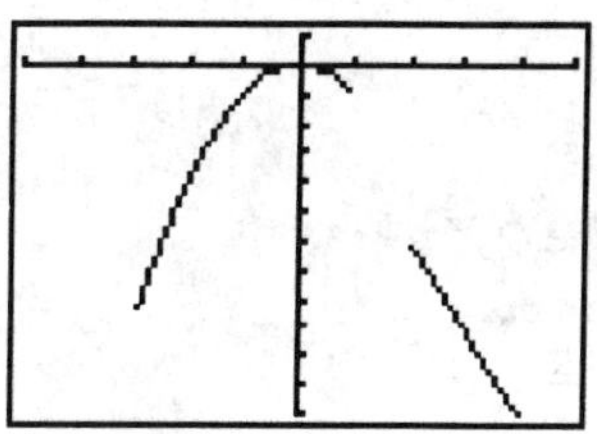

GRAPH $Y = \begin{cases} -X^2 & , \quad -3 < X < 1 \\ 0 & , \quad 1 \leq X \leq 2 \\ -3X & , \quad X > 2 \end{cases}$

by entering $Y_1 = (-X^2) \div (X > -3)(X < 1)$

$Y_2 = 0 \div (X \geq 1)(X \leq 2)$

$Y_3 = (-3X) \div (X > 2)$

increasing : $(-3, 0]$

decreasing : $[0, 1), (2, \infty)$

constant : $[1, 2]$

15.
XMIN = −5

XMAX = 5

XSCL = 1

YMIN = −5

YMAX = 1

YSCL = 1

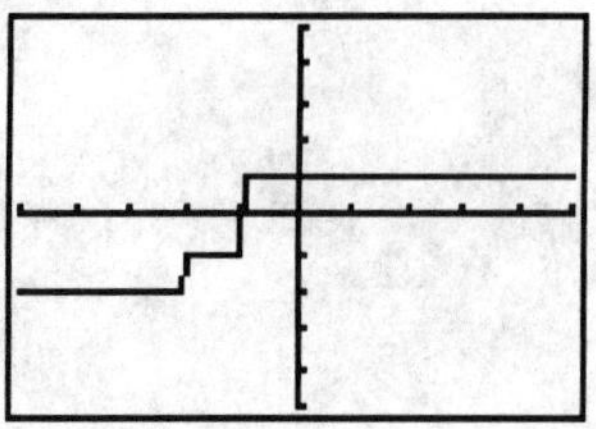

$$\text{GRAPH } Y = \begin{cases} -2 & , \quad X < -2 \\ -1 & , \quad -2 \le X \le -1 \\ 1 & , \quad X > -1 \end{cases}$$

by entering

$Y = -2(X < -2) - 1(X \ge -2)(X \le -1) + 1(X > -1)$

constant : $(-\infty, -2), [-2, -1], (-1, \infty)$

17.

XMIN = −5

XMAX = 5

XSCL = 1

YMIN = −2

YMAX = 10

YSCL = 1

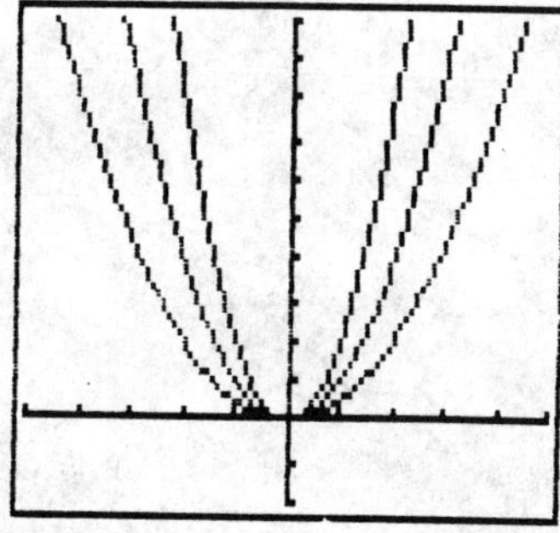

GRAPH $Y = X^2$

$Y = 2X^2$

$Y = \frac{1}{2}X^2$

19.
XMIN = −5
XMAX = 5
XSCL = 1
YMIN = −6
YMAX = 6
YSCL = 1

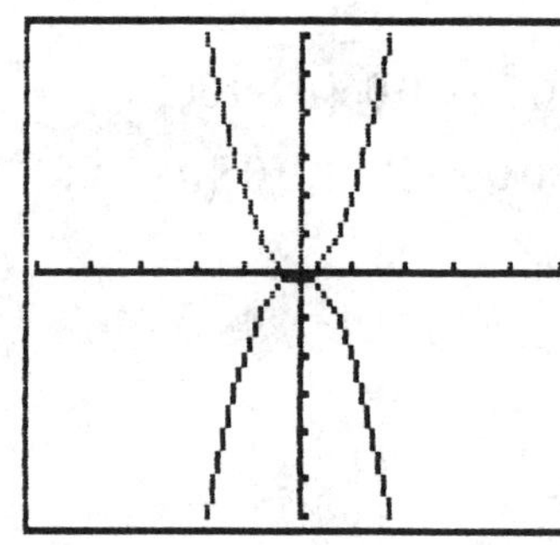

GRAPH $Y = 2X^2$
$Y = -2X^2$

21.
XMIN = −5
XMAX = 5
XSCL = 1
YMIN = −6
YMAX = 6
YSCL = 1

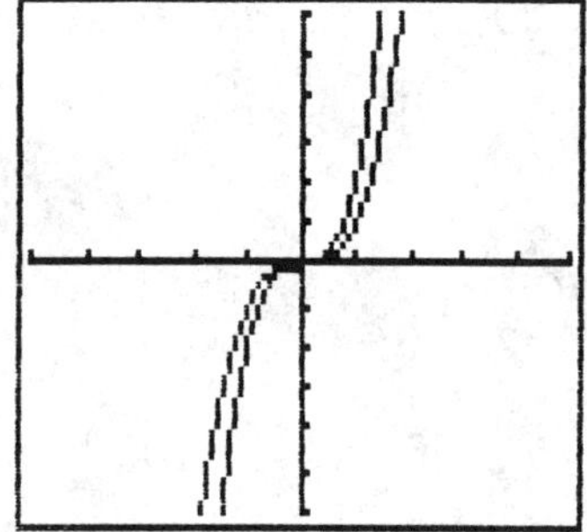

GRAPH $Y = X^3$
$Y = 2X^3$

23.
XMIN = −5
XMAX = 5
XSCL = 1
YMIN = −6
YMAX = 6
YSCL = 1

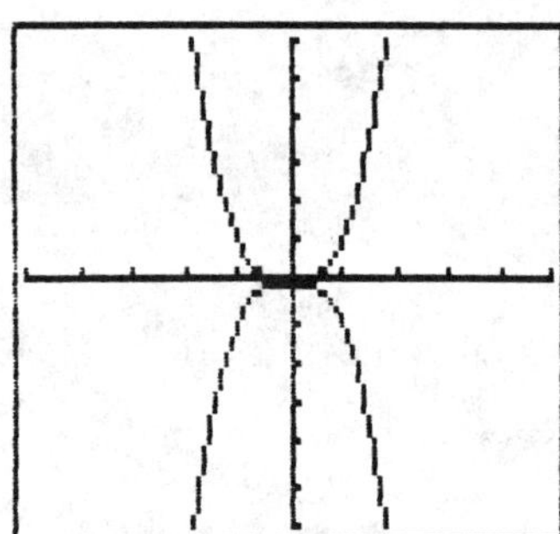

GRAPH $Y = X^3$
$Y = -X^3$

25.

$$C(u)=\begin{cases} 6.50 & , \text{ if } 0\le u\le 100 \\ 6.50+0.06(u-100) & , \text{ if } 100<u\le 200 \\ 6.50+0.06(100)+0.05(u-200) & , \text{ if } u>200 \end{cases}$$

which simpli fies to

$$C(u)=\begin{cases} 6.50 & , \text{ if } 0\le u\le 100 \\ 0.50+0.06u & , \text{ if } 100<u\le 200 \\ 2.50+0.05u & , \text{ if } u>200 \end{cases}$$

27.

$$R(x)=\begin{cases} 30{,}000 & , \text{ if } 0<x\le 100 \\ [300-1(x-100)]x & , \text{ if } x>100 \end{cases}$$

which simpli fies to

$$R(x)=\begin{cases} 30{,}000 & , \text{ if } 0<x\le 100 \\ 400x-x^2 & , \text{ if } x>100 \end{cases}$$

29a).

$C(m)=14+0.08m$

29b).

Domain : $\{m|m\ge 0\}$

29c).

$$\begin{aligned} C(100) &= 14+0.08(100) \\ &= 14+8 \\ &= 22 \end{aligned}$$

\$22

31a).

$$F(t)=\begin{cases} 2 & , \text{ if } 0<t\le \frac{1}{2} \\ 2+1(1.20) & , \text{ if } \frac{1}{2}<t\le 1 \\ 2+2(1.20) & , \text{ if } 1<t\le 1\frac{1}{2} \\ 2+3(1.20) & , \text{ if } 1\frac{1}{2}<t\le 2 \\ 2+4(1.20) & , \text{ if } 2<t\le 2\frac{1}{2} \\ 2+5(1.20) & , \text{ if } 2\frac{1}{2}<t\le 3 \end{cases}$$

which simplifies to

$$F(t)=\begin{cases} 2 & , \text{ if } 0<t\le \frac{1}{2} \\ 3.20 & , \text{ if } \frac{1}{2}<t\le 1 \\ 4.40 & , \text{ if } 1<t\le 1\frac{1}{2} \\ 5.60 & , \text{ if } 1\frac{1}{2}<t\le 2 \\ 6.80 & , \text{ if } 2<t\le 2\frac{1}{2} \\ 8 & , \text{ if } t>2\frac{1}{2} \end{cases}$$

31b).

11:20 A.M. until 1 :05 P.M. : $t=1\frac{3}{4}$

$F\left(1\frac{3}{4}\right)=\$5.60$

31c).

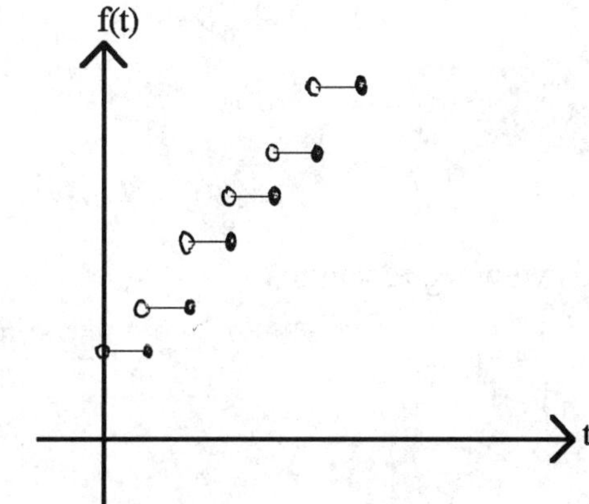

For Exercises 33-35:

a).

TENS viewing rectangle

XSCL = 1

YSCL = 1

GRAPH $Y=X^2+1$

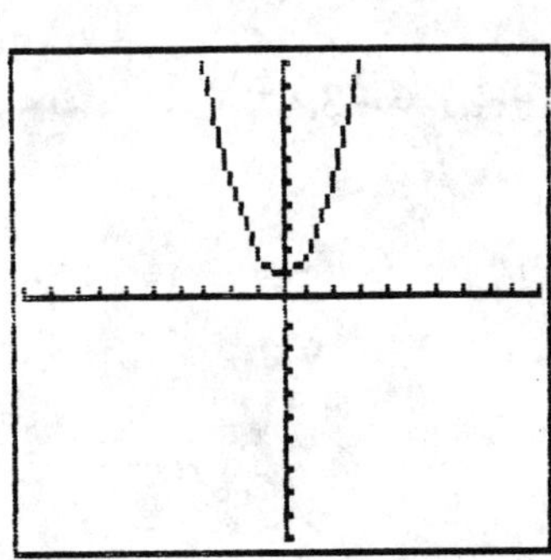

b).

EQUAL viewing rectangle

XSCL = 1

YSCL = 1

GRAPH $Y = 3X - \frac{1}{4}$

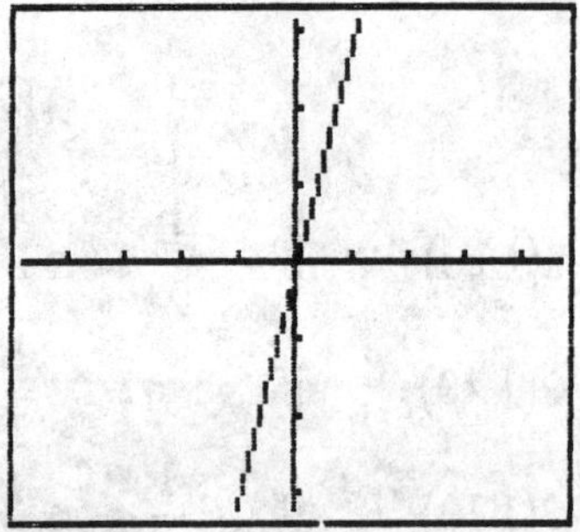

c).

EQUAL viewing rectangle

XSCL = 1

YSCL = 1

GRAPH $Y = -2X - 1$

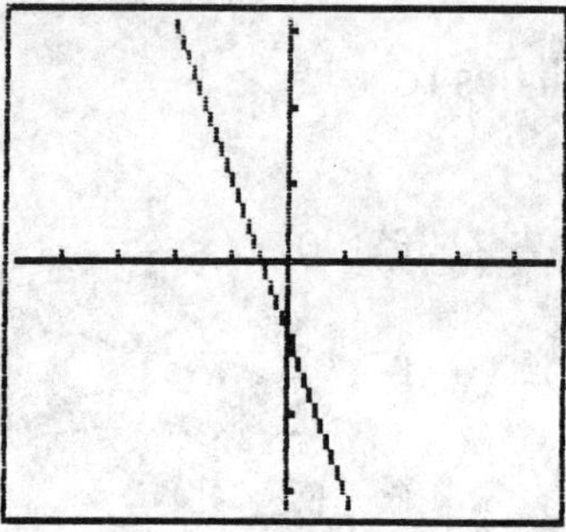

d).

EQUAL viewing rectangle

XSCL = 1

YSCL = 1

GRAPH $Y = -3X^2$

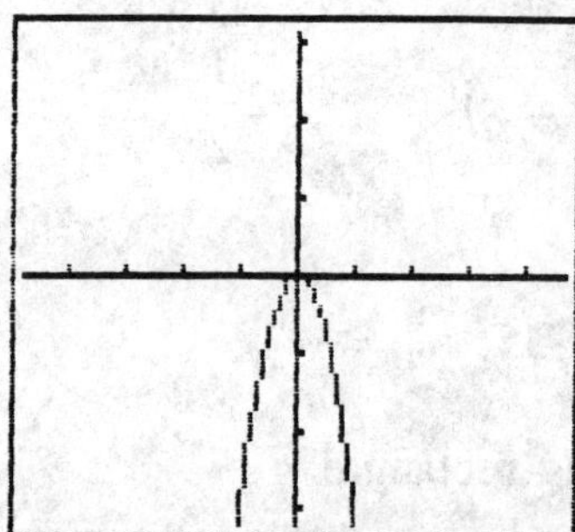

e).

XMIN = −9
XMAX = 9
XSCL = 1
YMIN = −4
YMAX = 8
YSCL = 1

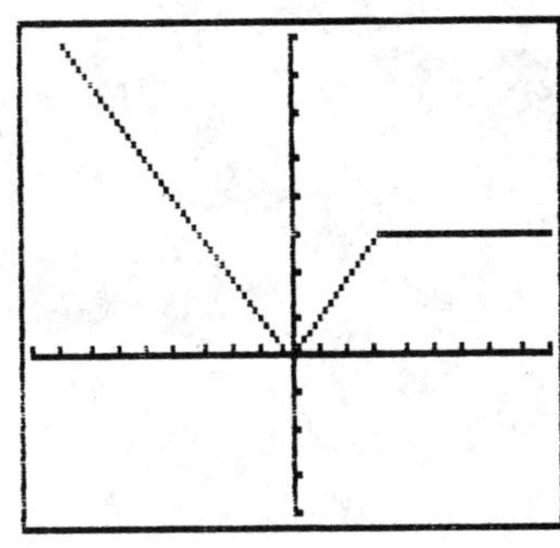

$$\text{GRAPH } Y = \begin{cases} -X & , \quad X \le 0 \\ X & , \quad 0 < X \le 3 \\ 3 & , \quad X > 3 \end{cases}$$

by entering $Y = (-X)(X \le 0) + X(X > 0)(X \le 3) + 3(X > 3)$

f).

EQUAL viewing rectangle
XSCL = 1
YSCL = 1

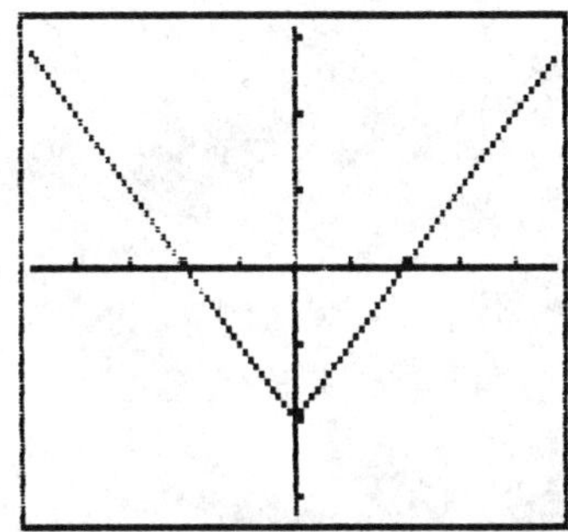

GRAPH $Y = ABS\ X - 2$

g).

XMIN = −10
XMAX = 10
XSCL = 1
YMIN = −5
YMAX = 15
YSCL = 1

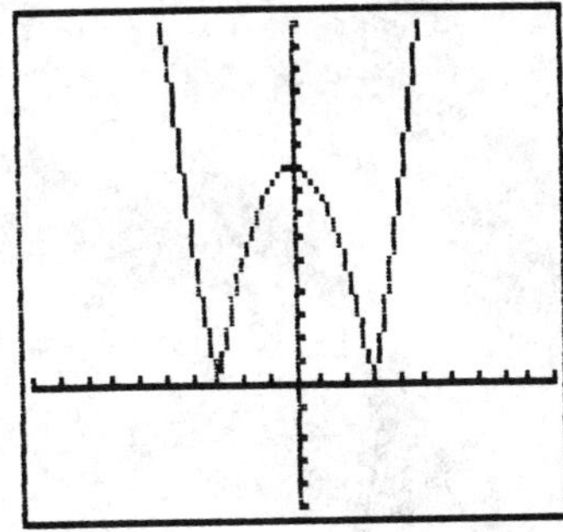

GRAPH $Y = ABS(X^2 - 9)$

h).

XMIN = −10
XMAX = 5
XSCL = 1
YMIN = −5
YMAX = 5
YSCL = 1

GRAPH $Y = \sqrt{\ }(1 - X)$

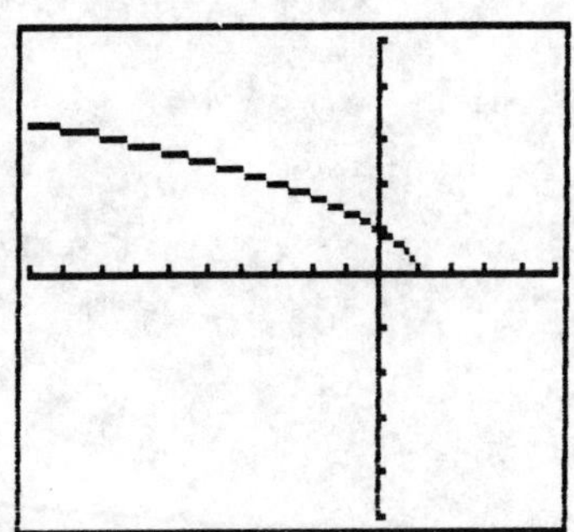

33a).
$\{y|y \geq 1\}$

33b).
$\{y|-\infty < y < \infty\}$

33c).
$\{y|-\infty < y < \infty\}$

33d).
$\{y|y \leq 0\}$

33e).
$\{y|y \geq 0\}$

33f).
$\{y|y \geq -2\}$

33g).
$\{y|y \geq 0\}$

33h).
$\{y|y \geq 0\}$

35.
$f(x)$ is decreasing where the graph is falling.

35a).
$(-\infty, 0)$

35b).
never

35c).
$(-\infty, \infty)$

35d).
$[0, \infty)$

35e).
$(-\infty, 0]$

35f).
$(-\infty, 0]$

35g).
$(-\infty, -3], [0, 3]$

35h).
$(-\infty, 1]$

37.
$y=-3$

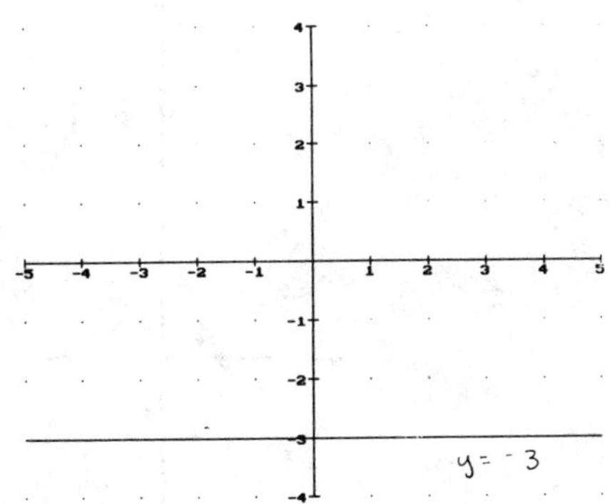

39.
$y=1+x^2$

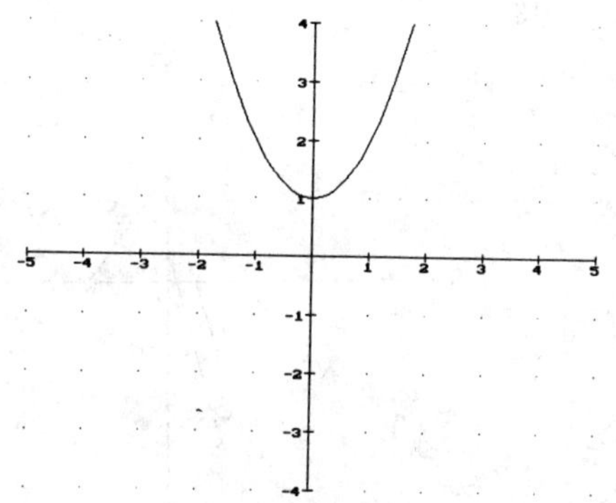

41.
$y=2+f(x)$
Shift the graph of $y=f(x)$ up two units.

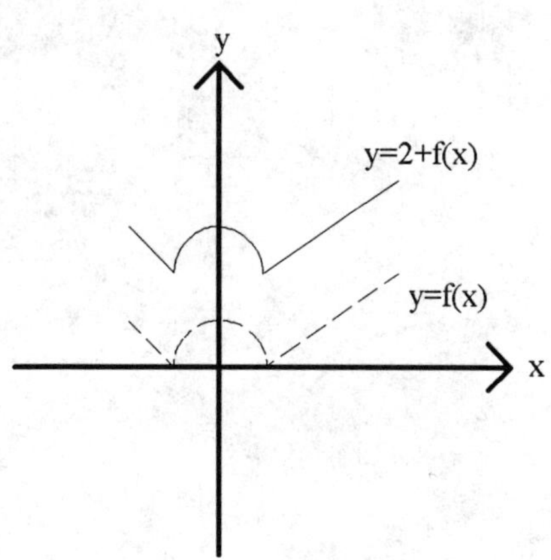

43.

$y = -f(x)$

Reflect the graph of $y = f(x)$ across the x-axis.

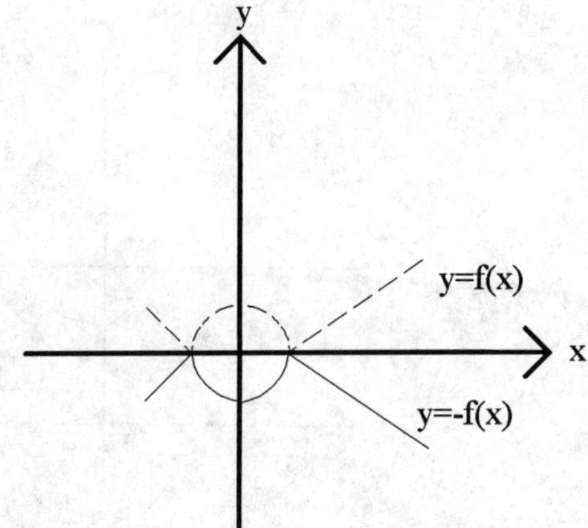

45.

$y = f(x) - 3$

Shift the graph of $y = f(x)$ down three units.

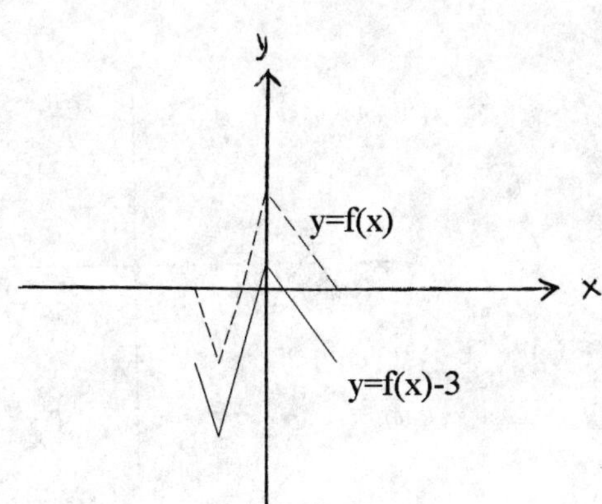

47.

$y = -f(x)$

Reflect the graph of $y = f(x)$ across the x-axis.

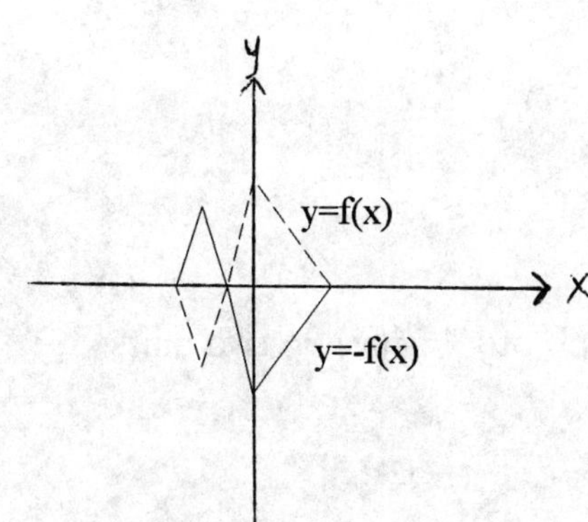

49.

$y = f(-x)$

Reflect the graph of $y = f(x)$ across the y-axis.

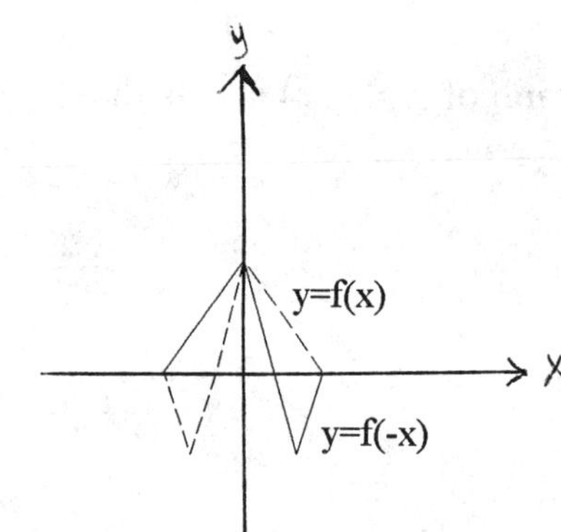

51.

$y = f(x+1)$

Shift the graph of $y = f(x)$ one unit to the left.

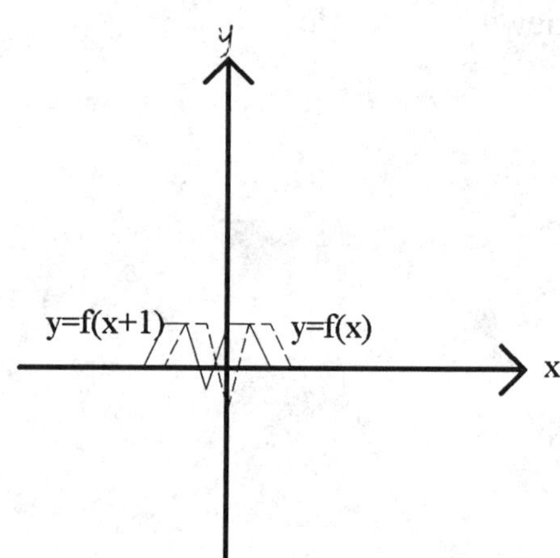

53.

$y = 5 - f(x)$

Reflect the graph of $y = f(x)$ across the x-axis, and then shift up five times.

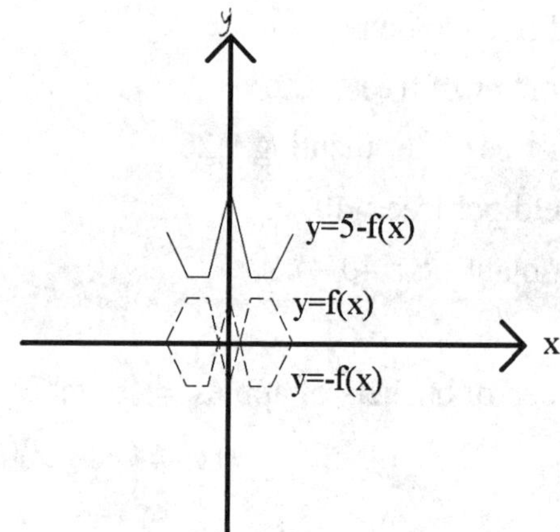

55.

$y = f(-x)$

Reflect the graph of $y = f(x)$ across the y-axis.

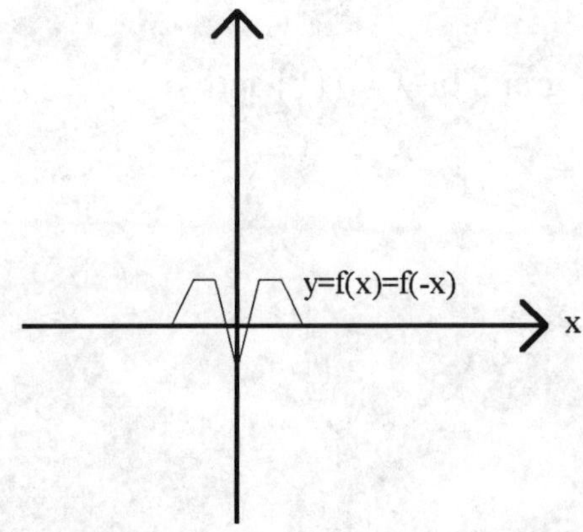

57.

EQUAL viewing rectangle

XSCL = 1

YSCL = 1

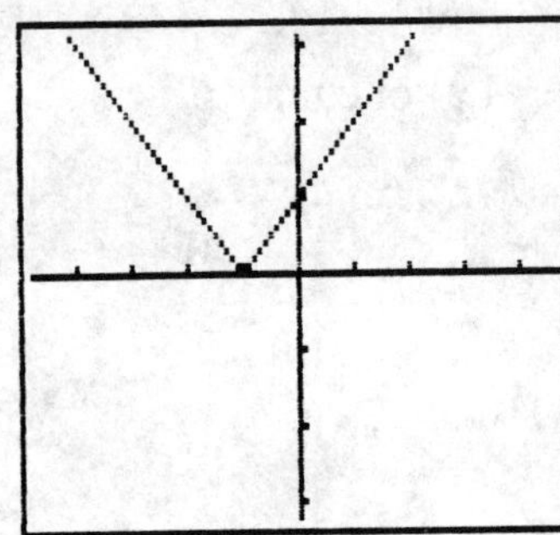

59a).

Number of trees to be planted : x

Current number of trees : 20

Number of trees after planting : $20 + x$

Current yield per tree : 40

Yield after planting : $40 - x$

Total number of bushels of apples = (number of trees)(yield per tree)

$$B = (20 + x)(40 - x)$$

59b).

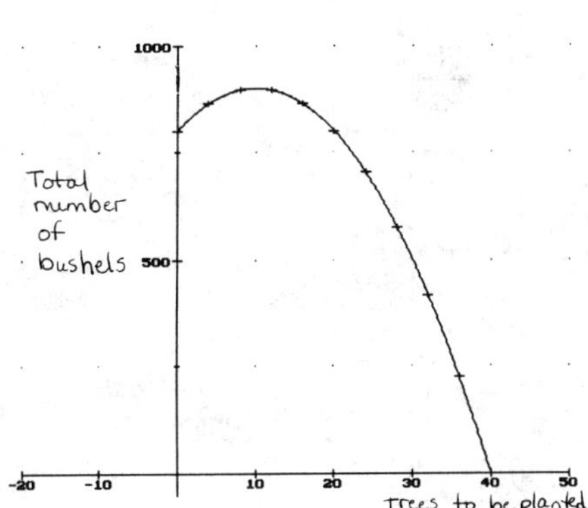

59c).
10 trees

59d).
900 bushels of apples

CHAPTER 3 SECTION 4

SECTION 3.4 PROGRESS CHECK

1. (Page 185)

$$m=\frac{0-(-5)}{-5-2}=\frac{5}{-7}=-\frac{5}{7}$$

$$y-y_1=m(x-x_1)$$
$$y-0=-\frac{5}{7}(x-(-5))$$
$$y=-\frac{5}{7}(x+5)$$
$$y=-\frac{5}{7}x-\frac{25}{7}$$

2. (Page 185)

$$2y+x-3=0$$
$$2y=-x+3$$
$$y=-\frac{1}{2}x+\frac{3}{2}$$
$$m=-\frac{1}{2},\ b=\frac{3}{2}$$

EXERCISE SET 3.4

1.

$$m=\frac{3-(-3)}{2-(-1)}=\frac{6}{3}=2$$

increasing function since $m>0$

3.

$$m=\frac{3-0}{-2-0}=\frac{3}{-2}=-\frac{3}{2}$$

decreasing function since $m<0$

5.

$$m=\frac{2-1}{\frac{1}{2}-\frac{3}{2}}=\frac{1}{-1}=-1$$

decreasing function since $m<0$

7.

$$m_{AB}=\frac{-5-(-1)}{-1-1}=\frac{-4}{-2}=2$$
$$m_{AC}=\frac{-5-3}{-1-3}=\frac{-8}{-4}=2$$
$$m_{BC}=\frac{-1-3}{1-3}=\frac{-4}{-2}=2$$

A, B, C are collinear since

$m_{AB}=m_{AC}=m_{BC}$

9.

$$\begin{aligned} y - y_1 &= m(x - x_1) \\ y - 3 &= 2(x - (-1)) \\ y - 3 &= 2(x + 1) \\ y - 3 &= 2x + 2 \\ y &= 2x + 5 \end{aligned}$$

11.

$$\begin{aligned} y &= mx + b \\ y &= 3x + 0 \\ y &= 3x \end{aligned}$$

13.

$$m = \frac{4 - (-6)}{2 - (-3)} = \frac{10}{5} = 2$$

$$\begin{aligned} y - y_1 &= m(x - x_1) \\ y - 4 &= 2(x - 2) \\ y - 4 &= 2x - 4 \\ y &= 2x \end{aligned}$$

15.

$$m = \frac{2 - 0}{3 - 0} = \frac{2}{3}$$

$$\begin{aligned} y &= mx + b \\ y &= \frac{2}{3}x + 0 \\ y &= \frac{2}{3}x \end{aligned}$$

17.

$$m = \frac{-1 - 1}{-\frac{1}{2} - \frac{1}{2}} = \frac{-2}{-1} = 2$$

$$\begin{aligned} y - y_1 &= m(x - x_1) \\ y - 1 &= 2\left(x - \frac{1}{2}\right) \\ y - 1 &= 2x - 1 \\ y &= 2x \end{aligned}$$

19.

$$\begin{aligned} y &= mx + b \\ y &= 3x + 2 \end{aligned}$$

21.

$$\begin{aligned} y &= mx + b \\ y &= 0x + 2 \\ y &= 2 \end{aligned}$$

23.

$$\begin{aligned} y &= mx + b \\ y &= \frac{1}{3}x - 5 \end{aligned}$$

25.

$$3x+4y=5$$
$$4y=-3x+5$$
$$y=-\frac{3}{4}x+\frac{5}{4}$$
$$m=-\frac{3}{4}\ ,\ b=\frac{5}{4}$$

27.

$$y-4=0$$
$$y=4$$
$$y=0x+4$$
$$m=0\ ,\ b=4$$

29.

$$3x+4y+2=0$$
$$3x+2=-4y$$
$$-\frac{3}{4}x-\frac{1}{2}=y$$
$$m=-\frac{3}{4}\ ,\ b=-\frac{1}{2}$$

31a).
$y=3$

31b).
$x=-6$

33a).
$y=0$

33b).
$x=-7$

35a).
$y=-9$

35b).
$x=9$

37.
$y=-3x+2$

37a).
$m=-3$

37b).
$$m_1=-\left(\frac{1}{-3}\right)=\frac{1}{3}$$

39.

$$3y=4x-1$$
$$y=\frac{4}{3}x-\frac{1}{3}$$

39a).
$$m=\frac{4}{3}$$

39b).
$$m_1=-\frac{3}{4}$$

41.

$$y = -3x + 2$$
$$m = -3$$

41a).

$$y - y_1 = m(x - x_1)$$
$$y - 3 = -3(x - 1)$$
$$y - 3 = -3x + 3$$
$$y = -3x + 6$$

41b).

$$m_1 = -\left(\frac{1}{-3}\right) = \frac{1}{3}$$

$$y - y_1 = m_1(x - x_1)$$
$$y - 3 = \frac{1}{3}(x - 1)$$
$$y - 3 = \frac{1}{3}x - \frac{1}{3}$$
$$y = \frac{1}{3}x + \frac{8}{3}$$

43.

$$3x + 5y = 2$$
$$5y = -3x + 2$$
$$y = -\frac{3}{5}x + \frac{2}{5}$$

$$m = -\frac{3}{5}$$

43a).

$$y - y_1 = m(x - x_1)$$
$$y - 2 = -\frac{3}{5}(x - (-3))$$
$$y - 2 = -\frac{3}{5}(x + 3)$$
$$y - 2 = -\frac{3}{5}x - \frac{9}{5}$$
$$y = -\frac{3}{5}x + \frac{1}{5}$$

43b).

$$m_1 = -\left(-\frac{5}{3}\right) = \frac{5}{3}$$

$$y - y_1 = m_1(x - x_1)$$
$$y - 2 = \frac{5}{3}(x - (-3))$$
$$y - 2 = \frac{5}{3}(x + 3)$$
$$y - 2 = \frac{5}{3}x + 5$$
$$y = \frac{5}{3}x + 7$$

45a).

C	F	
100	212	$(100, 212)$
0	32	$(0, 32)$

$$m = \frac{212-32}{100-0} = \frac{180}{100} = \frac{9}{5}$$

$$F - F_1 = m(C - C_1)$$
$$F - 32 = \frac{9}{5}(C - 0)$$
$$F - 32 = \frac{9}{5}C$$
$$F = \frac{9}{5}C + 32$$

45b).
$C = 20$

$$F = \frac{9}{5}(20) + 32$$
$$F = 68°$$

47.
In five years the growth was :

$600{,}000 - 200{,}000 = 400{,}000$

Sales in five years :

$600{,}000 + 400{,}000 = \$1{,}000{,}000$

49.
If $P(-2, 2)$ is in the line $3x + cy = 4$, its coordinates must satisfy the equation.

$$3(-2) + c(2) = 4$$
$$-6 + 2c = 4$$
$$2c = 10$$
$$c = 5$$

51.

$$m = \frac{-3-5}{-2-(-1)} = \frac{-8}{-1} = 8$$

$$y - y_1 = m(x - x_1)$$
$$y - 5 = 8(x - (-1))$$
$$y - 5 = 8(x + 1)$$
$$y = 8x + 13$$
$$f(x) = 8x + 13$$

53.

$(i.)$ Suppose $a > 0$

If $\quad x_2 > x_1$

then $\quad ax_2 > ax_1$

$$ax_2 + b > ax_1 + b$$
$$f(x_2) > f(x_1)$$

Thus $f(x)$ is an increasing function.

$(ii.)$Suppose $a < 0$

If $\quad x_2 > x_1$

then $\quad ax_2 < ax_1$

$$ax_2 + b < ax_1 + b$$
$$f(x_2) < f(x_1)$$

Thus $f(x)$ is a decreasing function.

$(iii.)$Suppose $a = 0$

then $\quad f(x) = ax + b$

$$= 0x + b$$
$$= b$$

Thus $f(x) = b$ which is a constant function.

55.

$y = mx + b_1$ and
$y = mx + b_2$ are
two lines with the same slope.
$b_1 \neq b_2$ so the two lines are different.
Solving simultaneously,
$mx + b_1 = mx + b_2$
Subtract mx from both sides
$b_1 = b_2$ ⊗
Thus the system of equations has no solution which means the lines are parallel.

57a).

$\angle$CAQ and $\angle$BQC are complements of $\angle$AQC

57b).

Their angles have equal measure.

57c).

Corresponding sides of similar triangles are proportional.

57d).

Definition of slope : $\dfrac{\text{rise}}{\text{run}}$

57e).

Using transitive property on (c) & (d) and definition of reciprocal, m_1 and m_2 are reciprocals. One line has positive slope, the other has negative slope, thus $m_2 = -\dfrac{1}{m_1}$.

59.

$$(x_1, f(x)), (x_2, f(x))$$

$$m = \frac{y_2 - y_1}{x_2 - x_1} = \frac{f(x_2) - f(x_1)}{x_2 - x_1}$$

In Exercises 61-63, the values you obtain when you TRACE on your graphics calculator are approximate. It is often necessary to ZOOM-IN to obtain more accurate results.

61.

$$(0, 1),\ (50, 1.01),\ (75, 1.02),\ (100, 1.04)$$

$$m = \frac{1.01 - 1}{50 - 0} = \frac{.01}{50} = 0.0002$$

$$m = \frac{1.02 - 1.01}{75 - 50} = \frac{0.01}{25} = 0.0004$$

$\neq$

Not linear

63.

$(2,-4),\ (10,20),\ (50,140),\ (85,245)$

$$m=\frac{20-(-4)}{10-2}=\frac{24}{8}=3$$

$$m=\frac{140-20}{50-10}=\frac{120}{40}=3$$

$$m=\frac{245-140}{85-50}=\frac{105}{35}=3$$

$$P-20=3(G-10)$$
$$P-20=3G-30$$
$$P=3G-10$$

XMIN = 0
XMAX = 100
XSCL = 10
YMIN = −50
YMAX = 250
YSCL = 50

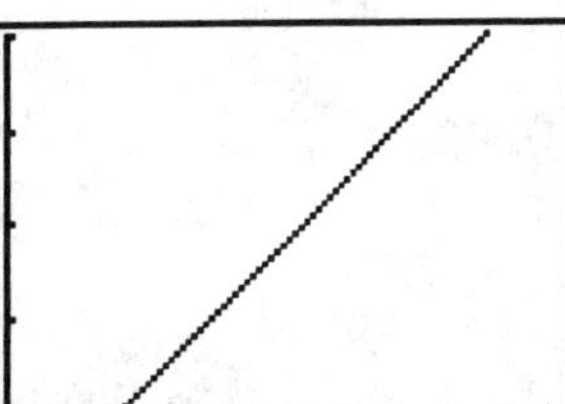

GRAPH $Y=3X-10$

65.

10:05 a.m. until 12:00 noon :

$$1\text{ hour }55\text{ minutes }=\frac{23}{12}$$

$$\text{Average speed }=\frac{\text{Distance}}{\text{Time}}$$

$$=\frac{683}{\frac{23}{12}}=356.35\text{ mph}$$

67.

Distance = 0.25

time = 5 minutes $=\frac{5}{60}=\frac{1}{12}$ hour

rate = 2.5 mph

$$\frac{\text{Distance}}{\text{rate}}=\text{time}$$

$$\frac{0.25}{2.5}\leq\frac{1}{12}\ ?$$

$$0.1\not\leq 0.08$$

The student will not complete the full distance.

69a).

$t=0$ to $t=4$ Time : $4-0=4$

$d=0$ to $d=4$ Distance : $4-0=4$

$$\text{Average speed} = \frac{D}{t} = \frac{4}{4} = 1 \text{ ft/sec}$$

69b).

$t=4$ to $t=7$ Time : $7-4=3$

$d=4$ to $d=4$ Distance : $4-4=0$

$$\text{Average speed} = \frac{D}{t} = \frac{0}{3} = 0 \text{ ft/sec}$$

69c).

$t=7$ to $t=10$ Time : $10-7=3$

$d=4$ to $d=7$ Distance : $7-4=3$

$$\text{Average speed} = \frac{D}{t} = \frac{3}{3} = 1 \text{ ft/sec}$$

69d).

$t=0$ to $t=7$ Time : $7-0=7$

$d=0$ to $d=4$ Distance : $4-0=4$

$$\text{Average speed} = \frac{D}{t} = \frac{4}{7} = 0.57 \text{ ft/sec}$$

69e).

$t=4$ to $t=10$ Time : $10-4=6$

$d=4$ to $d=7$ Distance : $7-4=3$

$$\text{Average speed} = \frac{D}{t} = \frac{3}{6} = \frac{1}{2} = 0.5 \text{ ft/sec}$$

69f).

$t=0$ to $t=10$ Time : $10-0=10$

$d=0$ to $d=7$ Distance : $7-0=7$

$$\text{Average speed} = \frac{D}{t} = \frac{7}{10} = 0.7 \text{ ft/sec}$$

69g).

Average speed $=0$, hence the particle was at rest between $t=4$ and $t=7$.

71a).

d	p
0	15
45	35

$$m = \frac{35-15}{45-0} = \frac{20}{45} = \frac{4}{9}$$

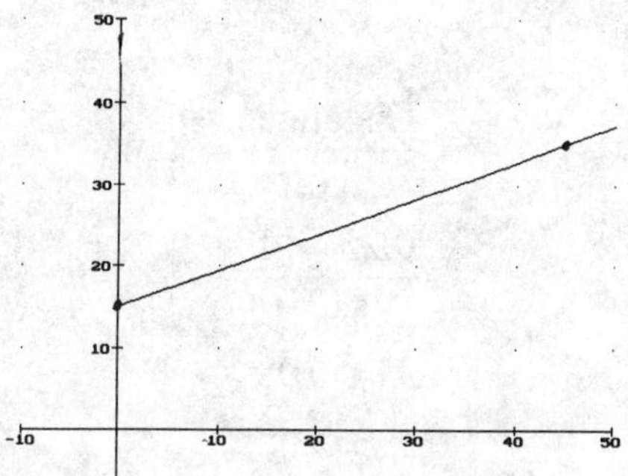

$$p - p_1 = m(d - d_1)$$

$$p - 15 = \frac{4}{9}(d - 0)$$

$$p - 15 = \frac{4}{9}d$$

$$p = \frac{4}{9}d + 15$$

71b).

Let $d = 10{,}800$

$$p = \frac{4}{9}(10800) + 15$$

$$p = 4800 + 15$$

$$p = 4815 \text{ lbs per in}^2.$$

71c).

Let $p = 99$

$$99 = \frac{4}{9}d + 15$$

$$84 = \frac{4}{9}d$$

$$189 = d$$

$$d = 189 \text{ feet}$$

73a).

The car decreased in value from $20500 to $14500 in the first three years.

decrease : $20500-14500=\$6000$

It will decrease another $6000 in value over the next 3 years.

Value after 6 years :

$20,500 - ($6000 + $6000) = $8500

(6, 8500)

After 8 years the value is:

$\$8500+\$900=\$9400$

(8, 9400)

$$m=\frac{9400-8500}{8-6}=\frac{900}{2}=450$$

value : v time in years : t

$$v-8500=450(t-6)$$
$$v=450t+5800$$

In the first 6 years, the car depreciates $2000 per year.

$$v(t)=\begin{cases}-2000t+20{,}500 & \text{if } 0<t\le 6\\ 450t+5800 & \text{if } t>6\end{cases}$$

73b).

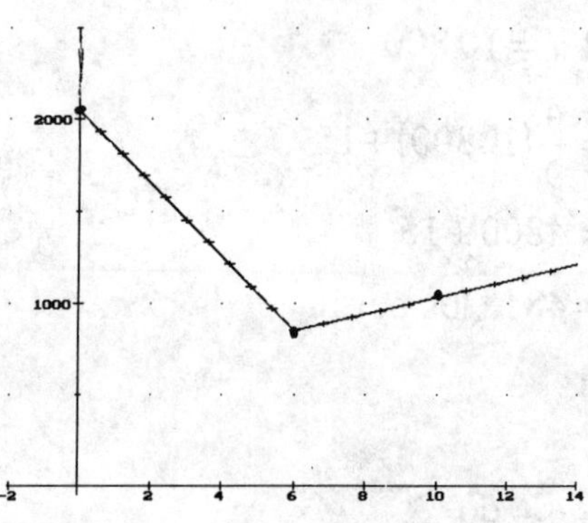

73c).

$$v(10)=450(10)+5800$$
$$=10{,}300$$

$10,300

CHAPTER 3 SECTION 5

SECTION 3.5 PROGRESS CHECK

1. (Page 195)
$f(x)=2x^2$ and $g(x)=x^2-5x+6$

1a).
$$(f+g)(x)=f(x)+g(x)$$
$$=2x^2+x^2-5x+6$$
$$=3x^2-5x+6$$

1b).
$$(f-g)(x)=f(x)-g(x)$$
$$=2x^2-(x^2-5x+6)$$
$$=2x^2-x^2+5x-6$$
$$=x^2+5x-6$$

1c).
$$(f\cdot g)(x)=f(x)g(x)$$
$$=2x^2(x^2-5x+6)$$
$$=2x^4-10x^3+12x^2$$

1d).
$$\left(\frac{f}{g}\right)(x)=\frac{f(x)}{g(x)}$$
$$=\frac{2x^2}{x^2-5x+6}$$

1e).
$$x^2-5x+6\neq 0$$
$$(x-3)(x-2)\neq 0$$
$$x-3\neq 0 \quad x-2\neq 0$$
$$x\neq 3 \quad x\neq 2$$
Domain : $\{x|x\neq 2, x\neq 3\}$

2. (Page 197)
$f(x)=x^2-2x$ and $g(x)=3x$

2a).
$$f[g(-1)]=f(3(-1))=f(-3)=(-3)^2-2(-3)$$
$$=9+6=15$$

2b).
$$g[f(-1)]=g[(-1)^2-2(-1)]=g(3)=3(3)=9$$

2c).
$$f[g(x)]=f(3x)=(3x)^2-2(3x)=9x^2-6x$$

2d).
$$g[f(x)]=g(x^2-2x)$$
$$=3(x^2-2x)=3x^2-6x$$

2e).
$$(f\circ g)(2)=f[g(2)]=9(2)^2-6(2)=24$$
(using (c))

2f)
$$(g\circ f)(2)=g[f(2)]=3(2)^2-6(2)=0$$
(using (d))

3. (Page 198)

Use denominator function for $g(x): g(x) = 7 - x$

Then $f(x) = \frac{2}{x}$

$$(f \circ g)(x) = f[g(x)] = f(7-x)$$
$$= \frac{2}{7-x} = h(x)$$

4. (Page 200)

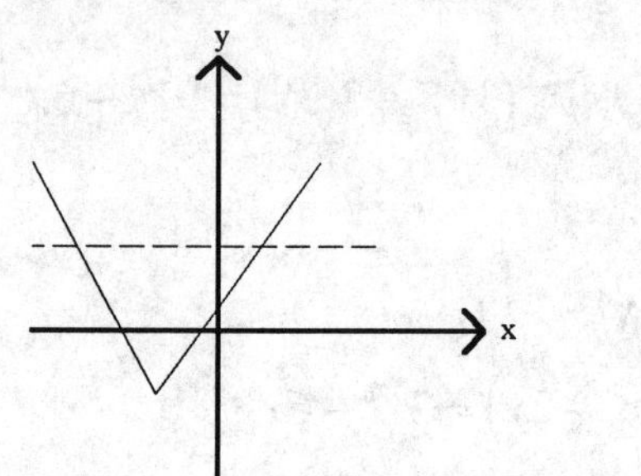

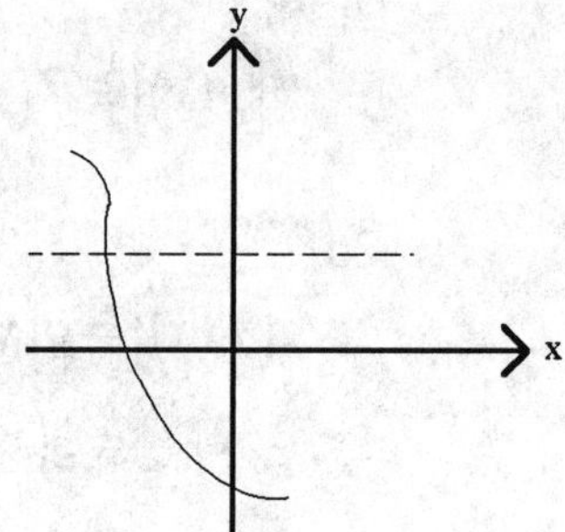

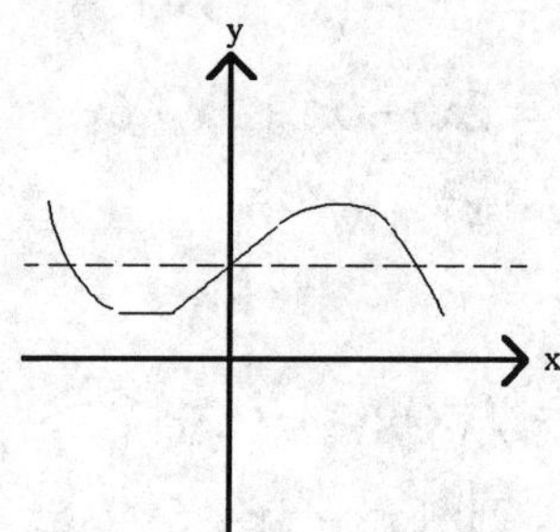

(b) Is the only graph that passes the horizontal line test.

5. (Page 202)

$f(x)=5x$ and $g(x)=\frac{x}{5}$

5a).

$(f\circ g)(x)=f[g(x)]=f\left(\frac{x}{5}\right)=5\left(\frac{x}{5}\right)=x$

5b).

$(g\circ f)(x)=g[f(x)]=g(5x)=\frac{5x}{5}=x$

5c).

yes since $(f\circ g)(x)=(g\circ f)(x)=x$

f and *g* are inverses of each other.

5d).

yes since $(f\circ g)(x)=(g\circ f)(x)=x$

f and *g* are inverses of each other.

6. (Page 205)

$f(x)=y=3x+5$

$y-5=3x$

$\frac{y-5}{3}=x$

$f^{-1}(x)=\frac{x-5}{3}$

EXERCISE SET 3.5

1.

$(f+g)(x)=f(x)+g(x)$

$=x^2+1+x-2$

$=x^2+x-1$

3.

$(f-g)(x)=f(x)-g(x)$

$=x^2+1-(x-2)$

$=x^2+1-x+2$

$=x^2-x+3$

5.

$(f\cdot g)(x)=f(x)\cdot g(x)$

$=(x^2+1)(x-2)$

$=x^3-2x^2+x-2$

7.

$\left(\frac{f}{g}\right)(x)=\frac{f(x)}{g(x)}$

$=\frac{x^2+1}{x-2}$

9.

dom_f : all real numbers

dom_g : all real numbers

11.

$$\begin{aligned}(f \circ g)(x) &= f[g(x)] \\ &= f(2x^2 + x) \\ &= 2(2x^2 + x) + 1 \\ &= 4x^2 + 2x + 1\end{aligned}$$

13.

$$\begin{aligned}(f \circ g)(2) &= f[g(2)] \\ &= g(2 \cdot 2^2 + 2) \\ &= f(10) \\ &= 2(10) + 1 \\ &= 21\end{aligned}$$

15.

$$\begin{aligned}(f \circ g)(x+1) &= f[g(x+1)] \\ &= f\left(2(x+1)^2 + x + 1\right) \\ &= f\left(2(x^2 + 2x + 1) + x + 1\right) \\ &= f(2x^2 + 4x + 2 + x + 1) \\ &= f(2x^2 + 5x + 3) \\ &= 2(2x^2 + 5x + 3) + 1 \\ &= 4x^2 + 10x + 6 + 1 \\ &= 4x^2 + 10x + 7\end{aligned}$$

17.

$$\begin{aligned}(g \circ f)(x-1) &= g[f(x-1)] \\ &= g\left(2(x-1) + 1\right) \\ &= g(2x - 2 + 1) \\ &= g(2x - 1) \\ &= 2(2x-1)^2 + (2x - 1) \\ &= 2(4x^2 - 4x + 1) + 2x - 1 \\ &= 8x^2 - 8x + 2 + 2x - 1 \\ &= 8x^2 - 6x + 1\end{aligned}$$

19.

$$\begin{aligned}(f \circ g)(x) &= f[g(x)] \\ &= f\left(\sqrt{x+2}\right) \\ &= \left(\sqrt{x+2}\right)^2 + 4 \\ &= x + 2 + 4 \\ &= x + 6,\ x \geq -2\end{aligned}$$

21.

$$\begin{aligned}(f \circ f)(-1) &= f[f(-1)] \\ &= f\left((-1)^2 + 4\right) \\ &= f(5) \\ &= 5^2 + 4 \\ &= 29\end{aligned}$$

23.

$$\begin{aligned}(g \circ f)(x) &= g[f(x)] \\ &= g\left(x^2 + 4\right) \\ &= \sqrt{x^2 + 4 + 2} \\ &= \sqrt{x^2 + 6}\end{aligned}$$

domain : $\{x \mid x \in \Re\}$

25.

$$\begin{aligned}(f \circ g)(x) &= f[g(x)] \\ &= f(x+2) \\ &= x+2-1 \\ &= x+1\end{aligned}$$

$$\begin{aligned}(g \circ f)(x) &= g[f(x)] \\ &= g(x-1) \\ &= x-1+2 \\ &= x+1\end{aligned}$$

27.

$$\begin{aligned}(f \circ g)(x) &= f[g(x)] \\ &= f\left(\frac{1}{x-1}\right) \\ &= \frac{1}{\frac{1}{x-1}+1} \\ &= \frac{x-1}{1+x-1} \\ &= \frac{x-1}{x}\end{aligned}$$

$$\begin{aligned}(g \circ f)(x) &= g[f(x)] \\ &= g\left(\frac{1}{x+1}\right) \\ &= \frac{1}{\frac{1}{x+1}-1} \\ &= \frac{x+1}{1-(x+1)} \\ &= \frac{x+1}{1-x-1} \\ &= \frac{x+1}{-x} \\ &= -\frac{x+1}{x}\end{aligned}$$

29.

$g(x)=x^2 \qquad f(x)=x+3$

$$\begin{aligned}(f\circ g)(x)&=f[g(x)]\\&=f(x^2)\\&=x^2+3\\&=h(x)\end{aligned}$$

31.

$g(x)=3x+2 \qquad f(x)=x^8$

$$\begin{aligned}(f\circ g)(x)&=f[g(x)]\\&=f(3x+2)\\&=(3x+2)^8\\&=h(x)\end{aligned}$$

33.

$g(x)=x^3-2x^2 \qquad f(x)=x^{\frac{1}{3}}$

$$\begin{aligned}(f\circ g)(x)&=f[g(x)]\\&=f(x^3-2x^2)\\&=(x^3-2x^2)^{\frac{1}{3}}\\&=h(x)\end{aligned}$$

35.

$g(x)=x^2-4 \qquad f(x)=|x|$

$$\begin{aligned}(f\circ g)(x)&=f[g(x)]\\&=f(x^2-4)\\&=|x^2-4|\\&=h(x)\end{aligned}$$

37.

$g(x)=4-x \qquad f(x)=\sqrt{x}$

$$\begin{aligned}(f\circ g)(x)&=f[g(x)]\\&=f(4-x)\\&=\sqrt{4-x}\\&=h(x)\end{aligned}$$

39.

$$\begin{aligned}f[g(x)]&=f\left(\frac{1}{2}x-2\right)\\&=2\left(\frac{1}{2}x-2\right)+4\\&=x-4+4\\&=x\end{aligned}$$

$$\begin{aligned}g[f(x)]&=g(2x+4)\\&=\frac{1}{2}(2x+4)-2\\&=x+2-2\\&=x\end{aligned}$$

41.

$$\begin{aligned} f[g(x)] &= f\left(-\frac{1}{3}x+\frac{2}{3}\right) \\ &= 2-3\left(-\frac{1}{3}x+\frac{2}{3}\right) \\ &= 2+x-2 \\ &= x \end{aligned}$$

$$\begin{aligned} g[f(x)] &= g(2-3x) \\ &= -\frac{1}{3}(2-3x)+\frac{2}{3} \\ &= -\frac{2}{3}+x+\frac{2}{3} \\ &= x \end{aligned}$$

43.

$$\begin{aligned} f[g(x)] &= f\left(\frac{1}{x}\right) \\ &= \frac{1}{\frac{1}{x}} \\ &= 1\cdot\frac{x}{1} \\ &= x \end{aligned}$$

$$\begin{aligned} g[f(x)] &= g\left(\frac{1}{x}\right) \\ &= \frac{1}{\frac{1}{x}} \\ &= 1\cdot\frac{x}{1} \\ &= x \end{aligned}$$

45.

$$\begin{aligned} f(x) &= y = 2x+3 \\ y-3 &= 2x \\ \frac{y-3}{2} &= x \\ f^{-1}(x) &= \frac{x-3}{2} \end{aligned}$$

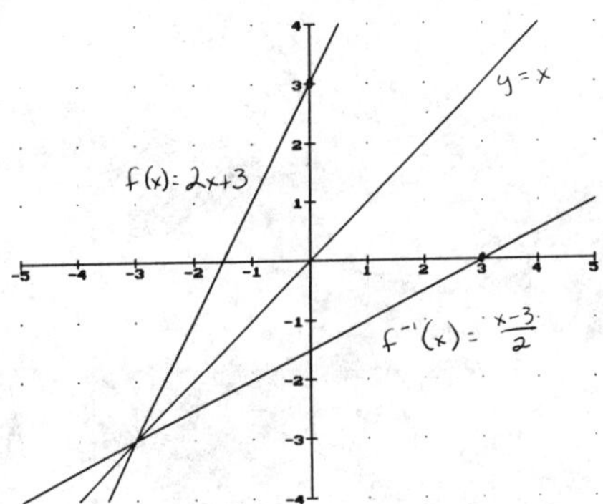

47.

$$f(x) = y = 3 - 2x$$
$$y - 3 = -2x$$
$$\frac{y-3}{-2} = x$$
$$f^{-1}(x) = \frac{-x+3}{2}$$

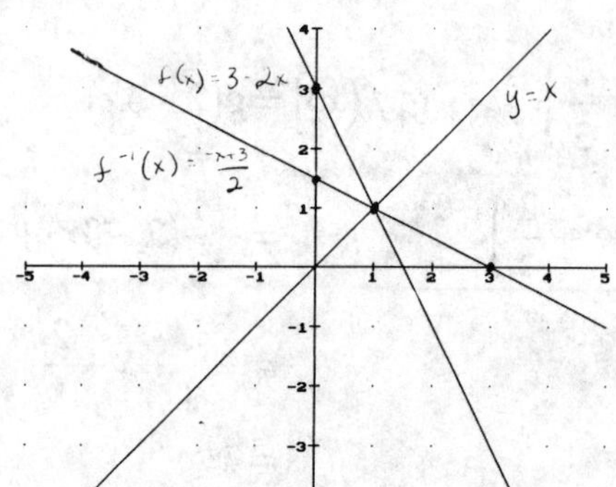

49.

$$f(x) = y = \frac{1}{3}x - 5$$
$$y + 5 = \frac{1}{3}x$$
$$3y + 15 = x$$
$$f^{-1}(x) = 3x + 15$$

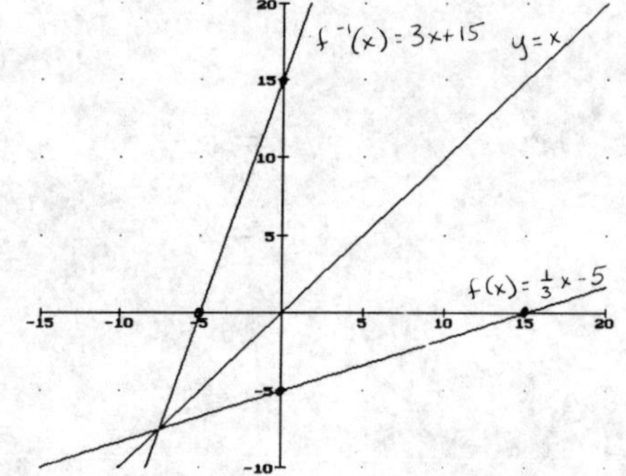

51.

$$f(x) = y = x^3 + 1$$
$$y - 1 = x^3$$
$$\sqrt[3]{y-1} = x$$
$$f^{-1}(x) = \sqrt[3]{x-1}$$

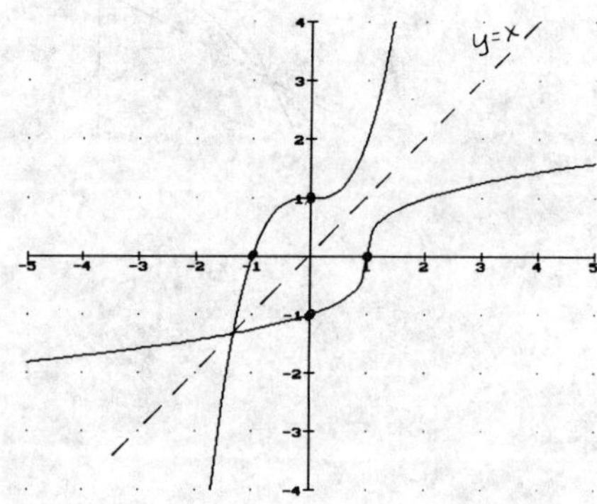

53.
Using the Horizontal Line Test, $(b), (d), (e), (h), (i)$ are one-to-one.

55.

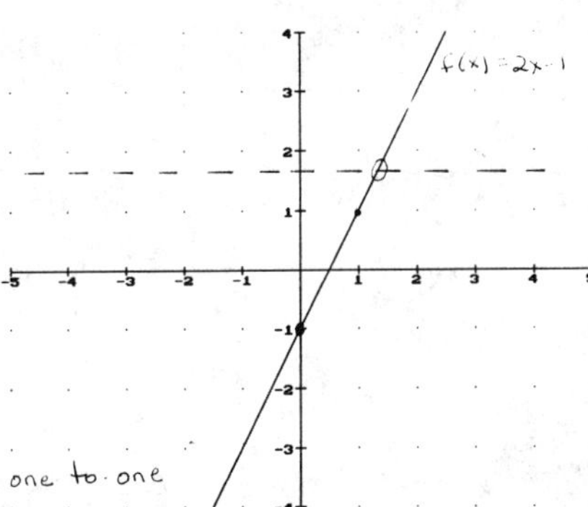

57.

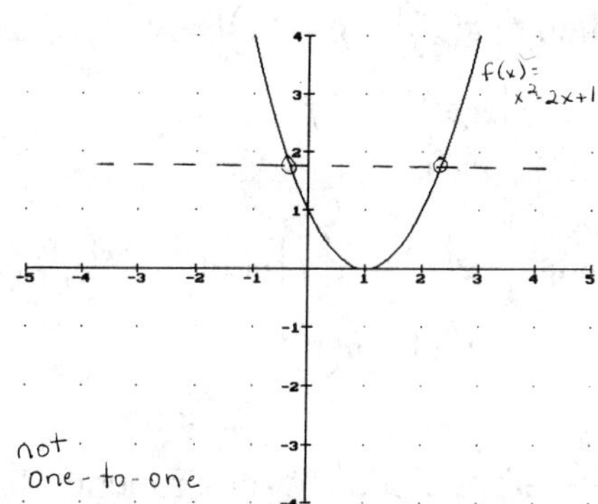

59.

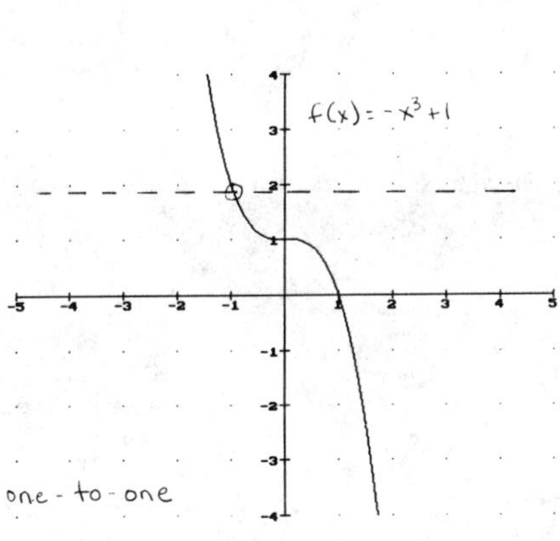

61.

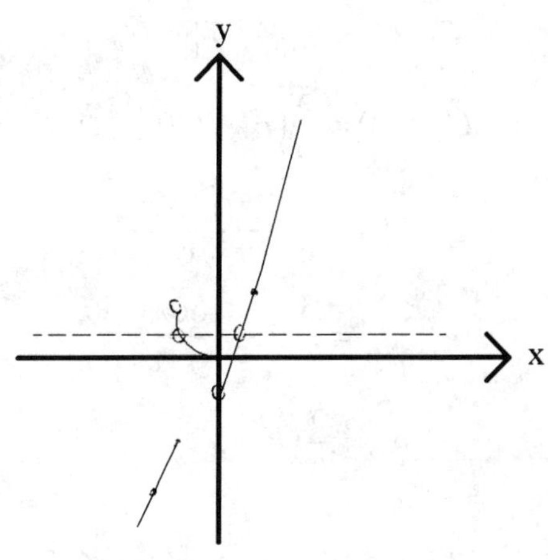

63.

Suppose $g(x)$ and $h(x)$ are both inverses of $f(x)$.

Then $f[g(x)] = g[f(x)] = x$
and $f[h(x)] = h[f(x)] = x$

Thus $g(x) = g[f[h(x)]] = f[g[h(x)]] = h(x)$
$g(x) = h(x)$

Therefore a one-to-one function has at most one inverse.

65.

$$f(x) = y = ax + b \quad , \quad a \neq 0$$
$$y - b = ax$$
$$\frac{y-b}{a} = x$$
$$f^{-1}(x) = \frac{x-b}{a}$$

67a).

F	C
230	110
−139	−95

$$m = \frac{110-(-95)}{230-(-139)} = \frac{205}{369} = \frac{5}{9}$$
$$C - C_1 = m(F - F_1)$$
$$C - 110 = \frac{5}{9}(F - 230)$$
$$C - 110 = \frac{5}{9}F - \frac{1150}{9}$$
$$C = \frac{5}{9}F - \frac{160}{9}$$
$$C = \frac{5}{9}(F - 32)$$

67b).

$$C = \frac{5}{9}(F - 32)$$
$$\frac{9}{5}C = F - 32$$
$$\frac{9}{5}C + 32 = F$$
$$F = \frac{9}{5}C + 32$$

69a).

$$\begin{aligned}(f \circ g)(P_m) &= f[g(P_m)] \\ &= f\left(\frac{4000}{P_m}\right) \\ &= 20\left(\frac{4000}{P_m}\right) - 5000 \\ &= \frac{80000}{P_m} - 5000\end{aligned}$$

69b).

It is the relationship between n the number of computers produced and *pm*, the price of microchips.

71a).

$$\begin{aligned}(g-f)(n) &= g(n) - f(n) \\ &= 22.5n - (5500 + 2.5n) \\ &= 20n - 5500\end{aligned}$$

71b).

Revenue − Cost = Profit

71c).

$$\begin{aligned}(g-f)(500) &= 20(500) - 5500 \\ &= 4500\end{aligned}$$

The profit earned by selling 500 items is $4,500.

$$\begin{aligned}(g-f)(250) &= 20(250) - 5500 \\ &= -500\end{aligned}$$

If only 250 items are sold there will be a loss of $500.

71d).

$$\begin{aligned}(g-f)(n) &= 0 \\ 20n - 5500 &= 0 \\ 20n &= 5500 \\ n &= 275\end{aligned}$$

71e).

They must sell 275 items to break even.

CHAPTER 3 SECTION 6

SECTION 3.6 PROGRESS CHECK

1a). (Page 210)

$$P = kV^2$$

$$64 = k(16)^2$$

$$k = \frac{64}{256} = \frac{1}{4}$$

1b).

$$C = kr$$

$$25.13 = k(4)$$

$$k = \frac{25.13}{4} = 6.2825$$

$$C = 6.2825r$$

2. (Page 212)

$$v = \frac{k}{w^3}$$

$$2 = \frac{k}{(-2)^3}$$

$$k = -16$$

EXERCISE SET 3.6

1a).

$$\frac{y}{x}=\frac{8}{2}$$
$$y=4x$$
$$k=4$$

1b).

$$y=4x$$

1c).

$$y=4(8)=32$$
$$y=4(12)=48$$

$$80=4x$$
$$20=x$$

$$120=4x$$
$$30=x$$

x	8	12	**20**	**30**
y	**32**	**48**	80	120

3a).

$$\frac{y}{x}=\frac{-\frac{1}{4}}{8}=-\frac{1}{4}\cdot\frac{1}{8}=-\frac{1}{32}$$

$$y=-\frac{1}{32}x$$
$$k=-\frac{1}{32}$$

3b).

$$y=-\frac{1}{32}(12)$$
$$=-\frac{3}{8}$$

5a).

$$S = kt^2$$
$$10 = k(10)^2$$
$$\frac{1}{10} = k$$

$$S = \frac{1}{10}t^2$$

5b).

$$S = \frac{1}{10}(5)^2$$
$$= \frac{5}{2}$$

7a).

$$y = \frac{k}{x}$$
$$-\frac{1}{2} = \frac{k}{6}$$

$$k = -3$$

$$y = -\frac{3}{x}$$

7b).

$$y = -\frac{3}{12}$$

$$y = -\frac{1}{4}$$

9a).

$$K = \frac{k}{r^3}$$
$$8 = \frac{k}{4^3}$$

$$k = 512$$
$$K = \frac{512}{r^3}$$

9b).

$$K = \frac{512}{5^2}$$
$$= \frac{512}{125}$$

11a).

$$M = \frac{kr^2}{s^2}$$
$$4 = \frac{k(4)^2}{(2)^2}$$

$$16 = 16k$$
$$k = 1$$
$$M = \frac{r^2}{s^2}$$

11b).

$$M = \frac{6^2}{5^2}$$
$$= \frac{36}{25}$$

13a).

$$T = \frac{kpv^3}{u^2}$$

$$24 = \frac{k(3)(2)^3}{4^2}$$

$$384 = 24k$$

$$k = 16$$

$$T = \frac{16pv^3}{u^2}$$

13b).

$$T = \frac{16(2)(3)^3}{(36)^2}$$

$$= \frac{864}{1296}$$

$$= \frac{2}{3}$$

15.

$$S = kt^2$$

$$144 = k(3)^2$$

$$k = 16$$

$$S = 16t^2$$

15a).

$$t = 5$$

$$S = 16(5)^2 = 400 \text{ ft}$$

15b).

$$S = 784$$

$$784 = 16t^2$$

$$49 = t^2$$

$$t = 7 \text{ sec}$$

17.

$$R = \frac{k}{A}$$

$$20 = \frac{k}{8}$$

$$k = 160$$

$$R = \frac{160}{A}$$

$$R = \frac{160}{12} = \frac{40}{3} \text{ ohms}$$

19.

$$I = \frac{k}{d^2}$$

$$200 = \frac{k}{(4)^2}$$

$$k = 3200$$

$$I = \frac{3200}{d^2}$$

19a).

$$d = 6$$

$$I = \frac{3200}{(6)^2} = \frac{800}{9} \text{ candlepower}$$

19b).

$$I = 50$$

$$50 = \frac{3200}{d^2}$$

$$50d^2 = 3200$$

$$d^2 = 64$$

$$d = 8 \text{ ft}$$

21.

C: equipment cost

p: number of presses

h: number of hours

$$C = kph$$
$$1200 = k(4)(6)$$
$$k = 50$$
$$C = 50ph$$
$$3600 = (50)p(12)$$
$$p = 6$$

6 presses

23.

L: illumi nation

I: intensity

d: distance

$$L = \frac{kI}{d^2}$$
$$50 = \frac{k(400)}{(2)^2}$$
$$k = \frac{1}{2}$$
$$L = \frac{\frac{1}{2}I}{d^2} = \frac{I}{2d^2}$$
$$L = \frac{3840}{2(4)^2}$$
$$= 120 \text{ candlepower per sq ft.}$$

25a).

$$I = \frac{k}{d^2}$$

25b).

$$3.80 = \frac{k}{(5.5)^2}$$
$$k = 114.95$$
$$I = \frac{114.95}{d^2}$$
$$I = \frac{114.95}{(4)^2}$$
$$= 7.18 \text{ watts per sq meter}$$

27.

A: Dow Jones Industrial Average

P: Oil price

$$A = \frac{k}{p}$$
$$2520 = \frac{k}{18.50}$$
$$k = 46620$$
$$A = \frac{46620}{p}$$

1990 : Price of oil $= 18.50 + 13.50 = 32$

$$A = \frac{46620}{32}$$
$$= 1456.88$$

29.

$$V = \frac{kT}{p}$$

$$245 = \frac{k(56)}{25}$$

$$k = 109.375$$

$$V = \frac{109.375T}{p}$$

29a).

$$210 = \frac{(109.375)(45)}{p}$$

$$210P = 4921.875$$

$$P = 23.4375 \text{ kg per cu.cm}$$

29b).

$$305 = \frac{109.375T}{40}$$

$$T = 111.54^{\circ} F$$

31a).

$$L = \frac{kwl^2}{d}$$

31b).

$$200 = \frac{k(3)(5)^2}{4}$$

$$k = 10.67$$

$$L = \frac{10.67wl^2}{d}$$

31c).

$$L = \frac{(10.67)(6)(5)^2}{2}$$

$$= 800.25 \text{ lb}$$

31d).

$$\frac{k(2w)l^2}{\frac{1}{2}d} = \frac{4kwl^2}{d} = 4L$$

The load can be multipli ed by 4.

31e).

$$\frac{kw(2l)^2}{d}$$

$$= \frac{4kwl^2}{d}$$

$$= 4L$$

The load can be multipli ed by 4.

CHAPTER 3 REVIEW EXERCISES

1.

$$\begin{aligned} d &= \sqrt{[-4-2]^2+[-6-(-1)]^2} \\ &= \sqrt{(-6)^2+(-5)^2} \\ &= \sqrt{61} \end{aligned}$$

2.

$$\overline{AB} = \sqrt{(-2-3)^2+[-6-(-4)]^2} = \sqrt{(-5)^2+(-2)^2} = \sqrt{29}$$

$$\overline{AC} = \sqrt{[3-(-1)]^2+[-4-2]^2} = \sqrt{4^2+(-6)^2} = \sqrt{52} = 2\sqrt{13}$$

$$\overline{BC} = \sqrt{[-2-(-1)]^2+[-6-2]^2} = \sqrt{(-1)^2+(-8)^2} = \sqrt{65}$$

The longest side is BC and its length is $\sqrt{65}$.

3.

x	y
−2	−1
−1	0
0	1
1	0
2	−1

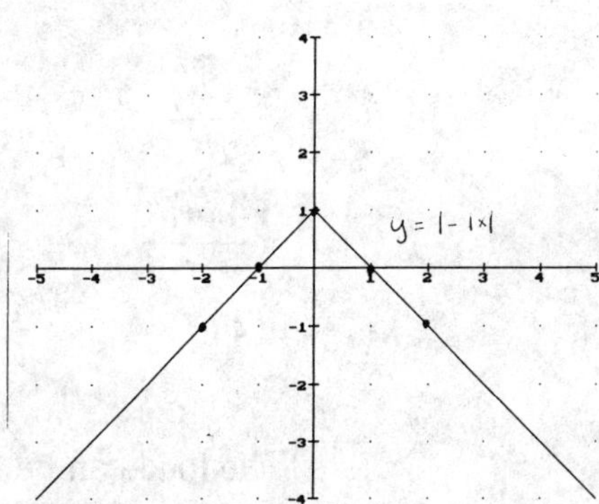

4.

x	y
2	0
3	1
6	2
11	3

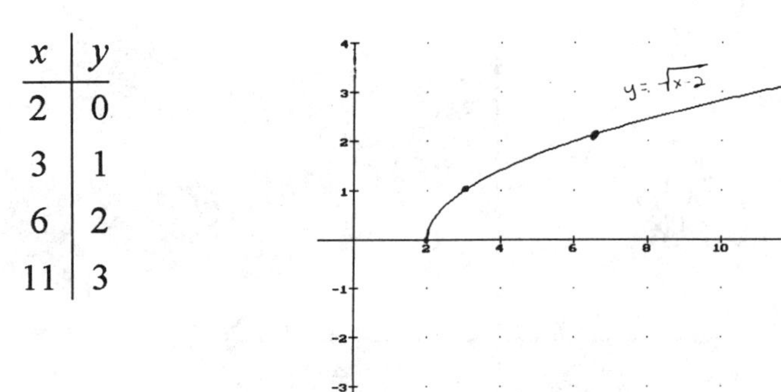

5.

$$(-y)^2 = 1 - x^3 \qquad y^2 = 1-(-x)^3 \qquad (-y)^2 = 1-(-x)^3$$

$$y^2 = 1 - x^3 \qquad y^2 = 1 + x^3 \qquad y^2 = 1 + x^3$$

yields original — doesn't yield original — doesn't yield original

Symmetric with respect to x-axis

6.

$$(-y)^2 = \frac{x^2}{x^2-5} \qquad y^2 = \frac{(-x)^2}{(-x)^2-5} \qquad (-y)^2 = \frac{(-x)^2}{(-x)^2-5}$$

$$y^2 = \frac{x^2}{x^2-5} \qquad y^2 = \frac{x^2}{x^2-5} \qquad y^2 = \frac{x^2}{x^2-5}$$

yields original — yields original — yields original

Symmetric with respect to x-axis, y-axis and origin.

7.

Since it passes the vertical line test, it is a function.

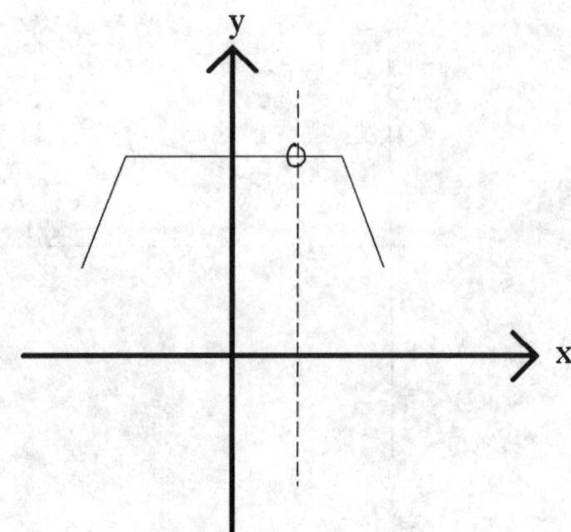

8.

Since it passes the vertical line test, it is a function.

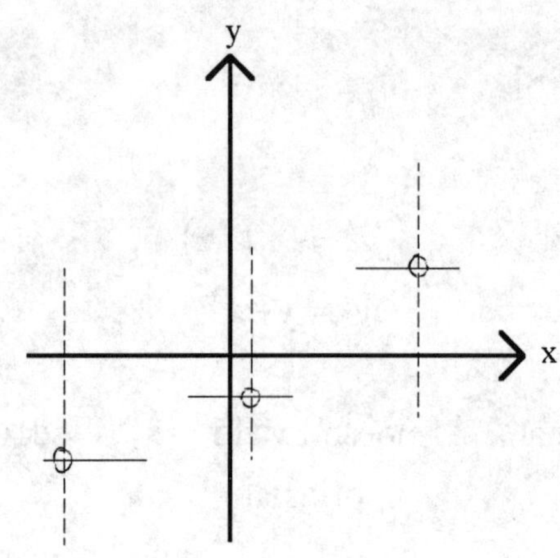

9.

Domain : $3x-5\geq 0$

$3x\geq 5$

$x\geq \frac{5}{3}$

$\left\{x \middle| x\geq \frac{5}{3}\right\}$

10.

Domain : $x^2+2x+1\neq 0$

$(x+1)(x+1)\neq 0$

$x+1\neq 0$

$x\neq -1$

$\{x|x\neq -1\}$

11.

Find the value of x for which $f(x)=15$

$15=\sqrt{x-1}$

$225=x-1$

$x=226$

12.

Find the value of t for which $f(t)=10$

$10=t^2+1$

$9=t^2$

$t=\pm 3$

In Exercises 13-22, the RANGE values given show important features of the functions. Other RANGE values are also acceptable. The display on your calculator may vary from that shown.

13.
TENS viewing rectangle
XSCL = 1
YSCL = 1

GRAPH $Y = 5X - 4$

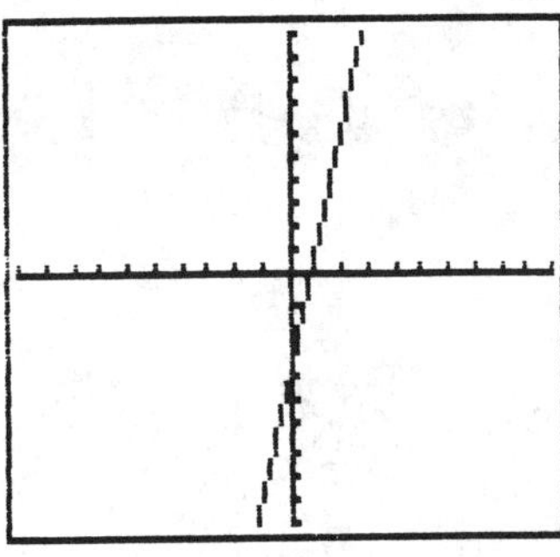

14.
TENS viewing rectangle
XSCL = 1
YSCL = 1

GRAPH $Y = 3X^3 + 2$

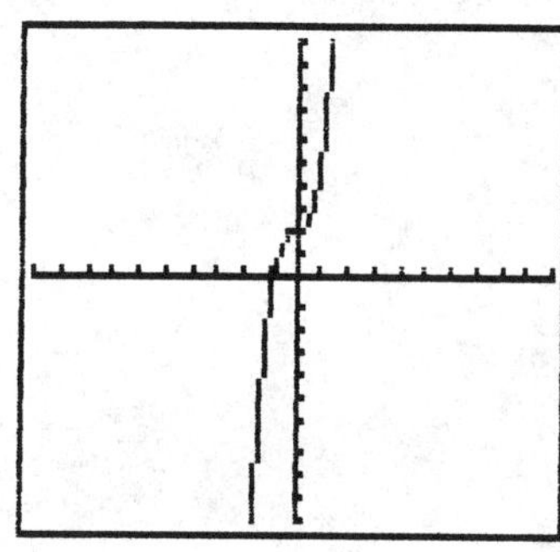

15.
XMIN = −5
XMAX = 5
XSCL = 1
YMIN = −6
YMAX = 1
YSCL = 1

GRAPH $Y = X - X^2$

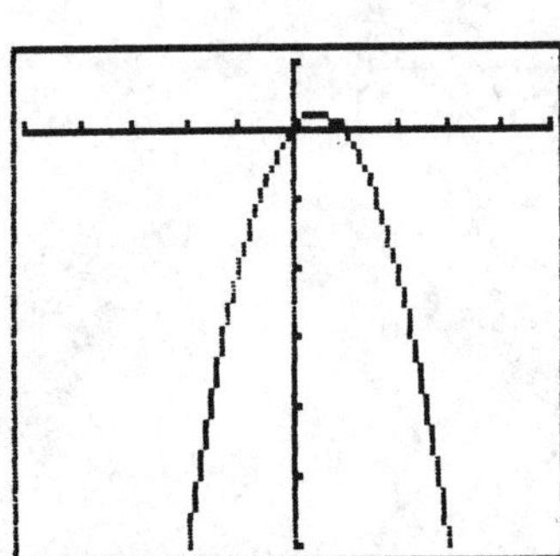

16.

XMIN = −5

XMAX = 5

XSCL = 1

YMIN = −1

YMAX = 5

YSCL = 1

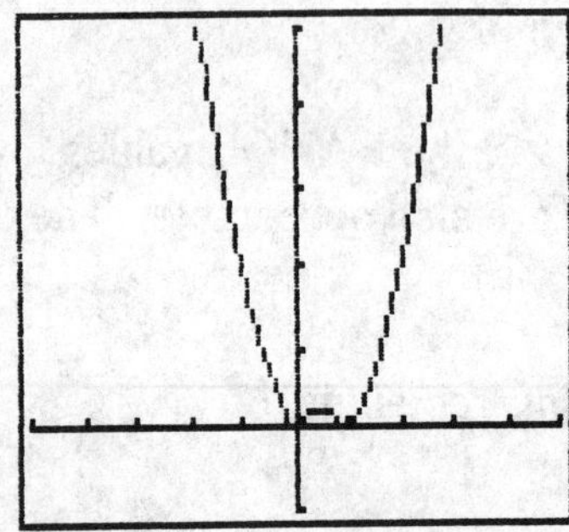

GRAPH $Y = ABS(X - X^2)$

17.

EQUAL viewing rectangle

XSCL = 1

YSCL = 1

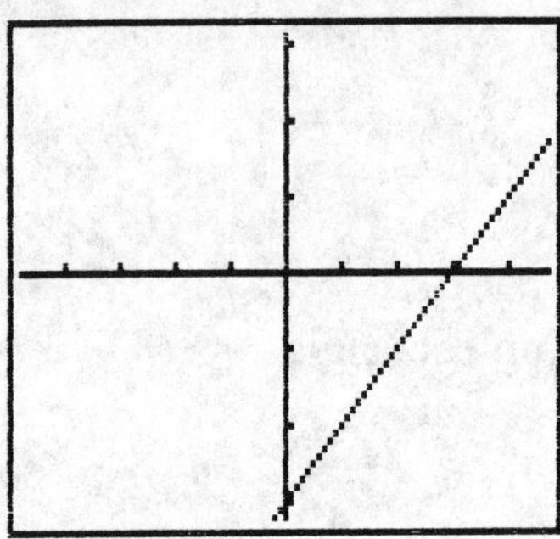

GRAPH $Y = X - 3$

18.

EQUAL viewing rectangle

XSCL = 1

YSCL = 1

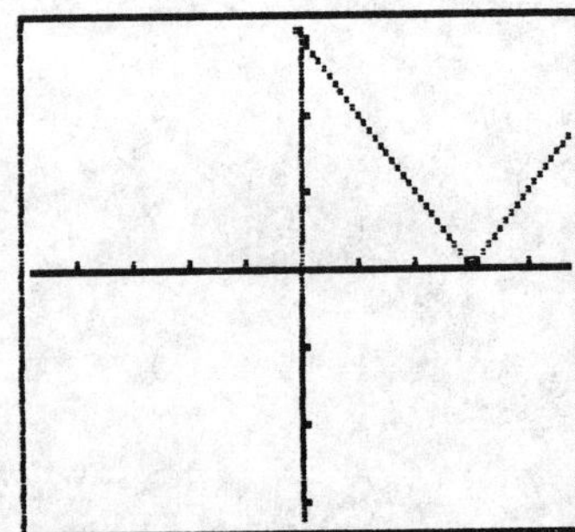

GRAPH $Y = ABS(X - 3)$

19.

EQUAL viewing rectangle

XSCL = 1

YSCL = 1

GRAPH $Y = (ABS(X-3)) \div (X-3)$

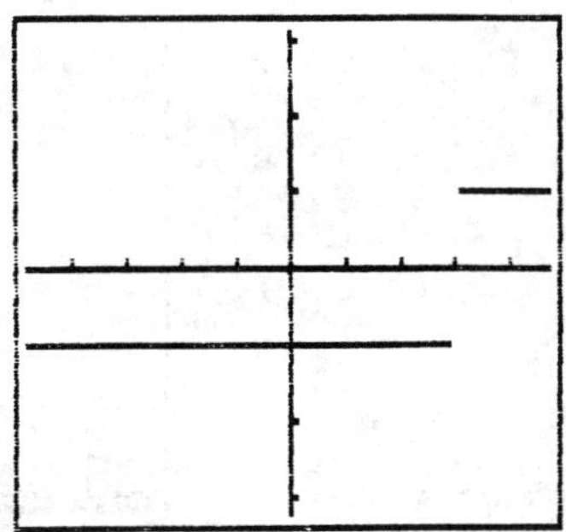

20.

XMIN = −2

XMAX = 10

XSCL = 1

YMIN = 0

YMAX = 14

YSCL = 1

GRAPH $Y = 2\sqrt{}X + 7$

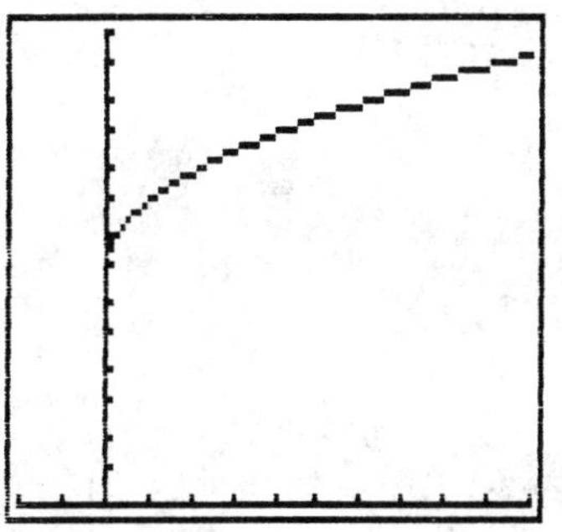

21.

XMIN = −10

XMAX = 2

XSCL = 1

YMIN = −7

YMAX = 7

YSCL = 1

GRAPH $Y = 2\sqrt{}(X + 7)$

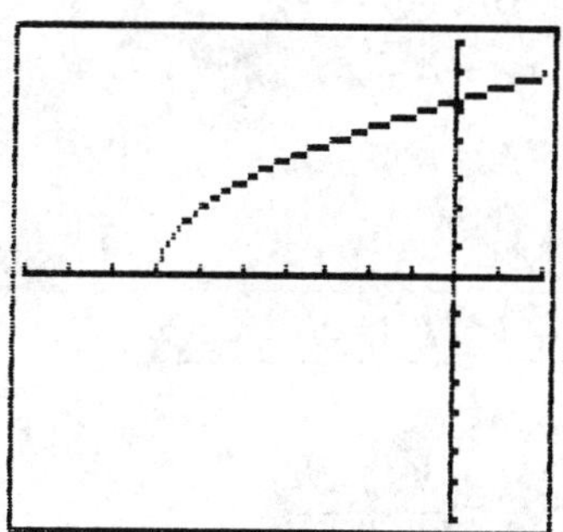

22.

XMIN = −4

XMAX = 4

XSCL = 1

YMIN = 0

YMAX = 6

YSCL = 1

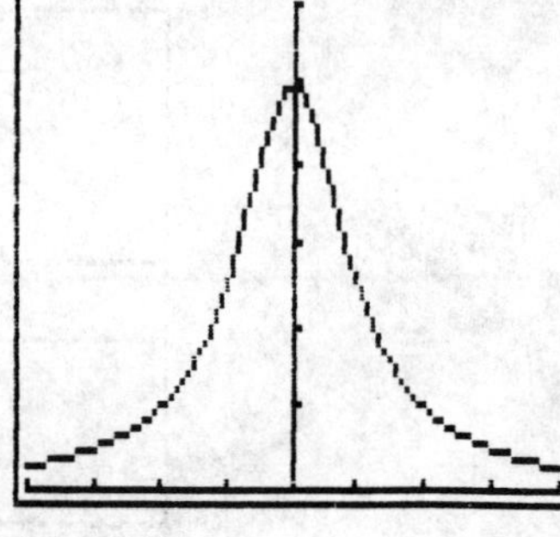

GRAPH $Y = 5 \div (X^2 + 1)$

23.

$$\begin{aligned} f(-3) &= (-3)^2 - (-3) \\ &= 9 + 3 \\ &= 12 \end{aligned}$$

24.

$$\begin{aligned} f(y-1) &= (y-1)^2 - (y-1) \\ &= y^2 - 2y + 1 - y + 1 \\ &= y^2 - 3y + 2 \end{aligned}$$

25.

$$\begin{aligned} f(2+h) &= (2+h)^2 - (2+h) \\ &= 4 + 4h + h^2 - 2 - h \\ &= h^2 + 3h + 2 \end{aligned}$$

$$\begin{aligned} f(2) &= (2)^2 - 2 \\ &= 4 - 2 \\ &= 2 \end{aligned}$$

$$\begin{aligned} \frac{f(2+h) - f(2)}{h} &= \frac{h^2 + 3h + 2 - 2}{h} \\ &= \frac{h^2 + 3h}{h} \\ &= \frac{h(h+3)}{h} \\ &= h + 3 \end{aligned}$$

26.

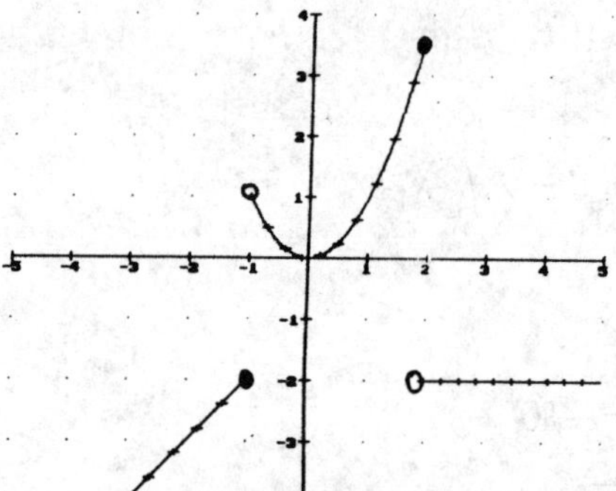

27.

$f(x)$ is increasing for $x \le -1, 0 \le x \le 2$ since the graph is rising.

$f(x)$ is decreasing for $-1 < x \le 0$ since the graph is falling

$f(x)$ is contant for $x > 2$

28.

$$f(-4) = (-4) - 1$$
$$= -5$$

29.

$$f(4) = -2$$

30.

Shift the graph of $y = f(x)$ up 3 units.

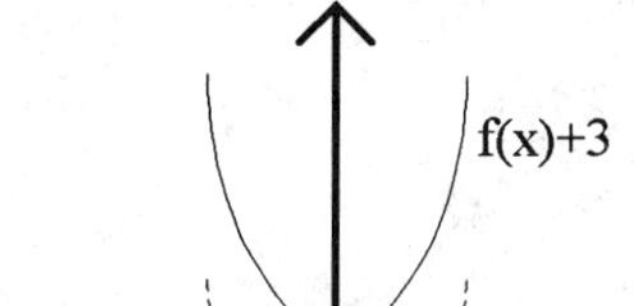
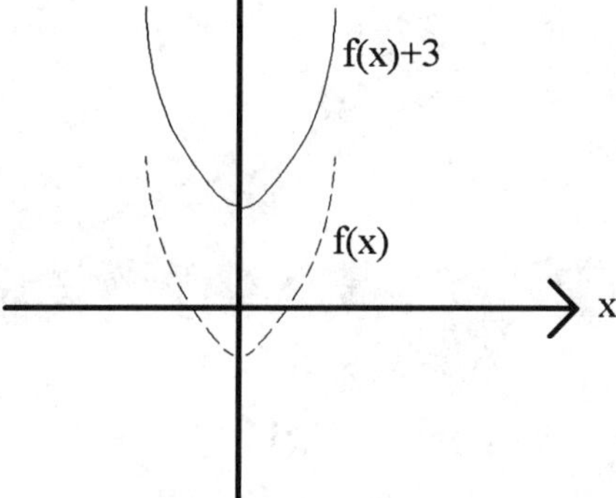

31.

Shift the graph of $y = f(x)$ one unit to the left.

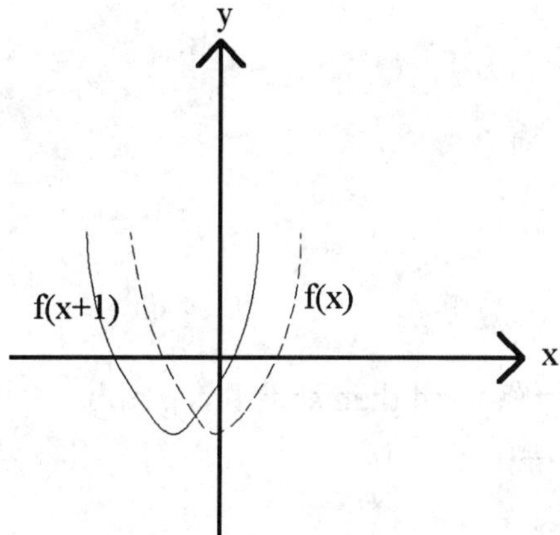

32.

$y = -f(x)$

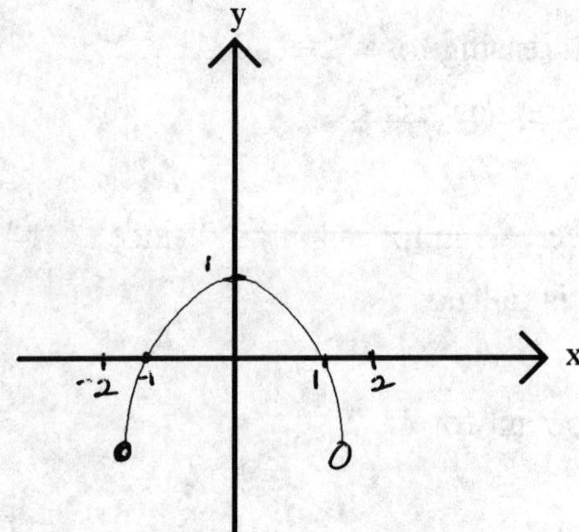

33.

$y = |f(x)|$

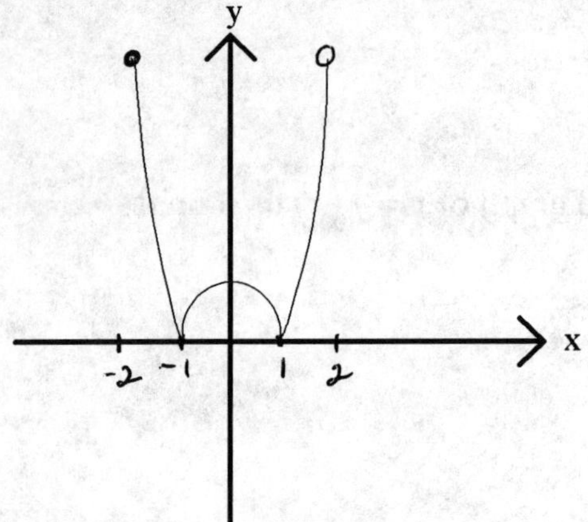

34.

Double the values of $y = f(x)$.

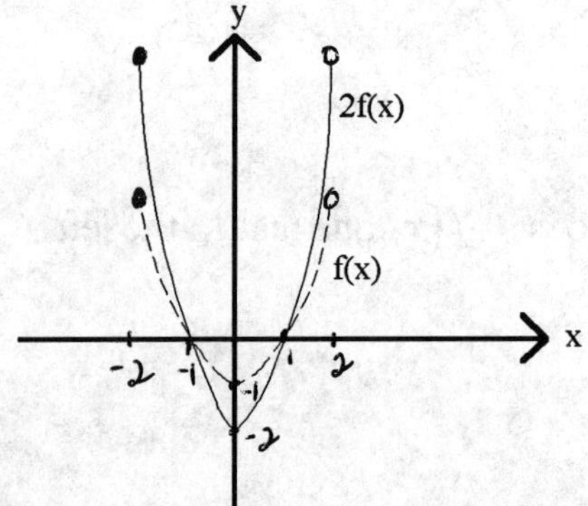

35.

Halve $y = f(x)$ and then shift this graph down one unit.

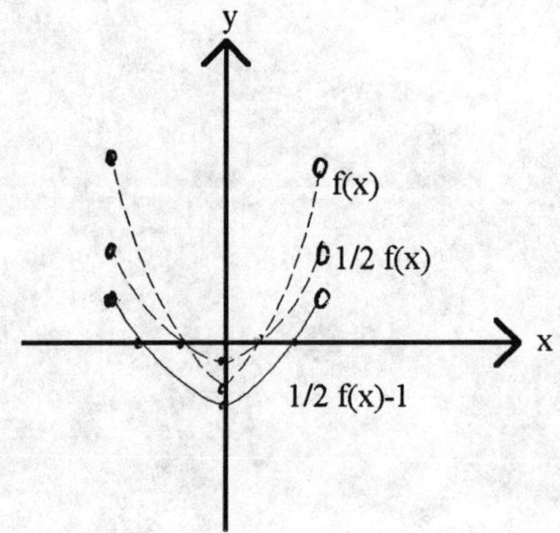

36.

$$m = \frac{-6-3}{-4-(-1)} = \frac{-9}{-3} = 3$$

37.

$$y - y_1 = m(x - x_1)$$
$$y - 3 = 3(x - (-1))$$
$$y - 3 = 3(x + 1)$$
$$y = 3x + 6$$

38.

Parallel to y-axis means the line is vertical.

$x = -4$

39.

$y = 3$

40.

$$4x - y - 3 = 0$$
$$4x - 3 = y$$
$$m = 4$$

$$y - y_1 = m(x - x_1)$$
$$y - (-6) = 4(x - (-4))$$
$$y + 6 = 4(x + 4)$$
$$y = 4x + 10$$

41.

$$2y + x - 5 = 0$$
$$x - 5 = -2y$$
$$-\frac{1}{2}x + \frac{5}{2} = y$$
$$m = -\frac{1}{2}$$
$$m_{\perp} = 2$$

$$y - y_1 = m_{\perp}(x - x_1)$$
$$y - 3 = 2(x - (-1))$$
$$y - 3 = 2(x + 1)$$
$$y = 2x + 5$$

42.

$$(f + g)(x) = f(x) + g(x)$$
$$= (x + 1) + (x^2 - 1)$$
$$= x^2 + x$$

43.

$$(f \cdot g)(-1) = f(-1)g(-1)$$
$$= (-1 + 1)\left[(-1)^2 - 1\right]$$
$$= (0)(0)$$
$$= 0$$

44.

$$\left(\frac{f}{g}\right)(x)=\frac{f(x)}{g(x)}$$
$$=\frac{x+1}{x^2-1}$$
$$=\frac{x+1}{(x+1)(x-1)}$$
$$=\frac{1}{x-1}\ ,\quad x\neq -1$$

45.

$$\left(\frac{f}{g}\right)(x)=\frac{f(x)}{g(x)}$$
$$=\frac{x+1}{x^2-1}$$

$$x^2-1\neq 0$$
$$x^2\neq 1$$
$$x\neq \pm 1$$

Domain $[x\,|\,x\neq \pm 1]$

46.

$$(g\circ f)(x)=g[f(x)]$$
$$=g(x+1)$$
$$=(x+1)^2-1$$
$$=x^2+2x+1-1$$
$$=x^2+2x$$

47.

$$(f\circ g)(2)=f[g(2)]$$
$$=f\left((2)^2-1\right)$$
$$=f(3)$$
$$=3+1$$
$$=4$$

48.

$$(f\circ g)(x)=f[g(x)]$$
$$=f(x^2)$$
$$=\sqrt{x^2}-2$$
$$=|x|-2$$

49.

$$(g\circ f)(x)=g[f(x)]$$
$$=g\left(\sqrt{x}-2\right)$$
$$=\left(\sqrt{x}-2\right)^2$$
$$=x-4\sqrt{x}+4$$

50.

$$(f\circ g)(-2)=|-2|-2$$
$$=2-2$$
$$=0$$

(Using results of #42)

51.

$$(g\circ f)(-2)=g[f(-2)]$$
$$=g\left(\sqrt{-2}-2\right)$$
not defined

-2 is not in the domain of $f(x)$.

52.

$$f[g(x)] = f\left(\frac{x}{2} - 2\right) = 2\left(\frac{x}{2} - 2\right) + 4$$
$$= x - 4 + 4 = x$$
$$g[f(x)] = g(2x + 4) = \frac{2x + 4}{2} - 2$$
$$= \frac{2x}{2} + \frac{4}{2} - 2 = x + 2 - 2 = x$$

53.

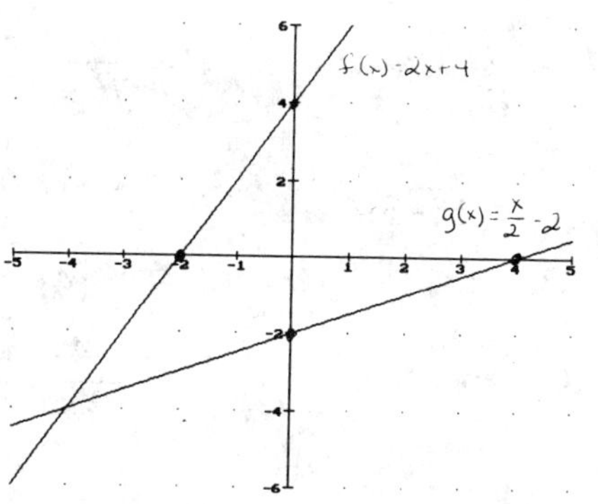

54.

$$R = kq \qquad R = 4q$$
$$20 = k(5) \qquad = 4(40)$$
$$k = 4 \qquad = 160$$

55.

$$S = \frac{k}{t^3}$$
$$8 = \frac{k}{(-1)^3}$$
$$k = -8$$
$$S = -\frac{8}{t^3}$$
$$S = -\frac{8}{(-2)^3}$$
$$= \frac{-8}{-8}$$
$$= 1$$

56.

$$P = \frac{kqr}{t^2}$$

$$-3 = \frac{k(2)(-3)}{(4)^2}$$

$$-48 = -6k$$

$$k = 8$$

$$P = \frac{8qr}{t^2}$$

$$= \frac{8(-1)\left(\frac{1}{2}\right)}{(4)^2}$$

$$= -\frac{4}{16}$$

$$= -\frac{1}{4}$$

57.

investment at 8.5% : x

investment at 9% : $3x$

investment at 5.25% : $10000-(x+3x)$

$$= 10000 - 4x$$

57a).

$$I = 0.085(x) + 0.09(3x) + 0.0525(10000-4x)$$
$$= 0.085x + 0.27x + 525 - 0.21x$$
$$= 0.145x + 525$$

57b).

$$3000 = 3x$$
$$x = 1000$$

$$I = 0.145(1000) + 525$$
$$= \$670$$

58.

$(950, 2), (975, 2.10), (1050, 2.40)$

58a).

$$m = \frac{2.10-2}{975-950} = \frac{0.10}{25} = 0.004$$

$$G - G_1 = m(S - S_1)$$
$$G - 2 = 0.004(S - 950)$$
$$G - 2 = 0.004S - 3.8$$
$$G = 0.004S - 1.8$$

58b).

$$G + 1.8 = 0.004S$$
$$\frac{G+1.8}{0.004} = S$$
$$G + 1.8 = 0.004S$$
$$1000G + 1800 = 4S$$
$$250G + 450 = S$$

58c).

Find S when $G = 4.00$

$$S = \frac{4.00+1.8}{0.004} = 1450$$

59.

Total parts $= 1$ part $H + 8$ parts $O = 9$ parts

59a).

$$H = \frac{1}{9}w$$

59b).

$$O = \frac{8}{9}w$$

59c).

$$w = 315$$
$$H = \frac{1}{9}(315)$$
$$= 35g$$

$$O = \frac{8}{9}(315)$$
$$= 280g$$

60a).

$$R = \frac{k_1 V}{I} \qquad P = k_2 IV$$

Let $k_1 = k_2 = 1$

Then $R = \dfrac{V}{I}$ $\qquad P = IV$

$V = 120$

$P = 1000$

Find I and R

$$\begin{aligned} P &= IV \\ 1000 &= I(120) \\ I &= 8.33 \text{ amps} \end{aligned}$$

$$\begin{aligned} R &= \frac{V}{I} \\ &= \frac{120}{8.33} \\ &= 14.41 \text{ ohms} \end{aligned}$$

60b).

$$R = \frac{k_1 V}{I} \qquad P = k_2 IV$$

Let $k_1 = k_2 = 1$

Then $R = \dfrac{V}{I}$ $\qquad P = IV$

$V = 240$

$P = 1000$

Find I

$$\begin{aligned} P &= IV \\ 1000 &= I(240) \\ I &= 4.17 \text{ amps} \end{aligned}$$

Find R

$$\begin{aligned} R &= \frac{V}{I} \\ &= \frac{240}{4.17} \\ &= 57.55 \text{ ohms} \end{aligned}$$

60c).

$$R = \frac{k_1 V}{I} \qquad P = k_2 IV$$

Let $k_1 = k_2 = 1$

Then $R = \dfrac{V}{I}$ $\qquad P = IV$

$R = 240$
$V = 120$

Find I and P

$$R = \frac{V}{I}$$
$$240 = \frac{120}{I}$$
$$I = \frac{120}{240} = 0.5 \text{ amps}$$
$$P = IV$$
$$= (0.5)(120)$$
$$= 60 \text{ watts}$$

CHAPTER 3 REVIEW TEST

1.

Label points : $A(2,5)$, $B(-3,1)$, $C(-3,4)$

$$\overline{AB}=\sqrt{(-3-2)^2+(1-5)^2}=\sqrt{(-5)^2+(-4)^2}=\sqrt{41}$$

$$\overline{AC}=\sqrt{(-3-2)^2+(4-5)^2}=\sqrt{(-5)^2+(-1)^2}=\sqrt{26}$$

$$\overline{BC}=\sqrt{[-3-(-3)]^2+(1-4)^2}=\sqrt{(0)^2+(-3)^2}=\sqrt{9}=3$$

$$P=\overline{AB}+\overline{AC}+\overline{BC}=\sqrt{41}+\sqrt{26}+3$$

2.

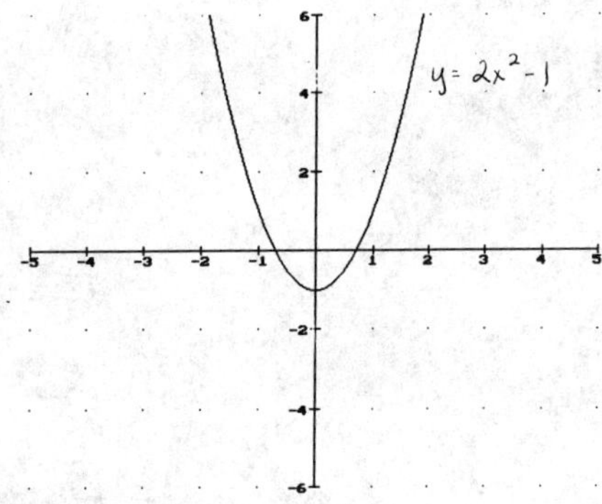

3.

$$(-y)=\frac{1}{x^3} \qquad y=\frac{1}{(-x)^3} \qquad (-y)=\frac{1}{(-x)^3}$$

$$-y=-\frac{1}{x^3}$$

$$-y=\frac{1}{x^3} \qquad y=-\frac{1}{x^3} \qquad y=\frac{1}{x^3}$$

doesn't yield original | doesn't yield original | yields original

Symmetric with respect to the origin.

4.

Domain : $x \geq 0$ for $\sqrt{x}$ to be defined

$$\sqrt{x}-1 \neq 0$$
$$\sqrt{x} \neq 1$$
$$x \neq 1$$

$$\{x | x \geq 0, x \neq 1\}$$

5.

Find the value of x for which $f(x)=4$

$$4=\sqrt{x-1}$$
$$16=x-1$$
$$17=x$$

6.

$$f(2t)=2(2t)^2+3$$
$$=8t^2+3$$

7.

f is increasing for $x \geq 0$ since the graph is rising.

f is decreasing for $-2 \leq x \leq 0$ since the graph is falling.

f is constant for $x < -2$.

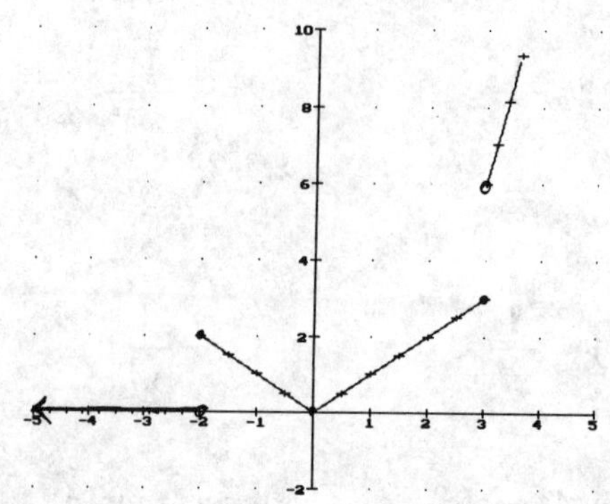

8.

$f(-5) = 0$ since $-5 < -2$ we use the first part of $f(x)$

9.

$f(-2) = |-2| = 2$ since $-2 \leq -2 \leq 3$ we use the second part of $f(x)$

10.

Shift the graph of $y = f(x)$ one unit to the left and halve the functional values.

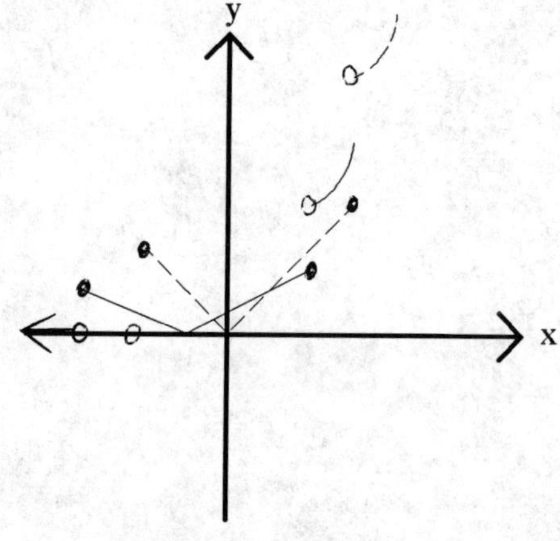

11.

$$m = \frac{5-2}{-3-(-5)} = \frac{3}{2}$$

$$\begin{aligned} y - y_1 &= m(x - x_1) \\ y - 5 &= \frac{3}{2}(x - (-3)) \\ y - 5 &= \frac{3}{2}(x + 3) \\ y - 5 &= \frac{3}{2}x + \frac{9}{2} \\ y &= \frac{3}{2}x + \frac{19}{2} \end{aligned}$$

12.

$$x = -3$$

13.

$$\begin{aligned} 2y - x &= 4 \\ 2y &= x + 4 \\ y &= \frac{1}{2}x + 2 \end{aligned}$$

$$m = \frac{1}{2};\ b = 2$$

14.

Parallel to the x-axis means the line is horizontal.

$$y = -1$$

15.

$$\begin{aligned} y - 3x - 2 &= 0 \\ y &= 3x + 2 \\ m &= 3 \\ m_1 &= -\frac{1}{3} \\ y - y_1 &= m_1(x - x_1) \\ y - 3 &= -\frac{1}{3}(x - (-2)) \\ y - 3 &= -\frac{1}{3}(x + 2) \\ y - 3 &= -\frac{1}{3}x - \frac{2}{3} \\ y &= -\frac{1}{3}x + \frac{7}{3} \end{aligned}$$

16.

$$\begin{aligned} (f - g)(2) &= f(2) - g(2) \\ &= \frac{1}{2-1} - (2)^2 \\ &= 1 - 4 \\ &= -3 \end{aligned}$$

17.

$$\left(\frac{g}{f}\right)(x) = \frac{g(x)}{f(x)}$$
$$= \frac{x^2}{\frac{1}{x-1}}$$
$$= x^2(x-1)$$
$$= x^3 - x^2$$

18.

$$(g \circ f)(3) = g[f(3)]$$
$$= g\left(\frac{1}{3-1}\right)$$
$$= g\left(\frac{1}{2}\right)$$
$$= \left(\frac{1}{2}\right)^2$$
$$= \frac{1}{4}$$

19.

$$f[g(x)] = f\left[-\frac{1}{3}(x-1)\right]$$
$$= -3\left[-\frac{1}{3}(x-1)\right]+1$$
$$= x-1+1$$
$$= x$$

$$g[f(x)] = g(-3x+1)$$
$$= -\frac{1}{3}(-3x+1-1)$$
$$= -\frac{1}{3}(-3x)$$
$$= x$$

20.

$$h = kr^3$$
$$2 = k\left(-\frac{1}{2}\right)^3$$
$$2 = -\frac{1}{8}k$$
$$k = -16$$
$$h = -16r^3$$
$$h = -16(4)^3$$
$$= -1024$$

21.

$$T = \frac{kab^2}{c^3}$$
$$64 = \frac{k(-1)\left(\frac{1}{2}\right)^2}{(2)^3}$$
$$512 = -\frac{1}{4}k$$
$$k = -2048$$
$$T = \frac{-2048ab^2}{c^3}$$
$$= \frac{-2048(2)(4)^2}{(-1)^3}$$
$$= 65{,}536$$

CUMULATIVE REVEIW EXERCISES: CHAPTERS 1-3

1.
Polynomial

2.
Not a polynomial :

$\frac{3}{x} = 3x^{-1}$ negative exponent

3.
Not a polynomial :

$3\sqrt{x} = 3x^{\frac{1}{2}}$ fractional exponent

4.
Polynomial

5.
Polynomial

6.
$$\sqrt[4]{a^8b^6} = \sqrt[4]{a^4 \cdot a^4 \cdot b^4 \cdot b^2} = a^2b\sqrt[4]{b^2} = a^2b\sqrt{b}$$

7.
$$\frac{3}{7-\sqrt{x}} = \frac{3}{7-\sqrt{x}} \cdot \frac{7+\sqrt{x}}{7+\sqrt{x}} = \frac{21+3\sqrt{x}}{49-x}$$

8.
$$\left(\frac{a^2}{b^6}\right)^{\frac{1}{2}}\left(\frac{a^5}{b^{\frac{5}{4}}}\right)^{\frac{2}{5}} = \frac{a}{b^3} \cdot \frac{a^2}{b^{\frac{1}{2}}} = \frac{a^3}{b^{\frac{7}{2}}}$$

9.
$$\frac{3-x}{x^3+x} - \frac{x^2}{x^2+1} = \frac{3-x}{x(x^2+1)} - \frac{(x^2)x}{(x^2+1)x} = \frac{(3-x)-x^3}{x(x^2+1)} = \frac{3-x-x^3}{x(x^2+1)}$$

10.
$$\left(\frac{x^2-9}{2x^2+3x-2}\right)\left(\frac{2x-1}{3-x}\right) = \frac{(x+3)(x-3)(2x-1)}{(2x-1)(x+2)(-1)(x-3)}$$
$$= \frac{(x+3)\cancel{(x-3)}\cancel{(2x-1)}}{\cancel{(2x-1)}(x+2)(-1)\cancel{(x-3)}} = -\frac{x+3}{x+2}$$

11.
$$5-u=10 \qquad 7v+2=9$$
$$-u=5 \qquad 7v=7$$
$$u=-5 \qquad v=1$$

12.
$$(3-2i)^2 = (3-2i)(3-2i) = 9-12i+4i^2$$
$$= 9-12i+4(-1) = 5-12i$$

13.

$$(2+4i)(3-5i) = 6+2i-20i^2 = 6+2i-20(-1) = 26+2i$$

14.

$$\begin{aligned}(2+i)^4(3-i)^2 &= \left[(2+i)^2\right]^2(3-i)(3-i)\\ &= \left[(2+i)(2+i)\right]^2(9-6i+i^2)\\ &= (4+4i+i^2)^2(8-6i)\\ &= (3+4i)^2(8-6i)\\ &= (3+4i)(3+4i)(8-6i)\\ &= (9+24i+16i^2)(8-6i)\\ &= (-7+24i)(8-6i)\\ &= -56+234i-144i^2\\ &= 88+234i\end{aligned}$$

15.

$$x^{\frac{5}{2}}\left(x^{-\frac{2}{3}}-1\right) = x^{\frac{5}{2}-\frac{2}{3}} - x^{\frac{5}{2}} = x^{\frac{11}{6}} - x^{\frac{5}{2}}$$

16.

$$\sqrt{12}-\sqrt{75}+2\sqrt{27} = 2\sqrt{3}-5\sqrt{3}+6\sqrt{3} = 3\sqrt{3}$$

17.

$$\begin{aligned}\frac{2}{x^2-x}-\frac{2x}{x-1} &= \frac{2}{x(x-1)}-\frac{(2x)x}{(x-1)x}\\ &= \frac{2-2x^2}{x(x-1)}\\ &= \frac{-2(x^2-1)}{x(x-1)}\\ &= \frac{-2(x+1)\cancel{(x-1)}}{x\cancel{(x-1)}} = \frac{-2(x+1)}{x}\end{aligned}$$

18.

$$\begin{aligned}\frac{x^{-2}+1}{x^{-1}-3} &= \frac{\frac{1}{x^2}+1}{\frac{1}{x}-3} = \frac{\frac{1}{x^2}+1}{\frac{1}{x}-3}\cdot\frac{x^2}{x^2}\\ &= \frac{1+x^2}{x-3x^2}\end{aligned}$$

19.

$$\begin{aligned}3x^2+x^2-2x &= x(3x^2+x-2)\\ &= x(3x-2)(x+1)\end{aligned}$$

20.

$$\begin{aligned}(x-1)^{\frac{2}{3}}+2(x-1)^{\frac{5}{3}} &= (x-1)^{\frac{2}{3}}\left[1+2(x-1)\right]\\ &= (x-1)^{\frac{2}{3}}\left[1+2x-2\right]\\ &= (x-1)^{\frac{2}{3}}\left[2x-1\right]\end{aligned}$$

21.

$$\frac{2}{\sqrt{x}-\sqrt{2}} = \frac{2}{\sqrt{x}-\sqrt{2}} \cdot \frac{\sqrt{x}+\sqrt{2}}{\sqrt{x}+\sqrt{2}} = \frac{2\sqrt{x}+2\sqrt{2}}{x-2}$$

22.

$$3x-2 \le x+3$$
$$2x-2 \le 3$$
$$2x \le 5$$
$$x \le \frac{5}{2}$$

Alternate method : $3x-2 \le x+3$

$3x-2 = x+3$

$x = \frac{5}{2}$ Critical value

Interval,

Critical Value	Test Point	Substitution	Verification
$x < \frac{5}{2}$	$x = 0$	$3(0)-2 \le 0+3$	True
$x = \frac{5}{2}$	$x = \frac{5}{2}$	$3\left(\frac{5}{2}\right)-2 \le \frac{5}{2}+3$	True
$x > \frac{5}{2}$	$x = 3$	$3(3)-2 \le 3+3$	False

$$x \le \frac{5}{2}$$

23.

$|1-2h|>2$

$1-2h>2 \quad \text{or} \quad 1-2h<-2$

$-2h>1 \quad \text{or} \quad -2h<-3$

$h<-\frac{1}{2} \quad \text{or} \quad h>\frac{3}{2}$

Alternate Method : $|1-2h|>2$

$|1-2h|=2$

$1-2h=2 \quad \text{or} \quad 1-2h=-2$

$h=-\frac{1}{2} \quad \text{or} \quad h=\frac{3}{2}$

Interval, Critical Value	Test Point	Substitution	Verification
$h<-\frac{1}{2}$	$h=-1$	$\lvert 1-2(-1)\rvert>2$	True
$h=-\frac{1}{2}$	$h=-\frac{1}{2}$	$\left\lvert 1-2\left(-\frac{1}{2}\right)\right\rvert>2$	False
$-\frac{1}{2}<h<\frac{3}{2}$	$h=0$	$\lvert 1-2(0)\rvert>2$	False
$h=\frac{3}{2}$	$h=\frac{3}{2}$	$\left\lvert 1-2\left(\frac{3}{2}\right)\right\rvert>2$	False
$h>\frac{3}{2}$	$h=2$	$\lvert 1-2(2)\rvert>2$	True

$h<-\frac{1}{2} \text{ or } h>\frac{3}{2}$

24.

$$2t^2 - 5t \geq 12$$
$$2t^2 - 5t = 12$$
$$2t^2 - 5t - 12 = 0$$
$$(2t+3)(t-4) = 0$$
$$(2t+3) = 0 \qquad t - 4 = 0$$
$$t = -\frac{3}{2} \qquad t = 4$$

Interval,

Critical Value	Test Point	Substitution	Verification
$t < -\frac{3}{2}$	$t = -2$	$2(-2)^2 - 5(-2) \geq 12$	True
$t = -\frac{3}{2}$	$t = -\frac{3}{2}$	$2\left(-\frac{3}{2}\right)^2 - 5\left(-\frac{3}{2}\right) \geq 12$	True
$-\frac{3}{2} < t < 4$	$t = 0$	$2(0)^2 - 5(0) \geq 12$	False
$t = 4$	$t = 4$	$2(4)^2 - 5(4) \geq 12$	True
$t > 4$	$t = 5$	$2(5)^2 - 5(5) \geq 12$	True

$$\left\{ t \,\middle|\, t \leq -\frac{3}{2} \text{ or } t \geq 4 \right\}$$

25.

$|x^2-3| \le 2$

$|x^2-3| = 2$

$x^2-3=2 \qquad x^2-3=-2$

$x^2=5 \qquad x^2=1$

$x=\pm\sqrt{5} \qquad x=\pm 1$

Interval, Critical Value	Test Point	Substitution	Verification
$x<-\sqrt{5}$	$x=-3$	$\lvert(-3)^2-3\rvert \le 2$	False
$x=-\sqrt{5}$	$x=-\sqrt{5}$	$\lvert(-\sqrt{5})^2-3\rvert \le 2$	True
$-\sqrt{5}<x<-1$	$x=-2$	$\lvert(-2)^2-3\rvert \le 2$	True
$x=-1$	$x=-1$	$\lvert(-1)^2-3\rvert \le 2$	True
$-1<x<1$	$x=0$	$\lvert(0)^2-3\rvert \le 2$	False
$x=1$	$x=1$	$\lvert(1)^2-3\rvert \le 2$	True
$1<x<\sqrt{5}$	$x=2$	$\lvert(2)^2-3\rvert \le 2$	True
$x=\sqrt{5}$	$x=\sqrt{5}$	$\lvert(\sqrt{5})^2-3\rvert \le 2$	True
$x>\sqrt{5}$	$x=3$	$\lvert(3)^2-3\rvert \le 2$	False

$$\left\{x \,\middle|\, -\sqrt{5} \le x \le -1, \quad 1 \le x \le \sqrt{5}\right\}$$

26.

$2x-1>0$

$2x>1$

$x>\frac{1}{2}$

$$\left\{x \,\middle|\, x>\frac{1}{2}\right\}$$

27.

$x^2-1 \ne 0$

$x^2 \ne 1$

$x \ne \pm 1$

$$\{x \mid x \ne \pm 1\}$$

28.
Label the points : $A(-2,2)$, $B(4,2)$ $C(-2,-3)$

$$\overline{AB}=\sqrt{(-2-4)^2+(2-2)^2}=\sqrt{(-6)^2+0^2}=\sqrt{36}=6$$

$$\overline{AC}=\sqrt{[-2-(-2)]^2+[2-(-3)]^2}=\sqrt{0^2+5^2}=\sqrt{25}=5$$

$$\overline{BC}=\sqrt{[4-(-2)]^2+[2-(-3)]^2}=\sqrt{6^2+5^2}=\sqrt{61}$$

The length of the hypotenuse is $\sqrt{61}$.

29a).

$$\begin{aligned} f(2a-1)&=1-(2a-1)^2\\ &=1-(4a^2-4a+1)\\ &=1-4a^2+4a-1\\ &=-4a^2+4a \end{aligned}$$

29b).
Find a value of t for which $f(x)=-15$

$$\begin{aligned} -15&=1-t^2\\ -16&=-t^2\\ 16&=t^2\\ t&=\pm 4 \end{aligned}$$

29c).

$$f(t+h)=1-(t+h)^2=1-(t^2+2th+h^2)=1-t^2-2th-h^2$$

$$\frac{f(t+h)-f(t)}{h}=\frac{1-t^2-2th-h^2-(1-t^2)}{h}$$

$$=\frac{1-t^2-2th-h^2-1+t^2}{h}=\frac{-2th-h^2}{h}=\frac{\cancel{h}(-2t-h)}{\cancel{h}}$$

$$=-2t-h$$

30a).
$G(0)=0$
Since $-2\le 0\le 2$ use the second part

30b).
$G(-3)=\frac{1}{2}$
Since $-3<-2$ use the first part

30c).

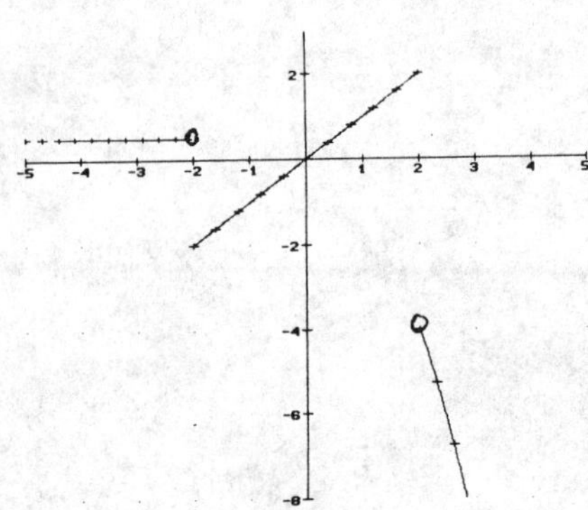

31.

$$L = 0.05x + 0.07y + 0.035z$$

32.

$$2x^2 - x - 3 = 0$$
$$(2x-3)(x+1) = 0$$
$$2x - 3 = 0 \qquad x + 1 = 0$$
$$2x = 3 \qquad x = -1$$
$$x = \frac{3}{2}$$

33.

$$2x^2 - x + 19 = 15$$
$$2x^2 - x + 4 = 0$$
$$x = \frac{-b \pm \sqrt{b^2 - 4ac}}{2a}$$
$$= \frac{-(-1) \pm \sqrt{(-1)^2 - 4(2)(4)}}{2(2)}$$
$$= \frac{1 \pm \sqrt{-31}}{4}$$
$$= \frac{1 \pm i\sqrt{31}}{4}$$

34.

$$\sqrt{x} + 12 = x$$
$$\sqrt{x} = x - 12$$
$$x = x^2 - 24x + 144$$
$$0 = x^2 - 25x + 144$$
$$0 = (x-16)(x-9)$$
$$x - 16 = 0 \qquad x - 9 = 0$$
$$x = 16 \qquad x = 9$$

Check :

$x = 16$

$\sqrt{16} + 12 = 16$?

$4 + 12 = 16$?

$16 = 16$ ✓

$x = 9$

$\sqrt{9} + 12 = 9$?

$3 + 12 = 9$?

$15 \neq 9$?

$x = 9$

Extraneous solution

$$\{16\}$$

35.

$$\frac{3-5x}{x+2}=4$$

$$3-5x=4x+8$$
$$-5-5x=4x$$
$$-5=9x$$
$$x=-\frac{5}{9}$$

36.

$$6x^2+5x-4=0$$
$$(3x+4)(2x-1)=0$$
$$3x+4=0 \qquad 2x-1=0$$
$$3x=-4 \qquad 2x=1$$
$$x=-\frac{4}{3} \qquad x=\frac{1}{2}$$

37.

$$(3-x)^2-16=0$$
$$(3-x)^2=16$$
$$(3-x)^2=\pm\sqrt{16}$$
$$3-x=\pm 4$$
$$3-x=4 \qquad 3-x=-4$$
$$-x=1 \qquad -x=-7$$
$$x=-1 \qquad x=7$$

38.

$$x-\sqrt{-1-5x}=-3$$
$$x+3=\sqrt{-1-5x}$$
$$x^2+6x+9=-1-5x$$
$$x^2+11x+10=0$$
$$(x+10)(x+1)=0$$
$$x+10=0 \qquad x+1=0$$
$$x=-10 \qquad x=-1$$

Check :

$$x=-10$$
$$-10-\sqrt{-1-5(-10)}=-3 \quad ?$$
$$-10-\sqrt{49}=-3 \quad ?$$
$$-10-7=-3 \quad ?$$
$$-17\neq -3$$

$x=-10$ Extraneous

$$x=-1$$
$$-1-\sqrt{-1-5(-1)}=-3 \quad ?$$
$$-1-\sqrt{4}=-3 \quad ?$$
$$-1-2=-3 \quad ?$$
$$-3=-3$$

Solution: $x=-1$

39.

$$-5 \le 1-4x \le 9$$

$$-5 \le 1-4x \text{ and } 1-4x \le 9$$

$$-5 = 1-4x \qquad 1-4x = 9$$

$$\frac{3}{2} = x \qquad x = -2$$

Interval, Critical Value	Test Point	Substitution	Verification
$x < -2$	$x = -3$	$-5 \le 1-4(-3) \le 9$	False
$x = -2$	$x = -2$	$-5 \le 1-4(-2) \le 9$	True
$-2 < x < \frac{3}{2}$	$x = 0$	$-5 \le 1-4(0) \le 9$	True
$x = \frac{3}{2}$	$x = \frac{3}{2}$	$-5 \le 1-4\left(\frac{3}{2}\right) \le 9$	True
$x > \frac{3}{2}$	$x = 2$	$-5 \le 1-4(2) \le 9$	False

$$-2 \le x \le \frac{3}{2}$$

40.

$$\left|\frac{x-3}{2}\right| \geq 10$$

$$\frac{x-3}{2}=10 \qquad \frac{x-3}{2}=-10$$

$$x=23 \qquad x=-17$$

Interval, Critical Value	Test Point	Substitution	Verification
$x<-17$	$x=-18$	$\left\|\frac{-18-3}{2}\right\| \geq 10$	True
$x=-17$	$x=-17$	$\left\|\frac{-17-3}{2}\right\| \geq 10$	True
$-17<x<23$	$x=0$	$\left\|\frac{0-3}{2}\right\| \geq 10$	False
$x=23$	$x=23$	$\left\|\frac{23-3}{2}\right\| \geq 10$	True
$x>23$	$x=24$	$\left\|\frac{24-3}{2}\right\| \geq 10$	True

$x \leq -17$ or $x \geq 23$

41.

$x^2 - 3x - 10 \le 0$

$x^2 - 3x - 10 = 0$

$(x-5)(x+2) = 0$

$x - 5 = 0 \qquad x + 2 = 0$

$x = 5 \qquad x = -2$

Interval, Critical Value	Test Point	Substitution	Verification
$x < -2$	$x = -3$	$(-3)^2 - 3(-3) - 10 \le 0$	False
$x = -2$	$x = -2$	$(-2)^2 - 3(-2) - 10 \le 0$	True
$-2 < x < 5$	$x = 0$	$(0)^2 - 3(0) - 10 \le 0$	True
$x = 5$	$x = 5$	$(5)^2 - 3(5) - 10 \le 0$	True
$x > 5$	$x = 6$	$(6)^2 - 3(6) - 10 \le 0$	False

$\{x \mid -2 \le x \le 5\}$

42.

$\frac{3}{x-2} \ge 2$ Not defined when $x = 2$

Interval, Critical Value	Test Point	Substitution	Verification
$x < 2$	$x = 0$	$\frac{3}{0-2} \ge 1$	False
$x = 2$	$x = 2$	$\frac{3}{2-2} \ge 1$	False
$x > 2$	$x = 3$	$\frac{3}{3-2} \ge 1$	True

$\{x \mid x > 2\}$

43.

$y=-1$

44.

$$x=4y+1$$
$$x-1=4y$$
$$\frac{1}{4}x-\frac{1}{4}=y$$

$$m=\frac{1}{4}$$

$$m_1=-4\ ;\ \left(\frac{1}{2},2\right)$$

$$y-y_1=m_1(x-x_1)$$

$$y-2=-4\left(x-\frac{1}{2}\right)$$
$$y-2=-4x+2$$
$$y=-4x+4$$

45a).

$$m=\frac{-2-2}{3-(-1)}=-\frac{4}{4}=-1$$

45b).

$$y-y_1=m(x-x_1)$$
$$y-2=-1(x-(-1))$$
$$y-2=-1(x+1)$$
$$y=-x+1$$

45c).

Parallel to the y-axis means the line is vertical : $x=-1$

45d).

$$m_1=-\left(\frac{1}{-1}\right)=1$$
$$y-y_1=m_1(x-x_1)$$
$$y-(-1)=1(x-(-4))$$
$$y+1=x+4$$
$$y=x+3$$

46.

length of shortest leg: x

length of other leg: $3x$

$$(x)^2 + (3x)^2 = (20)^2$$
$$x^2 + 9x^2 = 400$$
$$10x^2 = 400$$
$$x^2 = 40$$
$$x = \sqrt{40} = 2\sqrt{10}$$

$2\sqrt{10}$ in

47.

$f(x) = -3x^2 - x + 2$

x-intercepts : Let $y = 0$

$0 = -3x^2 - x + 2$
$0 = 3x^2 + x - 2$
$0 = (3x - 2)(x + 1)$

$3x - 2 = 0 \qquad x + 1 = 0$
$3x = 2 \qquad x = -1$
$x = \frac{2}{3}$

$y-$ intercept : Let $x = 0$

$y = -3(0)^2 - 0 + 2$
$y = 2$

48.

$(a,0)$, $(0,a)$

$$m=\frac{0-a}{a-0}=\frac{-a}{a}=-1$$

$$y=mx+b$$
$$y=-x+a$$

49.

Original number of payments : x

Monthly payment : $\dfrac{12000}{x}$

$$\frac{12000}{x}+50=\frac{12000}{x-8}$$

$$(x)(x-8)\left(\frac{12000}{x}\right)+x(x-8)(50)=x(x-8)\left(\frac{12000}{x-8}\right)$$

$$12000(x-8)+50x^2-400x=12000x$$
$$12000x-96000+50x^2-400x=12000x$$
$$50x^2-400x-96000=0$$
$$x^2-8x-1920=0$$
$$(x-48)(x+40)=0$$
$$x-48=0 \quad x+40=0$$
$$x=48 \qquad x=-40$$

Original number of payments : 48

New number of payments : $48-8=40$

40 months at $\dfrac{12000}{40}=\$300$ per month.

50.
Let $H =$ Hometown and $W =$ Workville

$$\overline{HA} = \sqrt{(10-0)^2 + (0-0)^2} = 10$$
$$\overline{AW} = \sqrt{(7-0)^2 + (25-0)^2} = \sqrt{7^2 + 25^2} = \sqrt{674} \approx 25.96$$

Distance using Route 1 : $10 + 25.96 = 35.96$ mi

$$t = \frac{d}{r}$$
$$t = \frac{35.96}{50} = 0.72 \text{ hr}$$

$$\overline{HB} = \sqrt{(10-5)^2 + (0-10)^2} = \sqrt{5^2 + (-10)^2} = \sqrt{125} \approx 11.18$$
$$\overline{BW} = \sqrt{(7-5)^2 + (25-10)^2} = \sqrt{2^2 + 15^2} = \sqrt{229} \approx 15.13$$

Distance using Route 2 : $11.18 + 15.13 = 26.31$

$$t = \frac{d}{r}$$
$$t = \frac{26.31}{30} = 0.88 \text{ hr}$$

He will arrive at Workville in a minimum amount of time by using Route 1.

51.

$$P(x)=\begin{cases} \dfrac{1}{10}x^2-20x & \text{if} \quad 0\le x\le 300 \\ -\dfrac{1}{10}x^2+120x-24{,}000 & \text{if} \quad 300<x\le 1000 \end{cases}$$

51a).

Domain : $\{x|0\le x\le 1000\}$

51b).

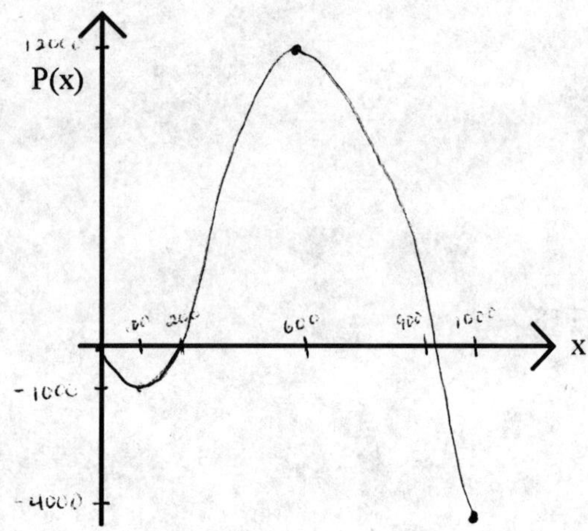

51c).

$i.)\,[-4000, 12000]$

$ii.)$ 600 books

$iii.)$ \$12,000

51d).

200 books

52.
L : Heat Loss

$$L = \frac{kAt_d}{T}$$

52a).

$$A = (2)(3) = 6$$

$$T = \frac{1}{24}$$

$$t_d = 68 - 38 = 30$$

$$2592 = \frac{k(6)(30)}{\frac{1}{24}}$$

$$k = 0.6$$

52b).

$$L = \frac{0.6At_d}{T}$$

$$3888 = \frac{0.6(6)t_d}{\frac{1}{24}}$$

$$t_d = 45^{\circ}$$

CHAPTER 3 ADDITIONAL PRACTICE EXERCISES

1. Find the distance between the points : $A(-6, 4)$ and $B(5, 2)$.

2. Given $f(x) = 2x^2 - x + 3$, evaluate $f(-2)$.

3. Given $f(x) = \sqrt{2x+3}$, find the domain of f.

4. Find the slope of the line through the points $A(8, -6)$ and $B(0, 5)$.

5. Find an equation of the line through $(-2, 1)$ that is perpendicular to the line $2x + 5y = 6$.

6. Find the inverse function of $f(x) = 3x + 2$.

7. Given $f(x) = 2x + 7$ and $g(x) = x^2 - 3$, find $(f \circ g)(x)$.

8. If T varies directly as r, and if $T = 15$ when $r = 30$, find T when $r = 62$.

9. Given $f(x) = 3x^2 + 2x - 1$ and $g(x) = 5x + 8$, find $(f - g)(x)$.

10.
Determine which of the following graphs determines y as a one-to-one function of x.

(a).

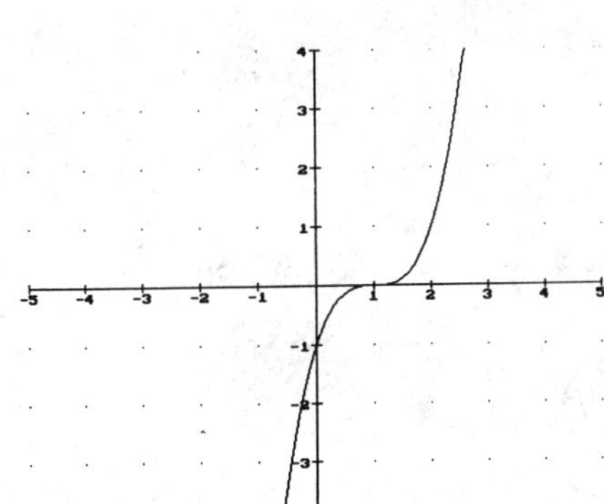

(b).

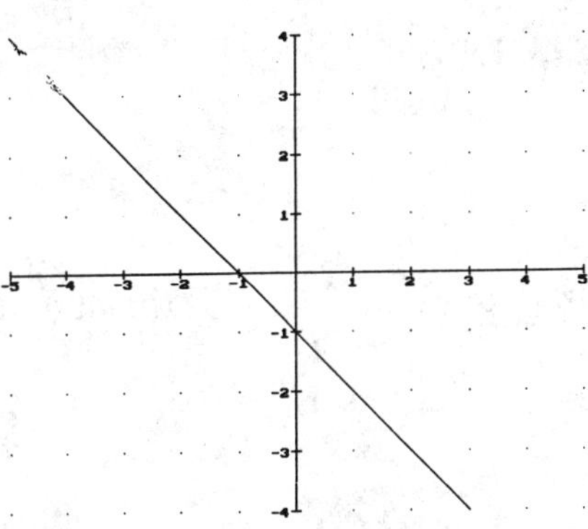

(c).

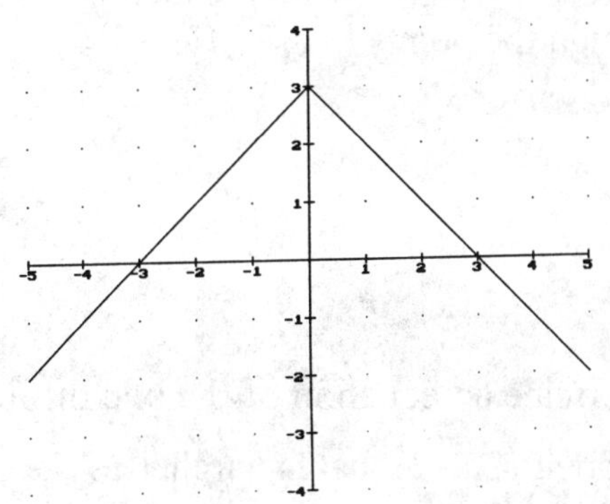

(d).

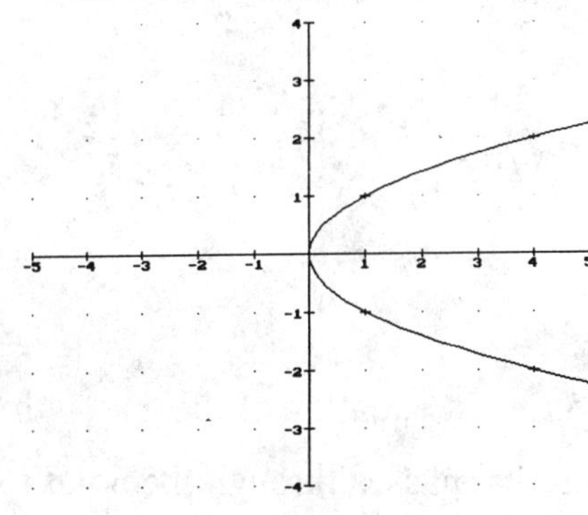

CHAPTER 3 PRACTICE TEST

1. Given $A(-2, 5)$ and $B(13, -7)$, find the midpoint of AB.

2. Sketch $y = -3x + 4$ and determine the intercepts.

3. Given $g(x) = \dfrac{2}{x+3}$, find the number whose image is -2.

4. Find the domain of the function $f(x) = \dfrac{1}{7x-6}$

5. Given the function g defined by $g(x) = x^2 - 3x + 1$, determine $g(a+1)$.

6. Sketch the graph of $y = \sqrt{x-1} + 2$.

7. Sketch the graph of f defined by

$$f(x) = \begin{cases} -x^2 & \text{if} \quad -1 \le x < 2 \\ x+1 & \text{if} \quad x \ge 2 \end{cases}$$

8. Determine the slope of the line passing through the points $(-7, 7), (3, -5)$.

9. Write an equation of the line with slope -3 that passes through the point $(0, 8)$.

10. Determine an equation of the line through the point $(2, -7)$ that is parallel to $5x - 3y + 1 = 0$.

11. Find the real number c such that $P(3, -1)$ is on the line $-2x + cy = 6$.

12. Given $f(x) = \dfrac{2}{3x+1}$ and $g(x) = x^2 + 3$ determine $(f \cdot g)(1)$.

13. Given $f(x) = \sqrt{x+2}$ and $g(x) = 2x + 1$ determine $(f \circ g)(x)$.

14. Given $f(x) = 6 - 5x$, find $f^{-1}(x)$.

15.
If y varies directly as the square of x, and $y = 10$ when $x = 1$, find y when $x = 6$.

16.
If B varies directly as C and inversely as the cube of D, and $B = 5$ when $C = 2$ and $D = 1$, find B when $C = 4$ and $D = 2$.

17.
Which of the following graphs determine y to be a function of x.

(a).

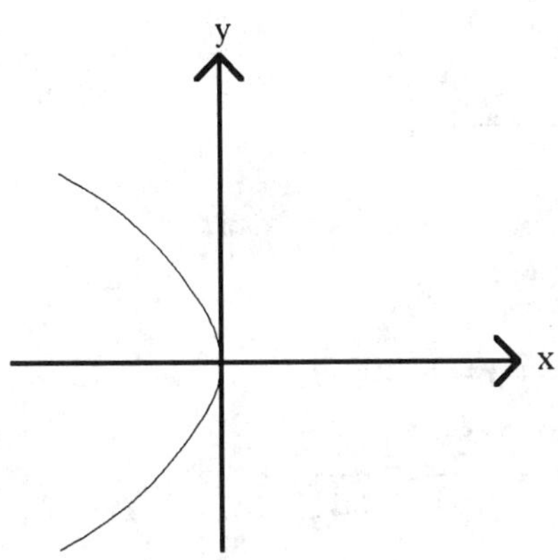

(b).

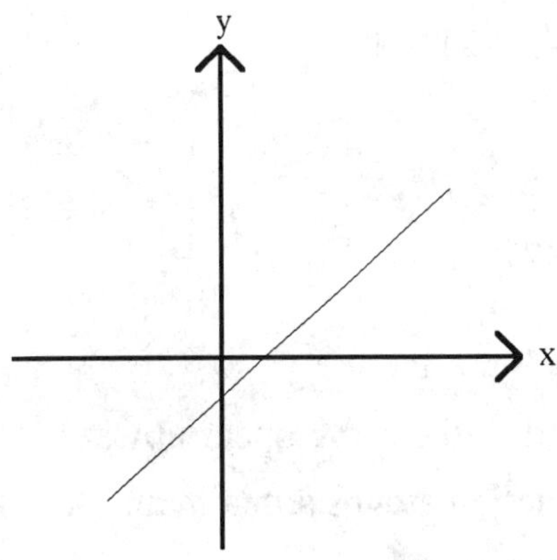

(c).

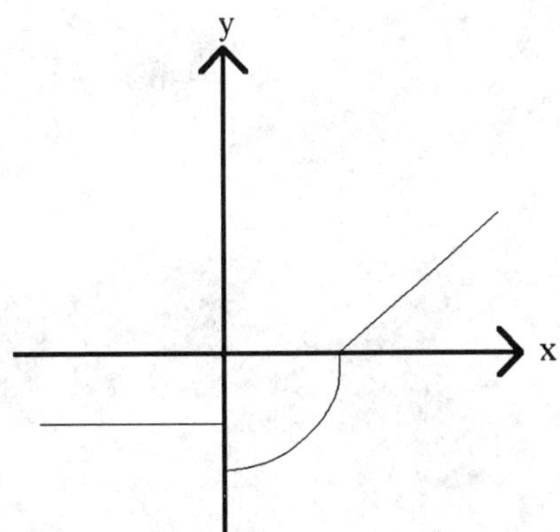

18.
Given $f(x) = 4x - 3$,

determine $\dfrac{f(x+h) - f(x)}{h}$

19.
Determine whether $f(x) = 5x - 4$ and $g(x) = \dfrac{x+4}{5}$ are inverses.

20.
Find the perimeter of the triangle whose vertices are $A(1, -3)$, $B(0, 6)$ and $C(-2, 4)$.

CHAPTER 4 SECTION 1

SECTION 4.1 PROGRESS CHECK

1a). (Page 227)

$$\begin{aligned} f(x) &= -3x^2 - 12x - 13 \\ &= -3(x^2 + 4x \qquad) - 13 \\ &= -3(x^2 + 4x + 4) - 13 + 12 \\ &= -3(x+2)^2 - 1 \end{aligned}$$

1b).

$$\begin{aligned} f(x) &= 2x^2 - 2x + 3 \\ &= 2(x^2 - x \qquad) + 3 \\ &= 2\left(x^2 - x + \frac{1}{4}\right) + 3 - \frac{1}{2} \\ &= 2\left(x - \frac{1}{2}\right)^2 + \frac{5}{2} \end{aligned}$$

2a.) (Page 232)

$f(x) = 2x^2 + x - 1$

$a = 2 > 0$, thus the curve opens upward and the function attains a minimum value at the vertex.

2b).

$$f(x) = 2x^2 + x - 1$$

$$x = -\frac{b}{2a} = -\frac{1}{2(2)} = -\frac{1}{4}$$

2c)

$$\begin{aligned} f(x) &= 2x^2 + x - 1 \\ f\left(-\frac{1}{4}\right) &= 2\left(-\frac{1}{4}\right)^2 + \left(-\frac{1}{4}\right) - 1 \\ &= 2\left(\frac{1}{16}\right) - \frac{1}{4} - 1 \\ &= -\frac{9}{8} \end{aligned}$$

EXERCISE SET 4.1

For the graphics calculator portion of the answer the RANGE values given show important features of the function. Other RANGE values are acceptable. The display on your calculator may vary from that shown.

1.

$$\begin{aligned} f(x) &= \left(x^2 - 6x \qquad\right) + 10 \\ &= \left(x^2 - 6x + 9\right) + 10 - 9 \\ &= (x-3)^2 + 1 \end{aligned}$$

XMIN = −1
XMAX = 7
XSCL = 1
YMIN = 0
YMAX = 12
YSCL = 1

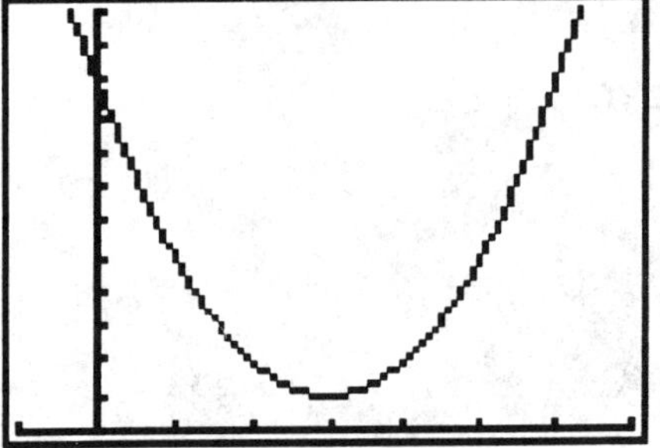

3.

$$\begin{aligned} f(x) &= \left(-2x^2 + 4x \qquad\right) - 5 \\ &= -2\left(x^2 - 2x \qquad\right) - 5 \\ &= -2\left(x^2 - 2x + 1\right) - 5 + 2 \\ &= -2(x-1)^2 - 3 \end{aligned}$$

XMIN = −5
XMAX = 5
XSCL = 1
YMIN = −10
YMAX = 0
YSCL = 1

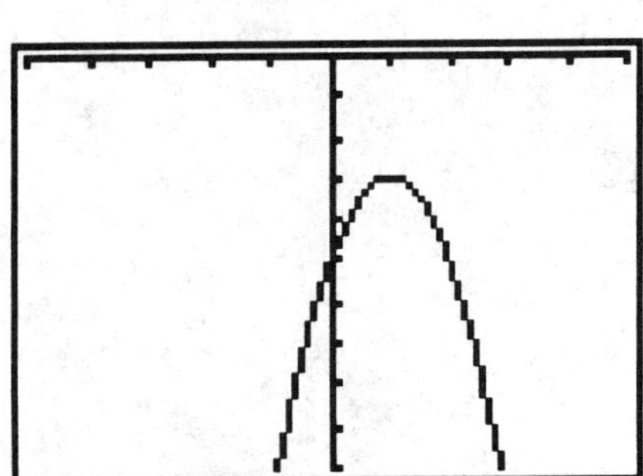

5.

$$\begin{aligned} f(x) &= (2x^2 + 6x \quad) + 5 \\ &= 2(x^2 + 3x \quad) + 5 \\ &= 2\left(x^2 + 3x + \frac{9}{4}\right) + 5 - \frac{9}{2} \\ &= 2\left(x + \frac{3}{2}\right)^2 + \frac{1}{2} \end{aligned}$$

XMIN = −5
XMAX = 5
XSCL = 1
YMIN = 0
YMAX = 10
YSCL = 1

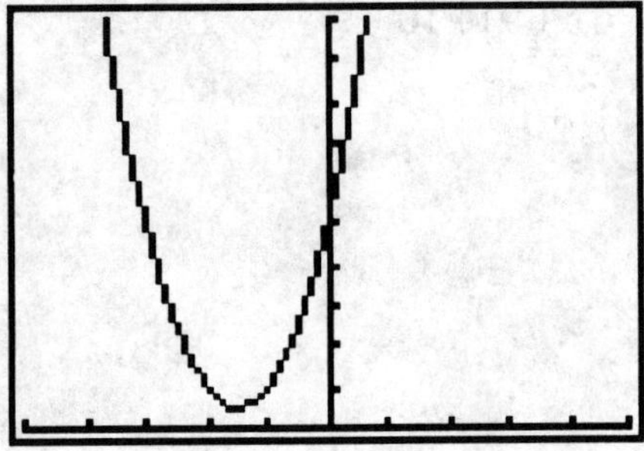

7.

$$\begin{aligned} f(x) &= -x^2 - x \\ &= -(x^2 + x \quad) \\ &= -\left(x^2 + x + \frac{1}{4}\right) + \frac{1}{4} \\ &= -\left(x + \frac{1}{2}\right)^2 + \frac{1}{4} \end{aligned}$$

XMIN = −5
XMAX = 5
XSCL = 1
YMIN = −10
YMAX = 1
YSCL = 1

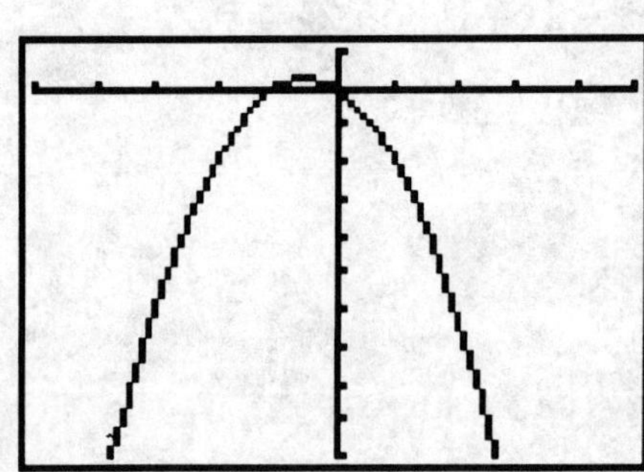

9.

$$f(x) = (-2x^2 \quad) + 5$$
$$= -2(x-0)^2 + 5$$

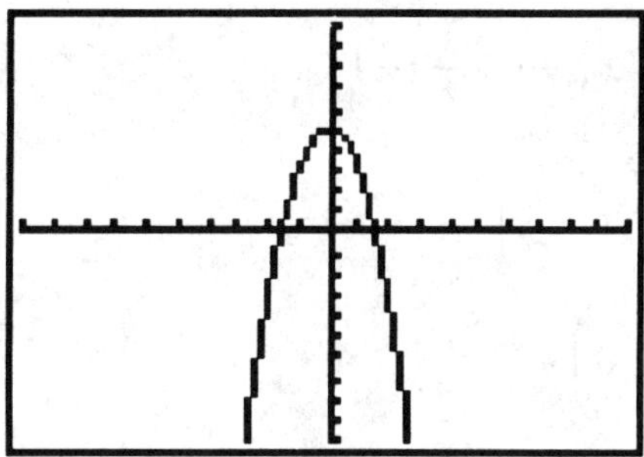

XMIN = −10
XMAX = 10
XSCL = 1
YMIN = −10
YMAX = 10
YSCL = 1

11.

$$f(x) = 2x^2 - 4x$$
$$= 2(x^2 - 2x \quad)$$
$$= 2(x^2 - 2x + 1) - 2$$
$$= 2(x-1)^2 - 2$$

Vertex: $(1, -2)$

$2x^2 - 4x = 0$

$2x(x-2) = 0$

x-intercepts: 0, 2

$f(0) = 2(0)^2 - 4(0) = 0$

y-intercept: 0

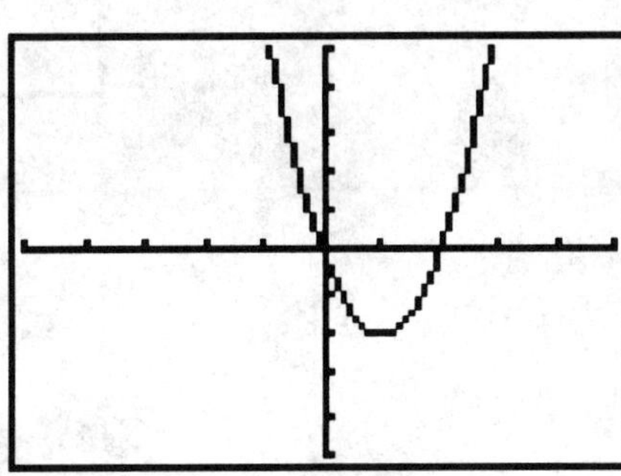

XMIN = −5
XMAX = 5
XSCL = 1
YMIN = −5
YMAX = 5
YSCL = 1

13.

$$\begin{aligned} f(x) &= \left(-4x^2+4x \quad\right)-1 \\ &= -4\left(x^2-x \quad\right)-1 \\ &= -4\left(x^2-x+\frac{1}{4}\right)-1+1 \\ &= -4\left(x-\frac{1}{2}\right)^2 \end{aligned}$$

Vertex: $\left(\frac{1}{2},\ 0\right)$

$$-4x^2+4x-1=0$$
$$(-2x+1)(2x-1)=0$$

x-intercept: $\frac{1}{2}$

$$f(0)=-4(0)^2+4(0)-1=-1$$

y-intercept: -1

XMIN = −5
XMAX = 5
XSCL = 1
YMIN = −10
YMAX = 1
YSCL = 1

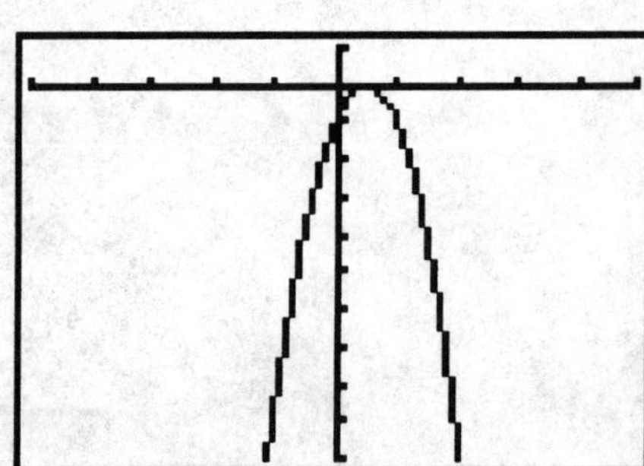

15.

$$f(x)=\left(\frac{1}{2}x^2+2x\quad\right)+4$$
$$=\frac{1}{2}\left(x^2+4x\quad\right)+4$$
$$=\frac{1}{2}\left(x^2+4x+4\right)+4-2$$
$$=\frac{1}{2}(x+2)^2+2$$

Vertex: $(-2, 2)$

$$\frac{1}{2}x^2+2x+4=0$$
$$x^2+4x+8=0$$
$$x=\frac{-4\pm\sqrt{4^2-4(1)(8)}}{2(1)}$$
$$=\frac{-4\pm\sqrt{-16}}{2}$$

No x-intercepts

$$f(0)=\frac{1}{2}(0)^2+2(0)+4=4$$

y-intercept: 4

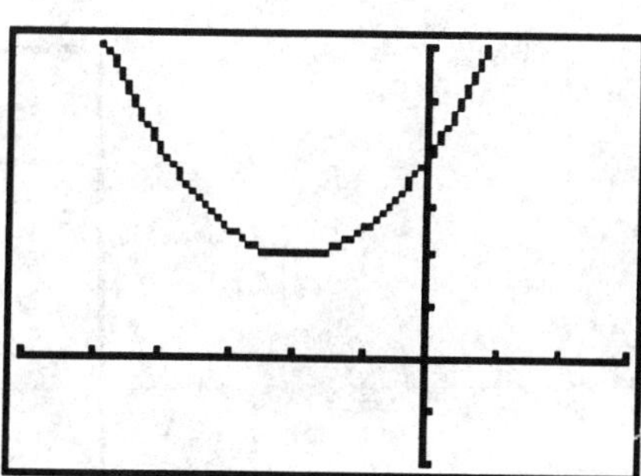

XMIN = −6
XMAX = 3
XSCL = 1
YMIN = −2
YMAX = 6
YSCL = 1

17.

$$\begin{aligned} f(x) &= \left(-\frac{1}{2}x^2 + 3x \quad\right) - 4 \\ &= -\frac{1}{2}\left(x^2 - 6x \quad\right) - 4 \\ &= -\frac{1}{2}\left(x^2 - 6x + 9\right) - 4 + \frac{9}{2} \\ &= -\frac{1}{2}(x-3)^2 + \frac{1}{2} \end{aligned}$$

Vertex: $\left(3, \frac{1}{2}\right)$

$$-\frac{1}{2}x^2 + 3x - 4 = 0$$
$$x^2 - 6x + 8 = 0$$
$$(x-4)(x-2) = 0$$

x-intercepts: 4, 2

$$f(0) = -\frac{1}{2}(0)^2 + 3(0) - 4 = -4$$

y-intercept: -4

XMIN = −5
XMAX = 10
XSCL = 1
YMIN = −10
YMAX = 3
YSCL = 1

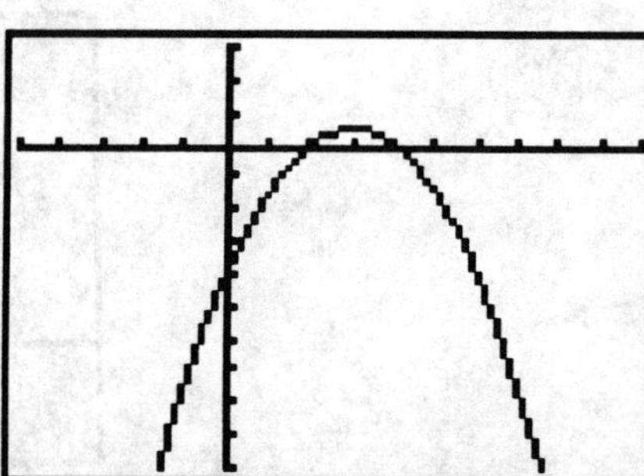

In Exercises 19-25, the following RANGE values are being used:

XMIN= -10
XMAX= 10
XSCL= 1
YMIN= -10
YMAX= 10
YSCL= 1

The linear factors determine the x-intercepts of the graph.
If $(x-r)$ is a factor then $x=r$ is an x-intercept.

19.
$Y=(X+1)(X-2)$

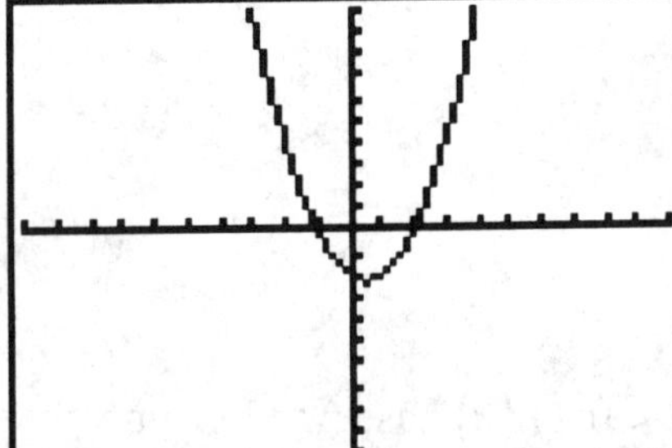

21.
$Y=X(X+5)$

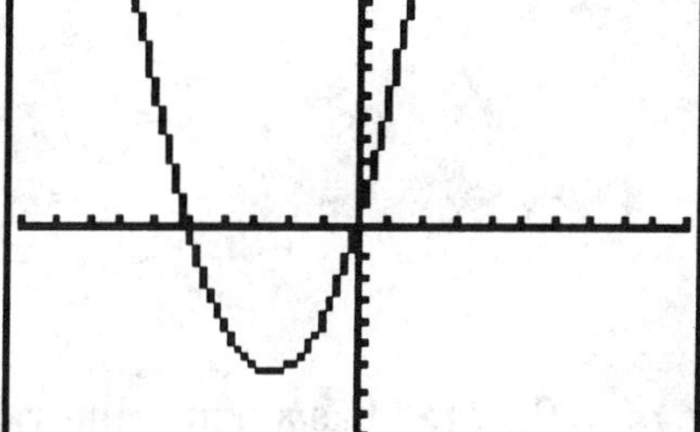

23.
$Y=(X-3)(X-7)$

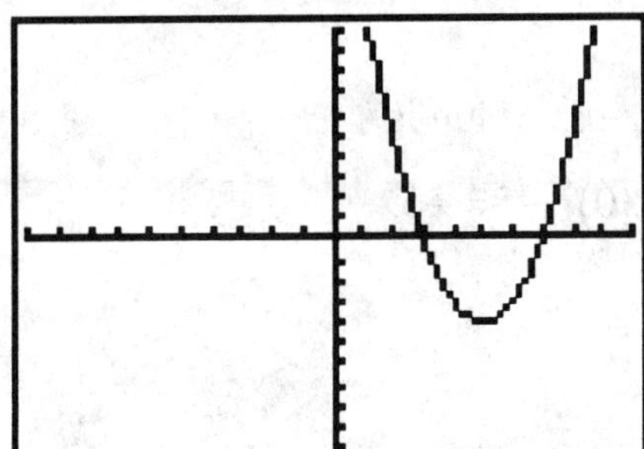

25.
$Y=0.3(X-3)(X-7)$

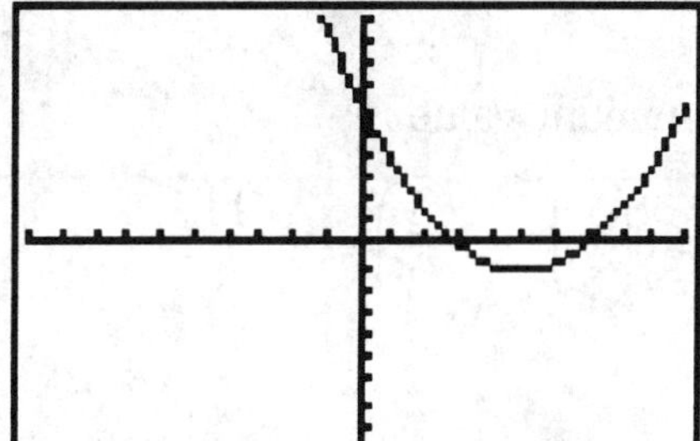

In Exercises 27 - 29, r_1, r_2 determine the x - intercepts.

27.

$r_1 = -4,\ r_2 = 5$

$f(x) = a(x+4)(x-5)$

Vertex: $(1, -20)$

$-20 = a(1+4)(1-5)$

$-20 = -20a$

$a = 1$

$f(x) = (x+4)(x-5)$

$= x^2 - x - 20$

29.

$r_1 = -3,\ r_2 = -8$

$f(x) = a(x+3)(x+8)$

y - intercept: $(0, -24)$

$-24 = a(0+3)(0+8)$

$-24 = 24a$

$a = -1$

$f(x) = -(x+3)(x+8)$

$= -x^2 - 11x - 24$

31.

$f(x) = 3x^2 - 2x + 4$

31a).

Since $a = 3 > 0$, $f(x)$ has a minimum value.

31b).

The minimum occurs at

$$x = -\frac{b}{2a} = \frac{-(-2)}{2(3)} = \frac{1}{3}.$$

31c).

The minimum value is

$$f\left(\frac{1}{3}\right) = 3\left(\frac{1}{3}\right)^2 - 2\left(\frac{1}{3}\right) + 4 = \frac{11}{3}.$$

33.

$f(x) = -2x^2 - 5$

33a).

Since $a = -2 < 0$, $f(x)$ has a maximum value.

33b).

The maximum occurs at

$$x = -\frac{b}{2a} = \frac{-0}{2(-2)} = 0.$$

33c).

The maximum value is

$$f(0) = -2(0)^2 - 5 = -5.$$

35.

$f(x) = x^2 + 5x$

35a).

Since $a = 1 > 0$, $f(x)$ has a minimum value.

35b).

The minimum occurs at

$x = -\frac{b}{2a} = \frac{-5}{2(1)} = -\frac{5}{2}$.

35c).

The minimum value is

$f\left(-\frac{5}{2}\right) = \left(-\frac{5}{2}\right)^2 + 5\left(-\frac{5}{2}\right) = -\frac{25}{4}$.

37.

$f(x) = 2x^2 - \frac{1}{2}x - \frac{3}{2}$

37a).

Since $a = 2 > 0$, $f(x)$ has a minimum value.

37b).

The minimum occurs at

$x = -\frac{b}{2a} = \frac{-\left(-\frac{1}{2}\right)}{2(2)} = \frac{1}{8}$.

37c).

The minimum value is

$f\left(\frac{1}{8}\right) = 2\left(\frac{1}{8}\right)^2 - \frac{1}{2}\left(\frac{1}{8}\right) - \frac{3}{2} = -\frac{49}{32}$.

39.

one number: x

other number: $20 - x$

$P(x) = x(20 - x) = -x^2 + 20x$

$x = -\frac{b}{2a} = \frac{-20}{2(-1)} = 10$

$x = 10$

$20 - x = 20 - 10 = 10$

The two numbers are 10 and 10.

41.

one number: x

other number: $50 - x$

$P(x) = (x)^2 + (50 - x)^2$

$= x^2 + 2500 - 100x + x^2$

$= 2x^2 - 100x + 2500$

$x = -\frac{b}{2a} = \frac{-(-100)}{2(2)} = 25$

$x = 25$

$50 - x = 50 - 25 = 25$

The two numbers are 25 and 25.

43.

Unknown point: $\left(x, \sqrt{x}\right)$

Given point: $(1, 0)$

$$d = \sqrt{(x-1)^2 + \left(\sqrt{x}-0\right)^2}$$
$$= \sqrt{x^2 - 2x + 1 + x}$$
$$= \sqrt{x^2 - x + 1}$$

Minimu m occurs where

$$x = -\frac{b}{2a} = \frac{-(-1)}{2(1)} = \frac{1}{2}.$$

$$f\left(\frac{1}{2}\right) = \sqrt{\frac{1}{2}} = \frac{\sqrt{2}}{2}$$

The point we seek is $\left(\frac{1}{2}, \frac{\sqrt{2}}{2}\right)$.

45.

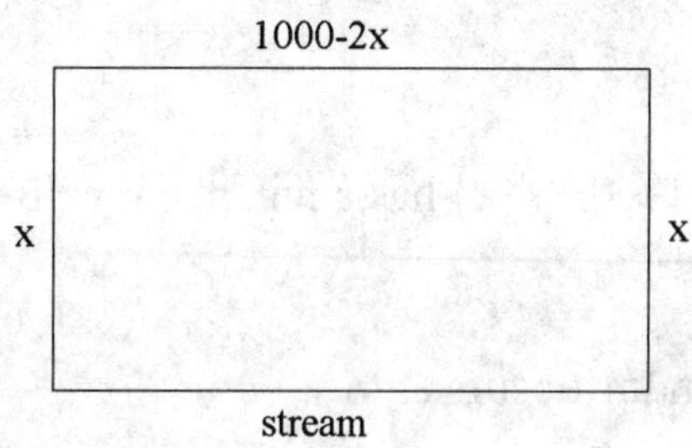

$$A(x) = x(1000 - 2x)$$
$$= -2x^2 + 1000x$$

A maxi mum is a ssumed when

$$x = -\frac{b}{2a} = -\frac{1000}{2(-2)} = 250.$$

$$x = 250$$
$$1000 - 2x = 1000 - 2(250) = 500$$

The dimensio ns are 250 ft × 500 ft.

47.

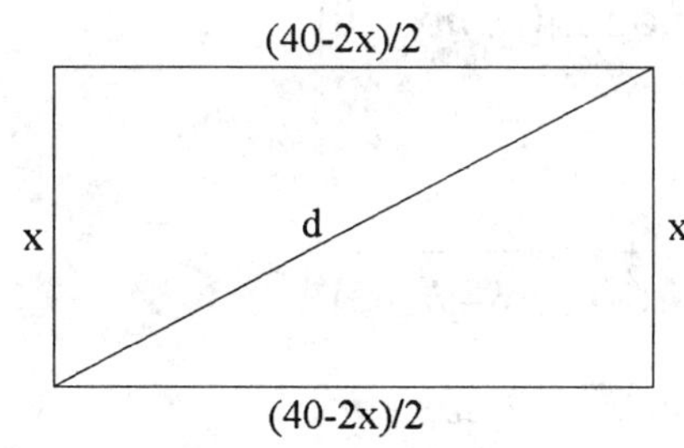

$$d^2 = x^2 + \left(\frac{40-2x}{2}\right)^2$$
$$= x^2 + (20-x)^2$$
$$= x^2 + 400 - 40x + x^2$$
$$= 2x^2 - 40x + 400$$

A mini mum is assumed if

$$x = -\frac{b}{2a} = \frac{-(-40)}{2(2)} = 10.$$

$$x = 10$$
$$\frac{40-2x}{2} = 20 - x = 20 - 10 = 10$$

The dimemsio ns are 10 ft $\times$ 10 ft.

49.

$$s(t) = 80t - 16t^2$$

Maximu m is attained when

$$t = -\frac{b}{2a} = \frac{-80}{2(-16)} = \frac{5}{2}.$$

The ball reaches its maximum h eight at 2.5 sec.

$$s(2.5) = 80(2.5) - 16(2.5)^2 = 100$$

The maximum height is 100 ft.

51.

$$R(x) = (4+.25x)(200-10x)$$
$$= 800 + 10x - 2.5x^2$$

$$x = -\frac{b}{2a} = \frac{-10}{2(-2.5)} = 2$$

An increase of $2(2.5) = \$0.50$ will maximi ze the gross revenue.

53.

unknown point: $(x, mx+b)$
origin: $(0, 0)$

$$d = \sqrt{(x-0)^2 + (mx+b-0)^2}$$
$$= \sqrt{x^2 + m^2x^2 + 2mbx + b^2}$$
$$= \sqrt{(1+m^2)x^2 + 2mbx + b^2}$$

A mini mum is attained if

$$x = -\frac{b_1}{2a} = \frac{-2mb}{2(1+m^2)} = -\frac{bm}{1+m^2}$$

(Note: b_1 has been used to distin guish between b in $y = mx+b$ and b in $ax^2 + bx + c$.)

$$y = m\left(\frac{-bm}{1+m^2}\right) + b$$
$$= \frac{-bm^2}{1+m^2} + b$$
$$= \frac{-bm^2 + b(1+m^2)}{1+m^2}$$
$$= \frac{b}{1+m^2}$$

The point closest to the origin is

$$\left(-\frac{bm}{1+m^2}, \frac{b}{1+m^2}\right).$$

55.

$$\begin{aligned} P(x) &= R(x) - C(x) \\ &= 50x - x^2 - (3x^2 - 750x + 100) \\ &= -4x^2 + 800x - 100 \end{aligned}$$

$$x = -\frac{b}{2a} = \frac{-800}{2(-4)} = 100$$

100 units should be produced to maximize profit.

CHAPTER 4 SECTION 2

EXERCISE SET 4.2

1.

$P(x) = 2x^4 - x^3 + 2x - 2$

$P(-2) = 34 > 0$

$P(-1) = -1 < 0$

$P(-2)$ and $P(-1)$ are of opposite sign, therefore, $P(x)$ has at least one root in the interval $[-2, -1]$.

3.

$P(x) = x^5 - 3x^3 + x^2 - 3$

$P(1) = -4 < 0$

$P(2) = 9 > 0$

$P(1)$ and $P(2)$ are of opposite sign, therefore, $P(x)$ has at least one root in the interval $[1, 2]$.

5.

$P(x) = x^6 - 3x^3 + x^2 - 2$

$P(1) = -3 < 0$

$P(2) = 42 > 0$

$P(1)$ and $P(2)$ are of opposite sign, therefore, $P(x)$ has at least one root in the interval $[1, 2]$.

7.

$P(x) = -2x^3 - x^2 + 3x - 4$

$P(-2) = 2 > 0$

$P(-1) = -6 < 0$

$P(-2)$ and $P(-1)$ are of opposite sign, therefore, $P(x)$ has at least one root in the interval $[-2, -1]$.

In Exercises 9-15, the RANGE settings for graphs (a), (b), and (c) are as follows:

(a)	XMIN=-10	(b)	XMIN=-10	(c)	XMIN=-10
	XMAX=10		XMAX=10		XMAX=10
	XSCL=1		XSCL=1		XSCL=1
	YMIN=-100		YMIN=-10,000		YMIN=-100,000
	YMAX=100		YMAX=10,000		YMAX=100,000
	YSCL=0		YSCL=0		YSCL=0

Graphs for Exercises 9-15 follow the chart.

For a polynomial, *P(x)*, if the degree of the leading term is odd, one end of the graph will extend upward and the other downward.

If the degree of the leading term is even, then both ends of the graph will extend upward or will both extend downward.

This together with the sign of the leading coefficient, a_n, determines how the graph extends.

degree of leading term	sign of leading coefficient	Large values of $\|x\|$, $x > 0$	Large values of $\|x\|$, $x < 0$
odd	$a_n > 0$	U	D
	$a_n < 0$	D	U
even	$a_n > 0$	U	U
	$a_n < 0$	D	D

	Polynomial function	Leading term	Sign of leading coefficient, a_n; degree: odd / even	Large values of $\|x\|$ $x > 0$	Large values of $\|x\|$, $x < 0$
9.	$P(x) = x^7 - 175x^3 + 23x^2$	x^7	$a_n > 0$ degree odd	U	D
11.	$P(x) = -8x^3 + 17x^2 - 15$	$-8x^3$	$a_n < 0$ degree odd	D	U
13.	$P(x) = -5x^{10} + 16x^7 + 5$	$-5x^{10}$	$a_n < 0$ degree even	D	D
15.	$P(x) = 4x^8 - 10x^6 + x^3 - 8$	$4x^8$	$a_n > 0$ degree even	U	U

Graphs for Exercises 9-15

9a).

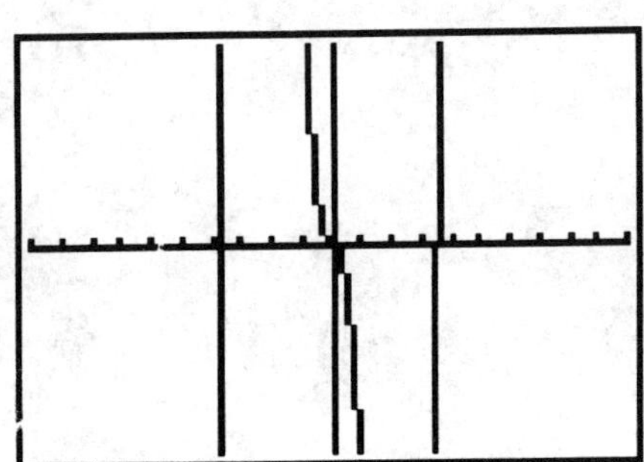

9b).

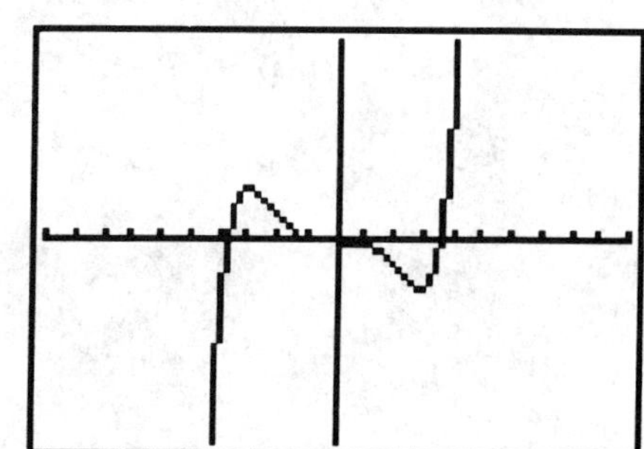

9c).

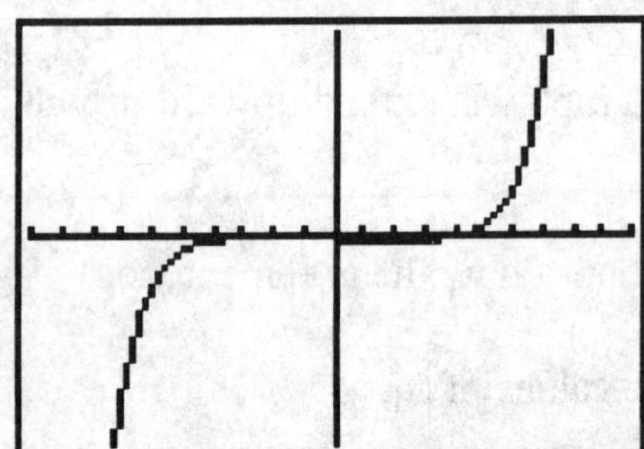

11a).

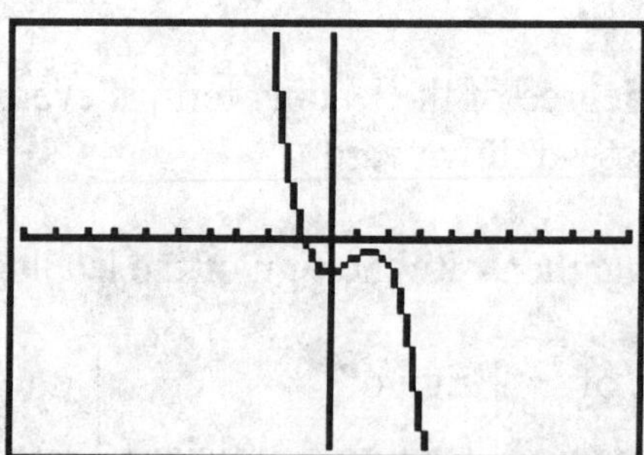

11b).

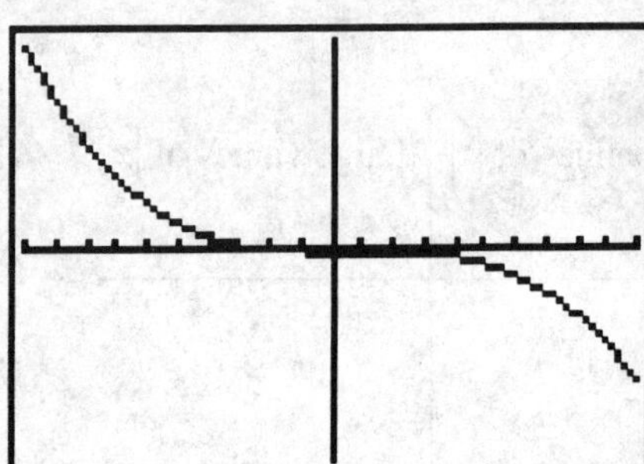

11c).

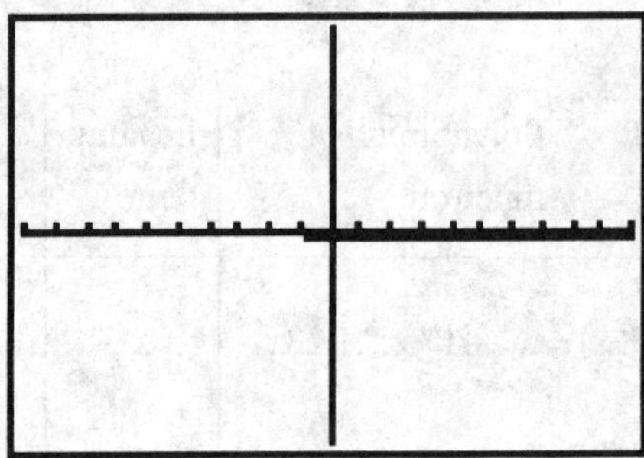

13a).

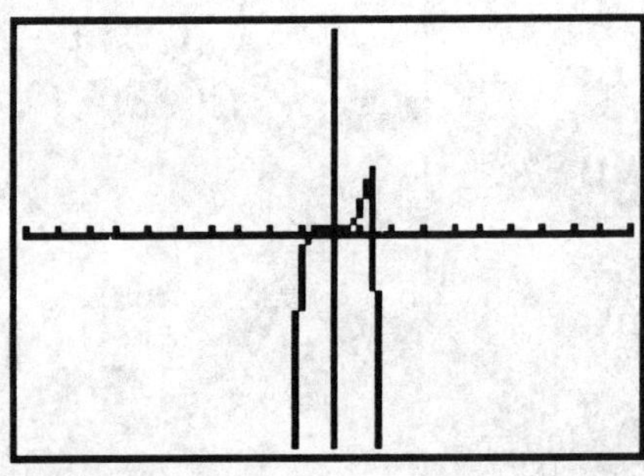

13b).

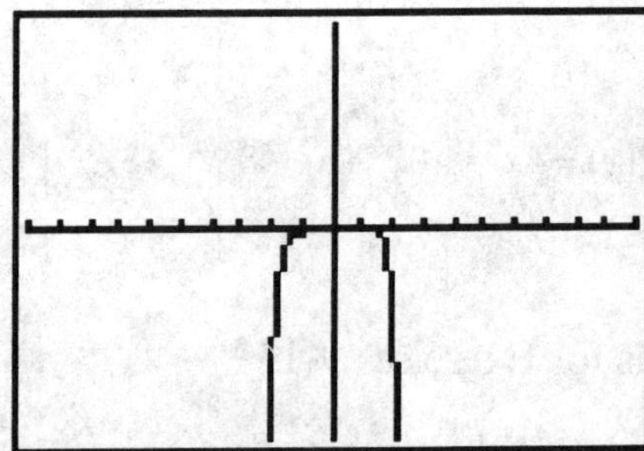

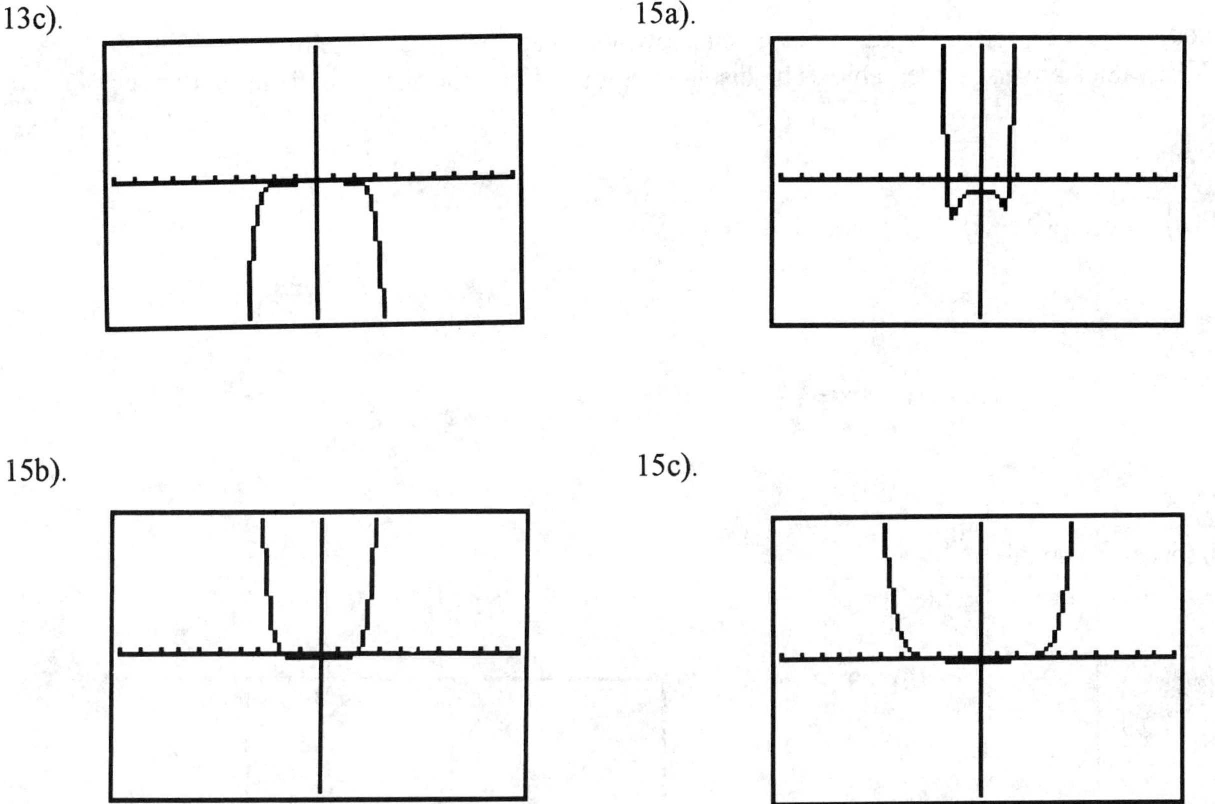

For each of the Exercises 9-15, viewing rectangle (c) most clearly shows the polynomial function behaving like its leading term.

In Exercises 17-21, the RANGE values given show important features of the function. Other RANGE values are also acceptable. The display on your calculator may vary from that shown.

17.
$P(x)=(x-3)(2x-1)(x+2)$

x-intercepts: $(x-3)(2x-1)(x+2)=0$

$$x=3,\ x=\frac{1}{2},\ x=-2$$

Test interval	Test point	Sign of $P(x)$
$(-\infty,-2)$	$x=-3$	$P(-3)<0$
$\left(-2,\frac{1}{2}\right)$	$x=0$	$P(0)>0$
$\left(\frac{1}{2},3\right)$	$x=1$	$P(1)<0$
$(3,\infty)$	$x=4$	$P(4)>0$

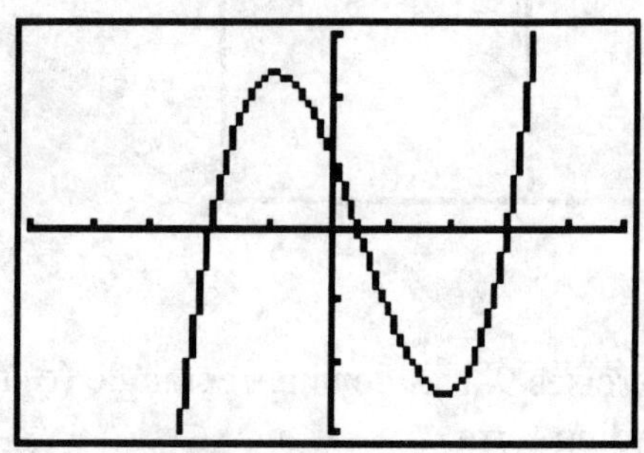

$P(x)>0$: $\left(-2,\frac{1}{2}\right)$, $(3,\infty)$

$P(x)<0$: $(-\infty,-2)$, $\left(\frac{1}{2},3\right)$

XMIN $=-5$
XMAX $=5$
XSCL $=1$
YMIN $=-15$
YMAX $=15$
YSCL $=5$

19.
$$P(x) = 2x^3 + 3x^2 - 5x$$

x-intercepts: $2x^3 + 3x^2 - 5x = 0$

$$x(x-1)(2x+5) = 0$$

$$x = 0,\ x = 1,\ x = -\frac{5}{2}$$

Test interval	Test point	Sign of $P(x)$
$\left(-\infty, -\frac{5}{2}\right)$	$x = -3$	$P(-3) < 0$
$\left(-\frac{5}{2}, 0\right)$	$x = -1$	$P(-1) > 0$
$(0, 1)$	$x = \frac{1}{2}$	$P\left(\frac{1}{2}\right) < 0$
$(1, \infty)$	$x = 2$	$P(2) > 0$

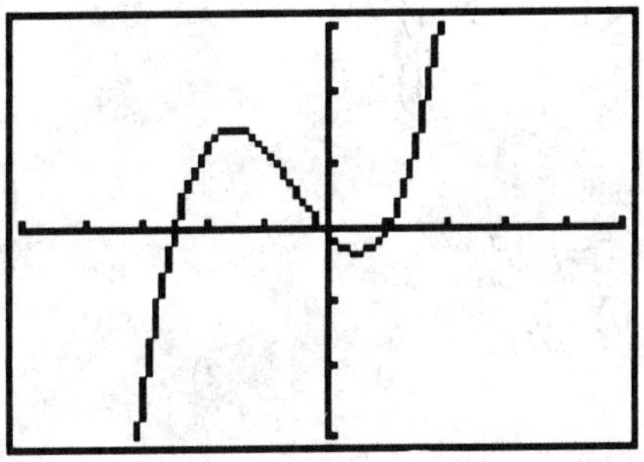

$P(x) > 0$: $\left(-\frac{5}{2}, 0\right)$, $(1, \infty)$

$P(x) < 0$: $\left(-\infty, -\frac{5}{2}\right)$, $(0, 1)$

XMIN = −5
XMAX = 5
XSCL = 1
YMIN = −15
YMAX = 15
YSCL = 5

21.

$P(x)=x^4-x^3-6x^2$

x-intercepts: $x^4-x^3-6x=0$

$x^2(x-3)(x+2)=0$

$x=0,\ x=3,\ x=-2$

Test interval	Test point	Sign of $P(x)$
$(-\infty,-2)$	$x=-3$	$P(-3)>0$
$(-2,0)$	$x=-1$	$P(-1)<0$
$(0,3)$	$x=1$	$P(1)<0$
$(3,\infty)$	$x=4$	$P(4)>0$

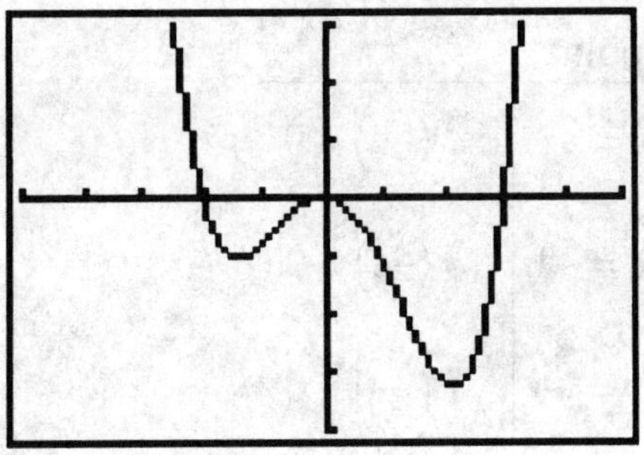

$P(x)>0$: $(-\infty,-2)$, $(3,\infty)$

$P(x)<0$: $(-2,0)$, $(0,3)$

XMIN = −5
XMAX = 5
XSCL = 1
YMIN = −20
YMAX = 15
YSCL = 5

23.

roots: 2, −4, 4

$P(x)=(x-2)(x+4)(x-4)$

$=x^3-2x^2-16x+32$

25.

roots: −1, −2, −3

$P(x)=(x+1)(x+2)(x+3)$

$=x^3+6x^2+11x+6$

27.

roots: 4, $1+\sqrt{3}$, $1-\sqrt{3}$

$P(x)=(x-4)\left[x-\left(1+\sqrt{3}\right)\right]\left[x-\left(1-\sqrt{3}\right)\right]$

$=x^3-6x^2+6x+8$

29.

$y = (x-1)(x+3)(x-5)$

$-10 \le x \le 10$ and $-30 \le y \le 30$

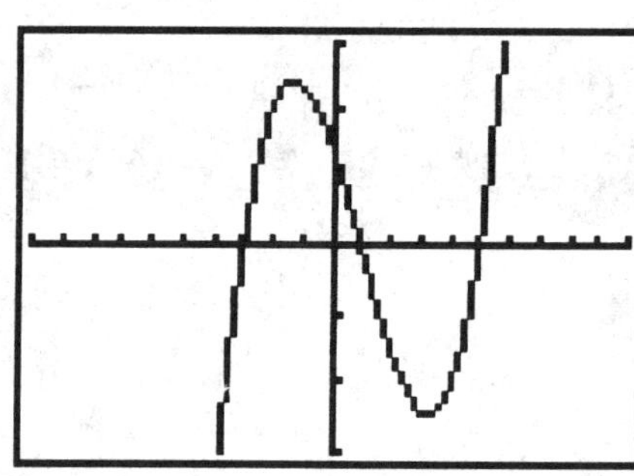

XSCL = 1

YSCL = 10

x-intercepts: $(x-1)(x+3)(x-5) = 0$

$x = 1,\ x = -3,\ x = 5$

y-intercept: $y(0) = (0-1)(0+3)(0-5) = 15$

31.

$y = -x(x-5)(x+5)(x+8)$

$-10 \le x \le 10$ and $-300 \le y \le 600$

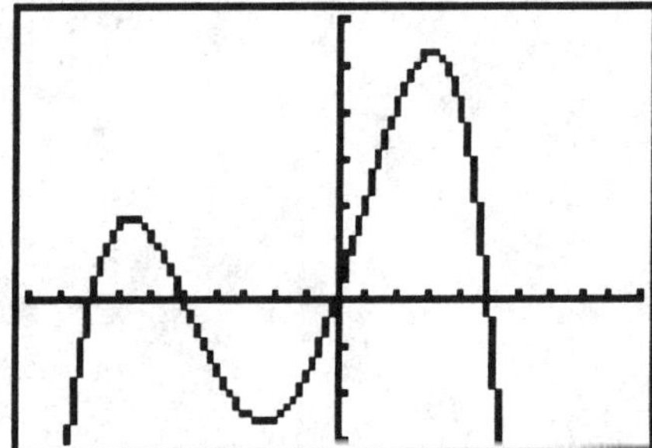

XSCL = 1

YSCL = 100

x-intercepts: $-x(x-5)(x+5)(x+8) = 0$

$x = 0,\ x = 5,\ x = -5,\ x = -8$

y-intercept: $y(0) = 0$

33.

$y = (x-10)(x+30)(x-50)$

$-100 \le x \le 100$ and $-30{,}000 \le y \le 30{,}000$

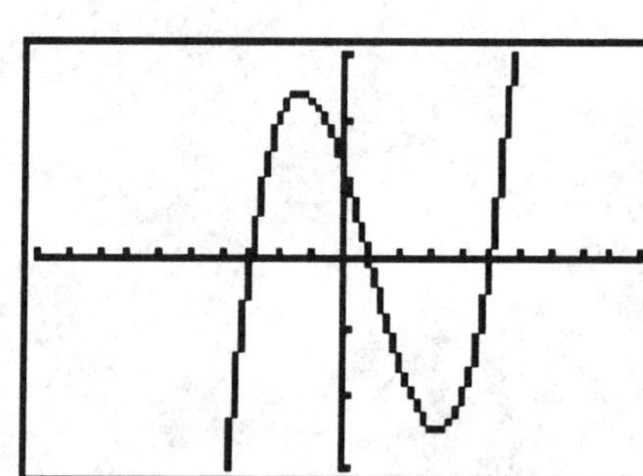

XSCL = 10

YSCL = 10,000

x-intercepts: $(x-10)(x+30)(x-50) = 0$

$x = 10,\ x = -30,\ x = 50$

y-intercept: $y(0) = (0-10)(0+30)(0-50) = 15{,}000$

35.

x-intercepts: $-9, -5, 1, 4$

(*i*) $y = (x+9)(x+5)(x-1)(x-4)$

$= x^4 + 9x^3 - 21x^2 - 169x + 180$

CHAPTER 4 SECTION 3

SECTION 4.3 PROGRESS CHECK

1. (Page 250)

$$\begin{array}{r} 4x-11 \\ x+2\overline{)4x^2-3x+6} \\ \underline{4x^2+8x} \\ -11x+6 \\ \underline{-11x-22} \\ 28 \end{array}$$

$$\frac{4x^2-3x+6}{x+2}=4x-11+\frac{28}{x+2}$$

2. (Page 252)

$$\begin{array}{r|rrrrr} 3 & 2 & 0 & -10 & -23 & 6 \\ & & 6 & 18 & 24 & 3 \\ \hline & 2 & 6 & 8 & 1 & |\,9 \end{array}$$

$Q(x)=2x^3+6x^2+8x+1$; $R=9$

EXERCISE SET 4.3

1.

$$\begin{array}{r} x-2 \\ x-5\overline{)x^2-7x+12} \\ \underline{x^2-5x} \\ -2x+12 \\ \underline{-2x+10} \\ 2 \end{array}$$

$Q(x)=x-2$; $R(x)=2$

3.

$$\begin{array}{r} 2x-4 \\ x^2+2x-1\overline{)2x^3+0x^2-2x} \\ \underline{2x^3+4x^2-2x} \\ -4x^2 \\ \underline{-4x^2-8x+4} \\ 8x-4 \end{array}$$

$Q(x)=2x-4$; $R(x)=8x-4$

5.

$$\begin{array}{r} 3x^3-9x^2+25x-75 \\ x+3\overline{)3x^4+0x^3-2x^2+0x+1} \\ \underline{3x^4+9x^3} \\ -9x^3-2x^2+0x+1 \\ \underline{-9x^3-27x^2} \\ 25x^2+0x+1 \\ \underline{25x^2+75x} \\ -75x+1 \\ \underline{-75x-225} \\ 226 \end{array}$$

$Q(x)=3x^3-9x^2+25x-75$; $R(x)=226$

7.

$$\begin{array}{r} 2x-3 \\ x^2+2\overline{)2x^3-3x^2+0x+0} \\ \underline{2x^3 \quad +4x} \\ -3x^2-4x \\ \underline{-3x^2 \quad -6} \\ -4x+6 \end{array}$$

$Q(x)=2x-3$; $R(x)=-4x+6$

9.

$$\begin{array}{r} x^2-x+1 \\ x^2+1\overline{)x^4-x^3+2x^2-x+1} \\ \underline{x^4 \quad +x^2} \\ -x^3+x^2-x+1 \\ \underline{-x^3 \quad -x} \\ x^2 \quad +1 \\ \underline{x^2 \quad +1} \\ 0 \end{array}$$

$Q(x)=x^2-x+1$; $R(x)=0$

11.

$$\begin{array}{r|rrrr} -2 & 1 & -1 & -6 & 5 \\ & & -2 & 6 & 0 \\ \hline & 1 & -3 & 0 & |\,5 \end{array}$$

$Q(x)=x^2-3x$; $R=5$

13.

$$\begin{array}{r|rrrrr} 3 & 1 & 0 & 0 & 0 & -81 \\ & & 3 & 9 & 27 & 81 \\ \hline & 1 & 3 & 9 & 27 & |\,0 \end{array}$$

$Q(x)=x^3+3x^2+9x+27$; $R=0$

15.

$$\begin{array}{r|rrrr} -1 & 3 & -1 & 0 & 8 \\ & & -3 & 4 & -4 \\ \hline & 3 & -4 & 4 & |\,4 \end{array}$$

$Q(x)=3x^2-4x+4$; $R=4$

17.

−2	1	0	0	0	0	32
		−2	4	−8	16	−32
	1	−2	4	−8	16	0

$Q(x) = x^4 - 2x^3 + 4x^2 - 8x + 16$; $R = 0$

19.

3	6	0	−1	0	4
		18	54	159	477
	6	18	53	159	481

$Q(x) = 6x^3 + 18x^2 + 53x + 159$; $R = 481$

CHAPTER 4 SECTION 4

SECTION 4.4 PROGRESS CHECK

1. (Page 254)

$$P(x) = 3x^2 - 2x - 6 \qquad x - r = x + 2$$
$$r = -2$$

Substitution:

$$R = P(-2) = 3(-2)^2 - 2(-2) - 6 = 10$$

Synthetic Division:

$$\begin{array}{r|rrr} -2 & 3 & -2 & -6 \\ & & -6 & 16 \\ \hline & 3 & -8 & |\,10 \end{array} \qquad R = 10$$

2. (Page 255)

$$P(x) = 3x^6 - 3x^5 - 4x^4 + 6x^3 - 2x^2 - x + 1$$
$$x - r = x - 1$$
$$r = 1$$

Substitution :

$$P(1) = 3(1)^6 - 3(1)^5 - 4(1)^4 + 6(1)^3 - 2(1)^2 - 1 + 1 = 0$$

Synthetic Division :

$$\begin{array}{r|rrrrrrr} 1 & 3 & -3 & -4 & 6 & -2 & -1 & 1 \\ & & 3 & 0 & -4 & 2 & 0 & -1 \\ \hline & 3 & 0 & -4 & 2 & 0 & -1 & |\,0 \end{array}$$

Since $R = 0$, we conclude $x - 1$ is a factor of $P(x)$.

EXERCISE SET 4.4

1.

$P(x) = x^3 - 4x^2 + 1$, $r = 2$

$P(2) = 2^3 - 4(2)^2 + 1 = -7$

$$\begin{array}{r|rrrr} 2 & 1 & -4 & 0 & 1 \\ & & 2 & -4 & -8 \\ \hline & 1 & -2 & -4 & |\,-7 \end{array}$$

$P(r) = -7$

3.

$P(x) = x^5 - 2$; $r = -2$

$P(-2) = (-2)^5 - 2 = -34$

$$\begin{array}{r|rrrrrr} -2 & 1 & 0 & 0 & 0 & 0 & -2 \\ & & -2 & 4 & -8 & 16 & -32 \\ \hline & 1 & -2 & 4 & -8 & 16 & |\,-34 \end{array}$$

$P(r) = -34$

5.

$P(x) = x^6 - 3x^4 + 2x^3 + 4; \quad r = -1$

$P(-1) = (-1)^6 - 3(-1)^4 + 2(-1)^3 + 4 = 0$

$$\begin{array}{r|rrrrrrr} -1 & 1 & 0 & -3 & 2 & 0 & 0 & 4 \\ & & -1 & 1 & 2 & -4 & 4 & -4 \\ \hline & 1 & -1 & -2 & 4 & -4 & 4 & |\,0 \end{array}$$

$P(r) = 0$

7.

$P(x) = x^3 - 2x^2 + x - 3 \qquad x - r = x - 2$

$r = 2$

$P(2) = 2^3 - 2(2)^2 + 2 - 3 = -1$

9.

$P(x) = -4x^3 + 6x - 2 \qquad x - r = x - 1$

$r = 1$

$P(1) = -4(1)^3 + 6(1) - 2 = 0$

11.

$P(x) = x^5 - 30 \qquad x - r = x + 2$

$r = -2$

$P(-2) = (-2)^5 - 30 = -62$

13.

$P(x) = x^3 - x^2 - 5x + 6 \qquad x - r = x - 2$

$r = 2$

$P(2) = 2^3 - 2^2 - 5(2) + 6 = 0$ Yes

15.

$P(x) = x^4 - 3x - 5 \qquad x - r = x + 2$

$r = -2$

$P(-2) = (-2)^4 - 3(-2) - 5 = 17 \neq 0$ No

17.

$P(x)=x^3+27$ $\quad x-r=x+3$

$r=-3$

$P(-3)=(-3)^3+27=0$ Yes

19.

$P(x)=x^4-16$ $\quad x-r=x+2$

$r=-2$

$P(-2)=(-2)^4-16=0$ Yes

21.

$P(x)=x^3-3x+2$ $\quad x=-2$

$P(-2)=(-2)^3-3(-2)+2=0$ Yes

23.

$P(x)=-2+x+2x^2-x^3$ $\quad x=-1$

$P(-1)=-2+(-1)+2(-1)^2-1(-1)^3=0$ Yes

25.

$P(x)=2x^2+4x-1$ $\quad x=\frac{3}{2}$

$P\left(\frac{3}{2}\right)=2\left(\frac{3}{2}\right)^2+4\left(\frac{3}{2}\right)-1=\frac{19}{2}\neq 0$ No

27.

$f(x)=(x+1)(x-2)$

$0=(x+1)(x-2)$

$x+1=0 \qquad x-2=0$

$x=-1 \qquad x=2$

29.

$f(x)=(1-x)(2x-1)$

$0=(1-x)(2x-1)$

$1-x=0 \qquad 2x-1=0$

$x=1 \qquad x=\frac{1}{2}$

31.

$f(x)=(1-2x)^2(1+2x)$

$0=(1-2x)(1-2x)(1+2x)$

$1-2x=0 \qquad 1-2x=0 \qquad 1+2x=0$

$x=\frac{1}{2} \qquad x=\frac{1}{2} \qquad x=-\frac{1}{2}$

33.

$$\begin{array}{r|rrr} r & 1 & -2 & -1 \\ & & r & r^2-2r \\ \hline & 1 & r-2 & r^2-2r-1 \end{array}$$

$r^2-2r-1=2$

$r^2-2r-3=0$

$(r-3)(r+1)=0$

$r-3=0 \qquad r+1=0$

$r=3 \qquad r=-1$

35.

$P(x)=x^3-3x^2+kx-1,$ $\quad x-r=x-2$

$r=2$

$$\begin{array}{r|rrrr} 2 & 1 & -3 & k & -1 \\ & & 2 & -2 & 2k-4 \\ \hline & 1 & -1 & k-2 & 2k-5 \end{array}$$

For $x-2$ to be a factor of $P(x)$, $P(r)=0$

$2k-5=0$

$k=\frac{5}{2}$

37.

$P(x) = x^8 - 256, \qquad x - r = x - 2$

$r = 2$

$P(2) = 2^8 - 256 = 0$

$x - 2$ is a factor of $P(x)$ since $P(2) = 0$.

39.

$P(x) = x^n - y^n, \qquad x - r = x - y$

$r = y$

$P(y) = y^n - y^n = 0$

$x - y$ is a factor of $P(x)$ since $P(y) = 0$.

CHAPTER 4 SECTION 5

SECTION 4.5 PROGRESS CHECK

1a). (Page 258)

$$z+w=(2+3i)+\left(\frac{1}{2}-2i\right)$$
$$=\frac{5}{2}+i$$
$$\overline{z+w}=\frac{5}{2}-i$$

$$\bar{z}=2-3i$$
$$\bar{w}=\frac{1}{2}+2i$$
$$\bar{z}+\bar{w}=(2-3i)+\left(\frac{1}{2}+2i\right)$$
$$=\frac{5}{2}-i$$

Thus $\overline{z+w}=\bar{z}+\bar{w}$.

1b).

$$z\cdot w=(2+3i)\left(\frac{1}{2}-2i\right)$$
$$=7-\frac{5}{2}i$$
$$\overline{z\cdot w}=7+\frac{5}{2}i$$

$$\bar{z}=2-3i$$
$$\bar{w}=\frac{1}{2}+2i$$
$$\bar{z}\cdot\bar{w}=(2-3i)\left(\frac{1}{2}+2i\right)$$
$$=7+\frac{5}{2}i$$

Thus $\overline{z\cdot w}=\bar{z}\cdot\bar{w}$.

1c).

$$z^2 = (2+3i)(2+3i) = -5+12i$$

$$\overline{z^2} = -5-12i$$

$$\overline{z} = 2-3i$$

$$\left(\overline{z}\right)^2 = (2-3i)(2-3i) = -5-12i$$

Thus $\overline{z^2} = \left(\overline{z}\right)^2$.

1d).

$$w^3 = \left(\frac{1}{2}-2i\right)\left(\frac{1}{2}-2i\right)\left(\frac{1}{2}-2i\right) = \left(-\frac{15}{4}-2i\right)\left(\frac{1}{2}-2i\right) = -\frac{47}{8}+\frac{13}{2}i$$

$$\overline{w^3} = -\frac{47}{8}-\frac{13}{2}i$$

$$\overline{w} = \frac{1}{2}+2i$$

$$\left(\overline{w}\right)^3 = \left(\frac{1}{2}+2i\right)\left(\frac{1}{2}+2i\right)\left(\frac{1}{2}+2i\right) = \left(-\frac{15}{4}+2i\right)\left(\frac{1}{2}+2i\right) = -\frac{47}{8}-\frac{13}{2}i$$

Thus $\overline{w^3} = \left(\overline{w}\right)^3$.

2. (Page 258)

zeros: 2, 4, -3

factors: $(x-2)$, $(x-4)$, $(x+3)$

$$\begin{aligned} P(x) &= (x-2)(x-4)(x+3) \\ &= (x^2-6x+8)(x+3) \\ &= x^3-3x^2-10x+24 \end{aligned}$$

3. (Page 261)

Since -2 is a zero of $P(x)$,

$x-(-2)=(x+2)$ is a factor of $P(x)$.

$$P(x)=(x+2)Q(x)$$

$$Q(x)=\frac{P(x)}{x+2}$$

$$\begin{array}{r|rrrr} -2 & 1 & 0 & -7 & -6 \\ & & -2 & 4 & 6 \\ \hline & 1 & -2 & -3 & |\,0 \end{array}$$

$$\begin{aligned} Q(x) &= x^2-2x-3 \\ &= (x-3)(x+1) \end{aligned}$$

$$0=(x-3)(x+1)$$

$$x-3=0 \qquad x+1=0$$

$$x=3 \qquad x=-1$$

Remaining zeros: -1, 3

4. (Page 262)

-2 is a zero of multiplicity 2

$$P(x) = (x+2)^2 Q(x) = (x^2+4x+4)Q(x)$$

$$Q(x) = \frac{P(x)}{x^2+4x+4}$$

$$\begin{array}{r|l} & x^2 + 1 \\ x^2+4x+4 & x^4+4x^3+5x^2+4x+4 \\ & \underline{x^4+4x^3+4x^2} \\ & \qquad\qquad x^2+4x+4 \\ & \qquad\qquad \underline{x^2+4x+4} \\ & \qquad\qquad\qquad 0 \end{array}$$

$$Q(x) = x^2+1 = (x+i)(x-i)$$
$$P(x) = (x+2)(x+2)(x+i)(x-i)$$

5. (Page 263)

Known zeros: $i,\ -3+i$

Zeros by Conjugate Zeros Theorem:

$-i,\ -3-i$

$$\begin{aligned} P(x) &= (x-i)[x-(-i)][x-(-3+i)][x-(-3-i)] \\ &= (x^2-i^2)[(x+3)-i][(x+3)+i] \\ &= (x^2+1)[(x+3)^2-i^2] \\ &= (x^2+1)(x^2+6x+10) \\ &= x^4+6x^3+11x^2+6x+10 \end{aligned}$$

EXERCISE SET 4.5

1.
$$(2-i)(2+i) = 4-i^2 = 5$$

3.
$$(3+4i)(3-4i) = 9-16i^2 = 25$$

5.
$$(-4-2i)(-4+2i) = 16-4i^2 = 20$$

7.
$$\begin{aligned} \frac{2+5i}{1-3i} &= \frac{2+5i}{1-3i}\cdot\frac{1+3i}{1+3i} \\ &= \frac{2+11i+15i^2}{1^2+3^2} \\ &= \frac{-13+11i}{10} \\ &= -\frac{13}{10}+\frac{11}{10}i \end{aligned}$$

9.

$$\frac{3-4i}{3+4i}=\frac{3-4i}{3+4i}\cdot\frac{3-4i}{3-4i}$$
$$=\frac{9-24i+16i^2}{3^2+4^2}$$
$$=\frac{-7-24i}{25}$$
$$=\frac{-7}{25}-\frac{24}{25}i$$

11.

$$\frac{3-2i}{2-i}=\frac{3-2i}{2-i}\cdot\frac{2+i}{2+i}$$
$$=\frac{6-i-2i^2}{2^2+1^2}$$
$$=\frac{8-i}{5}$$
$$=\frac{8}{5}-\frac{1}{5}i$$

13.

$$\frac{2+5i}{3i}=\frac{2+5i}{3i}\cdot\frac{-3i}{-3i}$$
$$=\frac{-6i-15i^2}{3^2}$$
$$=\frac{15-6i}{9}$$
$$=\frac{5}{3}-\frac{2}{3}i$$

15.

$$\frac{4i}{2+i}=\frac{4i}{2+i}\cdot\frac{2-i}{2-i}$$
$$=\frac{8i-4i^2}{2^2+1^2}$$
$$=\frac{4+8i}{5}$$
$$=\frac{4}{5}+\frac{8}{5}i$$

17.

$$\frac{1}{4+3i}=\frac{1}{4+3i}\cdot\frac{4-3i}{4-3i}$$
$$=\frac{4-3i}{4^2+3^2}$$
$$=\frac{4-3i}{25}$$
$$=\frac{4}{25}-\frac{3}{25}i$$

19.

$$\frac{1}{1-\frac{1}{3}i}=\frac{1}{1-\frac{1}{3}i}\cdot\frac{1+\frac{1}{3}i}{1+\frac{1}{3}i}$$
$$=\frac{1+\frac{1}{3}i}{1^2+\left(\frac{1}{3}\right)^2}$$
$$=\frac{1+\frac{1}{3}i}{\frac{10}{9}}$$
$$=\frac{9}{10}+\frac{3}{10}i$$

21.

$$\begin{aligned}\frac{1}{-5i} &= \frac{1}{-5i}\cdot\frac{5i}{5i}\\ &= \frac{5i}{5^2}\\ &= 0+\frac{1}{5}i\end{aligned}$$

23.

Let $z = a + bi$

$w = c + di$

$$\begin{aligned}z\cdot w &= (a+bi)(c+di)\\ &= ac+(bc+ad)i+bdi^2\\ &= (ac-bd)+(bc+ad)i\\ \overline{z\cdot w} &= (ac-bd)-(bc+ad)i\end{aligned}$$

$$\begin{aligned}\overline{z} &= a-bi\\ \overline{w} &= c-di\\ \overline{z}\cdot\overline{w} &= (a-bi)(c-di)\\ &= ac-adi-bci+bdi^2\\ &= (ac-bd)-(ad+bc)i\end{aligned}$$

Thus $\overline{z\cdot w} = \overline{z}\cdot\overline{w}$.

25.

$$\begin{aligned}P(x) &= (x-2)(x+4)(x-4)\\ &= x^3-2x^2-16x+32\end{aligned}$$

27.

$$\begin{aligned}P(x) &= (x+1)(x+2)(x+3)\\ &= x^3+6x^2+11x+6\end{aligned}$$

29.

$$\begin{aligned}P(x) &= (x-4)\left[x-\left(1+\sqrt{3}\right)\right]\left[x-\left(1-\sqrt{3}\right)\right]\\ &= (x-4)\left[(x-1)-\sqrt{3}\right]\left[(x-1)+\sqrt{3}\right]\\ &= (x-4)\left[(x-1)^2-3\right]\\ &= (x-4)\left(x^2-2x-2\right)\\ &= x^3-6x^2+6x+8\end{aligned}$$

31.

$$P(x) = a\left(x-\frac{1}{2}\right)\left(x-\frac{1}{2}\right)(x+2)$$

$$P(2) = 3 = a\left(2-\frac{1}{2}\right)\left(2-\frac{1}{2}\right)(2+2)$$

$$3 = 9a$$

$$a = \frac{1}{3}$$

$$\begin{aligned}P(x) &= \frac{1}{3}\left(x-\frac{1}{2}\right)\left(x-\frac{1}{2}\right)(x+2)\\ &= \frac{x^3}{3}+\frac{x^2}{3}-\frac{7x}{12}+\frac{1}{6}\end{aligned}$$

33.

$$P(x) = a(x - \sqrt{2})(x + \sqrt{2})(x - 4)$$

$$P(-1) = 5 = a(-1 - \sqrt{2})(-1 + \sqrt{2})(-1 - 4)$$

$$5 = 5a$$

$$a = 1$$

$$P(x) = (x - \sqrt{2})(x + \sqrt{2})(x - 4)$$

$$= x^3 - 4x^2 - 2x + 8$$

In Exercises 35-41, the RANGE values given show important features of the functions. There are other RANGE values that are acceptable. The display on your calculator may vary from that shown.

35.

$$(x - 3)(x + 1)(x - 2) = 0$$

roots: 3, -1, 2

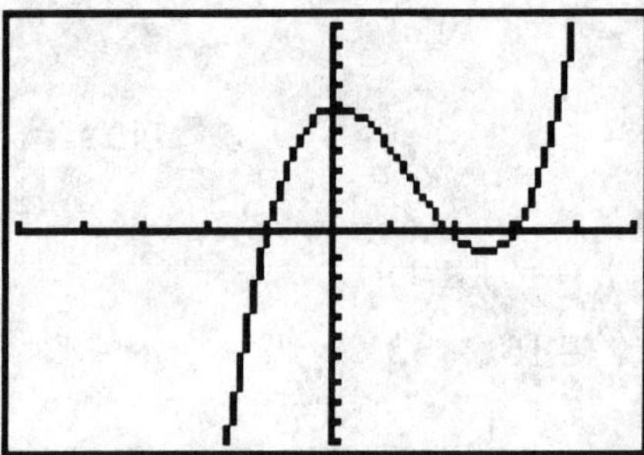

XMIN = -5

XMAX = 5

XSCL = 1

YMIN = -10

YMAX = 10

YSCL = 1

37.

$$(x+2)(x^2-16)=0$$
$$(x+2)(x+4)(x-4)=0$$

roots: $-2,\ -4,\ 4$

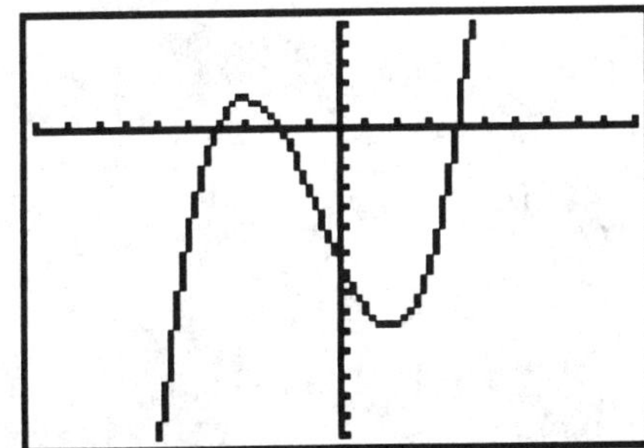

XMIN = − 10
XMAX = 10
XSCL = 1
YMIN = − 75
YMAX = 25
YSCL = 5

39.

$$(x^2+3x+2)(2x^2+x)=0$$
$$(x+2)(x+1)(x)(2x+1)=0$$

roots : $-2,\ -1,\ 0,\ -\frac{1}{2}$

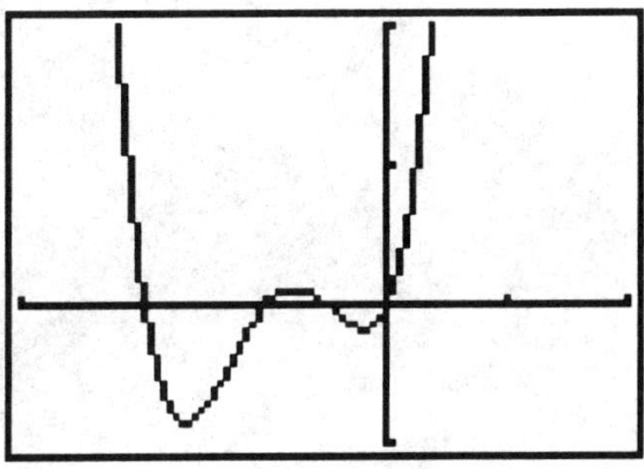

XMIN = − 3
XMAX = 2
XSCL = 1
YMIN = − 1
YMAX = 2
YSCL = 1

41.

$$(x-5)^3(x+5)^2=0$$

roots : 5, 5, 5, -5, -5

XMIN = -10
XMAX = 10
XSCL = 1
YMIN = -4000
YMAX = 1000
YSCL = 500

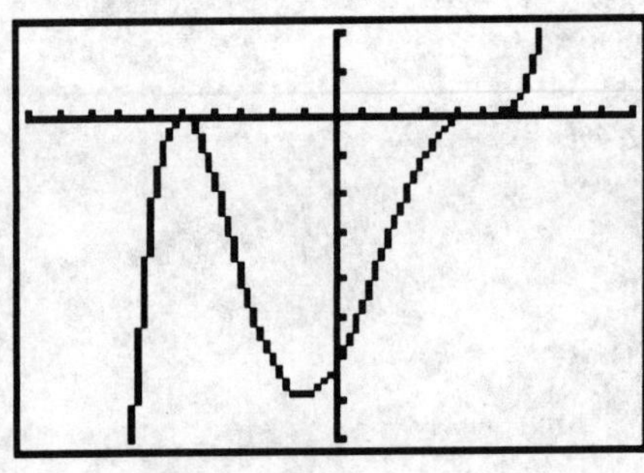

43.

$$P(x)=(x+2)^3$$
$$=x^3+6x^2+12x+8$$

45.

$$P(x)=\left(x-\frac{1}{2}\right)^2(x+1)^2$$
$$=x^4+x^3-\frac{3x^2}{4}-\frac{x}{2}+\frac{1}{4}$$

OR

$$=4\left(x^4+x^3-\frac{3x^2}{4}-\frac{x}{2}+\frac{1}{4}\right)$$
$$=4x^4+4x^3-3x^2-2x+1$$

In Exercises 47-51, use synthetic division to divide $P(x)$ by the given root to find the quotient polynomial, $Q(x)$. From the quotient polynomial, find the remaining roots of the equation.

47.

$$\begin{array}{r|rrrr} -1 & 1 & 0 & -3 & -2 \\ & & -1 & 1 & 2 \\ \hline & 1 & -1 & -2 & |\,0 \end{array}$$

$Q(x) = x^2 - x - 2$

$x^2 - x - 2 = 0$

$(x-2)(x+1) = 0$

remaining roots: 2, −1

49.

$$\begin{array}{r|rrrr} 5 & 1 & -8 & 18 & -15 \\ & & 5 & -15 & 15 \\ \hline & 1 & -3 & 3 & |\,0 \end{array}$$

$Q(x) = x^2 - 3x + 3$

$x^2 - 3x + 3 = 0$

$$x = \frac{3 \pm \sqrt{9-12}}{2}$$

$$= \frac{3 \pm i\sqrt{3}}{2}$$

remaining roots: $\frac{3}{2} + i\frac{\sqrt{3}}{2}, \frac{3}{2} - i\frac{\sqrt{3}}{2}$

51.

$$\begin{array}{r|rrrrr} -2 & 1 & 1 & -12 & -28 & -16 \\ & & -2 & 2 & 20 & 16 \\ \hline & 1 & -1 & -10 & -8 & |\,0 \end{array}$$

$Q(x) = x^3 - x^2 - 10x - 8$

$x^3 - x^2 - 10x - 8 = 0$

$(x+2)(x-4)(x+1) = 0$

remaining roots: −2, 4, −1

In Exercises 53-59, set the RANGE as follows:

XMIN = −10

XMAX = 10

XSCL = 1

YMIN = −100

YMAX = 100

YSCL = 10

53.
$y = (x+2)(x-3)$

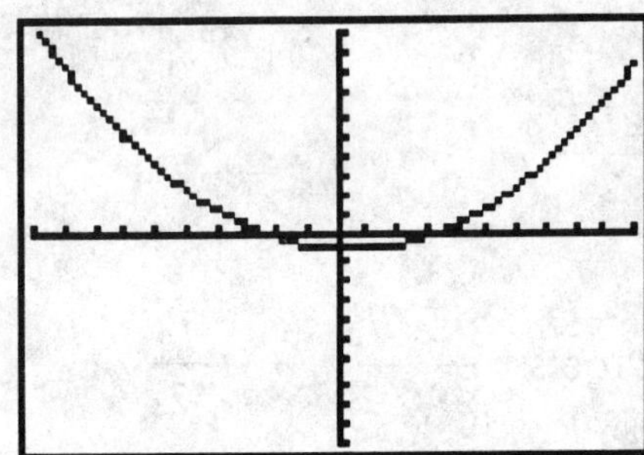

55.
$y = (x+2)(x-3)^3$

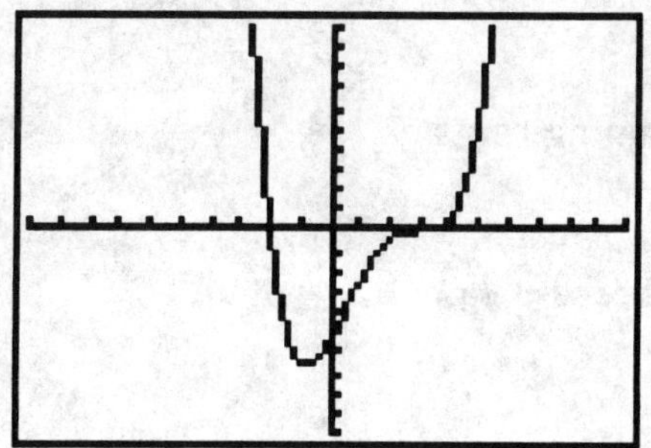

57.
$y = (x+2)^3(x-3)$

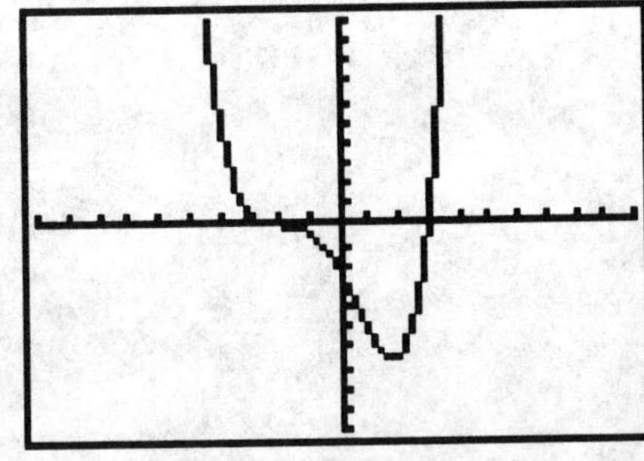

59.
$y = (x+2)^3(x-3)^2$

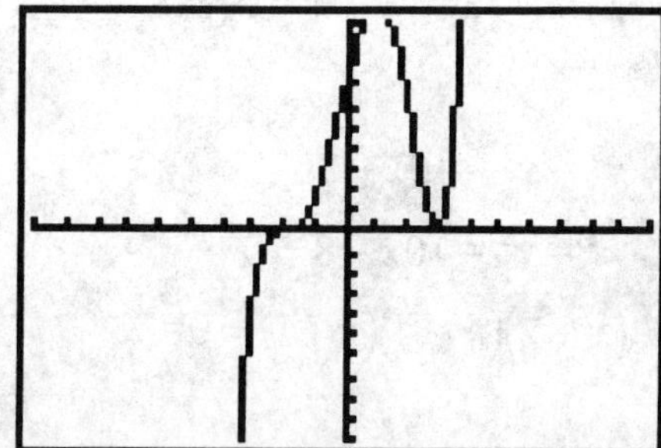

In Exercises 61-67, set the x-values in the RANGE as

XMIN = -10

XMAX = 10

61.

$$y=(x+2)(x-3)^4$$

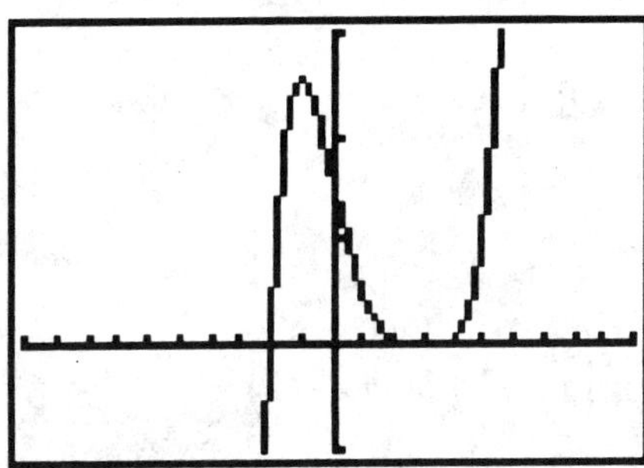

XSCL = 1

YMIN = -100

YMAX = 300

YSCL = 100

63.

$$y=(x+2)^4(x-3)$$

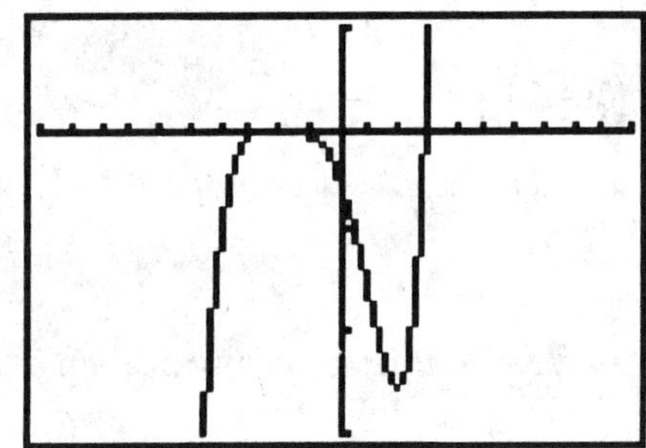

XSCL = 1

YMIN = -300

YMAX = 100

YSCL = 100

65.

$$y=(x+2)^4(x-3)^3$$

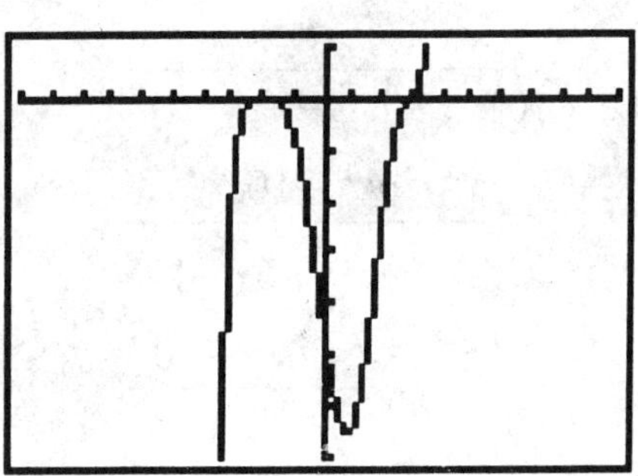

XSCL = 1

YMIN = -700

YMAX = 100

YSCL = 100

67.

$y=(x+2)^4(x-3)^5$

XSCL = 1

YMIN = −5000

YMAX = 1000

YSCL = 500

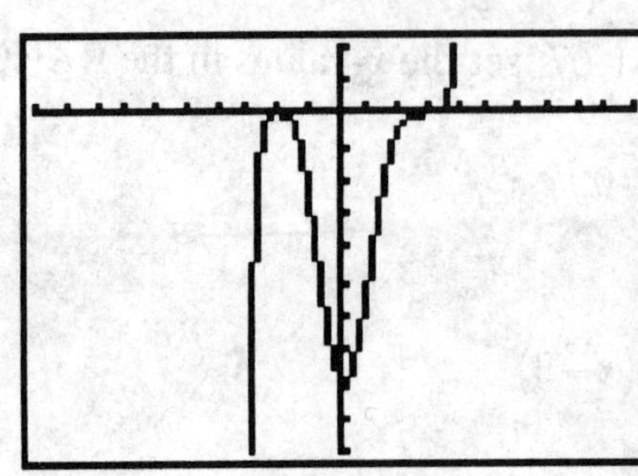

69.

$$P(x)=[x-(1+3i)](x+2)$$
$$=x^2+(1-3i)x-(2+6i)$$

71.

$$P(x)=[x-(1+i)][x-(2-i)]$$
$$=x^2-3x+(3+i)$$

73.

$$P(x)=(x+2)^2[x-(3-2i)]$$
$$=x^3+(1+2i)x^2+(-8+8i)x+(-12+8i)$$

In Exercises 75-79, if *a+bi* is a root then *a-bi* is also a root by the Conjugate Zeros Theorem.

75.

$$[x-(3-i)][x-(3+i)]=x^2-6x+10$$

$$\begin{array}{r|l} & x-1 \\ x^2-6x+10 & x^3-7x^2+16x-10 \\ & \underline{x^3-6x^2+10x} \\ & -x^2+6x-10 \\ & \underline{-x^2+6x-10} \\ & 0 \end{array}$$

$$(x^2-6x+10)(x-1)$$

77.

$$[x-(-1-2i)][x-(-1+2i)]=x^2+2x+5$$

$$\begin{array}{r|l} & x^2+2x+4 \\ x^2+2x+5 & x^4+4x^3+13x^2+18x+20 \\ & \underline{x^4+2x^3+5x^2} \\ & 2x^3+8x^2+18x+20 \\ & \underline{2x^3+4x^2+10x} \\ & 4x^2+8x+20 \\ & \underline{4x^2+8x+20} \\ & 0 \end{array}$$

$$(x^2+2x+5)(x^2+2x+4)$$

79.

$$[x-(-3-i)][x-(-3+i)](x+2)=(x^2+6x+10)(x+2)$$
$$=x^3+8x^2+22x+20$$

$$\begin{array}{r|l} & x^2-5x+6 \\ \hline x^3+8x^2+22x+20 & x^5+3x^4-12x^3-42x^2+32x+120 \\ & x^5+8x^4+22x^3+20x^2 \\ \hline & -5x^4-34x^3-62x^2+32x+120 \\ & -5x^4-40x^3-110x^2-100x \\ \hline & 6x^3+48x^2+132x+120 \\ & 6x^3+48x^2+132x+120 \\ \hline & 0 \end{array}$$

$$(x+2)(x^2+6x+10)(x^2-5x+6)=(x+2)(x^2+6x+10)(x-3)(x-2)$$

81.

$$x=a+bi$$
$$x-(a+bi)=0$$

$$P(x)=x-(a+bi)$$

83.

Suppose a polynomial with real coefficients has degree $2n+1$ for $n=0,1,2,...$ Then by the Fundamental Theorem of Algebra, Part II, it will have precisely $2n+1$ zeros.

All complex (non-real) roots come in conjugate pairs. Therefore, there must be an even number of complex (non-real) roots. The maximum number of complex roots would be $2n$ leaving $(2n+1)-(2n)=1$ real root. Hence there must be at least one real root to a polynomial of odd degree.

CHAPTER 4 SECTION 6

SECTION 4.6 PROGRESS CHECK

1. (Page 267)

$$P(x) = x^6 + 5x^4 - 4x^2 - 3$$

1 variation in sign $\Rightarrow$ 1 positive zero

$$P(-x) = (-x)^6 + 5(-x)^4 - 4(-x)^2 - 3$$
$$= x^6 + 5x^4 - 4x^2 - 3$$

1 variation in sign $\Rightarrow$ 1 negative zero

The degree of $P(x)$ is 6, thus there are
6 roots, 1 positive root, 1 negative root,
4 complex roots

2. (Page 269)

$P(x) = 9x^4 - 12x^3 + 13x^2 - 12x + 4$

If $\frac{p}{q}$ is a rational root reduced to lowest terms, then p is a factor of 4 and q is a factor of 9.

possible numerators: $\pm 1, \pm 2, \pm 4$ (the factors of 4)
possible denominators: $\pm 1, \pm 3, \pm 9$ (the factors of 9)
possible rational roots: $\pm 1, \pm 2, \pm 4, \pm\frac{1}{3}, \pm\frac{2}{3}, \pm\frac{4}{3}, \pm\frac{1}{9}, \pm\frac{2}{9}, \pm\frac{4}{9}$

$$\begin{array}{c|ccccc} \frac{2}{3} & 9 & -12 & 13 & -12 & 4 \\ & & 6 & -4 & 6 & -4 \\ \hline & 9 & -6 & 9 & -6 & |\,0 \end{array}$$

$$\begin{array}{c|cccc} \frac{2}{3} & 9 & -6 & 9 & -6 \\ & & 6 & 0 & 6 \\ \hline & 9 & 0 & 9 & |\,0 \end{array}$$

$\frac{2}{3}$ is a root two times. Using synthetic division on the remaining possibilities, they all have nonzero remainders. Thus the only rational roots are $\frac{2}{3}, \frac{2}{3}$.

3. (Page 271)

$P(x) = 9x^4 - 3x^3 + 16x^2 - 6x - 4$

possible numerators: $\pm 1, \pm 2, \pm 4$ (factors of 4)

possible denominators: $\pm 1, \pm 3, \pm 9$ (factors of 9)

possible rational roots: $\pm 1, \pm 2, \pm 4, \pm\frac{1}{3}, \pm\frac{2}{3}, \pm\frac{2}{9}, \pm\frac{1}{9}, \pm\frac{2}{9}, \pm\frac{4}{9}$

$$\begin{array}{r|rrrrr} \frac{2}{3} & 9 & -3 & 16 & -6 & -4 \\ & & 6 & 2 & 12 & 4 \\ \hline & 9 & 3 & 18 & 6 & |\,0 \end{array}$$

$$\begin{array}{r|rrrr} -\frac{1}{3} & 9 & 3 & 18 & 6 \\ & & -3 & 0 & -6 \\ \hline & 9 & 0 & 18 & |\,0 \end{array}$$

$$\begin{aligned} 9x^2 + 18 &= 0 \\ 9x^2 &= -18 \\ x^2 &= -2 \\ x &= \pm\sqrt{2}\,i \end{aligned}$$

zeros: $\frac{2}{3}, -\frac{1}{3}, \pm\sqrt{2}\,i$

EXERCISE SET 4.6

1.
$P(x) = 3x^4 - 2x^3 + 6x^2 + 5x - 2$

3 positive roots or 1 positive root

$P(-x) = 3x^4 + 2x^3 + 6x^2 - 5x - 2$

1 negative root

Possibilities:
3 positive roots, 1 negative root;
1 positive root, 1 negative root, 2 complex roots

3.
$P(x) = x^6 + 2x^4 + 4x^2 + 1$

0 positive roots

$P(-x) = x^6 + 2x^4 + 4x^2 + 1$

0 negative roots

Possibilities:
6 complex roots

5.

$P(x)=x^5-4x^3+7x-4$

3 positive roots or 1 positive root

$P(-x)=-x^5+4x^3-7x-4$

2 negative roots or 0 negative roots

Possibilities:

positive roots	negative roots	complex roots
3	2	0
3	0	2
1	2	2
1	0	4

7.

$P(x)=5x^3+2x^2+7x-1$

1 positive root

$P(-x)=-5x^3+2x^2-7x-1$

2 negative roots or 0 negative roots

Possibilities:

1 positive root, 2 negative roots;

1 positive root, 0 negative roots, 2 complex roots

9.

$P(x) = x^4 - 2x^3 + 5x^2 + 2$

2 positive roots or 0 positive roots

$P(-x) = x^4 + 2x^3 + 5x^2 + 2$

0 negative roots

Possibilities:

2 positive roots, 0 negative roots, 2 complex roots;

0 positive roots, 0 negative roots, 4 complex roots

11.

$P(x) = x^8 + 7x^3 + 3x - 5$

1 positive root

$P(-x) = x^8 - 7x^3 - 3x - 5$

1 negative root

Possibilities:

1 positive root, 1 negative root, 6 complex roots

In Exercises 13-21, only the synthetic division yielding a remainder of 0 is shown. All other possible rational roots leave a nonzero remainder and fail as roots.

13.

$P(x) = x^3 - 2x^2 - 5x + 6$

possible numerators: $\pm 1, \pm 2, \pm 3, \pm 6$ (factors of 6)

possible denominators: ± 1 (factors of 1)

possible rational roots: $\pm 1, \pm 2, \pm 3, \pm 6$

$$\begin{array}{r|rrrr} 1 & 1 & -2 & -5 & 6 \\ & & 1 & -1 & -6 \\ \hline & 1 & -1 & -6 & | \; 0 \end{array}$$

$$\begin{array}{r|rrrr} 3 & 1 & -2 & -5 & 6 \\ & & 3 & 3 & -6 \\ \hline & 1 & 1 & -2 & | \; 0 \end{array}$$

$$\begin{array}{r|rrrr} -2 & 1 & -2 & -5 & 6 \\ & & -2 & 8 & -6 \\ \hline & 1 & -4 & 3 & | \; 0 \end{array}$$

rational roots: $-2, 1, 3$

15.

$P(x) = 6x^4 - 7x^3 - 13x^2 + 4x + 4$

possible numerators: $\pm 1, \pm 2, \pm 4$ (factors of 4)

possible denominators: $\pm 1, \pm 2, \pm 3, \pm 6$ (factors of 6)

possible rational roots: $\pm 1, \pm 2, \pm 4, \pm \frac{1}{2}, \pm \frac{1}{3}, \pm \frac{2}{3}, \pm \frac{4}{3}, \pm \frac{1}{6}$

$$\begin{array}{r|rrrrr} 2 & 6 & -7 & -13 & 4 & 4 \\ & & 12 & 10 & -6 & -4 \\ \hline & 6 & 5 & -3 & -2 & | \, 0 \end{array}$$

$$\begin{array}{r|rrrrr} \frac{2}{3} & 6 & -7 & -13 & 4 & 4 \\ & & 4 & -2 & -10 & -4 \\ \hline & 6 & -3 & -15 & -6 & | \, 0 \end{array}$$

$$\begin{array}{r|rrrrr} -1 & 6 & -7 & -13 & 4 & 4 \\ & & -6 & 13 & 0 & -4 \\ \hline & 6 & -13 & 0 & 4 & | \, 0 \end{array}$$

$$\begin{array}{r|rrrrr} -\frac{1}{2} & 6 & -7 & -13 & 4 & 4 \\ & & -3 & 5 & 4 & -4 \\ \hline & 6 & -10 & -8 & 8 & | \, 0 \end{array}$$

rational roots: $-1, -\frac{1}{2}, \frac{2}{3}, 2$

17.

$P(x)=5x^6-x^5-5x^4+6x^3-x^2-5x+1$

possible numerators: ± 1 (factors of 1)

possible denominators: $\pm 1, \pm 5$ (factors of 5)

possible rational roots: $\pm 1, \pm\frac{1}{5}$

$$\begin{array}{r|rrrrrrr} 1 & 5 & -1 & -5 & 6 & -1 & -5 & 1 \\ & & 5 & 4 & -1 & 5 & 4 & -1 \\ \hline & 5 & 4 & -1 & 5 & 4 & -1 & \vert\, 0 \end{array}$$

$$\begin{array}{r|rrrrrrr} \frac{1}{5} & 5 & -1 & -5 & 6 & -1 & -5 & 1 \\ & & 1 & 0 & -1 & 1 & 0 & -1 \\ \hline & 5 & 0 & -5 & 5 & 0 & -5 & \vert\, 0 \end{array}$$

$$\begin{array}{r|rrrrrrr} -1 & 5 & -1 & -5 & 6 & -1 & -5 & 1 \\ & & -5 & 6 & -1 & -5 & 6 & -1 \\ \hline & 5 & -6 & 1 & 5 & -6 & 1 & \vert\, 0 \end{array}$$

$$\begin{array}{r|rrrrrr} -1 & 5 & -6 & 1 & 5 & -6 & 1 \\ & & -5 & 11 & -12 & 7 & -1 \\ \hline & 5 & -11 & 12 & -7 & 1 & \vert\, 0 \end{array}$$

rational roots: $-1, -1, \frac{1}{5}, 1$

19.

$P(x) = 4x^4 - x^3 + 5x^2 - 2x - 6$

possible numerators: $\pm 1, \pm 2, \pm 3, \pm 6$ (factors of 6)

possible denominators: $\pm 1, \pm 2, \pm 4$ (factors of 4)

possible rational roots: $\pm 1, \pm 2, \pm 3, \pm 6, \pm \frac{1}{2}, \pm \frac{3}{2}, \pm \frac{1}{4}, \pm \frac{3}{4}$

$$\begin{array}{r|rrrrr} 1 & 4 & -1 & 5 & -2 & -6 \\ & & 4 & 3 & 8 & 6 \\ \hline & 4 & 3 & 8 & 6 & | \; 0 \end{array}$$

$$\begin{array}{r|rrrrr} -\frac{3}{4} & 4 & -1 & 5 & -2 & -6 \\ & & -3 & 3 & -6 & 6 \\ \hline & 4 & -4 & 8 & -8 & | \; 0 \end{array}$$

rational roots: $-\frac{3}{4}, 1$

21.

$P(x) = 2x^5 - 13x^4 + 26x^3 - 22x^2 + 24x - 9$

possible numerators: $\pm 1, \pm 3, \pm 9$ (factors of 9)

possible denominators: $\pm 1, \pm 2$ (factors of 2)

possible rational roots: $\pm 1, \pm 3, \pm 9, \pm\frac{1}{2}, \pm\frac{3}{2}, \pm\frac{9}{2}$

$$\begin{array}{r|rrrrrr} \frac{1}{2} & 2 & -13 & 26 & -22 & 24 & -9 \\ & & 1 & -6 & 10 & -6 & 9 \\ \hline & 2 & -12 & 20 & -12 & 18 & \mid 0 \end{array}$$

$$\begin{array}{r|rrrrrr} 3 & 2 & -13 & 26 & -22 & 24 & -9 \\ & & 6 & -21 & 15 & -21 & 9 \\ \hline & 2 & -7 & 5 & -7 & 3 & \mid 0 \end{array}$$

$$\begin{array}{r|rrrrr} 3 & 2 & -7 & 5 & -7 & 3 \\ & & 6 & -3 & 6 & -3 \\ \hline & 2 & -1 & 2 & -1 & \mid 0 \end{array}$$

rational roots: $\frac{1}{2}, 3, 3$

23.

$P(x) = 4x^4 + x^3 + x^2 + x - 3$

possible numerators: $\pm 1, \pm 3$ (factors of 3)

possible denominators: $\pm 1, \pm 2, \pm 4$ (factors of 4)

possible rational roots: $\pm 1, \pm 3, \pm \frac{1}{2}, \pm \frac{3}{2}, \pm \frac{1}{4}, \pm \frac{3}{4}$

$$\begin{array}{r|rrrrr} \frac{3}{4} & 4 & 1 & 1 & 1 & -3 \\ & & 3 & 3 & 3 & 3 \\ \hline & 4 & 4 & 4 & 4 & |\,0 \end{array}$$

$4x^3 + 4x^2 + 4x + 4 = 0$ has the same roots as

$x^3 + x^2 + x + 1 = 0$

$$\begin{array}{r|rrrr} -1 & 1 & 1 & 1 & 1 \\ & & -1 & 0 & -1 \\ \hline & 1 & 0 & 1 & |\,0 \end{array}$$

$x^2 + 1 = 0$

$x^2 = -1$

$x = \pm i$

roots: $\frac{3}{4}, -1, \pm i$

25.

$P(x) = 5x^5 - 3x^4 - 10x^3 + 6x^2 - 40x + 24$

possible numerators: $\pm 1, \pm 2, \pm 3, \pm 4, \pm 6, \pm 8, \pm 12, \pm 24$ (factors of 24)

possible denominators: $\pm 1, \pm 5$ (factors of 5)

possible rational roots: $\pm 1, \pm 2, \pm 3, \pm 4, \pm 6, \pm 8, \pm 12, \pm 24, \pm \frac{1}{5}, \pm \frac{2}{5},$

$\pm \frac{3}{5}, \pm \frac{4}{5}, \pm \frac{6}{5}, \pm \frac{8}{5}, \pm \frac{12}{5}, \pm \frac{24}{5}$

$$\begin{array}{r|rrrrrr} 2 & 5 & -3 & -10 & 6 & -40 & 24 \\ & & 10 & 14 & 8 & 28 & -24 \\ \hline & 5 & 7 & 4 & 14 & -12 & \mid 0 \end{array}$$

$$\begin{array}{r|rrrrr} -2 & 5 & 7 & 4 & 14 & -12 \\ & & -10 & 6 & -20 & 12 \\ \hline & 5 & -3 & 10 & -6 & \mid 0 \end{array}$$

$$\begin{array}{r|rrrr} \frac{3}{5} & 5 & -3 & 10 & -6 \\ & & 3 & 0 & 6 \\ \hline & 5 & 0 & 10 & \mid 0 \end{array}$$

$5x^2 + 10 = 0$

$5x^2 = -10$

$x = \pm i\sqrt{2}$

roots: $-2, 2, \frac{3}{5}, \pm i\sqrt{2}$

27.

$P(x) = 6x^4 - x^3 - 5x^2 + 2x = x(6x^3 - x^2 - 5x + 2)$

Since x is a factor of $P(x)$, 0 is a root of $P(x) = 0$.

Consider $6x^3 - x^2 - 5x + 2 = 0$

possible numerators: $\pm 1, \pm 2$ (factors of 2)

possible denominators: $\pm 1, \pm 2, \pm 3, \pm 6$ (factors of 6)

possible rational roots: $\pm 1, \pm 2, \pm\frac{1}{2}, \pm\frac{1}{3}, \pm\frac{2}{3}, \pm\frac{1}{6}$

$$\begin{array}{r|rrrr} \frac{1}{2} & 6 & -1 & -5 & 2 \\ & & 3 & 1 & -2 \\ \hline & 6 & 2 & -4 & | \; 0 \end{array}$$

$6x^2 + 2x - 4 = 0$ has the same roots as

$3x^2 + x - 2 = 0$

$$\begin{array}{r|rrr} \frac{2}{3} & 3 & 1 & -2 \\ & & 2 & 2 \\ \hline & 3 & 3 & | \; 0 \end{array}$$

$3x + 3 = 0$

$x = -1$

roots: $-1, 0, \frac{1}{2}, \frac{2}{3}$

29.

$P(x) = 2x^4 - x^3 - 28x^2 + 30x - 8$

possible numerators: $\pm 1, \pm 2, \pm 4, \pm 8$ (factors of 8)

possible denominators: $\pm 1, \pm 2$ (factors of 2)

possible rational roots: $\pm 1, \pm 2, \pm 4, \pm 8, \pm \frac{1}{2}$

$$\begin{array}{r|rrrrr} \frac{1}{2} & 2 & -1 & -28 & 30 & -8 \\ & & 1 & 0 & -14 & 8 \\ \hline & 2 & 0 & -28 & 16 & \mid 0 \end{array}$$

$2x^3 - 28x + 16 = 0$ has the same roots as

$x^3 - 14x + 8 = 0$

$$\begin{array}{r|rrrr} -4 & 1 & 0 & -14 & 8 \\ & & -4 & 16 & -8 \\ \hline & 1 & -4 & 2 & \mid 0 \end{array}$$

$x^2 - 4x + 2 = 0$

$$x = \frac{-(-4) \pm \sqrt{(-4)^2 - 4(1)(2)}}{2(1)} = 2 \pm \sqrt{2}$$

roots: $\frac{1}{2}, -4, 2 \pm \sqrt{2}$

31.

$$P(x) = x^4 - 6x^3 + 10x^2 - 6x + 9$$

4 positive roots or 2 positive roots or 0 positive roots

$$P(-x) = x^4 + 6x^3 + 10x^2 + 6x + 9$$

0 negative roots

possible numerators: $\pm 1, \pm 3, \pm 9$ (factors of 9)

possible denominators: ± 1 (factors of 1)

possible rational roots: $\pm 1,\ \pm 3, \pm 9$

3	1	–6	10	–6	9
		3	–9	3	–9
	1	–3	1	–3	0

3	1	–3	1	–3
		3	0	3
	1	0	1	0

$$x^2 + 1 = 0$$
$$x^2 = -1$$
$$x = \pm i$$

roots: 3, 3, $\pm i$

33.

$P(x) = x^4 - 6x^2 + 8$

2 positive roots or 0 positive roots

$P(-x) = x^4 - 6x^2 + 8$

2 negative roots or 0 negative roots

possible numerators: $\pm 1, \pm 2, \pm 4, \pm 8$ (factors of 8)

possible denominators: ± 1 (factors of 1)

possible rational roots: $\pm 1, \pm 2, \pm 4, \pm 8$

$$\begin{array}{r|rrrrr} 2 & 1 & 0 & -6 & 0 & 8 \\ & & 2 & 4 & -4 & -8 \\ \hline & 1 & 2 & -2 & -4 & |\,0 \end{array}$$

$$\begin{array}{r|rrrr} -2 & 1 & 2 & -2 & -4 \\ & & -2 & 0 & 4 \\ \hline & 1 & 0 & -2 & |\,0 \end{array}$$

$x^2 - 2 = 0$

$x^2 = 2$

$x = \pm\sqrt{2}$

roots: $-2, 2, \pm\sqrt{2}$

35.

$P(x) = 4x^4 + 4x^3 - 3x^2 - 4x - 1$

1 positive root

$P(-x) = 4x^4 - 4x^3 - 3x^2 + 4x - 1$

3 negative roots or 1 negative root

possible numerators: ± 1 (factors of 1)

possible denominators: $\pm 1, \pm 2, \pm 4$ (factors of 4)

possible rational roots: $\pm 1, \pm\frac{1}{2}, \pm\frac{1}{4}$

$$\begin{array}{r|rrrrr} 1 & 4 & 4 & -3 & -4 & -1 \\ & & 4 & 8 & 5 & 1 \\ \hline & 4 & 8 & 5 & 1 & |\,0 \end{array}$$

$$\begin{array}{r|rrrr} -1 & 4 & 8 & 5 & 1 \\ & & -4 & -4 & -1 \\ \hline & 4 & 4 & 1 & |\,0 \end{array}$$

$$\begin{array}{r|rrr} -\frac{1}{2} & 4 & 4 & 1 \\ & & -2 & -1 \\ \hline & 4 & 2 & |\,0 \end{array}$$

$4x + 2 = 0$

$x = -\frac{1}{2}$

roots: $-\frac{1}{2}, -\frac{1}{2}, -1, 1$

37.

$$P(x) = x^3 + kx^2 + kx + 2$$

possible numerators: $\pm 1, \pm 2$ (factors of 2)
possible denominators: ± 1 (factors of 1)
possible rational roots: $\pm 1, \pm 2$

$$\begin{array}{r|rrrr} -2 & 1 & k & k & 2 \\ & & -2 & -2k+4 & 2k-8 \\ \hline & 1 & k-2 & -k+4 & 2k-6 \end{array}$$

$2k - 6 = 0$

$k = 3$

$k = 3, r = -2$

39.

$P(x) = x^4 - 3x^3 + kx^2 - 4x - 1$

possible numerators: ± 1 (factors of 1)
possible denominators: ± 1 (factors of 1)
possible rational roots: ± 1

$$\begin{array}{r|rrrrr} 1 & 1 & -3 & k & -4 & -1 \\ & & 1 & -2 & k-2 & k-6 \\ \hline & 1 & -2 & k-2 & k-6 & | \; k-7 \end{array}$$

$k - 7 = 0$
$k = 7$

$k = 7, r = 1$

$$\begin{array}{r|rrrrr} -1 & 1 & -3 & k & -4 & -1 \\ & & -1 & 4 & -k-4 & k+8 \\ \hline & 1 & -4 & k+4 & -k-8 & | \; k+7 \end{array}$$

$k + 7 = 0$
$k = -7$
$k = -7, r = -1$

41.

Suppose $P(x)$ is a polynomial with real coefficients that has one variation in sign. Suppose all of the real roots of $P(x)$ are negative. Represent these negative roots by $-r_1, -r_2, \ldots -r_n$ where $r_1, r_2, \ldots, r_n > 0$. Then $P(x) = (x+r_1)(x+r_2)\cdots(x+r_n)$ which when expanded has no variation in sign. Hence $P(x)$ must have at least one positive root. Suppose $P(x)$ must have at least one postive root, $r_1, r_2, \ldots, r_n$.
Then $P(x) = (x-r_1)(x-r_2)\cdots(x-r_n)$ which results in more than one variation in sign. Thus $P(x)$ must have exactly one positive zero.

43.

Let $x = \sqrt{5}$, then $x^2 = 5$ or $x^2 - 5 = 0$.
Let $P(x) = x^2 - 5$. By the Rational Zero Theorem, the only possible rational zeros of $P(x)$ are ± 1 and ± 5. Using substitution, we see that none of these numbers is a zero of $P(x)$, implying that $P(x)$ has no rational zeros. Since $P(\sqrt{5}) = 0$, $\sqrt{5}$ is a zero, hence $\sqrt{5}$ cannot be a rational number.

45.

Let $P(x)$ be a polynomial with real coefficients and let r be a positive zero of $P(x)$. The quotient polynomial $Q(x)$ can be obtained by synthetic division.

Let $a_0, a_1, a_2, \ldots, a_n$ be positive.

$$\begin{array}{r|ccccc} r & a_n & a_{n-1} & -a_{n-2} & -(r^3a_n + r^2a_{n-1} - ra_{n-2}) \\ & & ra_n & r^2a_n + ra_{n-1} & r^3a_n + r^2a_{n-1} - ra_{n-2} \\ \hline & a_n & ra_n + a_{n-1} & r^2a_n + ra_{n-1} - a_{n-2} & |\ 0 \end{array}$$

$-(r^3a_n + r^2a_{n-1} - ra_{n-2}) = -r(r^2a_n + ra_{n-1} - a_{n-2})$ has opposite sign of $r^2a_n + ra_{n-1} - a_{n-2}$ causing $P(x)$ to have at least one more sign variation than $Q(x)$.

The coefficients of $Q(x)$ remain positive at least until there is a variation in sign in $P(x)$. Once a sign in $Q(x)$ varies, the sign of $P(x)$ will have to vary an additional time in order to obtain a remainder of 0.

47.

Suppose $P(x) = (x - r_1)(x - r_2)\cdots(x - r_k)$ when $r_1, r_2, \ldots, r_k$ are positive numbers. Then $r_1, r_2, \ldots, r_k$ are positive roots of $P(x) = 0$. Since $P(x) = 0$ has k positive roots, $P(x)$ must have at least k variations in sign, (see #46). However when $P(x) = (x - r_1)(x - r_2)\cdots(x - r_k)$ is expanded there are only k possible sign variations. Thus each sign must vary making $P(x) = (x - r_1)(x - r_2)\cdots(x - r_k)$ have alternating signs.

49.

Let $P(x)=(x-r_1)(x-r_2)(x-r_3)(x+r_4)(x+r_4)(x^2+r_5)(x^2+r_6)\cdots(x^2+r_n)$

Note that the expansion of factors producing negative roots or complex roots do not cause a variation in sign. If there are $2n$ positive roots then the first and last terms of $P(x)$ have the same sign and thus an even number, $2k$, sign variations. Thus the difference between the number of sign variations and the number of positive roots is $2k-2n=2(k-n)$, an even number. If there are $2n+1$ positive roots then the first and last terms of $P(x)$ have different signs and thus an odd number, $2k+1$, sign variations. Therefore the difference between the number of sign variations and the number of positive roots is $(2k+1)-(2n+1)=2k-2n=2(k-n)$ an even number.

CHAPTER 4 SECTION 7

SECTION 4.7 PROGRESS CHECK

1. (Page 277)

$x^3 + x^2 - 3x - 3 = 0$

x		1.0	1.1	1.2	1.3	1.4	1.5	1.6	1.7	1.8	1.9	2.0
$P(x)$		–	–	–	–	–	–	–	–	+		

x		1.70	1.71	1.72	1.73	1.74	1.75	1.76	1.77	1.78	1.79	1.80
$P(x)$		–	–	–	–	+						

The first three digits of the root are 1.73.

2. (Page 279)

$P(x) = x^5 + x^4 + x + 2 = 0$

$P(-2) < 0$ $\quad P(-1.5) < 0$ $\quad P(-1) > 0$

$P(-1.5) < 0$ $\quad P(-1.25) > 0$ $\quad P(-1) > 0$

$P(-1.5) < 0$ $\quad P(-1.38) < 0$ $\quad P(-1.25) > 0$

$P(-1.38) < 0$ $\quad P(-1.32) < 0$ $\quad P(-1.25) > 0$

$P(-1.32) < 0$ $\quad P(-1.29) < 0$ $\quad P(-1.25) > 0$

$P(-1.29) < 0$ $\quad P(-1.27) > 0$ $\quad P(-1.25) > 0$

-1.27 is a root.

EXERCISE SET 4.7

1.

$$P(x) = 3x^3 - 2x^2 + 5x - 1$$
$$= [(3x-2)x+5]x - 1$$

3.

$$P(x) = x^5 + 2x^4 - 2x - 3$$
$$= \{\{[(x+2)x+0]x+0\}x-2\}x-3$$

5.

$$P(x) = 2x^4 - x^2 + x + 4$$
$$= \{[(2x+0)x-1]x+1\}x+4$$

7.

$2x^4 - x^3 + 2x - 2 = 0$

x		−2	−1.9	−1.8	−1.7	−1.6	−1.5	−1.4	−1.3	−1.2	−1.1	−1.0
$P(x)$		+	+	+	+	+	+	+	+	+	+	−

x		−1.10	−1.09	−1.08	−1.07	−1.06	−1.05	−1.04	−1.03	−1.02	−1.01	−1.0
$P(x)$		+	−									

−1.10

9.

$x^5 - 3x^3 + x^2 - 3 = 0$

x		1.0	1.1	1.2	1.3	1.4	1.5	1.6	1.7	1.8	1.9	2.0
$P(x)$		−	−	−	−	−	−	−	−	+		

x		1.70	1.71	1.72	1.73	1.74	1.75	1.76	1.77	1.78	1.79	1.8
$P(x)$		−	−	−	−	+						

1.73

11.

$x^6 - 3x^3 + x^2 - 2 = 0$

x		1.0	1.1	1.2	1.3	1.4	1.5	1.6	1.7	1.8	1.9	2.0
$P(x)$		−	−	−	−	−	+					

x		1.40	1.41	1.42	1.43	1.44	1.45	1.46	1.47	1.48	1.49	1.5
$P(x)$		−	−	−	−	+						

1.43

13.

$-2x^3 - x^2 + 3x - 4 = 0$

x		−2.0	−1.9	−1.8	−1.7	−1.6	−1.5	−1.4	−1.3	−1.2	−1.1	−1.0
$P(x)$		+	+	−								

x		−1.90	−1.89	−1.88	−1.87	−1.86	−1.85	−1.84	−1.83	−1.82	−1.81	−1.8
$P(x)$		+	+	+	−							

−1.87

15.

$2x^5 - 3x^2 - 5 = 0$

x		1.0	1.1	1.2	1.3	1.4	1.5	1.6	1.7	1.8	1.9	2.0
$P(x)$		−	−	−	−	−	+					

x		1.40	1.41	1.42	1.43	1.44	1.45	1.46	1.47	1.48	1.49	1.5
$P(x)$		−	+									

1.40

17.

$2x^5 - x^4 + x^2 - 3 = 0$

x		1.0	1.1	1.2	1.3	1.4	1.5	1.6	1.7	1.8	1.9	2.0
$P(x)$		−	−	+								

x		1.10	1.11	1.12	1.13	1.14	1.15	1.16	1.17	1.18	1.19	1.2
$P(x)$		−	+									

1.10

19.

$$P(x) = 2x^4 - x^3 + 2x - 2 = 0$$

$P(-2) > 0$	$P(-1.5) > 0$	$P(-1) < 0$
$P(-1.5) > 0$	$P(-1.25) > 0$	$P(-1) < 0$
$P(-1.25) > 0$	$P(-1.13) > 0$	$P(-1) < 0$
$P(-1.13) > 0$	$P(-1.07) < 0$	$P(-1) < 0$
$P(-1.13) > 0$	$P(-1.10) > 0$	$P(-1.07) < 0$

-1.10

21.

$$P(x) = x^5 - 3x^3 + x^2 - 3 = 0$$

$P(1) < 0$	$P(1.5) < 0$	$P(2) > 0$
$P(1.5) < 0$	$P(1.75) > 0$	$P(2) > 0$
$P(1.5) < 0$	$P(1.63) < 0$	$P(1.75) > 0$
$P(1.63) < 0$	$P(1.69) < 0$	$P(1.75) > 0$
$P(1.69) < 0$	$P(1.72) < 0$	$P(1.75) > 0$
$P(1.72) < 0$	$P(1.74) > 0$	$P(1.75) > 0$
$P(1.72) < 0$	$P(1.73) < 0$	$P(1.74) > 0$

1.73

23.

$$P(x) = x^6 - 3x^3 + x^2 - 2 = 0$$

$P(1) < 0$	$P(1.5) > 0$	$P(2) > 0$
$P(1) < 0$	$P(1.25) < 0$	$P(1.5) > 0$
$P(1.25) < 0$	$P(1.38) < 0$	$P(1.5) > 0$
$P(1.38) < 0$	$P(1.44) > 0$	$P(1.5) > 0$
$P(1.38) < 0$	$P(1.41) < 0$	$P(1.44) > 0$
$P(1.41) < 0$	$P(1.43) < 0$	$P(1.44) > 0$

1.43

25.

$P(x) = -2x^3 - x^2 + 3x - 4 = 0$

$P(-2) > 0$	$P(-1.5) < 0$	$P(-1) < 0$
$P(-2) > 0$	$P(-1.75) < 0$	$P(-1.5) < 0$
$P(-2) > 0$	$P(-1.88) > 0$	$P(-1.75) < 0$
$P(-1.88) > 0$	$P(-1.82) < 0$	$P(-1.75) < 0$
$P(-1.88) > 0$	$P(-1.85) < 0$	$P(-1.82) < 0$
$P(-1.88) > 0$	$P(-1.87) < 0$	$P(-1.85) < 0$

-1.87

27.

$P(x) = 2x^5 - 3x^2 - 5 = 0$

$P(1) < 0$	$P(1.5) > 0$	$P(2) > 0$
$P(1) < 0$	$P(1.25) < 0$	$P(1.5) > 0$
$P(1.25) < 0$	$P(1.38) < 0$	$P(1.5) > 0$
$P(1.38) < 0$	$P(1.44) > 0$	$P(1.5) > 0$
$P(1.38) < 0$	$P(1.41) > 0$	$P(1.44) > 0$
$P(1.38) < 0$	$P(1.40) < 0$	$P(1.41) > 0$

1.40

29.

$P(x) = 2x^5 - x^4 + x^2 - 3 = 0$

$P(1) < 0$	$P(1.5) > 0$	$P(2) > 0$
$P(1) < 0$	$P(1.25) > 0$	$P(1.5) > 0$
$P(1) < 0$	$P(1.13) > 0$	$P(1.25) > 0$
$P(1) < 0$	$P(1.07) < 0$	$P(1.13) > 0$
$P(1.07) < 0$	$P(1.10) < 0$	$P(1.13) > 0$
$P(1.10) < 0$	$P(1.12) > 0$	$P(1.13) > 0$
$P(1.10) < 0$	$P(1.11) > 0$	$P(1.12) > 0$

1.10

In Exercises 31-41, use the ZOOM-IN method on your graphics calculator to check the answers you found in Exercises 7-17.

CHAPTER 4 REVIEW EXERCISES

1.

$$\begin{aligned} f(x) &= -x^2 - 4x \\ &= -(x^2 + 4x \quad) \\ &= -(x^2 + 4x + 4) + 4 \\ &= -(x+2)^2 + 4 \end{aligned}$$

Vertex: $(-2, 4)$

x-intercepts

Let $y = f(x) = 0$

$0 = -x^2 - 4x$

$0 = -x(x+4)$

$x = 0, \quad x = -4$

y-intercept

Let $x = 0$

$y = f(0) = -0^2 - 4(0)$

$y = 0$

2.

$$\begin{aligned} f(x) &= x^2 - 5x + 7 \\ &= (x^2 - 5x \quad) + 7 \\ &= \left(x^2 - 5x + \frac{25}{4}\right) + 7 - \frac{25}{4} \\ &= \left(x - \frac{5}{2}\right)^2 + \frac{3}{4} \end{aligned}$$

Vertex: $\left(\frac{5}{2}, \frac{3}{4}\right)$

x-intercepts

Let $y = f(x) = 0$

$0 = x^2 - 5x + 7$

$$x = \frac{-(-5) \pm \sqrt{(-5)^2 - 4(1)(7)}}{2(1)}$$

$$= \frac{5 \pm \sqrt{-3}}{2}$$

No x-intercepts

y-intercept

Let $x = 0$

$y = f(0) = 0^2 - 5(0) + 7$

$y = 7$

3.

$f(x)=2x^2-x+1$

3a).

Since $a=2>0$, f has a minimum.

3b).

The minimum occurs at

$x=\dfrac{-b}{2a}=\dfrac{-(-1)}{2(2)}=\dfrac{1}{4}.$

3c).

The minimum value is

$f\left(\dfrac{1}{4}\right)=2\left(\dfrac{1}{4}\right)^2-\dfrac{1}{4}+1=\dfrac{7}{8}.$

4.

$f(x)=-x^2-3x-1$

4a).

Since $a=-1<0$, f has a maximum.

4b).

The maximum occurs at

$x=\dfrac{-b}{2a}=\dfrac{-(-3)}{2(-1)}=-\dfrac{3}{2}.$

4c).

The maximum value is

$f\left(-\dfrac{3}{2}\right)=-\left(-\dfrac{3}{2}\right)^2-3\left(-\dfrac{3}{2}\right)-1=\dfrac{5}{4}.$

	Polynomial function	Leading term	Sign of leading coefficients a_n, degree: odd / even	Large values of $\lvert x\rvert$, $x>0$	Large values of $\lvert x\rvert$, $x<0$
5.	$P(x)=-2x^5+27x^2+100$	$-2x^5$	$a_n<0$ degree odd	D	U
6.	$P(x)=4x^3-10{,}000$	$4x^3$	$a_n>0$ degree odd	U	D

7.

$$\begin{array}{r|rrrr} 1 & 2 & 0 & 6 & -4 \\ & & 2 & 2 & 8 \\ \hline & 2 & 2 & 8 & \mid 4 \end{array}$$

$Q(x)=2x^2+2x+8;\ R=4$

8.

$$\begin{array}{r|rrrrr} -2 & 1 & -3 & 0 & 2 & -5 \\ & & -2 & 10 & -20 & 36 \\ \hline & 1 & -5 & 10 & -18 & \mid 31 \end{array}$$

$Q(x)=x^3-5x^2+10x-18;\ R=31$

9.

$$\begin{array}{r|rrrr} 2 & 7 & -3 & 0 & 2 \\ & & 14 & 22 & 44 \\ \hline & 7 & 11 & 22 & |\,46 \end{array}$$

$P(2) = 46$

$$\begin{array}{r|rrrr} -1 & 7 & -3 & 0 & 2 \\ & & -7 & 10 & -10 \\ \hline & 7 & -10 & 10 & |\,-8 \end{array}$$

$P(-1) = -8$

10.

$$\begin{array}{r|rrrrrr} 2 & 1 & 0 & -4 & 0 & 2 & 0 \\ & & 2 & 4 & 0 & 0 & 4 \\ \hline & 1 & 2 & 0 & 0 & 2 & |\,4 \end{array}$$

$P(2) = 4$

$$\begin{array}{r|rrrrrr} -1 & 1 & 0 & -4 & 0 & 2 & 0 \\ & & -1 & 1 & 3 & -3 & 1 \\ \hline & 1 & -1 & -3 & 3 & -1 & |\,1 \end{array}$$

$P(-1) = 1$

11.

$$\begin{array}{r|rrrrr} -2 & 2 & 4 & 3 & 5 & -2 \\ & & -4 & 0 & -6 & 2 \\ \hline & 2 & 0 & 3 & -1 & |\,0 \end{array}$$

Since $P(-2) = 0$, $x+2$ is a factor.

12.

$$\begin{array}{r|rrrr} \frac{1}{2} & 2 & -5 & 6 & -2 \\ & & 1 & -2 & 2 \\ \hline & 2 & -4 & 4 & |\,0 \end{array}$$

Since $P\left(\frac{1}{2}\right) = 0$, $x - \frac{1}{2}$ is a factor.

13.

$$\begin{aligned} \frac{3-2i}{4+3i} &= \frac{3-2i}{4+3i} \cdot \frac{4-3i}{4-3i} \\ &= \frac{12-17i+6i^2}{4^2+3^2} \\ &= \frac{6-17i}{25} \\ &= \frac{6}{25} - \frac{17}{25}i \end{aligned}$$

14.

$$\begin{aligned} \frac{2+i}{-5i} &= \frac{2+i}{-5i} \cdot \frac{5i}{5i} \\ &= \frac{10i+5i^2}{-25i^2} \\ &= \frac{-5+10i}{25} \\ &= -\frac{1}{5} + \frac{2}{5}i \end{aligned}$$

15.

$$\frac{-5}{1+i}=\frac{-5}{1+i}\cdot\frac{1-i}{1-i}$$
$$=\frac{-5+5i}{1^2+1^2}$$
$$=\frac{-5+5i}{2}$$
$$=-\frac{5}{2}+\frac{5}{2}i$$

16.

$$\frac{1}{1+3i}=\frac{1}{1+3i}\cdot\frac{1-3i}{1-3i}$$
$$=\frac{1-3i}{1^2+3^2}$$
$$=\frac{1-3i}{10}$$
$$=\frac{1}{10}-\frac{3}{10}i$$

17.

$$\frac{1}{-4i}=\frac{1}{-4i}\cdot\frac{4i}{4i}$$
$$=\frac{4i}{-16i^2}$$
$$=\frac{4i}{16}$$
$$=0+\frac{1}{4}i$$

18.

$$\frac{1}{2-5i}=\frac{1}{2-5i}\cdot\frac{2+5i}{2+5i}$$
$$=\frac{2+5i}{2^2+5^2}$$
$$=\frac{2+5i}{29}$$
$$=\frac{2}{29}+\frac{5}{29}i$$

19.

$$P(x)=(x+3)(x+2)(x+1)$$
$$=x^3+6x^2+11x+6$$

20.

$$\pm\sqrt{-3}=\pm\sqrt{3}\,i$$
$$P(x)=(x-3)(x-\sqrt{3}\,i)(x+\sqrt{3}\,i)$$
$$=(x-3)(x^2+3)$$
$$=x^3-3x^2+3x-9$$

21.

$$P(x)=(x+2)(x-\sqrt{3})(x+\sqrt{3})(x-1)$$
$$=x^4+x^3-5x^2-3x+6$$

22.

The number $\frac{1}{2}$ is a zero of the linear factor $(2x-1)$, and -1 is a zero of the linear factor $x+1$.

$$P(x)=(2x-1)(2x-1)(x+1)(x+1)$$
$$=4x^4+4x^3-3x^2-2x+1$$

23.

$$P(x) = (x-i)(x-i)(x+i)(x+i)$$
$$= x^4 + 2x^2 + 1$$

24.

$$P(x) = (x+1)(x+1)(x+1)(x-3)$$
$$= x^4 - 6x^2 - 8x - 3$$

25.

$$\begin{array}{r|rrrr} -2 & 2 & -1 & -13 & -6 \\ & & -4 & 10 & 6 \\ \hline & 2 & -5 & -3 & |\,0 \end{array}$$

$$2x^2 - 5x - 3 = 0$$
$$(2x+1)(x-3) = 0$$
$$x = -\frac{1}{2},\ x = 3$$

26.

$$\begin{array}{r|rrrr} 4 & 1 & -2 & -9 & 4 \\ & & 4 & 8 & -4 \\ \hline & 1 & 2 & -1 & |\,0 \end{array}$$

$$x^2 + 2x - 1 = 0$$
$$x = \frac{-2 \pm \sqrt{2^2 - 4(1)(-1)}}{2(1)}$$
$$x = -1 \pm \sqrt{2}$$

27.

$$\begin{array}{r|rrrrr} -\frac{1}{2} & 2 & -15 & 34 & -19 & -20 \\ & & -1 & 8 & -21 & 20 \\ \hline & 2 & -16 & 42 & -40 & |\,0 \end{array}$$

$$2x^3 - 16x^2 + 42x - 40 = 0$$
$$x^3 - 8x^2 + 21x - 20 = 0$$
$$(x-4)(x^2 - 4x + 5) = 0$$
$$x = 4,\ x = \frac{-(-4) \pm \sqrt{(-4)^2 - 4(1)(5)}}{2(1)} = 2 \pm i$$

28.

$$P(x) = x^4 - 2x - 1$$

1 positive root

$$P(-x) = x^4 + 2x - 1$$

1 negative root

29.

$$P(x) = x^5 - x^4 + 3x^3 - 4x^2 + x - 5$$

maximu m of 5 positive roots

$$P(-x) = -x^5 - x^4 - 3x^3 - 4x^2 - x - 5$$

no sign variations, hence no negative roots

30.

$$P(x) = x^3 - 5$$

1 positive root

$$P(-x) = -x^3 - 5$$

no sign variations, hence no negative roots

31.

$P(x) = 3x^4 - 2x^2 + 1$

maximu m of 2 positive roots

$P(-x) = 3x^4 - 2x^2 + 1$

maximu m of 2 negative roots

32.

$P(x) = 6x^3 - 5x^2 - 33x - 18$

possible num erators: $\pm 1, \pm 2, \pm 3, \pm 6, \pm 9, \pm 18$ (factors of 18)

possible denominat ors: $\pm 1, \pm 2, \pm 3, \pm 6$ (factors of 6)

possible rational roots: $\pm 1, \pm 2, \pm 3, \pm 6, \pm 9, \pm 18, \pm\frac{1}{2}, \pm\frac{3}{2}, \pm\frac{9}{2}, \pm\frac{1}{3}, \pm\frac{2}{3}, \pm\frac{1}{6}$

$$\begin{array}{r|rrrr} 3 & 6 & -5 & -33 & -18 \\ & & 18 & 39 & 18 \\ \hline & 6 & 13 & 6 & 0 \end{array}$$

$$\begin{array}{r|rrr} -\frac{2}{3} & 6 & 13 & 6 \\ & & -4 & -6 \\ \hline & 6 & 9 & 0 \end{array}$$

$6x + 9 = 0$

$x = -\frac{3}{2}$

roots: $3, -\frac{2}{3}, -\frac{3}{2}$

33.

$$P(x) = 6x^4 - 7x^3 - 19x^2 + 32x - 12$$

possible numerators: $\pm 1, \pm 2, \pm 3, \pm 4, \pm 6, \pm 12$ (factors of 12)

possible denominators: $\pm 1, \pm 2, \pm 3, \pm 6$ (factors of 6)

possible rational roots: $\pm 1, \pm 2, \pm 3, \pm 4, \pm 6, \pm 12, \pm\frac{1}{2}, \pm\frac{1}{3}, \pm\frac{1}{6}, \pm\frac{2}{3}, \pm\frac{3}{2}, \pm\frac{4}{3}$

$$\begin{array}{r|rrrrr} 1 & 6 & -7 & -19 & 32 & -12 \\ & & 6 & -1 & -20 & 12 \\ \hline & 6 & -1 & -20 & 12 & |\,0 \end{array}$$

$$\begin{array}{r|rrrr} -2 & 6 & -1 & -20 & 12 \\ & & -12 & 26 & -12 \\ \hline & 6 & -13 & 6 & |\,0 \end{array}$$

$$\begin{array}{r|rrr} \frac{2}{3} & 6 & -13 & 6 \\ & & 4 & -6 \\ \hline & 6 & -9 & |\,0 \end{array}$$

$$6x - 9 = 0$$

$$x = \frac{3}{2}$$

roots: $1, -2, \frac{2}{3}, \frac{3}{2}$

34.
$P(x) = x^4 + 3x^3 + 2x^2 + x - 1$

possible numerators: ± 1 (factors of 1)
possible denominators: ± 1 (factors of 1)
possible rational roots: ± 1

$$\begin{array}{r|rrrrr} 1 & 1 & 3 & 2 & 1 & -1 \\ & & 1 & 4 & 6 & 7 \\ \hline & 1 & 4 & 6 & 7 & |\,6 \end{array}$$

$R \neq 0$, $x = 1$ is not a root

$$\begin{array}{r|rrrrr} -1 & 1 & 3 & 2 & 1 & -1 \\ & & -1 & -2 & 0 & -1 \\ \hline & 1 & 2 & 0 & 1 & |\,-2 \end{array}$$

$R \neq 0$, $x = -1$ is not a root

Hence, $P(x)$ has no rational roots.

35.

$P(x) = 6x^3 + 15x^2 - x - 10$

possible numerators: $\pm 1, \pm 2, \pm 5, \pm 10$ (factors of 10)

possible denominators: $\pm 1, \pm 2, \pm 3, \pm 6$ (factors of 6)

possible rational roots: $\pm 1, \pm 2, \pm 5, \pm 10, \pm \frac{1}{2}, \pm \frac{5}{2}, \pm \frac{1}{3}, \pm \frac{2}{3}, \pm \frac{5}{3}, \pm \frac{10}{3}, \pm \frac{1}{6}, \pm \frac{5}{6}$

$$\begin{array}{r|rrrr} -1 & 6 & 15 & -1 & -10 \\ & & -6 & -9 & 10 \\ \hline & 6 & 9 & -10 & \mid 0 \end{array}$$

$6x^2 + 9x - 10 = 0$

$$x = \frac{-9 \pm \sqrt{9^2 - 4(6)(-10)}}{2(6)} = \frac{-9 \pm \sqrt{321}}{12}$$

roots: $-1, \frac{-9 \pm \sqrt{321}}{12}$

36.

$P(x) = 2x^4 - 3x^3 - 10x^2 + 19x - 6$

possible numerators: $\pm 1, \pm 2, \pm 3, \pm 6$ (factors of 6)

possible denominators: $\pm 1, \pm 2$ (factors of 2)

possible rational roots: $\pm 1, \pm 2, \pm 3, \pm 6, \pm \frac{1}{2}, \pm \frac{3}{2}$

$$\begin{array}{r|rrrrr} 2 & 2 & -3 & -10 & 19 & -6 \\ & & 4 & 2 & -16 & 6 \\ \hline & 2 & 1 & -8 & 3 & |\ 0 \end{array}$$

$$\begin{array}{r|rrrr} \frac{3}{2} & 2 & 1 & -8 & 3 \\ & & 3 & 6 & -3 \\ \hline & 2 & 4 & -2 & |\ 0 \end{array}$$

$$2x^2 + 4x - 2 = 0$$
$$x^2 + 2x - 1 = 0$$
$$x = \frac{-2 \pm \sqrt{2^2 - 4(1)(-1)}}{2(1)} = -1 \pm \sqrt{2}$$

roots: $2, \frac{3}{2}, -1 \pm \sqrt{2}$

CHAPTER 4 REVIEW TEST

1.

$$
\begin{aligned}
y &= 3x^2 - 2x + 1 \\
&= 3\left(x^2 - \frac{2}{3}x \qquad\right) + 1 \\
&= 3\left(x^2 - \frac{2}{3}x + \frac{1}{9}\right) + 1 - \frac{1}{3} \\
&= 3\left(x - \frac{1}{3}\right)^2 + \frac{2}{3}
\end{aligned}
$$

Vertex: $\left(\frac{1}{3}, \frac{2}{3}\right)$

minimum value on graph: $\frac{2}{3}$

x-intercepts:

Let $y = 0$

$0 = 3x^2 - 2x + 1$

$$x = \frac{-(-2) \pm \sqrt{(-2)^2 - 4(3)(1)}}{2(3)}$$

$$x = \frac{2 \pm \sqrt{-8}}{6}$$

no x-intercepts

y-intercept

Let $x = 0$

$y = 3(0)^2 - 2(0) + 1 = 1$

$y = 1$

2.

$P(x) = -2x^9 + 3x^6 + 200$

$a_n = -2 < 0$

degree: 9

For large values of $|x|$, $x > 0$, the graph extends infinitely downward.

3.

$P(x) = -2x^9 + 3x^6 + 200$

$a_n = -2 < 0$

degree: 9

For large values of $|x|$, $x < 0$, the graph extends infinitely upward.

4.

$$\begin{array}{r} 2x^2 - 5 \\ x^2+2\,\overline{)\,2x^4 + 0x^3 - x^2 + 0x + 1} \\ \underline{2x^4 \qquad\quad + 4x^2 \qquad\qquad} \\ -5x^2 + 0x + 1 \\ \underline{-5x^2 \qquad - 10} \\ 11 \end{array}$$

$Q(x) = 2x^2 - 5, \;\; R = 11$

5.

$$\begin{array}{r|rrrrr} -2 & 3 & -1 & 0 & 0 & -2 \\ & & -6 & 14 & -28 & 56 \\ \hline & 3 & -7 & 14 & -28 & |\,54 \end{array}$$

$Q(x) = 3x^3 - 7x^2 + 14x - 28, \;\; R = 54$

6.

$$\begin{array}{r|rrrr} -2 & 1 & -2 & 7 & 5 \\ & & -2 & 8 & -30 \\ \hline & 1 & -4 & 15 & |\,-25 \end{array}$$

$P(-2) = -25$

7.

$$\begin{array}{r|rrrrrr} -2 & 4 & -2 & 0 & 0 & 0 & -5 \\ & & -8 & 20 & -40 & 80 & -160 \\ \hline & 4 & -10 & 20 & -40 & 80 & |\,-165 \end{array}$$

$R = -165$

8.

$$\begin{array}{r|rrrrr} 3 & 2 & -9 & 9 & 1 & -3 \\ & & 6 & -9 & 0 & 3 \\ \hline & 2 & -3 & 0 & 1 & |\,0 \end{array}$$

$P(3) = 0$, hence by the Factor Theorem $x - 3$ is a factor.

9.

$$\begin{aligned} P(x) &= (x+2)(x-1)(x-3) \\ &= x^3 - 2x^2 - 5x + 6 \end{aligned}$$

10.

$$\begin{aligned} P(x) &= (x+1)(x-1)\left[x-\left(3+\sqrt{2}\right)\right]\left[x-\left(3-\sqrt{2}\right)\right] \\ &= x^4 - 6x^3 + 6x^2 + 6x - 7 \end{aligned}$$

11.

$$(x^2+1)(x-2) = 0$$

$$\begin{array}{ll} x^2 + 1 = 0 & \qquad x - 2 = 0 \\ \quad x = \pm i & \qquad\quad x = 2 \end{array}$$

12.

$(x+1)^2(x^2-3x-2)=0$

$(x+1)^2=0 \qquad x^2-3x-2=0$

$x=-1,\ x=-1 \qquad x=\dfrac{-(-3)\pm\sqrt{(-3)^2-4(1)(-2)}}{2(1)}$

$=\dfrac{3\pm\sqrt{17}}{2}$

13.

$P(x)=(x+3)(x+3)(x-1)(x-1)(x-1)$

$=x^5+3x^4-6x^3-10x^2+21x-9$

14.

$-\dfrac{1}{4}$ is a zero of the linear factor $(4x+1)$

$P(x)=(4x+1)(4x+1)(x-i)(x+i)(x-1)$

$=16x^5-8x^4+9x^3-9x^2-7x-1$

15.

$P(x)=(x-i)[x-(1+i)]$

$=x^2-(1+2i)x+(i-1)$

16.

$\underline{-1}\rfloor$	4	0	−3	1
		−4	4	−1
	4	−4	1	| 0

$4x^2-4x+1=0$

$(2x-1)(2x-1)=0$

$x=\dfrac{1}{2},\quad x=\dfrac{1}{2}$

17.

$$\begin{array}{r|rrrrr} 1 & 1 & 0 & -1 & -2 & 2 \\ & & 1 & 1 & 0 & -2 \\ \hline & 1 & 1 & 0 & -2 & |\,0 \end{array}$$

$$\begin{array}{r|rrrr} 1 & 1 & 1 & 0 & -2 \\ & & 1 & 2 & 2 \\ \hline & 1 & 2 & 2 & |\,0 \end{array}$$

$x^2 + 2x + 2 = 0$

$$x = \frac{-2 \pm \sqrt{2^2 - 4(1)(2)}}{2(1)} = -1 \pm i$$

remaining roots: $1,\ -1 \pm i$

18.

If $2+i$ is a root then $2-i$ is a root.

There is one positive root.

$$[x-(2+i)][x-(2-i)] = x^2 - 4x + 5$$

$$\begin{array}{r} x-2 \\ x^2-4x+5 \overline{\big) x^3 - 6x^2 + 13x - 10} \\ \underline{x^3 - 4x^2 + 5x} \\ -2x^2 + 8x - 10 \\ \underline{-2x^2 + 8x - 10} \\ 0 \end{array}$$

$$P(x) = (x-2)(x^2 - 4x + 5)$$

19.

$P(x) = 2x^5 - 3x^4 + 1$

maximum number of positive real roots is 2.

20.

$P(-x) = 3x^4 - 2x^3 - 2x^2 - 1$

maximum number of negative real roots is 1.

21.

$P(x) = 6x^3 - 17x^2 + 14x + 3$

possible numerators: $\pm 1, \pm 3$ (factors of 3)

possible denominators: $\pm 1, \pm 2, \pm 3, \pm 6$ (factors of 6)

possible rational roots: $\pm 1, \pm 3, \pm \frac{1}{2}, \pm \frac{3}{2}, \pm \frac{1}{3}, \pm \frac{1}{6}$

Using synthetic division we see that none of the possible rational roots leave a remainder of 0, hence $P(x)$ has no rational roots.

22.

$P(x)=2x^5-x^4-4x^3+2x^2+2x-1$

possible numerators: ± 1 (factors of 1)

possible denominators: $\pm 1, \pm 2$ (factors of 2)

possible rational roots: $\pm 1, \pm\frac{1}{2}$

$$\begin{array}{r|rrrrrr} 1 & 2 & -1 & -4 & 2 & 2 & -1 \\ & & 2 & 1 & -3 & -1 & 1 \\ \hline & 2 & 1 & -3 & -1 & 1 & |\,0 \end{array}$$

$$\begin{array}{r|rrrrr} 1 & 2 & 1 & -3 & -1 & 1 \\ & & 2 & 3 & 0 & -1 \\ \hline & 2 & 3 & 0 & -1 & |\,0 \end{array}$$

$$\begin{array}{r|rrrr} -1 & 2 & 3 & 0 & -1 \\ & & -2 & -1 & 1 \\ \hline & 2 & 1 & -1 & |\,0 \end{array}$$

$$\begin{array}{r|rrr} -1 & 2 & 1 & -1 \\ & & -2 & 1 \\ \hline & 2 & -1 & |\,0 \end{array}$$

$2x-1=0$

$x=\frac{1}{2}$

roots: $1, 1, -1, -1, \frac{1}{2}$

23.

$P(x) = 3x^4 + 7x^3 - 3x^2 + 7x - 6$

possible numerators: $\pm 1, \pm 2, \pm 3, \pm 6$ (factors of 6)

possible denominators: $\pm 1, \pm 3$ (factors of 3)

possible rational roots: $\pm 1, \pm 2, \pm 3, \pm 6, \pm \frac{1}{3}, \pm \frac{2}{3}$

$$\begin{array}{r|rrrrr} -3 & 3 & 7 & -3 & 7 & -6 \\ & & -9 & 6 & -9 & 6 \\ \hline & 3 & -2 & 3 & -2 & \vert\, 0 \end{array}$$

$$\begin{array}{r|rrrr} \frac{2}{3} & 3 & -2 & 3 & -2 \\ & & 2 & 0 & 2 \\ \hline & 3 & 0 & 3 & \vert\, 0 \end{array}$$

$3x^2 + 3 = 0$

$x = \pm i$

roots: $-3, \frac{2}{3}, \pm i$

CHAPTER 4 ADDITIONAL PRACTICE EXERCISES

1. Write the reciprocal of $3-2i$ in the form $a+bi$.

2. Find the vertex and all intercepts of the parabola $f(x)=x^2+2x-15$.

3. Use synthetic division to find the quotient $Q(x)$ and the constant remainder R when $8x^4+3x^3-2x+6$ is divided by $x-2$.

4. Given $P(x)=3x^5-2x^3+7x+3$, use synthetic division to find $P(-2)$.

5. Use the Factor Theorem to show that $x-3$ is a factor of $2x^4-4x^3-x^2-45$.

6. Write $\dfrac{7-i}{-3i}$ in the form $a+bi$.

7. Find a polynomial of lowest degree that has zeros, 2, -4, 5.

8. Use Descartes' Rule of Signs to determine the maximum number of positive and negative real roots of $x^6-2x^5+13x^4+2x^2-6x-10=0$

9. Find all the rational roots of $x^5+2x^4-5x^3-10x^2+4x+8$.

10. Write $4x^5-3x^4+2x^2-x+7$ in nested form.

CHAPTER 4 PRACTICE TEST

1.
Find the vertex and intercepts of the parabola whose equation is $y = 2x^2 - 5x - 12$.

2.
Find two positive numbers such that their sum is 60 and their product is a maximum.

3.
$f(x) = -x^2 + 8x - 14$, determine

3a).
if f has a maximum value or a minimum value:

3b).
the value of x at which the maximum or minimum occurs:

3c).
the maximum or minimum value of f.

4.
Show that $2x^4 - 3x^3 + 3x - 1 = 0$ has a root in the interval $[0,1]$.

5.
Determine the behavior of the graph of $P(x) = -5x^7 - 4x^3 + 6x^2 - 1$ for large values of $|x|$, $x > 0$.

6.
Find the quotient and remainder when $5x^4 - 3x^3 + 4x^2 - 6$ is divided by $x^2 - 1$.

7.
Use synthetic division to find the quotient and remainder when $x^5 - 5x^4 - 4x^3 + 3x^2 - 115x - 1$ is divided by $x + 3$.

8.
Use synthetic division to find the quotient and remainder when $x^6 + 1$ is divided by $x - 1$.

9.
Use the Remainder Theorem and synthetic division to find $P(-3)$, given $P(x) = x^4 - 2x^3 + 3x^2 - 1$.

10.
Use the Factor Theorem to decide whether or not $x - 4$ is a factor of $x^3 - 2x^2 + x - 1$.

11.
Determine all zeros of
$f(x)=(x-3)(5-x)(3x-2)^3$.

12.
Determine the values of k for which
$x-3$ is a factor of x^3+kx^2+x+6.

13.
Perform the division $\frac{2-i}{6+5i}$ and write
the answer in the form $a+bi$.

14.
Find the reciprocal of $\frac{1}{3}+i$ and write
the answer in the form $a+bi$.

15.
Find a polynomial $P(x)$ of lowest degree
with zeros $2, -2, \sqrt{3}, -\sqrt{3}$.

16.
Use the given root, 5, to help in finding
the remaining roots of
$x^4-4x^3-14x^2+36x+45$.

17.
Use Descartes' Rule of Signs to determine
the maximum number of negative real
roots of $5x^6-7x^4+3x^3-2x+11$.

18.
Find all roots of the equation
$x^5+3x^4+8x^3+24x^2-9x-27=0$

19.
Write the polynomial $9x^4-8x^3+6x-7$
in nested form.

20.
Find a root of $x^3-3x^2+1=0$ in the interval
$[0, 1]$ to two decimal places.

CHAPTER 5 SECTION 1

SECTION 5.1 PROGRESS CHECK

1a). (Page 287)

$$S(x)=\frac{x-3}{2x^2-3x-2}$$

$$2x^2-3x-2=0$$
$$(2x+1)(x-2)=0$$
$$x=-\frac{1}{2},\quad x=2$$

Domain: all real numbers except $x=-\frac{1}{2}$, $x=2$

y – intercept: Let $x=0$.

$$y=S(0)=\frac{0-3}{2(0)^2-3(0)-2}=\frac{3}{2}$$
$$y=\frac{3}{2}$$

x – intercept: Let $y=S(x)=0$.

$$0=\frac{x-3}{2x^2-3x-2}$$
$$0=x-3$$
$$x=3$$

1b).

$$T(x)=\frac{5}{x^4+x^2+5}$$

$x^4+x^2+5\geq 5$ The denominator is never 0.

Domain: all real numbers

y-intercept: Let $x=0$.

$$y=T(0)=\frac{5}{0^4+0^2+5}=1$$

$y=1$

x-intercept: Let $y=T(x)=0$.

$$0=\frac{5}{x^4+x^2+5}$$

$0=5$

no x – intercepts

2a). (Page 293)

$$f(x)=\frac{x-1}{2x^2+1}$$

$$=\frac{x\left(1-\frac{1}{x}\right)}{x^2\left(2+\frac{1}{x^2}\right)}$$

$$=\frac{1-\frac{1}{x}}{x\left(2+\frac{1}{x^2}\right)}$$

$-\frac{1}{x}$, $\frac{1}{x^2}$ approach 0 as $|x|$ approaches ∞.

The factor of x in the denominator causes the denominator to approach ∞ as $|x|$ approaches ∞.

horizontal asymptote: $y=0$

2b).

$$g(x)=\frac{4x^2-3x+1}{-3x^2+1}$$

$$=\frac{x^2\left(4-\frac{3}{x}+\frac{1}{x^2}\right)}{x^2\left(-3+\frac{1}{x^2}\right)}$$

$$=\frac{4-\frac{3}{x}+\frac{1}{x^2}}{-3+\frac{1}{x^2}}$$

$-\frac{3}{x}$, $\frac{1}{x^2}$, $\frac{1}{x^2}$ approach 0 as $|x|$ approaches ∞.

horizontal asymptote: $y=\frac{4}{-3}$

$$y=-\frac{4}{3}$$

2c).

The degree of the numerator is 3 and the degree of the denominator is 2. Since the degree of numerator > degree of denominator, there is no horizontal asymptote.

3). (Page 295)

$$f(x)=\frac{x^2-x-6}{x^2-2x}$$

horizontal asymptotes: $f(x)=\dfrac{x^2\left(1-\frac{1}{x}-\frac{6}{x^2}\right)}{x^2\left(1-\frac{2}{x}\right)}$

$$=\frac{1-\frac{1}{x}-\frac{6}{x^2}}{1-\frac{2}{x}}$$

$-\dfrac{1}{x}$, $-\dfrac{6}{x^2}$, $-\dfrac{2}{x}$ approach 0 as $|x|$ approaches ∞.

horizontal asymptote: $y=\dfrac{1}{1}=1$

vertical asymptotes: $f(x)=\dfrac{(x-3)(x+2)}{x(x-2)}$

$$x=0,\ x-2=0$$

$$x=2$$

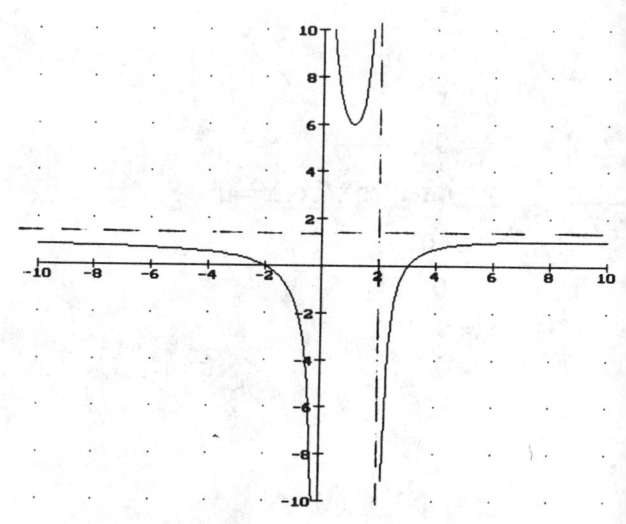

$y-$intercept: Let $x=0$.

$$y=\frac{0^2-0-6}{0^2-2(0)} \text{ undefined}$$

no y-intercept

$x-$intercepts: Let $y=f(x)=0$

$$0=\frac{x^2-x-6}{x^2-2x}$$

$$0=x^2-x-6$$

$$0=(x-3)(x+2)$$

$$x=3,\quad x=-2$$

4. (Page 296)

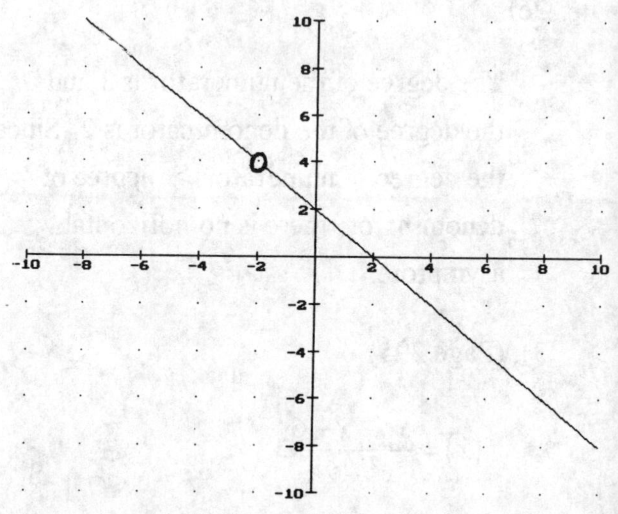

$$f(x)=\frac{4-x^2}{x+2}$$
$$=\frac{(2-x)(2+x)}{x+2}$$
$$=2-x,\quad x\neq -2$$

The graph of $f(x)$ coincides with the line $y=2-x$, with the exception that $f(x)$ is undefined at $x=-2$.

EXERCISE SET 5.1

1.

$$f(x)=\frac{x^2}{x-1}$$
$$x-1\neq 0$$
$$x\neq 1$$

Domain: $\{x|x\neq 1\}$

x-intercepts: Let $f(x)=0$

$$0=\frac{x^2}{x-1}$$
$$0=x^2$$
$$x=0$$

y-intercept: Let $x=0$

$$y=\frac{0^2}{0-1}$$
$$y=0$$

3.

$$g(x)=\frac{x^2+1}{x^2-2x}$$
$$x^2-2x\neq 0$$
$$x(x-2)\neq 0$$
$$x\neq 0 \quad x\neq 2$$

Domain: $\{x|\,x\neq 0,\ x\neq 2\}$

x-intercepts: Let $g(x)=0$

$$0=\frac{x^2+1}{x^2-2x}$$
$$0=x^2+1$$
$$x^2=-1$$

No solution - no x-intercepts

y-intercept: Since $x\neq 0$, there are no y-intercepts.

5.

$$f(x)=\frac{x^2-3}{x^2+3}$$

$$x^2+3\neq 0$$

$$x^2\neq -3$$

Domain: $\{x\,|-\infty<x<\infty\}$

$x-$intercepts: Let $f(x)=0$.

$$0=\frac{x^2-3}{x^2+3}$$

$$0=x^2-3$$

$$x^2=3$$

$$x=\pm\sqrt{3}$$

$y-$ intercept: Let $x=0$.

$$y=\frac{0^2-3}{0^2+3}$$

$$y=-1$$

In Exercises 7-27, the RANGE values given show important features of the functions. Other RANGE values are also acceptable. The display on your calculator may vary from that shown.

7.

vertical asymptote: $x-4=0$

$$x=4$$

$$f(x)=\frac{1}{x-4}$$

$$=\frac{x\left(\frac{1}{x}\right)}{x\left(1-\frac{4}{x}\right)}$$

$$=\frac{\frac{1}{x}}{1-\frac{4}{x}}$$

$\frac{1}{x}$, $-\frac{4}{x}$ approach 0 as $|x|$ approaches ∞.

horizontal asymptote: $y=\frac{0}{1-0}$

$$y=0$$

XMIN = –5
XMAX = 10
XSCL = 1
YMIN = –5
YMAX = 5
YSCL = 1
GRAPH $Y=1\div(X-4)$

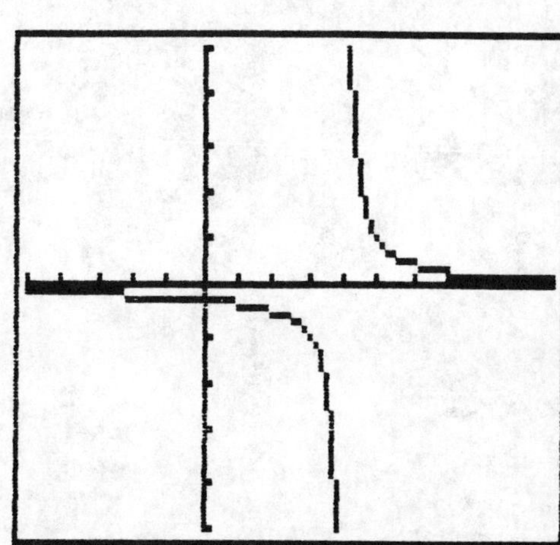

9.

vertical asymptote: $x+2=0$

$$x=-2$$

$$\begin{aligned} f(x) &= \frac{3}{x+2} \\ &= \frac{x\left(\frac{3}{x}\right)}{x\left(1+\frac{2}{x}\right)} \\ &= \frac{\frac{3}{x}}{1+\frac{2}{x}} \end{aligned}$$

$\frac{3}{x}, \frac{2}{x}$ approach 0 as $|x|$ approaches ∞

horizontal asymptote: $y=\frac{0}{1+0}=0$

XMIN = −6
XMAX = 4
XSCL = 1
YMIN = −5
YMAX = 5
YSCL = 1
GRAPH $Y=3\div(X+2)$

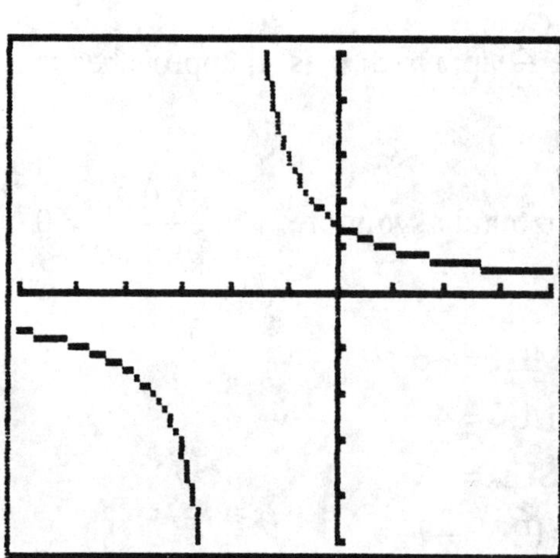

11.

vertical asymptote: $(x+1)^2 = 0$

$$x+1=0$$

$$x=-1$$

$$\begin{aligned} f(x) &= \frac{1}{(x+1)^2} \\ &= \frac{1}{x^2+2x+1} \\ &= \frac{x^2\left(\frac{1}{x^2}\right)}{x^2\left(1+\frac{2}{x}+\frac{1}{x^2}\right)} \\ &= \frac{\frac{1}{x^2}}{1+\frac{2}{x}+\frac{1}{x^2}} \end{aligned}$$

$\frac{1}{x^2}, \frac{2}{x}$ approach 0 as $|x|$ approaches ∞.

horizontal asymptote: $y = \frac{0}{1+0+0} = 0$

XMIN = −6
XMAX = 4
XSCL = 1
YMIN = −1
YMAX = 6
YSCL = 1
GRAPH $Y = 1 \div (X+1)^2$

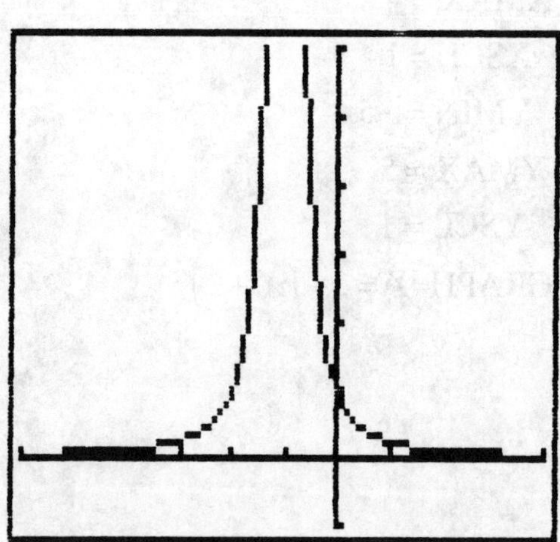

13.

vertical asymptote: $x - 2 = 0$

$$x = 2$$

$$f(x) = \frac{x+2}{x-2} = \frac{x\left(1+\frac{2}{x}\right)}{x\left(1-\frac{2}{x}\right)} = \frac{1+\frac{2}{x}}{1-\frac{2}{x}}$$

$\pm\frac{2}{x}$ approach 0 as $|x|$ approaches ∞.

horizontal asymptote: $y = \frac{1+0}{1-0} = 1$

XMIN = −10
XMAX = 10
XSCL = 1
YMIN = −5
YMAX = 5
YSCL = 1
GRAPH $Y = (X+2) \div (X-2)$

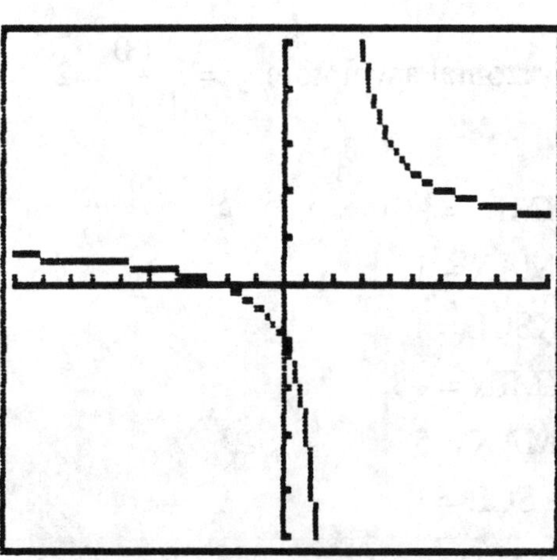

15.

vertical asymptotes: $x^2 - 4 = 0$

$$x^2 = 4$$

$$x = \pm 2$$

$$f(x) = \frac{2x^2 + 1}{x^2 - 4}$$

$$= \frac{x^2\left(2 + \frac{1}{x^2}\right)}{x^2\left(1 - \frac{4}{x^2}\right)}$$

$$= \frac{2 + \frac{1}{x^2}}{1 - \frac{4}{x^2}}$$

$\frac{1}{x^2}, -\frac{4}{x^2}$ approach 0 as $|x|$ approaches ∞.

horizontal asymptote: $y = \frac{2+0}{1-0} = 2$

XMIN = −10
XMAX = 10
XSCL = 1
YMIN = −5
YMAX = 5
YSCL = 1
GRAPH $Y = (2X^2 + 1) \div (X^2 - 4)$

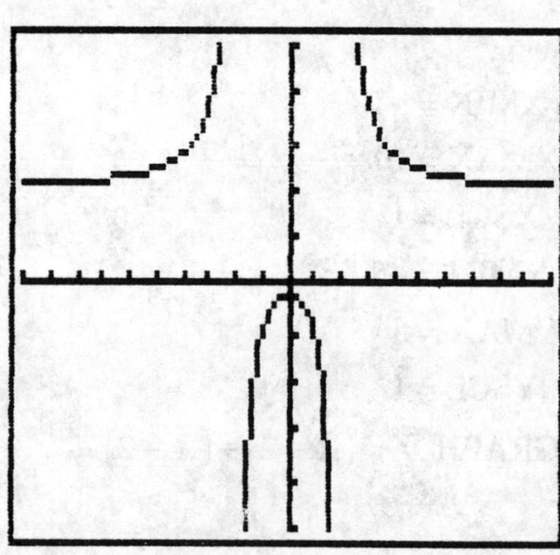

17.

vertical asymptotes: $2x^2 - x - 6 = 0$

$$(2x+3)(x-2) = 0$$

$$x = -\frac{3}{2} \qquad x = 2$$

$$f(x) = \frac{x^2+2}{2x^2-x-6}$$

$$= \frac{x^2\left(1+\frac{2}{x^2}\right)}{x^2\left(2-\frac{1}{x}-\frac{6}{x^2}\right)}$$

$$= \frac{1+\frac{2}{x^2}}{2-\frac{1}{x}-\frac{6}{x^2}}$$

$\frac{2}{x^2}, -\frac{1}{x}, -\frac{6}{x^2}$ approach 0 as $|x|$ approaches ∞.

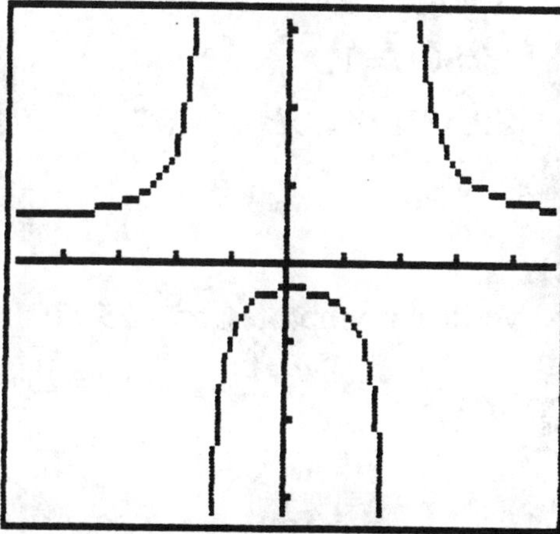

horizontal asymptote: $y = \frac{1+0}{2-0-0} = \frac{1}{2}$

EQUAL viewing rectangle

XSCL = 1

YSCL = 1

GRAPH $Y = (X^2+2) \div (2X^2 - X - 6)$

19.

vertical asymptote: $4x-4=0$

$$x=1$$

$$f(x)=\frac{x^2}{4x-4}$$

$$=\frac{1}{4}x+\frac{1}{4}+\frac{1}{4x-4}$$

oblique asymptote: $y=\frac{1}{4}x+\frac{1}{4}$

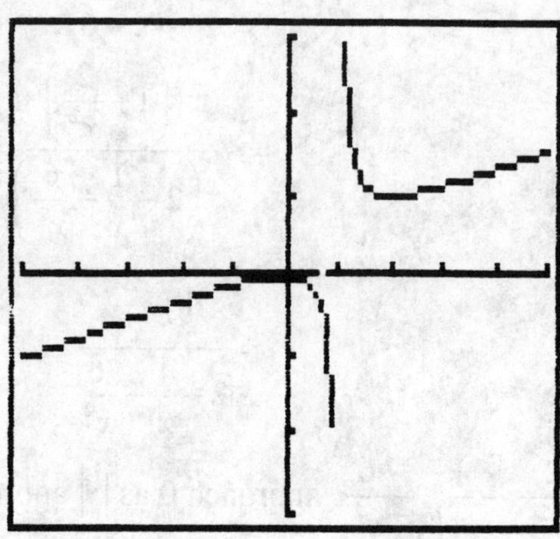

XMIN = −5

XMAX = 5

XSCL = 1

YMIN = −3

YMAX = 3

YSCL = 1

GRAPH $Y=X^2\div(4X-4)$

21.

vertical asymptotes: $x^2-25=0$

$$x^2=25$$

$$x=\pm 5$$

$$f(x)=\frac{x^3+4x^2+3x}{x^2-25}$$

$$=x+4+\frac{28x+100}{x^2-25}$$

oblique asymptote: $y=x+4$

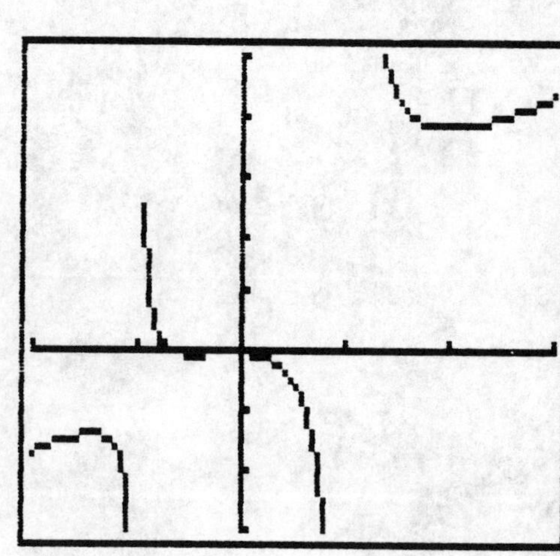

XMIN = −10

XMAX = 15

XSCL = 5

YMIN = −15

YMAX = 25

YSCL = 5

GRAPH $Y=(X^3+4X^2+3X)\div(X^2-25)$

23.

$x + 2 \neq 0$

$x \neq -2$

Domain: $\{x \mid x \neq -2\}$

XMIN, XMAX: EQUAL viewing rectangle

XSCL = 1

YMIN = −10

YMAX = 5

YSCL = 1

GRAPH $Y = (2X^2 - 8) \div (X + 2)$

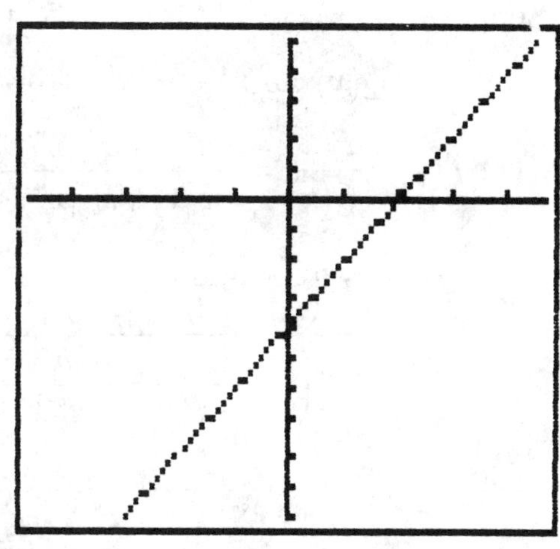

25.

$2x^2 - 8x + 8 \neq 0$

$x^2 - 4x + 4 \neq 0$

$(x - 2)^2 \neq 0$

$x \neq 2$

Domain: $\{x \mid x \neq 2\}$

XMIN, XMAX: four times the EQUAL viewing rectangle

XSCL = 5

YMIN = −5

YMAX = 5

YSCL = 1

GRAPH $Y = (X^2 + 2X - 8) \div (2X^2 - 8X + 8)$

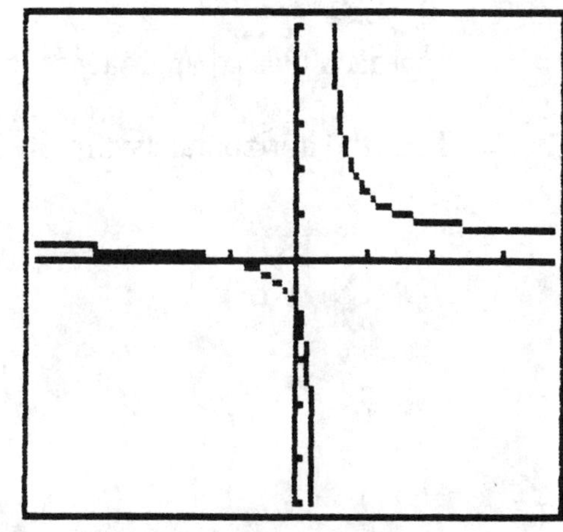

27.

$x^2 + x \neq 0$

$x(x + 1) \neq 0$

$x \neq 0 \quad x \neq -1$

Domain: $\{x \mid x \neq 0, x \neq -1\}$

EQUAL viewing rectangle

XSCL = 1

YSCL = 1

GRAPH $Y = 2X \div (X^2 + X)$

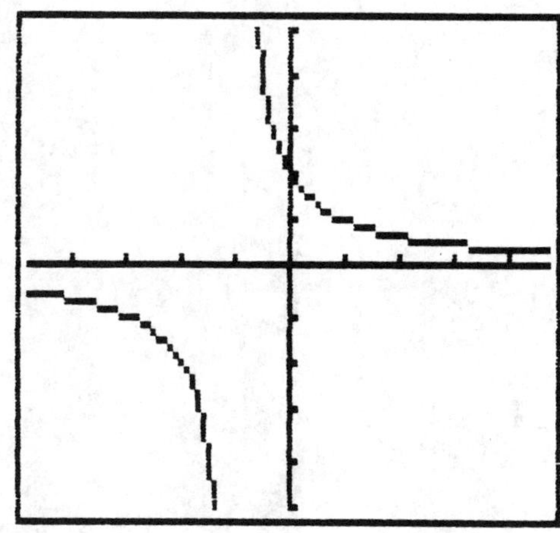

29.

Suppose $P(x) = a_n x^{n-k} + a_{n-1}x^{n-k-1} + \cdots + a_1 x + a_0,\ a_n \neq 0,\ k \geq 1$

$Q(x) = b_n x^n + b_{n-1}x^{n-1} + \cdots + b_1 x + b_0,\ b_n \neq 0$

$$\text{Then } f(x) = \frac{P(x)}{Q(x)} = \frac{a_n x^{n-k} + a_{n-1}x^{n-k-1} + \cdots + a_1 x + a_o}{b_n x^n + b_{n-1}x^{n-1} + \cdots + b_1 x + b_0}$$

$$= \frac{x^n\left(\frac{a_n}{x^k} + \frac{a_{n-1}}{x^{k+1}} + \cdots + \frac{a_1}{x^{n-1}} + \frac{a_o}{x^n}\right)}{x^n\left(b_n + \frac{b_{n-1}}{x} + \cdots + \frac{b_1}{x^{n-1}} + \frac{b_o}{x^n}\right)}$$

$$= \frac{\frac{a_n}{x^k} + \frac{a_{n-1}}{x^{k+1}} + \cdots + \frac{a_1}{x^{n-1}} + \frac{a_o}{x^n}}{b_n + \frac{b_{n-1}}{x} + \cdots + \frac{b_1}{x^{n-1}} + \frac{b_o}{x^n}}$$

$\frac{a_n}{x^k}, \frac{a_n - 1}{x^{k+1}}, \ldots, \frac{a_1}{x^{n-1}}, \frac{a_o}{x^n}, \frac{b_{n-1}}{x}, \ldots, \frac{b_1}{x^{n-1}} + \frac{b_o}{x^n}$

approach 0 as $|x|$ approaches ∞.

Thus the horizontal asymptote is $y = \frac{0 + 0 + \cdots + 0}{b_n + 0 + \cdots + 0} = 0$

CHAPTER 5 SECTION 2

SECTION 5.2 PROGRESS CHECK

1. (Page 302)

$$\left(x-\frac{1}{2}\right)^2+(y+5)^2=15$$
$$\left(x-\frac{1}{2}\right)^2+[y-(-5)]^2=\left(\sqrt{15}\right)^2$$

center: $\left(\frac{1}{2},-5\right)$

radius: $\sqrt{15}$

2. (Page 303)

$$4x^2+4y^2-8x+4y=103$$
$$4(x^2-2x)+4(y^2+y)=103$$
$$4(x^2-2x+1)+4\left(y^2+y+\frac{1}{4}\right)=103+4+1$$
$$4(x-1)^2+4\left(y+\frac{1}{2}\right)^2=108$$
$$(x-1)^2+\left(y+\frac{1}{2}\right)^2=27$$

center: $\left(1,-\frac{1}{2}\right)$

radius: $\sqrt{27}=3\sqrt{3}$

3. (Page 304)

$$x^2+y^2-12y+36=0$$
$$x^2+(y^2-12y)=-36$$
$$x^2+(y^2-12y+36)=-36+36$$
$$x^2+(y-6)^2=0$$

This equation represents a circle with center $(0, 6)$ and radius 0.
Since the radius is 0, this is actually just the point $(0, 6)$.

EXERCISE SET 5.2

1.

$$(x-2)^2+(y-3)^2=2^2$$
$$(x-2)^2+(y-3)^2=4$$

3.

$$[x-(-2)]^2+[y-(-3)]^2=\left(\sqrt{5}\right)^2$$
$$(x+2)^2+(y+3)^2=5$$

5.

$$(x-0)^2+(y-0)^2=3^2$$
$$x^2+y^2=9$$

7.

$$[x-(-1)]^2+(y-4)^2=\left(2\sqrt{2}\right)^2$$
$$(x+1)^2+(y-4)^2=8$$

In Exercises 9-23, the RANGE values given show important features of the functions. Other RANGE values are also acceptable. The display on your calculator may vary from that shown.

9.

$$(x-2)^2+(y-3)^2=4^2$$

center: $(2, 3)$

radius: $\sqrt{16}=4$

viewing rectangle: four times the EQUAL viewing rectangle

XSCL = 4

YSCL = 4

GRAPH $Y=\sqrt{\left(16-(X-2)^2\right)}+3$

and

$Y=-\sqrt{\left(16-(X-2)^2\right)}+3$

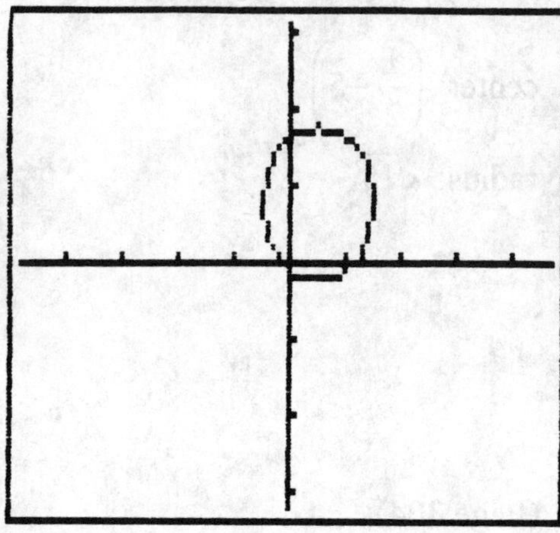

11.

$$(x-2)^2+[y-(-2)]^2=2^2$$

center: $(2, -2)$

radius: 2

viewing rectangle: two times the EQUAL viewing rectangle

XSCL = 2

YSCL = 2

GRAPH $Y=\sqrt{\left(4-(X-2)^2\right)}-2$

and

$Y=-\sqrt{\left(4-(X-2)^2\right)}-2$

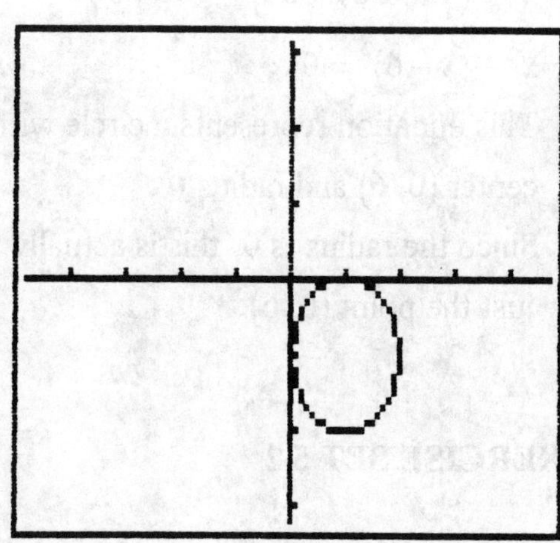

13.

$$[x-(-4)]^2+\left[y-\left(-\frac{3}{2}\right)\right]^2=(3\sqrt{2})^2$$

center: $\left(-4, -\frac{3}{2}\right)$

radius: $3\sqrt{2}$

viewing rectangle: two times the EQUAL viewing rectangle

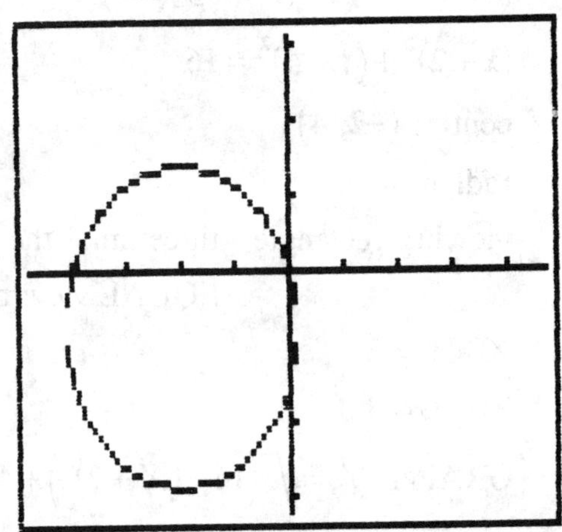

XSCL = 2

YSCL = 2

GRAPH $Y=\sqrt{\left(18-(X+4)^2\right)}-\frac{3}{2}$

and

$Y=-\sqrt{\left(18-(X+4)^2\right)}-\frac{3}{2}$

15.

$$\left(x-\frac{1}{3}\right)^2+(y-0)^2=-\frac{1}{9}$$

$r=\sqrt{-\frac{1}{9}}$ is not a real number, hence there is no circle.

17.

$$\left(x^2+4x\qquad\right)+\left(y^2-8y\qquad\right)=-4$$

$$\left(x^2+4x+4\right)+\left(y^2-8y+16\right)=-4+4+16$$

$$(x+2)^2+(y-4)^2=16$$

center: $(-2,\ 4)$

radius: 4

viewing rectangle: three times the EQUAL viewing rectangle

XSCL = 3

YSCL = 3

GRAPH $Y=\sqrt{\left(16-(X+2)^2\right)}+4$

and

$Y=-\sqrt{\left(16-(X+2)^2\right)}+4$

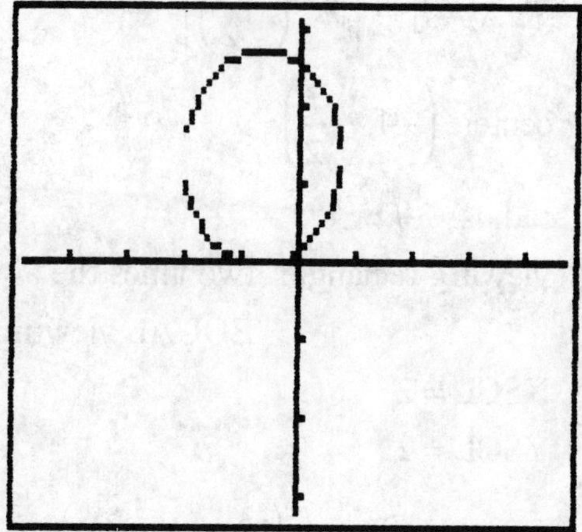

19.

$$(2x^2 - 6x \qquad) + (2y^2 - 10y \qquad) = -6$$
$$2(x^2 - 3x \qquad) + 2(y^2 - 5y \qquad) = -6$$
$$2\left(x^2 - 3x + \frac{9}{4}\right) + 2\left(y^2 - 5y + \frac{25}{4}\right) = -6 + \frac{9}{2} + \frac{25}{2}$$
$$2\left(x - \frac{3}{2}\right)^2 + 2\left(y - \frac{5}{2}\right)^2 = 11$$
$$\left(x - \frac{3}{2}\right)^2 + \left(y - \frac{5}{2}\right)^2 = \frac{11}{2}$$

center: $\left(\frac{3}{2}, \frac{5}{2}\right)$

radius: $\sqrt{\frac{11}{2}} = \frac{\sqrt{22}}{2}$

viewing rectangle: two times the EQUAL viewing rectangle

XSCL = 2

YSCL = 2

GRAPH $Y = \sqrt{\left(\frac{11}{2} - \left(X - \frac{3}{2}\right)^2\right)} + \frac{5}{2}$

and

$Y = -\sqrt{\left(\frac{11}{2} - \left(X - \frac{3}{2}\right)^2\right)} + \frac{5}{2}$

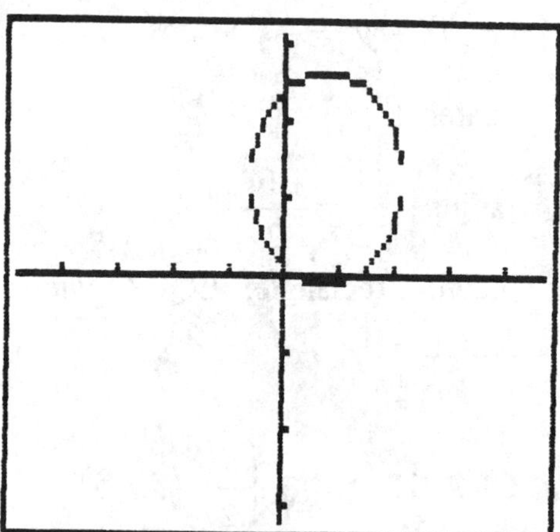

21.

$$\left(2x^2-4x\qquad\right)+\left(2y^2\qquad\right)=5$$
$$2\left(x^2-2x\qquad\right)+2\left(y^2\qquad\right)=5$$
$$2\left(x^2-2x+1\right)+2(y-0)^2=5+2$$
$$2(x-1)^2+2(y-0)^2=7$$
$$(x-1)^2+(y-0)^2=\frac{7}{2}$$
$$(x-1)^2+y^2=\frac{7}{2}$$

center: $(1, 0)$

radius: $\sqrt{\frac{7}{2}}=\frac{\sqrt{14}}{2}$

viewing rectangle: EQUAL viewing rectangle

XSCL = 1

YSCL = 1

GRAPH $Y=\sqrt{\left(\frac{7}{2}-(X-1)^2\right)}$

and

$Y=-\sqrt{\left(\frac{7}{2}-(X-1)^2\right)}$

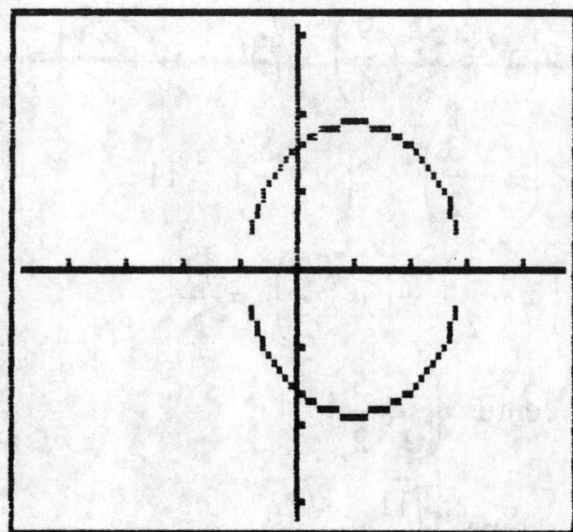

23.

$$(3x^2 - 12x \quad) + (3y^2 + 18y \quad) = -15$$
$$3(x^2 - 4x \quad) + 3(y^2 + 6y \quad) = -15$$
$$3(x^2 - 4x + 4) + 3(y^2 + 6y + 9) = -15 + 12 + 27$$
$$3(x-2)^2 + 3(y+3)^2 = 24$$
$$(x-2)^2 + (y+3)^2 = 8$$

center: $(2, -3)$

radius: $\sqrt{8} = 2\sqrt{2}$

viewing rectangle: three times the EQUAL viewing rectangle

XSCL = 3

YSCL = 3

GRAPH $Y = \sqrt{(8 - (X-2)^2)} - 3$

and

$Y = -\sqrt{(8 - (X-2)^2)} - 3$

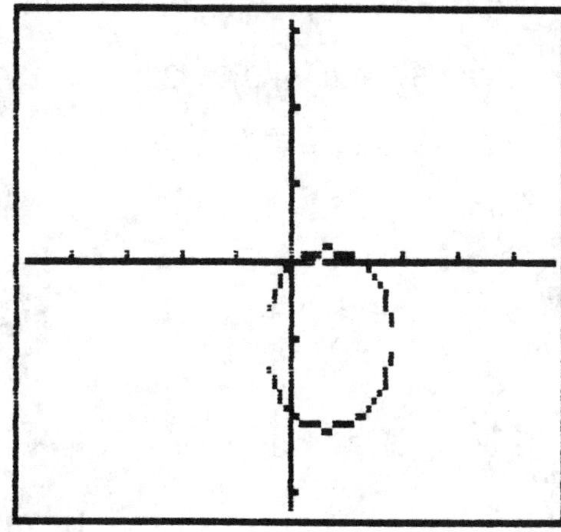

25.

$$(x^2 - 6x \quad) + (y^2 + 8y \quad) = -25$$
$$(x^2 - 6x + 9) + (y^2 + 8y + 16) = -25 + 9 + 16$$
$$(x-3)^2 + (y+4)^2 = 0$$

point at $(3, -4)$

27.

$$(x^2 + 3x \quad) + (y^2 - 5y \quad) = -7$$
$$\left(x^2 + 3x + \frac{9}{4}\right) + \left(y^2 - 5y + \frac{25}{4}\right) = -7 + \frac{9}{4} + \frac{25}{4}$$
$$\left(x + \frac{3}{2}\right)^2 + \left(y - \frac{5}{2}\right)^2 = \frac{3}{2}$$

circle

29.

$$(2x^2 - 12x \qquad) + (2y^2 \qquad) = 4$$
$$2(x^2 - 6x \qquad) + 2(y^2 \qquad) = 4$$
$$2(x^2 - 6x + 9) + 2(y - 0)^2 = 4 + 18$$
$$2(x - 3)^2 + 2(y - 0)^2 = 22$$
$$(x - 3)^2 + y^2 = 11$$

circle

31.

$$(2x^2 - 6x \qquad) + (2y^2 - 4y \qquad) = 2$$
$$2(x^2 - 3x \qquad) + 2(y^2 - 2y \qquad) = 2$$
$$2\left(x^2 - 3x + \frac{9}{4}\right) + 2(y^2 - 2y + 1) = 2 + \frac{9}{2} + 2$$
$$2\left(x - \frac{3}{2}\right)^2 + 2(y - 1)^2 = \frac{17}{2}$$
$$\left(x - \frac{3}{2}\right)^2 + (y - 1)^2 = \frac{17}{4}$$

circle

33.

$$(3x^2 + 12x \qquad) + (3y^2 - 4y \qquad) = 20$$
$$3(x^2 + 4x \qquad) + 3\left(y^2 - \frac{4}{3}y \qquad\right) = 20$$
$$3(x^2 + 4x + 4) + 3\left(y^2 - \frac{4}{3}y + \frac{4}{9}\right) = 20 + 12 + \frac{4}{3}$$
$$3(x + 2)^2 + 3\left(y - \frac{2}{3}\right)^2 = \frac{100}{3}$$
$$(x + 2)^2 + \left(y - \frac{2}{3}\right)^2 = \frac{100}{9}$$

circle

35.

$$(4x^2 - 12x \qquad) + (4y^2 - 20y \qquad) = -38$$
$$4(x^2 + 3x \qquad) + 4(y^2 - 5y \qquad) = -38$$
$$4\left(x^2 + 3x + \frac{9}{4}\right) + 4\left(y^2 - 5y + \frac{25}{4}\right) = -38 + 9 + 25$$
$$4\left(x + \frac{3}{2}\right)^2 + 4\left(y - \frac{5}{2}\right)^2 = -4$$
$$\left(x + \frac{3}{2}\right)^2 + \left(y - \frac{5}{2}\right)^2 = -1$$

neither

37.

$(x^2-2x \quad)+(y^2+4y \quad)=4$

$(x^2-2x+1)+(y^2+4y+4)=4+1+4$

$(x-1)^2+(y+2)^2=9$

radius $= 3$

area $= \pi r^2 = \pi(3)^2 = 9\pi$

39.

$(x^2-4x \quad)+(y^2+9y \quad)=3$

$(x^2-4x+4)+\left(y^2+9y+\frac{81}{4}\right)=3+4+\frac{81}{4}$

$(x-2)^2+\left(y+\frac{9}{2}\right)^2=\frac{109}{4}$

center: $\left(2,\ -\frac{9}{2}\right)$ radius $=\frac{\sqrt{109}}{2}$

$(3x^2-12x \quad)+(3y^2+27y \quad)=27$

$3(x^2-4x \quad)+3(y^2+9y \quad)=27$

$3(x^2-4x+4)+3\left(y^2+9y+\frac{81}{4}\right)=27+12+\frac{243}{4}$

$3(x-2)^2+3\left(y+\frac{9}{2}\right)^2=\frac{399}{4}$

$(x-2)^2+\left(y+\frac{9}{2}\right)^2=\frac{133}{4}$

center: $\left(2,\ -\frac{9}{2}\right)$ radius $=\frac{\sqrt{133}}{2}$

The two circles have the same center, but different radii, hence they are concentric.

41.

center: $(-5,\ 2)$

$[x-(-5)]^2+(y-2)^2=r^2$

$(x+5)^2+(y-2)^2=r^2$

$(-3,\ 4)$ lies on the circle, therefore the coordinates must satisfy the equation of the circle.

$(-3+5)^2+(4-2)^2=r^2$

$4+4=r^2$

$8=r^2$

$(x+5)^2+(y-2)^2=8$

43.

Given the two points $P(3,\ 5)$ and $Q(7,\ -3)$ find $\frac{1}{2}d(P,\ Q)$ to obtain the radius.

$\frac{1}{2}d(P,Q)=\frac{1}{2}\sqrt{(3-7)^2+(5+3)^2}$

$=\frac{1}{2}\sqrt{16+64}$

$=2\sqrt{5}$

The center is the midpoint of PQ.

midpoint: $\left(\frac{3+7}{2},\ \frac{5-3}{2}\right)=(5,\ 1)$

$(x-5)^2+(y-1)^2=20$

45.

The top and bottom halves of circles do not meet when points are not plotted at the "meeting points". In some instances this can be avoided by a careful choice of a viewing rectangle; x-values that are multiples of the EQUAL viewing rectangle values often work. In other cases, there is no way to define the viewing rectangle to insure that the halves of the circle will meet, for instance, $x^2 + y^2 = 7$.

47.

$(x-0)^2 + (y-2)^2 = 1^2$

$x^2 + (y-2)^2 = 1^2$

center: $(0, 2)$

radius: 1

CHAPTER 5 SECTION 3

SECTION 5.3 PROGRESS CHECK

1. (Page 309)

$$x^2 = 4py = -3y$$
$$4p = -3$$
$$p = -\frac{3}{4}$$
$$\text{focus: } (0,\ p) = \left(0,\ -\frac{3}{4}\right)$$
$$\text{directrix: } y = -p$$
$$y = \frac{3}{4}$$

2. (Page 310)

$$\text{focus: } (0,\ p) = (0,\ 3)$$
$$p = 3$$
$$x^2 = 4py$$
$$x^2 = 4(3)y$$
$$x^2 = 12y$$

3. (Page 312)

$$x^2 = 4py$$
$$(1)^2 = 4p(-2)$$
$$-\frac{1}{8} = p$$
$$x^2 = 4\left(-\frac{1}{8}\right)y$$
$$x^2 = -\frac{1}{2}y$$

4. (Page 313)

$$3[y-(-1)]^2 = 12\left(x - \frac{1}{3}\right)$$
$$[y-(-1)]^2 = 4\left(x - \frac{1}{3}\right)$$
$$(y-k)^2 = 4p(x-h)$$
$$h = \frac{1}{3},\quad k = -1,\quad 4p = 4$$
$$p = 1$$
$$\text{vertex: } \left(\frac{1}{3},\ -1\right)$$

axis: $y = -1$

parabola opens to the right

(y term squared, $p > 0$)

5a). (Page 314)

$$\left(y^2 - 2y \quad\right) = 2x + 5$$
$$\left(y^2 - 2y + 1\right) = 2x + 5 + 1$$
$$(y-1)^2 = 2(x+3)$$

$h = -3,\ k = 1$

vertex: $(-3,\ 1)$

axis: $y = 1$

5b).

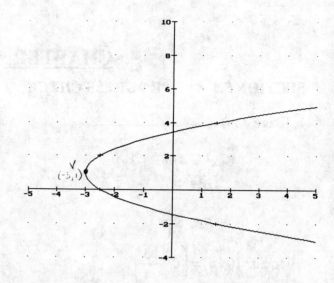

$$\left(x^2 - 2x \quad\right) = -2y + 1$$
$$\left(x^2 - 2x + 1\right) = -2y + 1 + 1$$
$$(x-1)^2 = -2(y-1)$$

$h = 1,\ k = 1$

vertex: $(1,\ 1)$

axis: $x = 1$

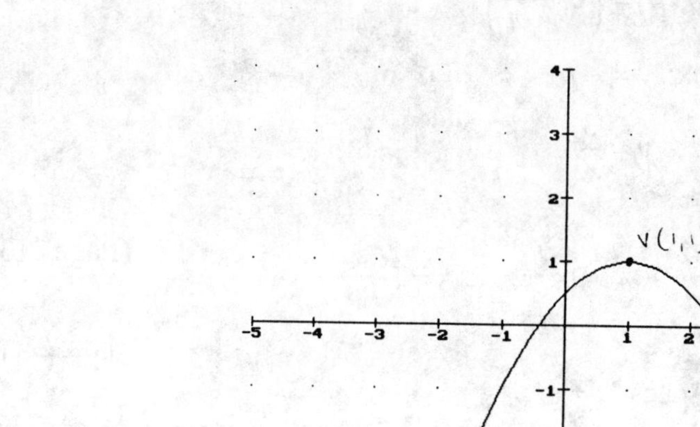

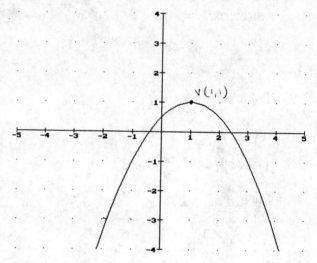

EXERCISE SET 5.3

1.

$$x^2 = 4y$$
$$4p = 4$$
$$p = 1$$

focus: $(0,\ p) = (0,\ 1)$

directrix: $y = -p$

$$y = -1$$

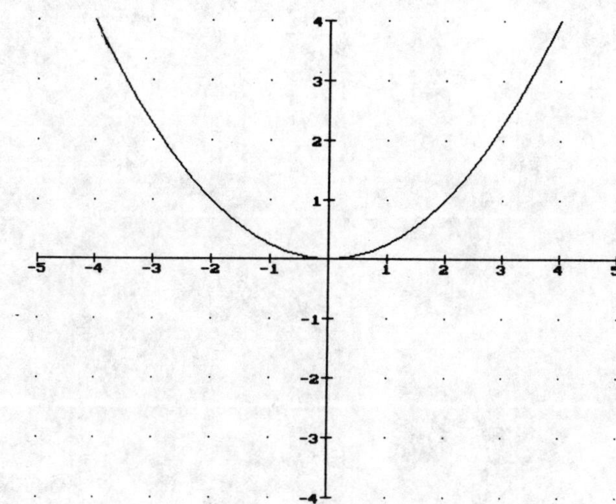

3.

$$y^2 = 2x$$
$$4p = 2$$
$$p = \frac{1}{2}$$
$$\text{focus: } (p,\ 0) = \left(\frac{1}{2},\ 0\right)$$
$$\text{directrix: } x = -p$$
$$x = -\frac{1}{2}$$

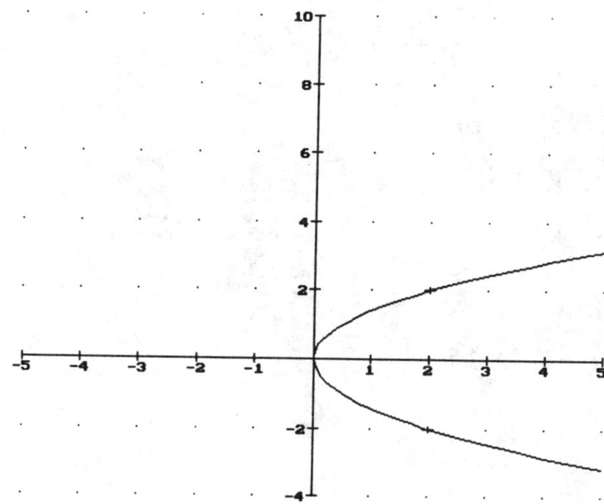

5.

$$x^2 + 5y = 0$$
$$x^2 = -5y$$
$$4p = -5$$
$$p = -\frac{5}{4}$$
$$\text{focus: } (0,\ p) = \left(0,\ -\frac{5}{4}\right)$$
$$\text{directrix: } y = -p$$
$$y = \frac{5}{4}$$

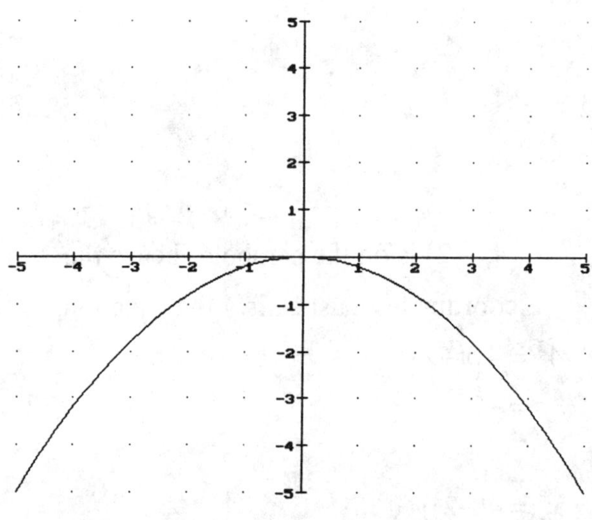

7.

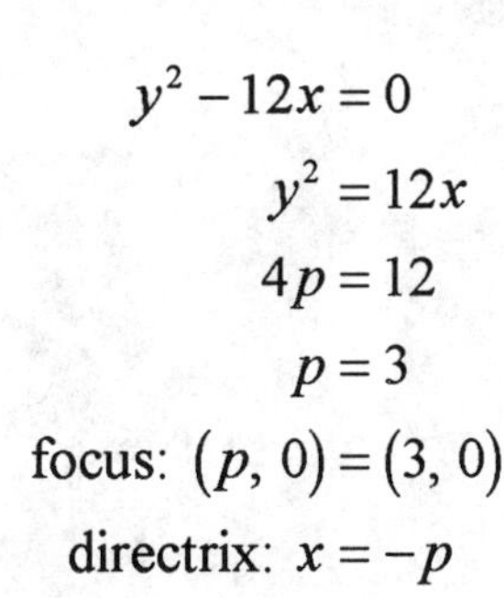

$$y^2 - 12x = 0$$
$$y^2 = 12x$$
$$4p = 12$$
$$p = 3$$
$$\text{focus: } (p,\ 0) = (3,\ 0)$$
$$\text{directrix: } x = -p$$
$$x = -3$$

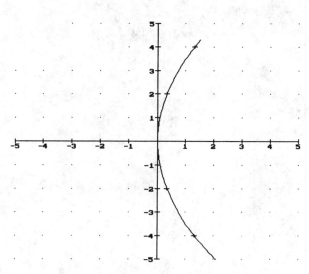

9.

focus: $(1, 0)$

$p = 1$

$y^2 = 4px$

$y^2 = 4(1)x$

$y^2 = 4x$

11.

directrix: $x = -\frac{3}{2}$

$-p = -\frac{3}{2}$

$p = \frac{3}{2}$

$y^2 = 4px$

$y^2 = 4\left(\frac{3}{2}\right)x$

$y^2 = 6x$

13.

$x^2 = 4py$

$(4, -2)$ is on the parabola, hence the coordinates must satisfy the equation.

$4^2 = 4p(-2)$

$p = -2$

$x^2 = 4(-2)y$

$x^2 = -8y$

15.

$y^2 = 4px$

$y^2 = 4\left(-\frac{5}{4}\right)x$

$y^2 = -5x$

17.

focus: $(-1, 0)$

$p = -1$

$y^2 = 4px$

$y^2 = 4(-1)x$

$y^2 = -4x$

19.

$y^2 = 4px$

$(4, 2)$ is on the parabola, hence the coordinates must satisfy the equation.

$2^2 = 4p(4)$

$p = \frac{1}{4}$

$y^2 = 4\left(\frac{1}{4}\right)x$

$y^2 = x$

21.

$4x^2 + y = 0$

$x^2 = -\frac{1}{4}y$

opens downward (x is squared, $p < 0$)

23.

$2x + y^2 = 0$

$y^2 = -2x$

opens to the left (y is squared, $p < 0$)

In Exercises 25-43, the RANGE values given show important features of the functions. Other RANGE values are also acceptable. The display on your calculator may vary from that shown.

25.

$$(x^2 - 2x \quad) = 3y - 7$$
$$(x^2 - 2x + 1) = 3y - 7 + 1$$
$$(x - 1)^2 = 3y - 6$$
$$(x - 1)^2 = 3(y - 2)$$
$$h = 1, \quad k = 2, \quad 4p = 3$$
$$p = \frac{3}{4}$$

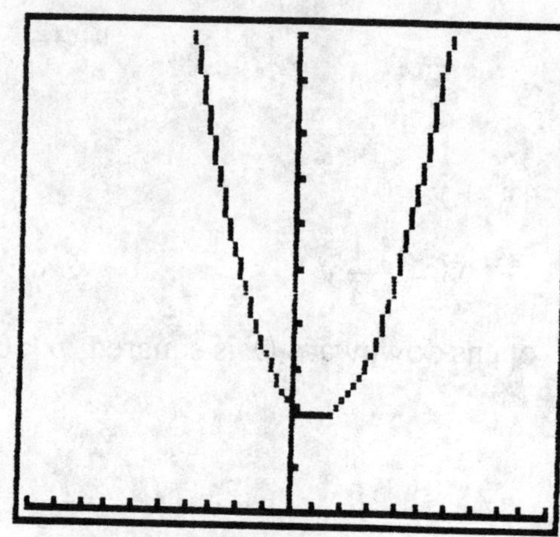

vertex: $(1, 2)$; axis: $x = 1$; opens upward

XMIN $= -10$

XMAX $= 10$

XSCL $= 1$

YMIN $= 0$

YMAX $= 10$

YSCL $= 1$

GRAPH $Y = (x - 1)^2 \div 3 + 2$

27.

$$\left(y^2-8y\qquad\right)=-2x-12$$
$$\left(y^2-8y+16\right)=-2x-12+16$$
$$(y-4)^2=-2x+4$$
$$(y-4)^2=-2(x-2)$$
$h=2,\ \ k=4,\ \ 4p=-2$

$$p=-\frac{1}{2}$$

vertex: $(2,\ 4)$; axis: $y=4$;
opens to the left
XMIN = −10
XMAX = 10
XSCL = 1
YMIN = −10
YMAX = 10
YSCL = 1
GRAPH $Y=4+\sqrt{\ }(-2(x-2))$
and
$Y=4-\sqrt{\ }(-2(x-2))$

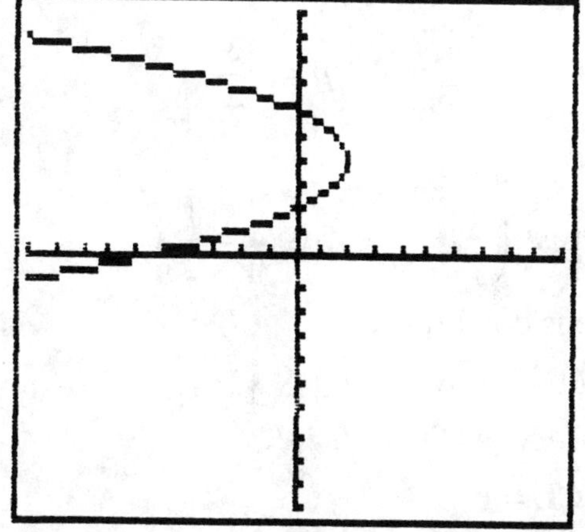

29.

$$\left(x^2 - x \qquad\right) = -3y - 1$$

$$\left(x^2 - x + \frac{1}{4}\right) = -3y - 1 + \frac{1}{4}$$

$$\left(x - \frac{1}{2}\right)^2 = -3y - \frac{3}{4}$$

$$\left(x - \frac{1}{2}\right)^2 = -3\left(y + \frac{1}{4}\right)$$

$$h = \frac{1}{2},\ \ k = -\frac{1}{4},\ \ 4p = -3$$

$$p = -\frac{3}{4}$$

vertex: $\left(\frac{1}{2}, -\frac{1}{4}\right)$; axis: $x = \frac{1}{2}$;

opens downward

XMIN = –4

XMAX = 4

XSCL = 1

YMIN = –4

YMAX = 4

YSCL = 1

GRAPH $Y = \left(x - \frac{1}{2}\right)^2 \div -3 - \frac{1}{4}$

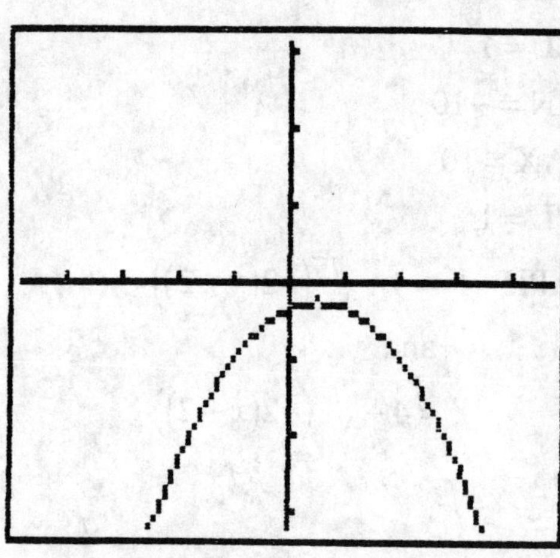

31.

$$\left(y^2 - 10y \qquad\right) = 3x - 24$$

$$\left(y^2 - 10y + 25\right) = 3x - 24 + 25$$

$$(y-5)^2 = 3x + 1$$

$$(y-5)^2 = 3\left(x + \frac{1}{3}\right)$$

$$h = -\frac{1}{3},\quad k = 5,\quad 4p = 3$$

$$p = \frac{3}{4}$$

vertex: $\left(-\frac{1}{3}, 5\right)$; axis: $y = 5$;

opens to the right

XMIN = −1

XMAX = 10

XSCL = 1

YMIN = −5

YMAX = 15

YSCL = 1

GRAPH $Y = \sqrt{\left(3\left(x + \frac{1}{3}\right)\right)} + 5$

and

$Y = -\sqrt{\left(3\left(x + \frac{1}{3}\right)\right)} + 5$

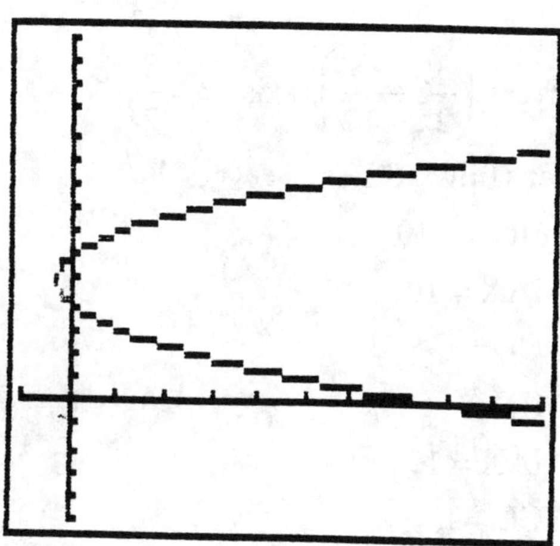

33.

$$\left(x^2 - 3x \qquad\right) = 3y - 1$$

$$\left(x^2 - 3x + \frac{9}{4}\right) = 3y - 1 + \frac{9}{4}$$

$$\left(x - \frac{3}{2}\right)^2 = 3y + \frac{5}{4}$$

$$\left(x - \frac{3}{2}\right)^2 = 3\left(y + \frac{5}{12}\right)$$

$$h = \frac{3}{2},\ \ k = -\frac{5}{12},\ \ 4p = 3$$

$$p = \frac{3}{4}$$

vertex: $\left(\frac{3}{2}, -\frac{5}{12}\right)$; axis: $x = \frac{3}{2}$;

opens upward

XMIN = −10

XMAX = 10

XSCL = 1

YMIN = −5

YMAX = 15

YSCL = 1

GRAPH $Y = \left(X - \frac{3}{2}\right)^2 \div 3 - \frac{5}{12}$

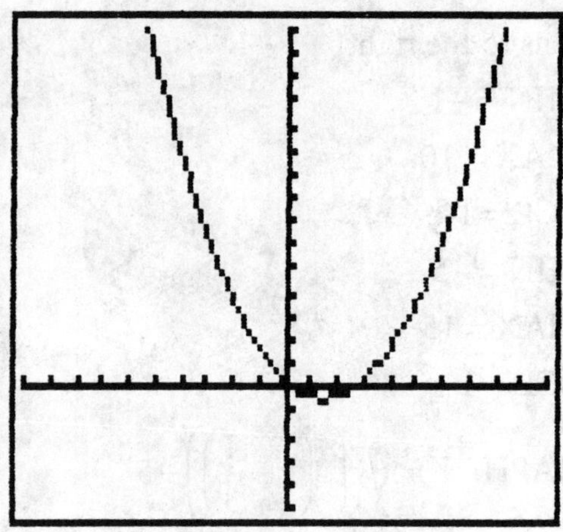

35.

$$\left(y^2+6y\quad\right)=-\frac{1}{2}x-7$$

$$\left(y^2+6y+9\right)=-\frac{1}{2}x-7+9$$

$$(y+3)^2=-\frac{1}{2}x+2$$

$$(y+3)^2=-\frac{1}{2}(x-4)$$

$$h=4,\quad k=-3,\quad 4p=-\frac{1}{2}$$

$$p=-\frac{1}{8}$$

vertex: $(4,-3)$; axis: $y=-3$;
opens to the left

XMIN = −30
XMAX = 5
XSCL = 5
YMIN = −10
YMAX = 5
YSCL = 1

GRAPH $Y=\sqrt{\left(-\frac{1}{2}(x-4)\right)}-3$

and

$Y=-\sqrt{\left(-\frac{1}{2}(x-4)\right)}-3$

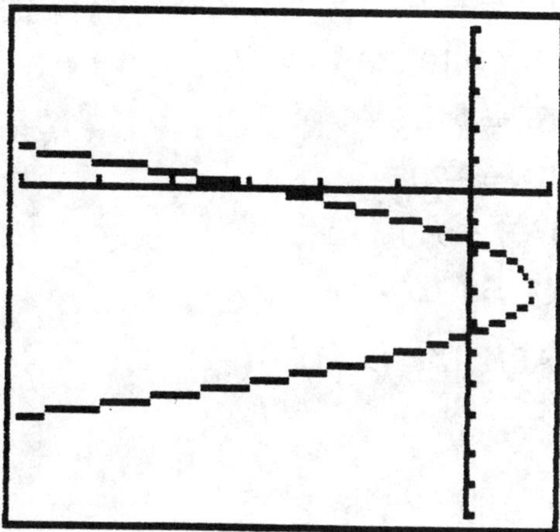

37.

$$\left(x^2+2x\quad\right)=-2y-3$$

$$\left(x^2+2x+1\right)=-2y-3+1$$

$$(x+1)^2=-2y-2$$

$$(x+1)^2=-2(y+1)$$

$h=-1,\ \ k=-1,\ \ 4p=-2$

$$p=-\frac{1}{2}$$

vertex: $(-1,-1)$; axis: $x=-1$;

opens downward

XMIN = −10

XMAX = 10

XSCL = 1

YMIN = −20

YMAX = 0

YSCL = 1

GRAPH $Y=(x+1)^2\div(-2)-1$

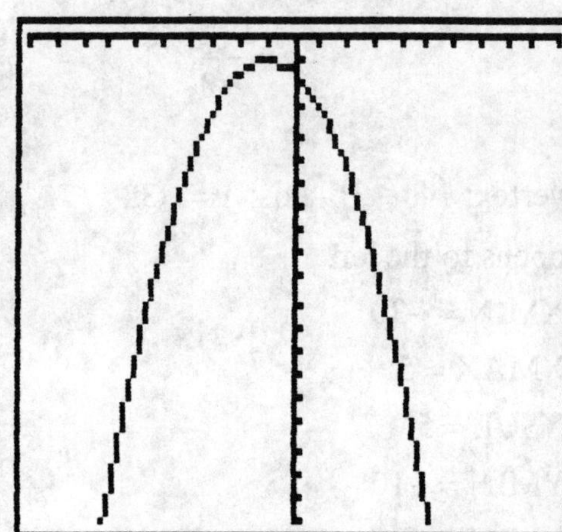

39.

$$\left(x^2-4x\quad\right)=2y-2$$
$$\left(x^2-4x+4\right)=2y-2+4$$
$$(x-2)^2=2y+2$$
$$(x-2)^2=2(y+1)$$
$$h=2,\ \ k=-1,\ \ 4p=2$$
$$p=\frac{1}{2}$$

vertex: $(2,-1)$; axis: $x=2$;
opens upward
XMIN $=-10$
XMAX $=10$
XSCL $=1$
YMIN $=-10$
YMAX $=10$
YSCL $=1$
GRAPH $Y=(x-2)^2\div 2-1$

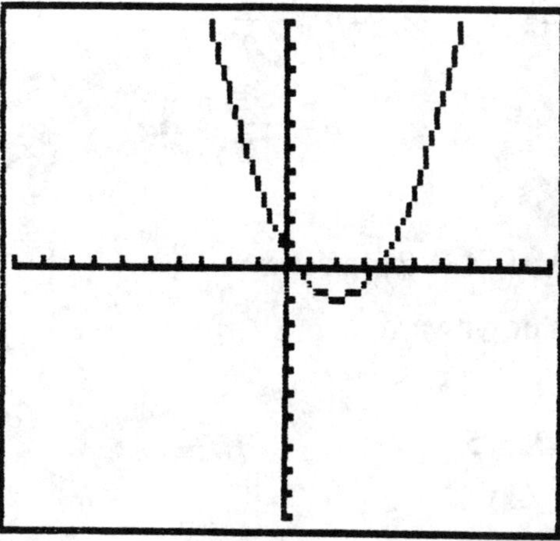

41.

$$\left(2x^2+16x \quad\right)=-y-34$$

$$2\left(x^2+8x \quad\right)=-y-34$$

$$2\left(x^2+8x+16\right)=-y-34+32$$

$$2(x+4)^2=-y-2$$

$$2(x+4)^2=-(y+2)$$

$$(x+4)^2=-\frac{1}{2}(y+2)$$

$$h=-4,\quad k=-2,\quad 4p=-\frac{1}{2}$$

$$p=-\frac{1}{8}$$

vertex: $(-4,-2)$; axis: $x=-4$;

opens downward

XMIN $=-15$

XMAX $=5$

XSCL $=1$

YMIN $=-50$

YMAX $=0$

YSCL $=10$

GRAPH $\quad Y=-2(x+4)^2-2$

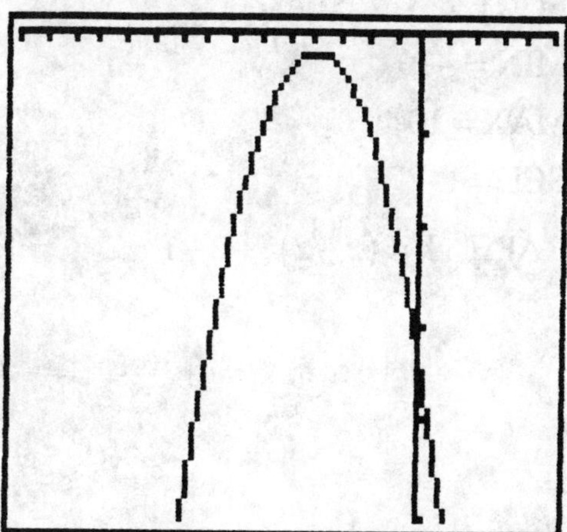

43.

$$y^2 = -2x - 2$$
$$y^2 = -2(x+1)$$
$$h = -1, \quad k = 0, \quad 4p = -2$$
$$p = -\frac{1}{2}$$

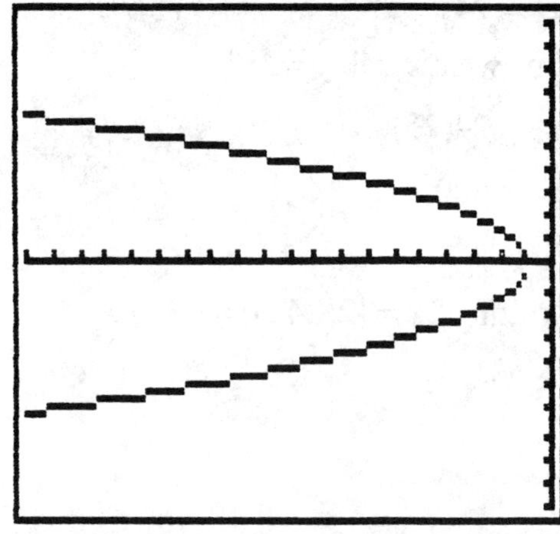

vertex: $(-1, 0)$; axis: $y = 0$;
opens to the left
XMIN = −10
XMAX = 10
XSCL = 1
YMIN = −20
YMAX = 0
YSCL = 1
GRAPH $Y = \sqrt{\ }(-2(x+1))$
and
$Y = -\sqrt{\ }(-2(x+1))$

45.

Area = 30

$$\pi r^2 = 30$$
$$r^2 = \frac{30}{\pi}$$
$$r \approx 3.1$$

The net extends from $45 - 3.1$ to $45 + 3.1$ or 41.9 to 48.1. The stuntman will land at 40 ft. missing the net.

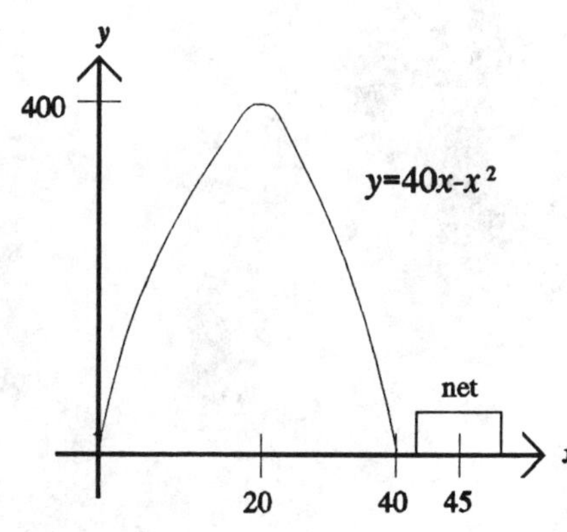

47a).

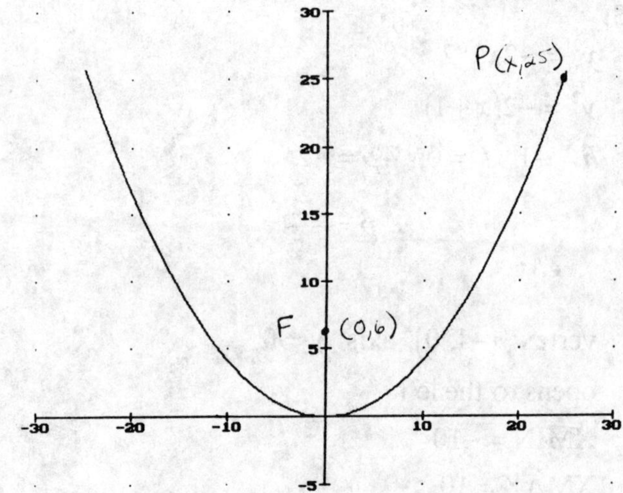

$p=6$

$x^2 = 4py$

$x^2 = 4(6)y$

$x^2 = 24y$

For what x will $y=25$?

$x^2 = 24(25)$

$x^2 = 600$

$x \approx 24.5$

width: $2x = (2(24.5)) = 49$ in.

47b).

$d(F,P)=\sqrt{(24.5-0)^2+(25-6)^2} \approx 31.0\text{in.}$

CHAPTER 5 SECTION 4

SECTION 5.4 PROGRESS CHECK

1a). (Page 321)

$$2x^2 + 3y^2 = 6$$

$$\frac{x^2}{3} + \frac{y^2}{2} = 1$$

The x-intercepts are $(\pm\sqrt{3},\ 0)$;

the y-intercepts are $(0,\ \pm\sqrt{2})$.

1b).

$$3x^2 + y^2 = 5$$

$$\frac{3x^2}{5} + \frac{y^2}{5} = 1$$

$$\frac{x^2}{\frac{5}{3}} + \frac{y^2}{5} = 1$$

The x-intercepts are $\left(\pm\sqrt{\frac{5}{3}},\ 0\right) = \left(\pm\frac{\sqrt{15}}{3},\ 0\right)$;

the y-intercepts are $(0,\ \pm\sqrt{5})$.

2a). (Page 323)

$$2x^2 - 5y^2 = 6$$

$$\frac{x^2}{3} - \frac{5y^2}{6} = 1$$

$$\frac{x^2}{3} - \frac{y^2}{\frac{6}{5}} = 1$$

The x-intercepts are $(\pm\sqrt{3},\ 0)$.

2b).

$$4y^2 - x^2 = 5$$

$$\frac{4y^2}{5} - \frac{x^2}{5} = 1$$

$$\frac{y^2}{\frac{5}{4}} - \frac{x^2}{5} = 1$$

The y-intercepts are $\left(0,\ \pm\frac{\sqrt{5}}{2}\right)$.

3. (Page 326)

$$4x^2 - 9y^2 = 144$$

$$\frac{x^2}{36} - \frac{y^2}{16} = 1$$

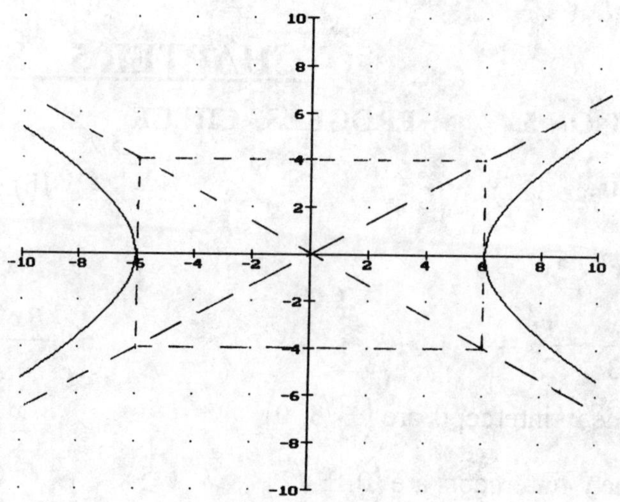

EXERCISE SET 5.4

In Exercises 1-33, it is a good idea to use RANGE values that are multiples of the EQUAL viewing rectangle to see the ellipses and hyperbolas in proper proportion. Other RANGE values are also acceptable. The display on your calculator may vary from that shown.

1.

$$\frac{x^2}{25} + \frac{y^2}{4} = 1$$

x-intercepts: $(\pm 5,\ 0)$

y-intercepts: $(0,\ \pm 2)$

viewing rectangle: 2 times the EQUAL viewing rectangle

XSCL = 2
YSCL = 2

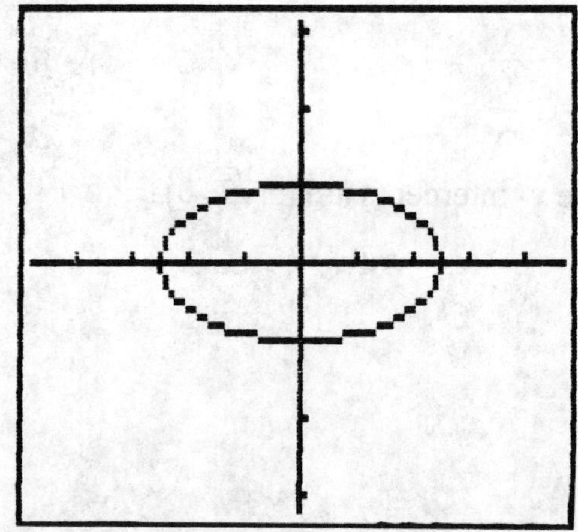

GRAPH: $Y = \sqrt{\ }(4 - 4X^2 \div 25)$

and

$Y = -\sqrt{\ }(4 - 4X^2 \div 25)$

3.

$$\frac{x^2}{8}+\frac{y^2}{4}=1$$

x-intercepts: $\left(\pm\sqrt{8},\,0\right)=\left(\pm2\sqrt{2},\,0\right)$

y-intercepts: $(0,\,\pm2)$

viewing rectangle: EQUAL viewing rectangle

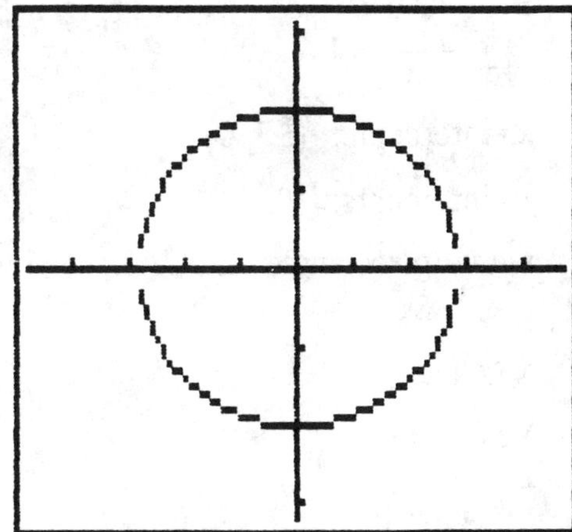

XSCL = 1

YSCL = 1

GRAPH: $Y=\sqrt{\ }\left(4-\left(X^2\div2\right)\right)$

and

$Y=-\sqrt{\ }\left(4-\left(X^2\div2\right)\right)$

5.

$$\frac{x^2}{16}+\frac{y^2}{25}=1$$

x-intercepts: $(\pm4,\,0)$

y-intercepts: $(0,\,\pm5)$

viewing rectangle: 2 times the EQUAL viewing rectangle

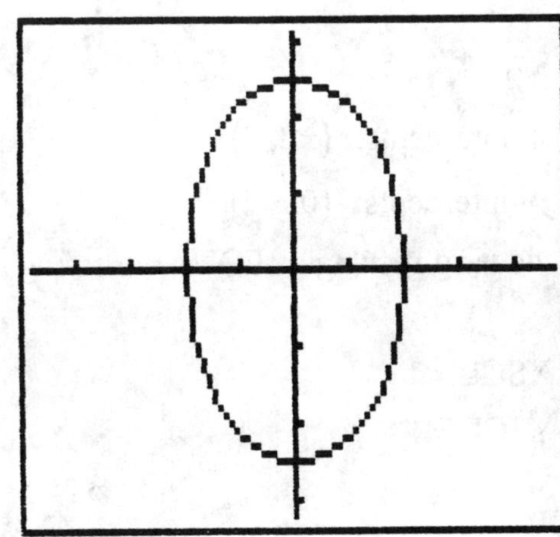

XSCL = 2

YSCL = 2

GRAPH: $Y=\sqrt{\ }\left(25-\left(25X^2\div16\right)\right)$

and

$Y=-\sqrt{\ }\left(25-\left(25X^2\div16\right)\right)$

7.

$4x^2 + 9y^2 = 36$

$\frac{x^2}{9} + \frac{y^2}{4} = 1$

x-intercepts: $(\pm 3, 0)$

y-intercepts: $(0, \pm 2)$

viewing rectangle: EQUAL viewing rectangle

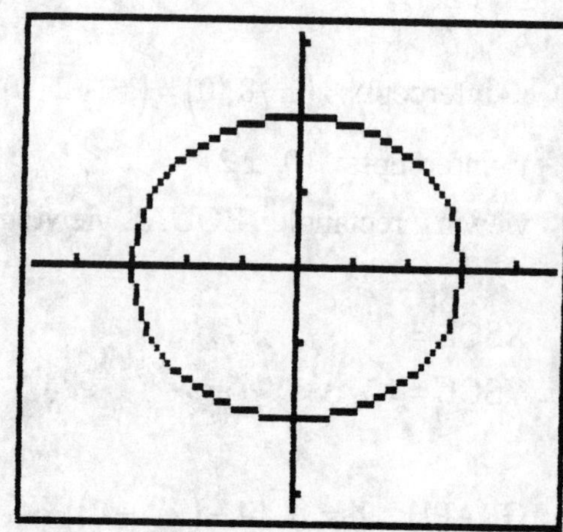

XSCL = 1

YSCL = 1

GRAPH: $Y = \sqrt{\ }(4 - (4X^2 \div 9))$

and

$Y = -\sqrt{\ }(4 - (4X^2 \div 9))$

9.

$4x^2 + 16y^2 = 16$

$\frac{x^2}{4} + \frac{y^2}{1} = 1$

x-intercepts: $(\pm 2, 0)$

y-intercepts: $(0, \pm 1)$

viewing rectangle: EQUAL viewing rectangle

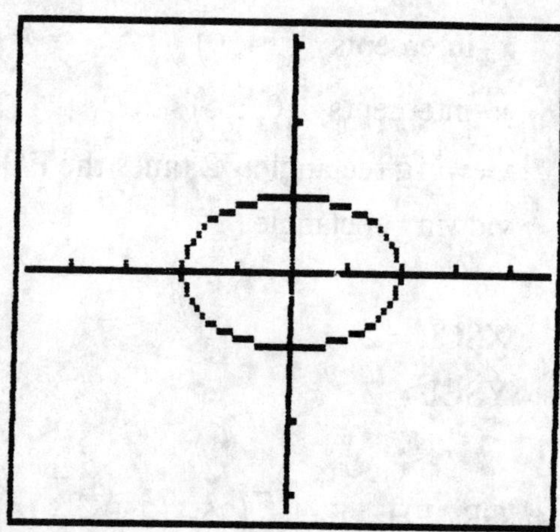

XSCL = 1

YSCL = 1

GRAPH: $Y = \sqrt{\ }(1 - (X^2 \div 4))$

and

$Y = -\sqrt{\ }(1 - (X^2 \div 4))$

11.

$$4x^2 + 16y^2 = 4$$

$$\frac{x^2}{1} + 4y^2 = 1$$

$$\frac{x^2}{1} + \frac{y^2}{\frac{1}{4}} = 1$$

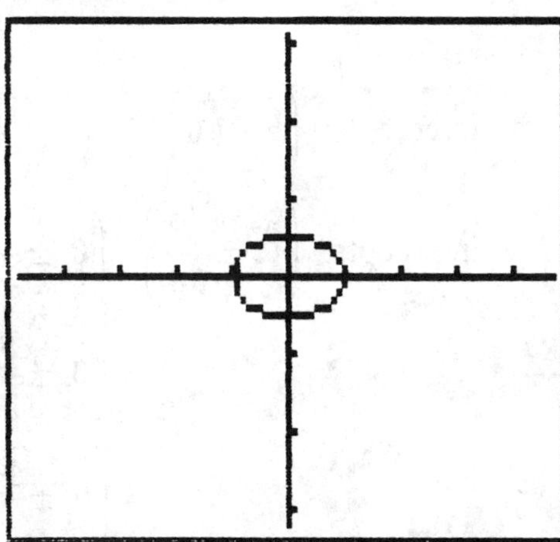

x-intercepts: $(\pm 1, 0)$

y-intercepts: $\left(0, \pm\frac{1}{2}\right)$

viewing rectangle: EQUAL viewing rectangle

XSCL = 1

YSCL = 1

GRAPH: $Y = \left(\sqrt{\ }(1 - X^2)\right) \div 2$

and

$Y = \left(-\sqrt{\ }(1 - X^2)\right) \div 2$

13.

$$8x^2 + 6y^2 = 24$$

$$\frac{x^2}{3} + \frac{y^2}{4} = 1$$

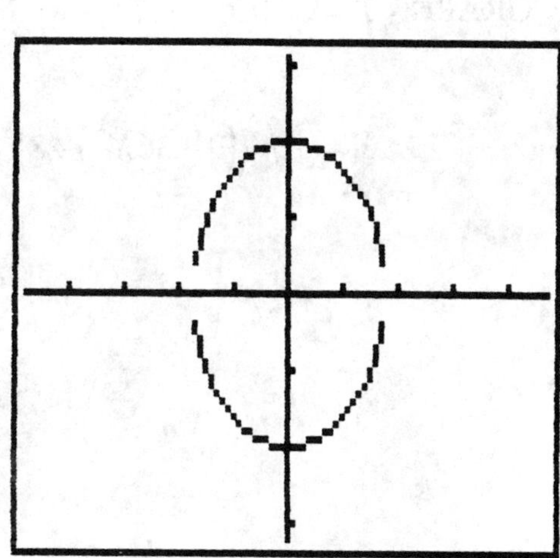

x-intercepts: $(\pm\sqrt{3}, 0)$

y-intercepts: $(0, \pm 2)$

viewing rectangle: EQUAL viewing rectangle

XSCL = 1

YSCL = 1

GRAPH: $Y = \sqrt{\ }\left(4 - (4X^2 \div 3)\right)$

and

$Y = -\sqrt{\ }\left(4 - (4X^2 \div 3)\right)$

15.

$$36x^2 + 8y^2 = 9$$

$$\frac{4x^2}{1} + \frac{8y^2}{9} = 1$$

$$\frac{x^2}{\frac{1}{4}} + \frac{y^2}{\frac{9}{8}} = 1$$

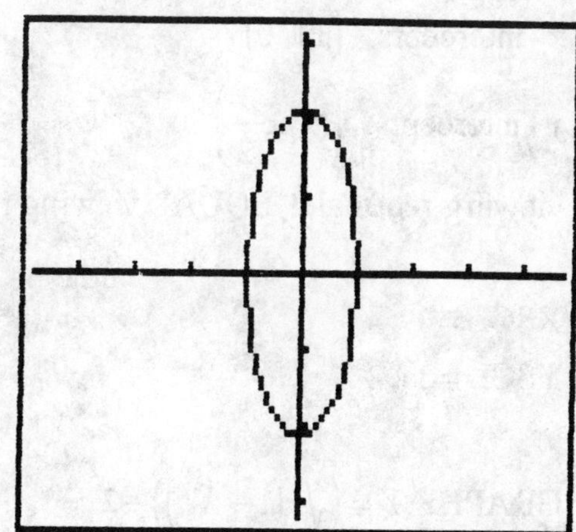

x-intercepts: $\left(\pm\frac{1}{2}, 0\right)$

y-intercepts: $\left(0, \pm\sqrt{\frac{9}{8}}\right) = \left(0, \pm\frac{3}{\sqrt{8}}\right)$

$$= \left(0, \pm\frac{3}{2\sqrt{2}}\right)$$

$$= \left(0, \pm\frac{3\sqrt{2}}{4}\right)$$

viewing rectangle: one-half the EQUAL viewing rectangle

XSCL = 0.5
YSCL = 0.5

GRAPH: $Y = \sqrt{\left((9 - 36X^2) \div 8\right)}$

and

$Y = -\sqrt{\left((9 - 36X^2) \div 8\right)}$

17.

$$\frac{x^2}{25} - \frac{y^2}{16} = -1$$

$$\frac{y^2}{16} - \frac{x^2}{25} = 1$$

y-intercepts: $(0, \pm 4)$

viewing rectangle: 4 times the EQUAL viewing rectangle

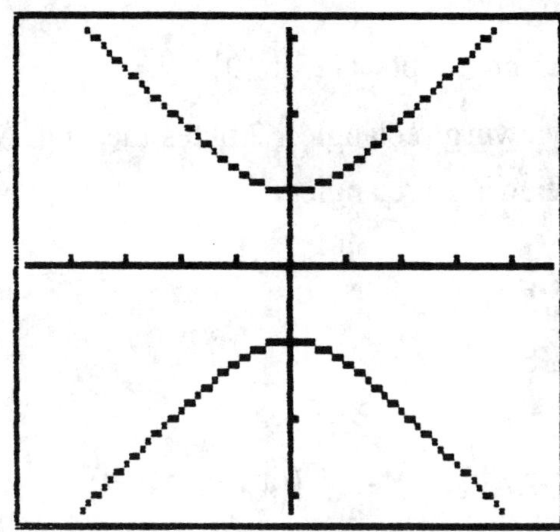

XSCL = 4

YSCL = 4

GRAPH: $Y = \sqrt{\ }\left(16 + \left(16X^2 \div 25\right)\right)$

and

$Y = -\sqrt{\ }\left(16 + \left(16X^2 \div 25\right)\right)$

19.

$$\frac{x^2}{36} - \frac{y^2}{1} = 1$$

x-intercepts: $(\pm 6, 0)$

viewing rectangle: 3 times the EQUAL viewing rectangle

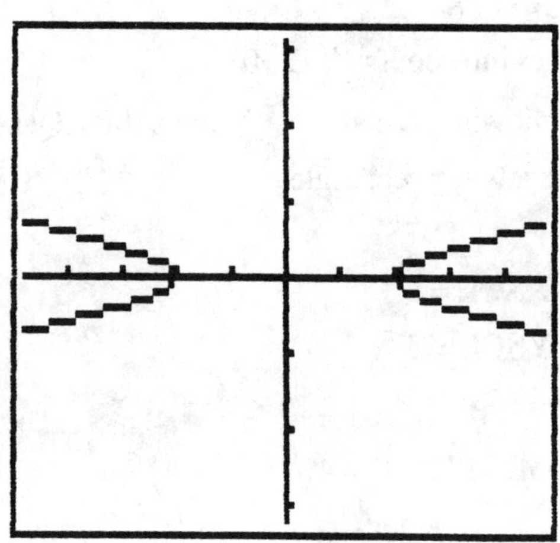

XSCL = 3

YSCL = 3

GRAPH: $Y = \sqrt{\ }\left(\left(X^2 \div 36\right) - 1\right)$

and

$Y = -\sqrt{\ }\left(\left(X^2 \div 36\right) - 1\right)$

21.

$$\frac{x^2}{6} - \frac{y^2}{8} = 1$$

x-intercepts: $\left(\pm\sqrt{6},\ 0\right)$

viewing rectangle: 2 times the EQUAL viewing rectangle

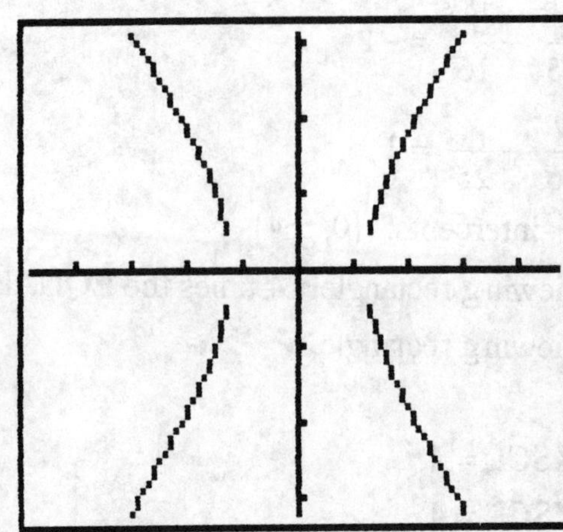

XSCL = 2
YSCL = 2

GRAPH: $Y = \sqrt{\ }\left(\left(4X^2 \div 3\right) - 8\right)$

and

$Y = -\sqrt{\ }\left(\left(4X^2 \div 3\right) - 8\right)$

23.

$$16x^2 - y^2 = 64$$

$$\frac{x^2}{4} - \frac{y^2}{64} = 1$$

x-intercepts: $(\pm 2,\ 0)$

viewing rectangle: 5 times the EQUAL viewing rectangle

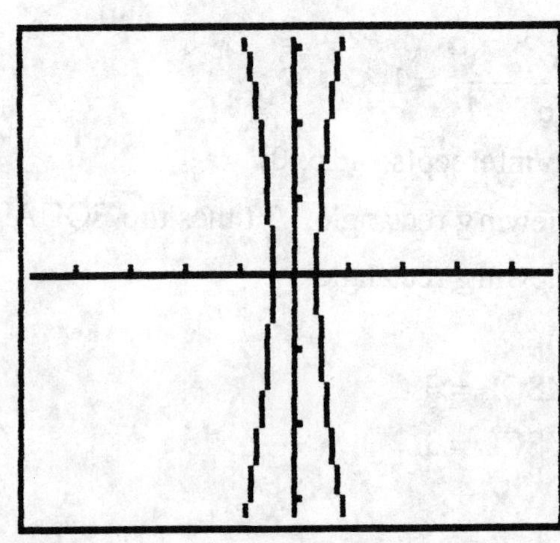

XSCL = 5
YSCL = 5

GRAPH: $Y = 4\sqrt{\ }\left(X^2 - 4\right)$

and

$Y = -4\sqrt{\ }\left(X^2 - 4\right)$

25.

$4y^2 - 4x^2 = 1$

$\dfrac{y^2}{\frac{1}{4}} - \dfrac{x^2}{\frac{1}{4}} = 1$

y-intercepts: $\left(0, \pm\dfrac{1}{2}\right)$

viewing rectangle: EQUAL viewing rectangle

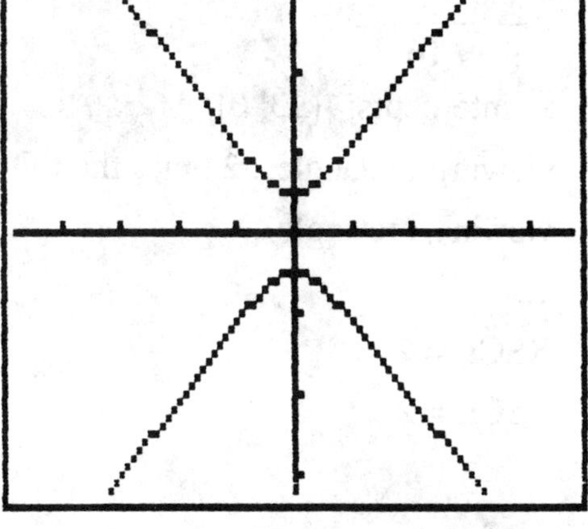

XSCL = 1

YSCL = 1

GRAPH: $Y = \sqrt{}\left(4X^2 + 1\right) \div 2$

and

$Y = -\sqrt{}\left(4X^2 + 1\right) \div 2$

27.

$4x^2 - 5y^2 = 20$

$\dfrac{x^2}{5} - \dfrac{y^2}{4} = 1$

x-intercepts: $\left(\pm\sqrt{5}, 0\right)$

viewing rectangle: EQUAL viewing rectangle

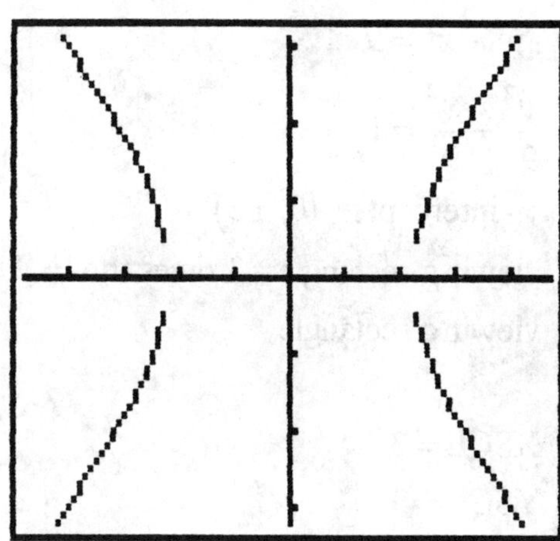

XSCL = 1

YSCL = 1

GRAPH: $Y = 2\sqrt{}\left(\left(X^2 - 5\right) \div 5\right)$

and

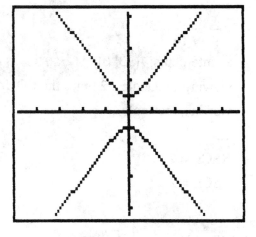

$Y = -2\sqrt{}\left(\left(X^2 - 5\right) \div 5\right)$

29.

$16x^2 - 9y^2 = 144$

$\frac{x^2}{9} - \frac{y^2}{16} = 1$

x-intercepts: $(\pm 3, 0)$

viewing rectangle: 2 times the EQUAL viewing rectangle

XSCL = 2

YSCL = 2

GRAPH: $Y = \frac{4}{3}\sqrt{(X^2 - 9)}$

and

$Y = -\frac{4}{3}\sqrt{(X^2 - 9)}$

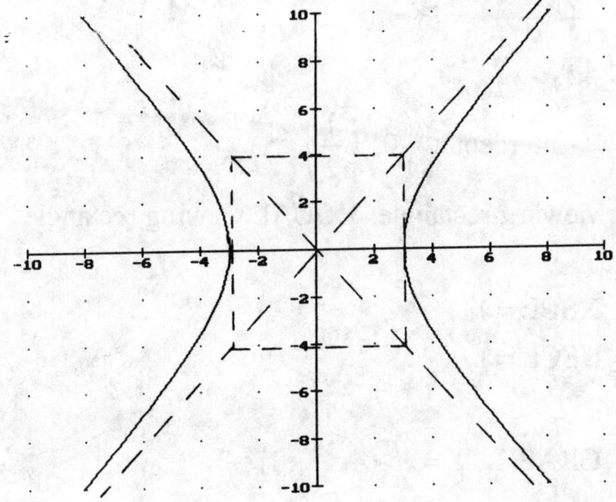

31.

$y^2 - x^2 = 9$

$\frac{y^2}{9} - \frac{x^2}{9} = 1$

y-intercepts: $(0, \pm 3)$

viewing rectangle: 3 times the EQUAL viewing rectangle

XSCL = 3

YSCL = 3

GRAPH: $Y = \sqrt{(X^2 + 9)}$

and

$Y = -\sqrt{(X^2 + 9)}$

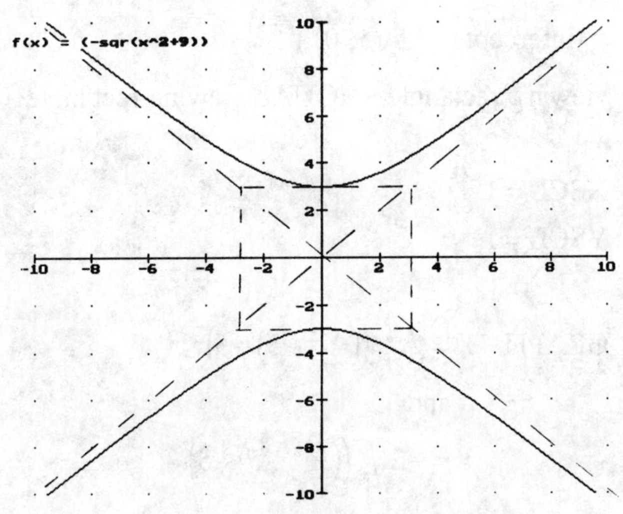

33.

$$\frac{x^2}{25} - \frac{y^2}{36} = 1$$

x-intercepts: $(\pm 5, 0)$

viewing rectangle: 3 times the EQUAL viewing rectangle

XSCL = 3

YSCL = 3

GRAPH: $Y = \frac{6}{5}\sqrt{\left(X^2 - 25\right)}$

and

$Y = -\frac{6}{5}\sqrt{\left(X^2 - 25\right)}$

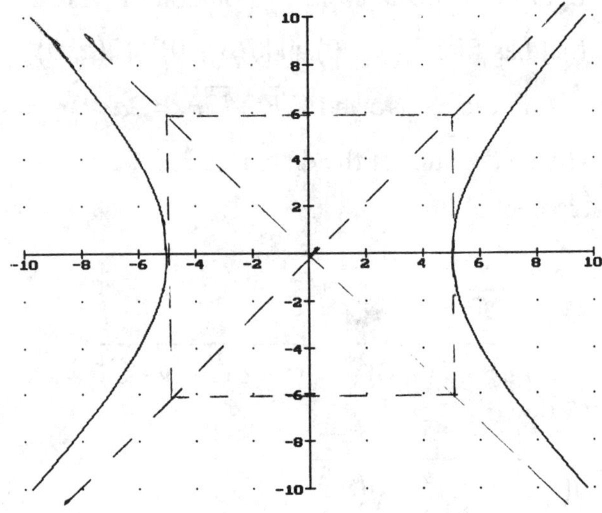

35.
(Refer to Figure 25 in your textbook.)

Let $P(x, y)$ be an arbitrary point on the ellipse that has foci $F_1(-c, 0)$ and $F_2(c, 0)$. $V_2(a, 0)$ lies on the ellipse and $\overline{V_2F_1}+\overline{V_2F_2}=2a.$

Hence the sum of the distances $\overline{PF_1}+\overline{PF_2}$ must also equal $2a$.

$$\overline{PF_1}+\overline{PF_2}=$$
$$\sqrt{(x+c)^2+(y-0)^2}+\sqrt{(x-c)^2+(y-0)^2}=2a$$

$$\sqrt{(x+c)^2+y^2}+\sqrt{(x-c)^2+y^2}=2a$$

$$\sqrt{(x+c)^2+y^2}=2a-\sqrt{(x-c)^2+y^2}$$

Square both sides.

$$(x+c)^2+y^2=$$
$$4a^2-4a\sqrt{(x-c)^2+y^2}+(x-c)^2+y^2$$

$$x^2+2xc+c^2=$$
$$4a^2-4a\sqrt{(x-c)^2+y^2}+x^2-2xc+c^2$$

$$4xc-4a^2=-4a\sqrt{(x-c)^2+y^2}$$

$$xc-a^2=-a\sqrt{(x-c)^2+y^2}$$

Square both sides.

$$(xc-a^2)^2=a^2\left[(x-c)^2+y^2\right]$$

35. (Continued)

$$x^2c^2-2xca^2+a^4=a^2\left(x^2-2cx+c^2+y^2\right)$$
$$x^2c^2-2xca^2+a^4=a^2x^2-2a^2cx+a^2c^2+a^2y^2$$
$$x^2c^2-a^2x^2-a^2y^2=a^2c^2-a^4$$
$$-x^2\left(a^2-c^2\right)-a^2y^2=-a^2\left(a^2-c^2\right)$$

Substitute $b^2=a^2-c^2$:

$$-x^2b^2-a^2y^2=-a^2b^2$$

Divide by $-a^2b^2$:

$$\frac{x^2}{a^2}+\frac{y^2}{b^2}=1$$

37.

$$\frac{x^2}{18^2}+\frac{y^2}{\left(\frac{9}{2}\right)^2}=1$$

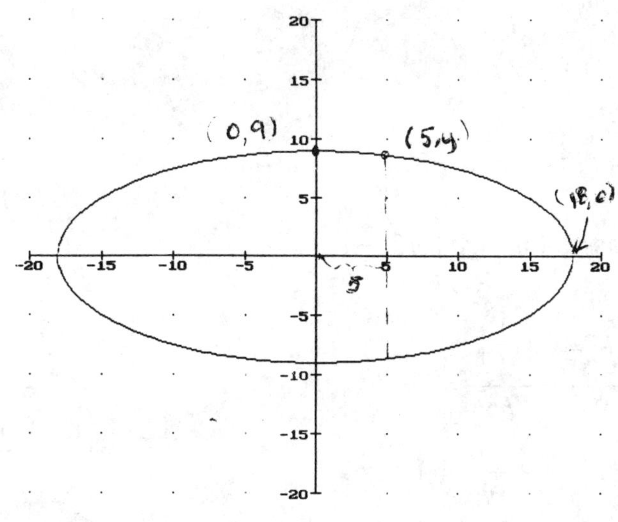

$$\frac{x^2}{324}+\frac{4y^2}{81}=1$$

Find y for $x = 5$.

$$\frac{5^2}{324}+\frac{4y^2}{81}=1$$

$$25+16y^2=324$$

$$16y^2=299$$

$$y^2=\frac{299}{16}$$

$$y=\frac{\sqrt{299}}{4}$$

The height is $2\left(\frac{\sqrt{299}}{4}\right)=\frac{\sqrt{299}}{2}$ in.

39.

Find y for $x = 125$.

$$\frac{x^2}{175^2}+\frac{y^2}{40^2}=1,\ y\geq 0$$

$$\frac{125^2}{175^2}+\frac{y^2}{40^2}=1$$

$$\frac{y^2}{40^2}=1-\frac{25}{49}$$

$$\frac{y^2}{40^2}=\frac{24}{49}$$

$$y^2=\frac{38400}{49}$$

$$y=\frac{80\sqrt{6}}{7}$$

The height is $\frac{80\sqrt{6}}{7}$ ft.

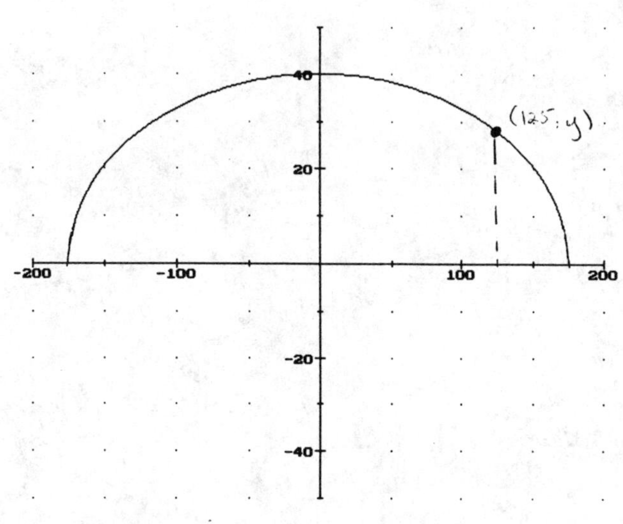

41.

Vertex: $(3, 0)$ Center: $(0, 0)$

Ellipse with horizontal major axis:

$$\frac{x^2}{a^2}+\frac{y^2}{b^2}=1$$

$$\frac{x^2}{3^2}+\frac{y^2}{b^2}=1$$

Passes through $\left(\sqrt{5}, \frac{4}{3}\right)$

$$\frac{\left(\sqrt{5}\right)^2}{9}+\frac{\left(\frac{4}{3}\right)^2}{b^2}=1$$

$$\frac{5}{9}+\frac{16}{9b^2}=1$$

$$\frac{16}{9b^2}=\frac{4}{9}$$

$$144=36b^2$$

$$b^2=4$$

$$\frac{x^2}{9}+\frac{y^2}{4}=1$$

43.

Center: $(0, 0)$ Focus: $(0, -4)$ Vertex: $(0, 5)$

$$-c=-4$$

$$c=4$$

$$b^2=a^2-c^2$$

$$b^2=25-16$$

$$b^2=9$$

Ellipse with vertical major axis:

$$\frac{x^2}{b^2}+\frac{y^2}{a^2}=1$$

$$\frac{x^2}{b^2}+\frac{y^2}{5^2}=1$$

$$\frac{x^2}{9}+\frac{y^2}{25}=1$$

45.

Center: $(0, 0)$ Vertex: $(3, 0)$

$$\frac{x^2}{a^2}-\frac{y^2}{b^2}=1$$

$$\frac{x^2}{3^2}-\frac{y^2}{b^2}=1$$

Passes through $\left(5, \frac{4}{3}\right)$

$$\frac{5^2}{9}-\frac{\left(\frac{4}{3}\right)^2}{b^2}=1$$

$$\frac{25}{9}-\frac{16}{9b^2}=1$$

$$\frac{16}{9}=\frac{16}{9b^2}$$

$$b^2=1$$

$$\frac{x^2}{9}-y^2=1$$

CHAPTER 5 SECTION 5

SECTION 5.5 PROGRESS CHECK

1a). (Page 329)

Substituting $h=-1$ and $k=-2$,

$$x' = x - h \qquad y' = y - k$$
$$x' = x + 1 \qquad y' = y + 2$$
$$x = x' - 1 \qquad y = y' - 2$$

1b).

$$(x', y') = (3, -3)$$
$$x = x' - 1 \qquad y = y' - 2$$
$$x = 3 - 1 \qquad y = -3 - 2$$
$$x = 2 \qquad y = -5$$
$$(x, y) = (2, -5)$$

2a). (Page 332)

$$\left(4x^2 + 16x \quad\right) + \left(y^2 + 2y \quad\right) = -13$$
$$4\left(x^2 + 4x \quad\right) + \left(y^2 + 2y \quad\right) = -13$$
$$4\left(x^2 + 4x + 4\right) + \left(y^2 + 2y + 1\right) = -13 + 16 + 1$$
$$4(x+2)^2 + (y+1)^2 = 4$$
$$\frac{(x+2)^2}{1} + \frac{(y+1)^2}{4} = 1$$
$$O' = (-2, -1)$$

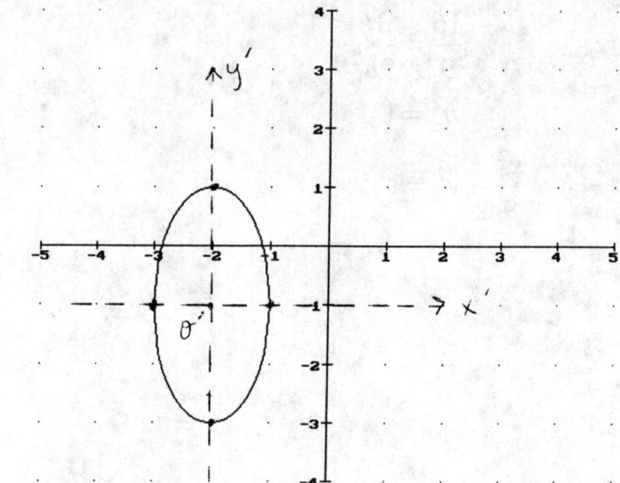

2b).

$$\left(4y^2 - 24y \quad\right) - \left(25x^2 + 50x \quad\right) = 89$$
$$4\left(y^2 - 6y \quad\right) - 25\left(x^2 + 2x \quad\right) = 89$$
$$4\left(y^2 - 6y + 9\right) - 25\left(x^2 + 2x + 1\right) = 89 + 36 - 25$$
$$4(y-3)^2 - 25(x+1)^2 = 100$$
$$\frac{(y-3)^2}{25} - \frac{(x+1)^2}{4} = 1$$
$$O' = (-1, 3)$$

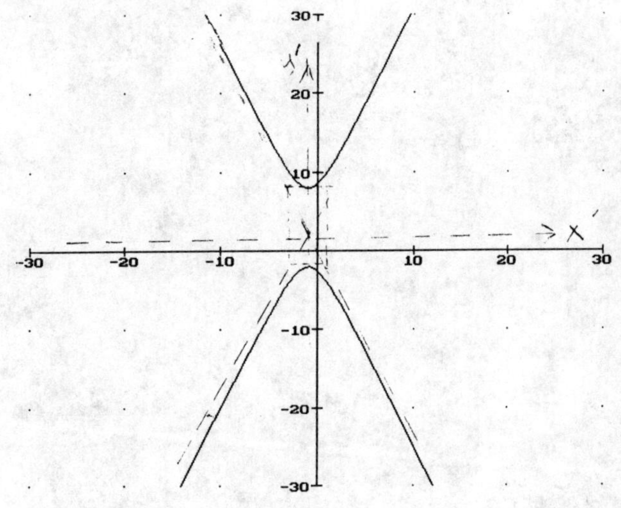

3. (Page 333)

$$y^2 + 4y \quad = 6x - 22$$
$$y^2 + 4y + 4 = 6x - 22 + 4$$
$$(y+2)^2 = 6x - 18$$
$$(y+2)^2 = 6(x-3)$$

Let $y' = y + 2,\ x' = x - 3$

$$(y')^2 = 6x'$$

Vertex: $(3, -2)$

$$4p = 6$$
$$p = \frac{3}{2}$$

4a). (Page 334)

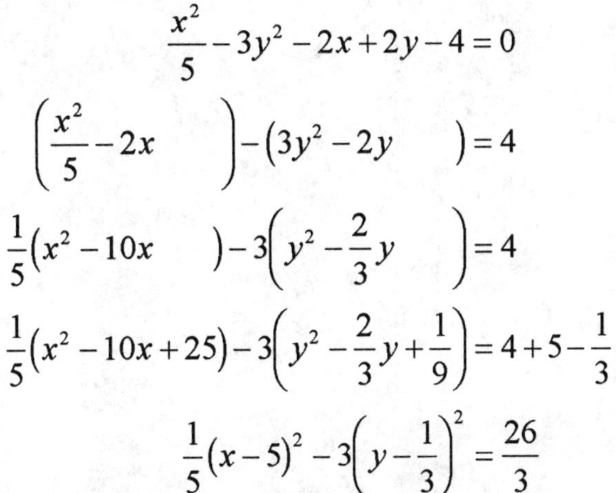

$$\frac{x^2}{5} - 3y^2 - 2x + 2y - 4 = 0$$
$$\left(\frac{x^2}{5} - 2x \qquad\right) - \left(3y^2 - 2y \qquad\right) = 4$$
$$\frac{1}{5}\left(x^2 - 10x \qquad\right) - 3\left(y^2 - \frac{2}{3}y \qquad\right) = 4$$
$$\frac{1}{5}\left(x^2 - 10x + 25\right) - 3\left(y^2 - \frac{2}{3}y + \frac{1}{9}\right) = 4 + 5 - \frac{1}{3}$$
$$\frac{1}{5}(x-5)^2 - 3\left(y - \frac{1}{3}\right)^2 = \frac{26}{3}$$

hyperbola

4b).

$$x^2 - 2y - 3x = 2$$
$$\left(x^2 - 3x \qquad\right) = 2y + 2$$
$$\left(x^2 - 3x + \frac{9}{4}\right) = 2y + 2 + \frac{9}{4}$$
$$\left(x - \frac{3}{2}\right)^2 = 2y + \frac{17}{4}$$
$$\left(x - \frac{3}{2}\right)^2 = 2\left(y + \frac{17}{8}\right)$$

parabola

4c).

$$x^2+y^2-4x-6y=-11$$
$$(x^2-4x\quad)+(y^2-6y\quad)=-11$$
$$(x^2-4x+4)+(y^2-6y+9)=-11+4+9$$
$$(x-2)^2+(y-3)^2=2$$

circle

4d).

$$4x^2+3y^2+6x-10=0$$
$$(4x^2+6x\quad)+(3y^2\quad)=10$$
$$4\left(x^2+\frac{3}{2}x\quad\right)+3(y^2\quad)=10$$
$$4\left(x^2+\frac{3}{2}x+\frac{9}{16}\right)+3(y-0)^2=10+\frac{9}{4}$$
$$4\left(x+\frac{3}{4}\right)^2+3(y-0)^2=\frac{49}{4}$$

ellipse

4e).

$$x^2-2x-3=0$$
$$(x-3)(x+1)=0$$
$$x-3=0 \quad x+1=0$$
$$x=3 \qquad x=-1$$

pair of lines

EXERCISE SET 5.5

In Exercises 1-3, $x'=x-h \quad y'=y-k$

$x'=x+1 \quad y'=y-4$

1.

$(x, y)=(0, 0)$

$x'=0+1=1 \quad y'=0-4=-4$

$(x', y')=(1, -4)$

3.

$(x, y)=(4, 3)$

$x'=4+1=5 \quad y'=3-4=-1$

$(x', y')=(5, -1)$

In Exercises 5 - 7, $x' = x - h \quad y' = y - k$

$x = x' + h \quad y = y' + k$

$x = x' - 3 \quad y = y' + 4$

5.

$(x', y') = (0, 0)$

$x = 0 - 3 = -3 \quad y = 0 + 4 = 4$

$(x, y) = (-3, 4)$

7.

$(x', y') = (4, 3)$

$x = 4 - 3 = 1 \quad y = 3 + 4 = 7$

$(x, y) = (1, 7)$

In Exercises 9-17, the RANGE values given show important features of the functions. Other RANGE values are also acceptable. The display on your calculator may vary from that shown.

9.

$$36x^2 - 100y^2 + 216x + 99 = 0$$

$$36(x^2 + 6x \quad\ ) - 100y^2 = -99$$

$$36(x^2 + 6x + 9) - 100y^2 = -99 + 324$$

$$36(x+3)^2 - 100(y-0)^2 = 225$$

$$\frac{4(x+3)^2}{25} - \frac{4(y-0)^2}{9} = 1$$

$$\frac{(x+3)^2}{\frac{25}{4}} - \frac{(y-0)^2}{\frac{9}{4}} = 1$$

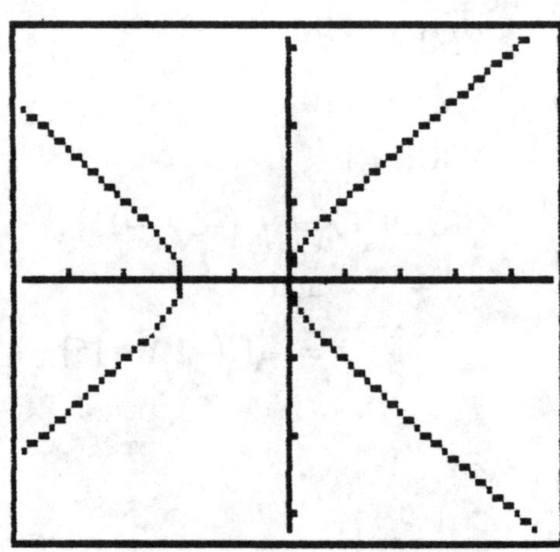

Viewing rectangle: 3 times the EQUAL viewing rectangle

XSCL = 3

YSCL = 3

GRAPH $Y = \dfrac{\sqrt{(36X^2 + 216X + 99)}}{10}$

and

$$Y = -\frac{\sqrt{(36X^2 + 216X + 99)}}{10}$$

11.

$$x^2+4x-y+5=0$$

$$\left(x^2+4x \qquad\right)=y-5$$

$$\left(x^2+4x+4\right)=y-5+4$$

$$(x+2)^2=y-1$$

Viewing rectangle: 3 times the EQUAL viewing rectangle

XSCL = 3

YSCL = 3

GRAPH $Y=X^2+4X+5$

13.

$$y^2+2x+15=0$$

$$y^2=-2x-15$$

$$y^2=-2\left(x+\frac{15}{2}\right)$$

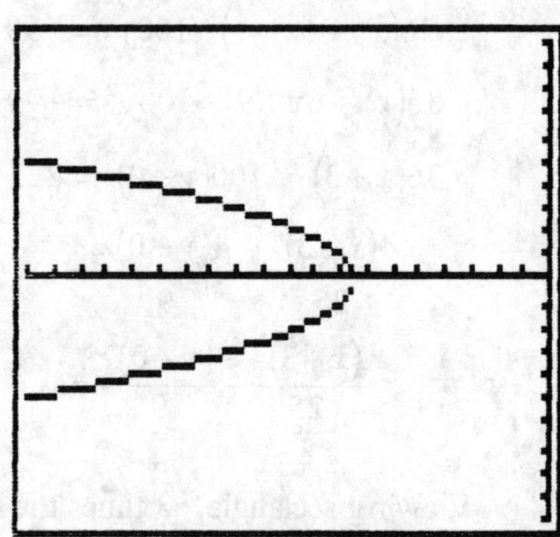

XMIN = −20

XMAX = 0

XSCL = 1

YMIN = −10

YMAX = 10

YSCL = 1

GRAPH $Y=\sqrt{(-2X-15)}$

and

$Y=-\sqrt{(-2X-15)}$

15.

$$x^2 + 4y^2 + 10x - 8y + 13 = 0$$
$$(x^2 + 10x + 25) + 4(y^2 - 2y + 1) = -13 + 25 + 4$$
$$(x+5)^2 + 4(y-1)^2 = 16$$
$$\frac{(x+5)^2}{16} + \frac{(y-1)^2}{4} = 1$$

Viewing rectangle: 2 times the EQUAL viewing rectangle

XSCL = 2

YSCL = 2

GRAPH $Y = \sqrt{\ }\left(4 - \left((X+5)^2 \div 4\right)\right) + 1$

and

$$Y = -\sqrt{\ }\left(4 - \left((X+5)^2 \div 4\right)\right) + 1$$

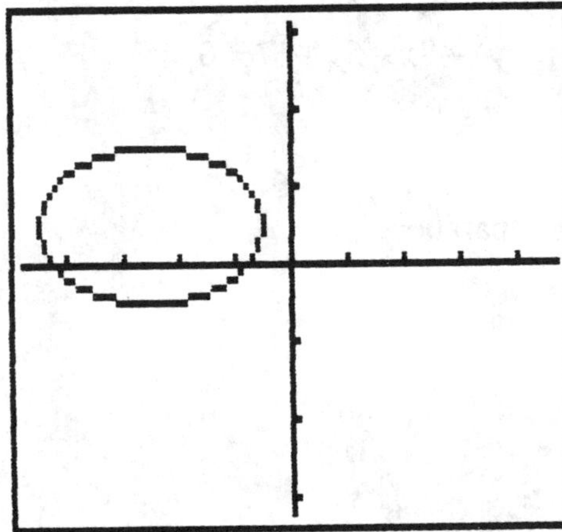

17.

$$x^2 + 9y^2 - 54y + 72 = 0$$
$$x^2 + 9(y^2 - 6y \quad\ ) = -72$$
$$x^2 + 9(y^2 - 6y + 9) = -72 + 81$$
$$(x-0)^2 + 9(y-3)^2 = 9$$
$$\frac{(x-0)^2}{9} + \frac{(y-3)^2}{1} = 1$$

Viewing rectangle: 2 times the EQUAL viewing rectangle

XSCL = 2

YSCL = 2

GRAPH $Y = \sqrt{\ }\left(1 - \left(X^2 \div 9\right)\right) + 3$

and

$$Y = -\sqrt{\ }\left(1 - \left(X^2 \div 9\right)\right) + 3$$

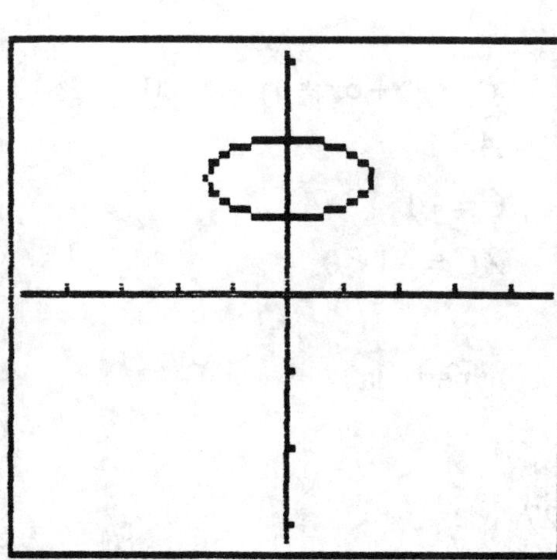

In Exercises 19 - 41, A and C refer to the general equation $Ax^2 + Cy^2 + Dx + Ey + F = 0$

19.

$y^2 - 8x + 6y + 17 = 0$

$A = 0$

parabola

21.

$4x^2 + y^2 + 24x - 4y + 24 = 0$

$A = 4$

$C = 1$

$A \neq C$

$AC = 4 > 0$

possible ellipse, check the constant term

$$4(x^2 + 6x \qquad) + (y^2 - 4y \qquad) = -24$$

$$4(x^2 + 6x + 9) + (y^2 - 4y + 4) = -24 + 36 + 4$$

$$4(x+3)^2 + (y-2)^2 = 16$$

$$\frac{(x+3)^2}{4} + \frac{(y-2)^2}{16} = 1$$

ellipse

23.

$x^2 - y^2 + 6x + 4y - 4 = 0$

$A = 1$

$C = -1$

$AC = -1 < 0$

hyperbola

25.

$25x^2 - 16y^2 + 210x + 96y + 656 = 0$

$A = 25$

$C = -16$

$AC = -400 < 0$

hyperbola

27.

$2x^2 + y - x + 3 = 0$

$C = 0$

parabola

29.

$$4x^2+4y^2-2x+3y-4=0$$
$$A=C=4\neq 0$$

possible circle, check that $r>0$

$$4\left(x^2-\frac{1}{2}x\qquad\right)+4\left(y^2+\frac{3}{4}y\qquad\right)=4$$
$$4\left(x^2-\frac{1}{2}x+\frac{1}{16}\right)+4\left(y^2+\frac{3}{4}y+\frac{9}{64}\right)=4+\frac{1}{4}+\frac{9}{16}$$
$$4\left(x-\frac{1}{4}\right)^2+4\left(y+\frac{3}{8}\right)^2=\frac{77}{16}$$
$$\left(x-\frac{1}{4}\right)^2+\left(y+\frac{3}{8}\right)^2=\frac{77}{64}$$
$$r=\frac{\sqrt{77}}{8}$$

circle

31.

$$36x^2-4y^2+x-y+2=0$$
$$A=36$$
$$C=-4$$
$$AC=-144<0$$

hyperbola

33.

$$16x^2+4y^2-2y+3=0$$
$$A=16$$
$$C=4$$
$$A\neq C$$
$$AC=64>0$$

possible ellipse, check the constant term

$$16x^2+4\left(y^2-\frac{1}{2}y\qquad\right)=-3$$
$$16x^2+4\left(y^2-\frac{1}{2}y+\frac{1}{16}\right)=-3+\frac{1}{4}=-\frac{11}{4}<0$$

no graph

35.

$x^2 + y^2 - 4x - 2y + 8 = 0$

$A = C = 1 \neq 0$

possible circle, check that $r > 0$

$(x^2 - 4x \quad) + (y^2 - 2y \quad) = -8$

$(x^2 - 4x + 4) + (y^2 - 2y + 1) = -8 + 4 + 1 = -3 < 0$

no graph

37.

$4x^2 + 9y^2 - x + 2 = 0$

$A = 4$

$C = 9$

$A \neq C$

$AC = 36 > 0$

possible ellipse, check the constant term

$4\left(x^2 - \frac{1}{4}x \quad\right) + 9y^2 = -2$

$4\left(x^2 - \frac{1}{4}x + \frac{1}{64}\right) + 9y^2 = -2 + \frac{1}{16} = -\frac{31}{16} < 0$

no graph

39.

$4x^2 - 9y^2 + 2x + y + 3 = 0$

$A = 4$

$C = -9$

$AC = -36 < 0$

hyperbola

41.

$x^2 + y^2 - 4x + 4 = 0$

$A = C = 1 \neq 0$

possible circle, check that $r > 0$

$(x^2 - 4x \quad) + y^2 = -4$

$(x^2 - 4x + 4) + y^2 = -4 + 4$

$(x - 2)^2 + (y - 0)^2 = 0$

point: $(2, 0)$

43.

XSCL = 0.5

YSCL = 0.5

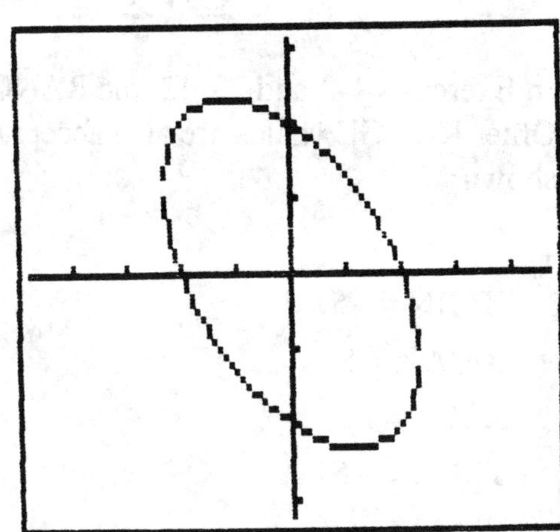

Graph: $Y=\left(-X+\sqrt{\left(X^2-4\left(X^2-1\right)\right)}\right)\div 2$

and

$Y=\left(-X-\sqrt{\left(X^2-4\left(X^2-1\right)\right)}\right)\div 2$

45.

XSCL = 0.5

YSCL = 0.5

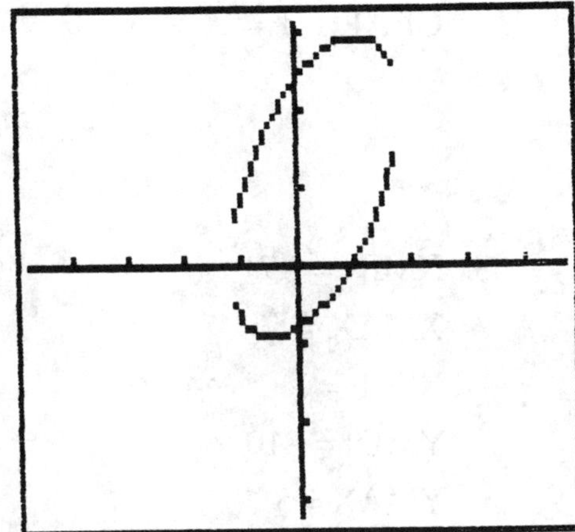

Graph:

$Y=\left(-(21X+12)+\sqrt{\left((21X+12)^2+60\left(-25X^2-3X+7\right)\right)}\right)\div -30$

and

$Y=\left(-(21X+12)-\sqrt{\left((21X+12)^2+60\left(-25X^2-3X+7\right)\right)}\right)\div -30$

47.

XSCL = 3

YSCL = 3

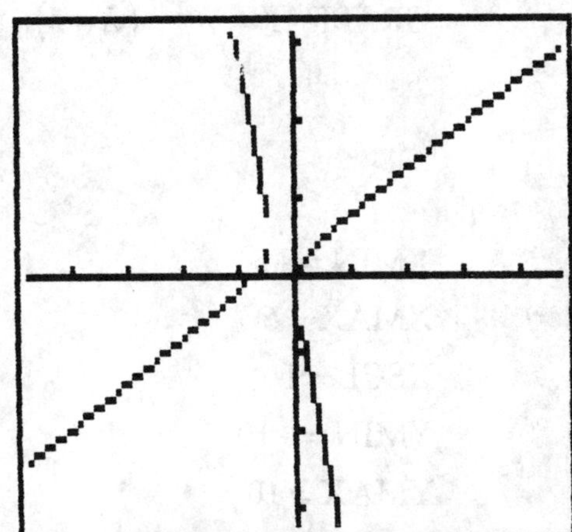

Graph:

$Y=\left(3X+1+\sqrt{\left((-3X-1)^2+4\left(2X^2+4X-1\right)\right)}\right)\div -2$

and

$Y=\left(3X+1-\sqrt{\left((-3X-1)^2+4\left(2X^2+4X-1\right)\right)}\right)\div -2$

CHAPTER 5 REVIEW EXERCISES

In Exercises 1-3 and 33-42, the RANGE values given show important features of the functions. Other RANGE values are also acceptable. The display on your calculator may vary from that shown.

1.
XMIN = −5
XMAX = 5
XSCL = 1
YMIN = −5
YMAX = 5
YSCL = 1
GRAPH $Y = X \div (X+1)$

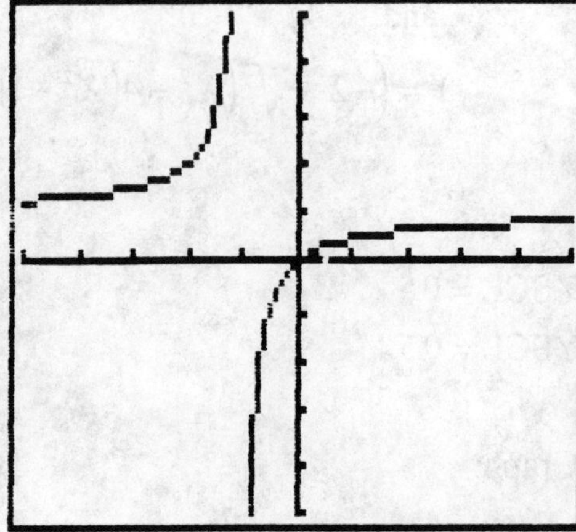

2.
XMIN = −5
XMAX = 5
XSCL = 1
YMIN = −10
YMAX = 5
YSCL = 1
GRAPH $Y = X^2 \div (X+1)$

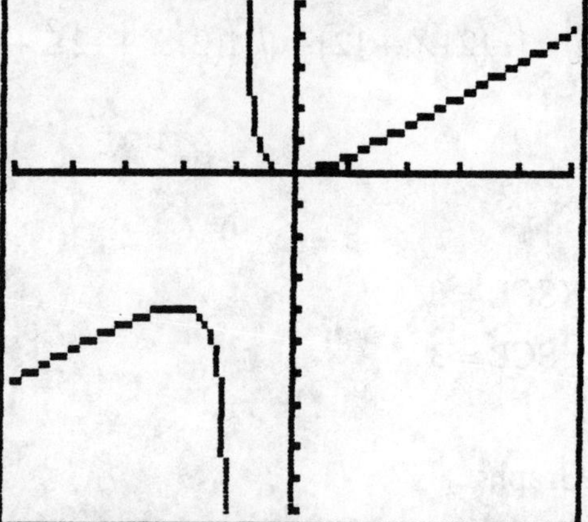

3.
XMIN = −5
XMAX = 5
XSCL = 1
YMIN = −10
YMAX = 10
YSCL = 1
GRAPH $Y = (X^2+2) \div (X^2-1)$

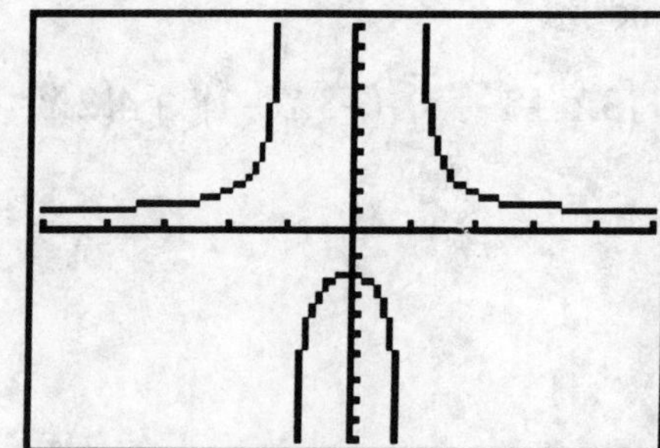

4.

$h=-5 \quad k=2 \quad r=4$

$$(x-h)^2+(y-k)^2=r^2$$
$$(x-(-5))^2+(y-2)^2=4^2$$
$$(x+5)^2+(y-2)^2=16$$

5.

$h=-3 \quad k=3 \quad r=2$

$$(x-h)^2+(y-k)^2=r^2$$
$$(x-(-3))^2+(y-3)^2=2^2$$
$$(x+3)^2+(y-3)^2=4$$

6.

$$(x-2)^2+(y+3)^2=9$$

$x-h=x-2 \qquad y-k=y+3 \qquad r^2=9$

$h=2 \qquad k=-3 \qquad r=3$

center: $(2,-3)$ radius: 3

7.

$$\left(x+\frac{1}{2}\right)^2+(y-4)^2=\frac{1}{9}$$

$x-h=x+\frac{1}{2} \qquad y-k=y-4 \qquad r^2=\frac{1}{9}$

$h=-\frac{1}{2} \qquad k=4 \qquad r=\frac{1}{3}$

center: $\left(-\frac{1}{2},4\right)$ radius: $\frac{1}{3}$

8.

$$x^2+y^2+4x-6y=-10$$
$$(x^2+4x\quad)+(y^2-6y\quad)=-10$$
$$(x^2+4x+4)+(y^2-6y+9)=-10+4+9$$
$$(x+2)^2+(y-3)^2=3$$

center: $(-2,3)$ radius: $\sqrt{3}$

9.

$$2x^2+2y^2-4x+4y=-3$$
$$(2x^2-4x\quad)+(2y^2+4y\quad)=-3$$
$$(x^2-2x\quad)+(y^2+2y\quad)=-\frac{3}{2}$$
$$(x^2-2x+1)+(y^2+2y+1)=-\frac{3}{2}+1+1$$
$$(x-1)^2+(y+1)^2=\frac{1}{2}$$

center: $(1,-1)$ radius: $\frac{\sqrt{2}}{2}$

10.

$$x^2 + y^2 - 6y + 3 = 0$$
$$x^2 + (y^2 - 6y \quad) = -3$$
$$x^2 + (y^2 - 6y + 9) = -3 + 9$$
$$(x-0)^2 + (y-3)^2 = 6$$

center: $(0, 3)$ radius: $\sqrt{6}$

11.

$$x^2 + y^2 - 2x - 2y = 8$$
$$(x^2 - 2x \quad) + (y^2 - 2y \quad) = 8$$
$$(x^2 - 2x + 1) + (y^2 - 2y + 1) = 8 + 1 + 1$$
$$(x-1)^2 + (y-1)^2 = 10$$

center: $(1, 1)$ radius: $\sqrt{10}$

12.

$$(y+5)^2 = 4\left(x - \frac{3}{2}\right)$$

$y - k = y + 5 \qquad x - h = x - \dfrac{3}{2}$

$k = -5 \qquad h = \dfrac{3}{2}$

vertex: $\left(\dfrac{3}{2}, -5\right)$

axis: $y = -5$

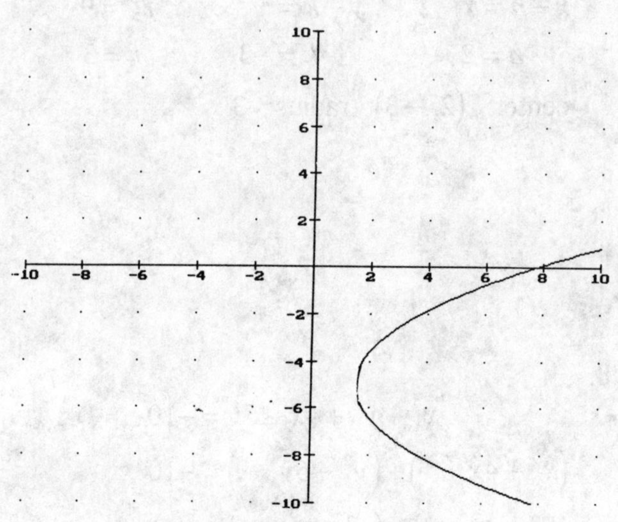

13.

$$(x-1)^2 = 2 - y = -1(y-2)$$

$x - h = x - 1 \qquad y - k = y - 2$

$h = 1 \qquad k = 2$

vertex: $(1, 2)$

axis: $x = 1$

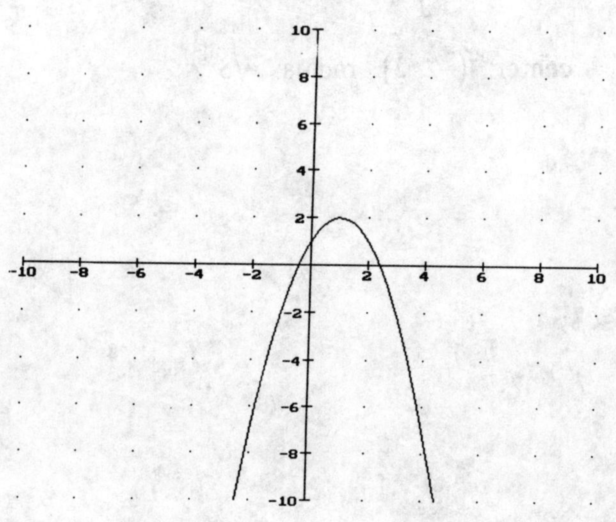

14.

$$y^2 + 3x + 9 = 0$$
$$y^2 = -3x - 9$$
$$y^2 = -3(x + 3)$$
$$x - h = x + 3 \qquad y - k = y - 0$$
$$h = -3 \qquad k = 0$$

vertex: $(-3, 0)$
axis: $y = 0$
opens left since $p < 0$

15.

$$y^2 + 4y + x + 2 = 0$$
$$y^2 + 4y \quad = -x - 2$$
$$y^2 + 4y + 4 = x - 2 + 4$$
$$(y + 2)^2 = -x + 2$$
$$(y + 2)^2 = -(x - 2)$$
$$y - k = y + 2 \qquad x - h = x - 2$$
$$k = -2 \qquad h = 2$$

vertex: $(2, -2)$
axis: $y = -2$
opens left since $p < 0$

16.

$$2x^2 - 12x - y + 16 = 0$$
$$2x^2 - 12x \quad = y - 16$$
$$x^2 - 6x \quad = \frac{1}{2}y - 8$$
$$x^2 - 6x + 9 = \frac{1}{2}y - 8 + 9$$
$$(x - 3)^2 = \frac{1}{2}y + 1$$
$$(x - 3)^2 = \frac{1}{2}(y + 2)$$
$$x - h = x - 3 \qquad y - k = y + 2$$
$$h = 3 \qquad k = -2$$

vertex: $(3, -2)$
axis: $x = 3$
opens up since $p > 0$

17.

$$x^2 + 4x + 2y + 5 = 0$$
$$x^2 + 4x \quad = -2y - 5$$
$$x^2 + 4x + 4 = -2y - 5 + 4$$
$$(x + 2)^2 = -2y - 1$$
$$(x + 2)^2 = -2\left(y + \frac{1}{2}\right)$$
$$x - h = x + 2 \qquad y - k = y + \frac{1}{2}$$
$$h = -2 \qquad k = -\frac{1}{2}$$

vertex: $\left(-2, -\frac{1}{2}\right)$
axis: $x = -2$
opens down since $p < 0$

18.

$$y^2-2y-4x+1=0$$
$$y^2-2y \quad =4x-1$$
$$y^2-2y+1=4x-1+1$$
$$(y-1)^2=4x$$
$$y-k=y-1 \qquad x-h=x$$
$$k=1 \qquad h=0$$

vertex: $(0, 1)$
axis: $y=1$
opens right since $p>0$

19.

$$x^2+6x+4y+9=0$$
$$x^2+6x \quad =-4y-9$$
$$x^2+6x+9=-4y-9+9$$
$$(x+3)^2=-4y$$
$$x-h=x+3 \qquad y-k=y$$
$$h=-3 \qquad k=0$$

vertex: $(-3, 0)$
axis: $x=-3$
opens down since $p<0$

20.

$$x^2=-\frac{2}{3}y$$
$$4p=-\frac{2}{3}$$
$$p=-\frac{1}{6}$$

focus: $\left(0, -\frac{1}{6}\right)$

directrix: $y=\frac{1}{6}$

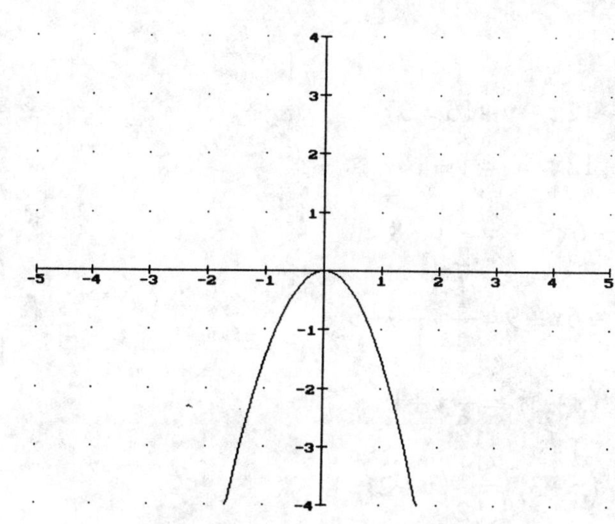

21.

$$3y^2+2x=0$$
$$y^2=-\frac{2}{3}x$$
$$4p=-\frac{2}{3}$$
$$p=-\frac{1}{6}$$

focus: $\left(-\frac{1}{6},0\right)$

directrix: $x=\frac{1}{6}$

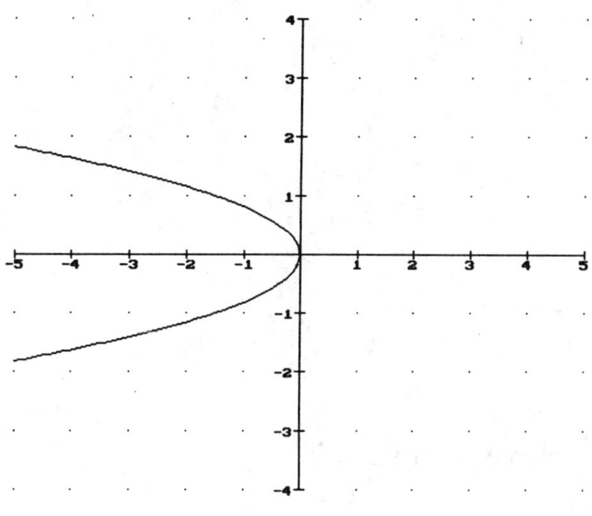

22.

directrix $y=\frac{7}{4}$
$$p=-\frac{7}{4}$$

$$x^2=4py$$
$$x^2=4\left(-\frac{7}{4}\right)y$$
$$x^2=-7y$$

23.

axis: $y-$axis

passes through $\left(1,\frac{5}{2}\right)$

$$x^2=4py$$
$$1^2=4p\left(\frac{5}{2}\right)$$
$$1=10p$$
$$p=\frac{1}{10}$$
$$x^2=4\left(\frac{1}{10}\right)y$$
$$x^2=\frac{2}{5}y$$

or

$$y=\frac{5}{2}x^2$$

24.

$9x^2 - 4y^2 = 36$

$\frac{x^2}{4} - \frac{y^2}{9} = 1$

x-intercepts: Let $y = 0$

$\frac{x^2}{4} = 1$

$x^2 = 4$

$x = \pm 2$

y-intercepts: Let $x = 0$

$\frac{-y^2}{9} = 1$

No solution

No y-intercepts

25.

$9x^2 + y^2 = 9$

$\frac{x^2}{1} + \frac{y^2}{9} = 1$

x-intercepts: Let $y = 0$

$x^2 = 1$

$x = \pm 1$

y-intercepts: Let $x = 0$

$\frac{y^2}{9} = 1$

$y^2 = 9$

$y = \pm 3$

26.

$5x^2 + 7y^2 = 35$

$\frac{x^2}{7} + \frac{y^2}{5} = 1$

x-intercepts: Let $y = 0$

$\frac{x^2}{7} = 1$

$x^2 = 7$

$x = \pm\sqrt{7}$

y-intercepts: Let $x = 0$

$\frac{y^2}{5} = 1$

$y^2 = 5$

$y = \pm\sqrt{5}$

27.

$9x^2 - 16y^2 = 144$

$\frac{x^2}{16} - \frac{y^2}{9} = 1$

x-intercepts: Let $y = 0$

$\frac{x^2}{16} = 1$

$x^2 = 16$

$x = \pm 4$

y-intercepts: Let $x = 0$

$-\frac{y^2}{9} = 1$

No solution

No y-intercepts

28.

$$3x^2 + 4y^2 = 9$$

$$\frac{x^2}{3} + \frac{4y^2}{9} = 1$$

$$\frac{x^2}{3} + \frac{y^2}{\frac{9}{4}} = 1$$

x-intercepts: Let $y = 0$

$$\frac{x^2}{3} = 1$$

$$x^2 = 3$$

$$x = \pm\sqrt{3}$$

y-intercepts: Let $x = 0$

$$\frac{y^2}{\frac{9}{4}} = 1$$

$$y^2 = \frac{9}{4}$$

$$y = \pm\frac{3}{2}$$

29.

$$3y^2 - 5x^2 = 20$$

$$\frac{3y^2}{20} - \frac{x^2}{4} = 1$$

$$\frac{y^2}{\frac{20}{3}} - \frac{x^2}{4} = 1$$

x-intercepts: Let $y = 0$

$$\frac{-x^2}{4} = 1$$

No solution

No x-intercepts

y-intercepts: Let $x = 0$

$$\frac{y^2}{\frac{20}{3}} = 1$$

$$y^2 = \frac{20}{3}$$

$$y = \pm\sqrt{\frac{20}{3}} = \pm\frac{2\sqrt{15}}{3}$$

30.

$$4x^2 - 4y^2 = 1$$

$$\frac{x^2}{\frac{1}{4}} - \frac{y^2}{\frac{1}{4}} = 1$$

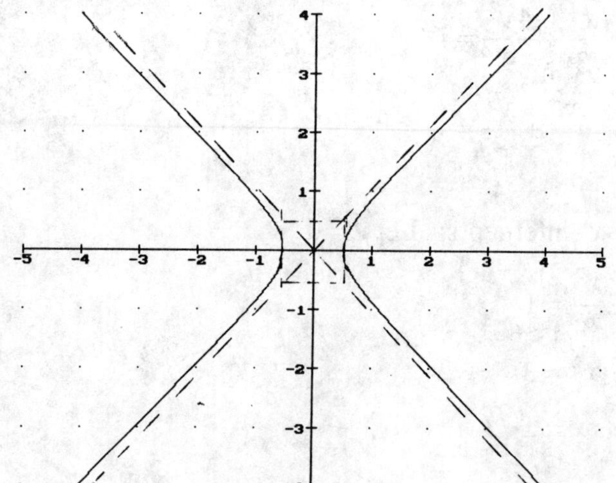

x-intercepts: Let $y = 0$

$$\frac{x^2}{\frac{1}{4}} = 1$$

$$x^2 = \frac{1}{4}$$

$$x = \pm\frac{1}{2}$$

$$a^2 = \frac{1}{4} \qquad b^2 = \frac{1}{4}$$

$$a = \pm\frac{1}{2} \qquad b = \pm\frac{1}{2}$$

asymptotes: $y = \pm\frac{b}{a}x$

$$y = \pm\frac{\frac{1}{2}}{\frac{1}{2}}x$$

$$y = \pm x$$

31.

$$9y^2 - 4x^2 = 36$$

$$\frac{y^2}{4} - \frac{x^2}{9} = 1$$

y-intercepts: Let $x = 0$

$$\frac{y^2}{4} = 1$$

$$y^2 = 4$$

$$y = \pm 2$$

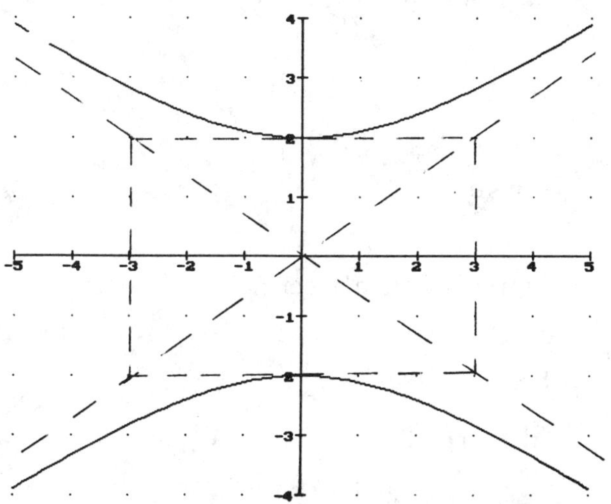

$a^2 = 4$ $\qquad$ $b^2 = 9$

$a = \pm 2$ $\qquad$ $b = \pm 3$

asymptotes: $y = \pm \frac{a}{b} x$

$$y = \pm \frac{2}{3} x$$

32.

vertex: $(2, 0)$ $\qquad$ center: $(0, 0)$

$$\frac{x^2}{a^2} - \frac{y^2}{b^2} = 1$$

$$\frac{x^2}{4} - \frac{y^2}{b^2} = 1$$

Passes through $(3, 1)$: $\frac{3^2}{4} - \frac{1^2}{b^2} = 1$

$$\frac{9}{4} - \frac{1}{b^2} = 1$$

$$\frac{5}{4} = \frac{1}{b^2}$$

$$b^2 = \frac{4}{5}$$

$$\frac{x^2}{4} - \frac{y^2}{\frac{4}{5}} = 1$$

$$\frac{x^2}{4} - \frac{5y^2}{4} = 1$$

33.

$2x^2 - 4x + y^2 = 0$

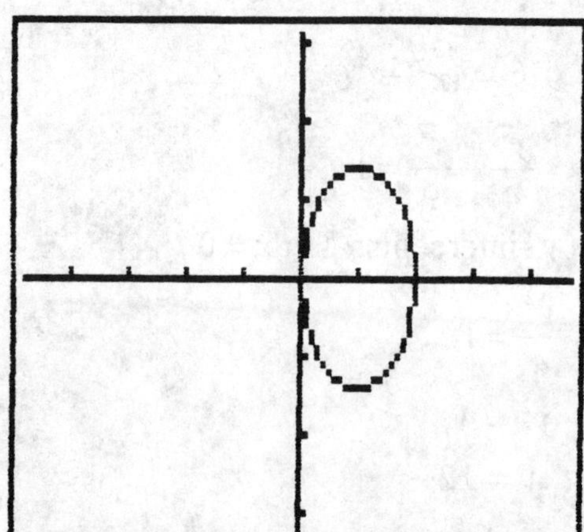

$A = 2 \qquad C = 1$

$A \neq C \qquad AC > 0$

ellipse

EQUAL viewing rectangle

XSCL = 1

YSCL = 1

GRAPH: $Y = \sqrt{}\left(4X - 2X^2\right)$

and

$Y = -\sqrt{}\left(4X - 2X^2\right)$

34.

$4y + x^2 - 2x = 1$

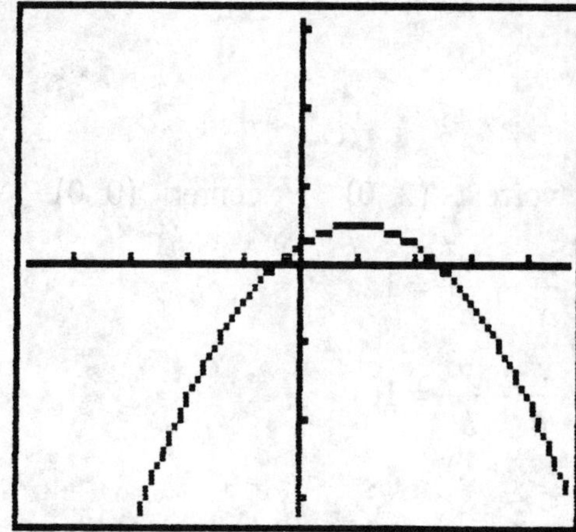

$C = 0$

parabola

EQUAL viewing rectangle

XSCL = 1

YSCL = 1

GRAPH: $Y = \left(-X^2 + 2X + 1\right) \div 4$

35.

$x^2 - 2x + y^2 - 4y = -6$

$A = C \neq 0$

$\left(x^2 - 2x + 1\right) + \left(y^2 - 4y + 4\right) = -6 + 1 + 4 = -1 < 0$

Not a conic section.

36.

$$-x^2 - 4x + y^2 + 4 = 0$$

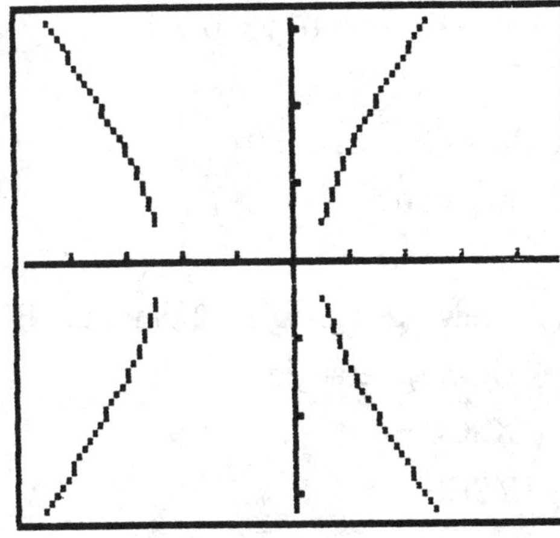

$A \neq C \qquad AC < 0$

hyperbola

Viewing rectangle: 2 times the EQUAL viewing retangle

XSCL = 2

YSCL = 2

GRAPH: $Y = \sqrt{\ }(X^2 + 4X - 4)$

and

$Y = -\sqrt{\ }(X^2 + 4X - 4)$

37.

$$y^2 + 2x - 4 = x^2 - 2x$$
$$y^2 = x^2 - 4x + 4$$
$$y^2 = (x-2)^2$$
$$y = \pm(x-2)$$

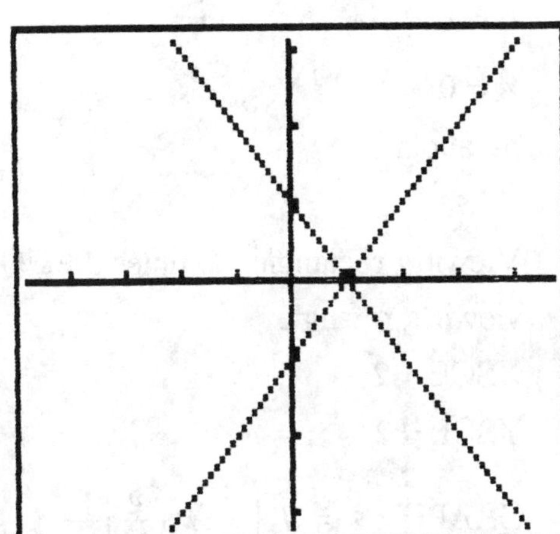

two lines

Viewing rectangle: 2 times the EQUAL viewing retangle

XSCL = 2

YSCL = 2

GRAPH: $Y = X - 2$

and

$Y = -X + 2$

38.

$$x^2 - 4x + 2y = 6 + y^2$$
$$x^2 - 4x - y^2 + 2y = 6$$

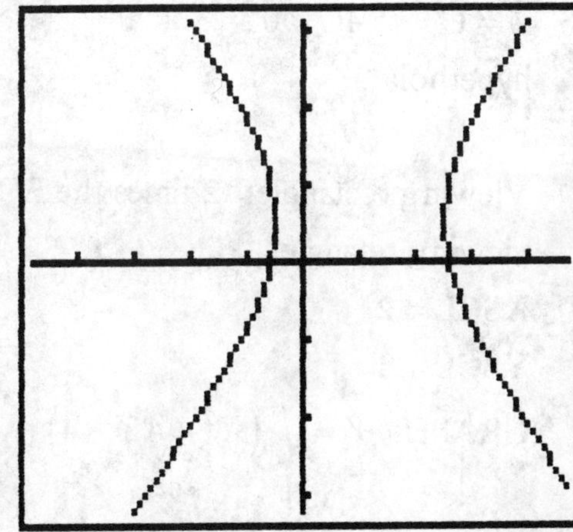

$A \neq C \qquad AC < 0$

hyperbola

Viewing rectangle: 2 times the EQUAL viewing retangle

XSCL = 2

YSCL = 2

GRAPH: $Y = \sqrt{\left(X^2 - 4X - 5\right)} + 1$

and

$Y = -\sqrt{\left(X^2 - 4X - 5\right)} + 1$

39.

$$2y^2 + 6y - 3x + 2 = 0$$

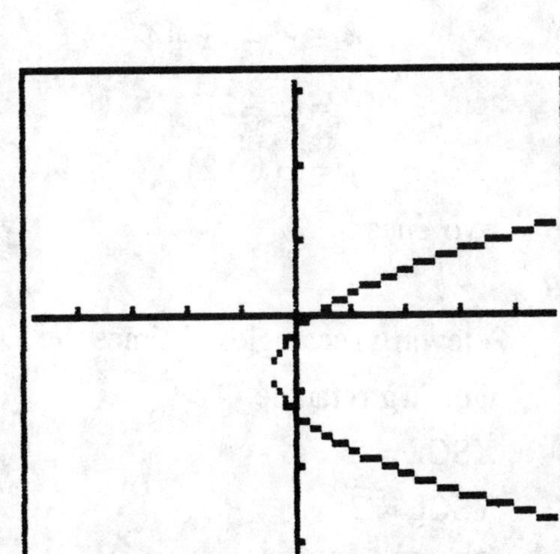

$A = 0$

parabola

Viewing rectangle: 2 times the EQUAL viewing retangle

XSCL = 2

YSCL = 2

GRAPH: $Y = \sqrt{\left(\frac{3}{2}X + \frac{5}{4}\right)} - \frac{3}{2}$

and

$Y = -\sqrt{\left(\frac{3}{2}X + \frac{5}{4}\right)} - \frac{3}{2}$

40.
$6x^2 - 7y^2 - 5x + 6y = 0$

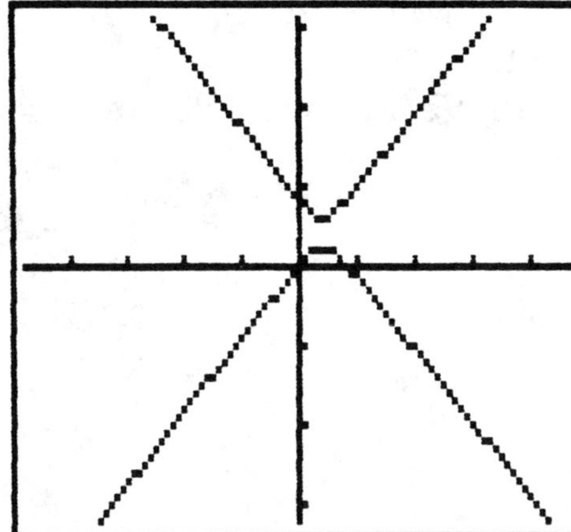

$A \neq C \qquad AC < 0$
hyperbola

Viewing rectangle: EQUAL viewing retangle
XSCL = 1
YSCL = 1

GRAPH: $Y = \sqrt{\left(\frac{6}{7}X^2 - \frac{5}{7}X + \frac{9}{49}\right)} - \frac{3}{7}$

and

$Y = -\sqrt{\left(\frac{6}{7}X^2 - \frac{5}{7}X + \frac{9}{49}\right)} - \frac{3}{7}$

41.
$2x^2 + y^2 + 12x - 2y + 17 = 0$

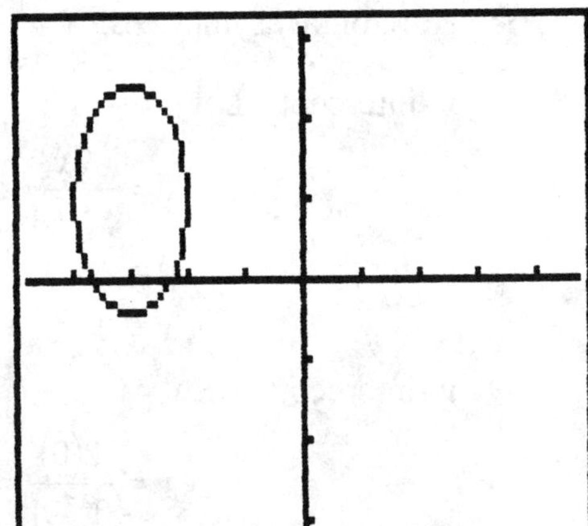

$A \neq C \qquad AC > 0$
ellipse

Viewing rectangle: EQUAL viewing retangle
XSCL = 1
YSCL = 1

GRAPH: $Y = \sqrt{\left(-2X^2 - 12X - 16\right)} + 1$

and

$Y = -\sqrt{\left(-2X^2 - 12X - 16\right)} + 1$

42.
$9x^2 + 4y^2 = -36$
$9x^2 + 4y^2 \geq 0 \qquad -36 < 0$
No graph

CHAPTER 5 REVIEW TEST

1.

Vertical asymptotes: $x^2-1=0$

$$x^2=1$$
$$x=\pm 1$$
$$f(x)=\frac{2x}{x^2-1}$$
$$=\frac{x^2\left(\frac{2}{x}\right)}{x^2\left(1-\frac{1}{x^2}\right)}$$
$$=\frac{\frac{2}{x}}{1-\frac{1}{x^2}}$$

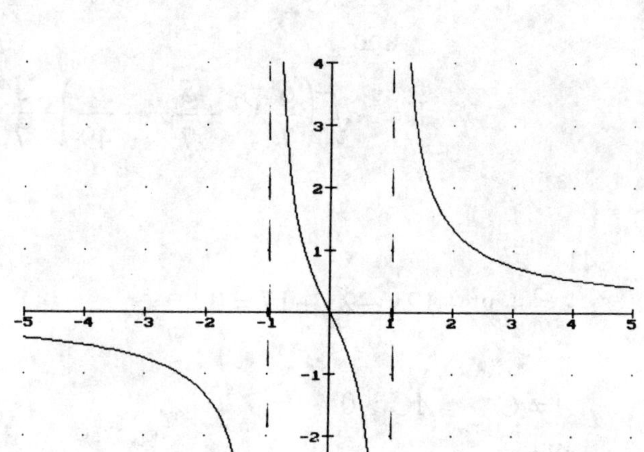

Horizontal asymptotes: $y=\frac{0}{1-0}=0$

x-intercept: Let $y=0$

$$0=\frac{2x}{x^2-1}$$
$$0=2x$$
$$x=0$$

y-intercept: Let $x=0$

$$y=\frac{2(0)}{0^2-1}=0$$

2.

$r=6 \qquad h=2 \qquad k=-3$

$$(x-h)^2+(y-k)^2=r^2$$
$$(x-2)^2+(y+3)^2=36$$

3.

$$x^2+y^2-2x+4y=-1$$
$$\left(x^2-2x \quad\right)+\left(y^2+4y \quad\right)=-1$$
$$\left(x^2-2x+1\right)+\left(y^2+4y+4\right)=-1+1+4$$
$$(x-1)^2+(y+2)^2=4$$

center: $(1,-2)$ radius: 2

4.

$$x^2 - 4x + y^2 = 1$$
$$(x^2 - 4x + 4) + y^2 = 1 + 4$$
$$(x-2)^2 + (y-0)^2 = 5$$

center: $(2, 0)$ radius: $\sqrt{5}$

5.

$$x^2 + 6x + 2y + 7 = 0$$
$$x^2 + 6x = -2y - 7$$
$$x^2 + 6x + 9 = -2y - 7 + 9$$
$$(x+3)^2 = -2y + 2$$
$$(x+3)^2 = -2(y-1)$$

$x - h = x + 3$ $y - k = y - 1$
$h = -3$ $k = 1$

vertex: $(-3, 1)$
axis: $x = -3$

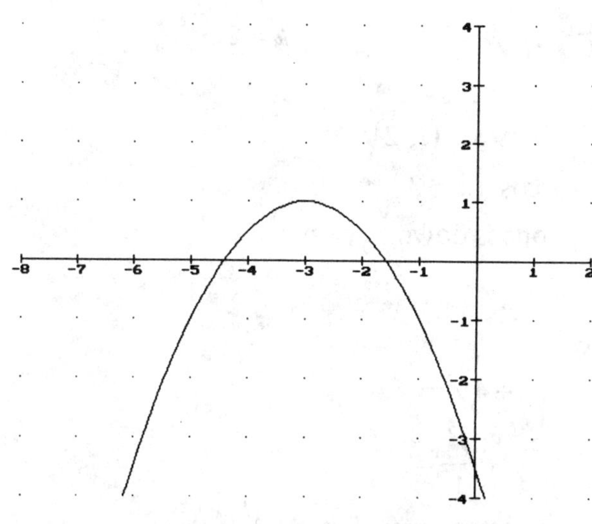

6.

$$y^2 - 4x - 4y + 8 = 0$$
$$y^2 - 4y = 4x - 8$$
$$y^2 - 4y + 4 = 4x - 8 + 4$$
$$(y-2)^2 = 4x - 4$$
$$(y-2)^2 = 4(x-1)$$

$y - k = y - 2$ $x - h = x - 1$
$k = 2$ $h = 1$

vertex: $(1, 2)$
axis: $y = 2$

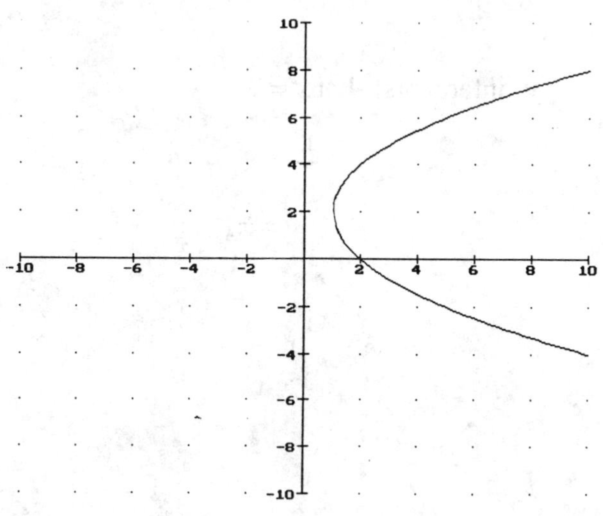

7.

$$x^2-6x+2y+5=0$$
$$x^2-6x \quad =-2y-5$$
$$x^2-6x+9=-2y-5+9$$
$$(x-3)^2=-2y+4$$
$$(x-3)^2=-2(y-2)$$

$x-h=x-3$ $\quad$ $y-k=y-2$

$h=3$ $\quad$ $k=2$

vertex: $(3, 2)$

axis: $x=3$

opens down since $p<0$

8.

$$y^2+8y-x+14=0$$
$$y^2+8y \quad =x-14$$
$$y^2+8y+16=x-14+16$$
$$(y+4)^2=x+2$$

$x-h=x+2$ $\quad$ $y-k=y+4$

$h=-2$ $\quad$ $k=-4$

vertex: $(-2, -4)$

axis: $y=-4$

opens right since $p>0$

9.

$$x^2+4y^2=4$$
$$\frac{x^2}{4}+\frac{y^2}{1}=1$$

x-intercepts: Let $y=0$

$$\frac{x^2}{4}=1$$
$$x^2=4$$
$$x=\pm 2$$

y-intercepts: Let $x=0$

$$\frac{y^2}{1}=1$$
$$y=\pm 1$$

10.

$$4y^2-9x^2=36$$
$$\frac{y^2}{9}-\frac{x^2}{4}=1$$

x-intercepts: Let $y=0$

$$\frac{-x^2}{4}=1$$
$$x^2=-4$$

No solution

No x-intercepts

y-intercepts: Let $x=0$

$$\frac{y^2}{9}=1$$
$$y^2=9$$
$$y=\pm 3$$

11.

$$4x^2 - 4y^2 = 1$$

$$\frac{x^2}{\frac{1}{4}} - \frac{y^2}{\frac{1}{4}} = 1$$

x-intercepts: Let $y = 0$

$$\frac{x^2}{\frac{1}{4}} = 0$$

$$x^2 = \frac{1}{4}$$

$$x = \pm\frac{1}{2}$$

y-intercepts: Let $x = 0$

$$\frac{-y^2}{\frac{1}{4}} = 1$$

$$y^2 = -\frac{1}{4}$$

No solution

No y-intercepts

12.

$$9x^2 - y^2 = 9$$

$$\frac{x^2}{1} - \frac{y^2}{9} = 1$$

x-intercepts: Let $y = 0$

$$x^2 = 1$$

$$x = \pm 1$$

asymptotes: $y = \pm\frac{b}{a}x$

$$y = \pm 3x$$

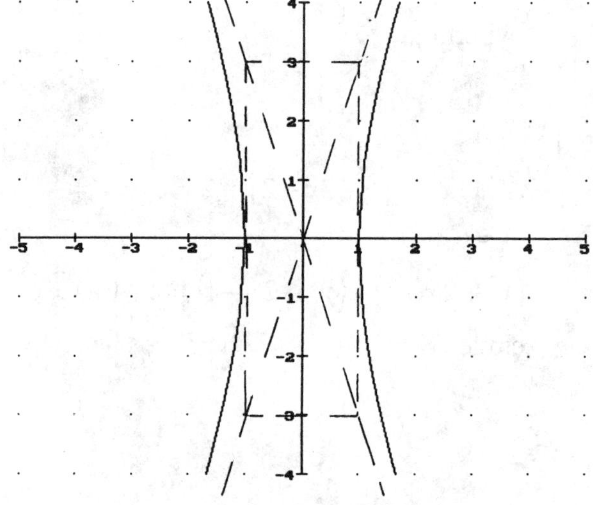

13.

$$4y^2 - x^2 = 1$$

$$\frac{y^2}{\frac{1}{4}} - \frac{x^2}{1} = 1$$

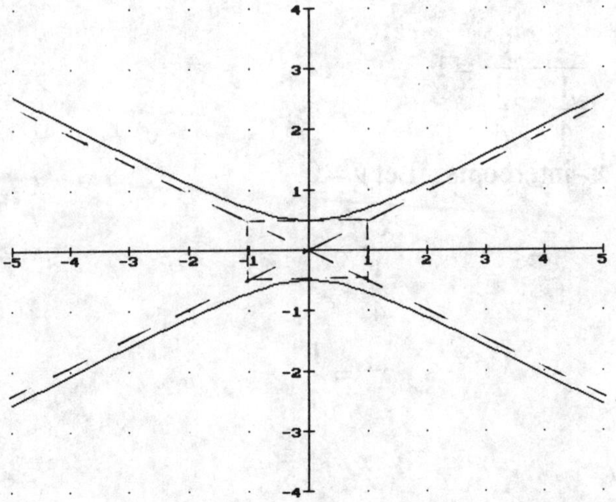

y-intercepts: Let $x = 0$

$$\frac{y^2}{\frac{1}{4}} = 1$$

$$y^2 = \frac{1}{4}$$

$$y = \pm\frac{1}{2}$$

asymptotes: $y = \pm\frac{a}{b}x$

$$y = \pm\frac{1}{4}x$$

14.

$$5y^2 - 4x^2 - 6x + 2 = 0$$

$A \neq C, \qquad AC < 0$

hyperbola

15.

$$3x^2 - 5x + 6y = 3$$

$C = 0$

parabola

16.

$$x^2 + y^2 + 2x - 2y - 2 = 0$$

$A = C \neq 0$

$$(x^2 + 2x + 1) + (y^2 - 2y + 1) = 2 + 1 + 1 > 0$$

circle

17.

$$x^2 + 9y^2 - 4x + 6y + 4 = 0$$

$A \neq C \qquad AC > 0$

$$(x^2 - 4x + 4) + 9\left(y^2 + \frac{2}{3}y + \frac{1}{9}\right) = 0 + 4 + 1 > 0$$

ellipse

CHAPTER 5 ADDITIONAL PRACTICE EXERCISES

1. Determine the domain and intercepts of $f(x)=\dfrac{2x-3}{x^2-25}$.

2. Find an equation of the circle with center $(-6, 4)$ and radius $\dfrac{1}{4}$.

3. Find the center and radius of the circle $x^2-2x+y^2+6y-9=0$.

4. Find an equation of the circle with center $(-1, -5)$ and passes through the point $(0, -4)$.

5. Determine the equation of the parabola that has its vertex at the origin and focus at $(-2, 0)$.

6. Determine the vertex and axis of the parabola $y^2+4y+2x-5=0$.

7. Find the intercepts and sketch the graph of $4x^2+y^2=16$.

8. Use the asymptotes and intercepts to sketch the graph of $16x^2-4y^2=64$.

9. Identify the conic section $3x^2-4y^2+8x-9y+5=0$.

10. Identify the conic section $9y^2-8y+7x=10$.

CHAPTER 5 PRACTICE TEST

1. Determine the domain and intercepts of $f(x)=\dfrac{x+7}{x^2+x-30}$.

2. Determine the focus and directrix of the parabola $x^2-6y=0$.

3. Determine the equation of the parabola with vertex at the origin and directrix $x=\frac{1}{2}$.

4. Determine the equation of the parabola with vertex at the origin, the axis is the x-axis and $p=3$.

5. Determine in which direction the parabola, $5x^2-6y=0$, opens.

6. Find an equation of the circle with center $(-2, 9)$ and radius 3.

7. Find an equation of the circle with center $(-2, -5)$ that passes through the point $(1, 3)$.

8. The two points $(4, -1)$ and $(8, 2)$ are the endpoints of the diameter of a circle. Write the equation of the circle in standard form.

9. Determine the center and radius of the circle $x^2-10x+y^2=8$.

10. Find the circumference of the circle whose equation is $x^2+y^2-4y+1=0$.

In Exercises 11-15, identify the conic section.

11. $5x^2+5y^2-3x+4y-9=0$.

12. $x^2-9x+4y+16=0$.

13. $4x^2-3y^2-2x+y=10$.

14. $2x^2+3y^2-2x=8$.

15.
$$\frac{x^2}{4}+y^2+6x+8y=4.$$

In Exercises 16-20, graph.

16.
$$x^2-y-6=0.$$

17.
$$x^2+y^2-4x-6y-3=0.$$

18.
$$2x^2+8y^2=32.$$

19.
$$x^2-4y^2-4x-12=0.$$

20.
$$x^2+y^2=12.$$

CHAPTER 6 SECTION 1

EXERCISE SET 6.1

1a).

$$F^{-1}[F(4)]=4$$

1b).

$$(F\circ F^{-1})(-5)=F[F^{-1}(-5)]=-5$$

1c).

$$F^{-1}[F(y+1)]=y+1$$

1d).

TENS viewing rectangle

XSCL = 1

YSCL = 1

GRAPH $X=T$ and $Y=2T-2$

GRAPH $Y=2T-2$ and $X=T$

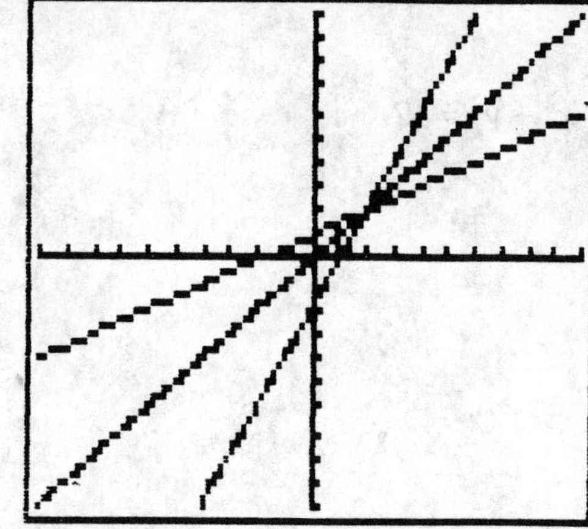

3.

$f(x)=\sqrt{x}$

passes horizontal line test,

is one-to-one

$y=\sqrt{x}$

$y^2=x$

$f^{-1}(x)=x^2,\ x\geq 0$

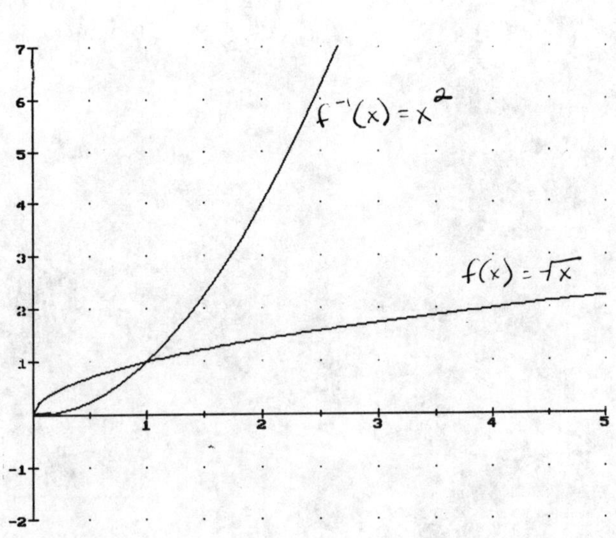

5.

$$f(x)=\frac{1}{x}$$

passes horizontal line test, is one-to-one

$$y=\frac{1}{x}$$

$$x=\frac{1}{y}$$

$$f^{-1}(x)=\frac{1}{x}$$

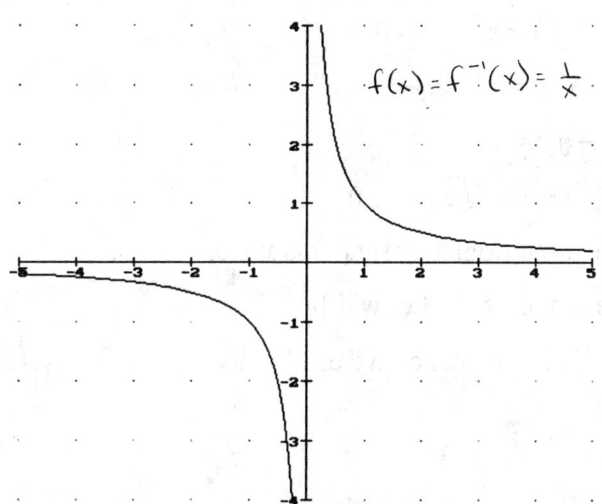

7.

$$f(x)=x^3+2x+3$$

$$3=x^3+2x+3$$

$$0=x^3+2x$$

$$0=x(x^2+2)$$

$$x=0$$

$$f^{-1}(3)=0$$

9a).

Yes, each x and each $G(x)$ are used once and only once.

9b).

$$G^{-1}(3)=5$$

9c).

$$G^{-1}(-4)=0$$

9d).

x	17	20	−4	3	10
$G^{-1}(x)$	−10	−5	0	5	10

11.

$$f(x)=x^2$$

11a).

$x\geq 0$

11b).

$x\leq 0$

13.

$$f(x) = x^4\ ,\quad x \le 0$$
$$y = x^4$$
$$-\sqrt[4]{y} = x$$
$$f^{-1}(x) = -\sqrt[4]{x}\ ,\quad x \ge 0$$

Since the domain of $f(x)$ is $x \le 0$,
the range of $f^{-1}(x)$ will be
$f^{-1}(x) \le 0$, hence we used $-\sqrt[4]{x}$.

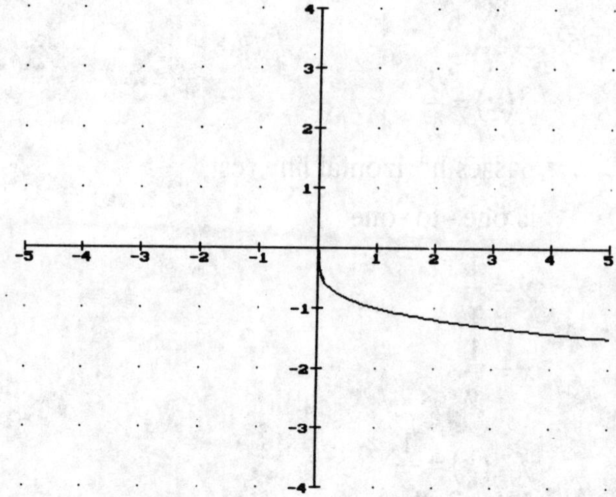

CHAPTER 6 SECTION 2

SECTION 6.2 PROGRESS CHECK

1a). (Page 348)

$$2^8 = 2^{x+1}$$
$$8 = x+1$$
$$7 = x$$

1b).

$$4^{2x+1} = 4^{11}$$
$$2x+1 = 11$$
$$2x = 10$$
$$x = 5$$

1c).

$$8^{x+1} = 2$$
$$\left(2^3\right)^{x+1} = 2$$
$$2^{3x+3} = 2^1$$
$$3x+3 = 1$$
$$3x = -2$$
$$x = -\frac{2}{3}$$

2. (Page 351)

$$A = P\left(1+\frac{r}{k}\right)^{kt}$$
$$A = 5000\left(1+\frac{0.06}{2}\right)^{2(12)}$$
$$= 10163.97$$
$$\$10,163.97$$

3. (Page 353)

$$A = Pe^{rt}$$
$$= 10000e^{0.10(6)}$$
$$= 18221.19$$
$$\$18,221.19$$

4. (Page 354)

$$65-35 = 30 \text{ years}$$
$$A = Pe^{rt}$$
$$20000 = Pe^{0.10(30)}$$
$$P = 995.74$$
$$\$995.74$$

5. (Page 355)

$$Q(t) = q_0 e^{0.005t}$$
$$Q(0) = 100$$
$$100 = q_0 e^{0.005(0)}$$
$$100 = q_0$$

$$Q(t) = 100e^{0.005t}$$

1 hour = 60 minutes $Q(60) = 100e^{0.005(60)}$

$$= 135$$

135 bacteria

6. (Page 357)

$$Q(t) = q_0 e^{-0.4t}$$
$$Q(0) = 200$$
$$200 = q_0 e^{-0.4(0)}$$
$$200 = q_0$$

$$Q(t) = 200e^{-0.4(t)}$$

0.1 minute $= 0.1(60)$ seconds $= 6$ seconds

$$Q(6) = 200e^{-0.4(6)}$$
$$= 18.14$$

18.14 g

EXERCISE SET 6.2

The graphs in Exercises 1-11 will match the graphs of 13-23 respectively. The RANGE values given show important features of the functions. Other RANGE values are also acceptable. The display on your calculator may vary from that shown.

1.

XMIN = −5
XMAX = 5
XSCL = 1
YMIN = −1
YMAX = 10
YSCL = 1
GRAPH $Y = 4^X$

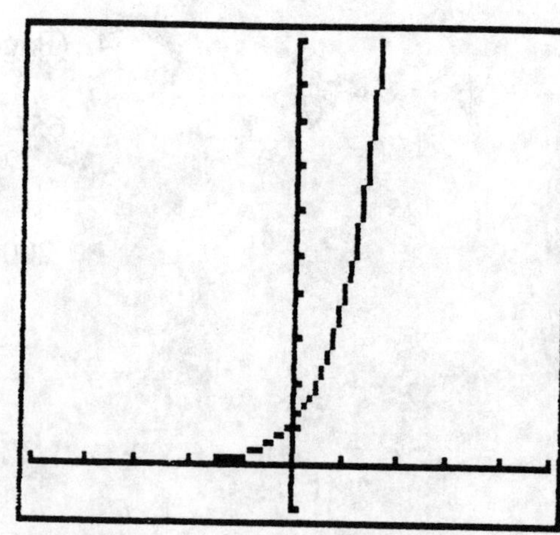

3\.

XMIN = −5
XMAX = 5
XSCL = 1
YMIN = −1
YMAX = 10
YSCL = 1
GRAPH $Y = 10^X$

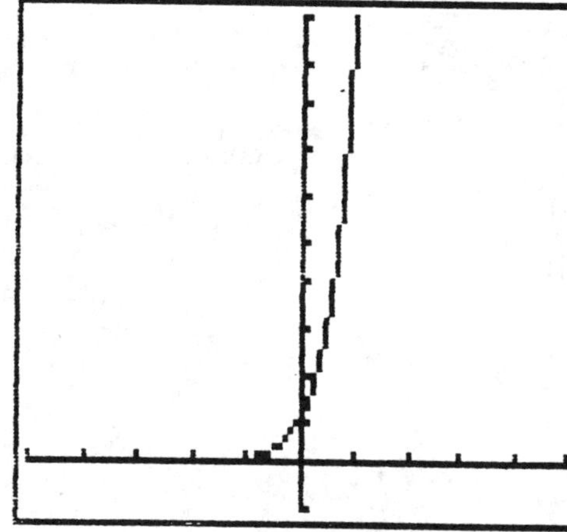

5\.

XMIN = −5
XMAX = 5
XSCL = 1
YMIN = −1
YMAX = 10
YSCL = 1
GRAPH $Y = 2^{X+1}$

7\.

XMIN = −5
XMAX = 5
XSCL = 1
YMIN = −1
YMAX = 10
YSCL = 1
GRAPH $Y = 2^{\text{ABS } X}$

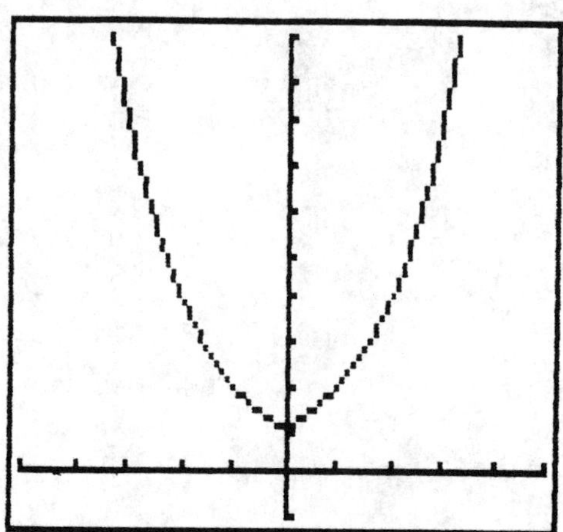

9.

XMIN = −5

XMAX = 5

XSCL = 1

YMIN = −1

YMAX = 10

YSCL = 1

GRAPH $Y = 2^{2X}$

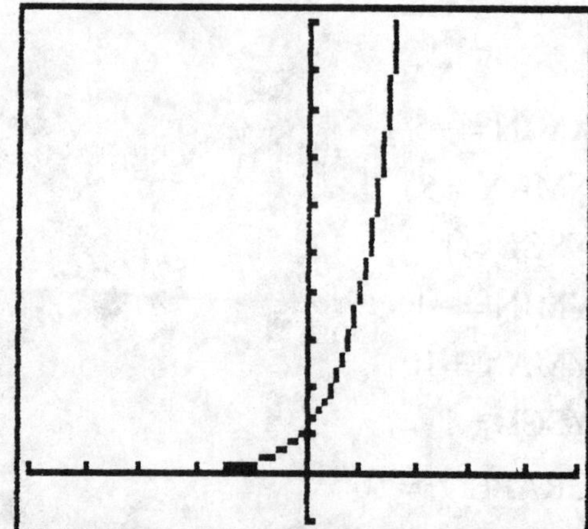

11.

XMIN = −5

XMAX = 5

XSCL = 1

YMIN = −1

YMAX = 10

YSCL = 1

GRAPH $Y = e^{X+1}$

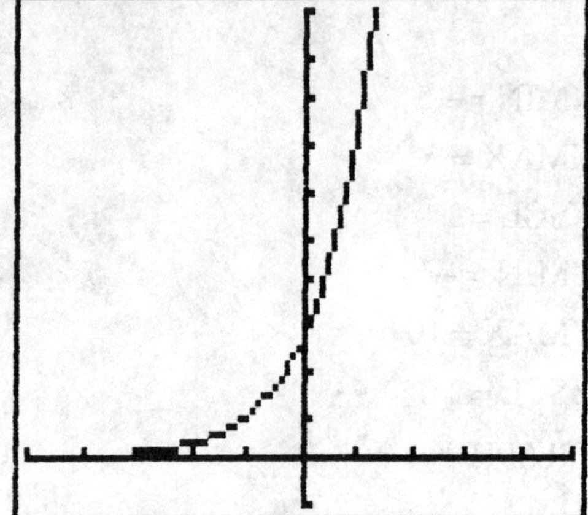

13.

XMIN = −5

XMAX = 5

XSCL = 1

YMIN = −1

YMAX = 10

YSCL = 1

GRAPH $Y = 4^{X}$

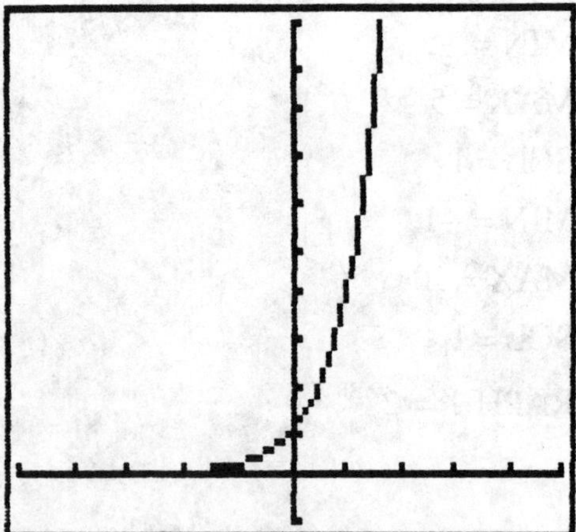

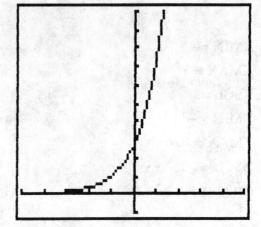

15.
XMIN = −5
XMAX = 5
XSCL = 1
YMIN = −1
YMAX = 10
YSCL = 1
GRAPH $Y = 10^X$

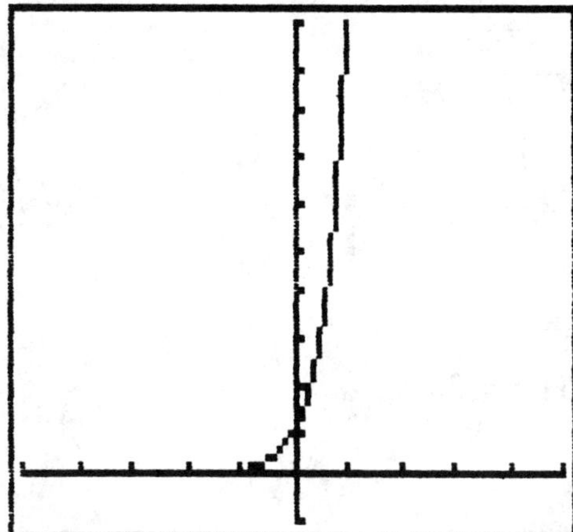

17.
XMIN = −5
XMAX = 5
XSCL = 1
YMIN = −1
YMAX = 10
YSCL = 1
GRAPH $Y = 2^{X+1}$

19.
XMIN = −5
XMAX = 5
XSCL = 1
YMIN = −1
YMAX = 10
YSCL = 1
GRAPH $Y = 2^{\text{ABS } X}$

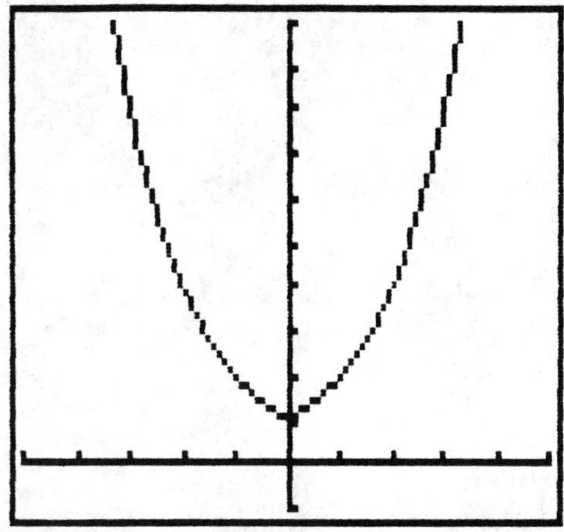

21.

XMIN = −5

XMAX = 5

XSCL = 1

YMIN = −1

YMAX = 10

YSCL = 1

GRAPH $Y = 2^{2X}$

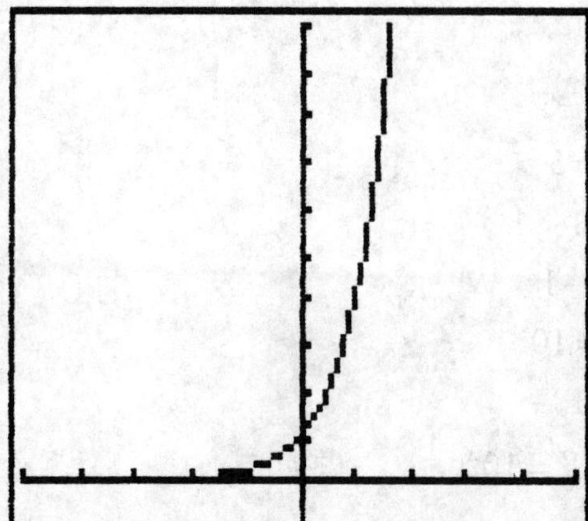

23.

XMIN = −5

XMAX = 5

XSCL = 1

YMIN = −1

YMAX = 10

YSCL = 1

GRAPH $Y = e^{X+1}$

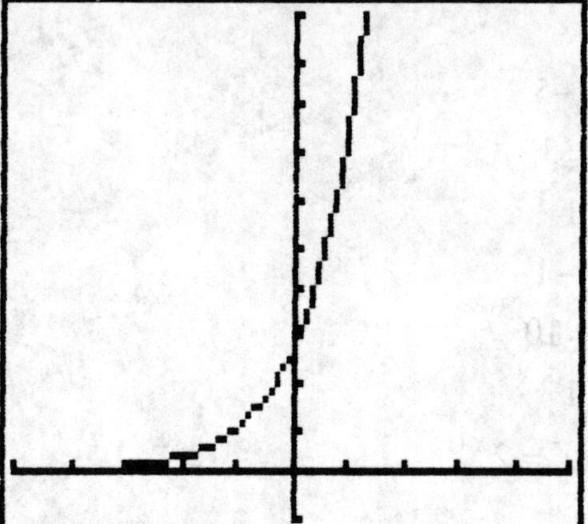

25.

$$2^x = 2^3$$
$$x = 3$$

27.

$$3^x = 9^{x-2}$$
$$3^x = \left(3^2\right)^{x-2}$$
$$3^x = 3^{2x-4}$$
$$x = 2x - 4$$
$$-x = -4$$
$$x = 4$$

29.

$$2^{3x} = 4^{x+1}$$
$$2^{3x} = \left(2^2\right)^{x+1}$$
$$2^{3x} = 2^{2x+2}$$
$$3x = 2x + 2$$
$$x = 2$$

31.

$$e^{x-1} = e^3$$
$$x - 1 = 3$$
$$x = 4$$

33.

$$(a+1)^x = (2a-1)^x$$
$$a+1 = 2a-1$$
$$2 = a$$

35.

$$(a+1)^x = (2a)^x$$
$$a+1 = 2a$$
$$1 = a$$

37.

$$e^2 = 7.389$$

39.

$$\frac{e^3 + e^{-3}}{2} = 10.068$$

41.

$$\frac{5}{6-3e^{-8}} = 0.833$$

43.

$$3^\pi \approx 31.54$$
$$\pi^3 \approx 31.01$$
$$3^\pi > \pi^3$$

45a).

$$500(1+0.075)^5 = 717.81$$

45b).

$$500\left(1+\frac{0.075}{2}\right)^{2(5)} = 722.52$$

45c).

$$500\left(1+\frac{0.075}{4}\right)^{4(5)} = 724.97$$

45d).

$$500\left(1+\frac{0.075}{12}\right)^{12(5)} = 726.65$$

45e).

$$500\left(1+\frac{0.075}{365}\right)^{365(5)} = 727.47$$

45f).

$$500\left(1+\frac{0.075}{365(24)}\right)^{365(24)(5)} = 727.49$$

45g).

$$500e^{0.075(5)} = 727.5$$

45h).

(a): \$500 compounded annually at 7.5% for 5 years

(b): \$500 compounded semi - annually at 7.5% for 5 years

(c): \$500 compounded quarterly at 7.5% for 5 years

(d): \$500 compounded monthly at 7.5% for 5 years

(e): \$500 compounded daily at 7.5% for 5 years

(f): \$500 compounded hourly at 7.5% for 5 years

(g): \$500 compounded continuously at 7.5% for 5 years

47a).

GRAPH $Y = (1+X)^{\frac{1}{X}}$

XSCL = 1

YSCL = 1

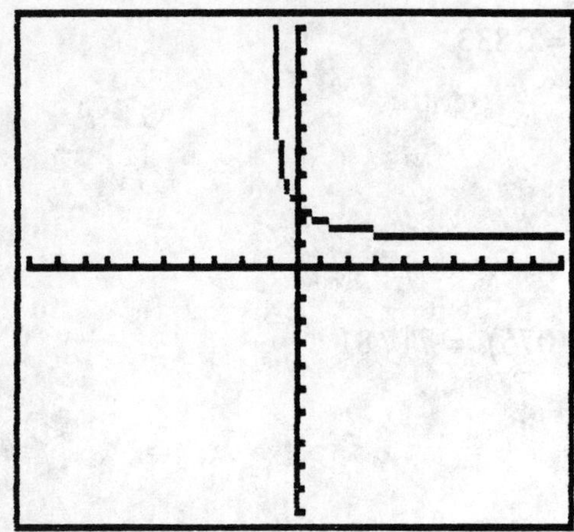

47b).

GRAPH $Y = (1+X)^{\frac{1}{X}}$

XSCL = 0.1

YSCL = 1

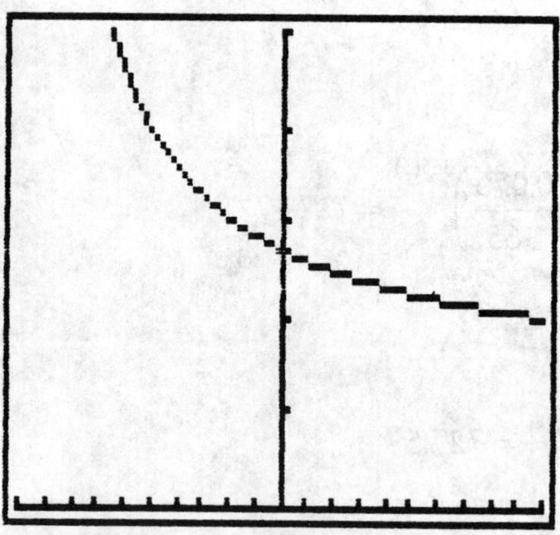

47c).

GRAPH $Y=(1+X)^{\frac{1}{X}}$

XSCL = 0.01

YSCL = 1

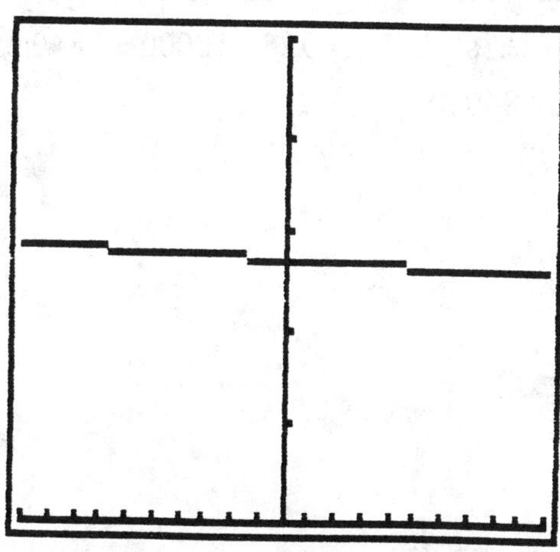

47d).

GRAPH $Y=(1+X)^{\frac{1}{X}}$

XSCL = 0.000001

YSCL = 1

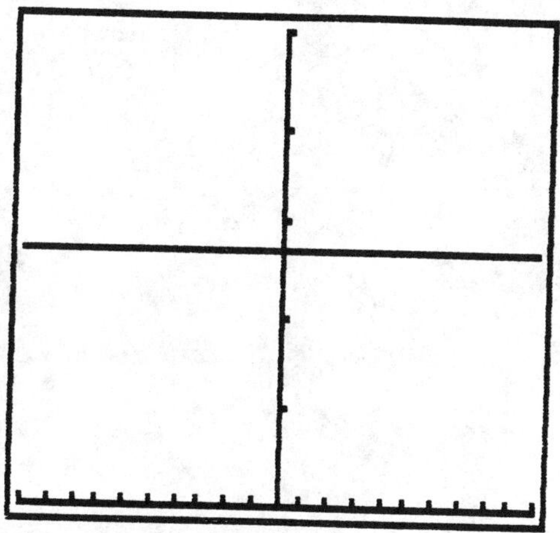

49.

Symmetric with respect to the y - axis.

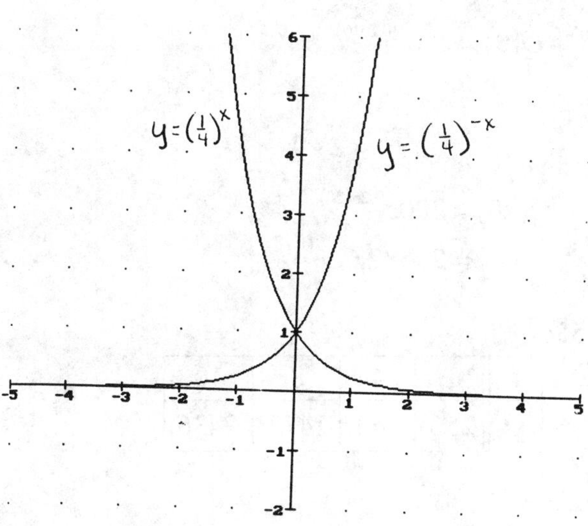

51.

$A = 10000e^{0.125(2)} = 12840.25$

Interest $= 12840.25 - 10000 = 2840.25$

\$2,840.25

53.

$Q(t) = q_0 2^{kt}$

$Q(2) = 10000$

$10000 = q_0 2^{2k}$

$Q(1) = 5000$

$Q(0) = 2500$

$2500 = q_0 2^{0k}$

$2500 = q_0$

$10000 = 2500(2^{2k})$

$4 = 2^{2k}$

$2^2 = 2^{2k}$

$2 = 2k$

$k = 1$

$Q(t) = 2500(2^t)$

$Q(7) = 2500(2^7)$

$= 320000$

$q_0 = 2500;\ Q(7) = 320{,}000$

55.

$Q(t) = 200e^{0.25t}$

55a).

$Q(0) = 200e^{0.25(0)}$

$= 200$

55b).

$Q(20) = 200e^{0.25(20)}$

$= 29682.6$

55c).

t	1	4	8	10
Q	256	543	1477	2436

57.

$Q(t) = q_0 e^{kt}$

$k = 2\% = 0.02;\ q_0 = 4$

$Q(t) = 4e^{0.02t}$

$t = 2050 - 1975 = 75$

$Q(75) = 4e^{0.02(75)}$

$= 17.93$

17.93 billi on people

59.

$$Q(t) = q_0 e^{-kt}$$
$$q_0 = 1000 \quad k = 4\% = 0.04$$
$$Q(t) = 1000e^{-0.04t}$$
$$Q(10) = 1000e^{-0.04(10)}$$
$$= 670.32$$
670.32 grams

61.

$$A = P\left(1+\frac{r}{k}\right)^{kt}$$
$$= 10000\left(1+\frac{0.08}{4}\right)^{4(18)}$$
$$= 41611.40$$
\$41,611.40

63.

$$A = P\left(1+\frac{r}{k}\right)^{kt}$$
$$= 25000\left(1+\frac{0.12}{12}\right)^{12(5)}$$
$$= 45417.42$$
\$45,417.42

65.

$$A = Pe^{rt}$$
$$= 100e^{0.055(10)}$$
$$= 173.33$$
\$173.33

67.

$$A = Pe^{rt} \qquad 64 - 40 = 25$$
$$50000 = Pe^{0.12(25)}$$
$$2489.35 = P$$
\$2,489.35

69.

$$A = Pe^{rt}$$
$$= 1000e^{0.09(8)}$$
$$= 2054.43$$
\$2,054.43

71a).

$$y(t) = (1.95)^{t/10}$$
$$1975 - 1958 = 17$$
$$y(17) = (1.95)^{17/10}$$
$$= 3.11$$
\$3.11

71b).

$$y(t) = (1.95)^{t/10}$$
$$1971 - 1958 = 13$$
$$1980 - 1958 = 22$$
$$y(22) - y(13) = (1.95)^{22/10} - (1.95)^{13/10}$$
$$= 4.35 - 2.38$$
$$= 1.97$$
\$1.97

CHAPTER 6 SECTION 3

SECTION 6.3 PROGRESS CHECK

1a). (Page 361)

$\log_a x = y$ is equivalent to $a^y = x$

$\log_4 64 = 3$

$4^3 = 64$

1b).

$\log_a x = y$ is equivalent to $a^y = x$

$\log_{10}\left(\frac{1}{10{,}000}\right) = -4$

$10^{-4} = \frac{1}{10{,}000}$

1c).

$\log_a x = y$ is equivalent to $a^y = x$

$\log_{25} 5 = \frac{1}{2}$

$25^{1/2} = 5$

1d).

$\log_a x = y$ is equivalent to $a^y = x$

$\ln 0.3679 \approx -1$

$\log_e 0.3679 \approx -1$

$e^{-1} \approx -0.3679$

2a). (Page 362)

$x = a^y$ is equivalent to $y = \log_a x$

$1000 = 10^3$

$\log_{10} 1000 = 3$

$\log 1000 = 3$

2b).

$x = a^y$ is equivalent to $y = \log_a x$

$6 = 36^{\frac{1}{2}}$

$\log_{36} 6 = \frac{1}{2}$

2c).

$x = a^y$ is equivalent to $y = \log_a x$

$\frac{1}{7} = 7^{-1}$

$\log_7 \frac{1}{7} = -1$

2d).

$x = a^y$ is equivalent to $y = \log_a x$

$20.09 \approx e^3$

$\log_e 20.09 \approx 3$

$\ln 20.09 \approx 3$

3a). (Page 365)

$$\log_x 1000 = 3$$
$$x^3 = 1000$$
$$x^3 = 10^3$$
$$x = 10$$

3b).

$$\log_2 x = 5$$
$$2^5 = x$$
$$x = 32$$

3c).

$$x = \log_7 \frac{1}{49}$$
$$7^x = \frac{1}{49}$$
$$7^x = 7^{-2}$$
$$x = -2$$

4a). (page 367)

$$\log_3 3^4 = 4$$

4b).

$$6^{\log_6 9} = 9$$

4c).

$$\log_5 1 = \log_5 5^0 = 0$$

4d).

$$\log_8 8 = \log_8 8^1 = 1$$

5a). (Page 369)

$$\log_2 x^2 = \log_2 9$$
$$x^2 = 9$$
$$x = \pm 3$$

5b).

$$\log_7 14 = \log_{2x} 14$$
$$7 = 2x$$
$$x = \frac{7}{2}$$

EXERCISE SET 6.3

In Exercises 1-11, $\log_a x = y$ is equivalent to $a^y = x$.

1.

$$\log_2 4 = 2$$
$$2^2 = 4$$

3.

$$\log_9 \frac{1}{81} = -2$$
$$9^{-2} = \frac{1}{81}$$

5.

$$\ln 20.09 \approx 3$$
$$\log_e 20.09 \approx 3$$
$$e^3 \approx 20.09$$

7.

$$\log_{10} 1000 = 3$$
$$10^3 = 1000$$

9.

$$\ln 1 = 0$$
$$\log_e 1 = 0$$
$$e^0 = 1$$

11.

$$\log_3 \frac{1}{27} = -3$$
$$3^{-3} = \frac{1}{27}$$

In Exercises 13 - 25, $x = a^y$ is equivalent to $y = \log_a x$.

13.

$$25 = 5^2$$
$$\log_5 25 = 2$$

15.

$$10000 = 10^4$$
$$\log_{10} 10000 = 4$$
$$\log 10000 = 4$$

17.

$$\frac{1}{8} = 2^{-3}$$
$$\log_2 \frac{1}{8} = -3$$

19.

$$1 = 2^0$$
$$\log_2 1 = 0$$

21.

$$6 = \sqrt{36}$$
$$6 = 36^{\frac{1}{2}}$$
$$\log_{36} 6 = \frac{1}{2}$$

23.

$$64 = 16^{\frac{3}{2}}$$
$$\log_{16} 64 = \frac{3}{2}$$

25.

$$\frac{1}{3} = 27^{-\frac{1}{3}}$$
$$\log_{27} \frac{1}{3} = -\frac{1}{3}$$

27.

$$\log_5 x = 2$$
$$5^2 = x$$
$$x = 25$$

29.

$$\log_{25} x = -\frac{1}{2}$$
$$25^{-\frac{1}{2}} = x$$
$$x = \frac{1}{5}$$

31.

$$\ln x = 2$$
$$\log_e x = 2$$
$$e^2 = x$$

33.

$$\ln x = -\frac{1}{2}$$
$$\log_e x = -\frac{1}{2}$$
$$e^{-\frac{1}{2}} = x$$

35.

$$\log_5 \frac{1}{25} = x$$
$$5^x = \frac{1}{25}$$
$$5^x = 5^{-2}$$
$$x = -2$$

37.

$$\log_x \frac{1}{8} = -\frac{1}{3}$$
$$x^{-\frac{1}{3}} = \frac{1}{8}$$
$$x = \left(\frac{1}{8}\right)^{-3}$$
$$x = 512$$

39.

$$\log_5(x+1) = 3$$
$$5^3 = x+1$$
$$125 = x+1$$
$$x = 124$$

41.

$$\log_{x+1} 24 = \log_3 24$$
$$x+1 = 3$$
$$x = 2$$

43.

$$\log_{x+1} 17 = \log_4 17$$
$$x+1 = 4$$
$$x = 3$$

45.

$$3^{\log_3 6} = 6$$

47.

$$e^{\ln 2} = 2$$

49.

$$\log_5 5^3 = 3$$

51.

$$\log_8 8^{\frac{1}{2}} = \frac{1}{2}$$

53.

$$\log_7 49 = \log_7 7^2 = 2$$

55.

$$\log_5 5 = \log_5 5^1 = 1$$

57.

$$\ln 1 = \log_e e^0 = 0$$

59.

$$\log_2 \frac{1}{4} = \log_2 2^{-2} = -2$$

61.

$$\log 10000 = \log_{10} 10^4 = 4$$

63.

$$\ln e^2 = 2$$

65.

$$\log \frac{8}{5} = 0.2041$$

67.

$$\frac{\ln 8}{\ln 5} = 1.2920$$

69.

$$\frac{\log\left(\frac{1}{2}\right)}{-0.0006} = 501.7167$$

71.

x	y
1	0
4	1
2	$\frac{1}{2}$
$\frac{1}{4}$	-1

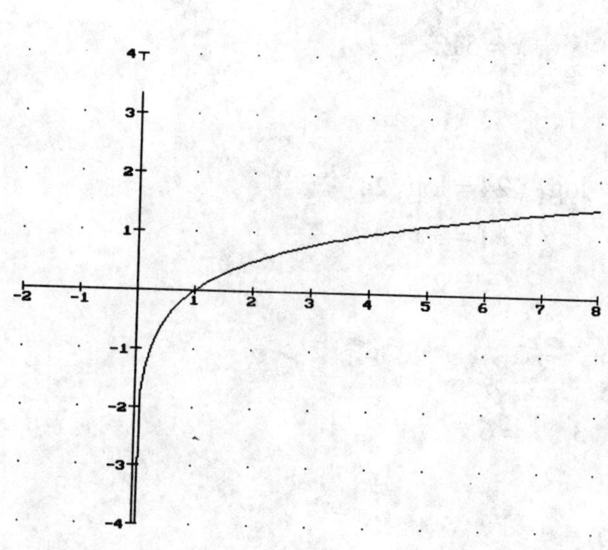

73.

x	y
$\frac{1}{20}$	-1
$\frac{1}{2}$	0
5	1
50	2

$$y = \log 2x = \log_{10} 2x$$
$$10^y = 2x$$
$$\frac{10^y}{2} = x$$

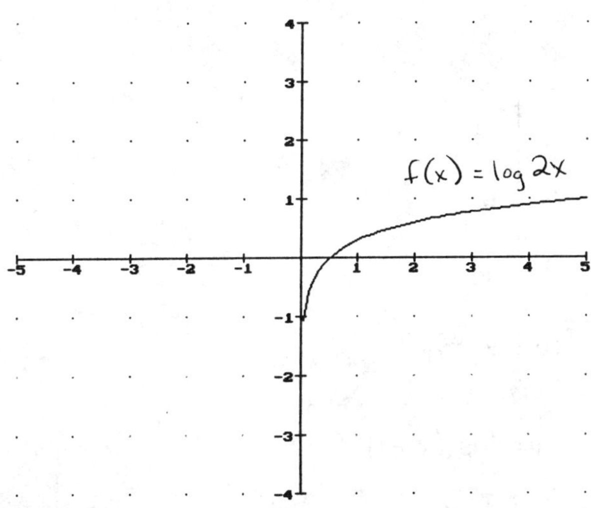

75.

x	y
0.7	-1
2	0
5.4	1
14.8	2

$$y = \ln\frac{x}{2} = \log_e \frac{x}{2}$$
$$e^y = \frac{x}{2}$$
$$x = 2e^y$$

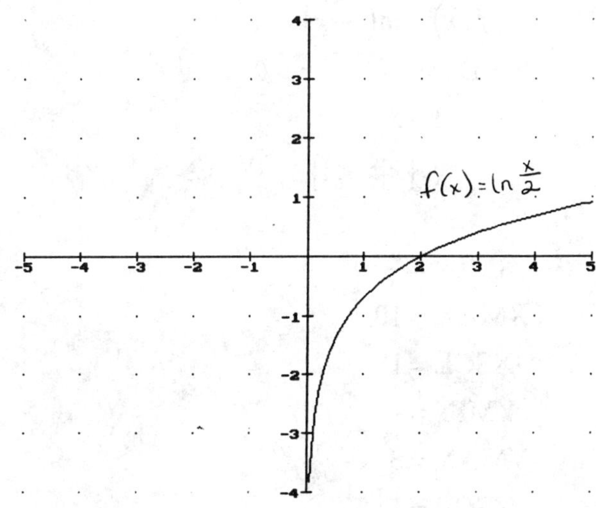

77.

x	y
$\frac{4}{3}$	-1
2	0
4	1
10	2

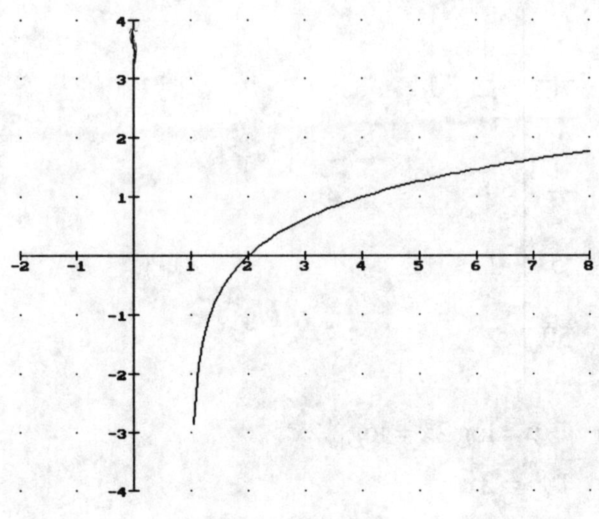

$$y=\log_3(x-1)$$
$$3^y=x-1$$
$$x=3^y+1$$

79.

$$f(x)=\ln(1-x)$$

Domain: $1-x>0$

$$1>x$$
$$\{x|x<1\}$$

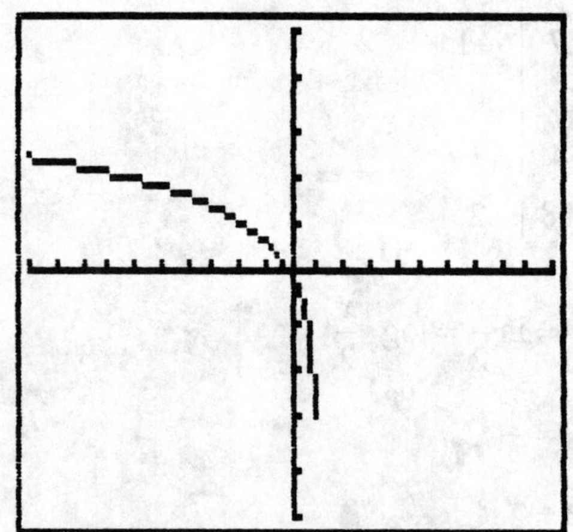

XMIN $=-10$
XMAX $=10$
XSCL $=1$
YMIN $=-5$
YMAX $=5$
YSCL $=1$
GRAPH $Y=\ln(1-X)$

81.

$f(x) = \log \dfrac{x}{x-1}$

Domain: $\dfrac{x}{x-1} > 0$

not defined when $x = 1$

$\dfrac{x}{x-1} = 0$

$x = 0$

Interval, Critical Value	Test Point	Substitution	Verification
$x < 0$	$x = -1$	$\dfrac{-1}{-1-1} > 0$	True
$x = 0$	$x = 0$	$\dfrac{0}{0-1} > 0$	False
$0 < x < 1$	$x = \dfrac{1}{2}$	$\dfrac{\frac{1}{2}}{\frac{1}{2}-1} > 0$	False
$x = 1$	$x = 1$	$\dfrac{1}{1-1} > 0$	False
$x > 1$	$x = 2$	$\dfrac{2}{2-1} > 0$	True

$\{x \mid x < 0 \text{ or } x > 1\}$

XMIN = −10

XMAX = 10

XSCL = 1

YMIN = −2

YMAX = 2

YSCL = 1

GRAPH $Y = \log[X \div (X-1)]$

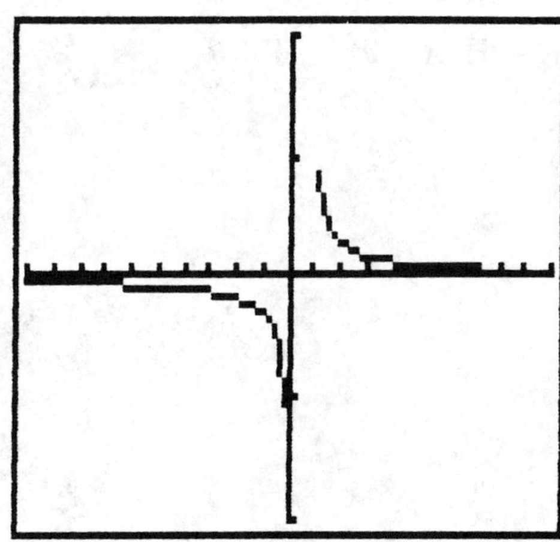

83.

$f(x) = \ln 2^x$

Domain: $2^x > 0$

$\{x \mid -\infty < x < \infty\}$

EQUAL viewing rectangle

XSCL = 1

YSCL = 1

GRAPH $Y = \ln 2^X$

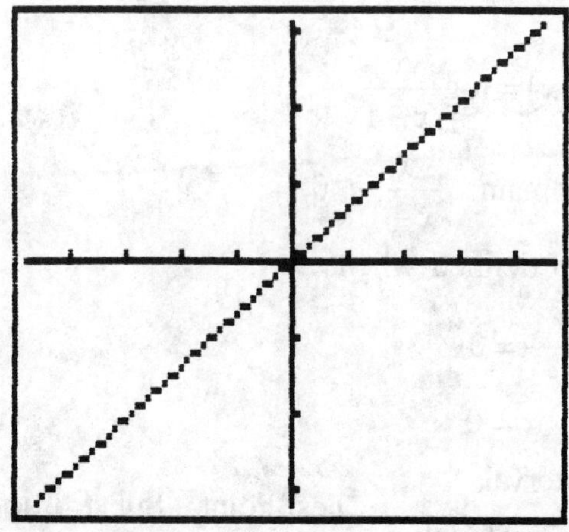

85.

$f(x) = \log\left(-\sqrt{x}\right)$

$-\sqrt{x} \le 0$ which is not in the domain of the logarithmic function, hence there is no graph.

87.

XSCL = 1

YSCL = 1

GRAPH $X = T \qquad Y = e^T$

GRAPH $X = e^T \qquad Y = T$

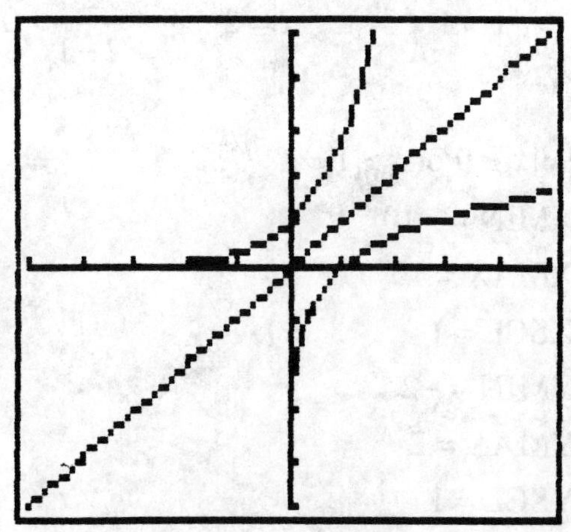

89.

$$A = P\left(1+\frac{r}{k}\right)^{kt}$$

$$3P_0 = P_0\left(1+\frac{0.06}{4}\right)^{4t}$$

$$3 = \left(1+\frac{0.06}{4}\right)^{4t}$$

$$3 = (1.015)^{4t}$$

$$\ln 3 = \ln(1.015)^{4t}$$

$$\ln 3 = 4t\ln(1.015)$$

$$t = 18.4$$

18.4 years

91.

$$A = Pe^{rt}$$

$$300 = 100e^{20r}$$

$$3 = e^{20r}$$

$$\ln 3 = \ln e^{20r}$$

$$\ln 3 = 20r$$

$$r = 0.055$$

5.5%

93.

x	$z = \ln y$
0	$\ln 2 = 0.69$
1	$\ln 6 = 1.79$
2	$\ln 18 = 2.89$
3	$\ln 54 = 3.99$

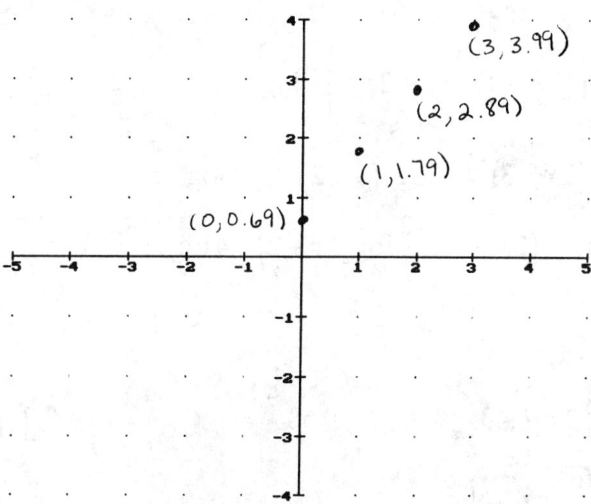

95.

$$y = ab^x$$

$(0, 2)$: $2 = ab^0$

$2 = a$

$$y = 2b^x$$

$(1, 6)$: $6 = 2b^1$

$3 = b$

$$y = 2(3^x)$$

CHAPTER 6 SECTION 4

SECTION 6.4 PROGRESS CHECK

1a). (Page 372)

$$\log_4[(1.47)(22.3)] = \log_4 1.47 + \log_4 22.3$$

1b).

$$\log_a \frac{x-1}{\sqrt{x}} = \log_a(x-1) - \log_a \sqrt{x}$$
$$= \log_a(x-1) - \log_a x^{\frac{1}{2}}$$
$$= \log_a(x-1) - \frac{1}{2}\log_a x$$

2. (Page 373)

$$\log_a \frac{(2x-3)^{-\frac{1}{2}}(y+2)^{-\frac{2}{3}}}{z^4}$$
$$= \log_a\left[(2x-3)^{-\frac{1}{2}}(y+2)^{-\frac{2}{3}}\right] - \log_a z^4$$
$$= \log_a(2x-3)^{-\frac{1}{2}} + \log_a(y+2)^{-\frac{2}{3}} - 4\log_a z$$
$$= -\frac{1}{2}\log_a(2x-3) - \frac{2}{3}\log_a(y+2) - 4\log_a z$$

3a). (Page 374)

$$\log_a 2 = 0.43, \quad \log_a 3 = 0.68$$
$$\log_a 18 = \log_a 2(3^2)$$
$$= \log_a 2 + \log_a 3^2$$
$$= \log_a 2 + 2\log_a 3$$
$$= 0.43 + 2(0.68)$$
$$= 1.79$$

3b).

$$\log_a 2 = 0.43, \quad \log_a 3 = 0.68$$
$$\log_a \sqrt[3]{\frac{9}{2}} = \log_a\left(\frac{9}{2}\right)^{\frac{1}{3}}$$
$$= \frac{1}{3}\log_a \frac{9}{2}$$
$$= \frac{1}{3}[\log_a 9 - \log_a 2]$$
$$= \frac{1}{3}[\log_a 3^2 - \log_a 2]$$
$$= \frac{1}{3}[2\log_a 3 - \log_a 2]$$
$$= \frac{1}{3}[2(0.68) - 0.43]$$
$$= 0.31$$

4. (Page 375)

$$\frac{1}{3}[\log_a(2x-1) - \log_a(2x-5)] + 4\log_a x$$
$$= \frac{1}{3}\log_a \frac{2x-1}{2x-5} + \log_a x^4$$
$$= \log_a\left(\frac{2x-1}{2x-5}\right)^{\frac{1}{3}} + \log_a x^4$$
$$= \log_a x^4 \sqrt[3]{\frac{2x-1}{2x-5}}$$

5a). (Page 377)

$$\log_5 16 = \frac{\ln 16}{\ln 5} = 1.7227$$

5b).

$$\log_{10} e = \frac{\ln e}{\ln 10} = \frac{1}{\ln 10} = 0.4343$$

EXERCISE SET 6.4

1.

$$\begin{aligned}\log_{10}[(120)(36)] &= \log_{10}(2^2 \cdot 3 \cdot 10 \cdot 2^2 \cdot 3^2)\\ &= \log_{10}(2^4 \cdot 3^3 \cdot 10)\\ &= \log_{10} 2^4 + \log_{10} 3^3 + \log_{10} 10\\ &= 4\log_{10} 2 + 3\log_{10} 3 + 1\end{aligned}$$

3.

$$\log_3(3^4) = 4\log_3 3 = 4$$

5.

$$\log_a(2xy) = \log_a 2 + \log_a x + \log_a y$$

7.

$$\begin{aligned}\log_a \frac{x}{yz} &= \log_a x - \log_a yz\\ &= \log_a x - [\log_a y + \log_a z]\\ &= \log_a x - \log_a y - \log_a z\end{aligned}$$

9.

$$\ln x^5 = 5\ln x$$

11.

$$\begin{aligned}\log_a(x^2y^3) &= \log_a x^2 + \log_a y^3\\ &= 2\log_a x + 3\log_a y\end{aligned}$$

13.

$$\begin{aligned}\log_a \sqrt{xy} &= \log_a(xy)^{\frac{1}{2}}\\ &= \frac{1}{2}\log_a(xy)\\ &= \frac{1}{2}[\log_a x + \log_a y]\\ &= \frac{1}{2}\log_a x + \frac{1}{2}\log_a y\end{aligned}$$

15.

$$\begin{aligned}\ln(x^2y^3z^4) &= \ln x^2 + \ln y^3 + \ln z^4\\ &= 2\ln x + 3\ln y + 4\ln z\end{aligned}$$

17.

$$\begin{aligned}\ln\left(\sqrt{x}\sqrt[3]{y}\right) &= \ln\sqrt{x}+\ln\sqrt[3]{y}\\ &= \ln x^{\frac{1}{2}}+\ln y^{\frac{1}{3}}\\ &= \frac{1}{2}\ln x+\frac{1}{3}\ln y\end{aligned}$$

19.

$$\begin{aligned}\log_a\left(\frac{x^2y^3}{z^4}\right) &= \log_a\left(x^2y^3\right)-\log_a z^4\\ &= \log_a x^2+\log_a y^3-\log_a z^4\\ &= 2\log_a x+3\log_a y-4\log_a z\end{aligned}$$

21.

$$\begin{aligned}\log 6 &= \log(2\cdot 3)=\log 2+\log 3\\ &= 0.30+0.47=0.77\end{aligned}$$

23.

$$\log 9=\log 3^2=2\log 3=2(0.47)=0.94$$

25.

$$\begin{aligned}\log 12 &= \log\left(2^2\cdot 3\right)=\log 2^2+\log 3\\ &= 2\log 2+\log 3\\ &= 2(0.30)+0.47\\ &= 1.07\end{aligned}$$

27.

$$\begin{aligned}\log\frac{15}{2} &= \log 15-\log 2=\log(3\cdot 5)-\log 2\\ &= \log 3+\log 5-\log 2\\ &= 0.47+0.70-0.30\\ &= 0.87\end{aligned}$$

29.

$$\begin{aligned}\log\sqrt{7.5} &= \log(7.5)^{\frac{1}{2}}=\frac{1}{2}\log 7.5\\ &= \frac{1}{2}\log\frac{15}{2}\\ &= \frac{1}{2}(\log 15-\log 2)\\ &= \frac{1}{2}[\log(3\cdot 5)-\log 2]\\ &= \frac{1}{2}(\log 3+\log 5-\log 2)\\ &= \frac{1}{2}(0.47+0.70-0.30)\\ &= 0.435\end{aligned}$$

31.

$$\begin{aligned}2\log x+\frac{1}{2}\log y &= \log x^2+\log y^{\frac{1}{2}}\\ &= \log x^2y^{\frac{1}{2}}\\ &= \log x^2\sqrt{y}\end{aligned}$$

33.

$$\begin{aligned}\frac{1}{3}\ln x+\frac{1}{3}\ln y&=\frac{1}{3}(\ln x+\ln y)\\&=\frac{1}{3}\ln(xy)\\&=\ln(xy)^{\frac{1}{3}}\\&=\ln\sqrt[3]{xy}\end{aligned}$$

35.

$$\begin{aligned}\frac{1}{3}\log_a x+2\log_a y-\frac{3}{2}\log_a z&=\log_a x^{\frac{1}{3}}+\log_a y^2-\log_a z^{\frac{3}{2}}\\&=\log_a\left(x^{\frac{1}{3}}y^2\right)-\log_a z^{\frac{3}{2}}\\&=\log_a\frac{x^{\frac{1}{3}}y^2}{z^{\frac{3}{2}}}\end{aligned}$$

37.

$$\begin{aligned}\frac{1}{2}(\log_a x+\log_a y)&=\frac{1}{2}\log_a(xy)\\&=\log_a(xy)^{\frac{1}{2}}\\&=\log_a\sqrt{xy}\end{aligned}$$

39.

$$\begin{aligned}\frac{1}{3}(2\ln x+4\ln y)-3\ln z&=\frac{1}{3}(\ln x^2+\ln y^4)-\ln z^3\\&=\frac{1}{3}\ln(x^2y^4)-\ln z^3\\&=\ln(x^2y^4)^{\frac{1}{3}}-\ln z^3\\&=\ln\frac{\sqrt[3]{x^2y^4}}{z^3}\end{aligned}$$

41.

$$\begin{aligned}\frac{1}{2}\log_a(x-1)-2\log_a(x+1)&=\log_a(x-1)^{\frac{1}{2}}-\log_a(x+1)^2\\&=\log_a\frac{\sqrt{x-1}}{(x+1)^2}\end{aligned}$$

43.

$$3\log_a x - 2\log_a(x-1) + \frac{1}{2}\log_a \sqrt[3]{x+1}$$
$$= \log_a x^3 - \log_a (x-1)^2 + \frac{1}{2}\log_a (x+1)^{\frac{1}{3}}$$
$$= \log_a \frac{x^3}{(x-1)^2} + \log_a (x+1)^{\frac{1}{6}}$$
$$= \log_a \frac{x^3(x+1)^{\frac{1}{6}}}{(x-1)^2}$$

45.

$$\log 17 = \log_{10} 17 = \frac{\ln 17}{\ln 10} \approx \frac{2.8332}{2.3026} = 1.2304$$

47.

$$\log_3 141 = \frac{\ln 141}{\ln 3} \approx \frac{4.9488}{1.0986} = 4.5046$$

49.

$$\log 245 = \log_{10} 245 = \frac{\ln 245}{\ln 10} \approx \frac{5.5013}{2.3026} = 2.3892$$

51a).

$$\log 18 = \log(2 \cdot 3^2)$$
$$= \log 2 + \log 3^2$$
$$= \log 2 + 2\log 3$$
$$= 0.301 + 2(0.477)$$
$$1.255$$

51b).

$$\log \frac{4}{9} = \log 4 - \log 9$$
$$= \log 2^2 - \log 3^2$$
$$= 2\log 2 - 2\log 3$$
$$= 2(0.301) - 2(0.477)$$
$$= -0.352$$

51c).

$$\log 5 = \log\left(\frac{1}{2} \cdot 10\right)$$
$$= \log \frac{1}{2} + \log 10$$
$$= \log 2^{-1} + \log 10$$
$$= -\log 2 + \log 10$$
$$= -0.301 + 1$$
$$= 0.699$$

53.

$$\log_5 10 = \frac{\ln 10}{\ln 5} = 1.4307$$

55.

$$\log_2 \sqrt{7} = \log_2 7^{\frac{1}{2}}$$
$$= \frac{1}{2}\log_2 7$$
$$= \frac{1}{2}\left(\frac{\ln 7}{\ln 2}\right)$$
$$= 1.4037$$

57.

XSCL = 1

YSCL = 1

GRAPH $Y = \ln X \div \ln 5$

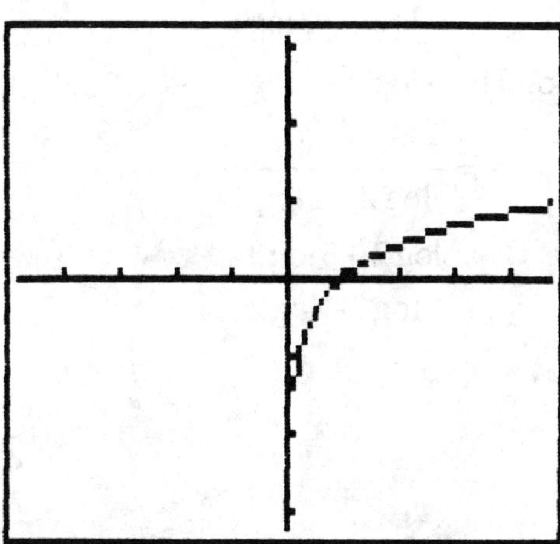

CHAPTER 6 SECTION 5

SECTION 6.5 PROGRESS CHECK

1. (Page 379)

$$\begin{aligned} 2^{x+1} &= 3^{2x-3} \\ \log 2^{x+1} &= \log 3^{2x-3} \\ (x+1)\log 2 &= (2x-3)\log 3 \\ x\log 2 + \log 2 &= 2x\log 3 - 3\log 3 \\ x\log 2 - 2x\log 3 &= -\log 2 - 3\log 3 \\ x(\log 2 - 2\log 3) &= -\log 2 - 3\log 3 \\ x &= \frac{-\log 2 - 3\log 3}{\log 2 - 2\log 3} \\ x &= \frac{\log 2 + 3\log 3}{2\log 3 - \log 2} \end{aligned}$$

2. (Page 380)

$$\begin{aligned} 5 - 2e^{3x-1} &= 0 \\ 5 &= 2e^{3x-1} \\ \frac{5}{2} &= e^{3x-1} \\ \ln\frac{5}{2} &= \ln e^{3x-1} \\ \ln\frac{5}{2} &= 3x-1 \\ \ln\frac{5}{2}+1 &= 3x \\ \frac{1}{3}\left(\ln\frac{5}{2}+1\right) &= x \end{aligned}$$

3. (Page 380)

$$\begin{aligned} \log x - \frac{1}{2} &= -\log 3 \\ \log x + \log 3 &= \frac{1}{2} \\ \log(3x) &= \frac{1}{2} \\ 10^{\log(3x)} &= 10^{\frac{1}{2}} \\ 3x &= 10^{\frac{1}{2}} \\ x &= \frac{10^{\frac{1}{2}}}{3} = \frac{\sqrt{10}}{3} \end{aligned}$$

4. (Page 381)

$$\begin{aligned} \log_3(x-8) &= 2 - \log_3 x \\ \log_3(x-8) + \log_3 x &= 2 \\ \log_3[x(x-8)] &= 2 \\ 3^{\log_3[x(x-8)]} &= 3^2 \\ x(x-8) &= 9 \\ x^2 - 8x - 9 &= 0 \\ (x-9)(x+1) &= 0 \\ x = 9 \qquad & \underbrace{x=-1}_{\text{not in domain of } \log_3 x} \end{aligned}$$

Solution: $x = 9$

5. (Page 383)

$$Q(t) = q_0 e^{0.1t}$$
$$2q_0 = q_0 e^{0.1t}$$
$$2 = e^{0.1t}$$
$$\ln 2 = \ln e^{0.1t}$$
$$\ln 2 = 0.1t$$
$$\frac{\ln 2}{0.1} = t$$
$$t \approx 6.93$$

6.93 minutes

EXERCISE SET 6.5

1.

$$5^x = 18$$
$$\log 5^x = \log 18$$
$$x \log 5 = \log 18$$
$$x = \frac{\log 18}{\log 5}$$

3.

$$2^{x-1} = 7$$
$$\log 2^{x-1} = \log 7$$
$$(x-1)\log 2 = \log 7$$
$$x - 1 = \frac{\log 7}{\log 2}$$
$$x = \frac{\log 7}{\log 2} + 1$$

5.

$$3^{2x} = 46$$
$$\log 3^{2x} = \log 46$$
$$2x \log 3 = \log 46$$
$$x = \frac{\log 46}{2\log 3}$$

7.

$$5^{2x-5} = 564$$
$$\log 5^{2x-5} = \log 564$$
$$(2x-5)\log 5 = \log 564$$
$$2x - 5 = \frac{\log 564}{\log 5}$$
$$2x = \frac{\log 564}{\log 5} + 5$$
$$x = \frac{\log 564}{2\log 5} + \frac{5}{2}$$

9.

$$3^{x-1} = 2^{2x-1}$$
$$\log 3^{x-1} = \log 2^{2x-1}$$
$$(x-1)\log 3 = (2x-1)\log 2$$
$$x\log 3 - \log 3 = 2x\log 2 - \log 2$$
$$x\log 3 - 2x\log 2 = \log 3 - \log 2$$
$$x(\log 3 - 2\log 2) = \log 3 - \log 2$$
$$x = \frac{\log 3 - \log 2}{\log 3 - 2\log 2}$$

11.

$$2^{-x} = 15$$
$$\log 2^{-x} = \log 15$$
$$-x\log 2 = \log 15$$
$$x = -\frac{\log 15}{\log 2}$$

13.

$$4^{-2x+1} = 12$$
$$\log 4^{-2x+1} = \log 12$$
$$(-2x+1)\log 4 = \log 12$$
$$-2x+1 = \frac{\log 12}{\log 4}$$
$$-2x = \frac{\log 12}{\log 4} - 1$$
$$x = -\frac{\log 12}{2\log 4} + \frac{1}{2}$$

15.

$$e^{x} = 18$$
$$\ln e^{x} = \ln 18$$
$$x = \ln 18$$

17.

$$e^{2x+3} = 30$$
$$\ln e^{2x+3} = \ln 30$$
$$2x+3 = \ln 30$$
$$2x = -3 + \ln 30$$
$$x = \frac{-3 + \ln 30}{2}$$

19.

$$\log x + \log 2 = 3$$
$$\log 2x = 3$$
$$10^{3} = 2x$$
$$\frac{1000}{2} = x$$
$$x = 500$$

21.

$$\log_x(3-5x)=1$$
$$x^1=3-5x$$
$$6x=3$$
$$x=\frac{1}{2}$$

23.

$$\log x+\log(x-3)=1$$
$$\log[x(x-3)]=1$$
$$10^1=x^2-3x$$
$$0=x^2-3x-10$$
$$0=(x-5)(x+2)$$
$$x=5 \qquad \underbrace{x=-2}_{\text{not in domain of } \log x}$$

Solution: $x=5$

25.

$$\log(3x+1)-\log(x-2)=1$$
$$\log\frac{3x+1}{x-2}=1$$
$$10^1=\frac{3x+1}{x-2}$$
$$10x-20=3x+1$$
$$7x=21$$
$$x=3$$

27.

$$\log_2 x=4-\log_2(x-6)$$
$$\log_2 x+\log_2(x-6)=4$$
$$\log_2[x(x-6)]=4$$
$$2^4=x(x-6)$$
$$16=x^2-6x$$
$$0=x^2-6x-16$$
$$0=(x-8)(x+2)$$
$$x=8 \qquad \underbrace{x=-2}_{\text{not in domain of } \log_2 x}$$

Solution: $x=8$

29.

$$\log_2(x+4) = 3 - \log_2(x-2)$$
$$\log_2(x+4) + \log_2(x-2) = 3$$
$$\log_2[(x+4)(x-2)] = 3$$
$$2^3 = (x+4)(x-2)$$
$$8 = x^2 + 2x - 8$$
$$0 = x^2 + 2x - 16$$
$$x = \frac{-2 \pm \sqrt{2^2 - 4(1)(-16)}}{2(1)}$$
$$= \frac{-2 \pm \sqrt{68}}{2}$$
$$= \frac{-2 \pm 2\sqrt{17}}{2}$$
$$x = -1 + \sqrt{17} \qquad \underbrace{x = -1 - \sqrt{17}}_{\text{not in domain of } \log_2(x-2)}$$

Solution: $x = -1 + \sqrt{17}$

31.

$$y=\frac{e^{x}-e^{-x}}{2}$$
$$2y=e^{x}-e^{-x}$$
$$e^{x}(2y)=e^{x}\left(e^{x}-e^{-x}\right)$$
$$2ye^{x}=e^{2x}-1$$
$$0=e^{2x}-2ye^{x}-1$$
Let $u=e^{x}$, $u>0$
$$0=u^{2}-2yu-1$$
$$u=\frac{-(-2y)\pm\sqrt{(-2y)^{2}-4(1)(-1)}}{2(1)}$$
$$u=\frac{2y\pm\sqrt{4y^{2}+4}}{2}$$
$$u=\frac{2y\pm2\sqrt{y^{2}+1}}{2}$$
$$u=y\pm\sqrt{y^{2}+1}$$
$u>0$, hence $u=y+\sqrt{y^{2}+1}$
$$u=e^{x}=y+\sqrt{y^{2}+1}$$
$$\ln e^{x}=\ln\left(y+\sqrt{y^{2}+1}\right)$$
$$x=\ln\left(y+\sqrt{y^{2}+1}\right)$$

33.

$$Q(t)=q_0e^{0.03t}$$
$$3q_0=q_0e^{0.03t}$$
$$3=e^{0.03t}$$
$$\ln 3=\ln e^{0.03t}$$
$$\ln 3=0.03t$$
$$t=\frac{\ln 3}{0.03}\approx 36.62$$
36.62 years

35.

$$Q(t)=q_0e^{-0.055t}$$
$$\frac{1}{2}q_0=q_0e^{-0.055t}$$
$$\frac{1}{2}=e^{-0.055t}$$
$$\ln\frac{1}{2}=\ln e^{-0.055t}$$
$$\ln\frac{1}{2}=-0.055t$$
$$t=\frac{\ln\frac{1}{2}}{-0.055}\approx 12.6$$
12.6 hours

37.

$$A=P\left(1+\frac{r}{k}\right)^{kt}$$
$$2P=P\left(1+\frac{0.08}{2}\right)^{2t}$$
$$2=(1.04)^{2t}$$
$$\log 2=\log 1.04^{2t}$$
$$\log 2=2t\log 1.04$$
$$t=\frac{\log 2}{2\log 1.04}\approx 8.84$$
8.84 years

39.

$$N = 60 - 60e^{-0.04t}$$
$$40 = 60 - 60e^{-0.04t}$$
$$-20 = -60e^{-0.04t}$$
$$\frac{1}{3} = e^{-0.04t}$$
$$\ln\frac{1}{3} = \ln e^{-0.04t}$$
$$\ln\frac{1}{3} = -0.04t$$
$$t = \frac{\ln\frac{1}{3}}{-0.04} \approx 27.47$$

27.47 days

41.

$$P = 400\left(1 - e^{-t}\right)$$
$$300 = 400\left(1 - e^{-t}\right)$$
$$\frac{3}{4} = 1 - e^{-t}$$
$$-\frac{1}{4} = -e^{-t}$$
$$\frac{1}{4} = e^{-t}$$
$$\ln\frac{1}{4} = \ln e^{-t}$$
$$\ln\frac{1}{4} = -t$$
$$t = -\ln\frac{1}{4} \approx 1.39$$

1.39 days

43.

$$x^2 + x = \ln(y+1)$$
$$e^{x^2+x} = e^{\ln(y+1)}$$
$$e^{x^2+x} = y + 1$$
$$y = e^{x^2+x} - 1$$

45.

$$\ln x + 3 = \ln(y+1) + \ln(y-1)$$
$$\ln x + 3 = \ln[(y+1)(y-1)]$$
$$e^{\ln x + 3} = e^{\ln[(y+1)(y-1)]}$$
$$e^{\ln x} \cdot e^3 = (y+1)(y-1)$$
$$xe^3 = y^2 - 1$$
$$y^2 = xe^3 + 1$$
$$y = \sqrt{xe^3 + 1}$$

CHAPTER 6 REVIEW EXERCISES

1.

$$f(x) = y = \frac{x}{3} - 2$$

$$y + 2 = \frac{x}{3}$$

$$3(y+2) = x$$

$$f^{-1}(x) = 3(x+2)$$

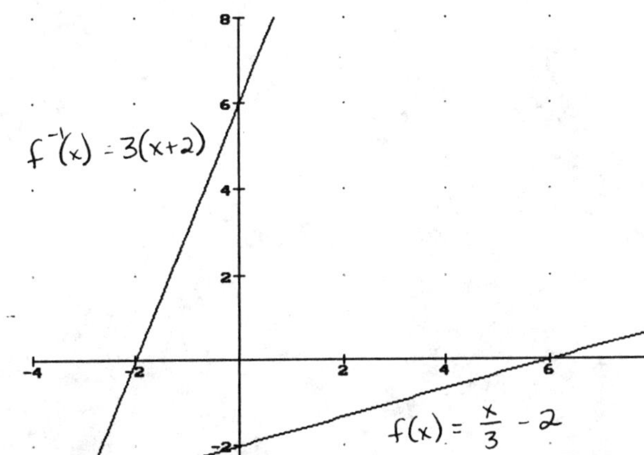

2.

$$f(x) = \frac{|x|}{3} - 2$$

$$f(1) = \frac{|1|}{3} - 2 = -\frac{5}{3}$$

$$f(-1) = \frac{|-1|}{3} - 2 = -\frac{5}{3}$$

$$f(1) = f(-1)$$

$$1 \neq -1$$

Not one-to-one

3.

$$f(x) = \left(\frac{1}{3}\right)^x$$

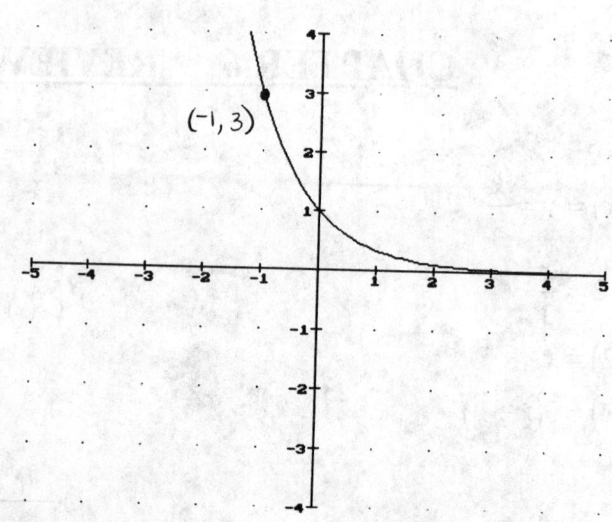

4.

$$2^{2x} = 8^{x-1}$$
$$2^{2x} = \left(2^3\right)^{x-1}$$
$$2^{2x} = 2^{3x-3}$$
$$2x = 2x - 3$$
$$-x = -3$$
$$x = 3$$

5.

$$(2a+1)^x = (3a-1)^x$$
$$2a+1 = 3a-1$$
$$2 = a$$

6.

$$A = P\left(1+\frac{r}{k}\right)^{kt}$$
$$A = 8000\left(1+\frac{0.12}{2}\right)^{2(4)}$$
$$A = 12750.78$$

\$12,750.78

7.

$$27 = 9^{\frac{3}{2}}$$
$$\log_9 27 = \frac{3}{2}$$

8.

$$\log_{64} 8 = \frac{1}{2}$$
$$64^{\frac{1}{2}} = 8$$

9.

$$\log_2 \frac{1}{8} = -3$$
$$2^{-3} = \frac{1}{8}$$

10.
$$6^0 = 1$$
$$\log_6 1 = 0$$

11
$$\log_x 16 = 4$$
$$x^4 = 16$$
$$x = \sqrt[4]{16}$$
$$x = 2$$

12.
$$\log_5 \frac{1}{125} = x - 1$$
$$5^{x-1} = \frac{1}{125}$$
$$5^{x-1} = 5^{-3}$$
$$x - 1 = -3$$
$$x = -2$$

13.
$$\ln x = -4$$
$$e^{\ln x} = e^{-4}$$
$$x = e^{-4}$$

14.
$$\log_3 (x+1) = \log_3 27$$
$$x + 1 = 27$$
$$x = 26$$

15.
$$\log_3 3^5 = 5$$

16.
$$\ln e^{-\frac{1}{3}} = -\frac{1}{3}$$

17.
$$\log_3 \frac{1}{3} = \log_3 3^{-1} = -1$$

18.
$$e^{\ln 3} = 3$$

19.

$$f(x) = y = \log_3 x + 1$$
$$y - 1 = \log_3 x$$
$$x = 3^{y-1}$$

x	y
$\frac{1}{9}$	-1
$\frac{1}{3}$	0
1	1
3	2

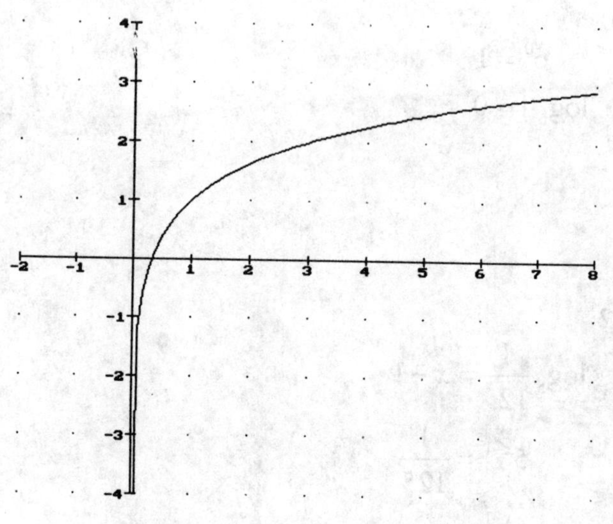

20.

$$\log_a \frac{\sqrt{x-1}}{2x} = \log_a (x-1)^{\frac{1}{2}} - \log_a 2x$$
$$= \frac{1}{2}\log_a (x-1) - [\log_a 2 + \log_a x]$$
$$= \frac{1}{2}\log_a (x-1) - \log_a 2 - \log_a x$$

21.

$$\log_a \frac{x(2-x)^2}{(y+1)^{\frac{1}{2}}} = \log_a \left[x(2-x)^2\right] - \log_a (y+1)^{\frac{1}{2}}$$
$$= \log_a x + \log_a (2-x)^2 - \frac{1}{2}\log_a (y+1)$$
$$= \log_a x + 2\log_a (2-x) - \frac{1}{2}\log_a (y+1)$$

22.

$$\ln\left[(x+1)^4(y-1)^2\right] = \ln(x+1)^4 + \ln(y-1)^2$$
$$= 4\ln(x+1) + 2\ln(y-1)$$

23.

$$\log\sqrt[5]{\frac{y^2z}{z+3}} = \log\left(\frac{y^2z}{z+3}\right)^{\frac{1}{5}}$$
$$= \frac{1}{5}\log\frac{y^2z}{z+3}$$
$$= \frac{1}{5}\left[\log(y^2z) - \log(z+3)\right]$$
$$= \frac{1}{5}\left[\log y^2 + \log z - \log(z+3)\right]$$
$$= \frac{1}{5}\left[2\log y + \log z - \log(z+3)\right]$$
$$= \frac{2}{5}\log y + \frac{1}{5}\log z - \frac{1}{5}\log(z+3)$$

24.

$$e^{-5} = 0.0067$$

25.

$$\ln\sqrt{2} = 0.3466$$

26.

$$\sqrt[3]{5-e^3} = -2.4709$$

27.

$$2500\left(1+\frac{0.08}{4}\right)^{80} = 12188.60$$

28.

$$2500e^{1.6} = 12382.581$$

29.

$$\frac{\log 5}{\log 2} = 2.3219$$

30.

$$\log_3 4 = \frac{\ln 4}{\ln 3} = 1.2619$$

31.

$$e^{5\ln 2} = e^{\ln 2^5} = 2^5 = 32$$

32.

$$2^e = 6.5809$$

33.

$$\log(5-\pi) = 0.2691$$

34.

Cliff: $21-12=9$ years

$$A=P\left(1+\frac{r}{k}\right)^{kt}$$
$$=8000\left(1+\frac{0.09}{4}\right)^{4(9)}$$
$$=17822.53$$

Joe: $23-14=9$ years

$$A=P\left(1+\frac{r}{k}\right)^{\mathrm{kt}}$$
$$=7000\left(1+\frac{0.11}{2}\right)^{2(9)}$$
$$=18350.26.$$

Joe can contribute more by \$527.73.

35.

$Q=q_0 50^t$ $\quad q_0$: initial blood count

$$5(10^6)=3(10^6)50^t$$
$$\frac{5}{3}=50^t$$
$$\ln\frac{5}{3}=\ln 50^t$$
$$\ln\frac{5}{3}=t\ln 50$$
$$\frac{\ln\frac{5}{3}}{\ln 50}=t$$
$$t=0.13$$

0.13 days $\approx$ 3.12 hours

36.

$Q=q_0e^{kt}$ $\quad q_0=320000$

$\mathrm{Q}(2)=640000$

$$Q=320000e^{kt}$$
$$640000=320000\mathrm{e}^{2\mathrm{k}}$$
$$2=e^{2k}$$
$$\ln 2=\ln e^{2k}=2k$$
$$k=\frac{\ln 2}{2}\approx 0.347$$
$$Q=320000e^{0.347t}$$

At 10 p.m., $t=6$

$$Q=320000e^{0.347(6)}$$
$$=2{,}566{,}558$$

37.

$Q=q_0e^{kt}$ $\quad q_0=150000$

$\mathrm{Q}(2)=225000$

$$Q=150000e^{kt}$$
$$225000=150000e^{k(2)}$$
$$1.5=e^{2k}$$
$$\ln 1.5=\ln e^{2k}$$
$$\ln 1.5=2k$$
$$k=\frac{\ln 1.5}{2}\approx 0.203$$
$$Q=150000e^{0.203t}$$

At 2 a.m., $t=10$

$$Q=150000e^{0.203(10)}$$
$$Q=1142113$$

1,142,113

38.

Bank A: $A = Pe^{rt}$

$= 1000e^{0.05(5)}$

$= 1284.03$

Bank B: $A = P\left(1+\frac{r}{k}\right)^{kt}$

$= 1000\left(1+\frac{.0505}{2}\right)^{2(5)}$

$= 1283.21$

Bank A is better

39a).

$C_F = C_0 e^{-kt}$

$1000 = 10000e^{-2t}$

$0.1 = e^{-2t}$

$\ln 0.1 = -2t$

$t = \frac{\ln 0.1}{-2} = 1.15$

39b).

$C_F = C_0 e^{-kt}$

$C^F = 10000e^{-t} \qquad k = 1$

$C^F = 10000e^{-2t} \qquad k = 2$

$C^F = 10000e^{-3t} \qquad k = 3$

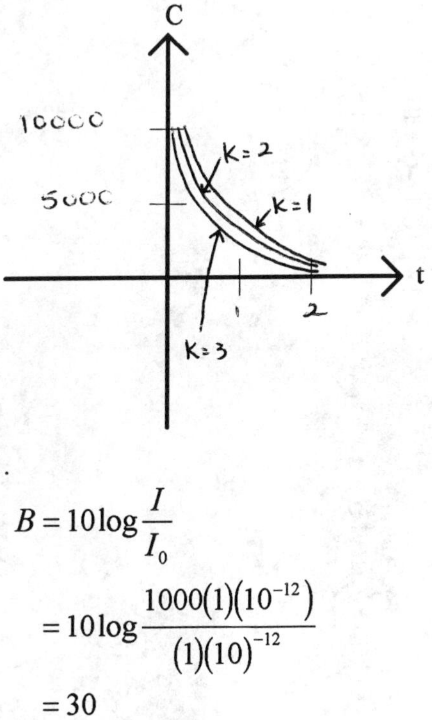

40.

$B = 10\log\frac{I}{I_0}$

$= 10\log\frac{(1.0)(10^{-8})}{(1.0)(10^{-12})}$

$= 40$

40 decibels

41.

$B = 10\log\frac{I}{I_0}$

$= 10\log\frac{1000(1)(10^{-12})}{(1)(10)^{-12}}$

$= 30$

30 decibels

CHAPTER 6 REVIEW TEST

1.

$$f(x) = y = -4x + 2$$
$$y - 2 = -4x$$
$$\frac{y-2}{-4} = x$$
$$f^{-1}(x) = -\frac{1}{4}x + \frac{1}{2}$$

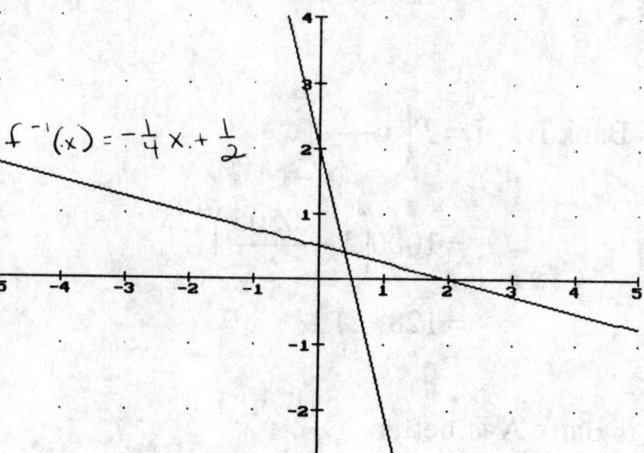

2.

$$f(x) = y = -\frac{1}{x}$$
$$x = -\frac{1}{y}$$
$$f^{-1}(x) = -\frac{1}{x}$$

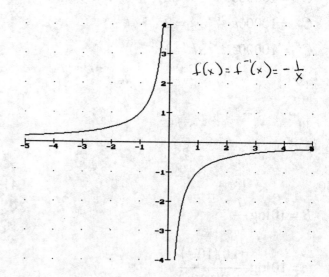

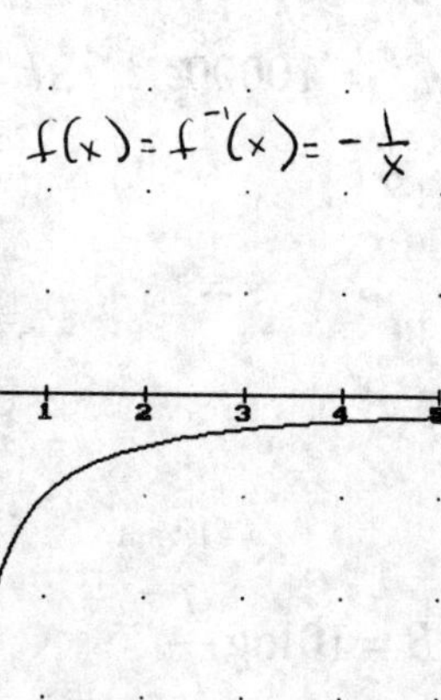

3.

$f(x)=2^{x+1}$

x	$f(x)$
-2	$\frac{1}{2}$
-1	1
0	2
1	4

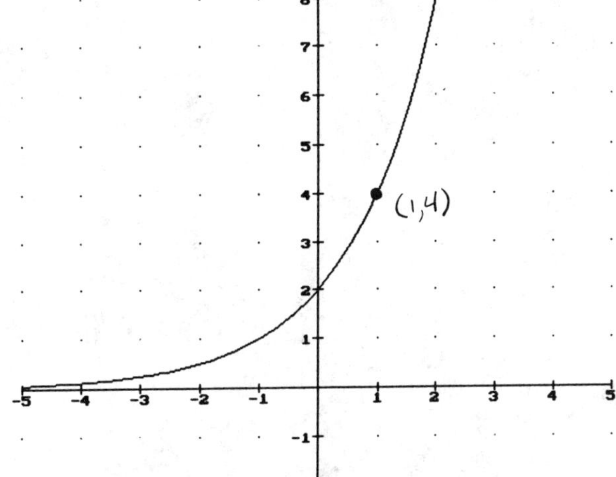

4.

$$\left(\frac{1}{2}\right)^x=\left(\frac{1}{4}\right)^{2x+1}$$
$$\left(\frac{1}{2}\right)^x=\left[\left(\frac{1}{2}\right)^2\right]^{2x+1}$$
$$\left(\frac{1}{2}\right)^x=\left(\frac{1}{2}\right)^{4x+2}$$
$$x=4x+2$$
$$-3x=2$$
$$x=-\frac{2}{3}$$

5.

$$\log_3\frac{1}{9}=-2$$
$$3^{-2}=\frac{1}{9}$$

6.

$$64=16^{\frac{3}{2}}$$
$$\log_{16}64=\frac{3}{2}$$

7.

$$\log_x 27=3$$
$$x^3=27$$
$$x=\sqrt[3]{27}$$
$$x=3$$

8.

$$\log_6\left(\frac{1}{36}\right) = 3x+1$$
$$6^{3x+1} = \frac{1}{36}$$
$$6^{3x+1} = 6^{-2}$$
$$3x+1 = -2$$
$$3x = -3$$
$$x = -1$$

9.

$$\ln e^{\frac{5}{2}} = \frac{5}{2}$$

10.

$$\log_5 \sqrt{5} = \log_5 5^{\frac{1}{2}} = \frac{1}{2}$$

11.

$$\log_a \frac{x^3}{y^2 z} = \log_a x^3 - \log_a y^2 z$$
$$= 3\log_a x - \left[\log_a y^2 + \log_a z\right]$$
$$= 3\log_a x - 2\log_a y - \log_a z$$

12.

$$\log \frac{x^2\sqrt{2y-1}}{y^3} = \log\left(x^2\sqrt{2y-1}\right) - \log y^3$$
$$= \log x^2 + \log(2y-1)^{\frac{1}{2}} - \log y^3$$
$$= 2\log x + \frac{1}{2}\log(2y-1) - 3\log y$$

13.

$$\log 5 = \log\left(\frac{1}{2}\cdot 10\right)$$
$$= \log\frac{1}{2} + \log 10$$
$$= \log 2^{-1} + 1$$
$$= -\log 2 + 1$$
$$= -0.3 + 1$$
$$= 0.7$$

14.

$$\log 2\sqrt{2} = \log 2^{\frac{3}{2}}$$
$$= \frac{3}{2}\log 2$$
$$= \frac{3}{2}(0.3)$$
$$= 0.45$$

15.

$$2\log x - 3\log(y+1) = \log x^2 - \log(y+1)^3$$
$$= \log \frac{x^2}{(y+1)^3}$$

16.

$$\frac{2}{3}\left[\log_a(x+3)-\log_a(x-3)\right]=\frac{2}{3}\log_a\frac{x+3}{x-3}$$
$$=\log_a\left(\frac{x+3}{x-3}\right)^{\frac{2}{3}}$$

17.

$$Q(t)=q_0e^{0.02t}$$
$$2q_0=q_0e^{0.02t}$$
$$2=e^{0.02t}$$
$$\ln 2=0.02t$$
$$t=\frac{\ln 2}{0.02}\approx 34.66$$

34.66 hours

18.

$$A=P\left(1+\frac{r}{k}\right)^{kt}$$
$$=500\left(1+\frac{0.12}{12}\right)^{12\left(\frac{1}{2}\right)}$$
$$=530.76$$

\$530.76

19.

$$\log x-\log 2=2$$
$$\log\frac{x}{2}=2$$
$$\frac{x}{2}=10^2$$
$$x=200$$

20.

$$\log_4(x-3)=1-\log_4 x$$
$$\log_4(x-3)+\log_4 x=1$$
$$\log_4\left[(x-3)x\right]=1$$
$$4^1=(x-3)x$$
$$4=x^2-3x$$
$$0=x^2-3x-4$$
$$0=(x-4)(x+1)$$
$$x=4 \qquad \underbrace{x=-1}_{\text{not in domain of }\log_4 x}$$

Solution: $x=4$

21.

$$\log 14=\log(2\cdot 7)$$
$$=\log 2+\log 7$$
$$=0.30+0.85$$
$$=1.15$$

22.

$$\begin{aligned}\log 3.5 &= \log\frac{7}{2}\\ &= \log 7 - \log 2\\ &= 0.85 - 0.30\\ &= 0.55\end{aligned}$$

23.

$$\begin{aligned}\log\sqrt{6} &= \log 6^{\frac{1}{2}}\\ &= \frac{1}{2}\log 6\\ &= \frac{1}{2}\log(2\cdot 3)\\ &= \frac{1}{2}[\log 2 + \log 3]\\ &= \frac{1}{2}[0.30 + 0.48]\\ &= 0.39\end{aligned}$$

24.

$$\begin{aligned}\log 0.7 &= \log\frac{7}{10}\\ &= \log 7 - \log 10\\ &= 0.85 - 1\\ &= -0.15\end{aligned}$$

25.

$$\begin{aligned}\frac{1}{3}\log_a x - \frac{1}{2}\log_a y &= \log_a x^{\frac{1}{3}} - \log_a y^{\frac{1}{2}}\\ &= \log_a \frac{x^{\frac{1}{3}}}{y^{\frac{1}{2}}}\end{aligned}$$

26.

$$\begin{aligned}\frac{4}{3}[\log x + \log(x-1)] &= \frac{4}{3}[\log[x(x-1)]]\\ &= \log(x^2 - x)^{\frac{4}{3}}\end{aligned}$$

27.

$$\begin{aligned}\ln 3x + 2\left(\ln y - \frac{1}{2}\ln z\right) &= \ln 3x + 2\left(\ln y - \ln z^{\frac{1}{2}}\right)\\ &= \ln 3x + 2\left(\ln\frac{y}{z^{\frac{1}{2}}}\right)\\ &= \ln 3x + \ln\left(\frac{y}{z^{\frac{1}{2}}}\right)^2\\ &= \ln\frac{3xy^2}{z}\end{aligned}$$

28.

$$\begin{aligned}2\log_a(x+2) - \frac{3}{2}\log_a(x+1) &= \log_a(x+2)^2 - \log_a(x+1)^{\frac{3}{2}}\\ &= \log_a\frac{(x+2)^2}{(x+1)^{\frac{3}{2}}}\end{aligned}$$

29.

$$\log_8 32 = \frac{\log 32}{\log 8} \approx \frac{1.5}{0.9} = 1.67$$

30.

$$\log_5 32 = \frac{\log 32}{\log 5} \approx \frac{1.5}{0.7} = 2.14$$

31.

original quantity: q_0

$$Q(t) = q_0 e^{-0.06t}$$

$$\frac{1}{2} q_0 = q_0 e^{-0.06t}$$

$$\frac{1}{2} = e^{-0.06t}$$

$$\ln \frac{1}{2} = -0.06t$$

$$t = \frac{\ln \frac{1}{2}}{-0.06} = 11.55$$

11.55 hours

32.

$$2^{3x-1} = 14$$

$$\log 2^{3x-1} = \log 14$$

$$(3x-1)\log 2 = \log 14$$

$$3x - 1 = \frac{\log 14}{\log 2}$$

$$3x = 1 + \frac{\log 14}{\log 2}$$

$$x = \frac{1}{3} + \frac{\log 14}{3\log 2}$$

33.

$$2\log x - \log 5 = 3$$

$$\log x^2 - \log 5 = 3$$

$$\log \frac{x^2}{5} = 3$$

$$10^3 = \frac{x^2}{5}$$

$$5000 = x^2$$

$$x = \sqrt{5000} = 50\sqrt{2}$$

34.

$$\log(2x-1) = 2 + \log(x-2)$$

$$\log(2x-1) - \log(x-2) = 2$$

$$\log \frac{2x-1}{x-2} = 2$$

$$10^2 = \frac{2x-1}{x-2}$$

$$100x - 200 = 2x - 1$$

$$98x = 199$$

$$x = \frac{199}{98}$$

CUMULATIVE REVIEW EXERCISES: CHAPTERS 4-6

1a).

$$P(x) = 2x^3 + 5x^2 + 3$$
$$P(-2) = 2(-2)^3 + 5(-2)^2 + 3 = 7$$

1b).

$$\begin{array}{r|rrrr} -1 & 2 & 5 & 0 & 3 \\ & & -2 & -3 & 3 \\ \hline & 2 & 3 & -3 & |\,6 \end{array}$$

$Q(x) = 2x^2 + 3x - 3;$ $\quad R = 6$

2.

$$5x^3 - x^2 + 20x - 4 = 0$$

p: $\pm 1, \pm 2, \pm 4$

q: $\pm 1, \pm 5$

$\frac{p}{q}$: $\pm 1, \pm 2, \pm 4, \pm\frac{1}{5}, \pm\frac{2}{5}, \pm\frac{4}{5}$

$$\begin{array}{r|rrrr} \frac{1}{5} & 5 & -1 & 20 & -4 \\ & & 1 & 0 & 4 \\ \hline & 5 & 0 & 20 & |\,0 \end{array}$$

$$5x^2 + 20 = 0$$
$$5x^2 = -20$$
$$x^2 = -4$$
$$x = \pm 2i$$

rational root: $x = \frac{1}{5}$

3.

$$\begin{array}{r|rrrr} \frac{1}{2} & 2 & -1 & -6 & 3 \\ & & 1 & 0 & -3 \\ \hline & 2 & 0 & -6 & |\,0 \end{array}$$

$$2x^2 - 6 = 0$$
$$2x^2 = 6$$
$$x^2 = 3$$
$$x = \pm\sqrt{3}$$

other x-intercepts: $(\sqrt{3}, 0), (-\sqrt{3}, 0)$

4a).

$$f(x)=\frac{2x^2+6x}{x^2-9}=\frac{2x(x+3)}{(x+3)(x-3)}=\frac{2x}{x-3}$$

$$x^2-9\neq 0$$
$$x^2\neq 9$$
$$x\neq \pm 3$$

Domain $\{x|x\neq \pm 3\}$

4b).

$$f(x)=\frac{2x^2+6x}{x^2-9}=\frac{2x(x+3)}{(x+3)(x-3)}=\frac{2x}{x-3}$$

vertical asymptote: $x=3$

$$\frac{2x}{x-3}=\frac{x(2)}{x\left(1-\frac{3}{x}\right)}=\frac{2}{1-\frac{3}{x}}$$

$-\frac{3}{x}$ approaches 0 as $|x|$ approaches ∞

horizontal asymptote: $y=\frac{2}{1-0}=2$

$y=2$

4c & d).

viewing rectangle: three times the EQUAL viewing rectangle

XSCL = 3

YSCL = 3

GRAPH $Y=(2X^2+6X)\div(X^2-9)$

5.

$$27^{x-1}=9^{3-x}$$
$$(3^3)^{x-1}=(3^2)^{3-x}$$
$$3^{3x-3}=3^{6-2x}$$

$$3x-3=6-2x$$
$$5x=9$$
$$x=\frac{9}{5}$$

6.

$$\log_x(2x+3)=2$$
$$x^2=2x+3$$
$$x^2-2x-3=0$$
$$(x-3)(x+1)=0$$

$x=3$ $\quad$ $\underbrace{x=-1}_{\text{not a base for log}}$

Solution: $x=3$

7.

$$\log_2(\log_3 x) = 1$$
$$2^1 = \log_3 x$$
$$3^2 = x$$
$$9 = x$$

8.

initial amount: q_0

$$Q(t) = q_0 e^{-kt}$$
$$\frac{1}{2} q_0 = q_0 e^{-k(5)}$$
$$\frac{1}{2} = e^{-5k}$$
$$\ln \frac{1}{2} = -5k$$
$$k = \frac{\ln \frac{1}{2}}{-5} = 0.139$$
$$Q(t) = q_0 e^{-0.139t}$$
$$3 = 9e^{-0.139t}$$
$$\frac{1}{3} = e^{-0.139t}$$
$$\ln \frac{1}{3} = -0.139t$$
$$t = \frac{\ln \frac{1}{3}}{-0.139} = 7.9$$

7.9 years

9.

$$f(x) = 2^x - 1$$

x	$f(x)$
-2	$-\frac{3}{4}$
-1	$-\frac{1}{2}$
0	0
1	1
2	3

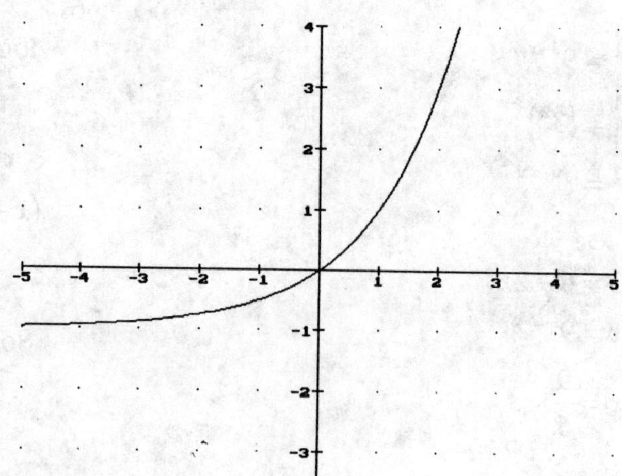

10a).

$$\log 18 = \log(2 \cdot 3^2)$$
$$= \log 2 + \log 3^2$$
$$= \log 2 + 2\log 3$$
$$= s + 2t$$

10b).

$$\log\frac{5}{4} = \log 5 - \log 4$$
$$= \log 5 - \log 2^2$$
$$= \log 5 - 2\log 2$$
$$= u - 2s$$

10c).

$$\log 1.5 = \log\frac{3}{2}$$
$$= \log 3 - \log 2$$
$$= t - s$$

10d).

$$\log\sqrt{7.5} = \log 7.5^{\frac{1}{2}}$$
$$= \frac{1}{2}\log 7.5$$
$$= \frac{1}{2}\log\frac{15}{2}$$
$$= \frac{1}{2}[\log 15 - \log 2]$$
$$= \frac{1}{2}[\log(3 \cdot 5) - \log 2]$$
$$= \frac{1}{2}[\log 3 + \log 5 - \log 2]$$
$$= \frac{1}{2}(t + u - s)$$

11.

$$A = Pe^{rt}$$
$$1000 = Pe^{0.08\left(\frac{1}{2}\right)}$$
$$P = 960.79$$
$$\$960.79$$

12.

$$\ln\frac{x^2\sqrt{x+1}}{x-1} = \ln\left(x^2\sqrt{x+1}\right) - \ln(x-1)$$
$$= \ln x^2 + \ln(x+1)^{\frac{1}{2}} - \ln(x-1)$$
$$= 2\ln x + \frac{1}{2}\ln(x+1) - \ln(x-1)$$

13.

$$A = Pe^{rt}$$
$$4P = Pe^{r(15)}$$
$$4 = e^{15r}$$
$$\ln 4 = 15r$$
$$r = \frac{\ln 4}{15} = 0.0924$$
$$9.24\%$$

14.

$$A = P\left(1 + \frac{r}{k}\right)^{kt}$$
$$= 6000\left(1 + \frac{0.07}{4}\right)^{4(4)}$$
$$= 7919.58$$

amount: \$7,919.58

interest = 7919.58 − 6000 = \$1,919.58

15.

$$A = Pe^{rt} \qquad 18 - 8 = 10 \text{ years}$$
$$20000 = Pe^{0.092(10)}$$
$$P = 7970.38$$
\$7,970.38

16.

$$A = Pe^{rt}$$
$$300000 = 75000e^{0.08t}$$
$$4 = e^{0.08t}$$
$$\ln 4 = 0.08t$$
$$t = 17.32$$
17.32 years

17.

Bond A: $A = Pe^{rt}$ $= 1000e^{0.07(3)}$ $= 1233.68$

Bond B: $A = Pe^{rt}$ $= 950e^{0.09(3)}$ $= 1244.47$

Bond B is the better investment.

18.

$$T = (T_0 - T_R)e^{-kt} + T_R$$
$$90 = (150 - 80)e^{-k(2)} + 80$$
$$10 = 70e^{-2k}$$
$$\frac{1}{7} = e^{-2k}$$
$$\ln\frac{1}{7} = -2k$$
$$k = \frac{\ln\frac{1}{7}}{-2} = 0.9730$$
$$k = 0.9730$$

CHAPTER 6 ADDITIONAL PRACTICE EXERCISES

1. $f(x)=\frac{3}{4}x-1$, find $f^{-1}(x)$

2. Evaluate $\frac{2}{3+e^{-2}}$

3. Sue purchases a $5,000 savings certificate paying 7% annual interest compounded continuously. Find the amount received when the savings certificate is redeemed at the end of 4 years.

4. Write $\log 0.001=-3$ in exponential form.

5. Write $216=6^3$ in logarithmic form.

6. Evaluate $\ln e^{3x+2}$

7. Solve for x: $\log_9 x=-3$

8. Solve for x: $\log_x 64=3$

9. Simplify $\ln\left(xy^3\sqrt{z}\right)$

10. Write as a single logarithm:
$$3\log x-\frac{1}{4}\log y$$

CHAPTER 6 PRACTICE TEST

1.
$f(x)=7x-9$, find $f^{-1}(x)$

2.
Find two restrictions of the function $f(x)=x^4$ that allow it to have an inverse.

3.
$f(x)=3x^3-1$, find $f\left[f^{-1}(-2)\right]$.

4.
Sketch the graph of $f(x)=2^{-x}$.

5.
Evaluate $\dfrac{e^2-e^{-2}}{4}$

6.
If \$500 is invested at an annual interest rate of 6.5% compounded continuously, how much is available after 8 years?

7.
If \$1000 is invested at an annual interest rate of 8% compounded quarterly, how much is available after 12 years?

8.
Write $\log_5 625=4$ in exponential form.

9.
Write $512=8^3$ in logarithmic form.

10.
Evaluate $\log_7 7^{\frac{1}{3}}$

11.
Evaluate $\log_5 \sqrt[3]{5}$

12.
Use your calculator to evaluate $\dfrac{\log 7}{\log 4}$.

13.
Solve for x: $\log_x 243=5$

14.
Solve for x: $\ln x=5$

15.
Solve for x: $\log_{2x} 30=\log_8 30$

16.
Evaluate $e^{5\ln x}$

17.

Write as a single logarithm:

$$4\log x - \frac{1}{3}\log y + \frac{1}{2}\log z$$

18.

Simplify $\log_a\left(x^3\sqrt{yz^5}\right)$

19.

Evaluate $\ln \frac{1}{e^{\frac{2}{3}}}$

20.

Evaluate $e^{\ln 5 - \ln 2}$

CHAPTER 7 SECTION 1

SECTION 7.1 PROGRESS CHECK

1a). (Page 395)

$$\frac{\theta}{\pi}=\frac{-210}{180}$$
$$\theta=\frac{-210\pi}{180}=-\frac{7\pi}{6}\text{ radians}$$

1b).

$$\frac{\theta}{\pi}=\frac{390}{180}$$
$$\theta=\frac{390\pi}{180}=\frac{13\pi}{6}\text{ radians}$$

2a). (Page 396)

$$\frac{\frac{9\pi}{2}}{\pi}=\frac{\theta}{180}$$
$$\theta=\frac{180\left(\frac{9\pi}{2}\right)}{\pi}=810°$$

2b).

$$\frac{\frac{-4\pi}{3}}{\pi}=\frac{\theta}{180}$$
$$\theta=\frac{\left(-\frac{4\pi}{3}\right)(180)}{\pi}=-240°$$

3a). (Page 398)

$95°-360°=-265°$

Hence $95°$ and $-265°$ are coterminal.

3b).

$$\frac{4\pi}{3}+2\pi(3)=\frac{4\pi}{3}+6\pi=\frac{22\pi}{3}$$

Hence $\frac{4\pi}{3}$ and $\frac{22\pi}{3}$ are coterminal.

4a). (Page 399)

$\theta=160°$ lies in the second quadrant.

$\theta'=180°-160°=20°$

4b).

$\theta=\frac{4\pi}{3}$ lies in the third quadrant.

$$\theta'=\frac{4\pi}{3}-\pi=\frac{\pi}{3}\text{ radians}$$

EXERCISE SET 7.1

1.
$270° < 313° < 360°$ Quadrant IV

3.
$0° < 14° < 90°$ Quadrant I

5.
$90° < 141° < 180°$ Quadrant II

7.
$-345° + 360° = 15°$

$0° < 15° < 90°$ Quadrant I

9.
$618° - 360° = 258°$

$180° < 258° < 270°$ Quadrant III

11.
$-195° + 360° = 165°$

$90° < 165° < 180°$ Quadrant II

13.
$\frac{\pi}{2} < \frac{7\pi}{8} < \pi$ Quadrant II

15.
$\frac{-8\pi}{3} + 2(2\pi) = \frac{4\pi}{3}$

$\pi < \frac{4\pi}{3} < \frac{3\pi}{2}$ Quadrant III

17.
$\frac{13\pi}{3} - 2(2\pi) = \frac{\pi}{3}$

$0 < \frac{\pi}{3} < \frac{\pi}{2}$ Quadrant I

19.
$\frac{\theta}{\pi} = \frac{30}{180}$

$\theta = \frac{30\pi}{180} = \frac{\pi}{6}$

21.
$\frac{\theta}{\pi} = \frac{-150}{180}$

$\theta = \frac{-150\pi}{180} = -\frac{5\pi}{6}$

23.
$\frac{\theta}{\pi} = \frac{75}{180}$

$\theta = \frac{75\pi}{180} = \frac{5\pi}{12}$

25.
$\frac{\theta}{\pi} = \frac{-450}{180}$

$\theta = \frac{-450\pi}{180} = -\frac{5\pi}{2}$

27.
$\frac{\theta}{\pi} = \frac{135}{180}$

$\theta = \frac{135\pi}{180} = \frac{3\pi}{4}$

29.

$$\frac{\theta}{\pi}=\frac{120}{180}$$

$$\theta=\frac{120\pi}{180}=\frac{2\pi}{3}$$

31.

$$\frac{\theta}{\pi}=\frac{45.22}{180}$$

$$\theta=\frac{45.22\pi}{180}=0.251\pi$$

33.

$$123°20'=123.33°$$

$$\frac{\theta}{\pi}=\frac{123.33}{180}$$

$$\theta=\frac{123.33\pi}{180}=0.685\pi\approx 2.15$$

35.

$$\frac{\frac{\pi}{4}}{\pi}=\frac{\theta}{180}$$

$$\theta=\frac{180\left(\frac{\pi}{4}\right)}{\pi}=45°$$

37.

$$\frac{\frac{3\pi}{2}}{\pi}=\frac{\theta}{180}$$

$$\theta=\frac{180\left(\frac{3\pi}{2}\right)}{\pi}=270°$$

39.

$$\frac{\frac{-\pi}{2}}{\pi}=\frac{\theta}{180}$$

$$\theta=\frac{180\left(\frac{-\pi}{2}\right)}{\pi}=-90°$$

41.

$$\frac{\frac{4\pi}{3}}{\pi}=\frac{\theta}{180}$$

$$\theta=\frac{180\left(\frac{4\pi}{3}\right)}{\pi}=240°$$

43.

$$\frac{\frac{5\pi}{2}}{\pi}=\frac{\theta}{180}$$

$$\theta=\frac{180\left(\frac{5\pi}{2}\right)}{\pi}=450°$$

45.

$$\frac{\frac{-5\pi}{3}}{\pi}=\frac{\theta}{180}$$

$$\theta=\frac{180\left(\frac{-5\pi}{3}\right)}{\pi}=-300°$$

47.

$$\frac{1.72}{\pi}=\frac{\theta}{180}$$

$$\theta=\frac{180(1.72)}{\pi}=98.55°\text{ or }98°33'$$

49.
$390° - 360° = 30°$ T

51.
$45° - (-45°) = 90°$ (not divisible by 360°) F

53.
$\frac{7\pi}{2} - \frac{\pi}{2} = 3\pi$ (not divisible by 2π) F

55.
$180° - 130° = 50°$

57.
$|-20°| = 20°$

59.
$-455° + 2(360°) = 265°$
$265° - 180° = 85°$

61.
$\frac{12\pi}{5} - 2\pi = \frac{2\pi}{5}$

63.
72° (own reference angle)

65.
$\frac{9\pi}{4} - 2\pi = \frac{\pi}{4}$

67.
$$\theta = \frac{\text{arc length}}{\text{radius}} = \frac{4}{7} \text{ radians}$$
$$\frac{\frac{4}{7}}{\pi} = \frac{\theta}{180}$$
$$\theta = \frac{180\left(\frac{4}{7}\right)}{\pi} \approx 32.74° = 32°44'$$

69.
$$\theta = \frac{\text{arc length}}{\text{radius}}$$
$$\frac{2\pi}{3} = \frac{4}{r}$$
$$r = \frac{4}{\frac{2\pi}{3}} = \frac{6}{\pi} \approx 1.9 \text{ m}$$

71.
1 rotation: arc = circumference of tire
$= 2\pi(13) = 26\pi$
≈ 81.7 in or 6.81 ft
1 mile = 5280 ft
$\frac{5280}{6.81} \approx 775.33$ rotations

73.

$r = 1.5$

$s = \dfrac{3\pi}{10}$

$\theta = \dfrac{s}{r} = \dfrac{\frac{3\pi}{10}}{1.5} = \dfrac{\pi}{5}$

$2\pi \div \dfrac{\pi}{5} = 10$ ribs

75.

$\dfrac{d}{180} = \dfrac{\theta}{\pi}$

$\theta = \dfrac{\pi d}{180}$ radians

$\dfrac{\pi d}{180} = \dfrac{s}{r}$

$s = \dfrac{\pi d r}{180}$

77.

$d = 2$

circumference: 2π

$\dfrac{30}{2\pi} = \dfrac{15}{\pi} \approx 4.8$ rotations

CHAPTER 7 SECTION 2

SECTION 7.2 PROGRESS CHECK

1. (Page 409)

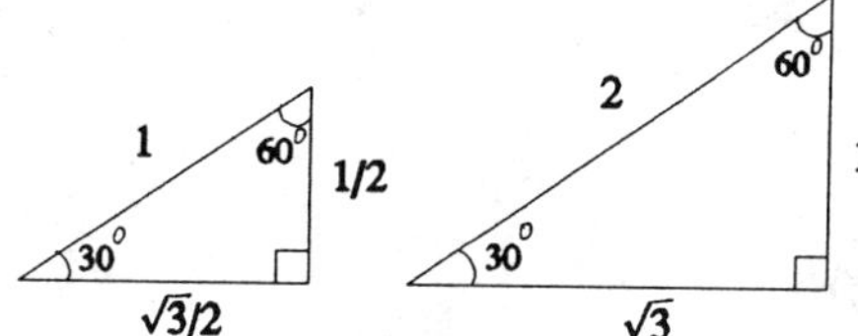

$$\frac{1}{2} = \frac{\frac{1}{2}}{1} = \frac{\frac{\sqrt{3}}{2}}{\sqrt{3}} \ ?$$

Multiply by $2\sqrt{3}$: $\sqrt{3} = \sqrt{3} = \sqrt{3}$

All three ratios are equal, hence the sides are proportional making the triangles similar.

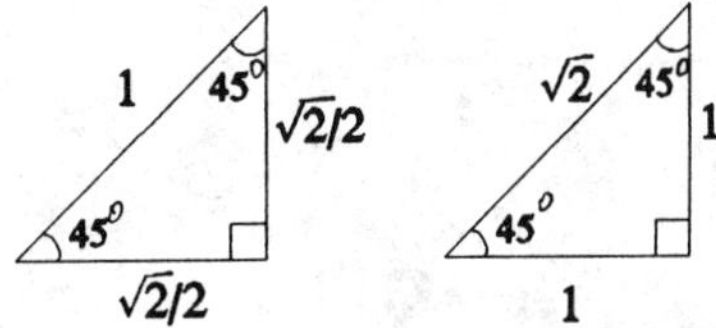

$$\frac{1}{\sqrt{2}} = \frac{\frac{\sqrt{2}}{2}}{1} = \frac{\frac{\sqrt{2}}{2}}{1} \ ?$$

Multiply by $2\sqrt{2}$: $2 = 2 = 2$

All three ratios are equal, hence the sides are proportional making the triangles similar.

	$\sin\theta$	$\cos\theta$
30°	$\frac{\frac{1}{2}}{1} = \frac{1}{2}$	$\frac{\frac{\sqrt{3}}{2}}{1} = \frac{\sqrt{3}}{2}$
45°	$\frac{\frac{\sqrt{2}}{2}}{1} = \frac{\sqrt{2}}{2}$	$\frac{\frac{\sqrt{2}}{2}}{1} = \frac{\sqrt{2}}{2}$
60°	$\frac{\frac{\sqrt{3}}{2}}{1} = \frac{\sqrt{3}}{2}$	$\frac{\frac{1}{2}}{1} = \frac{1}{2}$

2. (Page 412)

$$\beta = 90° - \alpha$$
$$\beta = 90° - 64° = 26°$$

$$\sin\ B = \frac{b}{c}$$
$$\sin\ 26° = \frac{24.7}{c}$$
$$c = \frac{24.7}{\text{Sin}\ 26°}$$
$$c = 56.34$$
$$\tan\ A = \frac{a}{b}$$
$$\tan 64° = \frac{a}{24.7}$$
$$a = 24.7 \tan 64°$$
$$a = 50.64$$

EXERCISE SET 7.2

1.

$$12^2 + (\text{opposite})^2 = 15^2$$

$$\text{opposite} = \sqrt{15^2 - 12^2} = 9$$

$$\sin\theta = \frac{\text{opposite}}{\text{hypotenuse}} = \frac{9}{15} = \frac{3}{5}$$

$$\cos\theta = \frac{\text{adjacent}}{\text{hypotenuse}} = \frac{12}{15} = \frac{4}{5}$$

$$\tan\theta = \frac{\text{opposite}}{\text{adjacent}} = \frac{9}{12} = \frac{3}{4}$$

$$\csc\theta = \frac{1}{\sin\theta} = \frac{1}{\frac{3}{5}} = \frac{5}{3}$$

$$\sec\theta = \frac{1}{\cos\theta} = \frac{1}{\frac{4}{5}} = \frac{5}{4}$$

$$\cot\theta = \frac{1}{\tan\theta} = \frac{1}{\frac{3}{4}} = \frac{4}{3}$$

3.

$$(\text{opposite})^2 + 6^2 = 10^2$$

$$\text{opposite} = \sqrt{10^2 - 6^2} = 8$$

$$\sin\theta = \frac{\text{opposite}}{\text{hypotenuse}} = \frac{8}{10} = \frac{4}{5}$$

$$\cos\theta = \frac{\text{adjacent}}{\text{hypotenuse}} = \frac{6}{10} = \frac{3}{5}$$

$$\tan\theta = \frac{\text{opposite}}{\text{adjacent}} = \frac{8}{6} = \frac{4}{3}$$

$$\csc\theta = \frac{1}{\sin\theta} = \frac{1}{\frac{4}{5}} = \frac{5}{4}$$

$$\sec\theta = \frac{1}{\cos\theta} = \frac{1}{\frac{3}{5}} = \frac{5}{3}$$

$$\cot\theta = \frac{1}{\tan\theta} = \frac{1}{\frac{4}{3}} = \frac{3}{4}$$

5.

$$(\text{opposite})^2 + 1^2 = \left(\sqrt{5}\right)^2$$

$$\text{opposite} = \sqrt{\left(\sqrt{5}\right)^2 - 1^2} = 2$$

$$\sin\theta = \frac{\text{opposite}}{\text{hypotenuse}} = \frac{2}{\sqrt{5}} = \frac{2\sqrt{5}}{5}$$

$$\cos\theta = \frac{\text{adjacent}}{\text{hypotenuse}} = \frac{1}{\sqrt{5}} = \frac{\sqrt{5}}{5}$$

$$\tan\theta = \frac{\text{opposite}}{\text{adjacent}} = \frac{2}{1} = 2$$

$$\csc\theta = \frac{1}{\sin\theta} = \frac{1}{\frac{2}{\sqrt{5}}} = \frac{\sqrt{5}}{2}$$

$$\sec\theta = \frac{1}{\cos\theta} = \frac{1}{\frac{1}{\sqrt{5}}} = \sqrt{5}$$

$$\cot\theta = \frac{1}{\tan\theta} = \frac{1}{2}$$

7.

$$1^2 + x^2 = (\text{hypotenuse})^2$$

$$\text{hypotenuse} = \sqrt{1+x^2}$$

$$\sin\theta = \frac{\text{opposite}}{\text{hypotenuse}} = \frac{1}{\sqrt{1+x^2}} = \frac{\sqrt{1+x^2}}{1+x^2}$$

$$\cos\theta = \frac{\text{adjacent}}{\text{hypotenuse}} = \frac{x}{\sqrt{1+x^2}} = \frac{x\sqrt{1+x^2}}{1+x^2}$$

$$\tan\theta = \frac{\text{opposite}}{\text{adjacent}} = \frac{1}{x}$$

$$\csc\theta = \frac{1}{\sin\theta} = \frac{1}{\frac{1}{\sqrt{1+x^2}}} = \sqrt{1+x^2}$$

$$\sec\theta = \frac{1}{\cos\theta} = \frac{1}{\frac{x}{\sqrt{1+x^2}}} = \frac{\sqrt{1+x^2}}{x}$$

$$\cot\theta = \frac{1}{\tan\theta} = \frac{1}{\frac{1}{x}} = x$$

9.

$$\sin\theta = \frac{h}{5}$$

$$h = 5\sin\theta$$

11.

$$\cot\theta = \frac{h}{6.5}$$

$$h = 6.5\cot\theta$$

13.

$$\csc\theta = \frac{h}{3.7}$$

$$h = 3.7\csc\theta$$

15.

$$\sin 40^\circ = \frac{40}{c}$$

$$c = \frac{40}{\sin 40^\circ} = 62.23$$

17.

$$\tan 22^\circ = \frac{b}{75}$$

$$b = 75\tan 22^\circ = 30.30$$

19.

$$\cos 42^\circ 30' = \frac{25}{c}$$

$$c = \frac{25}{\cos 42.5^\circ} = 33.91$$

CHAPTER 7 SECTION 3

SECTION 7.3 PROGRESS CHECK

1a). (Page 422)

$$t = \frac{-15\pi}{2} = -3(2\pi) - \frac{3\pi}{2}$$

Sweep out an arc of length $\frac{3\pi}{2}$ in a clockwise direction. $P\left(-\frac{15\pi}{2}\right) = P(0, 1)$

1b).

$$t = -3\pi = -1(2\pi) - \pi$$

Sweep out an arc of length π in a clockwise direction. $P(-3\pi) = P(-1, 0)$

1c).

$$t = \frac{-\pi}{2}$$

Sweep out an arc of length $\frac{\pi}{2}$ in a clockwise direction. $P\left(-\frac{\pi}{2}\right) = P(0, -1)$

EXERCISE SET 7.3

1.

$$\pi < \frac{4\pi}{3} < \frac{3\pi}{2}$$ Quadrant III

3.

$$\frac{\pi}{2} < \frac{5\pi}{6} < \pi$$ Quadrant II

5.

$$\frac{-19\pi}{18} + 2\pi = \frac{17\pi}{18}$$

$$\frac{\pi}{2} < \frac{17\pi}{18} < \pi$$ Quadrant II

7.

$$\frac{-7\pi}{6} + 2\pi = \frac{5\pi}{6}$$

$$\frac{\pi}{2} < \frac{5\pi}{6} < \pi$$ Quadrant II

9.

$$4\pi - 2(2\pi) = 0$$

11.

$$\frac{15\pi}{7} - 2\pi = \frac{\pi}{7}$$

13.

$$\frac{-21\pi}{2}+6(2\pi)=\frac{3\pi}{2}$$

15.

$$\frac{41\pi}{6}-3(2\pi)=\frac{5\pi}{6}$$

17.

$$-9\pi+5(2\pi)=\pi$$

19.

$$\frac{27\pi}{5}-2(2\pi)=\frac{7\pi}{5}$$

21.

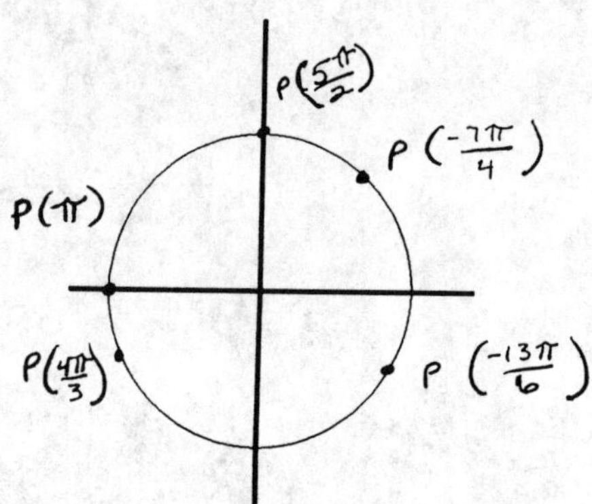

23.

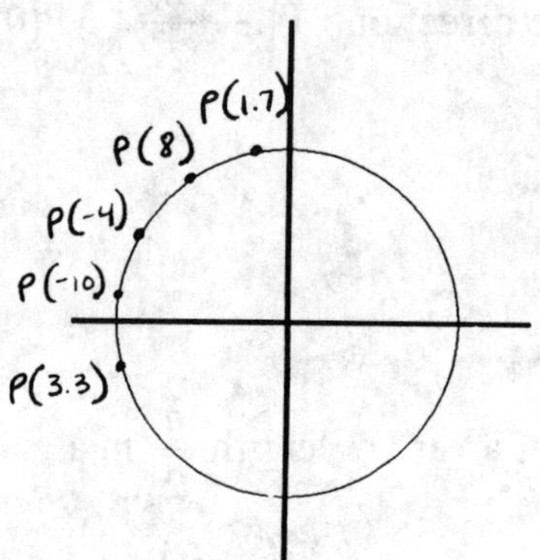

25.

$$3\pi-2\pi=\pi \qquad (-1,\ 0)$$

27.

$$\frac{7\pi}{2}-2\pi=\frac{3\pi}{2} \qquad (0,\ -1)$$

29.

$$\frac{23\pi}{2}-5(2\pi)=\frac{3\pi}{2} \qquad (0,\ -1)$$

31.

$$-18\pi+9(2\pi)=0 \qquad (1,\ 0)$$

33.

$$\frac{-15\pi}{2}+4(2\pi)=\frac{\pi}{2} \qquad (0,\ 1)$$

35.

$$\frac{17\pi}{2}-4(2\pi)=\frac{\pi}{2} \qquad (0,\ 1)$$

37.

$$\cos t = x = -\frac{3}{5}$$

$$\sin t = y = \frac{4}{5}$$

39.

$$\cos t = x = \frac{\sqrt{3}}{2}$$

$$\sin t = y = -\frac{1}{2}$$

41.

$$\cos t = x = -\frac{\sqrt{2}}{2}$$

$$\sin t = y = \frac{\sqrt{2}}{2}$$

43.

$$\cos t = x = \frac{\sqrt{15}}{4}$$

$$\sin t = y = -\frac{1}{4}$$

	$\tan t = \frac{\sin t}{\cos t}$	$\cot t = \frac{1}{\tan t}$	$\sec t = \frac{1}{\cos t}$	$\csc t = \frac{1}{\sin t}$
45.	$\frac{\frac{4}{5}}{-\frac{3}{5}} = -\frac{4}{3}$	$\frac{1}{-\frac{4}{3}} = -\frac{3}{4}$	$\frac{1}{-\frac{3}{5}} = -\frac{5}{3}$	$\frac{1}{\frac{4}{5}} = \frac{5}{4}$
47.	$\frac{-\frac{1}{2}}{\frac{\sqrt{3}}{2}} = -\frac{\sqrt{3}}{3}$	$\frac{1}{-\frac{\sqrt{3}}{3}} = -\sqrt{3}$	$\frac{1}{\frac{\sqrt{3}}{2}} = \frac{2\sqrt{3}}{3}$	$\frac{1}{-\frac{1}{2}} = -2$
49.	$\frac{\frac{\sqrt{2}}{2}}{-\frac{\sqrt{2}}{2}} = -1$	$\frac{1}{-1} = -1$	$\frac{1}{-\frac{\sqrt{2}}{2}} = -\sqrt{2}$	$\frac{1}{\frac{\sqrt{2}}{2}} = \sqrt{2}$
51.	$\frac{-\frac{1}{4}}{\frac{\sqrt{15}}{4}} = -\frac{\sqrt{15}}{15}$	$\frac{1}{-\frac{\sqrt{15}}{15}} = -\sqrt{15}$	$\frac{1}{\frac{\sqrt{15}}{4}} = \frac{4\sqrt{15}}{15}$	$\frac{1}{-\frac{1}{4}} = -4$

CHAPTER 7 SECTION 4

SECTION 7.4 PROGRESS CHECK

1. (Page 427)

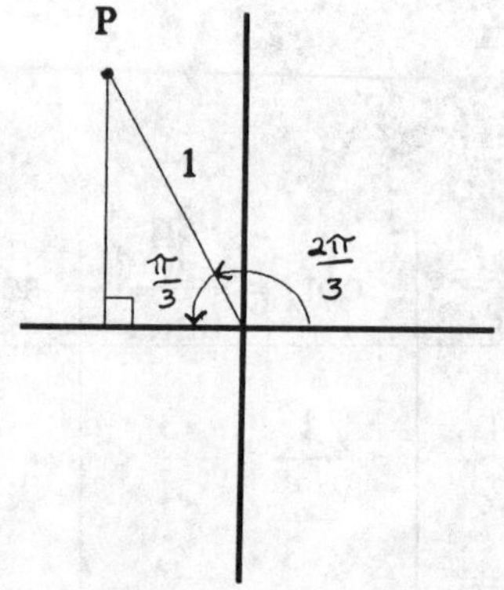

$$\frac{-4\pi}{3}+2\pi=\frac{2\pi}{3}$$

$$P\left(-\frac{1}{2},\frac{\sqrt{3}}{2}\right)$$

$$\sin\left(-\frac{4\pi}{3}\right)=\frac{\sqrt{3}}{2}$$

$$\csc\left(-\frac{4\pi}{3}\right)=\frac{1}{\sin\left(-\frac{4\pi}{3}\right)}=\frac{1}{\frac{\sqrt{3}}{2}}=\frac{2\sqrt{3}}{3}$$

2a). (Page 433)

$$\left.\begin{array}{l}\cos t<0\text{: Q II and III}\\ \tan t>0\text{: Q I and III}\end{array}\right\}\text{Quadrant III}$$

2b).

$$\left.\begin{array}{l}\cos t<0\text{: Q II and III}\\ \sin t>0\text{: Q I and II}\end{array}\right\}\text{Quadrant II}$$

2c).

$$\left.\begin{array}{l}\tan t<0\text{: Q II and IV}\\ \csc t>0\text{: Q I and II}\end{array}\right\}\text{Quadrant II}$$

3. (Page 434)

$$\cos\left(-\frac{3\pi}{4}\right)=\cos\left(\frac{3\pi}{4}\right)=-\frac{\sqrt{2}}{2}$$

$$\sin\left(-\frac{3\pi}{4}\right)=-\sin\left(\frac{3\pi}{4}\right)=-\frac{\sqrt{2}}{2}$$

$$\csc\left(-\frac{4\pi}{3}\right)=-\csc\left(\frac{4\pi}{3}\right)=-\left(-\frac{2\sqrt{3}}{3}\right)=\frac{2\sqrt{3}}{3}$$

4a). (Page 436)

$$\sin^2 t+\cos^2 t=1$$

$$\left(\frac{12}{13}\right)^2+\cos^2 t=1$$

$$\cos^2 t=1-\frac{144}{169}=\frac{25}{169}$$

$$\cos t=-\frac{5}{13}\ \text{(Quadrant II)}$$

4b).

$$\tan t=\frac{\sin t}{\cos t}=\frac{\frac{12}{13}}{-\frac{5}{13}}=-\frac{12}{5}$$

5. (Page 436)

$$\tan t\cos t+\sin t+\frac{1}{\csc t}=\left(\frac{\sin t}{\cos t}\right)\cos t+\sin t+\sin t$$
$$=\sin t+\sin t+\sin t$$
$$=3\sin t$$

EXERCISE SET 7.4

1.
$250°-180°=70°$

3.
$-330°+360°=30°$

5.
$\frac{6\pi}{5}-\pi=\frac{\pi}{5}$

7.
$360°-335°=25°$

9.
$|-47°|=47°$

11.
$\frac{15\pi}{7}-2\pi=\frac{\pi}{7}$

In Exercises 13-23, refer to the following reference triangles:

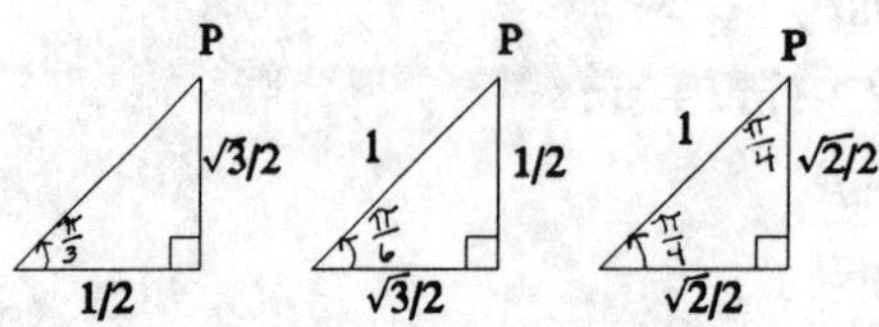

13.

$t = \frac{5\pi}{3}$ Quadrant IV

reference angle: $2\pi - \frac{5\pi}{3} = \frac{\pi}{3}$

$P(t) = \left(\frac{1}{2}, -\frac{\sqrt{3}}{2}\right)$

$\sin t = y = \frac{-\sqrt{3}}{2}$

$\cos t = x = \frac{1}{2}$

15.

$t = -5\pi$

$-5\pi + 3(2\pi) = \pi$ Quadrantal angle

$P(t) = (-1, 0)$

$\sin t = y = 0$

$\cos t = x = -1$

17.

$t = \frac{7\pi}{4}$ Quadrant IV

reference angle: $2\pi - \frac{7\pi}{4} = \frac{\pi}{4}$

$P(t) = \left(\frac{\sqrt{2}}{2}, -\frac{\sqrt{2}}{2}\right)$

$\sin t = y = -\frac{\sqrt{2}}{2}$

$\cos t = x = \frac{\sqrt{2}}{2}$

19.

$t = \frac{2\pi}{3}$ Quadrant II

reference angle: $\pi - \frac{2\pi}{3} = \frac{\pi}{3}$

$P(t) = \left(-\frac{1}{2}, \frac{\sqrt{3}}{2}\right)$

$\sin t = y = \frac{\sqrt{3}}{2}$

$\cos t = x = -\frac{1}{2}$

21.

$t=-\frac{\pi}{3}$ Quadrant IV

reference angle: $\left|-\frac{\pi}{3}\right|=\frac{\pi}{3}$

$P(t)=\left(\frac{1}{2},-\frac{\sqrt{3}}{2}\right)$

$\sin t=y=-\frac{\sqrt{3}}{2}$

$\cos t=x=\frac{1}{2}$

23.

$t=\frac{5\pi}{4}$ Quadrant III

reference angle: $\frac{5\pi}{4}-\pi=\frac{\pi}{4}$

$P(t)=\left(-\frac{\sqrt{2}}{2},-\frac{\sqrt{2}}{2}\right)$

$\sin t=y=-\frac{\sqrt{2}}{2}$

$\cos t=x=-\frac{\sqrt{2}}{2}$

In Exercises 25-39, use the identities

$\tan t=\frac{\sin t}{\cos t},\quad \sec t=\frac{1}{\cos t},\quad \csc t=\frac{1}{\sin t},\quad \cot t=\frac{1}{\tan t}$

25.

$t=135°$ Quadrant II

reference angle: $180°-135°=45°$

$P(t)=\left(-\frac{\sqrt{2}}{2},\frac{\sqrt{2}}{2}\right)$

$\sin 135°=\frac{\sqrt{2}}{2}$

$\cos 135°=-\frac{\sqrt{2}}{2}$

$\tan 135°=\frac{\frac{\sqrt{2}}{2}}{-\frac{\sqrt{2}}{2}}=-1$

$\csc 135°=\frac{1}{\frac{\sqrt{2}}{2}}=\sqrt{2}$

$\sec 135°=\frac{1}{-\frac{\sqrt{2}}{2}}=-\sqrt{2}$

$\cot 135°=\frac{1}{-1}=-1$

27.

$t=-30°$ Quadrant IV

reference angle: $|-30°|=30°$

$P(t)=\left(\frac{\sqrt{3}}{2},-\frac{1}{2}\right)$

$\sin(-30°)=-\frac{1}{2}$

$\cos(-30°)=\frac{\sqrt{3}}{2}$

$\tan(-30°)=\frac{-\frac{1}{2}}{\frac{\sqrt{3}}{2}}=-\frac{\sqrt{3}}{3}$

$\csc(-30°)=\frac{1}{-\frac{1}{2}}=-2$

$\sec(-30°)=\frac{1}{\frac{\sqrt{3}}{2}}=\frac{2\sqrt{3}}{3}$

$\cot(-30°)=\frac{1}{-\frac{\sqrt{3}}{3}}=-\sqrt{3}$

29.

$t = \frac{\pi}{3}$ Quadrant I

$P(t) = \left(\frac{1}{2}, \frac{\sqrt{3}}{2}\right)$

$\sin\frac{\pi}{3} = \frac{\sqrt{3}}{2}$

$\cos\frac{\pi}{3} = \frac{1}{2}$

$\tan\frac{\pi}{3} = \frac{\frac{\sqrt{3}}{2}}{\frac{1}{2}} = \sqrt{3}$

$\csc\frac{\pi}{3} = \frac{1}{\frac{\sqrt{3}}{2}} = \frac{2\sqrt{3}}{3}$

$\sec\frac{\pi}{3} = \frac{1}{\frac{1}{2}} = 2$

$\cot\frac{\pi}{3} = \frac{1}{\sqrt{3}} = \frac{\sqrt{3}}{3}$

31.

$t = \frac{\pi}{4}$ Quadrant I

$P(t) = \left(\frac{\sqrt{2}}{2}, \frac{\sqrt{2}}{2}\right)$

$\sin\frac{\pi}{4} = \frac{\sqrt{2}}{2}$

$\cos\frac{\pi}{4} = \frac{\sqrt{2}}{2}$

$\tan\frac{\pi}{4} = \frac{\frac{\sqrt{2}}{2}}{\frac{\sqrt{2}}{2}} = 1$

$\csc\frac{\pi}{4} = \frac{1}{\frac{\sqrt{2}}{2}} = \sqrt{2}$

$\sec\frac{\pi}{4} = \frac{1}{\frac{\sqrt{2}}{2}} = \sqrt{2}$

$\cot\frac{\pi}{4} = \frac{1}{1} = 1$

33.

$t=\frac{5\pi}{6}$ Quadrant II

reference angle: $\pi-\frac{5\pi}{6}=\frac{\pi}{6}$

$P(t)=\left(-\frac{\sqrt{3}}{2},\frac{1}{2}\right)$

$\sin\frac{5\pi}{6}=\frac{1}{2}$

$\cos\frac{5\pi}{6}=-\frac{\sqrt{3}}{2}$

$\tan\frac{5\pi}{6}=\frac{\frac{1}{2}}{-\frac{\sqrt{3}}{2}}=-\frac{\sqrt{3}}{3}$

$\csc\frac{5\pi}{6}=\frac{1}{\frac{1}{2}}=2$

$\sec\frac{5\pi}{6}=\frac{1}{-\frac{\sqrt{3}}{2}}=-\frac{2\sqrt{3}}{3}$

$\cot\frac{5\pi}{6}=\frac{1}{-\frac{\sqrt{3}}{3}}=-\sqrt{3}$

35.

$t=\frac{3\pi}{2}$ Quadrantal angle

$P(t)=(0,-1)$

$\sin\frac{3\pi}{2}=-1$

$\cos\frac{3\pi}{2}=0$

$\tan\frac{3\pi}{2}=-\frac{1}{0}$ undefined

$\csc\frac{3\pi}{2}=-\frac{1}{1}=-1$

$\sec\frac{3\pi}{2}=\frac{1}{0}$ undefined

$\cot\frac{3\pi}{2}=-\frac{0}{1}=0$

37.

$t = \frac{3\pi}{4}$ Quadrant II

reference angle: $\pi - \frac{3\pi}{4} = \frac{\pi}{4}$

$P(t) = \left(-\frac{\sqrt{2}}{2}, \frac{\sqrt{2}}{2}\right)$

$\sin\frac{3\pi}{4} = \frac{\sqrt{2}}{2}$

$\cos\frac{3\pi}{4} = -\frac{\sqrt{2}}{2}$

$\tan\frac{3\pi}{4} = \frac{\frac{\sqrt{2}}{2}}{-\frac{\sqrt{2}}{2}} = -1$

$\csc\frac{3\pi}{4} = \frac{1}{\frac{\sqrt{2}}{2}} = \sqrt{2}$

$\sec\frac{3\pi}{4} = \frac{1}{-\frac{\sqrt{2}}{2}} = -\sqrt{2}$

$\cot\frac{3\pi}{4} = -\frac{1}{1} = -1$

39.

$t = -\frac{5\pi}{4}$

$-\frac{5\pi}{4} + 2\pi = \frac{3\pi}{4}$ Quadrant II

reference angle: $\pi - \frac{3\pi}{4} = \frac{\pi}{4}$

$P(t) = \left(-\frac{\sqrt{2}}{2}, \frac{\sqrt{2}}{2}\right)$

$\sin\left(-\frac{5\pi}{4}\right) = \frac{\sqrt{2}}{2}$

$\cos\left(-\frac{5\pi}{4}\right) = -\frac{\sqrt{2}}{2}$

$\tan\left(-\frac{5\pi}{4}\right) = \frac{\frac{\sqrt{2}}{2}}{-\frac{\sqrt{2}}{2}} = -1$

$\csc\left(-\frac{5\pi}{4}\right) = \frac{1}{\frac{\sqrt{2}}{2}} = \sqrt{2}$

$\sec\left(-\frac{5\pi}{4}\right) = \frac{1}{-\frac{\sqrt{2}}{2}} = -\sqrt{2}$

$\cot\left(-\frac{5\pi}{4}\right) = -\frac{1}{1} = -1$

41.

$5\pi - 2(2\pi) = \pi$

$P(-1, 0)$

43.

$t = -\frac{\pi}{4}$ Quadrant IV

$P\left(\frac{\sqrt{2}}{2}, -\frac{\sqrt{2}}{2}\right)$

45.

$t=\frac{5\pi}{4}$ Quadrant III

reference angle: $\frac{5\pi}{4}-\pi=\frac{\pi}{4}$

$P\left(-\frac{\sqrt{2}}{2},-\frac{\sqrt{2}}{2}\right)$

47.

$t=\frac{4\pi}{3}$ Quadrant III

reference angle: $\frac{4\pi}{3}-\pi=\frac{\pi}{3}$

$P\left(-\frac{1}{2},-\frac{\sqrt{3}}{2}\right)$

49.

$-\frac{2\pi}{3}+2\pi=\frac{4\pi}{3}$ Quadrant III

reference angle: $\frac{4\pi}{3}-\pi=\frac{\pi}{3}$

$P\left(-\frac{1}{2},-\frac{\sqrt{3}}{2}\right)$

51.

$\frac{19\pi}{6}-2\pi=\frac{7\pi}{6}$ Quadrant III

reference angle: $\frac{7\pi}{6}-\pi=\frac{\pi}{6}$

$P\left(-\frac{\sqrt{3}}{2},-\frac{1}{2}\right)$

53.

$-\frac{5\pi}{6}+2\pi=\frac{7\pi}{6}$ Quadrant III

reference angle: $\frac{7\pi}{6}\pi=\frac{\pi}{6}$

$P\left(-\frac{\sqrt{3}}{2},-\frac{1}{2}\right)$

55.

$\frac{19\pi}{3}-3(2\pi)=\frac{\pi}{3}$ Quadrant I

$P\left(\frac{1}{2},\frac{\sqrt{3}}{2}\right)$

57.

$(-1, 0)$

negative x-axis

$t=\pi$ or $t=-\pi$

59.

$\left(-\frac{\sqrt{2}}{2},\frac{\sqrt{2}}{2}\right)$ Quadrant II with reference angle $\frac{\pi}{4}$

$t=\pi-\frac{\pi}{4}=\frac{3\pi}{4}$ or $t=\frac{3\pi}{4}-2\pi=-\frac{5\pi}{4}$

61.

$\left(-\frac{\sqrt{3}}{2}, \frac{1}{2}\right)$ Quadrant II with reference angle $\frac{\pi}{6}$

$t = \pi - \frac{\pi}{6} = \frac{5\pi}{6}$ or $t = \frac{5\pi}{6} - 2\pi = -\frac{7\pi}{6}$

63.

$\left(\frac{1}{2}, -\frac{\sqrt{3}}{2}\right)$ Quadrant IV with reference angle $\frac{\pi}{3}$

$t = 2\pi - \frac{\pi}{3} = \frac{5\pi}{3}$ or $t = -\frac{\pi}{3}$

65a).

$P(t) = \left(\frac{3}{5}, \frac{4}{5}\right)$

$P(t+\pi) = (-x, -y) = \left(-\frac{3}{5}, -\frac{4}{5}\right)$

65b).

$P\left(t - \frac{\pi}{2}\right) = (y, -x) = \left(\frac{4}{5}, -\frac{3}{5}\right)$

65c).

$P(-t) = (x, -y) = \left(\frac{3}{5}, -\frac{4}{5}\right)$

65d).

$P(-t-\pi) = (-x, y) = \left(-\frac{3}{5}, \frac{4}{5}\right)$

67.

unit circle: $x^2 + y^2 = 1$

Since (a, b) lies on the unit circle, $a^2 + b^2 = 1$

Does $(a, -b)$ satisfy $x^2 + y^2 = 1$?

$a^2 + (-b^2) = 1$?

$a^2 + b^2 = 1$ yes

Does $(-a, b)$ satisfy $x^2 + y^2 = 1$?

$(-a)^2 + b^2 = 1$?

$a^2 + b^2 = 1$ yes

Does $(-a, -b)$ satisfy $x^2 + y^2 = 1$?

$(-a)^2 + (-b)^2 = 1$?

$a^2 + b^2 = 1$ yes

69.

$$\left.\begin{array}{ll}\sin t > 0: & \text{Q I and II} \\ \cos t < 0: & \text{QII and III}\end{array}\right\}\text{Quadrant II}$$

71.

$$\left.\begin{array}{ll}\cos t < 0: & \text{Q II and III} \\ \tan t > 0: & \text{Q I and III}\end{array}\right\}\text{Quadrant III}$$

73.

$$\left.\begin{array}{ll}\sin t < 0: & \text{Q III and IV} \\ \cos t < 0: & \text{Q II and III}\end{array}\right\}\text{Quadrant III}$$

75.

$$\left.\begin{array}{ll}\sec t < 0: & \text{Q II and III} \\ \sin t < 0: & \text{Q III and IV}\end{array}\right\}\text{Quadrant III}$$

77.

$$\left.\begin{array}{ll}\csc t > 0: & \text{Q I and II} \\ \sec t < 0: & \text{Q II and III}\end{array}\right\}\text{Quadrant II}$$

79.

$$\left.\begin{array}{ll}\sec t < 0: & \text{Q II and III} \\ \cot t > 0: & \text{Q I and III}\end{array}\right\}\text{Quadrant III}$$

81.

$$\left.\begin{array}{ll}\sec t < 0: & \text{Q II and III} \\ \csc t < 0: & \text{Q III and IV}\end{array}\right\}\text{Quadrant III}$$

83.

$$\tan t = \frac{3}{2}$$

$$\tan(-t) = -\tan t = -\frac{3}{2}$$

85.

$$\tan t = 1$$

$$\tan(-t) = -\tan t = -1$$

87.

$$\tan t = \frac{\sqrt{2}}{2}$$

$$\tan(-t) = -\tan t = -\frac{\sqrt{2}}{2}$$

89.

$$\cos t = -\frac{\sqrt{3}}{2}$$

$$\cos(-t) = \cos t = -\frac{\sqrt{3}}{2}$$

91.

$$\tan t = \sqrt{3}$$

$$\tan(-t) = -\tan t = -\sqrt{3}$$

93.

$$\sin t = \frac{\sqrt{3}}{2}$$

$$\sin(-t) = -\sin t = -\frac{\sqrt{3}}{2}$$

95.

$$P(t) = (x,\ y) = \left(\frac{12}{13},\ \frac{5}{13}\right)$$

95a).

$$P(t+\pi) = (-x,\ -y) = \left(-\frac{12}{13},\ -\frac{5}{13}\right)$$

95b).

$$P\left(t-\frac{\pi}{2}\right) = (y,\ -x) = \left(\frac{5}{13},\ -\frac{12}{13}\right)$$

95c).

$$P(-t) = (x,\ -y) = \left(\frac{12}{13},\ -\frac{5}{13}\right)$$

95d).

$$P(-t-\pi) = (-x,\ y) = \left(-\frac{12}{13},\ \frac{5}{13}\right)$$

97.

$$P(t) = (a,\ b)$$

$$P\left(t+\frac{\pi}{2}\right) = (-b,\ a)$$

97a).

$$\sin\left(t+\frac{\pi}{2}\right) = a$$

97b).

$$\cos\left(t+\frac{\pi}{2}\right) = -b$$

99.

OP and OP′ are radii of the same circle, hence their lengths are equal.

$$\sqrt{(a-0)^2+(b-0)^2}=\sqrt{(a'-0)^2+(b'-0)^2}$$

$$\sqrt{a^2+b^2}=\sqrt{(a')^2+(b')^2}$$

$$a^2+b^2=(a')^2+(b')^2$$

From exercise 98, $\frac{b'}{a'}=-\frac{a}{b}$

$$a'=-\frac{bb'}{a}$$

Substituting, $a^2+b^2=\left(-\frac{bb'}{a}\right)^2+(b')^2$

$$a^2+b^2=\frac{b^2(b')^2}{a^2}+(b')^2$$

$$a^2+b^2=(b')^2\left[\frac{b^2}{a^2}+1\right]$$

$$a^2+b^2=(b')^2\left(\frac{b^2+a^2}{a^2}\right)$$

$$\left(\frac{a^2}{b^2+a^2}\right)(a^2+b^2)=(b')^2$$

$$a^2=(b')^2$$

$$b'=\pm a$$

99. (Continued)

Similarly, from exercise 98, $\frac{b'}{a'}=-\frac{a}{b}$

$$b'=-\frac{aa'}{b}$$

$$a^2+b^2=(a')^2+(b')^2$$

Substituting, $a^2+b^2=(a')^2+\left(-\frac{aa'}{b}\right)^2$

$$a^2+b^2=(a')^2+\frac{a^2(a')^2}{b^2}$$

$$a^2+b^2=(a')^2\left[1+\frac{a^2}{b^2}\right]$$

$$a^2+b^2=(a')^2\left[\frac{b^2+a^2}{b^2}\right]$$

$$\left(\frac{b^2}{b^2+a^2}\right)(a^2+b^2)=(a')^2$$

$$b^2=(a')^2$$

$$a'=\pm b$$

101.

$$\cos^2 t=1-\sin^2 t$$

$$=1-\left(\frac{3}{5}\right)^2$$

$$=\frac{16}{25}$$

$\cos t=-\frac{4}{5}$ (Quadrant II)

$$\tan t=\frac{\sin t}{\cos t}=\frac{\frac{3}{5}}{-\frac{4}{5}}=-\frac{3}{4}$$

103.

$$\sin^2 t=1-\cos^2 t$$

$$=1-\left(-\frac{5}{13}\right)^2$$

$$=\frac{144}{169}$$

$\sin t=-\frac{12}{13}$ (Quadrant III)

105.

$$\left.\begin{array}{l}\cos t > 0: \text{ Q I and IV} \\ \sin t < 0: \text{ Q III and IV}\end{array}\right\} \text{Quadrant IV}$$

$$\begin{aligned}\sin^2 t &= 1 - \cos^2 t \\ &= 1 - \left(\frac{4}{5}\right)^2 \\ &= \frac{9}{25} \\ \sin t &= -\frac{3}{5}\end{aligned}$$

107.

$$\left.\begin{array}{l}\sin t < 0: \text{ Q III and IV} \\ \tan t < 0: \text{ Q II and IV}\end{array}\right\} \text{Quadrant IV}$$

$$\begin{aligned}\cos^2 t &= 1 - \sin^2 t \\ &= 1 - \left(-\frac{3}{5}\right)^2 \\ &= \frac{16}{25} \\ \cos t &= \frac{4}{5}\end{aligned}$$

109.

$P(-5, 12)$

$$\begin{aligned}r &= \sqrt{x^2 + y^2} \\ &= \sqrt{(-5)^2 + 12^2} \\ &= 13\end{aligned}$$

$$\sin\theta = \frac{y}{r} = \frac{12}{13}$$

$$\cos\theta = \frac{x}{r} = -\frac{5}{13}$$

$$\tan\theta = \frac{\sin\theta}{\cos\theta} = \frac{\frac{12}{13}}{-\frac{5}{13}} = -\frac{12}{5}$$

$$\csc\theta = \frac{1}{\sin\theta} = \frac{1}{\frac{12}{13}} = \frac{13}{12}$$

$$\sec\theta = \frac{1}{\cos\theta} = \frac{1}{-\frac{5}{13}} = -\frac{13}{5}$$

$$\cot\theta = \frac{1}{\tan\theta} = \frac{1}{-\frac{12}{5}} = -\frac{5}{12}$$

111.

$P(-1, -1)$

$$\begin{aligned}r &= \sqrt{x^2 + y^2} \\ &= \sqrt{(-1)^2 + (-1)^2} \\ &= \sqrt{2}\end{aligned}$$

$$\sin\theta = \frac{y}{r} = -\frac{1}{\sqrt{2}} = -\frac{\sqrt{2}}{2}$$

$$\cos\theta = \frac{x}{r} = -\frac{1}{\sqrt{2}} = -\frac{\sqrt{2}}{2}$$

$$\tan\theta = \frac{\sin\theta}{\cos\theta} = \frac{-\frac{\sqrt{2}}{2}}{-\frac{\sqrt{2}}{2}} = 1$$

$$\csc\theta = \frac{1}{\sin\theta} = \frac{1}{-\frac{\sqrt{2}}{2}} = -\sqrt{2}$$

$$\sec\theta = \frac{1}{\cos\theta} = \frac{1}{-\frac{\sqrt{2}}{2}} = -\sqrt{2}$$

$$\cot\theta = \frac{1}{\tan\theta} = \frac{1}{1} = 1$$

113.

$P(-8, 6)$

$$r = \sqrt{x^2 + y^2}$$
$$= \sqrt{(-8)^2 + 6^2}$$
$$= 10$$

$$\sin\theta = \frac{y}{r} = \frac{6}{10} = \frac{3}{5}$$

$$\cos\theta = \frac{x}{r} = -\frac{8}{10} = -\frac{4}{5}$$

$$\tan\theta = \frac{\sin\theta}{\cos\theta} = \frac{\frac{3}{5}}{-\frac{4}{5}} = -\frac{3}{4}$$

$$\csc\theta = \frac{1}{\sin\theta} = \frac{1}{\frac{3}{5}} = \frac{5}{3}$$

$$\sec\theta = \frac{1}{\cos\theta} = \frac{1}{-\frac{4}{5}} = -\frac{5}{4}$$

$$\cot\theta = \frac{1}{\tan\theta} = \frac{1}{-\frac{3}{4}} = -\frac{4}{3}$$

115.

$P(12, -5)$

$$r = \sqrt{x^2 + y^2}$$
$$= \sqrt{12^2 + (-5)^2}$$
$$= 13$$

$$\sin\theta = \frac{y}{r} = -\frac{5}{13}$$

$$\cos\theta = \frac{x}{r} = \frac{12}{13}$$

$$\tan\theta = \frac{\sin\theta}{\cos\theta} = \frac{-\frac{5}{13}}{\frac{12}{13}} = -\frac{5}{12}$$

$$\csc\theta = \frac{1}{\sin\theta} = \frac{1}{-\frac{5}{13}} = -\frac{13}{5}$$

$$\sec\theta = \frac{1}{\cos\theta} = \frac{1}{\frac{12}{13}} = \frac{13}{12}$$

$$\cot\theta = \frac{1}{\tan\theta} = \frac{1}{-\frac{5}{12}} = -\frac{12}{5}$$

117.

$P(-12, -5)$

$$r = \sqrt{x^2 + y^2} = \sqrt{(-12)^2 + (-5)^2} = 13$$

$$\sin\theta = \frac{y}{r} = -\frac{5}{13}$$

$$\cos\theta = \frac{x}{r} = -\frac{12}{13}$$

$$\tan\theta = \frac{\sin\theta}{\cos\theta} = \frac{-\frac{5}{13}}{-\frac{12}{13}} = \frac{5}{12}$$

$$\csc\theta = \frac{1}{\sin\theta} = \frac{1}{-\frac{5}{13}} = -\frac{13}{5}$$

$$\sec\theta = \frac{1}{\cos\theta} = \frac{1}{-\frac{12}{13}} = -\frac{13}{12}$$

$$\cot\theta = \frac{1}{\tan\theta} = \frac{1}{\frac{5}{12}} = \frac{12}{5}$$

119.

$P(-2, 1)$

$$r = \sqrt{x^2 + y^2} = \sqrt{(-2)^2 + 1^2} = \sqrt{5}$$

$$\sin\theta = \frac{y}{r} = \frac{1}{\sqrt{5}} = \frac{\sqrt{5}}{5}$$

$$\cos\theta = \frac{x}{r} = -\frac{2}{\sqrt{5}} = -\frac{2\sqrt{5}}{5}$$

$$\tan\theta = \frac{\sin\theta}{\cos\theta} = \frac{\frac{1}{\sqrt{5}}}{-\frac{2}{\sqrt{5}}} = -\frac{1}{2}$$

$$\csc\theta = \frac{1}{\sin\theta} = \frac{1}{\frac{1}{\sqrt{5}}} = \sqrt{5}$$

$$\sec\theta = \frac{1}{\cos\theta} = \frac{1}{-\frac{2}{\sqrt{5}}} = -\frac{\sqrt{5}}{2}$$

$$\cot\theta = \frac{1}{\tan\theta} = \frac{1}{-\frac{1}{2}} = -2$$

121.

$$\sin 0.80 \approx 0.80 - \frac{(0.80)^3}{6} + \frac{(0.80)^5}{120} - \frac{(0.80)^7}{5040} = 0.7174$$

$$\sin 0.80 = 0.7174$$

123.

$$\sin(-0.20) \approx -0.20 - \frac{(-0.20)^3}{6} + \frac{(-0.20)^5}{120} - \frac{(-0.20)^7}{5040} = -0.1987$$

$$\sin(-0.20) = -0.1987$$

125.

$$\tan(0.1)=\frac{\sin 0.1}{\cos 0.1}$$

$$\sin(0.1)\approx 0.1-\frac{0.1^3}{6}+\frac{0.1^5}{120}-\frac{0.1^7}{5040}=0.0998$$

$$\cos(0.1)\approx 1-\frac{0.1^2}{2}+\frac{0.1^4}{24}-\frac{0.1^6}{720}=0.9950$$

$$\tan(0.1)\approx\frac{0.0998}{0.9950}=0.1003$$

$$\tan(0.1)=0.1003$$

127.

$$\begin{aligned}\sin(-t)&\approx -t-\frac{(-t)^3}{6}+\frac{(-t)^5}{120}-\frac{(-t)^7}{5040}\\&=-t+\frac{t^3}{6}-\frac{t^5}{120}+\frac{t^7}{5040}\\&=-\left[t-\frac{t^3}{6}+\frac{t^5}{120}-\frac{t^7}{5040}\right]\\&=-\sin t\end{aligned}$$

129.

$$\tan t\cos t=\left(\frac{\sin t}{\cos t}\right)(\cos t)=\sin t$$

131.

$$\frac{1-\sin^2 t}{\sin t}=\frac{\cos^2 t}{\sin t}=\frac{\cos^2 t}{\sin t}\cdot\frac{\frac{1}{\cos t}}{\frac{1}{\cos t}}=\frac{\cos t}{\frac{\sin t}{\cos t}}=\frac{\cos t}{\tan t}$$

133.

$$\begin{aligned}\cos t\left(\frac{1}{\cos t}-\cos t\right)&=\frac{\cos t}{\cos t}-\cos^2 t\\&=1-\cos^2 t\\&=\sin^2 t\end{aligned}$$

135.

$$\frac{1-\cos^2 t}{\cos^2 t}=\frac{\sin^2 t}{\cos^2 t}=\tan^2 t$$

137.

$$\begin{aligned}(\sin t-\cos t)^2&=\sin^2 t-2\sin t\cos t+\cos^2 t\\&=(\sin^2 t+\cos^2 t)-2\sin t\cos t\\&=1-2\sin t\cos t\end{aligned}$$

CHAPTER 7 SECTION 5

EXERCISE SET 7.5

1.
$\cos t = 0$

$f(t) = \cos t$

$f(t) = 0$ at $t = \dfrac{\pi}{2}, \dfrac{3\pi}{2}$

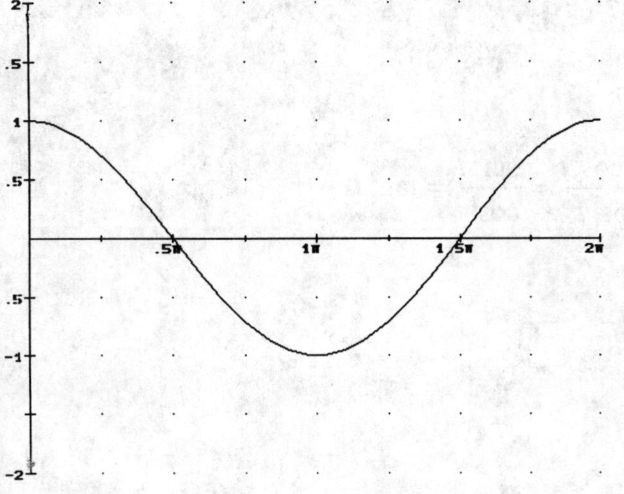

3.
$\tan t = 1$

$f(t) = \tan t - 1$

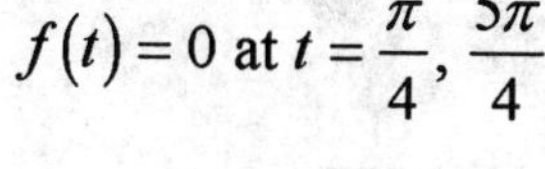
$f(t) = 0$ at $t = \dfrac{\pi}{4}, \dfrac{5\pi}{4}$

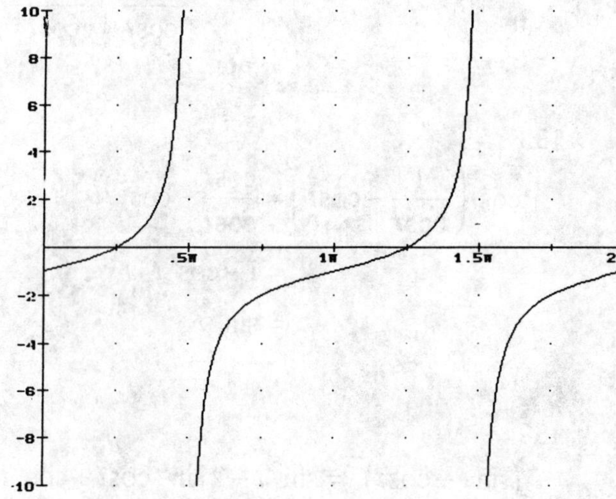

5.

$$\sin t = \frac{\sqrt{2}}{2}$$

$$f(t) = \sin t - \frac{\sqrt{2}}{2}$$

$$f(t) = 0 \text{ at } t = \frac{\pi}{4}, \frac{3\pi}{4}$$

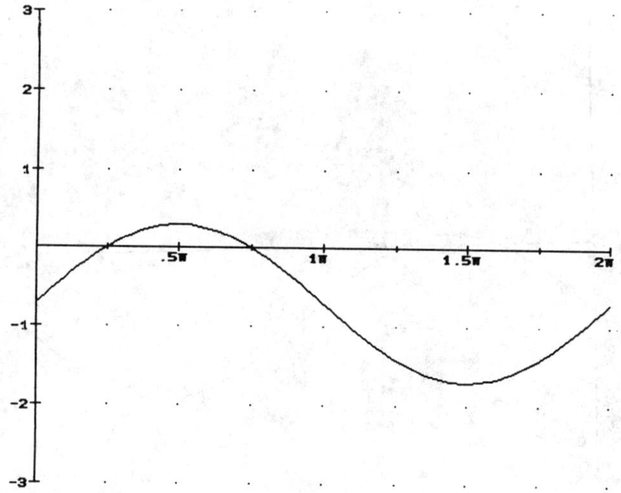

7.

$$\cos t = -\frac{\sqrt{3}}{2}$$

$$f(t) = \cos t + \frac{\sqrt{3}}{2}$$

$$f(t) = 0 \text{ at } t = \frac{5\pi}{6}, \frac{7\pi}{6}$$

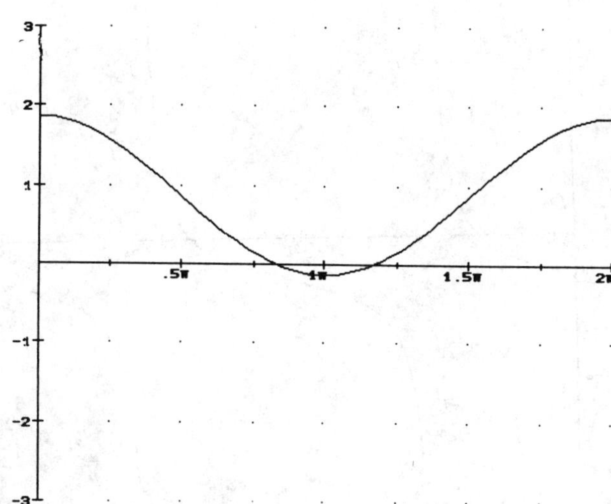

9.

$$\tan t = \sqrt{3}$$

$$f(t) = \tan t - \sqrt{3}$$

$$f(t) = 0 \text{ at } t = \frac{\pi}{3}, \frac{4\pi}{3}$$

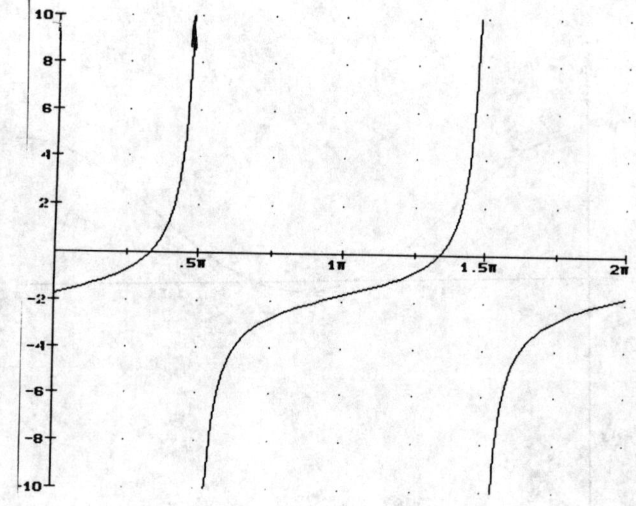

11.

$$\sin t = \frac{\sqrt{3}}{2}$$

$$f(t) = \sin t - \frac{\sqrt{3}}{2}$$

$$f(t) = 0 \text{ at } t = \frac{\pi}{3}, \frac{2\pi}{3}$$

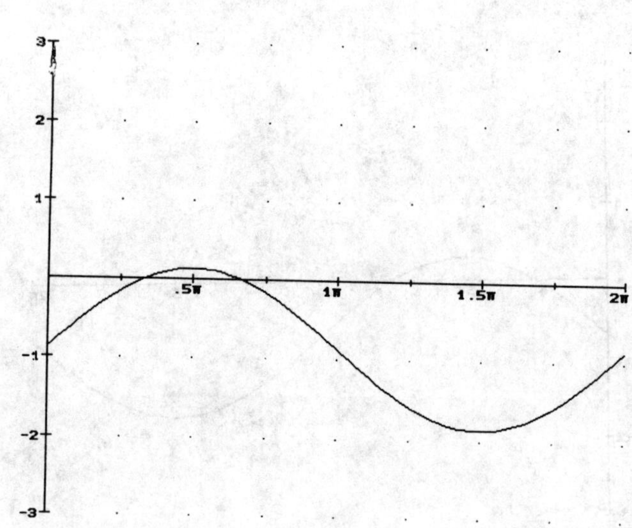

13.

$$\sin t = -\frac{1}{2}$$

$$f(t) = \sin t + \frac{1}{2}$$

$f(t) = 0$ at $t = \frac{7\pi}{6}, \frac{11\pi}{6}$

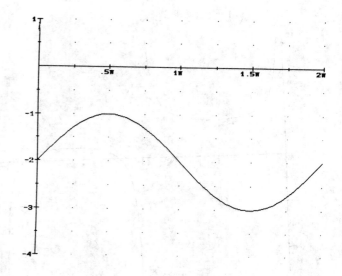

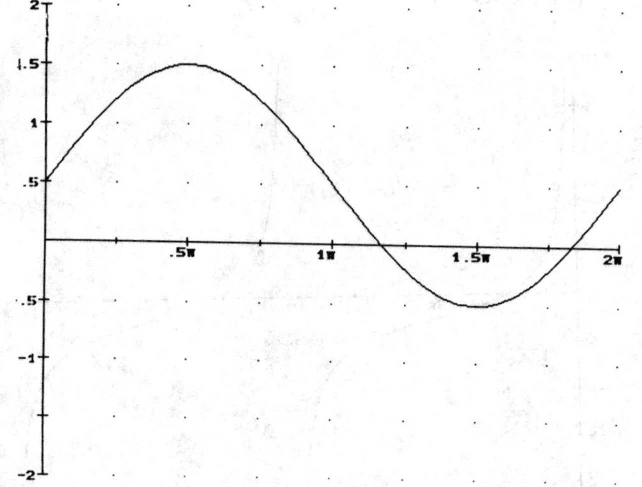

15.

$$\sin t = 2$$

$$f(t) = \sin t - 2$$

$f(t) = 0$ No solution

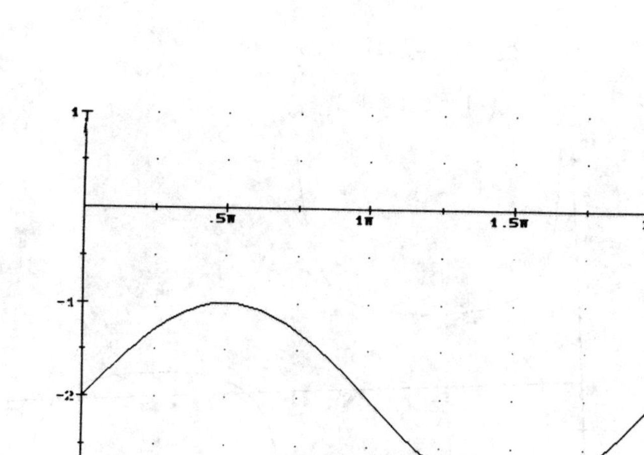

17.

$$\sin t = \frac{1}{2}$$

$$f(t) = \sin t - \frac{1}{2}$$

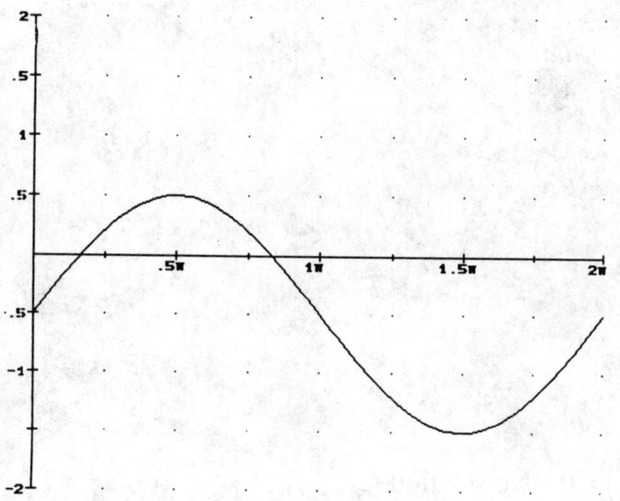

$f(t) = 0$ at $t = \frac{\pi}{6}, \frac{5\pi}{6}$

19.

$$\sec t = 1$$

$$f(t) = \sec t - 1$$

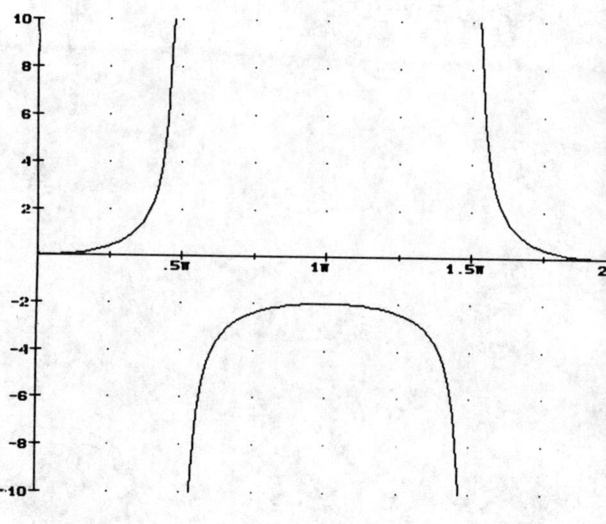

$f(t) = 0$ at $t = 0$

21.

$$\csc t = -2$$

$$f(t) = \csc t + 2$$

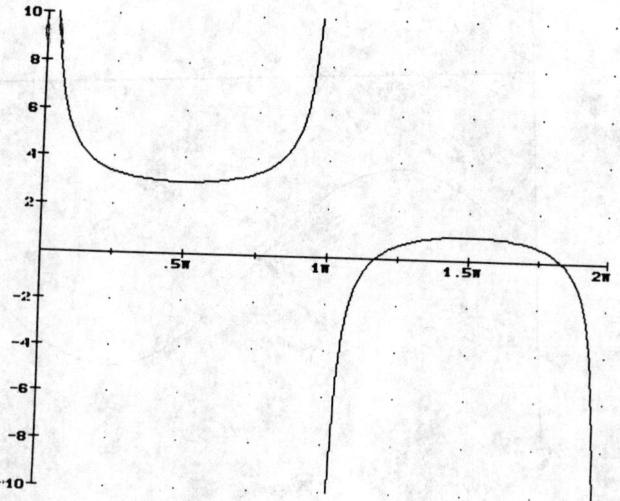

$f(t) = 0$ at $t = \frac{7\pi}{6}, \frac{11\pi}{6}$

23.

$$\cot t = 1$$

$$f(t) = \cot t - 1$$

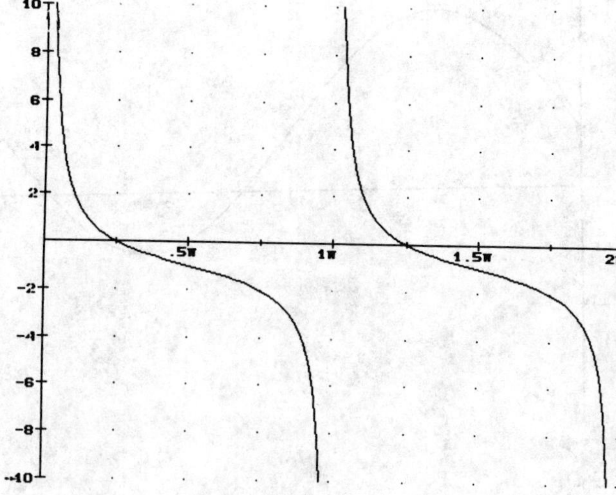

$f(t) = 0$ at $t = \frac{\pi}{4}, \frac{5\pi}{4}$

25.

$\cot t = -1$

$f(t) = \cot t + 1$

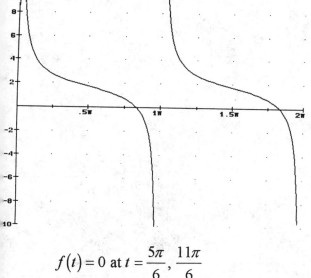

$f(t) = 0$ at $t = \dfrac{3\pi}{4}, \dfrac{7\pi}{4}$

27.

$\sec t = \sqrt{2}$

$f(t) = \sec t - \sqrt{2}$

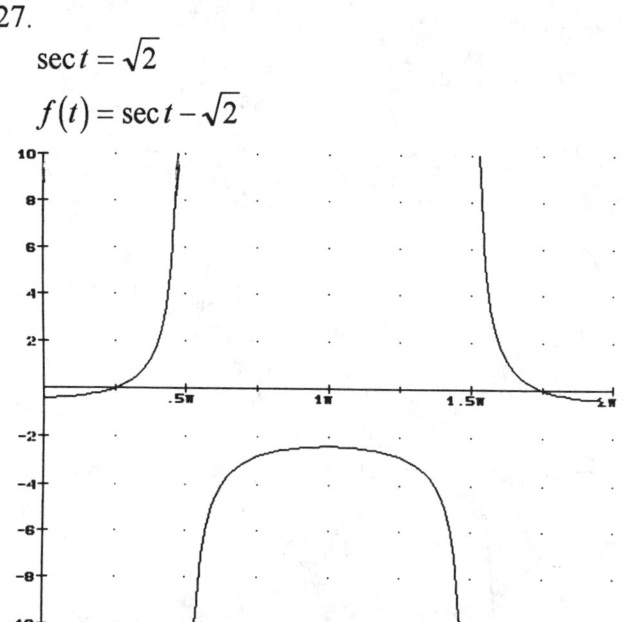

$f(t) = 0$ at $t = \dfrac{\pi}{4}, \dfrac{7\pi}{4}$

29.

$\cot t = -\sqrt{3}$

$f(t) = \cot t + \sqrt{3}$

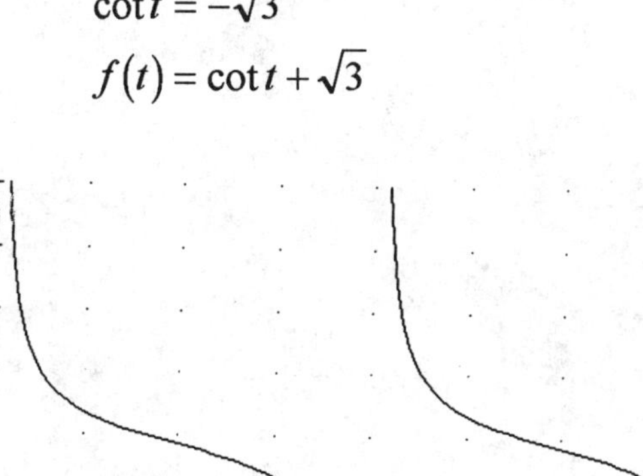

$f(t) = 0$ at $t = \dfrac{5\pi}{6}, \dfrac{11\pi}{6}$

31.

$\sin t = \dfrac{1}{2}$, $\sec t < 0 \quad t \in \left(\dfrac{\pi}{2}, \dfrac{3\pi}{2}\right)$

$f(t) = \sin t - \dfrac{1}{2}$

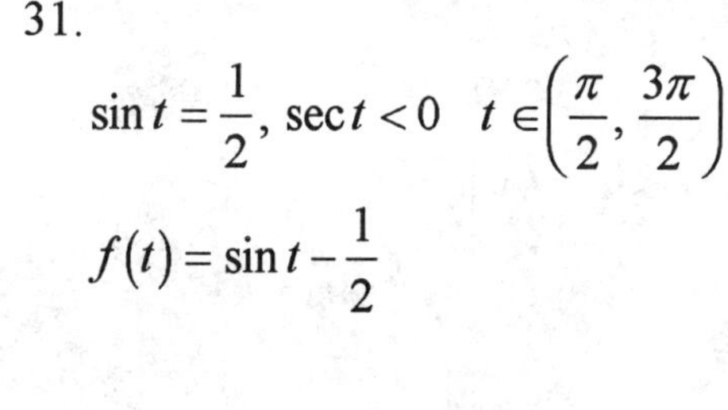

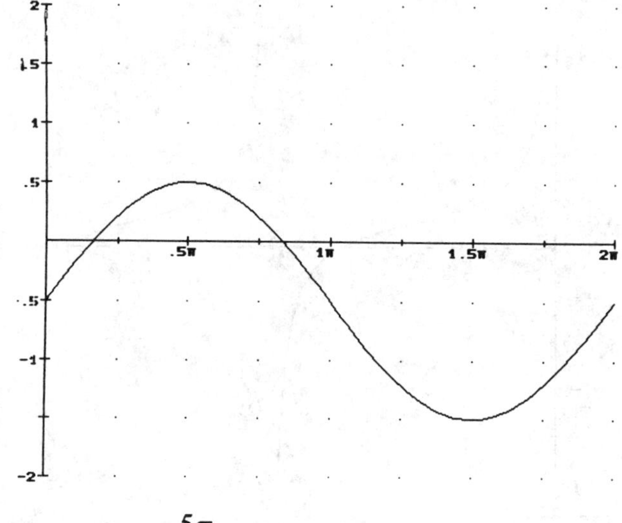

$f(t) = 0$ at $t = \dfrac{5\pi}{6}$

33.

$\sec t = -2,\ \csc t > 0 \quad t \in \left(\frac{\pi}{2}, \pi\right)$

$f(t) = \sec t + 2$

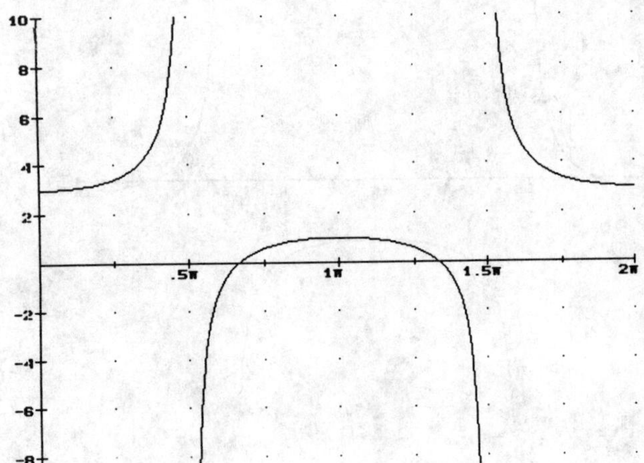

$f(t) = 0$ at $t = \frac{2\pi}{3}$

35.

$\csc t = -\sqrt{2},\ \sec t < 0 \quad t \in \left(\pi, \frac{3\pi}{2}\right)$

$f(t) = \csc t + \sqrt{2}$

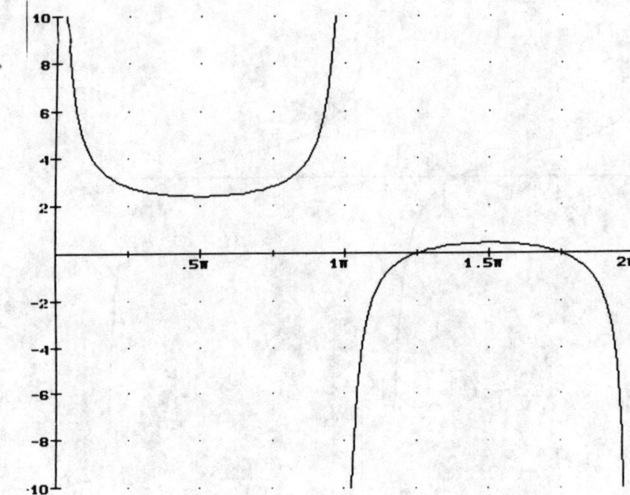

$f(t) = 0$ at $t = \frac{5\pi}{4}$

37.

$\cot t = -1,\ \sec t < 0 \quad t \in \left(\frac{\pi}{2}, \pi\right)$

$f(t) = \cot t + 1$

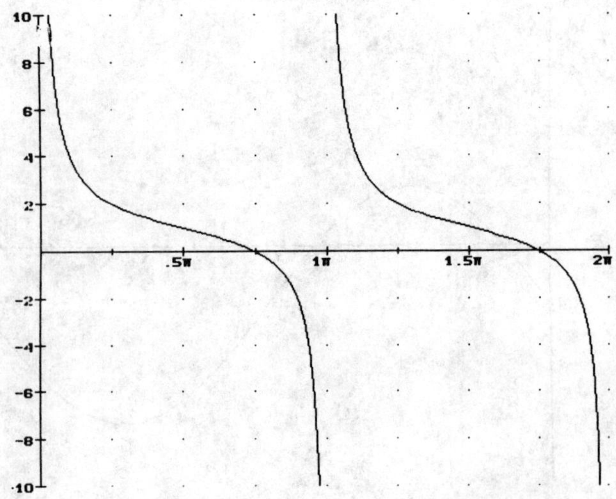

$f(t) = 0$ at $t = \frac{3\pi}{4}$

In Exercises 39-47, the RANGE values given show important features of the functions. Other RANGE values are also acceptable. The display on your calculator may vary from that shown.

39.
XMIN $= -2\pi$
XMAX $= 2\pi$
XSCL $= \frac{\pi}{2}$
YMIN $= -3$
YMAX $= 3$
YSCL $= 1$
GRAPH $Y = 1 + COS\ X$

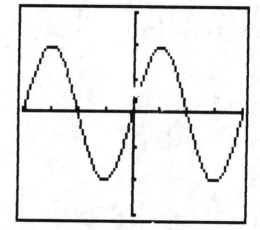

41.
XMIN $= -2\pi$
XMAX $= 2\pi$
XSCL $= \frac{\pi}{2}$
YMIN $= -3$
YMAX $= 3$
YSCL $= 1$
GRAPH $Y = 2\,SIN\ X$

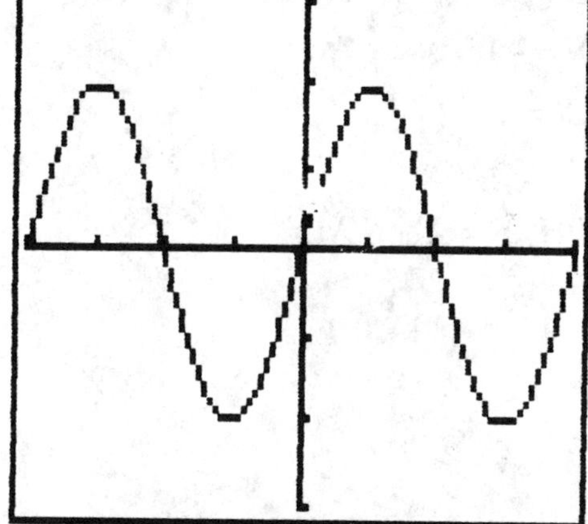

43.
XMIN $= -2\pi$
XMAX $= 2\pi$
XSCL $= \frac{\pi}{2}$
YMIN $= -3$
YMAX $= 3$
YSCL $= 1$
GRAPH $Y = SIN\ X + 0.5\,COS\ X$

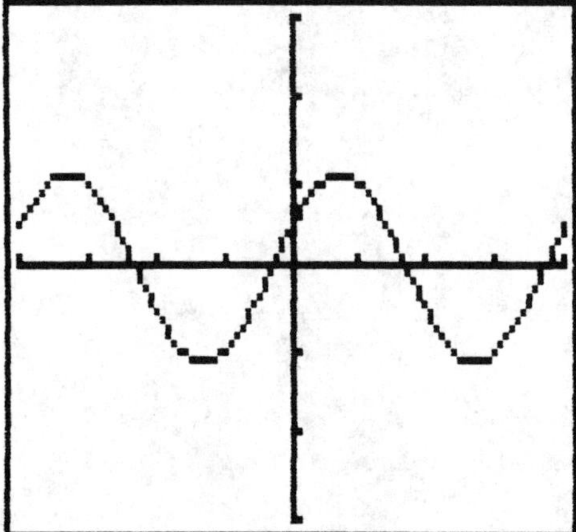

45.

XMIN $= -2\pi$

XMAX $= 2\pi$

XSCL $= \dfrac{\pi}{2}$

YMIN $= -3$

YMAX $= 3$

YSCL $= 1$

GRAPH $Y = SIN\ X - COS\ X$

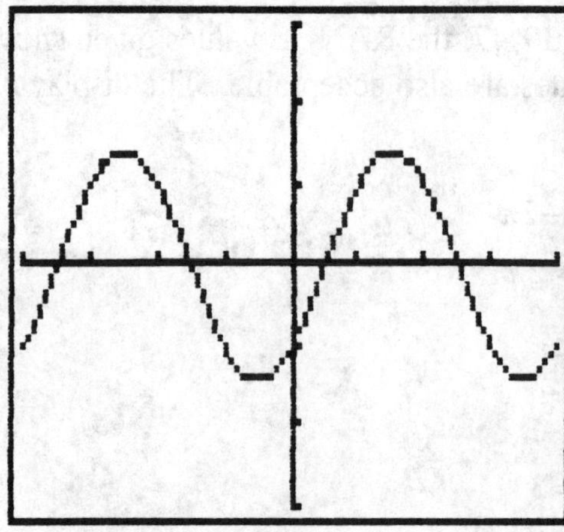

47.

XMIN $= -15$

XMAX $= 15$

XSCL $= 5$

YMIN $= -15$

YMAX $= 15$

YSCL $= 5$

GRAPH $Y = X + SIN\ X$

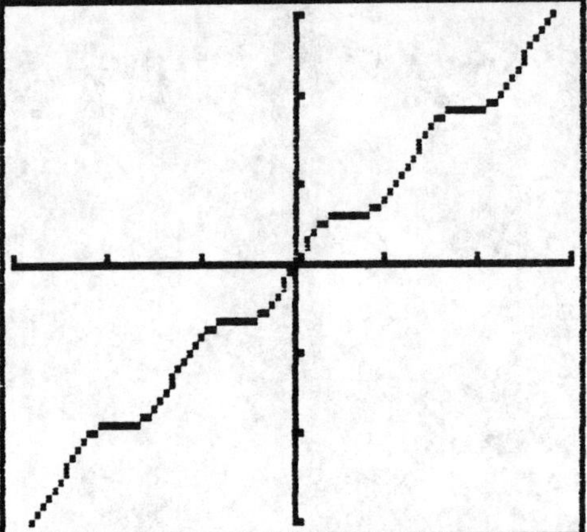

49.

Suppose $\sin(t+c) = \sin t$, $0 < c < 2\pi$ for all t.

Let $t = 0$

$\sin(0+c) = \sin 0$

$\sin c = 0$

Then $c = 0$ or $c = \pi$

However $0 < c < 2\pi$, thus $c = \pi$.

$\sin(t+\pi) = \sin t$

Let $t = \frac{\pi}{2}$

$\sin\left(\frac{\pi}{2}+\pi\right) = \sin\frac{\pi}{2}$?

$\sin\frac{3\pi}{2} = \sin\frac{\pi}{2}$?

$-1 \neq 1$

Hence, there is no $c \in (0, 2\pi)$ such that $\sin(t+c) = \sin t$. Therefore the period of $\sin t = 2\pi$.

51.

Suppose $\tan(t+c) = \tan t$, $0 < c < \pi$, for all t.

Let $t = 0$.

$\tan(0+c) = \tan 0$

$\tan c = 0$

No solution for $c \in (0, \pi)$

Hence, there is no value of $c \in (0, \pi)$ such that $\tan(t+c) = \tan t$. The period of $\tan t = \pi$.

53.

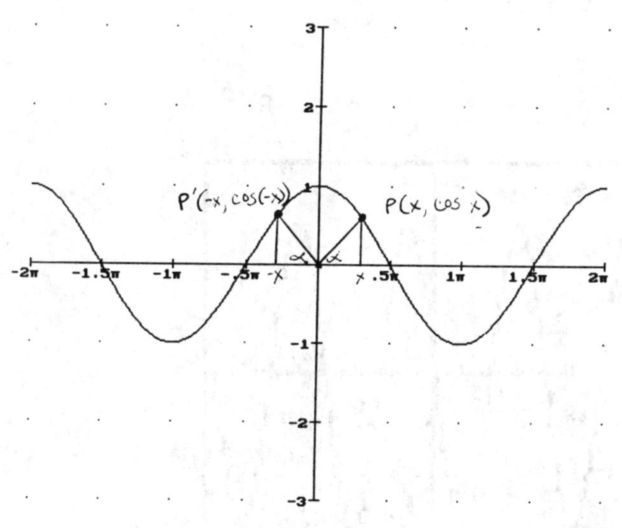

Note that $\cos(-x) = \cos x$

55.

$\tan t = \frac{\sin t}{\cos t}$

$\tan t$ will have vertical asymptotes when $\cos t = 0$.

$\cos t = 0$

$t = \frac{(2k+1)\pi}{2}$ for k an integer

CHAPTER 7 SECTION 6

SECTION 7.6 PROGRESS CHECK

1. (Page 452)

$f(x) = 2\cos\left(2x + \frac{\pi}{2}\right)$

$A = 2 \;\; B = 2 \;\; C = \frac{\pi}{2}$

amplitude $= |2| = 2$

period $= \frac{2\pi}{|B|} = \frac{2\pi}{|2|} = \pi$

phase shift $= -\frac{C}{B} = -\frac{\frac{\pi}{2}}{2} = -\frac{\pi}{4}$

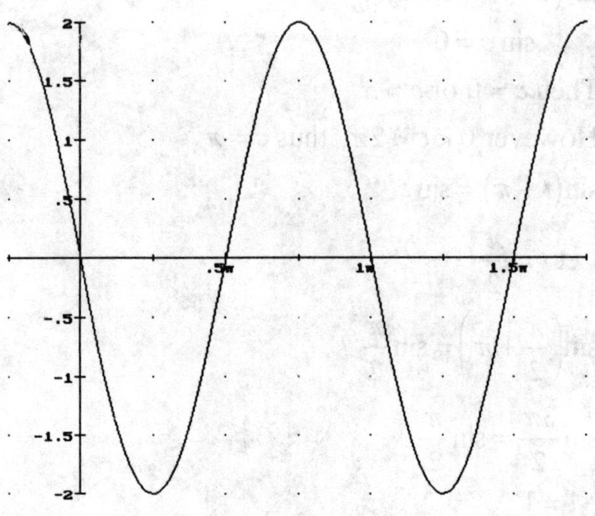

EXERCISE SET 7.6

The RANGE values given show important features of the functions. Other RANGE values are also acceptable. The display on your calculator may vary from that shown.

1.
$f(x) = 3\sin x$

$A = 3 \;\; B = 1$

amplitude $= |A| = |3| = 3$

period $= \frac{2\pi}{|B|} = \frac{2\pi}{1} = 2\pi$

XMIN $= -4\pi$

XMAX $= 4\pi$

XSCL $= \frac{\pi}{2}$

YMIN $= -4$

YMAX $= 4$

YSCL $= 1$

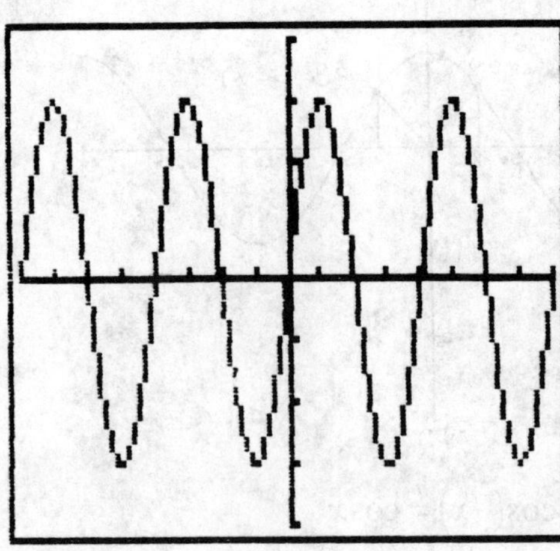

3.

$f(x) = \cos 4x$

$A = 1 \quad B = 4$

$\text{amplitude } = |A| = |1| = 1$

$\text{period } = \frac{2\pi}{|B|} = \frac{2\pi}{|4|} = \frac{\pi}{2}$

$\text{XMIN} = -2\pi$

$\text{XMAX} = 2\pi$

$\text{XSCL} = \frac{\pi}{2}$

$\text{YMIN} = -2$

$\text{YMAX} = 2$

$\text{YSCL} = 1$

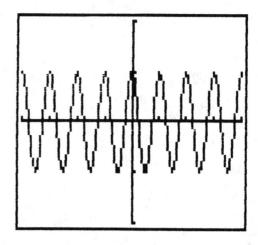

5.

$f(x) = -2\sin 4x$

$A = -2 \quad B = 4$

$\text{amplitude } = |A| = |-2| = 2$

$\text{period } = \frac{2\pi}{|B|} = \frac{2\pi}{|4|} = \frac{\pi}{2}$

$\text{XMIN} = -2\pi$

$\text{XMAX} = 2\pi$

$\text{XSCL} = \frac{\pi}{2}$

$\text{YMIN} = -3$

$\text{YMAX} = 3$

$\text{YSCL} = 1$

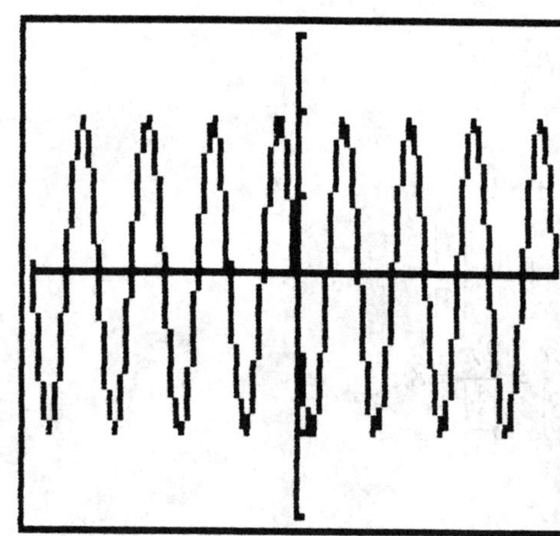

7.

$f(x)=2\cos\frac{x}{3}$

$A=2 \;\; B=\frac{1}{3}$

amplitude $=|A|=|2|=2$

period $=\frac{2\pi}{|B|}=\frac{2\pi}{\left|\frac{1}{3}\right|}=6\pi$

XMIN $=-3\pi$

XMAX $=3\pi$

XSCL $=\frac{\pi}{2}$

YMIN $=-3$

YMAX $=3$

YSCL $=1$

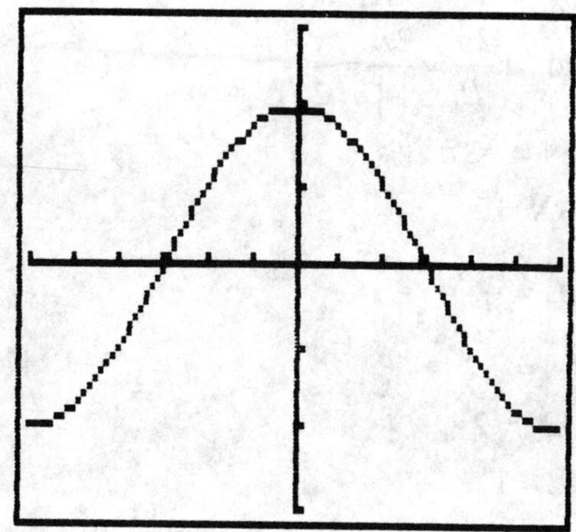

9.

$f(x)=\frac{1}{4}\sin\frac{x}{4}$

$A=\frac{1}{4} \;\; B=\frac{1}{4}$

amplitude $=|A|=\left|\frac{1}{4}\right|=\frac{1}{4}$

period $=\frac{2\pi}{|B|}=\frac{2\pi}{\left|\frac{1}{4}\right|}=8\pi$

XMIN $=-4\pi$

XMAX $=4\pi$

XSCL $=\frac{\pi}{2}$

YMIN $=-1$

YMAX $=1$

YSCL $=1$

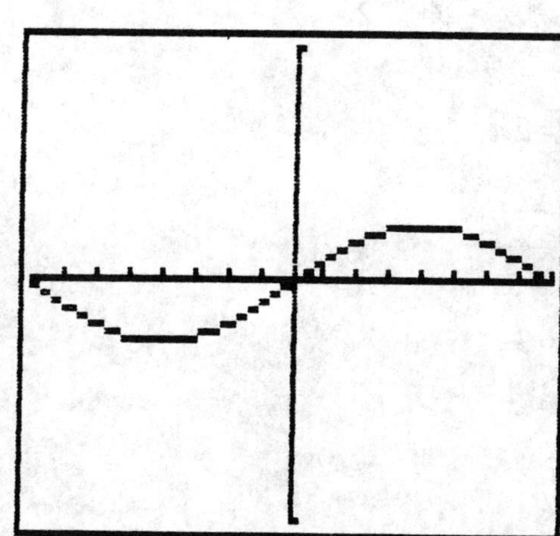

11.

$f(x) = -3\cos 3x$

$A = -3 \quad B = 3$

amplitude $= |A| = |-3| = 3$

period $= \frac{2\pi}{|B|} = \frac{2\pi}{|3|} = \frac{2\pi}{3}$

$\text{XMIN} = -2\pi$

$\text{XMAX} = 2\pi$

$\text{XSCL} = \frac{\pi}{2}$

$\text{YMIN} = -4$

$\text{YMAX} = 4$

$\text{YSCL} = 1$

13.

$f(x) = 2\sin(x - \pi)$

$A = 2 \quad B = 1 \quad C = -\pi$

amplitude $= |A| = |2| = 2$

period $= \frac{2\pi}{|B|} = \frac{2\pi}{|1|} = 2\pi$

phase shift $= -\frac{C}{B} = -\frac{-\pi}{1} = \pi$

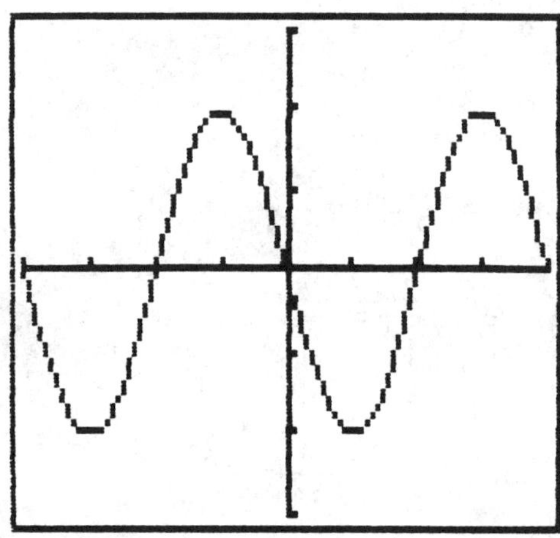

$\text{XMIN} = -2\pi$

$\text{XMAX} = 2\pi$

$\text{XSCL} = \frac{\pi}{2}$

$\text{YMIN} = -3$

$\text{YMAX} = 3$

$\text{YSCL} = 1$

15.

$f(x) = 3\cos(2x - \pi)$

$A = 3 \quad B = 2 \quad C = -\pi$

amplitude $= |A| = |3| = 3$

period $= \dfrac{2\pi}{|B|} = \dfrac{2\pi}{|2|} = \pi$

phase shift $= \dfrac{-C}{B} = -\dfrac{-\pi}{2} = \dfrac{\pi}{2}$

$\text{XMIN} = -2\pi$

$\text{XMAX} = 2\pi$

$\text{XSCL} = \dfrac{\pi}{2}$

$\text{YMIN} = -4$

$\text{YMAX} = 4$

$\text{YSCL} = 1$

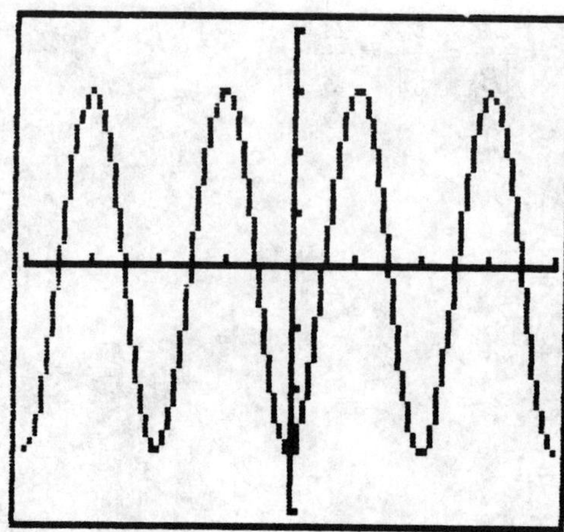

17.

$f(x) = \dfrac{1}{3}\sin\left(3x + \dfrac{3\pi}{4}\right)$

$A = \dfrac{1}{3} \quad B = 3 \quad C = \dfrac{3\pi}{4}$

amplitude $= |A| = \left|\dfrac{1}{3}\right| = \dfrac{1}{3}$

period $= \dfrac{2\pi}{|B|} = \dfrac{2\pi}{|3|} = \dfrac{2\pi}{3}$

phase shift $= \dfrac{-C}{B} = \dfrac{-\dfrac{3\pi}{4}}{3} = -\dfrac{\pi}{4}$

$\text{XMIN} = -2\pi$

$\text{XMAX} = 2\pi$

$\text{XSCL} = \dfrac{\pi}{2}$

$\text{YMIN} = -1$

$\text{YMAX} = 1$

$\text{YSCL} = 1$

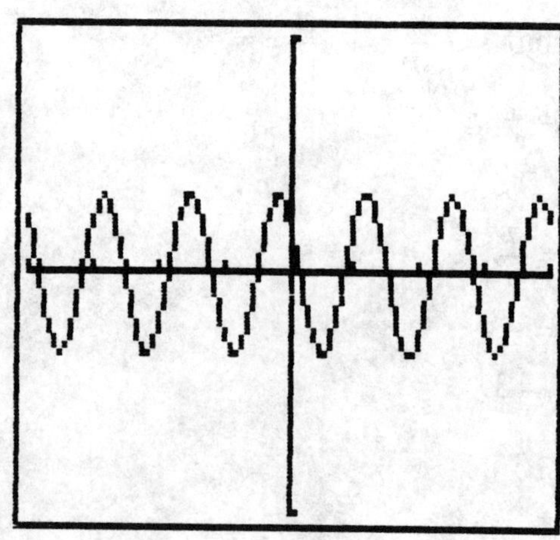

19.

$f(x)=2\cos\left(\frac{x}{4}-\pi\right)$

$A=2\ \ B=\frac{1}{4}\ \ C=-\pi$

amplitude $=|A|=|2|=2$

period $=\frac{2\pi}{|B|}=\frac{2\pi}{\left|\frac{1}{4}\right|}=8\pi$

phase shift $=-\frac{C}{B}=-\frac{-\pi}{\frac{1}{4}}=4\pi$

XMIN $=-4\pi$

XMAX $=4\pi$

XSCL $=\frac{\pi}{2}$

YMIN $=-3$

YMAX $=3$

YSCL $=1$

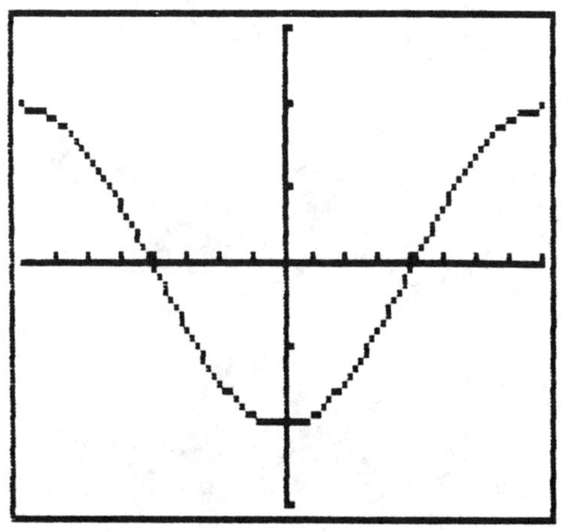

21.

amplitude $=3$

period $=2\pi$

$f(x)=3\sin x$

23.

amplitude $=5$

period $=\frac{\pi}{2}\qquad \frac{\pi}{2}=\frac{2\pi}{B}$

$B=4$

$f(x)=5\sin 4x$

25.

amplitude $=2=A$

period $=\frac{2\pi}{3}=\frac{2\pi}{B}$

$B=3$

phase shift $=0=-\frac{C}{B}=-\frac{C}{3}$

$C=0$

$y=2\sin 3x$

27.

$A=1$

period $=\pi$: $B=1$

phase shift $=\frac{\pi}{3}=-\frac{C}{B}=-\frac{C}{1}$

$C=-\frac{\pi}{3}$

$y=\tan\left(x-\frac{\pi}{3}\right)$

29.

TENS viewing rectangle

XSCL = 1

YSCL = 1

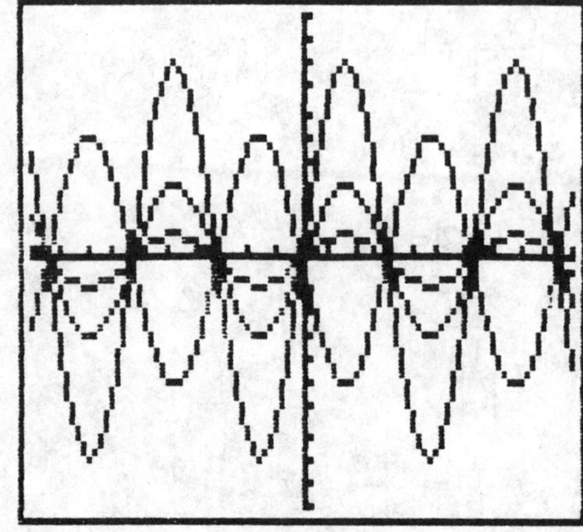

31.a).

$\text{XSCL} = \frac{\pi}{2}$

YSCL = 1

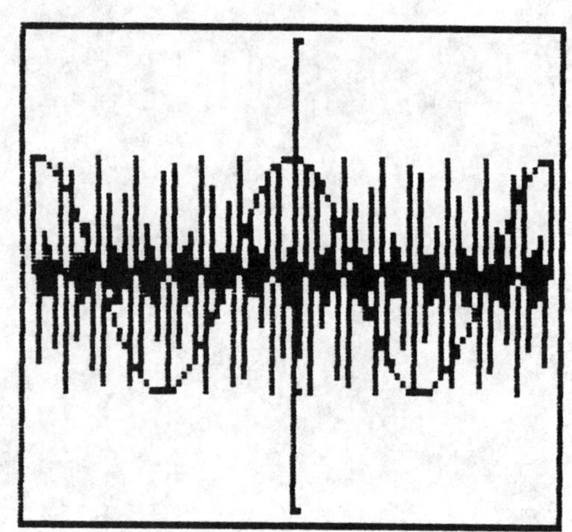

31b).

$\text{XSCL} = \frac{\pi}{2}$

YSCL = 1

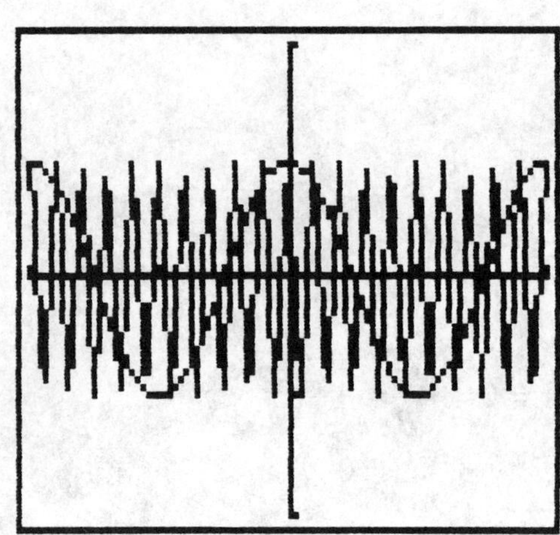

33.

$$\text{XSCL} = \frac{\pi}{2}$$
$$\text{YSCL} = 1$$

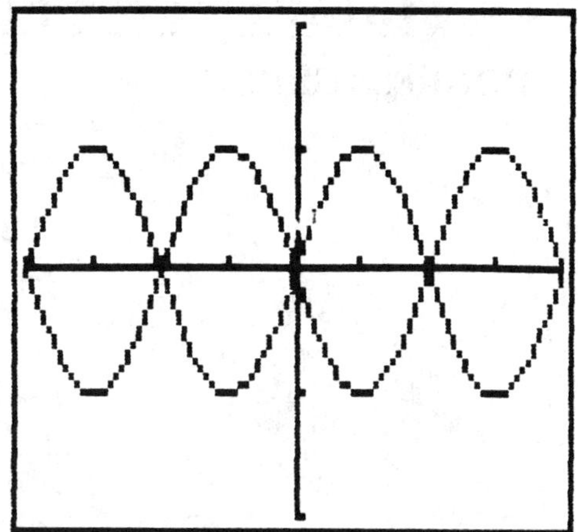

35.

$$\text{XSCL} = \frac{\pi}{2}$$
$$\text{YSCL} = 1$$

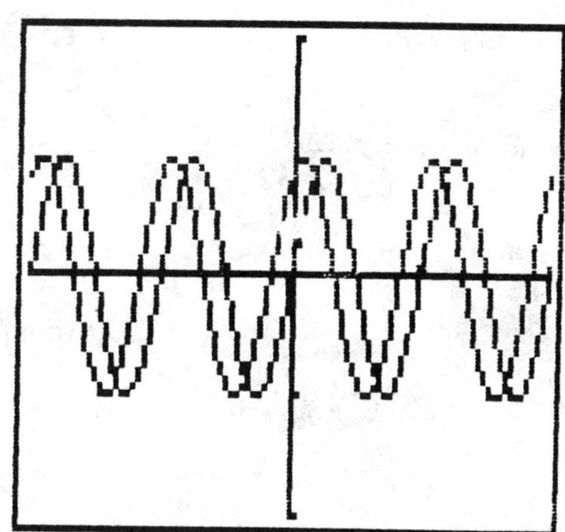

37.

$$y = x - \frac{x^3}{3!} + \frac{x^5}{5!} - \frac{x^7}{7!} + \frac{x^9}{9!} - \frac{x^{11}}{11!} + \frac{x^{13}}{13!} - \frac{x^{15}}{15!} + \frac{x^{17}}{17!} - \frac{x^{19}}{19!} + \frac{x^{21}}{21!}$$

CHAPTER 7 SECTION 7

SECTION 7.7 PROGRESS CHECK

1a). (Page 457)

$$\sin^{-1}\left(-\frac{\sqrt{3}}{2}\right)=y$$

$$\frac{-\sqrt{3}}{2}=\sin y \quad y\in\left[-\frac{\pi}{2},\frac{\pi}{2}\right]$$

$$y=-\frac{\pi}{3}$$

1b).

$$\arcsin\left(\tan\frac{5\pi}{4}\right)=\arcsin 1=y$$

$$1=\sin y \quad y\in\left[-\frac{\pi}{2},\frac{\pi}{2}\right]$$

$$y=\frac{\pi}{2}$$

2. (Page 461)

$$\cot\left[\sin^{-1}\left(-\frac{5}{13}\right)\right]$$

Let $t=\sin^{-1}\left(-\frac{5}{13}\right)$, find $\cot t$.

$$\sin t=-\frac{5}{13} \quad t\in\left[-\frac{\pi}{2},0\right]$$

$$\cos^2 t=1-\sin^2 t$$

$$\cos^2 t=1-\left(-\frac{5}{13}\right)^2$$

$$=\frac{144}{169}$$

$$\cos t=\frac{12}{13} \quad \text{(Quadrant IV)}$$

$$\cot t=\frac{\cos t}{\sin t}=\frac{\frac{12}{13}}{-\frac{5}{13}}=-\frac{12}{5}$$

3a). (Page 461)

$$\sin^{-1}(-0.725)=-0.8110344$$

3b).

$$\sec(\arcsin(-0.429))=1.1070464$$

4. (Page 463)

$$2\sin^2 x + 2\sin x - 1 = 0$$

Let $u = \sin x$

$$2u^2 + 2u - 1 = 0$$

$$u = \frac{-2 \pm \sqrt{2^2 - 4(2)(-1)}}{2(2)}$$

$$= \frac{-2 \pm \sqrt{12}}{4}$$

$$= \frac{-1 \pm \sqrt{3}}{2}$$

$$u = \frac{-1+\sqrt{3}}{2} \approx 0.366 \qquad u = \frac{-1-\sqrt{3}}{2} \approx -1.366$$

$$\sin x = \frac{-1+\sqrt{3}}{2} \qquad \sin x = -1.366$$

$$x = \sin^{-1}\left(\frac{-1+\sqrt{3}}{2}\right)$$

$$x \approx 0.3747344$$

EXERCISE SET 7.7

1.

$$\sin^{-1}\left(-\frac{1}{2}\right) = y$$

$$-\frac{1}{2} = \sin y \quad y \in \left[-\frac{\pi}{2}, \frac{\pi}{2}\right]$$

$$y = -\frac{\pi}{6}$$

3.

$$\arctan\sqrt{3} = y$$

$$\sqrt{3} = \tan y \quad y \in \left[-\frac{\pi}{2}, \frac{\pi}{2}\right]$$

$$y = \frac{\pi}{3}$$

5.

$$\arcsin\left(-\frac{\sqrt{2}}{2}\right) = y$$

$$-\frac{\sqrt{2}}{2} = \sin y \quad y \in \left[-\frac{\pi}{2}, \frac{\pi}{2}\right]$$

$$y = -\frac{\pi}{4}$$

7.

$$\arccos\left(-\frac{\sqrt{3}}{2}\right) = y$$

$$-\frac{\sqrt{3}}{2} = \cos y \quad y \in [0, \pi]$$

$$y = \frac{5\pi}{6}$$

9.

$$\sin^{-1}(-1) = y$$

$$-1 = \sin y \quad y \in \left[-\frac{\pi}{2}, \frac{\pi}{2}\right]$$

$$y = -\frac{\pi}{2}$$

11.

$$\cos^{-1} 0 = y$$

$$0 = \cos y \quad y \in [0, \pi]$$

$$y = \frac{\pi}{2}$$

13.

$$\cos^{-1} 1 = y$$

$$1 = \cos y \quad y \in [0, \pi]$$

$$y = 0$$

15.

$$\arctan(-1) = y$$

$$-1 = \tan y \quad y \in \left[-\frac{\pi}{2}, \frac{\pi}{2}\right]$$

$$y = -\frac{\pi}{4}$$

17.

$$\cos^{-1}\left(-\frac{1}{2}\right) = y$$

$$-\frac{1}{2} = \cos y \quad y \in [0, \pi]$$

$$y = \frac{2\pi}{3}$$

19.

$$\sin^{-1} 0.3709 = 0.3800$$

21.

$$\cos^{-1}(-0.7648) = 2.4415$$

23.

$$\arcsin(0.9636) = 1.3002$$

25.

$$\arctan 1 = y$$

$$1 = \tan y \quad y \in \left(-\frac{\pi}{2}, \frac{\pi}{2}\right)$$

$$y = \frac{\pi}{4}$$

$$\sin\frac{\pi}{4} = \frac{\sqrt{2}}{2}$$

27.

$$\cos\frac{\pi}{2} = 0$$

$$\tan^{-1} 0 = y$$

$$0 = \tan y \quad y \in \left(-\frac{\pi}{2}, \frac{\pi}{2}\right)$$

$$y = 0$$

29.

$$\sin\frac{9\pi}{4} = \sin\frac{\pi}{4} = \frac{\sqrt{2}}{2}$$

$$\cos^{-1}\frac{\sqrt{2}}{2} = y$$

$$\frac{\sqrt{2}}{2} = \cos y \quad y \in [0, \pi]$$

$$y = \frac{\pi}{4}$$

31.

$$\cos^{-1}\left(\cos\frac{2\pi}{3}\right) = \frac{2\pi}{3}$$

33.

$$\text{Let } t = \sin^{-1}\left(-\frac{5}{13}\right)$$

$$\sin t = \frac{-5}{13} \quad t \in \left[-\frac{\pi}{2}, 0\right]$$

$$\cos^2 t = 1 - \sin^2 t$$

$$= -1\left(-\frac{5}{13}\right)^2$$

$$= \frac{144}{169}$$

$$\cos t = \frac{12}{13} \quad \text{(Quadrant IV)}$$

$$\tan\left[\sin^{-1}\left(-\frac{5}{13}\right)\right] = \tan t = \frac{\sin t}{\cos t} = \frac{\frac{-5}{13}}{\frac{12}{13}} = -\frac{5}{12}$$

35.

$$\text{Let } t = \sin^{-1}\left(\frac{4}{5}\right)$$

$$\sin t = \frac{4}{5} \quad t \in \left[0, \frac{\pi}{2}\right]$$

$$\cos^2 t = 1 - \sin^2 t$$

$$= 1 - \left(\frac{4}{5}\right)^2$$

$$= \frac{9}{25}$$

$$\cos t = \frac{3}{5} \quad \text{(Quadrant I)}$$

$$\cos\left[\sin^{-1}\left(\frac{4}{5}\right)\right] = \frac{3}{5}$$

37.

$$\sin x = \frac{\sqrt{3}}{2}, \quad x \in \left[0, \frac{\pi}{2}\right]$$

$$x = \frac{\pi}{3}$$

39.

$$\tan x = \frac{1}{\sqrt{3}}, \quad x \in \left[0, \frac{\pi}{2}\right]$$

$$x = \frac{\pi}{6}$$

41.

$$2\sin x = -1, \quad x \in [\pi, 2\pi]$$

$$\sin x = -\frac{1}{2}$$

Reference angle of x is $\frac{\pi}{6}$. Since $\sin x < 0$, x is in quadrant III or IV.

$$x = \pi + \frac{\pi}{6} = \frac{7\pi}{6}$$

$$x = 2\pi - \frac{\pi}{6} = \frac{11\pi}{6}$$

43.

$$\tan x = \sqrt{3}, \quad x \in \left[\frac{\pi}{2}, \frac{3\pi}{2}\right]$$

Reference angle of x is $\frac{\pi}{3}$. Since $\tan x > 0$, x is in quadrant III.

$$x = \pi + \frac{\pi}{3} = \frac{4\pi}{3}$$

45.

$$7\sin^2 x - 1 = 0, \quad x \in \left[-\frac{\pi}{2}, \frac{\pi}{2}\right]$$

$$\sin^2 x = \frac{1}{7}$$

$$\sin x = \pm\frac{\sqrt{7}}{7}$$

$$x = \sin^{-1}\left(\pm\frac{\sqrt{7}}{7}\right)$$

47.

$$12\cos^2 x - \cos x - 1 = 0, \quad x \in [0, \pi]$$

$$(3\cos x - 1)(4\cos x + 1) = 0$$

$$3\cos x - 1 = 0 \qquad 4\cos x + 1 = 0$$

$$\cos x = \frac{1}{3} \qquad \cos x = -\frac{1}{4}$$

$$x = \cos^{-1}\left(\frac{1}{3}\right) \qquad x = \cos^{-1}\left(-\frac{1}{4}\right)$$

49.

$$9\sin^2 t - 12\sin t + 4 = 0, \quad t \in \left[-\frac{\pi}{2}, \frac{\pi}{2}\right]$$

$$(3\sin t - 2)^2 = 0$$

$$3\sin t - 2 = 0$$

$$\sin t = \frac{2}{3}$$

$$t = \sin^{-1}\left(\frac{2}{3}\right)$$

51.

$$\sin^{-1} x = \frac{1}{\sin x} \; ?$$

Let $x = 0$

$$\sin^{-1} 0 = \frac{1}{\sin 0} \; ?$$

$$0 \neq \frac{1}{0}$$

53.

$\sin^{-1}(\sin x) = x$?

Let $x = \pi$

$\sin^{-1}(\sin \pi) = \pi$?

$\sin^{-1}(0) = \pi$?

$0 \neq \pi$

55.

$2\cos^2 x + \cos x = 2, \quad [0, \pi]$

Let $u = \cos x$

$2u^2 + u = 2$

$2u^2 + u - 2 = 0$

$$u = \frac{-1 \pm \sqrt{1^2 - 4(2)(-2)}}{2(2)}$$

$$= \frac{-1 \pm \sqrt{17}}{4}$$

$u = \frac{-1+\sqrt{17}}{4} \approx 0.7808$ $\qquad$ $u = \frac{-1-\sqrt{17}}{4} \approx -1.2808$

$\cos x = 0.7808$ $\qquad$ ~~$\cos x = -1.2808$~~

$$x = \cos^{-1}\left(\frac{-1+\sqrt{17}}{4}\right)$$

$x = 0.6749$

57.

$\sin^2 x - 2\sin x - 2 = 0, \quad \left[-\frac{\pi}{2}, \frac{\pi}{2}\right]$

Let $u = \sin x$

$u^2 - 2u - 2 = 0$

$$u = \frac{-(-2) \pm \sqrt{(-2)^2 - 4(1)(-2)}}{2(1)}$$

$$= \frac{2 \pm \sqrt{12}}{2}$$

$$= 1 \pm \sqrt{3}$$

$u = 1 + \sqrt{3} \approx 2.7321$ $\qquad$ $u = 1 - \sqrt{3} \approx -0.7321$

$\sin x \neq 2.7321$ $\qquad$ $\sin x = -0.7321$

$x = \sin^{-1}\left(1 - \sqrt{3}\right) = -0.8213$

59.

XSCL = 1

YSCL = 1

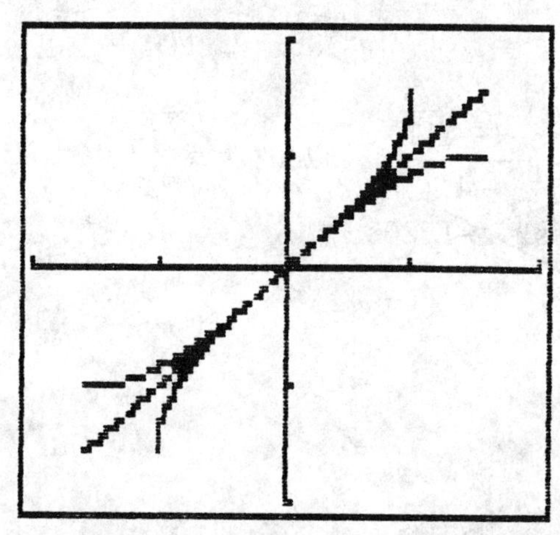

61.

XSCL = 1

YSCL = 1

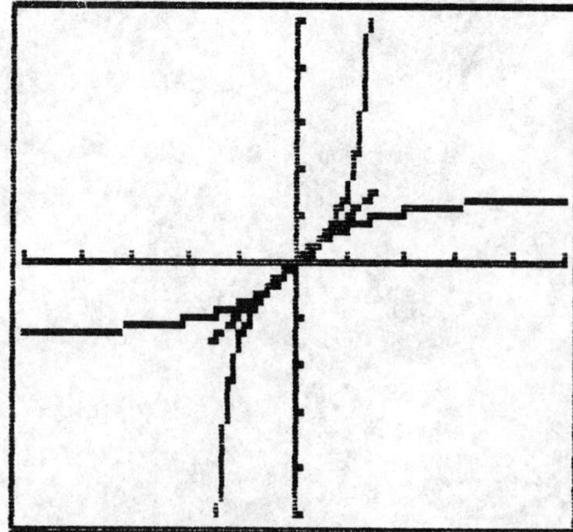

CHAPTER 7 SECTION 8

SECTION 7.8 PROGRESS CHECK

1. (Page 465)

$$\tan\alpha = \frac{17.4}{38.2} \approx 0.4555$$
$$\alpha = \tan^{-1} 0.4555 \approx 24.49 = 24°29'$$
$$\beta = 90° - 24°29' = 65°31'$$
$$c = \sqrt{a^2 + b^2}$$
$$= \sqrt{(17.4)^2 + (38.2)^2}$$
$$\approx 42.0$$

2. (Page 466)

$$\sin 32°30' = \frac{y}{125}$$
$$y = 125 \sin 32°30' \approx 67$$
$$67\,\text{m}$$

EXERCISE SET 7.8

1.

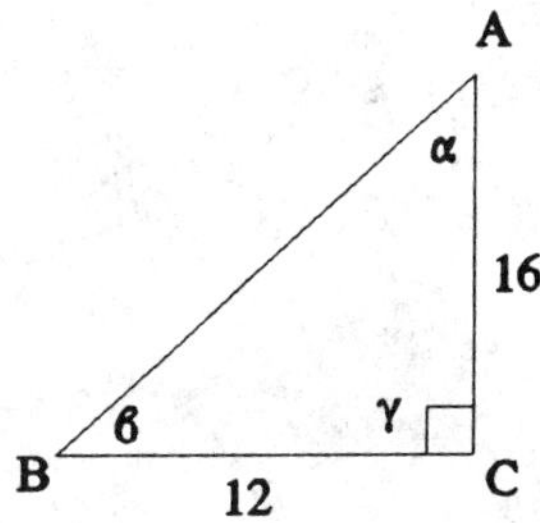

$$\tan\alpha = \frac{12}{16}$$
$$\alpha = \tan^{-1}\frac{12}{16} \approx 36.87° = 36°52'12''$$

3.

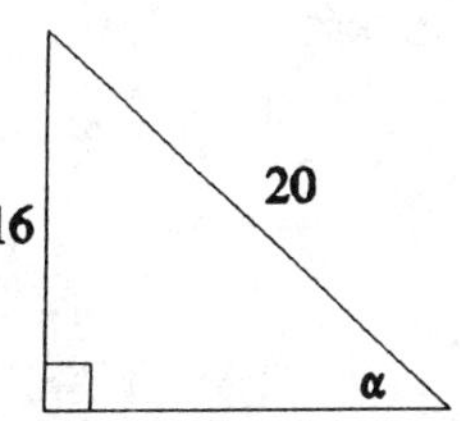

$$\sin\alpha = \frac{16}{20} = \frac{4}{5}$$
$$\alpha = \sin^{-1}\frac{4}{5} \approx 53.13° = 53°7'48''$$

5.

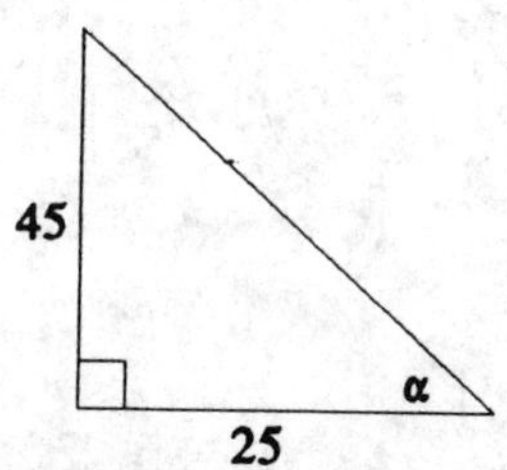

$$\tan\alpha = \frac{45}{25} = \frac{9}{5}$$

$$\alpha = \tan^{-1}\frac{9}{5} \approx 60.95° = 60°57'$$

7.

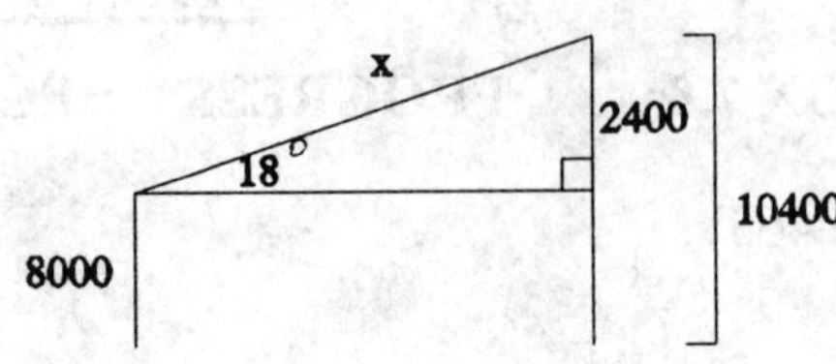

$$\sin 18° = \frac{2400}{x}$$

$$x = \frac{2400}{\sin 18°} \approx 7766.56$$

7766.56 feet

9.

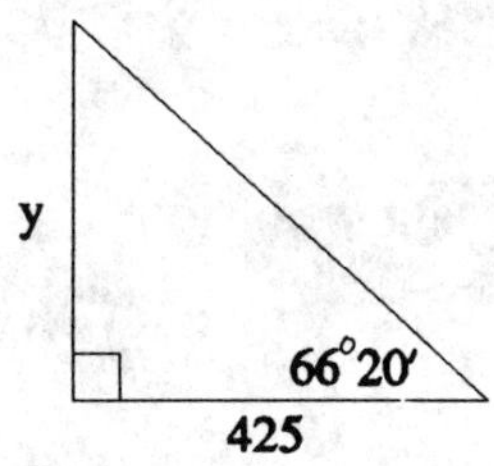

$$\tan 66°20' = \frac{y}{425}$$

$$y = 425\tan 66°20' \approx 969.71$$

969.71 meters

11.

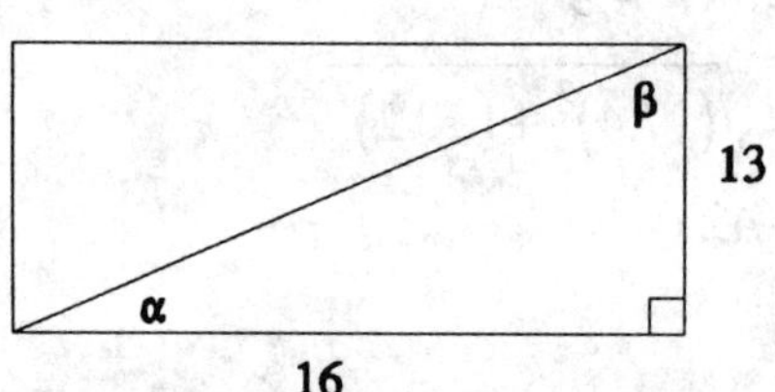

$$\tan\alpha = \frac{13}{16}$$

$$\alpha = \tan^{-1}\frac{13}{16} \approx 39.09° \approx 39°5'24''$$

$$\beta = 90° - 39°5'24'' = 50°54'36''$$

13.

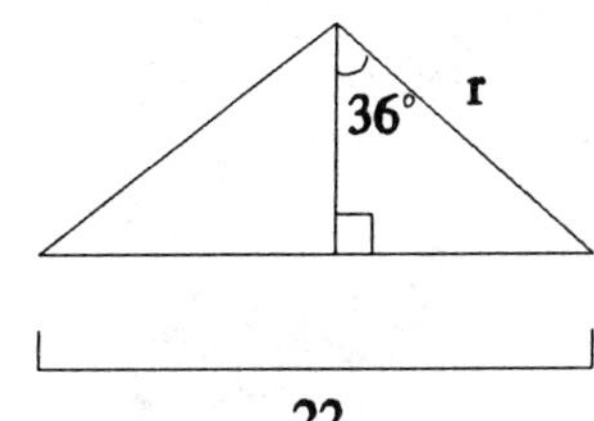

5 central angles: $\frac{360°}{5} = 72°$

$\sin 36° = \frac{11}{r}$

$r = \frac{11}{\sin 36°} \approx 18.7$

18.7 cm

15.

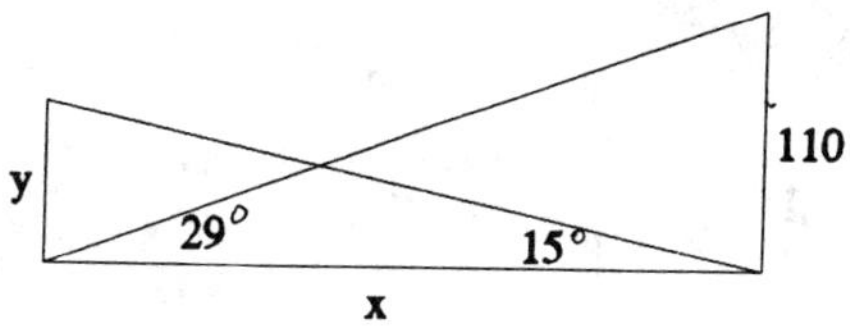

$\tan 29° = \frac{110}{x}$

$x = \frac{110}{\tan 29°} \approx 198.4$

$\tan 15° = \frac{y}{x} = \frac{y}{198.4}$

$y = 198.4 \tan 15° \approx 53.2$

53.2 feet

17.

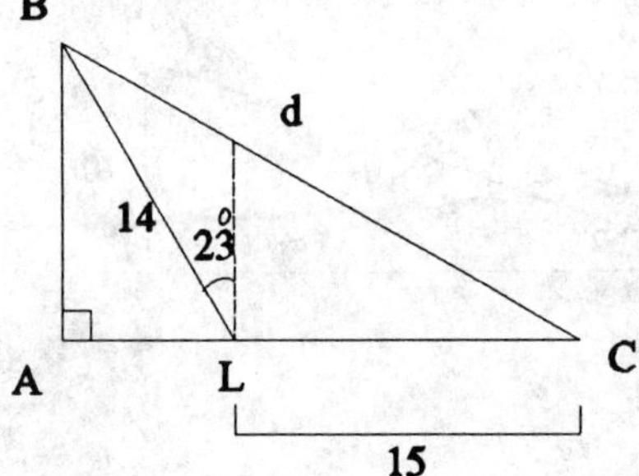

$\angle ALB = 90° - 23° = 67°$

$\cos 67° = \frac{\overline{AL}}{14}$

$\overline{AL} = 14\cos 67° \approx 5.5$

$\overline{AC} = 5.5 + 15 = 20.5$

$\sin 67° = \frac{\overline{AB}}{14}$

$\overline{AB} = 14\sin 67° \approx 12.9$

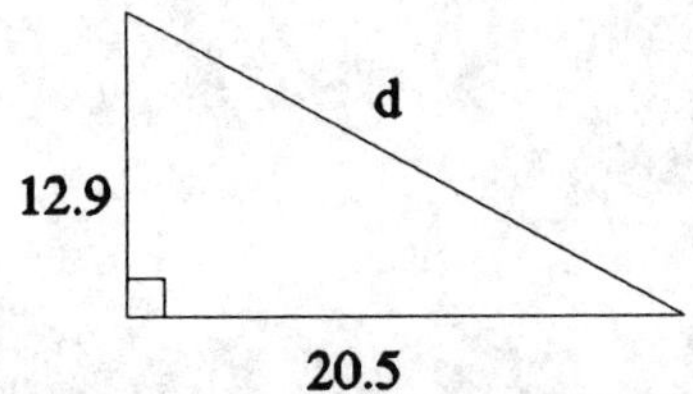

$$d = \sqrt{(12.9)^2 + (20.5)^2}$$
$$\approx 24.2$$

The distance is 24.2 miles.

CHAPTER 7 REVIEW EXERCISES

1.

$$\frac{\theta}{\pi}=\frac{-60}{180}$$

$$\theta=\frac{-60\pi}{180}=-\frac{\pi}{3}\text{ radians}$$

2.

$$\frac{\frac{3\pi}{2}}{\pi}=\frac{\theta}{180}$$

$$\theta=\frac{180\left(\frac{3\pi}{2}\right)}{\pi}=270°$$

3.

$$\frac{\frac{-5\pi}{12}}{\pi}=\frac{\theta}{180}$$

$$\theta=\frac{180\left(-\frac{5\pi}{12}\right)}{\pi}=-75°$$

4.

$$\frac{\theta}{\pi}=\frac{45}{180}$$

$$\theta=\frac{45\pi}{180}=\frac{\pi}{4}\text{ radians}$$

5.

$$\frac{\frac{5\pi}{9}}{\pi}=\frac{\theta}{180}$$

$$\theta=\frac{180\left(\frac{5\pi}{9}\right)}{\pi}=100°$$

coterminal

6.

$$\frac{\frac{4\pi}{3}}{\pi}=\frac{\theta}{180}$$

$$\theta=\frac{180\left(\frac{4\pi}{3}\right)}{\pi}=240°$$

$480°-360°=120°$

Not coterminal

7.

$$\frac{\frac{5\pi}{4}}{\pi}=\frac{\theta}{180}$$

$$\theta=\frac{180\left(\frac{5\pi}{4}\right)}{\pi}=225°$$

$-135°+360°=225°$

Coterminal

8.

$$\frac{3\pi}{2}<\frac{11\pi}{6}<2\pi$$

Quadrant IV

9.
$$-220°+360°=140°$$
$$90°<140°<180°$$
Quadrant II

10.
$$490°-360°=130°$$
$$90°<130°<180°$$
Quadrant II

11.
$$\frac{-11\pi}{3}+2(2\pi)=\frac{\pi}{3}$$
$$0<\frac{\pi}{3}<\frac{\pi}{2}$$
Quadrant I

12.
$$360°-310°=50°$$

13.
$$-185°+360°=175°$$
$$180°-175°=5°$$

14.
$$405°-360°=45°$$

15.
$$\text{radian measure} = \frac{\text{intercepted arc}}{\text{radius}}$$
$$=\frac{14}{10}=\frac{7}{5}=1.4\text{ radians}$$

16.
$$\frac{2\pi}{3}=\frac{\frac{5\pi}{2}}{r}$$
$$r=\frac{3\left(\frac{5\pi}{2}\right)}{2\pi}=\frac{15}{4}\text{cm}$$

17.
$$\frac{9\pi}{2}-2(2\pi)=\frac{\pi}{2}$$

18.
$$-\frac{15\pi}{2}+4(2\pi)=\frac{\pi}{2}$$

19.
$$-6\pi+3(2\pi)=0$$

20.
$$\frac{23\pi}{3}-3(2\pi)=\frac{5\pi}{3}$$

21.

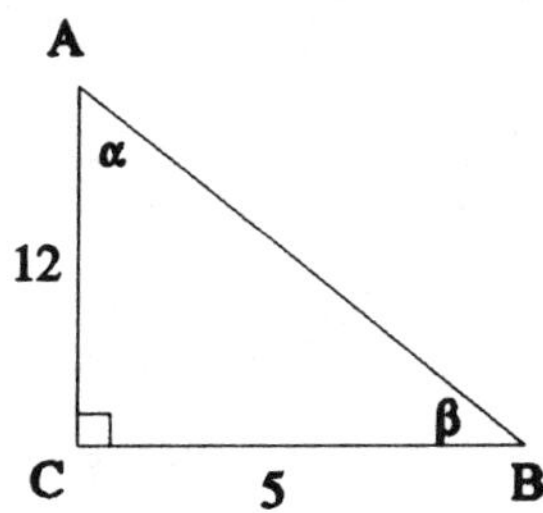

$$c = \sqrt{5^2 + 12^2} = 13$$

$$\sin\alpha = \frac{\text{opposite}}{\text{hypotenuse}} = \frac{5}{13}$$

22.

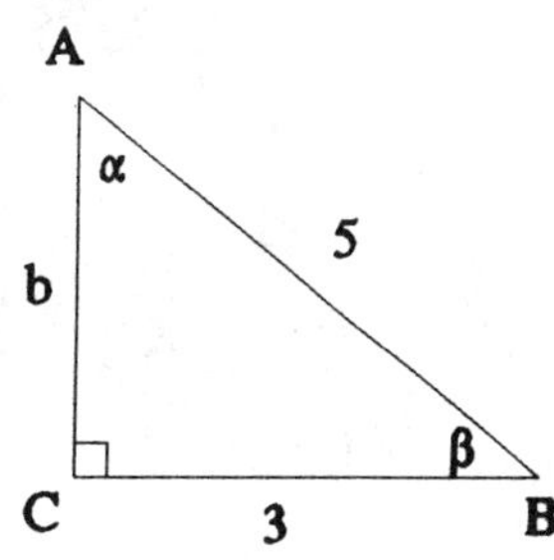

$$b = \sqrt{5^2 - 3^2} = 4$$

$$\tan\beta = \frac{\text{opposite}}{\text{adjacent}} = \frac{4}{3}$$

23.

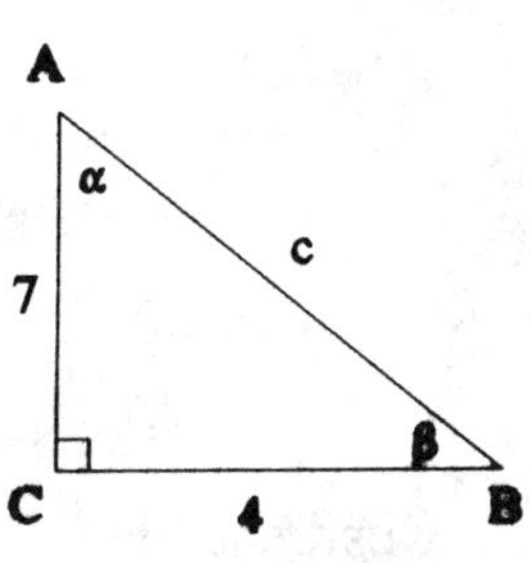

$$c = \sqrt{4^2 + 7^2} = \sqrt{65}$$

$$\sec\alpha = \frac{\text{hypotenuse}}{\text{adjacent}} = \frac{\sqrt{65}}{7}$$

24.

$P\left(-\sqrt{3}, 1\right)$

$x = -\sqrt{3}$

$y = 1$

$$r = \sqrt{x^2 + y^2} = \sqrt{\left(-\sqrt{3}\right)^2 + 1^2} = 2$$

$$\csc\theta = \frac{r}{y} = \frac{2}{1} = 2$$

25.

$P\left(\sqrt{2}, -\sqrt{2}\right)$

$x = \sqrt{2}$

$y = -\sqrt{2}$

$$\cot\theta = \frac{x}{y} = \frac{\sqrt{2}}{-\sqrt{2}} = -1$$

26.

$P\left(-1, -\sqrt{3}\right)$

$x = -1$

$y = -\sqrt{3}$

$$r = \sqrt{x^2 + y^2} = \sqrt{(-1)^2 + \left(-\sqrt{3}\right)^2} = 2$$

$$\cos\theta = \frac{x}{r} = -\frac{1}{2}$$

27.

$P\left(\sqrt{2}, \sqrt{2}\right)$

$x = \sqrt{2}$

$y = \sqrt{2}$

$r = \sqrt{x^2 + y^2} = \sqrt{\left(\sqrt{2}\right)^2 + \left(\sqrt{2}\right)^2} = 2$

$\sin\theta = \dfrac{y}{r} = \dfrac{\sqrt{2}}{2}$

In Exercises 28-36, use the following reference triangles.

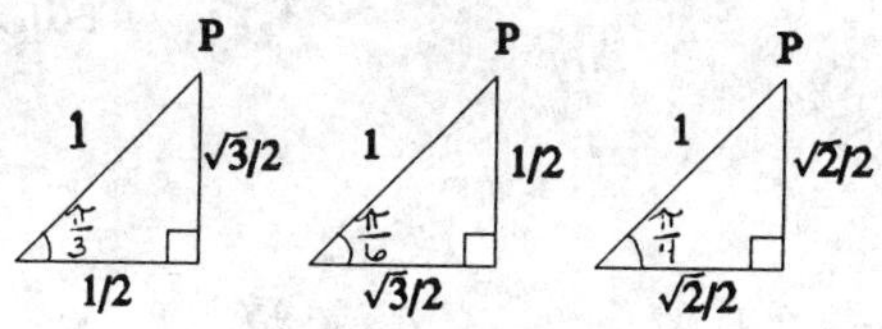

28.

$P\left(\dfrac{7\pi}{6}\right)$ Quadrant III

reference angle: $\dfrac{7\pi}{6} - \pi = \dfrac{\pi}{6}$

$\left(-\dfrac{\sqrt{3}}{2}, -\dfrac{1}{2}\right)$

29.

$P\left(-\dfrac{8\pi}{3}\right)$

$-\dfrac{8\pi}{3} + 2(2\pi) = \dfrac{4\pi}{3}$ Quadrant III

reference angle: $\dfrac{4\pi}{3} - \pi = \dfrac{\pi}{3}$

$\left(-\dfrac{1}{2}, -\dfrac{\sqrt{3}}{2}\right)$

30.

$P\left(\dfrac{5\pi}{6}\right)$ Quadrant II

reference angle: $\pi - \dfrac{5\pi}{6} = \dfrac{\pi}{6}$

$\left(-\dfrac{\sqrt{3}}{2}, \dfrac{1}{2}\right)$

31.

$P\left(-\dfrac{7\pi}{4}\right)$

$-\dfrac{7\pi}{4} + 2\pi = \dfrac{\pi}{4}$ Quadrant I

$\left(\dfrac{\sqrt{2}}{2}, \dfrac{\sqrt{2}}{2}\right)$

32.

$P\left(\frac{11\pi}{6}\right)$ Quadrant IV

reference angle: $2\pi - \frac{11\pi}{6} = \frac{\pi}{6}$

$\left(\frac{\sqrt{3}}{2}, -\frac{1}{2}\right)$

33.

reference angle: $\pi - \frac{2\pi}{3} = \frac{\pi}{3}$ Quadrant II

$P\left(-\frac{1}{2}, \frac{\sqrt{3}}{2}\right)$

$\sin\frac{2\pi}{3} = y = \frac{\sqrt{3}}{2}$

34.

$-\frac{5\pi}{4} + 2\pi = \frac{3\pi}{4}$ Quadrant II

reference angle: $\pi - \frac{3\pi}{4} = \frac{\pi}{4}$

$P\left(-\frac{\sqrt{2}}{2}, \frac{\sqrt{2}}{2}\right)$

$\cos\left(-\frac{5\pi}{4}\right) = x = -\frac{\sqrt{2}}{2}$

$\sec\left(-\frac{5\pi}{4}\right) = \frac{1}{\cos\left(-\frac{5\pi}{4}\right)} = \frac{1}{-\frac{\sqrt{2}}{2}} = -\sqrt{2}$

35.

reference angle: $\pi - \frac{5\pi}{6} = \frac{\pi}{6}$ Quadrant II

$P\left(-\frac{\sqrt{3}}{2}, \frac{1}{2}\right)$

$\tan = \frac{5\pi}{6} = \frac{y}{x} = \frac{\frac{1}{2}}{-\frac{\sqrt{3}}{2}} = -\frac{\sqrt{3}}{3}$

36.

$-\frac{\pi}{6} + 2\pi = \frac{11\pi}{6}$ Quadrant IV

reference angle: $\left|-\frac{\pi}{6}\right| = \frac{\pi}{6}$

$P\left(\frac{\sqrt{3}}{2}, -\frac{1}{2}\right)$

$\sin\left(-\frac{\pi}{6}\right) = y = -\frac{1}{2}$

$\csc\left(-\frac{\pi}{6}\right) = \frac{1}{\sin\left(-\frac{\pi}{6}\right)} = \frac{1}{-\frac{1}{2}} = -2$

In Exercises 37-41, $x = \frac{4}{5},\ y = -\frac{3}{5}$.

37.

$$P(t-\pi) = (-x, -y) = \left(-\frac{4}{5}, \frac{3}{5}\right)$$

38.

$$P\left(t+\frac{\pi}{2}\right) = (-y, x) = \left(\frac{3}{5}, \frac{4}{5}\right)$$

39.

$$P(-t) = (x, -y) = \left(\frac{4}{5}, \frac{3}{5}\right)$$

40.

$$P\left(t-\frac{\pi}{2}\right) = (y, -x) = \left(-\frac{3}{5}, -\frac{4}{5}\right)$$

41.

From Exercise 39, $P(-t) = (x, -y)$.

Hence $P(-t-\pi) = (-x, -(-y)) = \left(-\frac{4}{5}, -\frac{3}{5}\right)$.

In Exercises 42-45, use the reference triangles given for Exercises 28-36.

42.

$\sin t = -\frac{\sqrt{2}}{2}$

reference angle: $\frac{\pi}{4}$

Quadrant III: $\pi + \frac{\pi}{4} = \frac{5\pi}{4}$

43.

$\cos t = \frac{\sqrt{3}}{2}$

reference angle: $\frac{\pi}{6}$

Quadrant IV: $2\pi - \frac{\pi}{6} = \frac{11\pi}{6}$

44.

$\cot t = \frac{\sqrt{3}}{3}$

$\tan t = \frac{1}{\frac{\sqrt{3}}{3}} = \sqrt{3}$

reference angle: $\frac{\pi}{3}$

Quadrant I: $\frac{\pi}{3}$

45.

$\sec t = -2$

$\cos t = \frac{1}{-2}$

reference angle: $\frac{\pi}{3}$

Quadrant II: $\pi - \frac{\pi}{3} = \frac{2\pi}{3}$

46.

$$\left.\begin{array}{ll}\sin t<0\text{:} & \text{Q III and IV}\\ \cos t>0\text{:} & \text{Q I and IV}\end{array}\right\}\text{ Quadrant IV}$$

47.

$$\left.\begin{array}{rl}\sin(-t)=-\sin t>0\text{:} & \text{Q III and IV}\\ \tan t>0\text{:} & \text{Q I and III}\end{array}\right\}\text{ Quadrant III}$$

48.

$$\begin{aligned}\sin^2 t&=1-\cos^2 t\\&=1-\left(\frac{3}{5}\right)^2\\&=\frac{16}{25}\end{aligned}$$

$$\sin t=-\frac{4}{5}\quad\text{(Quadrant IV)}$$

$$\cot t=\frac{1}{\tan t}=\frac{1}{\frac{\sin t}{\cos t}}=\frac{\cos t}{\sin t}=\frac{\frac{3}{5}}{\frac{-4}{5}}=-\frac{3}{4}$$

49.

$$\left.\begin{array}{ll}\tan t>0\text{:} & \text{Q I and III}\\ \sin t<0\text{:} & \text{Q III and IV}\end{array}\right\}\text{ Quadrant III}$$

$$\begin{aligned}\cos^2 t&=1-\sin^2 t\\&=1-\left(-\frac{4}{5}\right)^2\\&=\frac{9}{25}\end{aligned}$$

$$\cos t=-\frac{3}{5}\quad\text{(Quadrant III)}$$

$$\sec t=\frac{1}{\cos t}=\frac{1}{-\frac{3}{5}}=-\frac{5}{3}$$

50.

$$\left.\begin{array}{ll}\cos t<0\text{:} & \text{Q II and III}\\ \sin t>0\text{:} & \text{Q I and II}\end{array}\right\}\text{ Quadrant II}$$

$$\begin{aligned}\cos^2 t&=1-\sin^2 t\\&=1-\left(\frac{12}{13}\right)^2\\&=\frac{25}{169}\end{aligned}$$

$$\cos t=-\frac{5}{13}\quad\text{(Quadrant II)}$$

$$\tan t=\frac{\sin t}{\cos t}=\frac{\frac{12}{13}}{\frac{-5}{13}}=-\frac{12}{5}$$

51.

$$\left.\begin{array}{ll}\tan t<0\text{:} & \text{Q II and IV}\\ \cos t<0\text{:} & \text{Q II and III}\end{array}\right\}\text{ Quadrant II}$$

$$\begin{aligned}\sin^2 t&=1-\cos^2 t\\&=1-\left(-\frac{5}{13}\right)^2\\&=\frac{144}{169}\end{aligned}$$

$$\sin t=\frac{12}{13}\quad\text{(Quadrant II)}$$

$$\csc t=\frac{1}{\sin t}=\frac{1}{\frac{12}{13}}=\frac{13}{12}$$

52.

$$\sin t\sec t=\sin t\left(\frac{1}{\cos t}\right)=\frac{\sin t}{\cos t}=\tan t$$

53.

$$\frac{\sin t}{\cos^2 t}=\frac{\sin t}{\cos t}\cdot\frac{1}{\cos t}=(\tan t)(\sec t)$$

54.

$$\cos 3.71 - \sin 1.44 \approx -0.8428 - 0.9915$$
$$= -1.8343$$

55.

$$\tan(-2.74) \approx 0.4247$$

56.

$$f(x) = 1 - \sin x$$

57.

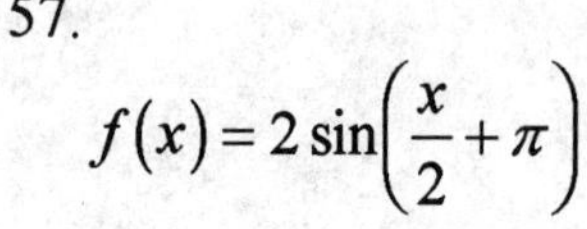

$$f(x) = 2\sin\left(\frac{x}{2} + \pi\right)$$

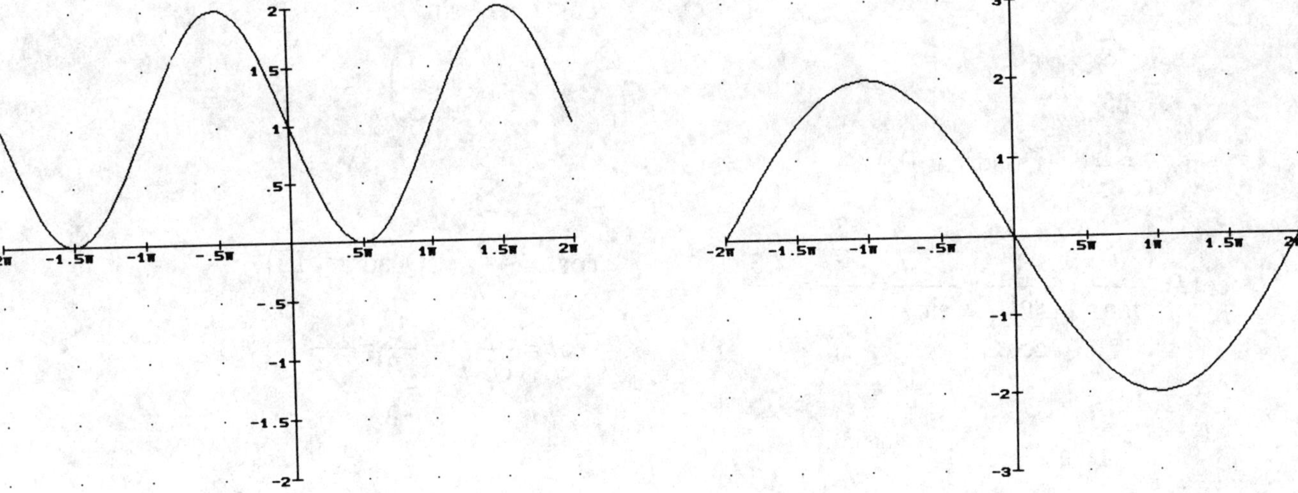

58.

$$f(x) = -\cos(2x - \pi)$$

$A = -1 \quad B = 2 \quad C = -\pi$

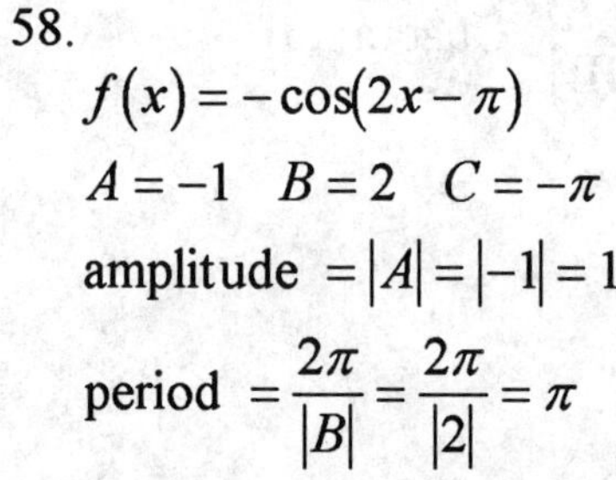

amplitude $= |A| = |-1| = 1$

period $= \dfrac{2\pi}{|B|} = \dfrac{2\pi}{|2|} = \pi$

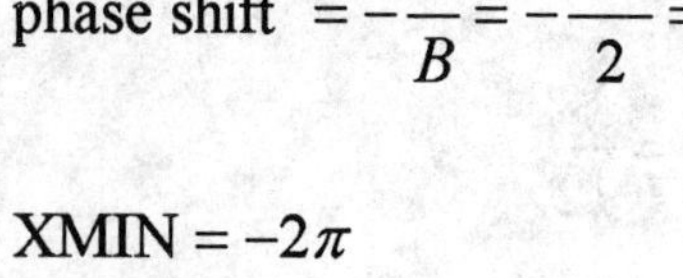

phase shift $= -\dfrac{C}{B} = -\dfrac{-\pi}{2} = \dfrac{\pi}{2}$

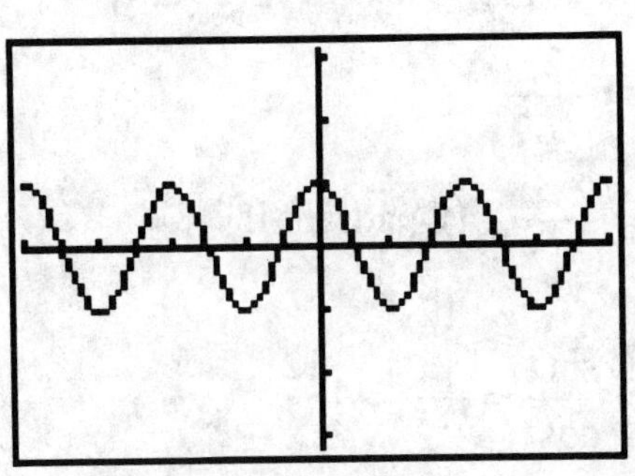

XMIN $= -2\pi$

XMAX $= 2\pi$

XSCL $= \dfrac{\pi}{2}$

YMIN $= -2$

YMAX $= 2$

YSCL $= 1$

59.

$$f(x) = 4\sin\left(-x + \frac{\pi}{2}\right)$$

$A = 4 \quad B = -1 \quad C = \dfrac{\pi}{2}$

amplitude $= |A| = |4| = 4$

period $= \dfrac{2\pi}{|B|} = \dfrac{2\pi}{|-1|} = 2\pi$

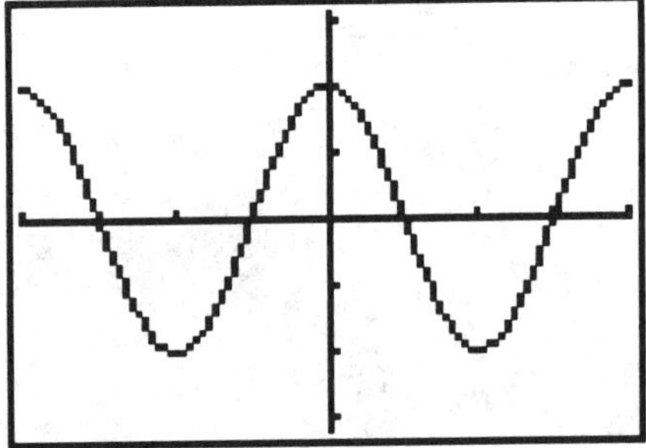

phase shift $= -\dfrac{C}{B} = -\dfrac{\frac{\pi}{2}}{-1} = \dfrac{\pi}{2}$

XMIN $= -2\pi$

XMAX $= 2\pi$

XSCL $= \dfrac{\pi}{2}$

YMIN $= -5$

YMAX $= 5$

YSCL $= 1$

60.

$f(x) = -2\sin\left(\frac{x}{3} + \frac{\pi}{3}\right)$

$A = -2 \quad B = \frac{1}{3} \quad C = \frac{\pi}{3}$

amplitude $= |A| = |-2| = 2$

period $= \frac{2\pi}{|B|} = \frac{2\pi}{\left|\frac{1}{3}\right|} = 6\pi$

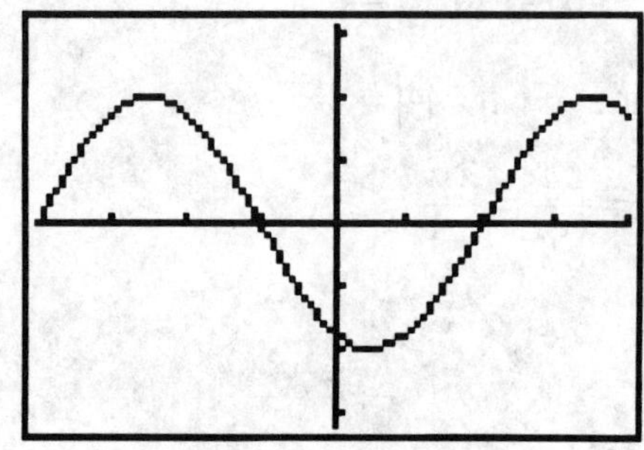

phase shift $= -\frac{C}{B} = -\frac{\frac{\pi}{3}}{\frac{1}{3}} = -\pi$

XMIN $= -3\pi$

XMAX $= 3\pi$

XSCL $= \frac{\pi}{2}$

YMIN $= -3$

YMAX $= 3$

YSCL $= 1$

61.

$\arcsin\left(-\frac{1}{2}\right) = y$

$-\frac{1}{2} = \sin y \quad y \in \left[-\frac{\pi}{2}, \frac{\pi}{2}\right]$

$y = -\frac{\pi}{6}$

62.

$\cos^{-1} 1 = y$

$1 = \cos y \quad y \in [0, \pi]$

$y = 0$

$\tan 0 = 0$

63.

$$\tan(\tan^{-1} 5) = 5$$

64.

$$5\cos^2 x - 4 = 0$$

$$\cos^2 x = \frac{4}{5}$$

$$\cos x = \pm\frac{2\sqrt{5}}{5}$$

$$x = \cos^{-1}\left(\pm\frac{2\sqrt{5}}{5}\right)$$

65.

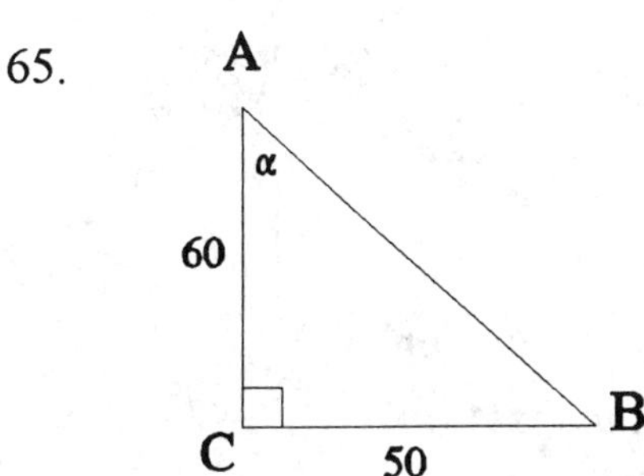

$$\tan\alpha = \frac{50}{60}$$

$$\alpha = \tan^{-1}\frac{5}{6} \approx 39.81° = 39°48'36''$$

66.

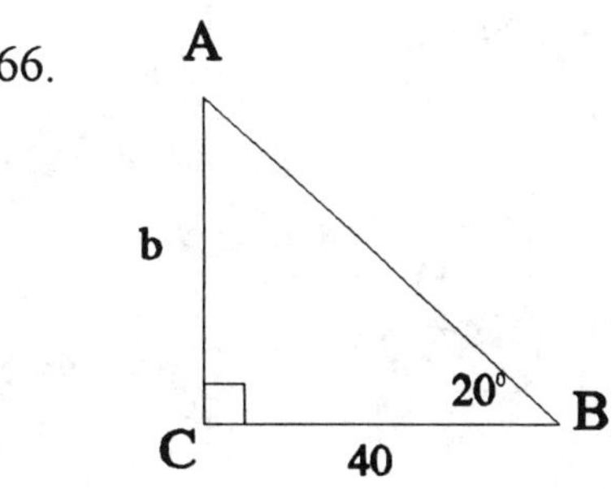

$$\tan 20° = \frac{b}{40}$$

$$b = 40\tan 20° \approx 14.56$$

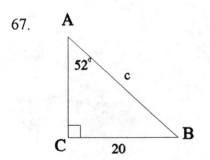

67.

A

52°

c

C

20

B

$$\sin 52° = \frac{20}{c}$$

$$c = \frac{20}{\sin 52°} \approx 25.38$$

68.

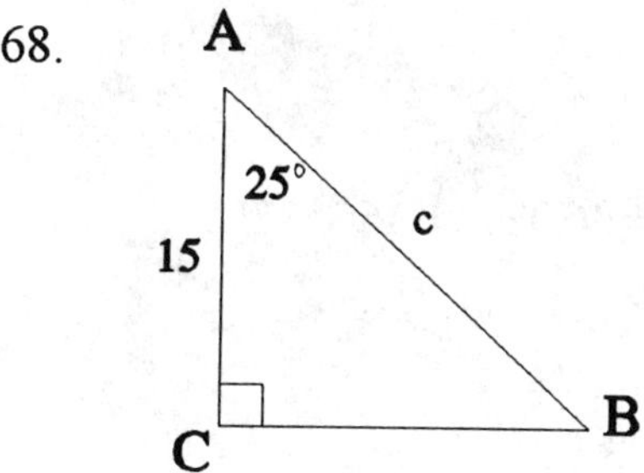

$$\cos 25° = \frac{15}{c}$$

$$c = \frac{15}{\cos 25°} \approx 16.55$$

69.

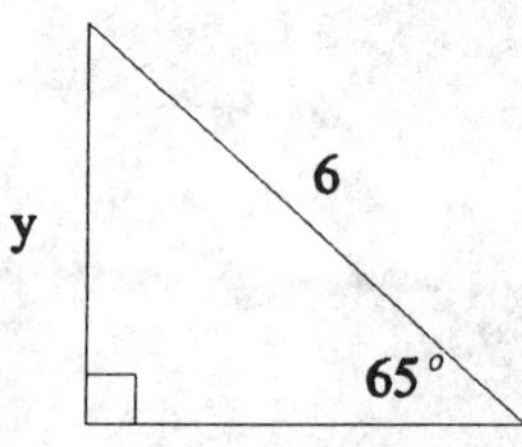

$$\sin 65° = \frac{y}{6}$$

$$y = 6\sin 65° \approx 5.44$$

5.44 meters

70.

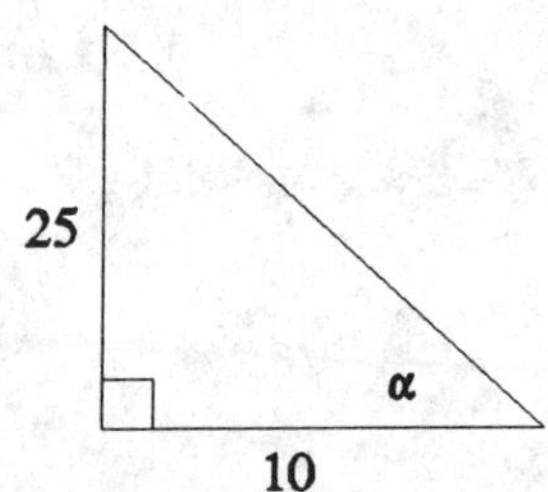

$$\tan\alpha = \frac{25}{10}$$

$$\alpha = \tan^{-1}\left(\frac{5}{2}\right) = 68.199° = 68°11'56''$$

71.

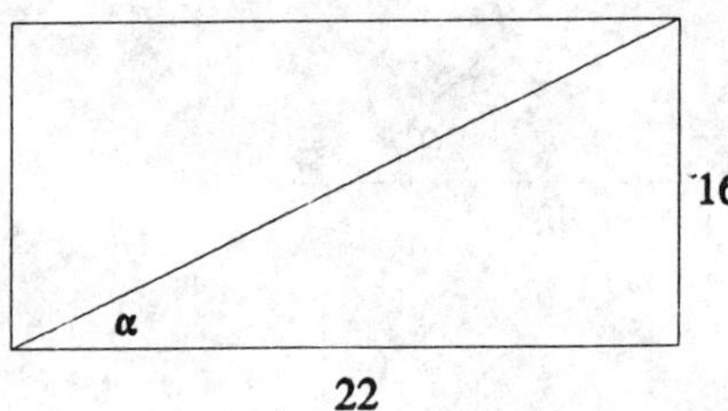

The smaller angle will be opposite the shorter side.

$$\tan\alpha = \frac{16}{22}$$

$$\alpha = \tan^{-1}\frac{8}{11} = 36.027° \approx 36°1'37''$$

72.

The graph is shifted up 2 units.

amplitude $= 1$ implies $A = 1$

period $= \dfrac{2\pi}{3} = \dfrac{2\pi}{B}$ implies $B = 3$

$$y = 2 + \cos 3x$$

73.

amplitude $= 1$ implies $A = 1$

period $= 2\pi$ implies $B = 1$

phase shift $= -\dfrac{3\pi}{4} = -\dfrac{C}{B}$

$$-\frac{3\pi}{4} = -\frac{C}{1}$$

$$C = \frac{3\pi}{4}$$

$$y = \cos\left(x + \frac{3\pi}{4}\right)$$

CHAPTER 7 REVIEW TEST

1.

$$\frac{\frac{5\pi}{3}}{\pi} = \frac{\theta}{180}$$

$$\theta = \frac{180\left(\frac{5\pi}{3}\right)}{\pi} = 300°$$

2.

$$\frac{\theta}{\pi} = -\frac{200}{180}$$

$$\theta = -\frac{200\pi}{180} = -\frac{10\pi}{9}$$

3.

$$\frac{\theta}{\pi} = \frac{75}{180}$$

$$\theta = \frac{75\pi}{180} = \frac{5\pi}{12}$$

4.

$$-25° + 360° = 335°$$

5.

$$\frac{17\pi}{4} - 2(2\pi) = \frac{\pi}{4} = 45°$$

6.

$$180° - 160° = 20°$$

7.

$$2\pi - \frac{7\pi}{4} = \frac{\pi}{4}$$

8.

$$\text{radian measure} = \frac{\text{intercepted arc}}{\text{radius}}$$

$$= \frac{12}{15} = \frac{4}{5} = 0.8 \text{ radian}$$

9.

$$\frac{19\pi}{3} - 3(2\pi) = \frac{\pi}{3}$$

10.

$$-22\pi + 11(2\pi) = 0$$

11.

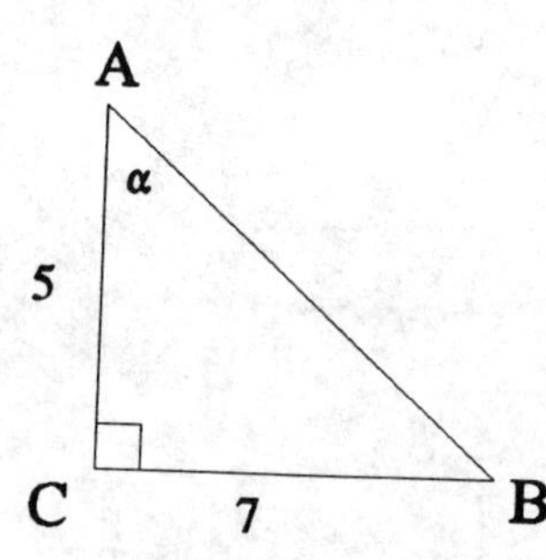

$$\tan\alpha = \frac{\text{opposite}}{\text{adjacent}} = \frac{7}{5}$$

12.

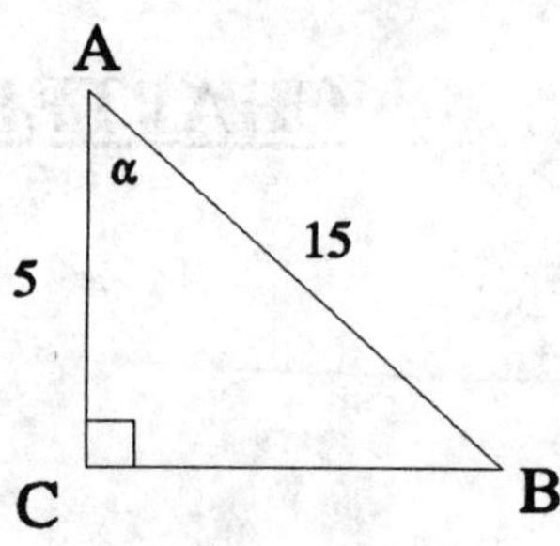

$$\sec\alpha = \frac{\text{hypotenuse}}{\text{adjacent}} = \frac{15}{5} = 3$$

13.

$P\left(-\sqrt{2}, \sqrt{2}\right)$

$$\cot\theta = \frac{x}{y} = \frac{-\sqrt{2}}{\sqrt{2}} = -1$$

14.

$P(0, -5)$

$$\begin{aligned} r &= \sqrt{x^2 + y^2} \\ &= \sqrt{0^2 + (-5)^2} \\ &= 5 \end{aligned}$$

$$\sin\theta = \frac{y}{r} = \frac{-5}{5} = -1$$

15.

$P\left(2, 2\sqrt{3}\right)$

$$\begin{aligned} r &= \sqrt{x^2 + y^2} \\ &= \sqrt{2^2 + \left(2\sqrt{3}\right)^2} \\ &= 4 \end{aligned}$$

$$\sec\theta = \frac{r}{x} = \frac{4}{2} = 2$$

16.

$$\frac{29\pi}{6} - 2(2\pi) = \frac{5\pi}{6} \quad \text{(Quadrant II)}$$

reference angle: $\pi - \dfrac{5\pi}{6} = \dfrac{\pi}{6}$

$$\left(-\frac{\sqrt{3}}{2}, \frac{1}{2}\right)$$

17.

$$-\frac{\pi}{3} + 2\pi = \frac{5\pi}{3} \quad \text{(Quadrant IV)}$$

reference angle: $\left|-\dfrac{\pi}{3}\right| = \dfrac{\pi}{3}$

$$\left(\frac{1}{2}, -\frac{\sqrt{3}}{2}\right)$$

In Exercises 18-20, $P(t)=\left(-\frac{5}{13},\frac{12}{13}\right)$, hence $x=-\frac{5}{13},\ y=\frac{12}{13}$.

18.

$$P(t+\pi)=(-x,-y)=\left(\frac{5}{13},-\frac{12}{13}\right)$$

19.

$$P\left(t-\frac{\pi}{2}\right)=(y,-x)=\left(\frac{12}{13},\frac{5}{13}\right)$$

20.

$$P(-t)=(x,-y)=\left(-\frac{5}{13},-\frac{12}{13}\right)$$

21.

$$\frac{7\pi}{3}-2\pi=\frac{\pi}{3}$$

$$\cos\frac{7\pi}{3}=\frac{1}{2}$$

22.

$$\frac{-2\pi}{3}+2\pi=\frac{4\pi}{3}$$

reference angle: $\frac{4\pi}{3}-\pi=\frac{\pi}{3}$

$$\sin\left(-\frac{2\pi}{3}\right)=-\frac{\sqrt{3}}{2}\quad\text{(Quadrant III)}$$

$$\csc\left(-\frac{2\pi}{3}\right)=\frac{1}{\sin\left(-\frac{2\pi}{3}\right)}=\frac{1}{-\frac{\sqrt{3}}{2}}=-\frac{2\sqrt{3}}{3}$$

23.

$$\tan t=1$$

reference angle: $\frac{\pi}{4}$

$$\pi+\frac{\pi}{4}=\frac{5\pi}{4}\quad\text{(Quadrant III)}$$

24.

$$\sec t=\sqrt{2}$$

$$\cos t=\frac{1}{\sqrt{2}}=\frac{\sqrt{2}}{2}$$

reference angle: $\frac{\pi}{4}$

$$2\pi-\frac{\pi}{4}=\frac{7\pi}{4}\quad\text{(Quadrant IV)}$$

25.

$$\left.\begin{array}{l}\tan t>0\text{: Q I and III}\\ \cos t<0\text{: Q II and III}\end{array}\right\}\text{ Quadrant III}$$

$$\begin{aligned}\sin^2 t&=1-\cos^2 t\\&=1-\left(-\frac{12}{13}\right)^2\\&=\frac{25}{169}\end{aligned}$$

$$\sin t=-\frac{5}{13}$$

26.

$$\cos^2 t = 1 - \sin^2 t$$
$$= 1 - \left(\frac{3}{5}\right)^2$$
$$= \frac{16}{25}$$
$$\cos t = -\frac{4}{5} \quad \text{(Quadrant II)}$$
$$\sec t = \frac{1}{\cos t} = \frac{1}{-\frac{4}{5}} = -\frac{5}{4}$$

27.

$$1 - \tan x = 1 - \frac{\sin x}{\cos x}$$
$$= \frac{\cos x}{\cos x} - \frac{\sin x}{\cos x}$$
$$= \frac{\cos x - \sin x}{\cos x}$$

28.

$$\tan(-3.68) = -0.5973$$

29.

$$\cos 1.15 - \sin 0.72 = 0.4085 - 0.6594$$
$$= -0.2509$$

30.

$$f(x) = x + \cos x$$

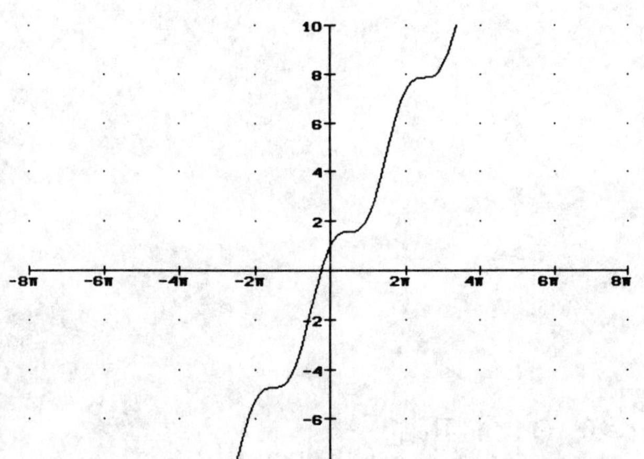

31.

$$f(x) = -2\cos(\pi - x)$$
$$A = -2 \quad B = -1 \quad C = \pi$$
$$\text{amplitude } = |A| = |-2| = 2$$
$$\text{period } = \frac{2\pi}{|B|} = \frac{2\pi}{|-1|} = 2\pi$$
$$\text{phase shift } = -\frac{C}{B} = -\frac{\pi}{-1} = \pi$$

32.

$$f(x) = 2\sin\left(\frac{x}{2} - \frac{\pi}{2}\right)$$

$$A = 2 \quad B = \frac{1}{2} \quad C = -\frac{\pi}{2}$$

$$\text{amplitude } = |A| = |2| = 2$$

$$\text{period } = \frac{2\pi}{|B|} = \frac{2\pi}{\left|\frac{1}{2}\right|} = 4\pi$$

$$\text{phase shift } = -\frac{C}{B} = -\frac{\frac{-\pi}{2}}{\frac{1}{2}} = \pi$$

33.

$$\tan^{-1}\left(-\sqrt{3}\right) = y$$

$$-\sqrt{3} = \tan y \quad y \in \left(-\frac{\pi}{2}, \frac{\pi}{2}\right)$$

$$y = -\frac{\pi}{3}$$

34.

$$\text{Let } u = \sin^{-1}\frac{\sqrt{3}}{2}$$

$$\sin u = \frac{\sqrt{3}}{2} \quad u \in \left(-\frac{\pi}{2}, \frac{\pi}{2}\right)$$

$$u = \frac{\pi}{3}$$

$$\cos\left(\sin^{-1}\frac{\sqrt{3}}{2}\right) = \cos u = \cos\frac{\pi}{3} = \frac{1}{2}$$

35.

$$6\tan^2 x - 13\tan x + 6 = 0 \quad x \in \left(-\frac{\pi}{2}, \frac{\pi}{2}\right)$$

$$(2\tan x - 3)(3\tan x - 2) = 0$$

$$2\tan x - 3 = 0 \qquad 3\tan x - 2 = 0$$

$$\tan x = \frac{3}{2} \qquad \tan x = \frac{2}{3}$$

$$x = \tan^{-1}\frac{3}{2} \qquad x = \tan^{-1}\frac{2}{3}$$

36.

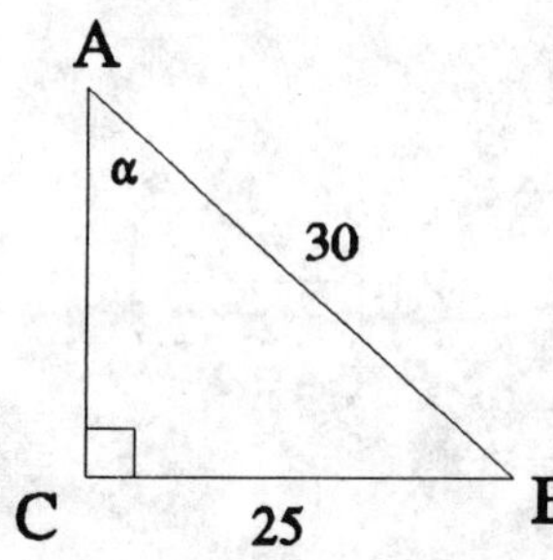

$$\sin\alpha = \frac{25}{30}$$

$$\alpha = \sin^{-1}\frac{5}{6} \approx 56.44° = 56°26'35''$$

37.

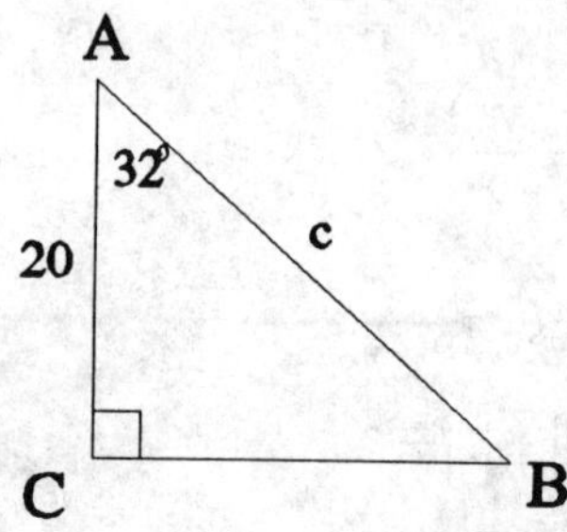

$$\cos 32° = \frac{20}{c}$$

$$c = \frac{20}{\cos 32°} \approx 23.58$$

38.

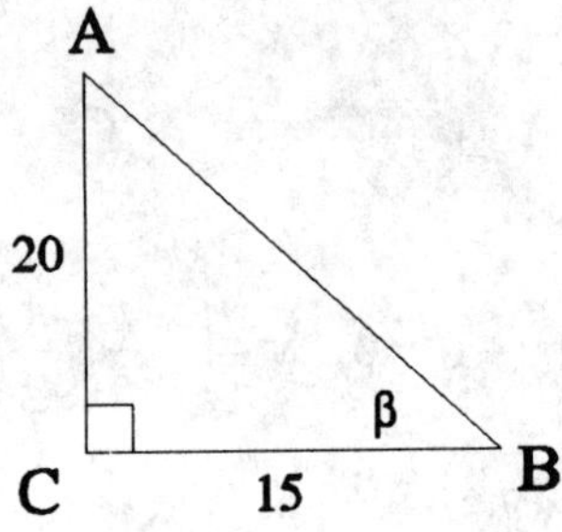

$$\tan\beta = \frac{20}{15}$$

$$\beta = \tan^{-1}\frac{4}{3} \approx 53.13° = 53°7'48''$$

39.

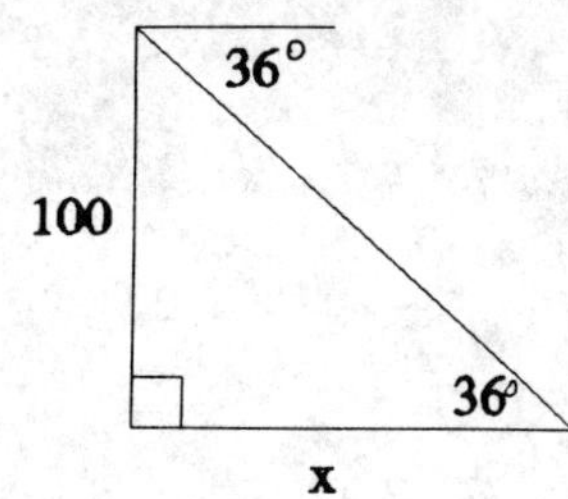

$$\tan 36° = \frac{100}{x}$$

$$x = \frac{100}{\tan 36°} \approx 137.64$$

137.64 meters

CHAPTER 7 ADDITIONAL PRACTICE EXERCISES

1. Determine the quadrant in which $\theta = 655°$ lies.

2. Convert 65° to radian measure.

3. Determine if the pair of angles $\frac{2\pi}{7}, \frac{16\pi}{7}$ are conterminal.

4. Find the reference angle for $\frac{5\pi}{8}$.

5. Given the triangle ABC with $\gamma = 90°$, $\beta = 36°$, $a = 37$ find b.

6. Find the point on the unit circle that corresponds to $t = -\frac{21\pi}{2}$.

7 Find the quadrant in which $\mathrm{P}(t)$ lies if $\cot < 0$, $\csc t > 0$.

8. Given $\sin t = -\frac{3}{4}$, find $\sin(-t)$.

9. Determine the amplitude and period of $f(x) = 5\sin 6x$.

10. Evaluate $\operatorname{arccsc}(-1)$.

CHAPTER 7 PRACTICE TEST

1. Determine the quadrant in which $\theta = -\dfrac{7\pi}{12}$ lies.

2. Convert $\dfrac{2\pi}{9}$ radians to degree measure.

3. Find the reference angle for $268°$.

4. Determine if $-62°$ and $655°$ are coterminal.

5. Find the length of an arc subtended by a central angle $\dfrac{\pi}{8}$ radians on a circle of radius 16 cm.

6. Find $\sin\theta$.

5

θ

2

7. Find a real number t in the interval $[0, 2\pi]$ so that $P\left(\dfrac{18\pi}{5}\right) = P(t)$.

8. $P(t) = P\left(\dfrac{4}{5}, -\dfrac{3}{5}\right)$ is the point on the unit circle. Find $\cot t$.

9. Evaluate $\tan\dfrac{5\pi}{3}$.

10. Find the point on the unit circle that corresponds to $t = \dfrac{33\pi}{2}$.

11. Evaluate $\csc\left(-\dfrac{2\pi}{3}\right)$.

12. Find the quadrant in which $P(t)$ lies if $\sin t < 0$ and $\sec t > 0$.

13.
Given $\sec t = 1.269$, find the $\sec(-t)$.

14.
If $\cos t = \frac{3}{5}$ and t is in quadrant IV, find $\cot t$.

15.
Given $P(1, -2)$ lies on the terminal side of θ, find $\csc\theta$.

16.
Evaluate $\sec 10.62$.

17.
Given $f(x) = \frac{1}{3}\cos\left(\frac{x}{4} - \frac{\pi}{3}\right)$, determine the amplitude, period and phase shift.

18.
Evaluate $\operatorname{arccot}\sqrt{3}$.

19.
Evaluate $\sin^{-1} 0.788$.

20.
Given triangle ABC with $\gamma = 90°$, $a = 4$, $b = 12$, find α.

CHAPTER 8 SECTION 1

SECTION 8.1 PROGRESS CHECK

1. (Page 481)

$$\frac{\csc\theta}{1+\cot^2\theta}=\frac{\dfrac{1}{\sin\theta}}{1+\dfrac{\cos^2\theta}{\sin^2\theta}}=\frac{\sin\theta}{\sin^2\theta+\cos^2\theta}=\frac{\sin\theta}{1}=\sin\theta$$

2. (Page 482)

$$\frac{\sin^2 y-1}{1-\sin y}=\frac{(\sin y+1)(\sin y-1)}{-1(\sin y-1)}=-\sin y-1$$

3. (Page 482)

$$\sin x\sec x=\sin x\left(\frac{1}{\cos x}\right)=\frac{\sin x}{\cos x}=\tan x$$

4. (Page 482)

$$\cos x+\tan x\sin x=\cos x+\frac{\sin x}{\cos x}(\sin x)=\cos x+\frac{\sin^2 x}{\cos x}$$

$$=\frac{\cos^2+\sin^2 x}{\cos x}=\frac{1}{\cos x}=\sec x$$

5. (Page 483)

$$\frac{1+\cos t}{\sin t}+\frac{\sin t}{1+\cos t}=\frac{(1+\cos t)(1+\cos t)+(\sin t)(\sin t)}{\sin t(1+\cos t)}$$

$$=\frac{1+2\cos t+\cos^2 t+\sin^2 t}{\sin t(1+\cos t)}=\frac{2+2\cos t}{\sin t(1+\cos t)}=\frac{2(1+\cos t)}{\sin t(1+\cos t)}$$

$$=\frac{2}{\sin t}=2\csc t$$

6. (Page 484)

$$\frac{\sin x+\cos x}{\tan^2 x-1}=\frac{\sin x+\cos x}{\dfrac{\sin^2 x}{\cos^2 x}-1}=\frac{\cos^2 x(\sin x+\cos x)}{\sin^2 x-\cos^2 x}$$

$$=\frac{\cos^2 x(\sin x+\cos x)}{(\sin x+\cos x)(\sin x-\cos x)}=\frac{\cos^2 x}{\sin x-\cos x}$$

EXERCISE SET 8.1

1.

$$\csc\gamma - \cos\gamma\cot\gamma = \frac{1}{\sin\gamma} - \cos\gamma\left(\frac{\cos\gamma}{\sin\gamma}\right)$$

$$= \frac{1-\cos^2\gamma}{\sin\gamma} = \frac{\sin^2\gamma}{\sin\gamma} = \sin\gamma$$

3.

$$\sec v + \tan v = \frac{1}{\cos v} + \frac{\sin v}{\cos v} = \frac{1+\sin v}{\cos v}$$

5.

$$\sin\alpha\sec\alpha = \sin\alpha\left(\frac{1}{\cos\alpha}\right) = \frac{\sin\alpha}{\cos\alpha} = \tan\alpha$$

7.

$$3-\sec^2 x = 2+1-\sec^2 x$$
$$= 2-\left(\sec^2 x - 1\right)$$
$$= 2-\tan^2 x$$

9.

$$\frac{\sec^2 y}{\tan y} = \frac{\tan^2 y+1}{\tan y}$$
$$= \frac{\tan^2 y}{\tan y} + \frac{1}{\tan y}$$
$$= \tan y + \cot y$$

11.

$$\frac{\sin u}{\csc u} + \frac{\cos u}{\sec u} = \frac{\sin u}{\frac{1}{\sin u}} + \frac{\cos u}{\frac{1}{\cos u}}$$
$$= \sin^2 u + \cos^2 u = 1$$

13.

$$\frac{\sec^2\theta - 1}{\sec^2\theta} = \frac{\sec^2\theta}{\sec^2\theta} - \frac{1}{\sec^2\theta}$$
$$= 1-\cos^2\theta$$
$$= \sin^2\theta$$

15.

$$\cos\gamma + \cos\gamma\tan^2\gamma = \cos\gamma\left(1+\tan^2\gamma\right)$$
$$= \cos\gamma\left(\sec^2\gamma\right)$$
$$= \cos\gamma\left(\frac{1}{\cos^2\gamma}\right)$$
$$= \frac{1}{\cos\gamma}$$
$$= \sec\gamma$$

17.

$$\frac{\sec w \sin w}{\tan w+\cot w}=\frac{\left(\frac{1}{\cos w}\right)(\sin w)}{\frac{\sin w}{\cos w}+\frac{\cos w}{\sin w}}$$
$$=\frac{\sin^2 w}{\sin^2 w+\cos^2 w}$$
$$=\frac{\sin^2 w}{1}$$
$$=\sin^2 w$$

19.

$$(\sin\alpha+\cos\alpha)^2+(\sin\alpha-\cos\alpha)^2=\sin^2\alpha+2\sin\alpha\cos\alpha+\cos^2\alpha+\sin^2\alpha-2\sin\alpha\cos\alpha+\cos^2\alpha$$
$$=\left(\sin^2\alpha+\cos^2\alpha\right)+\left(\sin^2\alpha+\cos^2\alpha\right)$$
$$=1+1$$
$$=2$$

21.

$$\sec^2 v+\cos^2 v=\sec^2 v+\frac{1}{\sec^2 v}=\frac{\sec^4 v+1}{\sec^2 v}$$

23.

$$\frac{\sin^2\alpha}{1+\cos\alpha}=\left(\frac{\sin^2\alpha}{1+\cos\alpha}\right)\left(\frac{1-\cos\alpha}{1-\cos\alpha}\right)$$
$$=\frac{\sin^2\alpha(1-\cos\alpha)}{1-\cos^2\alpha}$$
$$=\frac{\sin^2\alpha(1-\cos\alpha)}{\sin^2\alpha}$$
$$=1-\cos\alpha$$

25.

$$\frac{\cos t}{1+\sin t}=\left(\frac{\cos t}{1+\sin t}\right)\left(\frac{1-\sin t}{1-\sin t}\right)$$
$$=\frac{\cos t(1-\sin t)}{1-\sin^2 t}$$
$$=\frac{\cos t(1-\sin t)}{\cos^2 t}$$
$$=\frac{1-\sin t}{\cos t}$$

27.

$$\csc^2\theta-\frac{\cos^2\theta}{\sin^2\theta}=\csc^2\theta-\cot^2\theta=1$$

EXERCISE SET 8.1

1.

$$\csc\gamma - \cos\gamma \cot\gamma = \frac{1}{\sin\gamma} - \cos\gamma\left(\frac{\cos\gamma}{\sin\gamma}\right)$$

$$= \frac{1-\cos^2\gamma}{\sin\gamma} = \frac{\sin^2\gamma}{\sin\gamma} = \sin\gamma$$

3.

$$\sec v + \tan v = \frac{1}{\cos v} + \frac{\sin v}{\cos v} = \frac{1+\sin v}{\cos v}$$

5.

$$\sin\alpha \sec\alpha = \sin\alpha\left(\frac{1}{\cos\alpha}\right) = \frac{\sin\alpha}{\cos\alpha} = \tan\alpha$$

7.

$$3-\sec^2 x = 2+1-\sec^2 x$$
$$= 2-\left(\sec^2 x - 1\right)$$
$$= 2-\tan^2 x$$

9.

$$\frac{\sec^2 y}{\tan y} = \frac{\tan^2 y+1}{\tan y}$$
$$= \frac{\tan^2 y}{\tan y} + \frac{1}{\tan y}$$
$$= \tan y + \cot y$$

11.

$$\frac{\sin u}{\csc u} + \frac{\cos u}{\sec u} = \frac{\sin u}{\frac{1}{\sin u}} + \frac{\cos u}{\frac{1}{\cos u}}$$
$$= \sin^2 u + \cos^2 u = 1$$

13.

$$\frac{\sec^2\theta - 1}{\sec^2\theta} = \frac{\sec^2\theta}{\sec^2\theta} - \frac{1}{\sec^2\theta}$$
$$= 1-\cos^2\theta$$
$$= \sin^2\theta$$

15.

$$\cos\gamma + \cos\gamma \tan^2\gamma = \cos\gamma\left(1+\tan^2\gamma\right)$$
$$= \cos\gamma\left(\sec^2\gamma\right)$$
$$= \cos\gamma\left(\frac{1}{\cos^2\gamma}\right)$$
$$= \frac{1}{\cos\gamma}$$
$$= \sec\gamma$$

17.

$$\begin{aligned}\frac{\sec w \sin w}{\tan w + \cot w} &= \frac{\left(\frac{1}{\cos w}\right)(\sin w)}{\frac{\sin w}{\cos w} + \frac{\cos w}{\sin w}} \\ &= \frac{\sin^2 w}{\sin^2 w + \cos^2 w} \\ &= \frac{\sin^2 w}{1} \\ &= \sin^2 w\end{aligned}$$

19.

$$\begin{aligned}(\sin\alpha + \cos\alpha)^2 + (\sin\alpha - \cos\alpha)^2 &= \sin^2\alpha + 2\sin\alpha\cos\alpha + \cos^2\alpha + \sin^2\alpha - 2\sin\alpha\cos\alpha + \cos^2\alpha \\ &= (\sin^2\alpha + \cos^2\alpha) + (\sin^2\alpha + \cos^2\alpha) \\ &= 1 + 1 \\ &= 2\end{aligned}$$

21.

$$\sec^2 v + \cos^2 v = \sec^2 v + \frac{1}{\sec^2 v} = \frac{\sec^4 v + 1}{\sec^2 v}$$

23.

$$\begin{aligned}\frac{\sin^2\alpha}{1+\cos\alpha} &= \left(\frac{\sin^2\alpha}{1+\cos\alpha}\right)\left(\frac{1-\cos\alpha}{1-\cos\alpha}\right) \\ &= \frac{\sin^2\alpha(1-\cos\alpha)}{1-\cos^2\alpha} \\ &= \frac{\sin^2\alpha(1-\cos\alpha)}{\sin^2\alpha} \\ &= 1-\cos\alpha\end{aligned}$$

25.

$$\begin{aligned}\frac{\cos t}{1+\sin t} &= \left(\frac{\cos t}{1+\sin t}\right)\left(\frac{1-\sin t}{1-\sin t}\right) \\ &= \frac{\cos t(1-\sin t)}{1-\sin^2 t} \\ &= \frac{\cos t(1-\sin t)}{\cos^2 t} \\ &= \frac{1-\sin t}{\cos t}\end{aligned}$$

27.

$$\csc^2\theta - \frac{\cos^2\theta}{\sin^2\theta} = \csc^2\theta - \cot^2\theta = 1$$

47.

$$\sin x = \sqrt{1-\cos^2 x}$$

$$\text{Let } x = \frac{3\pi}{2}$$

$$\sin\frac{3\pi}{2} = \sqrt{1-\cos^2\left(\frac{3\pi}{2}\right)} \quad ?$$

$$-1 = \sqrt{1-0} \quad ?$$

$$-1 \neq 1$$

49.

$$(\sin t + \cos t)^2 = \sin^2 t + \cos^2 t$$

$$\text{Let } t = \frac{\pi}{4}$$

$$\left(\sin\frac{\pi}{4} + \cos\frac{\pi}{4}\right)^2 = \sin^2\frac{\pi}{4} + \cos^2\frac{\pi}{4} \quad ?$$

$$\left(\frac{\sqrt{2}}{2} + \frac{\sqrt{2}}{2}\right)^2 = \left(\frac{\sqrt{2}}{2}\right)^2 + \left(\frac{\sqrt{2}}{2}\right)^2 \quad ?$$

$$2 = \frac{1}{2} + \frac{1}{2} \quad ?$$

$$2 \neq 1$$

51.

$$\sqrt{\cos^2 x} = \cos x$$

$$\text{Let } x = \pi$$

$$\sqrt{\cos^2 \pi} = \cos\pi \quad ?$$

$$\sqrt{(-1)^2} = -1 \quad ?$$

$$1 \neq -1$$

CHAPTER 8 SECTION 2

SECTION 8.2 PROGRESS CHECK

1. (Page 487)

$$\begin{aligned}\cos 15° &= \cos(60° - 45°)\\ &= \cos 60° \cos 45° + \sin 60° \sin 45°\\ &= \left(\frac{1}{2}\right)\left(\frac{\sqrt{2}}{2}\right) + \left(\frac{\sqrt{3}}{2}\right)\left(\frac{\sqrt{2}}{2}\right)\\ &= \frac{\sqrt{2}+\sqrt{6}}{4}\end{aligned}$$

2. (Page 487)

$$\begin{aligned}\cos\left(\frac{5\pi}{12}\right) &= \cos\left(\frac{9\pi}{12} - \frac{4\pi}{12}\right)\\ &= \cos\frac{9\pi}{12}\cos\frac{4\pi}{12} + \sin\frac{9\pi}{12}\sin\frac{4\pi}{12}\\ &= \left(-\frac{\sqrt{2}}{2}\right)\left(\frac{1}{2}\right) + \left(\frac{\sqrt{2}}{2}\right)\left(\frac{\sqrt{3}}{2}\right)\\ &= \frac{\sqrt{6}-\sqrt{2}}{4}\end{aligned}$$

3. (Page 489)

$$\begin{aligned}\tan(x - \pi) &= \frac{\tan x - \tan \pi}{1 + \tan x \tan \pi}\\ &= \frac{\tan x - 0}{1 + (\tan x)(0)}\\ &= \tan x\end{aligned}$$

4. (Page 490)

$$\sin^2\alpha = 1-\cos^2\alpha$$

$$\sin^2\alpha = 1-\left(-\frac{4}{5}\right)^2$$

$$\sin^2\alpha = \frac{9}{25}$$

$$\sin^2\alpha = -\frac{3}{5} \qquad \text{(Quadrant III)}$$

$$\sin^2\beta = 1-\cos^2\beta$$

$$\sin^2\beta = 1-\left(\frac{3}{5}\right)^2$$

$$\sin^2\beta = \frac{16}{25}$$

$$\sin\beta = \frac{4}{5} \qquad \text{(Quadrant I)}$$

$$\begin{aligned}\cos(\alpha-\beta) &= \cos\alpha\cos\beta+\sin\alpha\sin\beta\\ &= \left(-\frac{4}{5}\right)\left(\frac{3}{5}\right)+\left(-\frac{3}{5}\right)\left(\frac{4}{5}\right)\\ &= -\frac{24}{25}\end{aligned}$$

Since $\cos(\alpha-\beta)$ is negative, $\alpha-\beta$ lies in either Quadrant II or Quadrant III.

Consider $\sin(\alpha-\beta) = \left(-\frac{3}{5}\right)\left(\frac{3}{5}\right)-\left(-\frac{4}{5}\right)\left(\frac{4}{5}\right) = \frac{7}{25} > 0.$

Hence $\alpha-\beta$ lies in Quadrant II.

EXERCISE SET 8.2

1.

Let $s=\frac{\pi}{2}$ and $t=\frac{\pi}{4}$

$$\cos\left(\frac{\pi}{2}-\frac{\pi}{4}\right)=\cos\frac{\pi}{2}-\cos\frac{\pi}{4} \quad ?$$

$$\cos\frac{\pi}{4}=0-\frac{\sqrt{2}}{2} \quad ?$$

$$\frac{\sqrt{2}}{2}\neq-\frac{\sqrt{2}}{2}$$

3.

Let $s=\frac{\pi}{2}$ and $t=\pi$

$$\sin\left(\frac{\pi}{2}-\pi\right)=\sin\frac{\pi}{2}-\sin\pi \quad ?$$

$$\sin\left(-\frac{\pi}{2}\right)=1-0 \quad ?$$

$$-1\neq 1$$

5.

Let $s=\frac{\pi}{3}$ and $t=\frac{\pi}{4}$

$$\tan\left(\frac{\pi}{3}+\frac{\pi}{4}\right)=\tan\frac{\pi}{3}+\tan\frac{\pi}{4} \quad ?$$

$$\tan\frac{7\pi}{12}=\sqrt{3}+1 \quad ?$$

$$-3.73\neq 2.73$$

7.

$$\cos\left(\frac{\pi}{6}+\frac{\pi}{4}\right)=\cos\frac{\pi}{6}\cos\frac{\pi}{4}-\sin\frac{\pi}{6}\sin\frac{\pi}{4}$$

$$=\left(\frac{\sqrt{3}}{2}\right)\left(\frac{\sqrt{2}}{2}\right)-\left(\frac{1}{2}\right)\left(\frac{\sqrt{2}}{2}\right)$$

$$=\frac{\sqrt{6}-\sqrt{2}}{4}$$

9.

$$\sin\left(\frac{\pi}{4}+\frac{\pi}{3}\right)=\sin\frac{\pi}{4}\cos\frac{\pi}{3}+\cos\frac{\pi}{4}\sin\frac{\pi}{3}$$

$$=\left(\frac{\sqrt{2}}{2}\right)\left(\frac{1}{2}\right)+\left(\frac{\sqrt{2}}{2}\right)\left(\frac{\sqrt{3}}{2}\right)$$

$$=\frac{\sqrt{2}+\sqrt{6}}{4}$$

11.

$$\begin{aligned}\cos(30°+180°) &= \cos 30° \cos 180° - \sin 30° \sin 180° \\ &= \left(\frac{\sqrt{3}}{2}\right)(-1) - \left(\frac{1}{2}\right)(0) \\ &= -\frac{\sqrt{3}}{2}\end{aligned}$$

13.

$$\begin{aligned}\tan(300°-60°) &= \frac{\tan 300° - \tan 60°}{1+\tan 300° \tan 60°} \\ &= \frac{-\sqrt{3}-\sqrt{3}}{1+\left(-\sqrt{3}\right)\left(\sqrt{3}\right)} \\ &= \sqrt{3}\end{aligned}$$

15.

$$\begin{aligned}\sin\frac{11\pi}{12} &= \sin\left(\frac{\pi}{6}+\frac{3\pi}{4}\right) \\ &= \sin\frac{\pi}{6}\cos\frac{3\pi}{4} + \cos\frac{\pi}{6}\sin\frac{3\pi}{4} \\ &= \left(\frac{1}{2}\right)\left(-\frac{\sqrt{2}}{2}\right) + \left(\frac{\sqrt{3}}{2}\right)\left(\frac{\sqrt{2}}{2}\right) \\ &= \frac{-\sqrt{2}+\sqrt{6}}{4}\end{aligned}$$

17.

$$\begin{aligned}\cos\frac{7\pi}{12} &= \cos\left(\frac{5\pi}{6}-\frac{\pi}{4}\right) \\ &= \cos\frac{5\pi}{6}\cos\frac{\pi}{4} + \sin\frac{5\pi}{6}\sin\frac{\pi}{4} \\ &= \left(-\frac{\sqrt{3}}{2}\right)\left(\frac{\sqrt{2}}{2}\right) + \left(\frac{1}{2}\right)\left(\frac{\sqrt{2}}{2}\right) \\ &= \frac{-\sqrt{6}+\sqrt{2}}{4}\end{aligned}$$

19.

$$\begin{aligned}\sin\frac{7\pi}{6} &= \sin\left(\pi+\frac{\pi}{6}\right) \\ &= \sin\pi\cos\frac{\pi}{6} + \cos\pi\sin\frac{\pi}{6} \\ &= (0)\left(\frac{\sqrt{3}}{2}\right) + (-1)\left(\frac{1}{2}\right) \\ &= -\frac{1}{2}\end{aligned}$$

21.

$$\tan 15° = \tan(45° - 30°)$$
$$= \frac{\tan 45° - \tan 30°}{1 + \tan 45° \tan 30°}$$
$$= \frac{1 - \frac{\sqrt{3}}{3}}{1 + (1)\left(\frac{\sqrt{3}}{3}\right)}$$
$$= \frac{3 - \sqrt{3}}{3 + \sqrt{3}}$$
$$= 2 - \sqrt{3}$$

23.

$$\sin 47° = \cos(90° - 47°) = \cos 43°$$

25.

$$\tan\frac{\pi}{6} = \cot\left(\frac{\pi}{2} - \frac{\pi}{6}\right) = \cot\frac{\pi}{3}$$

27.

$$\cos\frac{\pi}{3} = \sin\left(\frac{\pi}{2} - \frac{\pi}{3}\right) = \sin\frac{\pi}{6}$$

29.

$$\sin\left(\frac{\pi}{2} - t\right) = \cos t$$
$$\cos^2 t = 1 - \sin^2 t$$
$$= 1 - \left(-\frac{3}{5}\right)^2$$
$$= \frac{16}{25}$$
$$\cos t = -\frac{4}{5} \qquad \text{(Quadrant II)}$$

Hence $\sin\left(\frac{\pi}{2} - t\right) = -\frac{4}{5}$

31.

$$\tan\left(\theta + \frac{\pi}{4}\right) = \frac{\tan\theta + \tan\frac{\pi}{4}}{1 - (\tan\theta)\left(\tan\frac{\pi}{4}\right)}$$
$$= \frac{\frac{4}{3} + 1}{1 - \left(\frac{4}{3}\right)(1)}$$
$$= -7$$

33.

$$\begin{aligned}
\sin^2 t &= 1-\cos^2 t \\
&= 1-(0.4)^2 \\
&= 0.84
\end{aligned}$$

$$\sin t = -0.92 \qquad \text{(Quadrant IV)}$$

$$\tan t = \frac{\sin t}{\cos t} = \frac{-0.92}{0.4} = -2.3$$

$$\begin{aligned}
\tan(t+\pi) &= \frac{\tan t + \tan\pi}{1-\tan t\tan\pi} \\
&= \frac{-2.3+0}{1-(-2.3)(0)} \\
&= -2.3
\end{aligned}$$

35.

$$\begin{aligned}
\cos^2 s &= 1-\sin^2 s \\
&= 1-\left(\frac{3}{5}\right)^2 \\
&= \frac{16}{25}
\end{aligned}$$

$$\cos s = -\frac{4}{5} \qquad \text{(Quadrant II)}$$

$$\begin{aligned}
\sin^2 t &= 1-\cos^2 t \\
&= 1-\left(-\frac{12}{13}\right)^2 \\
&= \frac{25}{169}
\end{aligned}$$

$$\sin t = -\frac{5}{13} \qquad \text{(Quadrant III)}$$

$$\begin{aligned}
\sin(s+t) &= \sin s\cos t + \cos s\,\sin t \\
&= \left(\frac{3}{5}\right)\left(-\frac{12}{13}\right)+\left(-\frac{4}{5}\right)\left(-\frac{5}{13}\right) \\
&= -\frac{16}{65}
\end{aligned}$$

37.

$$\sec\alpha = \frac{1}{\cos\alpha} = \frac{1}{\frac{5}{13}} = \frac{13}{5}$$

$$\begin{aligned}\tan^2\alpha &= \sec^2\alpha - 1\\ &= \left(\frac{13}{5}\right)^2 - 1\\ &= \frac{144}{25}\end{aligned}$$

$$\tan\alpha = \frac{12}{5} \qquad \text{(Quadrant I)}$$

$$\begin{aligned}\tan(\alpha+\beta) &= \frac{\tan\alpha + \tan\beta}{1 - \tan\alpha\tan\beta}\\ &= \frac{\frac{12}{5} + (-2)}{1 - \left(\frac{12}{5}\right)(-2)}\\ &= \frac{2}{29}\end{aligned}$$

39.

$$\begin{aligned}\sin 2\alpha &= \sin(\alpha+\alpha)\\ &= \sin\alpha\cos\alpha + \cos\alpha\sin\alpha\\ &= 2\sin\alpha\cos\alpha\end{aligned}$$

41.

$$\begin{aligned}\tan 2\alpha = \tan(\alpha+\alpha) &= \frac{\tan\alpha + \tan\alpha}{1 - \tan\alpha\tan\alpha}\\ &= \frac{2\tan\alpha}{1 - \tan^2\alpha}\end{aligned}$$

43.

$$\begin{aligned}\cos(x-y)\cos(x+y) &= (\cos x\cos y + \sin x\sin y)(\cos x\cos y - \sin x\sin y)\\ &= \cos^2 x\cos^2 y - \sin^2 x\sin^2 y\end{aligned}$$

45.

$$\begin{aligned}\csc\left(t+\frac{\pi}{2}\right)&=\frac{1}{\sin\left(t+\frac{\pi}{2}\right)}\\&=\frac{1}{\sin t\cos\frac{\pi}{2}+\cos t\sin\frac{\pi}{2}}\\&=\frac{1}{\cos t}\\&=\sec t\end{aligned}$$

47.

$$\begin{aligned}\tan\left(x+\frac{\pi}{4}\right)&=\frac{\tan x+\tan\frac{\pi}{4}}{1-\tan x\tan\frac{\pi}{4}}\\&=\frac{\tan x+1}{1-\tan x}\\&=\frac{1+\tan x}{1-\tan x}\end{aligned}$$

49.

$$\begin{aligned}\cot(s-t)&=\frac{1}{\tan(s-t)}\\&=\frac{1}{\frac{\tan s-\tan t}{1+\tan s\tan t}}\\&=\frac{1+\tan s\tan t}{\tan s-\tan t}\end{aligned}$$

51.

$$\begin{aligned}\sin(s+t)+\sin(s-t)&=\sin s\cos t+\cos s\sin t+\sin s\cos t-\cos s\sin t\\&=2\sin s\cos t\end{aligned}$$

53.

$$\begin{aligned}\frac{\sin(x+h)-\sin x}{h}&=\frac{\sin x\cos h+\cos x\sin h-\sin x}{h}\\&=\frac{(\sin x\cos h-\sin x)}{h}+\frac{\cos x\sin h}{h}\\&=\frac{\sin x(\cos h-1)}{h}+\frac{\cos x\sin h}{h}\\&=\sin x\left(\frac{\cos h-1}{h}\right)+\cos x\left(\frac{\sin h}{h}\right)\end{aligned}$$

CHAPTER 8 SECTION 3

SECTION 8.3 PROGRESS CHECK

1. (Page 493)

$$\cos^2\theta = 1-\sin^2\theta$$
$$= 1-\left(\frac{5}{13}\right)^2$$
$$= \frac{144}{169}$$
$$\cos\theta = \frac{12}{13} \quad \text{(Quadrant I)}$$

$$\sin 2\theta = 2\sin\theta\cos\theta$$
$$= 2\left(\frac{5}{13}\right)\left(\frac{12}{13}\right)$$
$$= \frac{120}{169}$$

$$\cos 2\theta = \cos^2\theta - \sin^2\theta$$
$$= \left(\frac{12}{13}\right)^2 - \left(\frac{5}{13}\right)^2$$
$$= \frac{119}{169}$$

$$\tan 2\theta = \frac{\sin 2\theta}{\cos 2\theta} = \frac{\frac{120}{169}}{\frac{119}{169}} = \frac{120}{119}$$

2. (Page 493)

$$\cos 3t = \cos(2t+t)$$
$$= \cos 2t\cos t - \sin 2t\sin t$$
$$= (\cos^2 t - \sin^2 t)\cos t - (2\sin t\cos t)\sin t$$
$$= \cos^3 t - \sin^2 t\cos t - 2\sin^2 t\cos t$$
$$= \cos^3 t - 3\sin^2 t\cos t$$
$$= \cos^3 t - 3(1-\cos^2 t)\cos t$$
$$= \cos^3 t - 3\cos t + 3\cos^3 t$$
$$= 4\cos^3 t - 3\cos t$$

3. (Page 494)

$$\frac{1+\cos 2\theta}{\sin 2\theta} = \frac{1+2\cos^2\theta - 1}{2\sin\theta\cos\theta} = \frac{2\cos^2 t}{2\sin\theta\cos\theta} = \frac{\cos\theta}{\sin\theta} = \cot\theta$$

4. (Page 496)

$$\tan\frac{3\pi}{8} = \tan\frac{\frac{3\pi}{4}}{2}$$

$$= \sqrt{\frac{1-\cos\frac{3\pi}{4}}{1+\cos\frac{3\pi}{4}}} \quad \text{(Quadrant I)}$$

$$= \sqrt{\frac{1-\left(-\frac{1}{\sqrt{2}}\right)}{1+\left(-\frac{1}{\sqrt{2}}\right)}}$$

$$= \frac{\sqrt{1+\frac{1}{\sqrt{2}}}}{\sqrt{1-\frac{1}{\sqrt{2}}}} \cdot \frac{\sqrt{1+\frac{1}{\sqrt{2}}}}{\sqrt{1+\frac{1}{\sqrt{2}}}}$$

$$= \frac{1+\frac{1}{\sqrt{2}}}{\sqrt{1-\frac{1}{2}}}$$

$$= \frac{1+\frac{1}{\sqrt{2}}}{\frac{1}{\sqrt{2}}}$$

$$= \sqrt{2}+1$$

5. (Page 497)

$$\sec^2\alpha = 1+\tan^2\alpha$$

$$= 1+\left(\frac{3}{4}\right)^2$$

$$= \frac{25}{16}$$

$$\sec\alpha = -\frac{5}{4} \quad \text{(Quadrant III)}$$

$$\cos\alpha = \frac{1}{\sec\alpha} = \frac{1}{-\frac{5}{4}} = -\frac{4}{5}$$

$$\tan\frac{\alpha}{2} = -\sqrt{\frac{1-\cos\alpha}{1+\cos\alpha}}$$

$$= -\sqrt{\frac{1-\left(-\frac{4}{5}\right)}{1+\left(-\frac{4}{5}\right)}} = -\sqrt{9} = -3$$

(Quadrant II)

EXERCISE SET 8.3

1.

$$\cos^2 u = 1 - \sin^2 u$$
$$= 1 - \left(\frac{3}{5}\right)^2$$
$$= \frac{16}{25}$$

$$\cos 2u = \cos^2 u - \sin^2 u$$
$$= \frac{16}{25} - \left(\frac{3}{5}\right)^2$$
$$= \frac{7}{25}$$

3.

$$\cos\alpha = \frac{1}{\sec\alpha} = \frac{1}{-2} = -\frac{1}{2}$$

$$\sin^2 \alpha = 1 - \cos^2 \alpha$$
$$= 1 - \left(-\frac{1}{2}\right)^2$$
$$= \frac{3}{4}$$
$$\sin\alpha = \frac{\sqrt{3}}{2} \qquad \text{(Quadrant II)}$$

$$\sin 2\alpha = 2\sin\alpha\cos\alpha$$
$$= 2\left(\frac{\sqrt{3}}{2}\right)\left(-\frac{1}{2}\right)$$
$$= -\frac{\sqrt{3}}{2}$$

5.

$$\sin t = \frac{1}{\csc t} = \frac{1}{-\frac{17}{8}} = -\frac{8}{17}$$

$$\begin{aligned}\cos^2 t &= 1 - \sin^2 t \\ &= 1 - \left(-\frac{8}{17}\right)^2 \\ &= \frac{225}{289} \\ \cos t &= \frac{15}{17} \quad \text{(Quadrant IV)}\end{aligned}$$

$$\tan t = \frac{\sin t}{\cos t} = \frac{-\frac{8}{17}}{\frac{15}{17}} = -\frac{8}{15}$$

$$\begin{aligned}\tan 2t &= \frac{2\tan t}{1 - \tan^2 t} \\ &= \frac{2\left(-\frac{8}{15}\right)}{1 - \left(-\frac{8}{15}\right)^2} \\ &= -\frac{240}{161}\end{aligned}$$

7.

$$\begin{aligned}\cos^2 2\alpha &= 1 - \sin^2 2\alpha \\ &= 1 - \left(-\frac{4}{5}\right)^2 \\ &= \frac{9}{25} \\ \cos 2\alpha &= \frac{3}{5} \quad \text{(Quadrant IV)}\end{aligned}$$

$$\begin{aligned}\sin 4\alpha &= 2\sin 2\alpha \cos 2\alpha \\ &= 2\left(-\frac{4}{5}\right)\left(\frac{3}{5}\right) \\ &= -\frac{24}{25}\end{aligned}$$

9.

$$\begin{aligned}\cos\frac{\theta}{2} &= \sqrt{\frac{1+\cos\theta}{2}} \\ \frac{8}{17} &= \sqrt{\frac{1+\cos\theta}{2}} \\ \frac{64}{289} &= \frac{1+\cos\theta}{2} \\ 128 &= 289 + 289\cos\theta \\ \cos\theta &= -\frac{161}{289}\end{aligned}$$

11.

$$\sin 42° = \sin\frac{84°}{2} = \sqrt{\frac{1-\cos 84°}{2}} \quad \text{(Quadrant I)}$$

$$\begin{aligned}0.67 &= \sqrt{\frac{1-\cos 84°}{2}} \\ 0.4489 &= \frac{1-\cos 84°}{2} \\ 0.8978 &= 1 - \cos 84° \\ \cos 84° &= 0.1022\end{aligned}$$

13.

$$\sin 15° = \sin\frac{30°}{2} = \sqrt{\frac{1-\cos 30°}{2}}$$

$$= \sqrt{\frac{1-\frac{\sqrt{3}}{2}}{2}}$$

$$= \sqrt{\frac{2-\sqrt{3}}{4}}$$

$$= \frac{\sqrt{2-\sqrt{3}}}{2}$$

15.

$$\tan\frac{\pi}{8} = \tan\frac{\frac{\pi}{4}}{2} = \sqrt{\frac{1-\cos\frac{\pi}{4}}{1+\cos\frac{\pi}{4}}}$$

$$= \sqrt{\frac{1-\frac{1}{\sqrt{2}}}{1+\frac{1}{\sqrt{2}}}}$$

$$= \sqrt{\frac{\sqrt{2}-1}{\sqrt{2}+1}}$$

$$= \frac{\sqrt{\sqrt{2}-1}}{\sqrt{\sqrt{2}+1}}$$

$$= \frac{\sqrt{\sqrt{2}-1}\cdot\sqrt{\sqrt{2}-1}}{\sqrt{\sqrt{2}+1}\cdot\sqrt{\sqrt{2}-1}}$$

$$= \frac{\left(\sqrt{\sqrt{2}-1}\right)^2}{\sqrt{\left(\sqrt{2}\right)^2-1^2}}$$

$$= \frac{\sqrt{2}-1}{\sqrt{2-1}}$$

$$= \frac{\sqrt{2}-1}{\sqrt{1}}$$

$$= \sqrt{2}-1$$

17.

$$\begin{aligned}
\csc 165^\circ &= \frac{1}{\sin 165^\circ} \\
&= \frac{1}{\sin \frac{330^\circ}{2}} \\
&= \frac{1}{\sqrt{\frac{1-\cos 330^\circ}{2}}} \qquad \text{(Quadrant II)} \\
&= \frac{1}{\sqrt{\frac{1-\frac{\sqrt{3}}{2}}{2}}} \\
&= \frac{1}{\sqrt{\frac{2-\sqrt{3}}{4}}} \\
&= \frac{2}{\sqrt{2-\sqrt{3}}} \\
&= \frac{2\left(\sqrt{2+\sqrt{3}}\right)}{\left(\sqrt{2-\sqrt{3}}\right)\left(\sqrt{2+\sqrt{3}}\right)} \\
&= \frac{2\left(\sqrt{2+\sqrt{3}}\right)}{\sqrt{\left(2-\sqrt{3}\right)\left(2+\sqrt{3}\right)}} \\
&= \frac{2\left(\sqrt{2+\sqrt{3}}\right)}{\sqrt{2^2-\left(\sqrt{3}\right)^2}} \\
&= \frac{2\left(\sqrt{2+\sqrt{3}}\right)}{\sqrt{4-3}} \\
&= \frac{2\left(\sqrt{2+\sqrt{3}}\right)}{\sqrt{1}} \\
&= 2\sqrt{2+\sqrt{3}}
\end{aligned}$$

19.

$$\begin{aligned}
\cos^2\theta &= 1-\sin^2\theta \\
&= 1-\left(-\frac{4}{5}\right)^2 \\
&= \frac{9}{25} \\
\cos\theta &= \frac{3}{5} \qquad \text{(Quadrant IV)} \\
\cos\frac{\theta}{2} &= \sqrt{\frac{1+\cos\theta}{2}} \\
&= -\sqrt{\frac{1+\frac{3}{5}}{2}} \qquad \text{(Quadrant II)} \\
&= -\frac{2\sqrt{5}}{5}
\end{aligned}$$

29.

$$\begin{aligned}\tan 2y &= \frac{2\tan y}{1-\tan^2 y}\\ &= \frac{\dfrac{2}{\cot y}}{1-\dfrac{1}{\cot^2 y}}\cdot\frac{\cot^2 y}{\cot^2 y}\\ &= \frac{2\cot y}{\cot^2 y-1}\\ &= \frac{2\cot y}{\left(\csc^2 y-1\right)-1}\\ &= \frac{2\cot y}{\csc^2 y-2}\end{aligned}$$

31.

$$\begin{aligned}4\sin\alpha\cos^3\alpha-4\sin^3\alpha\cos\alpha &= 4\sin\alpha\cos\alpha\left(\cos^2\alpha-\sin^2\alpha\right)\\ &= 2\sin 2\alpha(\cos 2\alpha)\\ &= \sin 4\alpha\end{aligned}$$

33.

$$\begin{aligned}\frac{1-\tan^2 u}{1+\tan^2 u} &= \frac{1-\dfrac{\sin^2 u}{\cos^2 u}}{1+\dfrac{\sin^2 u}{\cos^2 u}}\\ &= \frac{\cos^2 u-\sin^2 u}{\cos^2 u+\sin^2 u}\\ &= \frac{\cos 2u}{1}\\ &= \cos 2u\end{aligned}$$

35.

$$\begin{aligned}\frac{\sin t}{2} &= \frac{1}{2}\sin\left(\frac{t}{2}+\frac{t}{2}\right)\\ &= \frac{1}{2}\left(\sin\frac{t}{2}\cos\frac{t}{2}+\cos\frac{t}{2}\sin\frac{t}{2}\right)\\ &= \sin\frac{t}{2}\cos\frac{t}{2}\end{aligned}$$

37.

$$\sin\alpha - \cos\alpha\tan\frac{\alpha}{2} = \sin\left(2\left(\frac{\alpha}{2}\right)\right) - \cos\left(2\left(\frac{\alpha}{2}\right)\right)\tan\frac{\alpha}{2}$$

$$= 2\sin\frac{\alpha}{2}\cos\frac{\alpha}{2} - \left(\cos^2\frac{\alpha}{2} - \sin^2\frac{\alpha}{2}\right)\frac{\sin\frac{\alpha}{2}}{\cos\frac{\alpha}{2}}$$

$$= 2\sin\frac{\alpha}{2}\cos\frac{\alpha}{2} - \sin\frac{\alpha}{2}\cos\frac{\alpha}{2} + \frac{\sin^3\frac{\alpha}{2}}{\cos\frac{\alpha}{2}}$$

$$= \sin\frac{\alpha}{2}\cos\frac{\alpha}{2} + \frac{\sin^3\frac{\alpha}{2}}{\cos\frac{\alpha}{2}}$$

$$= \frac{\sin\frac{\alpha}{2}\cos^2\frac{\alpha}{2} + \sin^3\frac{\alpha}{2}}{\cos\frac{\alpha}{2}}$$

$$= \frac{\sin\frac{\alpha}{2}\left(\cos^2\frac{\alpha}{2} + \sin^2\frac{\alpha}{2}\right)}{\cos\frac{\alpha}{2}}$$

$$= \frac{\sin\frac{\alpha}{2}(1)}{\cos\frac{\alpha}{2}}$$

$$= \tan\frac{\alpha}{2}$$

39.

$$\cos^4 x - \sin^4 x = \left(\cos^2 x - \sin^2 x\right)\left(\cos^2 x + \sin^2 x\right)$$

$$= (\cos 2x)(1)$$

$$= \cos 2x$$

41.

$$\frac{2\tan\alpha}{1+\tan^2\alpha}=\frac{\frac{2\sin\alpha}{\cos\alpha}}{1+\frac{\sin^2\alpha}{\cos^2\alpha}}\cdot\frac{\cos^2\alpha}{\cos^2\alpha}$$

$$=\frac{2\sin\alpha\cos\alpha}{\cos^2\alpha+\sin^2\alpha}$$

$$=\frac{\sin 2\alpha}{1}$$

$$=\sin 2\alpha$$

43.

$$\frac{\sec^2 t}{2-\sec^2 t}=\frac{\frac{1}{\cos^2 t}}{2-\frac{1}{\cos^2 t}}$$

$$=\frac{1}{2\cos^2 t-1}$$

$$=\frac{1}{2\cos^2 t-(\sin^2 t+\cos^2 t)}$$

$$=\frac{1}{\cos^2 t-\sin^2 t}$$

$$=\frac{1}{\cos 2t}$$

$$=\sec 2t$$

45.

$$\frac{1-\cos t}{\sin t}=\frac{2\sin^2\frac{t}{2}}{2\sin\frac{t}{2}\cos\frac{t}{2}}$$

$$=\frac{\sin\frac{t}{2}}{\cos\frac{t}{2}}$$

$$=\tan\frac{t}{2}$$

47.

Let $\theta=\text{arc}\cos\frac{3}{5}$, then $\cos\theta=\frac{3}{5}$

$$\sin^2\theta=1-\cos^2\theta$$

$$=1-\left(\frac{3}{5}\right)^2$$

$$=\frac{16}{25}$$

$$\sin\theta=\frac{4}{5}$$

$$\sin 2\theta=2\sin\theta\cos\theta$$

$$=2\left(\frac{4}{5}\right)\left(\frac{3}{5}\right)$$

$$=\frac{24}{25}$$

49.

Let $\theta = \arcsin\frac{5}{13}$, then $\sin\theta = \frac{5}{13}$

$$\begin{aligned}\cos^2\theta &= 1-\sin^2\theta \\ &= 1-\left(\frac{5}{13}\right)^2 \\ &= \frac{144}{169}\end{aligned}$$

$$\cos\theta = \frac{12}{13}$$

$$\tan\theta = \frac{\sin\theta}{\cos\theta} = \frac{\frac{5}{13}}{\frac{12}{13}} = \frac{5}{12}$$

$$\begin{aligned}\tan 2\theta &= \frac{2\tan\theta}{1-\tan^2\theta} \\ &= \frac{2\left(\frac{5}{12}\right)}{1-\left(\frac{5}{12}\right)^2} \\ &= \frac{120}{119}\end{aligned}$$

CHAPTER 8 SECTION 4

SECTION 8.4 PROGRESS CHECK

1. (Page 499)

$$\sin 5x \sin 2x = \frac{\cos(5x-2x)-\cos(5x+2x)}{2} = \frac{\cos 3x - \cos 7x}{2}$$

2. (Page 500)

$$\cos\frac{\pi}{3}\sin\frac{\pi}{6} = \frac{\sin\left(\frac{\pi}{3}+\frac{\pi}{6}\right)-\sin\left(\frac{\pi}{3}-\frac{\pi}{6}\right)}{2} = \frac{\sin\frac{\pi}{2}-\sin\frac{\pi}{6}}{2} = \frac{1-\frac{1}{2}}{2} = \frac{1}{4}$$

3. (Page 500)

$$\cos 6x + \cos 2x = 2\cos\frac{6x+2x}{2}\cos\frac{6x-2x}{2} = 2\cos 4x \cos 2x$$

4. (Page 501)

$$\sin\frac{11\pi}{12} - \sin\frac{5\pi}{12} = 2\cos\frac{\frac{11\pi}{12}+\frac{5\pi}{12}}{2}\sin\frac{\frac{11\pi}{12}-\frac{5\pi}{12}}{2} = 2\cos\frac{2\pi}{3}\sin\frac{\pi}{4} = 2\left(-\frac{1}{2}\right)\left(\frac{\sqrt{2}}{2}\right) = -\frac{\sqrt{2}}{2}$$

EXERCISE SET 8.4

1.

$$2\sin 5\alpha \cos\alpha = 2\left[\frac{\sin(5\alpha+\alpha)+\sin(5\alpha-\alpha)}{2}\right] = \sin 6\alpha + \sin 4\alpha$$

3.

$$\sin 3x \sin(-2x) = \frac{\cos(3x-(-2x))-\cos(3x-2x)}{2}$$
$$= \frac{\cos 5x - \cos x}{2}$$

5.

$$-2\cos 2\theta \cos 5\theta = -2\left[\frac{\cos(2\theta+5\theta)+\cos(2\theta-5\theta)}{2}\right]$$
$$= -(\cos 7\theta + \cos(-3\theta))$$
$$= -(\cos 7\theta + \cos 3\theta)$$

7.

$$\cos(\alpha+\beta)\cos(\alpha-\beta) = \frac{\cos(\alpha+\beta+\alpha-\beta)+\cos(\alpha+\beta-(\alpha-\beta))}{2}$$
$$= \frac{\cos 2\alpha + \cos 2\beta}{2}$$

9.

$$\cos\frac{7\pi}{8}\sin\frac{5\pi}{8} = \frac{\sin\left(\frac{7\pi}{8}+\frac{5\pi}{8}\right)-\sin\left(\frac{7\pi}{8}-\frac{5\pi}{8}\right)}{2}$$
$$= \frac{\sin\frac{3\pi}{2}-\sin\frac{\pi}{4}}{2}$$
$$= \frac{-1-\frac{\sqrt{2}}{2}}{2}$$
$$= \frac{-2-\sqrt{2}}{4}$$

11.

$$\begin{aligned}\sin 120^\circ \cos 60^\circ &= \frac{\sin(120^\circ+60^\circ)+\sin(120^\circ-60^\circ)}{2}\\ &= \frac{\sin 180^\circ + \sin 60^\circ}{2}\\ &= \frac{0+\frac{\sqrt{3}}{2}}{2}\\ &= \frac{\sqrt{3}}{4}\end{aligned}$$

13.

$$\begin{aligned}\sin 5x + \sin x &= 2\sin\frac{5x+x}{2}\cos\frac{5x-x}{2}\\ &= 2\sin 3x\cos 2x\end{aligned}$$

15.

$$\begin{aligned}\cos 2\theta + \cos 6\theta &= 2\cos\frac{2\theta+6\theta}{2}\cos\frac{2\theta-6\theta}{2}\\ &= 2\cos 4\theta\cos(-2\theta)\\ &= 2\cos 4\theta\cos 2\theta\end{aligned}$$

17.

$$\begin{aligned}\sin(\alpha+\beta)+\sin(\alpha-\beta) &= 2\sin\frac{\alpha+\beta+\alpha-\beta}{2}\cos\frac{\alpha+\beta-(\alpha-\beta)}{2}\\ &= 2\sin\alpha\cos\beta\end{aligned}$$

19.

$$\begin{aligned}\sin(7x)-\sin(3x) &= 2\cos\frac{7x+3x}{2}\sin\frac{7x-3x}{2}\\ &= 2\cos(5x)\sin(2x)\end{aligned}$$

21.

$$\begin{aligned}\cos 75^\circ + \cos 15^\circ &= 2\cos\frac{75^\circ+15^\circ}{2}\cos\frac{75^\circ-15^\circ}{2}\\ &= 2\cos 45^\circ\cos 30^\circ\\ &= 2\left(\frac{\sqrt{2}}{2}\right)\left(\frac{\sqrt{3}}{2}\right)\\ &= \frac{\sqrt{6}}{2}\end{aligned}$$

23.

$$\cos\frac{3\pi}{4} - \cos\frac{\pi}{4} = -2\sin\frac{\frac{3\pi}{4}+\frac{\pi}{4}}{2}\sin\frac{\frac{3\pi}{4}-\frac{\pi}{4}}{2}$$

$$= -2\sin\frac{\pi}{2}\sin\frac{\pi}{4}$$

$$= -2(1)\left(\frac{\sqrt{2}}{2}\right)$$

$$= -\sqrt{2}$$

25.

$$\sin 40° + \sin 20° = 2\sin\frac{40°+20°}{2}\cos\frac{40°-20°}{2}$$

$$= 2\sin 30°\cos 10°$$

$$= 2\left(\frac{1}{2}\right)\cos 10°$$

$$= \cos 10°$$

27.

$$\frac{\sin 5\theta - \sin 3\theta}{\cos 3\theta - \cos 5\theta} = \frac{2\cos\frac{5\theta+3\theta}{2}\cdot\sin\frac{5\theta-3\theta}{2}}{-2\sin\frac{3\theta+5\theta}{2}\cdot\sin\frac{3\theta-5\theta}{2}}$$

$$= -\frac{\cos 4\theta\sin\theta}{\sin 4\theta\sin(-\theta)}$$

$$= \frac{\cos 4\theta\sin\theta}{\sin 4\theta\sin\theta}$$

$$= \cot 4\theta$$

29.

$$\frac{\sin t - \sin s}{\cos t - \cos s} = \frac{2\cos\frac{t+s}{2}\cdot\sin\frac{t-s}{2}}{-2\sin\frac{t+s}{2}\cdot\sin\frac{t-s}{2}}$$

$$= -\cot\frac{t+s}{2}$$

31.

$$\frac{\sin 50° - \sin 10°}{\cos 50° - \cos 10°} = \frac{2\cos\frac{50°+10°}{2}\cdot\sin\frac{50°-10°}{2}}{-2\sin\frac{50°+10°}{2}\cdot\sin\frac{50°-10°}{2}}$$

$$= -\frac{\cos 30°}{\sin 30°}$$

$$= -\frac{\frac{\sqrt{3}}{2}}{\frac{1}{2}}$$

$$= -\sqrt{3}$$

33.

$$\frac{\cot x - \tan x}{\cot x + \tan x} = \frac{\frac{1}{\tan x} - \tan x}{\frac{1}{\tan x} + \tan x}$$

$$= \frac{1-\tan^2 x}{1+\tan^2 x}$$

$$= \frac{1-\tan^2 x}{\sec^2 x}$$

$$= \cos^2 x\left(1-\frac{\sin^2 x}{\cos^2 x}\right)$$

$$= \cos^2 x - \sin^2 x$$

$$= \cos 2x$$

35.

$$\sin ax \cos bx = \frac{\sin(ax+bx)+\sin(ax-bx)}{2}$$
$$= \frac{\sin(a+b)x+\sin(a-b)x}{2}$$

37.

Proof of (2):

$$\frac{\sin(s+t)-\sin(s-t)}{2} = \frac{\sin s\cos t+\cos s\sin t-(\sin s\cos t-\cos s\sin t)}{2}$$
$$= \frac{\sin s\cos t+\cos s\sin t-\sin s\cos t+\cos s\sin t}{2}$$
$$= \cos s\sin t$$

Proof of (3):

$$\frac{\cos(s+t)+\cos(s-t)}{2} = \frac{\cos s\cos t-\sin s\sin t+\cos s\cos t+\sin s\sin t}{2}$$
$$= \cos s\cos t$$

Proof of (4):

$$\frac{\cos(s-t)-\cos(s+t)}{2} = \frac{\cos s\cos t+\sin s\sin t-(\cos s\cos t-\sin s\sin t)}{2}$$
$$= \frac{\cos s\cos t+\sin s\sin t-\cos s\cos t+\sin s\sin t}{2}$$
$$= \sin s\sin t$$

CHAPTER 8 SECTION 5

SECTION 8.5 PROGRESS CHECK

1. (Page 503)

$$2\sin^2 t - 3\sin t + 1 = 0 \qquad [0, 2\pi]$$

Let $x - \sin t$

$$2x^2 - 3x + 1 = 0$$

$$(2x-1)(x-1) = 0$$

$$x = \frac{1}{2} \qquad x = 1$$

$$\sin t = \frac{1}{2} \quad \sin t = 1$$

$$t = \frac{\pi}{6} \qquad t = \frac{\pi}{2}$$

$$t = \pi - \frac{\pi}{6} = \frac{5\pi}{6}$$

$$t = \frac{\pi}{6}, \frac{5\pi}{6}, \frac{\pi}{2}$$

2. (Page 504)

$$\cos 2\theta + \cos\theta = 0$$

$$2\cos^2\theta - 1 + \cos\theta = 0$$

Let $x - \cos\theta$

$$2x^2 - 1 + x = 0$$

$$2x^2 + x - 1 = 0$$

$$(2x-1)(x+1) = 0$$

$$x = \frac{1}{2} \qquad x = -1$$

$$\cos\theta = \frac{1}{2} \quad \cos\theta = -1$$

$$\theta = \frac{\pi}{3} + 2\pi n \qquad \theta = \pi + 2\pi n$$

$$\theta = \left(2\pi - \frac{\pi}{3}\right)2\pi n = \frac{5\pi}{3} + 2\pi n$$

$$\theta = \frac{\pi}{3} + 2\pi n, \ \frac{5\pi}{3} + 2\pi n, \ \pi + 2\pi n$$

or $\theta = 60° + 360° n, \ 300° + 360° n, \ 180° + 360° n$

EXERCISE SET 8.5

1.

$$2\sin\theta - 1 = 0$$

$$\sin\theta = \frac{1}{2}$$

$$\theta = \frac{\pi}{6} \text{ or } \theta = \pi - \frac{\pi}{6} = \frac{5\pi}{6}$$

$$\theta = \frac{\pi}{6}, \frac{5\pi}{6}$$

or $\theta = 30°, \ 150°$

3.

$$\cos\alpha + 1 = 0$$

$$\cos\alpha = -1$$

$$\alpha = \pi$$

or $\alpha = 180°$

5.

$$4\cos^2\alpha = 3$$
$$\cos^2\alpha = \frac{3}{4}$$
$$\cos\alpha = \pm\frac{\sqrt{3}}{2}$$
$$\alpha = \frac{\pi}{6}, \frac{5\pi}{6}, \frac{7\pi}{6}, \frac{11\pi}{6}$$
or $\alpha = 30°,\ 150°,\ 210°,\ 330°$

7.

$$3\tan^2\alpha = 1$$
$$\tan^2\alpha = \frac{1}{3}$$
$$\tan\alpha = \pm\frac{1}{\sqrt{3}}$$
$$\alpha = \frac{\pi}{6}, \frac{5\pi}{6}, \frac{7\pi}{6}, \frac{11\pi}{6}$$
or $\alpha = 30°,\ 150°,\ 210°,\ 330°$

9.

$$2\sin^2\beta = \sin\beta$$
$$2\sin^2\beta - \sin\beta = 0$$
$$\sin\beta(2\sin\beta - 1) = 0$$

$\sin\beta = 0$ $2\sin\beta - 1 = 0$

$\beta = 0, \pi$ $\sin\beta = \frac{1}{2}$

or $\beta = 0°,\ 180°$ $\beta = \frac{\pi}{6}, \frac{5\pi}{6}$

or $\beta = 30°,\ 150°$

11.

$$2\cos^2\theta - 3\cos\theta + 1 = 0$$
Let $x = \cos\theta$
$$2x^2 - 3x + 1 = 0$$
$$(2x - 1)(x - 1) = 0$$

$x = \frac{1}{2}$ $x = 1$

$\cos\theta = \frac{1}{2}$ $\cos\theta = 1$

$\theta = \frac{\pi}{3}, \frac{5\pi}{3}$ $\theta = 0$

or $\theta = 60°, 300°$ or $\theta = 0°$

13.

$$\sin 5\theta = 1$$

$$5\theta = \frac{\pi}{2}, \frac{5\pi}{2}, \frac{9\pi}{2}, \frac{13\pi}{2}, \frac{17\pi}{2}$$

$$\theta = \frac{\pi}{10}, \frac{\pi}{2}, \frac{9\pi}{10}, \frac{13\pi}{10}, \frac{17\pi}{10}$$

or $\theta = 18°, 90°, 162°, 234°, 306°$

15.

$$2\sin^2\alpha - 3\cos\alpha = 0$$

$$2(1-\cos^2\alpha) - 3\cos\alpha = 0$$

$$-2\cos^2\alpha + 2 - 3\cos\alpha = 0$$

$$2\cos^2\alpha + 3\cos\alpha - 2 = 0$$

Let $x = \cos\alpha$

$$2x^2 + 3x - 2 = 0$$

$$(2x-1)(x+2) = 0$$

$x = \frac{1}{2}$ $x = -2$

$\cos\alpha = \frac{1}{2}$ ~~$\cos\alpha = -2$~~

$$\alpha = \frac{\pi}{3}, \frac{5\pi}{3}$$

or $\alpha = 60°, 300°$

17.

$$2\cos^2\theta - 1 = \sin\theta$$

$$2(1-\sin^2\theta) - 1 = \sin\theta$$

$$-2\sin^2\theta + 1 = \sin\theta$$

$$0 = 2\sin^2\theta + \sin\theta - 1$$

Let $x = \sin\theta$

$$0 = 2x^2 + x - 1$$

$$0 = (2x-1)(x+1)$$

$x = \frac{1}{2}$ $x = -1$

$\sin\theta = \frac{1}{2}$ $\sin\theta = -1$

$\theta = \frac{\pi}{6}, \frac{5\pi}{6}$ $\theta = \frac{3\pi}{2}$

or $\theta = 30°, 150°$ or $\theta = 270°$

19.

$$\sin^2\beta + 3\cos\beta - 3 = 0$$

$$1 - \cos^2\beta + 3\cos\beta - 3 = 0$$

$$-\cos^2\beta + 3\cos\beta - 2 = 0$$

Let $x = \cos\beta$

$$-x^2 + 3x - 2 = 0$$

$$x^2 - 3x + 2 = 0$$

$$(x-2)(x-1) = 0$$

$x = 2$ $x = 1$

~~$\cos\beta = 2$~~ $\cos\beta = 1$

$$\beta = 0$$

or $\beta = 0°$

21.

$$3\tan^2 x - 1 = 0$$

$$\tan^2 x = \frac{1}{3}$$

$$\tan x = \pm\frac{1}{\sqrt{3}}$$

$$x = \frac{\pi}{6} \pm \pi n \ , \ \frac{5\pi}{6} \pm \pi n$$

23.

$$3\cot^2\theta - 1 = 0$$

$$\cot^2\theta = \frac{1}{3}$$

$$\cot\theta = \pm\frac{1}{\sqrt{3}}$$

$$\theta = \frac{\pi}{3} + \pi n \ , \ \frac{2\pi}{3} + \pi n$$

25.

$$\sec 2u - 2 = 0$$

$$\sec 2u = 2$$

$$2u = \frac{\pi}{3} + 2\pi n \ , \ \frac{5\pi}{3} + 2\pi n$$

$$u = \frac{\pi}{6} + \pi n \ , \ \frac{5\pi}{6} + \pi n$$

27.

$$\sin 4x = 0$$

$$4x = 0 + \pi n$$

$$x = \frac{\pi n}{4}$$

29.

$$4\cos^2 2t - 3 = 0$$

$$\cos^2 2t = \frac{3}{4}$$

$$\cos 2t = \pm\frac{\sqrt{3}}{2}$$

$$2t = \frac{\pi}{6} + \pi n \ , \ \frac{5\pi}{6} + \pi n$$

$$t = \frac{\pi}{12} + \frac{\pi n}{2} \ , \ \frac{5\pi}{12} + \frac{\pi n}{2}$$

31.

$$\sin 2t + 2\cos t = 0$$
$$2\cos t \sin t + 2\cos t = 0$$
$$2\cos t(\sin t + 1) = 0$$

$2\cos t = 0$ $\qquad$ $\sin t + 1 = 0$

$\cos t = 0$ $\qquad$ $\sin t = -1$

$t = \dfrac{\pi}{2} + \pi n$ $\qquad$ $t = \dfrac{3\pi}{2} + 2\pi n$ (These values are also generated by $t = \dfrac{\pi}{2} + \pi n$.)

$$t = \frac{\pi}{2} + \pi n$$

33.

$$\cos 2t + \sin t = 0$$
$$1 - 2\sin^2 t + \sin t = 0$$

Let $x = \sin t$

$$1 - 2x^2 + x = 0$$
$$2x^2 - x - 1 = 0$$
$$(2x+1)(x-1) = 0$$

$x = -\dfrac{1}{2}$ $\qquad$ $x = 1$

$\sin t = -\dfrac{1}{2}$ $\qquad$ $\sin t = 1$

$t = \dfrac{7\pi}{6} + 2\pi n, \dfrac{11\pi}{6} + 2\pi n$ $\qquad$ $t = \dfrac{\pi}{2} + 2\pi n$

35.

$$\tan^2 x - \tan x = 0$$
$$\tan x(\tan x - 1) = 0$$

$\tan x = 0$ $\qquad$ $\tan x = 1$

$x = 0 + \pi n$ $\qquad$ $x = \dfrac{\pi}{4} + \pi n$

$$x = \pi n \ , \ \frac{\pi}{4} + \pi n$$

37.

$$2\sin^2 x + 3\sin x - 2 = 0$$

Let $u = \sin x$

$$2u^2 + 3u - 2 = 0$$

$$(2u-1)(u+2) = 0$$

$$u = \frac{1}{2} \qquad u = -2$$

$$\sin x = \frac{1}{2} \qquad \sin x = -2 \text{ (crossed out)}$$

$$x = \frac{\pi}{6} + 2\pi n,\ \frac{5\pi}{6} + 2\pi n$$

39.

$$5\sin^2 x - \sin x - 2 = 0$$

Let $u = \sin x$

$$5u^2 - u - 2 = 0$$

$$u = \frac{-(-1) \pm \sqrt{(-1)^2 - 4(5)(-2)}}{2(5)}$$

$$= \frac{1 \pm \sqrt{41}}{10}$$

$$u = \frac{1+\sqrt{41}}{10} = 0.74 \qquad u = \frac{1-\sqrt{41}}{10} = -0.54$$

$\sin x = 0.74$ | $\sin x = -0.54$

$x = 0.83$ | $x = 5.71$

or $x = 2.31$ | or $x = 3.71$

41.

$$3\tan^2 u + 5\tan u + 1 = 0$$

Let $x = \tan u$

$$3x^2 + 5x + 1 = 0$$

$$x = \frac{-5 \pm \sqrt{5^2 - 4(3)(1)}}{2(3)}$$

$$= \frac{-5 \pm \sqrt{13}}{6}$$

$$x = \frac{-5+\sqrt{13}}{6} = -0.2324 \qquad x = \frac{-5-\sqrt{13}}{6} = -1.4343$$

$\tan u = -0.2324$ | $\tan u = -1.4343$

$u \approx 6.05$ or $u \approx 2.91$ | $u \approx 5.32$ or $u \approx 2.18$

43.

$x \approx 0.739085113$

45.

$x \approx 1.428492216$

$x \approx 3.801193214$

$x \approx 4.944776791$

CHAPTER 8 REVIEW EXERCISES

1.

$$\sin\theta\sec\theta+\tan\theta=\sin\theta\left(\frac{1}{\cos\theta}\right)+\tan\theta$$
$$=\tan\theta+\tan\theta$$
$$2\tan\theta$$

2.

$$\frac{\cos^2 x}{1-\sin x}=\frac{\cos^2 x}{1-\sin x}\cdot\frac{1+\sin x}{1+\sin x}$$
$$=\frac{\cos^2 x(1+\sin x)}{1-\sin^2 x}$$
$$=\frac{\cos^2 x(1+\sin x)}{\cos^2 x}$$
$$=1+\sin x$$

3.

$$\sin\alpha+\sin\alpha\cot^2\alpha=\sin\alpha(1+\cot^2\alpha)$$
$$=\sin\alpha(\csc^2\alpha)$$
$$=\sin\alpha\left(\frac{1}{\sin^2\alpha}\right)$$
$$=\frac{1}{\sin\alpha}$$
$$=\csc\alpha$$

4.

$$\sin\left(\frac{\pi}{6}+\frac{\pi}{4}\right)=\sin\frac{\pi}{6}\cos\frac{\pi}{4}+\cos\frac{\pi}{6}\sin\frac{\pi}{4}$$
$$=\left(\frac{1}{2}\right)\left(\frac{\sqrt{2}}{2}\right)+\left(\frac{\sqrt{3}}{2}\right)\left(\frac{\sqrt{2}}{2}\right)$$
$$=\frac{\sqrt{2}+\sqrt{6}}{4}$$

5.

$$\cos(45°+90°)=\cos 45°\cos 90°-\sin 45°\sin 90°$$
$$=\left(\frac{\sqrt{2}}{2}\right)(0)-\left(\frac{\sqrt{2}}{2}\right)(1)$$
$$=-\frac{\sqrt{2}}{2}$$

6.

$$\tan\left(\frac{\pi}{3}+\frac{\pi}{4}\right)=\frac{\tan\frac{\pi}{3}+\tan\frac{\pi}{4}}{1-\tan\frac{\pi}{3}\tan\frac{\pi}{4}}$$
$$=\frac{\sqrt{3}+1}{1-(\sqrt{3})(1)}$$
$$=\frac{\sqrt{3}+1}{1-\sqrt{3}}$$
$$=-2-\sqrt{3}$$

7.

$$\begin{aligned}\sin\frac{7\pi}{12} &= \sin\left(\frac{\pi}{4}+\frac{\pi}{3}\right)\\ &= \sin\frac{\pi}{4}\cos\frac{\pi}{3}+\cos\frac{\pi}{4}\sin\frac{\pi}{3}\\ &= \left(\frac{\sqrt{2}}{2}\right)\left(\frac{1}{2}\right)+\left(\frac{\sqrt{2}}{2}\right)\left(\frac{\sqrt{3}}{2}\right)\\ &= \frac{\sqrt{2}+\sqrt{6}}{4}\end{aligned}$$

8.

$$\csc 15° = \sec(90°-15°) = \sec 75°$$

9.

$$\cos 23° = \sin(90°-23°) = \sin 67°$$

10.

$$\sin\frac{\pi}{8} = \cos\left(\frac{\pi}{2}-\frac{\pi}{8}\right) = \cos\frac{3\pi}{8}$$

11.

$$\tan\frac{2\pi}{7} = \cot\left(\frac{\pi}{2}-\frac{2\pi}{7}\right) = \cot\frac{3\pi}{14}$$

12.

$$\begin{aligned}\sin^2\theta &= 1-\cos^2\theta\\ &= 1-\left(-\frac{12}{13}\right)^2\\ &= \frac{25}{169}\\ \sin\theta &= \frac{5}{13} \qquad \text{(Quadrant II)}\end{aligned}$$

$$\begin{aligned}\sin(\pi-\theta) &= \sin\pi\cos\theta-\cos\pi\sin\theta\\ &= (0)\left(-\frac{12}{13}\right)-(-1)\left(\frac{5}{13}\right)\\ &= \frac{5}{13}\end{aligned}$$

13.

$$\cos\alpha = \frac{1}{\sec\alpha} = \frac{1}{\frac{5}{4}} = \frac{4}{5}$$

$$\sin^2\alpha = 1-\cos^2\alpha$$
$$= 1-\left(\frac{4}{5}\right)^2$$
$$= \frac{9}{25}$$

$$\sin\alpha = -\frac{3}{5} \qquad \text{(Quadrant IV)}$$

$$\csc\left(\alpha+\frac{\pi}{3}\right) = \frac{1}{\sin\left(\alpha+\frac{\pi}{3}\right)}$$
$$= \frac{1}{\sin\alpha\cos\frac{\pi}{3}+\cos\alpha\sin\frac{\pi}{3}}$$
$$= \frac{1}{\left(-\frac{3}{5}\right)\left(\frac{1}{2}\right)+\left(\frac{4}{5}\right)\left(\frac{\sqrt{3}}{2}\right)}$$
$$= \frac{1}{-\frac{3}{10}+\frac{2\sqrt{3}}{5}}$$
$$= \frac{10}{4\sqrt{3}-3}$$
$$= \frac{10\left(4\sqrt{3}+3\right)}{39}$$

14.

$$\cos^2 t = 1-\sin^2 t$$
$$= 1-\left(-\frac{3}{5}\right)^2$$
$$= \frac{16}{25}$$

$$\cos t = -\frac{4}{5} \qquad \text{(Quadrant III)}$$

$$\tan t = \frac{\sin t}{\cos t} = \frac{-\frac{3}{5}}{-\frac{4}{5}} = \frac{3}{4}$$

$$\tan(t+\pi) = \frac{\tan t+\tan\pi}{1-\tan t\tan\pi}$$
$$= \frac{\frac{3}{4}+0}{1-\left(\frac{3}{4}\right)(0)}$$
$$= \frac{3}{4}$$

15.

$$\sin^2\alpha = 1-\cos^2\alpha$$
$$= 1-\left(-\frac{12}{13}\right)^2$$
$$= \frac{25}{169}$$
$$\sin\alpha = \frac{5}{13} \qquad \text{(Quadrant II)}$$
$$\tan\alpha = \frac{\sin\alpha}{\cos\alpha} = \frac{\frac{5}{13}}{-\frac{12}{13}} = -\frac{5}{12}$$
$$\tan(\alpha+\beta) = \frac{\tan\alpha+\tan\beta}{1-\tan\alpha\tan\beta}$$
$$= \frac{-\frac{5}{12}-\frac{5}{2}}{1-\left(-\frac{5}{12}\right)\left(-\frac{5}{2}\right)}$$
$$= 70$$

16.

$$\cos^2 x = 1-\sin^2 x$$
$$= 1-\left(\frac{3}{5}\right)^2$$
$$= \frac{16}{25}$$
$$\cos x = -\frac{4}{5} \qquad \text{(Quadrant II)}$$
$$\sin y = \frac{1}{\csc y} = \frac{1}{\frac{13}{12}} = \frac{12}{13}$$
$$\cos^2 y = 1-\sin^2 y$$
$$= 1-\left(\frac{12}{13}\right)^2$$
$$= \frac{25}{169}$$
$$\cos y = \frac{5}{13} \qquad \text{(Quadrant I)}$$
$$\cos(x-y) = \cos x\cos y+\sin x\sin y$$
$$= \left(-\frac{4}{5}\right)\left(\frac{5}{13}\right)+\left(\frac{3}{5}\right)\left(\frac{12}{13}\right)$$
$$= \frac{16}{65}$$

17.

$$\sin u = \frac{1}{\csc u} = \frac{1}{-\frac{5}{4}} = -\frac{4}{5}$$

$$\begin{aligned}\cos 2u &= 1 - 2\sin^2 u \\ &= 1 - 2\left(-\frac{4}{5}\right)^2 \\ &= -\frac{7}{25}\end{aligned}$$

18.

$$\begin{aligned}\sec^2\alpha &= \tan^2\alpha + 1 \\ &= \left(-\frac{3}{4}\right)^2 + 1 \\ &= \frac{25}{16}\end{aligned}$$

$$\sec\alpha = -\frac{5}{4} \qquad \text{(Quadrant II)}$$

$$\cos\alpha = \frac{1}{\sec\alpha} = \frac{1}{-\frac{5}{4}} = -\frac{4}{5}$$

$$\tan\alpha = \frac{\sin\alpha}{\cos\alpha}$$

$$-\frac{3}{4} = \frac{\sin\alpha}{-\frac{4}{5}}$$

$$\sin\alpha = \frac{3}{5}$$

$$\begin{aligned}\sin 2\alpha &= 2\sin\alpha\cos\alpha \\ &= 2\left(\frac{3}{5}\right)\left(-\frac{4}{5}\right) \\ &= -\frac{24}{25}\end{aligned}$$

19.

$$\begin{aligned}\cos^2 2t &= 1 - \sin^2 2t \\ &= 1 - \left(\frac{3}{5}\right)^2 \\ &= \frac{16}{25}\end{aligned}$$

$$\cos 2t = \frac{4}{5} \qquad \text{(Quadrant I)}$$

$$\begin{aligned}\sin 4t &= 2\sin 2t\cos 2t \\ &= 2\left(\frac{3}{5}\right)\left(\frac{4}{5}\right) \\ &= \frac{24}{25}\end{aligned}$$

20.

$$\begin{aligned}\cos^2\theta &= 1 - \sin^2\theta \\ &= 1 - (0.5)^2 \\ &= 0.75 = \frac{3}{4}\end{aligned}$$

$$\cos\theta = -\frac{\sqrt{3}}{2} \qquad \text{(Quadrant II)}$$

$$\begin{aligned}\sin 2\theta &= 2\sin\theta\cos\theta \\ &= 2(0.5)\left(-\frac{\sqrt{3}}{2}\right) \\ &= -\frac{\sqrt{3}}{2}\end{aligned}$$

21.

$$\sin^2\frac{\theta}{2} = 1-\cos^2\frac{\theta}{2}$$

$$= 1-\left(\frac{12}{13}\right)^2$$

$$= \frac{25}{169}$$

$$\sin\frac{\theta}{2} = \frac{5}{13} \qquad \text{(Quadrant I)}$$

$$\sin\theta = 2\sin\frac{\theta}{2}\cos\frac{\theta}{2}$$

$$= 2\left(\frac{5}{13}\right)\left(\frac{12}{13}\right)$$

$$= \frac{120}{169}$$

22.

$$\cos^2\alpha = 1-\sin^2\alpha$$

$$= 1-\left(-\frac{3}{5}\right)^2$$

$$= \frac{16}{25}$$

$$\cos\alpha = -\frac{4}{5} \qquad \text{(Quadrant III)}$$

$$\cos\frac{\alpha}{2} = -\sqrt{\frac{1-\frac{4}{5}}{2}} \qquad \text{(Quadrant II)}$$

$$= -\frac{\sqrt{10}}{10}$$

23.

$$\tan t = \frac{1}{\cot t} = \frac{1}{-\frac{4}{3}} = -\frac{3}{4}$$

$$\sec^2 t = 1+\tan^2 t$$

$$= 1+\left(-\frac{3}{4}\right)^2$$

$$= \frac{25}{16}$$

$$\sec t = \frac{5}{4}$$

$$\cos t = \frac{1}{\sec t} = \frac{1}{\frac{5}{4}} = \frac{4}{5}$$

$$\tan\frac{t}{2} = -\sqrt{\frac{1-\cos t}{1+\cos t}} \qquad \text{(Quadrant II)}$$

$$= -\sqrt{\frac{1-\frac{4}{5}}{1+\frac{4}{5}}}$$

$$= -\frac{1}{3}$$

24.

$$\cos 2x = -\sqrt{\frac{1+\cos 4x}{2}} \qquad \text{(Quadrant II)}$$

$$= -\sqrt{\frac{1+\frac{2}{3}}{2}}$$

$$= -\frac{\sqrt{30}}{6}$$

25.

$$\cos 15° = \cos\left(\frac{30°}{2}\right)$$
$$= \sqrt{\frac{1+\cos 30°}{2}} \quad \text{(Quadrant I)}$$
$$= \sqrt{\frac{1+\frac{\sqrt{3}}{2}}{2}}$$
$$= \frac{\sqrt{2+\sqrt{3}}}{2}$$

26.

$$\sin\frac{\pi}{8} = \sin\left(\frac{\frac{\pi}{4}}{2}\right)$$
$$= \sqrt{\frac{1-\cos\frac{\pi}{4}}{2}} \quad \text{(Quadrant I)}$$
$$= \sqrt{\frac{1-\frac{\sqrt{2}}{2}}{2}}$$
$$= \frac{\sqrt{2-\sqrt{2}}}{2}$$

27.

$$\tan 112.5° = \tan\frac{225°}{2}$$
$$= -\sqrt{\frac{1-\cos 225°}{1+\cos 225°}} \quad \text{(Quadrant II)}$$
$$= -\sqrt{\frac{1-\left(-\frac{1}{\sqrt{2}}\right)}{1+\left(-\frac{1}{\sqrt{2}}\right)}}$$
$$= -\sqrt{\frac{\sqrt{2}+1}{\sqrt{2}-1}} = -\frac{\sqrt{\sqrt{2}+1}}{\sqrt{\sqrt{2}-1}}$$
$$= -\frac{\sqrt{\sqrt{2}+1}}{\sqrt{\sqrt{2}-1}} \cdot \frac{\sqrt{\sqrt{2}+1}}{\sqrt{\sqrt{2}+1}}$$
$$= -\frac{\left(\sqrt{\sqrt{2}+1}\right)^2}{\sqrt{\left(\sqrt{2}\right)^2 - 1^2}} = -\frac{\sqrt{2}+1}{\sqrt{2-1}}$$
$$= -\frac{\sqrt{2}+1}{\sqrt{1}} = -\sqrt{2}-1$$
$$= -1-\sqrt{2}$$

28.

$$\cos 30x = \cos^2 15x - \sin^2 15x$$
$$= 1 - \sin^2 15x - \sin^2 15x$$
$$= 1 - 2\sin^2 15x$$

29.

$$\frac{1}{2}\sin 2y = \frac{1}{2}(2\sin y\cos y)$$
$$= \sin y\left(\frac{1}{\sec y}\right)$$
$$= \frac{\sin y}{\sec y}$$

30.

$$\frac{1-\cos\alpha}{\sin\alpha} = \frac{1-\cos\alpha}{\pm\sqrt{1-\cos^2\alpha}}$$
$$= \pm\frac{1-\cos\alpha}{\sqrt{1-\cos^2\alpha}}$$
$$= \pm\frac{\sqrt{(1-\cos\alpha)^2}}{\sqrt{1-\cos^2\alpha}}$$
$$= \pm\sqrt{\frac{(1-\cos\alpha)^2}{(1-\cos\alpha)(1+\cos\alpha)}}$$
$$= \pm\sqrt{\frac{1-\cos\alpha}{1+\cos\alpha}}$$
$$= \tan\frac{\alpha}{2}$$

31.

$$\sin\frac{3\alpha}{2}\sin\frac{\alpha}{2} = \frac{\cos\left(\frac{3\alpha}{2}-\frac{\alpha}{2}\right)-\cos\left(\frac{3\alpha}{2}+\frac{\alpha}{2}\right)}{2}$$
$$= \frac{\cos\alpha - \cos 2\alpha}{2}$$

32.

$$\cos 3x - \cos x = -2\sin\frac{3x+x}{2}\sin\frac{3x-x}{2}$$
$$= -2\sin 2x\sin x$$

33.

$$\sin 75°\sin 15° = \frac{\cos(75°-15°)-\cos(75°+15°)}{2}$$
$$= \frac{\cos 60° - \cos 90°}{2}$$
$$= \frac{\frac{1}{2}-0}{2}$$
$$= \frac{1}{4}$$

34.

$$\cos\frac{3\pi}{4}+\cos\frac{\pi}{4} = 2\cos\left(\frac{\frac{3\pi}{4}+\frac{\pi}{4}}{2}\right)\cos\left(\frac{\frac{3\pi}{4}-\frac{\pi}{4}}{2}\right)$$
$$= 2\cos\frac{\pi}{2}\cos\frac{\pi}{4}$$
$$2(0)\left(\frac{\sqrt{2}}{2}\right)$$
$$= 0$$

35.

$$2\cos^2\alpha - 1 = 0$$
$$\cos^2\alpha = \frac{1}{2}$$
$$\cos\alpha = \pm\frac{1}{\sqrt{2}}$$
$$\alpha = \frac{\pi}{4},\ \frac{3\pi}{4},\ \frac{5\pi}{4},\ \frac{7\pi}{4}$$

36.

$$2\sin\theta\cos\theta = 0$$
$$\sin\theta = 0 \qquad \cos\theta = 0$$
$$\theta = 0,\ \pi \qquad \theta = \frac{\pi}{2},\ \frac{3\pi}{2}$$

37.

$$\sin 2t - \sin t = 0$$
$$2\sin t\cos t - \sin t = 0$$
$$\sin t(2\cos t - 1) = 0$$
$$\sin t = 0 \qquad 2\cos t - 1 = 0$$
$$t = 0,\ \pi \qquad \cos t = \frac{1}{2}$$
$$t = \frac{\pi}{3},\ \frac{5\pi}{3}$$

38.

$$\cos^2\alpha - 2\cos\alpha = 0$$
$$\cos\alpha(\cos\alpha - 2) = 0$$
$\cos\alpha = 0$ ~~$\cos\alpha = 2$~~
$$\alpha = 90° + 180° n$$

39.

$$\tan 3x + 1 = 0$$
$$\tan 3x = -1$$
$$3x = 135° + 180° n$$
$$x = 45° + 60° n$$

40.

$$4\sin^2 2t = 3$$
$$\sin^2 2t = \frac{3}{4}$$
$$\sin 2t = \pm\frac{\sqrt{3}}{2}$$

$$2t = 60° + 180° n\ ,\ 120° + 180° n$$
$$t = 30° + 90° n\ ,\ 60° + 90° n$$

41.

$$x \approx 3.183086798$$

CHAPTER 8 REVIEW TEST

1.

$$\begin{aligned} 4-\tan^2 x &= 4-\left(\sec^2 x-1\right) \\ &= 5-\sec^2 x \end{aligned}$$

2.

$$\begin{aligned} \cos\left(270°+30°\right) &= \cos 270° \cos 30° - \sin 270° \sin 30° \\ &= (0)\left(\frac{\sqrt{3}}{2}\right)-(-1)\left(\frac{1}{2}\right) \\ &= \frac{1}{2} \end{aligned}$$

3.

$$\begin{aligned} \tan\left(\frac{\pi}{4}-\frac{\pi}{3}\right) &= \frac{\tan\frac{\pi}{4}-\tan\frac{\pi}{3}}{1+\tan\frac{\pi}{4}\tan\frac{\pi}{3}} \\ &= \frac{1-\sqrt{3}}{1+(1)\left(\sqrt{3}\right)} \\ &= -2+\sqrt{3} \end{aligned}$$

4.

$$\begin{aligned} \sin 47° &= \cos\left(90°-47°\right) \\ &= \cos 43° \end{aligned}$$

5.

$$\sin^2\theta = 1-\cos^2\theta$$
$$= 1-\left(\frac{4}{5}\right)^2$$
$$= \frac{9}{25}$$
$$\sin\theta = -\frac{3}{5} \qquad \text{(Quadrant IV)}$$

$$\sin(\theta-\pi) = \sin\theta\cos\pi - \cos\theta\sin\pi$$
$$= \left(-\frac{3}{5}\right)(-1)-\left(\frac{4}{5}\right)(0)$$
$$= \frac{3}{5}$$

6.

$$\cos^2 x = 1-\sin^2 x$$
$$= 1-\left(-\frac{5}{13}\right)^2$$
$$= \frac{144}{169}$$
$$\cos x = -\frac{12}{13} \qquad \text{(Quadrant III)}$$

$$\tan x = \frac{\sin x}{\cos x} = \frac{-\frac{5}{13}}{-\frac{12}{13}} = \frac{5}{12}$$

$$\tan(x-y) = \frac{\tan x - \tan y}{1+\tan x\tan y}$$
$$= \frac{\frac{5}{12}-\frac{8}{3}}{1+\left(\frac{5}{12}\right)\left(\frac{8}{3}\right)}$$
$$= -\frac{81}{76}$$

7.

$$\cos 2v = 1-2\sin^2 v$$
$$= 1-2\left(-\frac{12}{13}\right)^2$$
$$= -\frac{119}{169}$$

8.

$$\cos 4\alpha = 2\cos^2 2\alpha - 1$$
$$= 2\left(-\frac{4}{5}\right)^2 - 1$$
$$= \frac{7}{25}$$

9.

$$\sin\alpha = \frac{1}{\csc\alpha} = \frac{1}{-2}$$

$$\cos^2\alpha = 1 - \sin^2\alpha$$
$$= 1 - \left(-\frac{1}{2}\right)^2$$
$$= \frac{3}{4}$$

$$\cos\alpha = -\frac{\sqrt{3}}{2} \qquad \text{(Quadrant III)}$$

$$\cos\frac{\alpha}{2} = -\sqrt{\frac{1+\cos\alpha}{2}} \qquad \text{(Quadrant II)}$$
$$= -\sqrt{\frac{1-\frac{\sqrt{3}}{2}}{2}}$$
$$= -\frac{\sqrt{2-\sqrt{3}}}{2}$$

10.

$$\tan 15° = \tan\left(\frac{30°}{2}\right)$$
$$= \sqrt{\frac{1-\cos 30°}{1+\cos 30°}} \qquad \text{(Quadrant I)}$$
$$= \sqrt{\frac{1-\frac{\sqrt{3}}{2}}{1+\frac{\sqrt{3}}{2}}}$$
$$= \frac{\sqrt{2-\sqrt{3}}}{\sqrt{2+\sqrt{3}}}$$

11.

$$\sin\frac{x}{4} = \sin\left(\frac{x}{8}+\frac{x}{8}\right)$$
$$= \sin\frac{x}{8}\cos\frac{x}{8} + \cos\frac{x}{8}\sin\frac{x}{8}$$
$$= 2\sin\frac{x}{8}\cos\frac{x}{8}$$

12.

$$\sin 2x + \sin 3x = 2\sin\frac{2x+3x}{2}\cos\frac{2x-3x}{2}$$
$$= 2\sin\frac{5x}{2}\cos\left(-\frac{x}{2}\right)$$
$$= 2\sin\frac{5x}{2}\cos\frac{x}{2}$$

13.

$$\sin 150° - \sin 30° = 2\cos\frac{150°+30°}{2}\sin\frac{150°-30°}{2}$$
$$= 2\cos 90°\sin 60°$$
$$= 2(0)\left(\frac{\sqrt{3}}{2}\right)$$
$$= 0$$

14.

$$4\sin\alpha = 3$$
$$\sin^2\alpha = \frac{3}{4}$$
$$\sin\alpha = \pm\frac{\sqrt{3}}{2}$$
$$\alpha = \frac{\pi}{3}, \frac{2\pi}{3}, \frac{4\pi}{3}, \frac{5\pi}{3}$$

15.

$$\sin^2\theta - \cos^2\theta = 0$$
$$\cos^2\theta - \sin^2\theta = 0 \quad \text{(Multiply by } -1\text{)}$$
$$\cos 2\theta = 0$$
$$2\theta = 90° + 180°n$$
$$\theta = 45° + 90°n$$

CHAPTER 8 ADDITIONAL PRACTICE EXERCISES

1. Verify $\tan x \csc x = \sec x$

2. Verify $\dfrac{\sin x + \cos x}{\sin x} = 1 + \cot x$

3. Write $\tan 39°$ in terms of its cofunction.

4. Use the addition formula to find the exact value of $\sin(30° + 180°)$.

5. If $\cos\alpha = \dfrac{4}{5}$, with α in quadrant IV, find $\sin(\alpha - \pi)$

6. If $\sin\theta = -\dfrac{4}{5}$ and θ is in quadrant III, find $\cos 2\theta$.

7. Use the half-angle formula to find the exact value of $\cos\dfrac{5\pi}{8}$.

8. Express $3\sin 4\alpha \cos 5\alpha$ as a sum or difference.

9. Express $\cos 2\theta + \cos 6\theta$ as a product.

10. Find all solutions of $4\cos^2\alpha - 2 = 0$.

CHAPTER 8 PRACTICE TEST

1. Verify $\dfrac{\tan^2\theta}{\sec\theta - 1} = 1 + \sec\theta$

2. Verify $\sin(-\theta)\csc\theta = -1$

3. Verify $\dfrac{\tan x - 1}{1 - \cot x} = \sin x \sec x$

4. Use the addition formula to find the exact value of $\sin\dfrac{7\pi}{12}$.

5. Use the subtraction formula to find the exact value of $\cot 15°$.

6. Write $\sec 68°$ in terms of its cofunction.

7. Write $\sin\dfrac{\pi}{9}$ in terms of its cofunction.

8. If $\cos\theta = -\dfrac{1}{3}$ and θ is in quadrant III, find $\sin\dfrac{\theta}{2}$.

9. If $\cot\alpha = \dfrac{3}{4}$ and α is in quadrant I, find $\cos\dfrac{\alpha}{2}$.

10. Verify $4\sin^2 2x + 2\cos 4x = 2$

11. Use the half-angle formula to find the exact value of $\sin\dfrac{5\pi}{8}$.

12. Use the half-angle formula to find the exact value of $\tan 105°$.

13. Express $4\sin 3\theta \cos 2\theta$ as a sum or difference.

14. Express $\cos 5x \cos(-4x)$ as a sum or difference.

15.

Express $\sin\frac{3x}{2} - \sin\frac{5x}{2}$ as a product.

16.

Express $\cos 9\alpha + \cos 2\alpha$ as a product.

17.

Find all solutions of $\cos^2 x - 1 = 0$ in the interval $[0, 2\pi)$.

18.

Find all solutions of $2\cos^2\theta + \cos\theta = 1$ in the interval $[0, 2\pi)$.

19.

Find all solutions of $1 - 2\cos^2 x = 0$. Express answer in radian measure.

20.

Find all solutions of $\csc^2 x - 3\csc x + 2 = 0$. Express answer in degree measure.

CHAPTER 9 SECTION 1

EXERCISE SET 9.1

1.

$$\frac{a}{\sin\alpha}=\frac{b}{\sin\beta}$$

$$\frac{12.4}{\sin 25°}=\frac{b}{\sin 82°}$$

$$b=\frac{12.4\sin 82°}{\sin 25°}$$

$$b=29.06$$

3.

$$\alpha=180°-(\beta+\gamma)$$
$$=180°-(23°+47°)$$
$$=110°$$

$$\frac{a}{\sin\alpha}=\frac{c}{\sin\gamma}$$

$$\frac{9.3}{\sin 110°}=\frac{c}{\sin 47°}$$

$$c=\frac{9.3\sin 47°}{\sin 110°}$$

$$c=7.24$$

5.

$$\beta=180°-(\alpha+\gamma)$$
$$=180°-(42°20'+78°40')$$
$$=59°$$

$$\frac{a}{\sin\alpha}=\frac{b}{\sin\beta}$$

$$\frac{a}{\sin 42°20'}=\frac{20}{\sin 59°}$$

$$a=\frac{20\sin 42°20'}{\sin 59°}$$

$$a=15.71$$

7.

$$\frac{a}{\sin\alpha}=\frac{b}{\sin\beta}$$

$$\frac{25}{\sin 65°}=\frac{30}{\sin\beta}$$

$$\sin\beta=\frac{30\sin 65°}{25}$$

$$\sin\beta=1.0876$$

No solution

9.

$$\frac{a}{\sin\alpha}=\frac{c}{\sin\gamma}$$

$$\frac{12.6}{\sin\alpha}=\frac{6.3}{\sin 30°}$$

$$\sin\alpha=\frac{12.6\sin 30°}{6.3}$$

$$\sin\alpha=1$$

$$\alpha=\sin^{-1}1$$

$$\alpha=90°$$

$$\beta=180°-(\alpha+\gamma)$$

$$\beta=180°-(90°+30°)$$

$$\beta=60°$$

$$\frac{b}{\sin\beta}=\frac{c}{\sin\gamma}$$

$$\frac{b}{\sin 60°}=\frac{6.3}{\sin 30°}$$

$$b=\frac{6.3\sin 60°}{\sin 30°}$$

$$b=10.91$$

11.

$$\frac{c}{\sin\gamma}=\frac{b}{\sin\beta}$$

$$\frac{6}{\sin 45°}=\frac{7}{\sin\beta}$$

$$\sin\beta=\frac{7\sin 45°}{6}$$

$$\sin\beta=0.8250$$

$$\beta=\sin^{-1}(0.8250)$$

$\beta=55.59°$ or $\beta=124.41°$

$$\alpha=180°-(\beta+\gamma) \qquad \alpha=180°-(\beta+\gamma)$$

$$=180°-(55.59°+45°) \qquad =180°-(124.41°+45°)$$

$$=79.41° \qquad =10.59°$$

$$\frac{a}{\sin\alpha}=\frac{c}{\sin\gamma} \qquad \frac{a}{\sin\alpha}=\frac{c}{\sin\gamma}$$

$$\frac{a}{\sin 79.41°}=\frac{6}{\sin 45°} \qquad \frac{a}{\sin 10.59°}=\frac{6}{\sin 45°}$$

$$a=\frac{6\sin 79.41°}{\sin 45°} \qquad a=\frac{6\sin 10.59°}{\sin 45°}$$

$a=8.34$ or $a=1.56$

13.

Find c.

$$\angle BCA=180°-(95°+47°)=38°$$

$$\frac{c}{\sin 38°}=\frac{160}{\sin 95°}$$

$$c=\frac{160\sin 38°}{\sin 95°}$$

$$c=98.88$$

98.88 meters

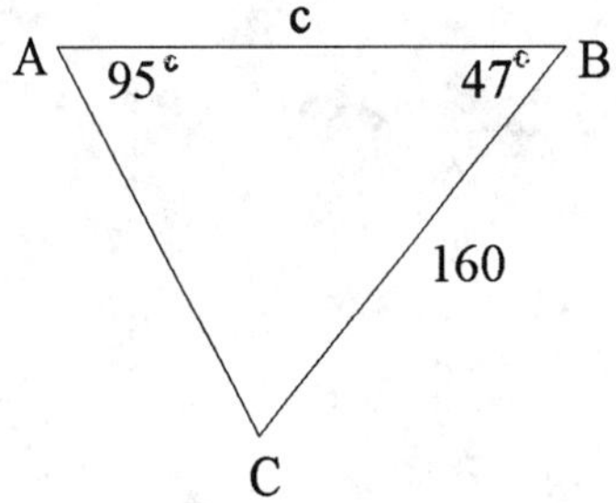

15.

$$\alpha = 180° - 45° = 135°$$

$$\frac{750}{\sin 135°} = \frac{c}{\sin 40°}$$

$$c = \frac{750 \sin 40°}{\sin 135°}$$

$$c = 681.78$$

681.78 meters

17.

point A: position at 2 p.m.

point B: position at 4 p.m.

4 p.m. − 2 p.m. = 2 hours

2 hours at 22 mph = 44 miles

$\overline{AB} = 44$

Find b.

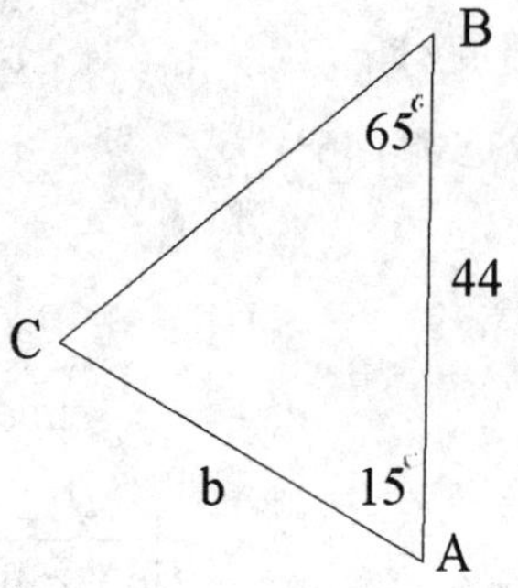

$$\gamma = 180° - (\alpha + \beta)$$
$$= 180° - (15° + 65°)$$
$$= 100°$$

$$\frac{44}{\sin 100°} = \frac{b}{\sin 65°}$$

$$b = \frac{44 \sin 65°}{\sin 100°}$$

$$b = 40.49$$

40.49 miles

19.

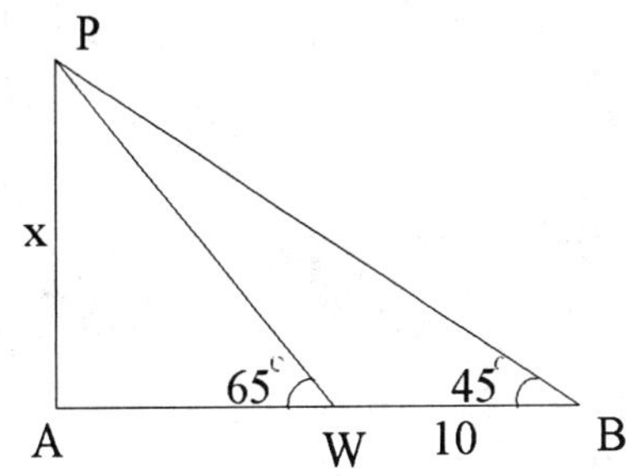

$$\angle PWB = 180° - 65° = 115°$$

$$\angle WPB = 180° - (115° + 45°) = 20°$$

$$\frac{\overline{PW}}{\sin 45°} = \frac{10}{\sin 20°}$$

$$\overline{PW} = \frac{10 \sin 45°}{\sin 20°}$$

$$\overline{PW} = 20.67$$

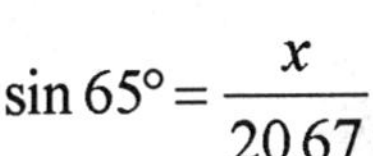

$$\sin 65° = \frac{x}{20.67}$$

$$x = 20.67 \sin 65°$$

$$x = 18.73$$

18.73 meters

21.

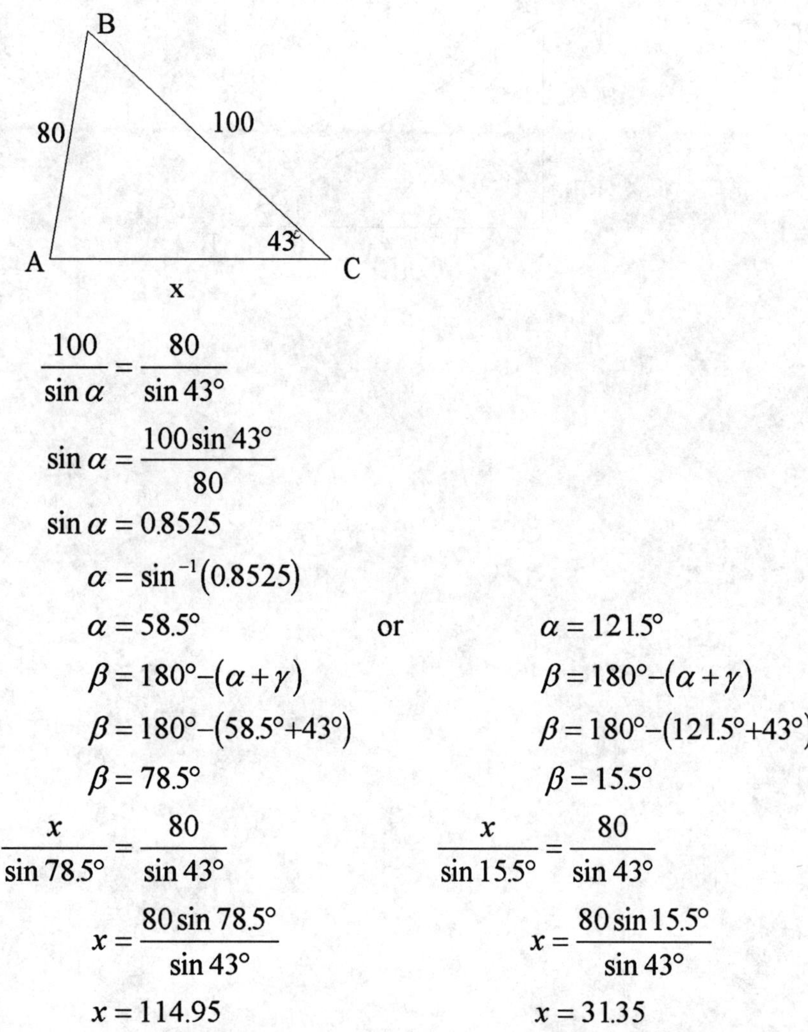

$$\frac{100}{\sin\alpha}=\frac{80}{\sin 43°}$$

$$\sin\alpha=\frac{100\sin 43°}{80}$$

$$\sin\alpha=0.8525$$

$$\alpha=\sin^{-1}(0.8525)$$

$\alpha=58.5°$ or $\alpha=121.5°$

$$\beta=180°-(\alpha+\gamma) \qquad \beta=180°-(\alpha+\gamma)$$

$$\beta=180°-(58.5°+43°) \qquad \beta=180°-(121.5°+43°)$$

$$\beta=78.5° \qquad \beta=15.5°$$

$$\frac{x}{\sin 78.5°}=\frac{80}{\sin 43°} \qquad \frac{x}{\sin 15.5°}=\frac{80}{\sin 43°}$$

$$x=\frac{80\sin 78.5°}{\sin 43°} \qquad x=\frac{80\sin 15.5°}{\sin 43°}$$

$$x=114.95 \qquad x=31.35$$

x needs to be larger than the shorter side (80 cm), hence $x=114.95$ cm.

23.

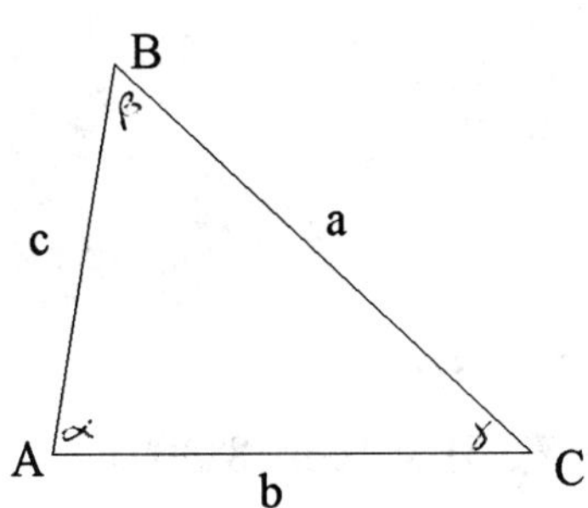

23a).

Suppose $b \sin\alpha > a$.

From the Law of Sines, $\dfrac{a}{\sin\alpha} = \dfrac{b}{\sin\beta}$

$$b \sin\alpha = a \sin\beta$$

Hence $a \sin\beta > a$

$$\sin\beta > 1$$

No such angle β.

23b).

Suppose $b \sin\alpha = a$.

$$b \sin\alpha = a \sin\beta$$

$$a = a \sin\beta$$

$$1 = \sin\beta$$

$$\beta = 90°$$

CHAPTER 9 SECTION 2

EXERCISE SET 9.2

1.

$$b^2 = a^2 + c^2 - 2ac\cos\beta$$
$$15^2 = 10^2 + 21^2 - 2(10)(21)\cos\beta$$
$$0.7524 = \cos\beta$$
$$\beta = \cos^{-1}(0.7524)$$
$$\beta = 41.2° = 41°12'$$

3.

$$b^2 = a^2 + c^2 - 2ac\cos\beta$$
$$b^2 = 25^2 + 30^2 - 2(25)(30)\cos 28°30'$$
$$b^2 = 206.77$$
$$b = 14.38$$

5.

$$c^2 = a^2 + b^2 - 2ac\cos\gamma$$
$$c^2 = 10^2 + 12^2 - 2(10)(12)\cos 108°$$
$$c^2 = 318.16$$
$$c = 17.84$$

7.

$$c^2 = a^2 + b^2 - 2ab\cos\gamma$$
$$c^2 = 7^2 + 6^2 - 2(7)(6)\cos 68°$$
$$c^2 = 53.53$$
$$c = 7.32$$

$$a^2 = b^2 + c^2 - 2bc\cos\alpha$$
$$7^2 = 6^2 + 7.32^2 - 2(6)(7.32)\cos\alpha$$
$$0.4620 = \cos\alpha$$
$$\alpha = \cos^{-1}(0.4620)$$
$$\alpha = 62.48° = 62°29'$$

9.

$$c^2 = a^2 + b^2 - 2ab\cos\gamma$$
$$15^2 = 9^2 + 12^2 - 2(9)(12)\cos\gamma$$
$$0 = \cos\gamma$$
$$\gamma = \cos^{-1}0$$
$$\gamma = 90°$$

11.

$$50^2 = 25^2 + 40^2 - 2(25)(40)\cos\alpha$$

$$\cos\alpha = -0.1375$$

$$\alpha = \cos^{-1}(-0.1375)$$

$$\alpha = 97.9°$$

smaller angle: $180° - 97.9° = 82.1° = 82°6'$

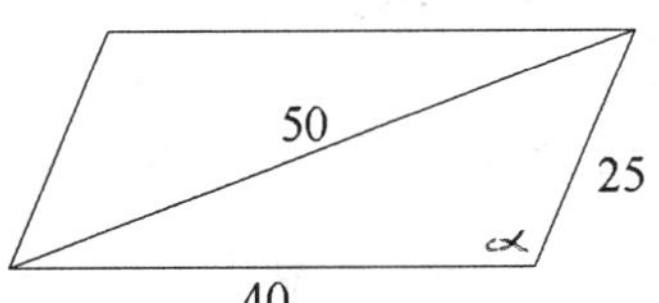

13.

$11 \text{ a.m.} - 9 \text{ a.m.} = 2 \text{ hours}$

2 hours at 15 mph = 30 miles = $\overline{BC}$

Find $\overline{AC}$.

$1 \text{ p.m.} - 11 \text{ a.m.} = 2 \text{ hours}$

2 hours at 15 mph = 30 miles = $\overline{BA}$

$\angle ABC = 32° + 90° = 122°$

$$\left(\overline{AC}\right)^2 = 30^2 + 30^2 - 2(30)(30)\cos 122°$$

$$\left(\overline{AC}\right)^2 = 2753.85$$

$$\overline{AC} = 52.48$$

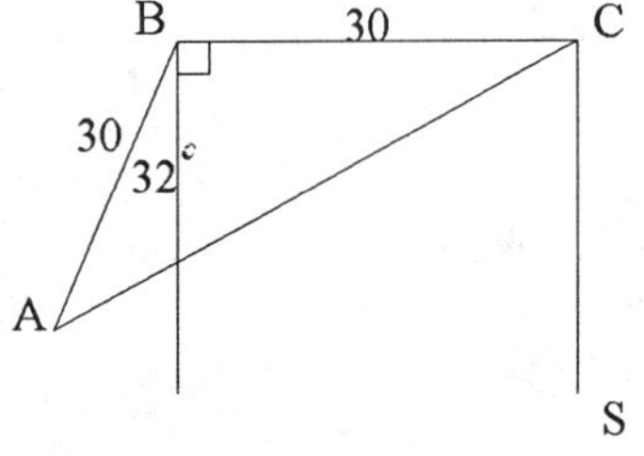

Since ΔABC is isosceles,

$$\angle BCA = \frac{180° - 122°}{2} = 29°.$$

$$\angle ACS = 90° - 29° = 61°.$$

The ship is 52.48 miles from port with a bearing of S 61° W.

15.

$2{:}30 \text{ p.m.} - 2{:}00 \text{ p.m.} = \frac{1}{2} \text{ hour}$

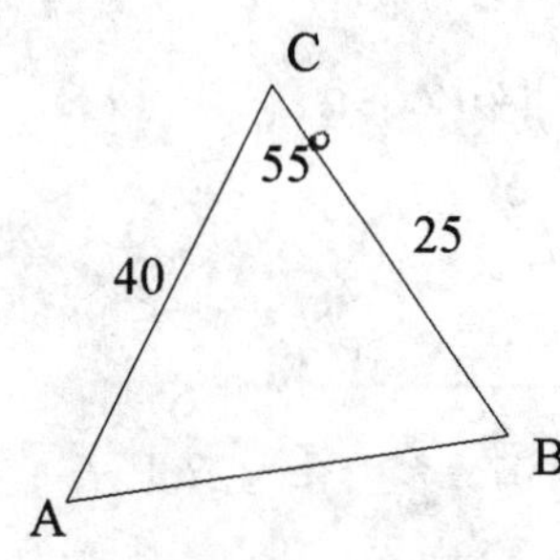

$\frac{1}{2}$ hour at 50 mph = 25 miles

$\frac{1}{2}$ hour at 80 mph = 40 miles

Find c.

$$c^2 = 40^2 + 25^2 - 2(40)(25)\cos 55°$$
$$c^2 = 1077.85$$
$$c = 32.83$$

32.83 miles

17.

$$c^2 = a^2 + b^2 - 2ab\cos\gamma$$
$$c^2 = 20^2 + 30^2 - 2(20)(30)\cos 37°$$
$$c^2 = 341.64$$
$$c = 18.48$$

$$\begin{aligned}\text{Perimeter} &= a + b + c \\ &= 20 + 30 + 18.48 \\ &= 68.48\end{aligned}$$

19a).

$$a^2 = b^2 + c^2 - 2bc\cos\alpha$$
$$b^2 = a^2 + c^2 - 2ac\cos\beta$$
$$c^2 = b^2 + a^2 - 2ab\cos\gamma$$

Adding these three equations:

$$a^2 + b^2 + c^2 = 2a^2 + 2b^2 + 2c^2 - 2bc\cos\alpha - 2ac\cos\beta - 2ab\cos\gamma$$
$$2bc\cos\alpha + 2ac\cos\beta + 2ab\cos\gamma = a^2 + b^2 + c^2$$
$$2(bc\cos\alpha + ac\cos\beta + ab\cos\gamma) = a^2 + b^2 + c^2$$

19b).

$$2bc\cos\alpha + 2ac\cos\beta + 2ab\cos\gamma = a^2 + b^2 + c^2$$

Divide by $2abc$:

$$\frac{2bc\cos\alpha}{2abc} + \frac{2ac\cos\beta}{2abc} + \frac{2ab\cos\gamma}{2abc} = \frac{a^2 + b^2 + c^2}{2abc}$$

$$\frac{\cos\alpha}{a} + \frac{\cos\beta}{b} + \frac{\cos\gamma}{c} = \frac{a^2 + b^2 + c^2}{2abc}$$

21.

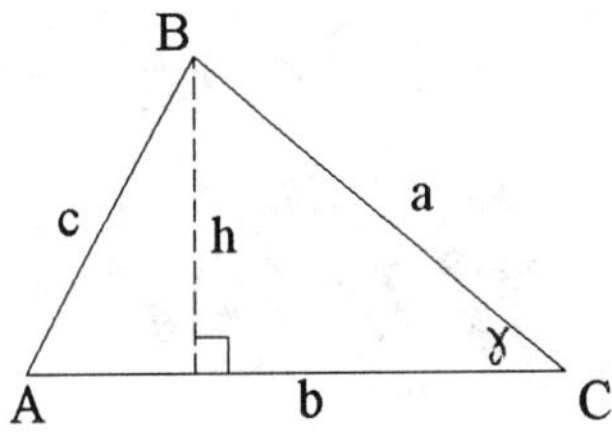

$$\sin\gamma = \frac{h}{a}$$

$$h = a\sin\gamma$$

$$\text{Area} = \frac{1}{2}\ \text{base} \cdot \text{height}$$

$$= \frac{1}{2}(b)(a\sin\gamma)$$

$$= \frac{1}{2}ab\sin\gamma$$

CHAPTER 9 SECTION 3

SECTION 9.3 PROGRESS CHECK

1. (Page 524)

$1-\sqrt{3}i$ (Quadrant IV)

$a=1 \qquad b=-\sqrt{3}$

$$r=\sqrt{a^2+b^2}$$

$$=\sqrt{1^2+\left(-\sqrt{3}\right)^2}$$

$$\tan\theta=\frac{b}{a}=\frac{-\sqrt{3}}{1}$$

$$\theta=300^\circ$$

$$1-\sqrt{3}i=r(\cos\theta+i\sin\theta)$$

$$=2(\cos 300^\circ+i\sin 300^\circ)$$

2. (Page 525)

$$\sqrt{2}\left(\cos\frac{\pi}{4}+\sin\frac{\pi}{4}\right)=\sqrt{2}\left(\frac{\sqrt{2}}{2}+\frac{\sqrt{2}}{2}i\right)$$

$$=1+i$$

3. (Page 526)

$1+\sqrt{3}i$ (Quadrant I)

$$r=\sqrt{1^2+\left(\sqrt{3}\right)^2}=2$$

$$\tan\theta=\frac{\sqrt{3}}{1}$$

$$\theta=60^\circ$$

$$1+\sqrt{3}i=2(\cos 60^\circ+i\sin 60^\circ)$$

$$1-\sqrt{3}i=2(\cos 300^\circ+i\sin 300^\circ)$$

(See Progress Check #1)

$$[2(\cos 60^\circ+i\sin 60^\circ)][2(\cos 300^\circ+i\sin 300^\circ)]$$

$$=(2)(2)[\cos(60^\circ+300^\circ)+i\sin(60^\circ+300^\circ)]$$

$$=4(\cos 360^\circ+i\sin 360^\circ)$$

$$=4(1+0)$$

$$=4$$

4. (Page 527)

$$1+\sqrt{3}i=2(\cos 60^\circ+i\sin 60^\circ)$$

(See Progress Check #3)

$$1-\sqrt{3}i=2(\cos 300^\circ+i\sin 300^\circ)$$

$$\frac{1+\sqrt{3}i}{1-\sqrt{3}i}=\frac{2}{2}[\cos(60^\circ-300^\circ)+i\sin(60^\circ-300^\circ)]$$

$$=\cos(-240^\circ)+i\sin(-240^\circ)$$

$$=\cos 240^\circ-i\sin 240^\circ$$

$$=-\frac{1}{2}-i\left(-\frac{\sqrt{3}}{2}\right)$$

$$=-\frac{1}{2}+\frac{\sqrt{3}}{2}i$$

5. (Page 528)

$\sqrt{3}+i$ (Quadrant I)

$a=\sqrt{3} \qquad b=1$

$r=\sqrt{\left(\sqrt{3}\right)^2+1^2}=2$

$\tan\theta=\dfrac{1}{\sqrt{3}}$

$\theta=30°$

$\sqrt{3}+i=2(\cos 30°+i\sin 30°)$

$\left(\sqrt{3}+i\right)^6=2^6\left[\cos(6\cdot 30°)+i\sin(6\cdot 30°)\right]$

$=64(\cos 180°+i\sin 180°)$

$=64(-1+0)$

$=-64$

6. (Page 530)

$\dfrac{\sqrt{3}}{2}-\dfrac{1}{2}i$ (Quadrant IV)

$a=\dfrac{\sqrt{3}}{2} \qquad b=-\dfrac{1}{2}$

$r=\sqrt{\left(\dfrac{\sqrt{3}}{2}\right)^2+\left(\dfrac{1}{2}\right)^2}=\sqrt{1}=1$

$\tan\theta=\dfrac{-\frac{1}{2}}{\frac{\sqrt{3}}{2}}=-\dfrac{1}{\sqrt{3}}$

$\theta=330°$

$z=\cos 330°+i\sin 330°$

$u_0=\sqrt{1}\left[\cos\left(\dfrac{330°}{2}\right)+i\sin\left(\dfrac{330°}{2}\right)\right]$

$=\cos 165°+i\sin 165°$

$u_1=\sqrt{1}\left[\cos\left(\dfrac{330°+360°}{2}\right)+i\sin\left(\dfrac{330°+360°}{2}\right)\right]$

$=\cos 345°+i\sin 345°$

EXERCISE SET 9.3

1.

$|3-2i|=\sqrt{3^2+(-2)^2}=\sqrt{13}$

3.

$|1+i|=\sqrt{1^2+1^2}=\sqrt{2}$

5.

$$|-6-2i| = \sqrt{(-6)^2+(-2)^2} = 2\sqrt{10}$$

7.

$3-3i$ (Quadrant IV)

$a = 3 \quad b = -3$

$$r = \sqrt{3^2+(-3)^2} = 3\sqrt{2}$$

$$\tan\theta = -\frac{3}{3} = -1$$

$$\theta = \frac{7\pi}{4}$$

$$3-3i = 3\sqrt{2}\left(\cos\frac{7\pi}{4} + i\sin\frac{7\pi}{4}\right)$$

9.

$\sqrt{3}-i$ (Quadrant IV)

$a = \sqrt{3} \quad b = -1$

$$r = \sqrt{\left(\sqrt{3}\right)^2+(-1)^2} = 2$$

$$\tan\theta = -\frac{1}{\sqrt{3}}$$

$$\theta = \frac{11\pi}{6}$$

$$\sqrt{3}-i = 2\left(\cos\frac{11\pi}{6} + i\sin\frac{11\pi}{6}\right)$$

11.

$-1+i$ (Quadrant II)

$a = -1 \quad b = 1$

$$r = \sqrt{(-1)^2-1^2} = \sqrt{2}$$

$$\tan\theta = \frac{1}{-1} = -1$$

$$\theta = \frac{3\pi}{4}$$

$$-1+i = \sqrt{2}\left(\cos\frac{3\pi}{4} + i\sin\frac{3\pi}{4}\right)$$

13.

-4 (negative x-axis)

$a = -4 \quad b = 0$

$$r = \sqrt{(-4)^2+0^2} = 4$$

$$\theta = \pi$$

$$-4 = 4(\cos\pi + i\sin\pi)$$

15.

$$4(\cos 180° + i\sin 180°) = 4(-1+0i) = -4$$

17.

$$\sqrt{2}(\cos 135° + i\sin 135°) = \sqrt{2}\left(-\frac{\sqrt{2}}{2} + \frac{\sqrt{2}}{2}i\right)$$
$$= -1+i$$

19.

$$5\left(\cos\frac{3\pi}{2} + i\sin\frac{3\pi}{2}\right) = 5(0-1i) = -5i$$

21.

$$2(\cos 150°+i\sin 150°)\cdot 3(\cos 210°+i\sin 210°)$$
$$=2\cdot 3[\cos(150°+210°)+i\sin(150°+210°)]$$
$$=6(\cos 360°+i\sin 360°)$$

23.

$$2\left(\cos\frac{\pi}{5}+i\sin\frac{\pi}{5}\right)\cdot\left(\cos\frac{\pi}{4}+i\sin\frac{\pi}{4}\right)$$
$$=2\cdot 1\left[\cos\left(\frac{\pi}{5}+\frac{\pi}{4}\right)+i\sin\left(\frac{\pi}{5}+\frac{\pi}{4}\right)\right]$$
$$=2\left(\cos\frac{9\pi}{20}+i\sin\frac{9\pi}{20}\right)$$

25.

$$1-i$$
$$a=1 \qquad b=-1$$
$$r=\sqrt{1^2+(-1)^2}=\sqrt{2}$$
$$\tan\theta=-\frac{1}{1}=-1$$
$$\theta=\frac{7\pi}{4}$$
$$1-i=\sqrt{2}\left(\cos\frac{7\pi}{4}+i\sin\frac{7\pi}{4}\right)$$

$$2i$$
$$a=0 \qquad b=2$$
$$r=\sqrt{0^2+2^2}=2$$
$$\theta=\frac{\pi}{2}$$
$$2i=2\left(\cos\frac{\pi}{2}+i\sin\frac{\pi}{2}\right)$$

$$(i-1)(2i)=\sqrt{2}\left(\cos\frac{7\pi}{4}+i\sin\frac{7\pi}{4}\right)\cdot 2\left(\cos\frac{\pi}{2}+i\sin\frac{\pi}{2}\right)$$
$$=2\sqrt{2}\left[\cos\left(\frac{7\pi}{4}+\frac{\pi}{2}\right)+i\sin\left(\frac{7\pi}{4}+\frac{\pi}{2}\right)\right]$$
$$=2\sqrt{2}\left(\cos\frac{9\pi}{4}+i\sin\frac{9\pi}{4}\right)$$
$$=2\sqrt{2}\left(\frac{\sqrt{2}}{2}+\frac{\sqrt{2}}{2}i\right)$$
$$=2+2i$$

27.

$-2+2\sqrt{3}i$

$a=-2 \qquad b=2\sqrt{3}$

$r=\sqrt{(-2)^2+\left(2\sqrt{3}\right)^2}=4$

$\tan\theta=\dfrac{2\sqrt{3}}{-2}=-\sqrt{3}$

$\theta=\dfrac{2\pi}{3}$

$-2+2\sqrt{3}i=4\left(\cos\dfrac{2\pi}{3}+i\sin\dfrac{2\pi}{3}\right)$

$3+3i$

$a=3 \qquad b=3$

$r=\sqrt{3^2+3^2}=3\sqrt{2}$

$\tan\theta=\dfrac{3}{3}=1$

$\theta=\dfrac{\pi}{4}$

$3+3i=3\sqrt{2}\left(\cos\dfrac{\pi}{4}+i\sin\dfrac{\pi}{4}\right)$

$$
\begin{aligned}
\left(-2+2\sqrt{3}i\right)(3+3i) &= 4\left(\cos\frac{2\pi}{3}+i\sin\frac{2\pi}{3}\right)\cdot 3\sqrt{2}\left(\cos\frac{\pi}{4}+i\sin\frac{\pi}{4}\right)\\
&= 4\cdot 3\sqrt{2}\left[\cos\left(\frac{2\pi}{3}+\frac{\pi}{4}\right)+i\sin\left(\frac{2\pi}{3}+\frac{\pi}{4}\right)\right]\\
&= 12\sqrt{2}\left[\cos\frac{2\pi}{3}\cos\frac{\pi}{4}-\sin\frac{2\pi}{3}\sin\frac{\pi}{4}+i\left(\sin\frac{2\pi}{3}\cos\frac{\pi}{4}+\cos\frac{2\pi}{3}\sin\frac{\pi}{4}\right)\right]\\
&= 12\sqrt{2}\left[-\frac{1}{2}\cdot\frac{\sqrt{2}}{2}-\frac{\sqrt{3}}{2}\cdot\frac{\sqrt{2}}{2}+i\left(\frac{\sqrt{3}}{2}\cdot\frac{\sqrt{2}}{2}+\left(-\frac{1}{2}\right)\left(\frac{\sqrt{2}}{2}\right)\right)\right]\\
&= 12\sqrt{2}\left[\left(-\frac{\sqrt{2}}{4}-\frac{\sqrt{6}}{4}\right)+i\left(\frac{\sqrt{6}}{4}-\frac{\sqrt{2}}{4}\right)\right]\\
&= \left(-6\sqrt{3}-6\right)+\left(-6+6\sqrt{3}\right)i
\end{aligned}
$$

29.

5

$a = 5 \qquad b = 0$

$r = \sqrt{5^2 + 0^2} = 5$

$\theta = 0$

$5 = 5(\cos 0 + i \sin 0)$

$-2 - 2i$

$a = -2 \qquad b = -2$

$r = \sqrt{(-2)^2 + (-2)^2} = 2\sqrt{2}$

$\tan\theta = \frac{-2}{-2} = 1$

$\theta = \frac{5\pi}{4}$

$-1 - 2i = 2\sqrt{2}\left(\cos\frac{5\pi}{4} + i\sin\frac{5\pi}{4}\right)$

$$
\begin{aligned}
5(-2-2i) &= 5(\cos 0 + i\sin 0)\cdot 2\sqrt{2}\left(\cos\frac{5\pi}{4} + i\sin\frac{5\pi}{4}\right)\\
&= 5\cdot 2\sqrt{2}\left[\cos\left(0+\frac{5\pi}{4}\right) + i\sin\left(0+\frac{5\pi}{4}\right)\right]\\
&= 10\sqrt{2}\left(\cos\frac{5\pi}{4} + i\sin\frac{5\pi}{4}\right)\\
&= 10\sqrt{2}\left(-\frac{\sqrt{2}}{2} - \frac{\sqrt{2}}{2}i\right)\\
&= -10 - 10i
\end{aligned}
$$

31a).

5

$a = 5 \qquad b = 0$

$r = \sqrt{5^2 + 0^2} = 5$

$\theta = 0$

$5 = 5(\cos 0 + i \sin 0)$

$$-2 - 2i = 2\sqrt{2}\left(\cos\frac{5\pi}{4} + i\sin\frac{5\pi}{4}\right) \qquad \text{(See Exercise 29)}$$

$$\begin{aligned}
\frac{5}{(-2-2i)} &= \frac{5(\cos 0 + i\sin 0)}{2\sqrt{2}\left(\cos\frac{5\pi}{4} + i\sin\frac{5\pi}{4}\right)} \\
&= \frac{5}{2\sqrt{2}}\left[\cos\left(0 - \frac{5\pi}{4}\right) + i\sin\left(0 - \frac{5\pi}{4}\right)\right] \\
&= \frac{5\sqrt{2}}{4}\left(\cos\left(-\frac{5\pi}{4}\right) + i\sin\left(-\frac{5\pi}{4}\right)\right) \\
&= \frac{5\sqrt{2}}{4}\left(\cos\frac{3\pi}{4} + i\sin\frac{3\pi}{4}\right) \\
&= \frac{5\sqrt{2}}{4}\left(-\frac{\sqrt{2}}{2} + \frac{\sqrt{2}}{2}i\right) \\
&= -\frac{5}{4} + \frac{5}{4}i
\end{aligned}$$

31b).

$$\frac{5}{-2-2i}\cdot\frac{(-2+2i)}{(-2+2i)} = \frac{-10+10i}{4+4} = \frac{-10+10i}{8} = \frac{-5}{4} + \frac{5}{4}i$$

33a).

$$3+3i = 3\sqrt{2}\left(\cos\frac{\pi}{4}+i\sin\frac{\pi}{4}\right) \quad \text{(See Exercise 27)}$$

$$-2+2\sqrt{3}i = 4\left(\cos\frac{2\pi}{3}+i\sin\frac{2\pi}{3}\right) \quad \text{(See Exercise 27)}$$

$$\frac{3+3i}{-2+2\sqrt{3}i} = \frac{3\sqrt{2}\left(\cos\frac{\pi}{4}+i\sin\frac{\pi}{4}\right)}{4\left(\cos\frac{2\pi}{3}+i\sin\frac{2\pi}{3}\right)}$$

$$= \frac{3\sqrt{2}}{4}\left[\cos\left(\frac{\pi}{4}-\frac{2\pi}{3}\right)+i\sin\left(\frac{\pi}{4}-\frac{2\pi}{3}\right)\right]$$

$$= \frac{3\sqrt{2}}{4}\left[\cos\left(-\frac{5\pi}{12}\right)+i\sin\left(-\frac{5\pi}{12}\right)\right]$$

$$= \frac{3\sqrt{2}}{4}\left(\cos\frac{19\pi}{12}+i\sin\frac{19\pi}{12}\right)$$

$$= \frac{3\sqrt{2}}{4}\left[\left(\frac{\sqrt{6}}{4}-\frac{\sqrt{2}}{4}\right)-i\left(\frac{\sqrt{2}}{4}+\frac{\sqrt{6}}{4}\right)\right]$$

$$= \left(\frac{3\sqrt{3}}{8}-\frac{3}{8}\right)+\left(-\frac{3}{8}-\frac{3\sqrt{3}}{8}i\right)$$

33b).

$$\frac{3+3i}{-2+2\sqrt{3}i}\cdot\frac{-2-2\sqrt{3}i}{-2-2\sqrt{3}i} = \frac{-6-6i-6\sqrt{3}i+6\sqrt{3}}{4+12}$$

$$= \frac{\left(-6+6\sqrt{3}\right)}{16}+\frac{\left(-6-6\sqrt{3}\right)}{16}i$$

$$= \frac{-3+3\sqrt{3}}{8}+\frac{\left(-3-3\sqrt{3}\right)}{8}i$$

35a).

$$2i = 2\left(\cos\frac{\pi}{2} + i\sin\frac{\pi}{2}\right)$$

$$1 - i = \sqrt{2}\left(\cos\frac{7\pi}{4} + i\sin\frac{7\pi}{4}\right) \quad \text{(See Exercise 25)}$$

$$\frac{2i}{1-i} = \frac{2\left(\cos\frac{\pi}{2} + i\sin\frac{\pi}{2}\right)}{\sqrt{2}\left(\cos\frac{7\pi}{4} + i\sin\frac{7\pi}{4}\right)}$$

$$= \frac{2}{\sqrt{2}}\left[\cos\left(\frac{\pi}{2} - \frac{7\pi}{4}\right) + i\sin\left(\frac{\pi}{2} - \frac{7\pi}{4}\right)\right]$$

$$= \frac{2}{\sqrt{2}}\left[\cos\left(\frac{-5\pi}{4}\right) + i\sin\left(\frac{-5\pi}{4}\right)\right]$$

$$= \frac{2}{\sqrt{2}}\left(\cos\frac{3\pi}{4} + i\sin\frac{3\pi}{4}\right)$$

$$= \frac{2}{\sqrt{2}}\left(-\frac{\sqrt{2}}{2} + \frac{\sqrt{2}}{2}i\right)$$

$$= -1 + i$$

35b).

$$\frac{2i}{1-i} \cdot \frac{1+i}{1+i} = \frac{2i-2}{1+1} = \frac{2i-2}{2} = -1 + i$$

37.

$$-2+2i$$

$$a=-2 \qquad b=2$$

$$r=\sqrt{(-2)^2+2^2}=2\sqrt{2}$$

$$\tan\theta=\frac{2}{-2}=-1$$

$$\theta=\frac{3\pi}{4}$$

$$-2+2i=2\sqrt{2}\left(\cos\frac{3\pi}{4}+i\sin\frac{3\pi}{4}\right)$$

$$(-2+2i)^6=\left[2\sqrt{2}\left(\cos\frac{3\pi}{4}+i\sin\frac{3\pi}{4}\right)\right]^6$$

$$=\left(2\sqrt{2}\right)^6\left[\cos\left(6\cdot\frac{3\pi}{4}\right)+i\sin\left(6\cdot\frac{3\pi}{4}\right)\right]$$

$$=512\left(\cos\frac{9\pi}{2}+i\sin\frac{9\pi}{2}\right)$$

$$=512(0+i)$$

$$=0+512i$$

39.

$$1-i=\sqrt{2}\left(\cos\frac{7\pi}{4}+i\sin\frac{7\pi}{4}\right) \quad \text{(See Exercise 25)}$$

$$(1-i)^9=\left[\sqrt{2}\left(\cos\frac{7\pi}{4}+i\sin\frac{7\pi}{4}\right)\right]^9$$

$$=\left(\sqrt{2}\right)^9\left[\cos\left(\frac{7\pi}{4}\cdot 9\right)+i\sin\left(\frac{7\pi}{4}\cdot 9\right)\right]$$

$$=16\sqrt{2}\left(\cos\frac{63\pi}{4}+i\sin\frac{63\pi}{4}\right)$$

$$=16\sqrt{2}\left(\cos\frac{7\pi}{4}+i\sin\frac{7\pi}{4}\right)$$

$$=16\sqrt{2}\left(\frac{\sqrt{2}}{2}-\frac{\sqrt{2}}{2}i\right)$$

$$=16-16i$$

41.

$$-1-i$$

$$a=-1 \qquad b=-1$$

$$r=\sqrt{(-1)^2+(-1)^2}=\sqrt{2}$$

$$\tan\theta=\frac{-1}{-1}=1$$

$$\theta=\frac{5\pi}{4}$$

$$-1-i=\sqrt{2}\left(\cos\frac{5\pi}{4}+i\sin\frac{5\pi}{4}\right)$$

$$\begin{aligned}(-1-i)^7&=\left[\sqrt{2}\left(\cos\frac{5\pi}{4}+i\sin\frac{5\pi}{4}\right)\right]^7\\&=\left(\sqrt{2}\right)^7\left[\cos\left(\frac{5\pi}{4}\cdot 7\right)+i\sin\left(\frac{5\pi}{4}\cdot 7\right)\right]\\&=8\sqrt{2}\left(\cos\frac{35\pi}{4}+i\sin\frac{35\pi}{4}\right)\\&=8\sqrt{2}\left(\cos\frac{3\pi}{4}+i\sin\frac{3\pi}{4}\right)\\&=8\sqrt{2}\left(-\frac{\sqrt{2}}{2}+\frac{\sqrt{2}}{2}i\right)\\&=-8+8i\end{aligned}$$

43.

$$-16 = 16(\cos\pi + i\sin\pi)$$

$$u_0 = \sqrt[4]{16}\left[\cos\frac{\pi}{4} + i\sin\frac{\pi}{4}\right] = 2\left(\frac{\sqrt{2}}{2} + \frac{\sqrt{2}}{2}i\right) = \sqrt{2} + \sqrt{2}i$$

$$u_1 = \sqrt[4]{16}\left[\cos\frac{\pi + 2\pi}{4} + i\sin\frac{\pi + 2\pi}{4}\right]$$

$$= 2\left(\cos\frac{3\pi}{4} + i\sin\frac{3\pi}{4}\right)$$

$$= 2\left(-\frac{\sqrt{2}}{2} + \frac{\sqrt{2}}{2}i\right)$$

$$= -\sqrt{2} + \sqrt{2}i$$

$$u_2 = \sqrt[4]{16}\left[\cos\frac{\pi + 2\pi(2)}{4} + i\sin\frac{\pi + 2\pi(2)}{4}\right]$$

$$= 2\left(\cos\frac{5\pi}{4} + i\sin\frac{5\pi}{4}\right)$$

$$= 2\left(-\frac{\sqrt{2}}{2} - \frac{\sqrt{2}}{2}i\right)$$

$$= -\sqrt{2} - \sqrt{2}i$$

$$u_3 = \sqrt[4]{16}\left[\cos\frac{\pi + 2\pi(3)}{4} + i\sin\frac{\pi + 2\pi(3)}{4}\right]$$

$$= 2\left(\cos\frac{7\pi}{4} + i\sin\frac{7\pi}{4}\right)$$

$$= 2\left(\frac{\sqrt{2}}{2} - \frac{\sqrt{2}}{2}i\right)$$

$$= \sqrt{2} - \sqrt{2}i$$

The roots are $\sqrt{2} + \sqrt{2}i$; $-\sqrt{2} + \sqrt{2}i$; $-\sqrt{2} - \sqrt{2}i$; $\sqrt{2} - \sqrt{2}i$

45.

$$1-\sqrt{3}i = 2\left(\cos\frac{5\pi}{3}+i\sin\frac{5\pi}{3}\right) \quad \text{(See Exercise 28)}$$

$$u_0 = \sqrt{2}\left(\cos\frac{\frac{5\pi}{3}}{2}+i\sin\frac{\frac{5\pi}{3}}{2}\right)$$

$$= \sqrt{2}\left(\cos\frac{5\pi}{6}+i\sin\frac{5\pi}{6}\right) = \sqrt{2}(\cos 150° + i\sin 150°)$$

$$u_1 = \sqrt{2}\left(\cos\frac{\frac{5\pi}{3}+2\pi}{2}+i\sin\frac{\frac{5\pi}{3}+2\pi}{2}\right)$$

$$= \sqrt{2}\left(\cos\frac{11\pi}{6}+i\sin\frac{11\pi}{6}\right) = \sqrt{2}(\cos 330° + i\sin 330°)$$

47.

$$z^3+8=0$$

$$(z+2)(z^2-2z+4)=0$$

$$z+2=0 \qquad z^2-2z+4=0$$

$$z=-2 \qquad z=\frac{-(-2)\pm\sqrt{(-2)^2-4(1)(4)}}{2(1)}$$

$$=\frac{2\pm\sqrt{-12}}{2}$$

$$=\frac{2\pm2\sqrt{3}i}{2}$$

$$=1\pm\sqrt{3}i$$

$-2,\ 1\pm\sqrt{3}i$

49.

$$z^4-16=0$$

$$(z^2+4)(z^2-4)=0$$

$$z^2+4=0 \qquad z^2-4=0$$

$$z^2=-4 \qquad z^2=4$$

$$z=\pm2i \qquad z=\pm2$$

51.

$$\frac{r_1(\cos\theta_1 + i\sin\theta_1)}{r_2(\cos\theta_2 + i\sin\theta_2)} = \frac{r_1}{r_2}\cdot\frac{(\cos\theta_1 + i\sin\theta_1)}{(\cos\theta_2 + i\sin\theta_2)}\cdot\frac{(\cos\theta_2 - i\sin\theta_2)}{(\cos\theta_2 - i\sin\theta_2)}$$

$$= \frac{r_1}{r_2}\cdot\frac{\cos\theta_1\cos\theta_2 + \sin\theta_1\sin\theta_2 + \sin\theta_1\cos\theta_2 i - \cos\theta_1\sin\theta_2 i}{\cos^2\theta_2 + i\sin^2\theta_2}$$

$$= \frac{r_1}{r_2}\cdot\frac{\cos(\theta_1 - \theta_2) + i\sin(\theta_1 - \theta_2)}{1}$$

$$= \frac{r_1}{r_2}\left[\cos(\theta_1 - \theta_2) + i\sin(\theta_1 - \theta_2)\right]$$

CHAPTER 9 SECTION 4

SECTION 9.4 PROGRESS CHECK

1. (Page 532)

$$(3, -30^\circ),\ (3, -30^\circ+360^\circ) = (3, 330^\circ)$$
$$(3, -30^\circ+2\cdot 360^\circ) = (3, 690^\circ)$$
$$(3, -30^\circ-360^\circ) = (3, -390^\circ)$$

2. (Page 533)

$$(-3, -45^\circ) = (3, -45^\circ+180^\circ) = (3, 135^\circ)$$
$$(-3, -45^\circ) = (-3, -45^\circ+360^\circ) = (-3, 315^\circ)$$

(b) and (c)

3a). (Page 535)

$$\left(2, \frac{\pi}{3}\right) \quad r = 2, \quad \theta = \frac{\pi}{3}$$

$$x = r\cos\theta = 2\cos\frac{\pi}{3} = 2\left(\frac{1}{2}\right) = 1$$

$$y = r\sin\theta = 2\sin\frac{\pi}{3} = 2\left(\frac{\sqrt{3}}{2}\right) = \sqrt{3}$$

$$(x, y) = \left(1, \sqrt{3}\right)$$

3b).

$$(-1, -1) \quad x = -1, \quad y = -1$$

$$r = \sqrt{x^2 + y^2} = \sqrt{(-1)^2 + (-1)^2} = \sqrt{2}$$

$$\tan\theta = \frac{y}{x} = \frac{-1}{-1} = 1$$

Since θ lies in quadrant III, $\theta = \frac{5\pi}{4}$

$(r, \theta) = \left(\sqrt{2}, \frac{5\pi}{4}\right)$ (not unique)

4a). (Page 538)

$r = 2\sin\theta$

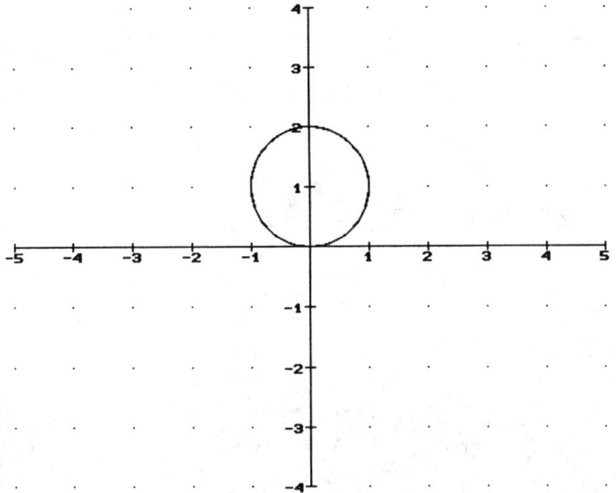

4b).

$r = 1 + 2\cos\theta$

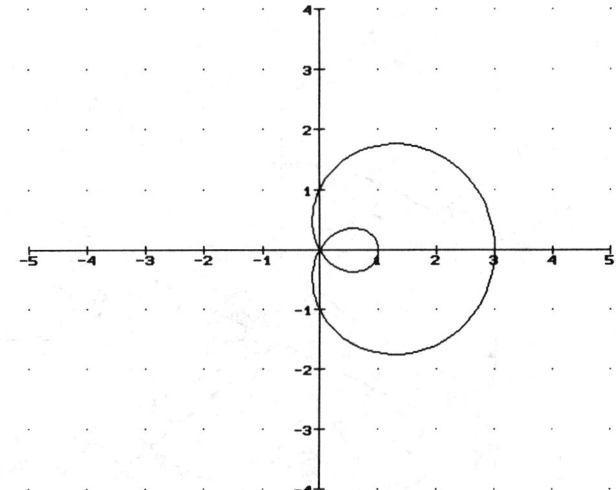

4c).

$r = \cos 3\theta$

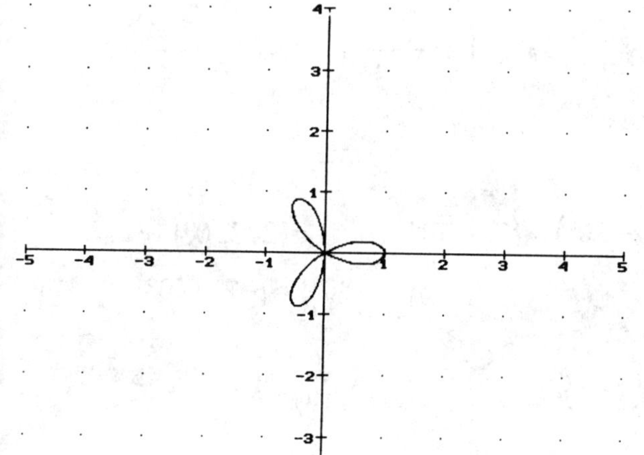

5. (Page 538)

$x = 5$

$x = r\cos\theta$

$r\cos\theta = 5$

EXERCISE SET 9.4

1.

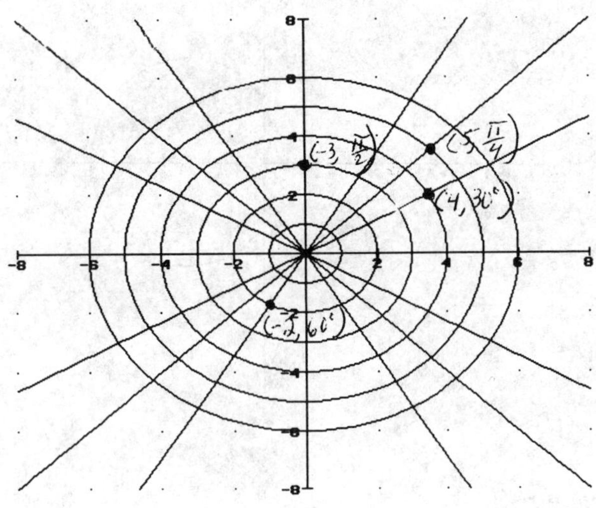

3a).
$$(6,\ 135°),\ (6,\ 135°+360°) = (6,\ 495°),$$
$$(6,\ 135°-360°) = (6,\ -225°)$$

3b).
$$(-2,\ 120°),\ (-2,\ 120°+360°) = (-2,\ 480°),$$
$$(2,\ 120°+180°) = (2,\ 300°)$$

3c).
$$\left(4,\ \frac{5\pi}{6}\right),\ \left(4,\ \frac{5\pi}{6}+2\pi\right) = \left(4,\ \frac{17\pi}{6}\right),$$
$$\left(-4,\ \frac{5\pi}{6}+\pi\right) = \left(-4,\ \frac{11\pi}{6}\right)$$

3d).
$$\left(-4,\ -\frac{7\pi}{4}\right),\ \left(-4,\ -\frac{7\pi}{4}+2\pi\right) = \left(-4,\ \frac{\pi}{4}\right),$$
$$\left(4,\ -\frac{7\pi}{4}+\pi\right) = \left(4,\ -\frac{3\pi}{4}\right)$$

5a).
$$(-2,\ 30°) = (2,\ 30°+180°) = (2,\ 210°)$$

5b).
$$(-4,\ -60°) = (4,\ -60°+180°) = (4,\ 120°)$$

5c).
$$\left(-3,\ \frac{2\pi}{3}\right) = \left(3,\ \frac{2\pi}{3}+\pi\right) = \left(3,\ \frac{5\pi}{3}\right)$$

5d).
$$\left(-1,\ -\frac{7\pi}{6}\right) = \left(1,\ -\frac{7\pi}{6}+\pi\right) = \left(1,\ -\frac{\pi}{6}\right)$$

7.
$$(2,\ -30°) = (2,\ -30°+360°) = (2,\ 330°)$$
$$(-2,\ -30°+180°) = (-2,\ 150°)$$

(a), (c) yes (b), (d) no

9a).

$(5,\ 330°)$

$r=5\quad \theta=330°$

$x=r\cos\theta=5\cos 330°=5\left(\frac{\sqrt{3}}{2}\right)=\frac{5\sqrt{3}}{2}$

$y=r\sin\theta=5\sin 330°=5\left(-\frac{1}{2}\right)=-\frac{5}{2}$

$(x,\ y)=\left(\frac{5\sqrt{3}}{2},\ -\frac{5}{2}\right)$

9b).

$(2,\ 270°)$

$r=2\quad \theta=270°$

$x=r\cos\theta=2\cos 270°=0$

$y=r\sin\theta=2\sin 270°=-2$

$(x,\ y)=(0,\ -2)$

9c).

$\left(4,\ \frac{\pi}{6}\right)$

$r=4\quad \theta=\frac{\pi}{6}$

$x=r\cos\theta=4\cos\frac{\pi}{6}=4\left(\frac{\sqrt{3}}{2}\right)=2\sqrt{3}$

$y=r\sin\theta=4\sin\frac{\pi}{6}=4\left(\frac{1}{2}\right)=2$

$(x,\ y)=(2\sqrt{3},\ 2)$

9d).

$\left(-3,\ -\frac{2\pi}{3}\right)$

$r=-3\quad \theta=-\frac{2\pi}{3}$

$x=r\cos\theta=-3\cos\left(-\frac{2\pi}{3}\right)=-3\left(-\frac{1}{2}\right)=\frac{3}{2}$

$y=r\sin\theta=-3\sin\left(-\frac{2\pi}{3}\right)=-3\left(-\frac{\sqrt{3}}{2}\right)=\frac{3\sqrt{3}}{2}$

$(x,\ y)=\left(\frac{3}{2},\ \frac{3\sqrt{3}}{2}\right)$

11a).

$(-2, 2)$ (Quadrant II)

$x=-2 \quad y=2$

$r=\sqrt{x^2+y^2}=\sqrt{(-2)^2+2^2}=2\sqrt{2}$

$\tan\theta=\dfrac{y}{x}=\dfrac{2}{-2}=-1$

$\theta=\dfrac{3\pi}{4}$

$(r, \theta)=\left(2\sqrt{2}, \dfrac{3\pi}{4}\right)$

11b).

$(1, -\sqrt{3})$ (Quadrant IV)

$x=1 \quad y=-\sqrt{3}$

$r=\sqrt{x^2+y^2}=\sqrt{(1)^2+(-\sqrt{3})^2}=2$

$\tan\theta=\dfrac{y}{x}=\dfrac{-\sqrt{3}}{1}$

$\theta=\dfrac{5\pi}{3}$

$(r, \theta)=\left(2, \dfrac{5\pi}{3}\right)$

11c).

$(\sqrt{3}, 1)$ (Quadrant I)

$x=\sqrt{3}, \quad y=1$

$r=\sqrt{x^2+y^2}=\sqrt{(\sqrt{3})^2+(1)^2}=2$

$\tan\theta=\dfrac{y}{x}=\dfrac{1}{\sqrt{3}}$

$\theta=\dfrac{\pi}{6}$

$(r, \theta)=\left(2, \dfrac{\pi}{6}\right)$

11d).

$(-4, -4)$ (Quadrant III)

$x=-4, \quad y=-4$

$r=\sqrt{x^2+y^2}=\sqrt{(-4)^2+(-4)^2}=4\sqrt{2}$

$\tan\theta=\dfrac{y}{x}=\dfrac{-4}{-4}=1$

$\theta=\dfrac{5\pi}{4}$

$(r, \theta)=\left(4\sqrt{2}, \dfrac{5\pi}{4}\right)$

13.
$\theta = 45°$

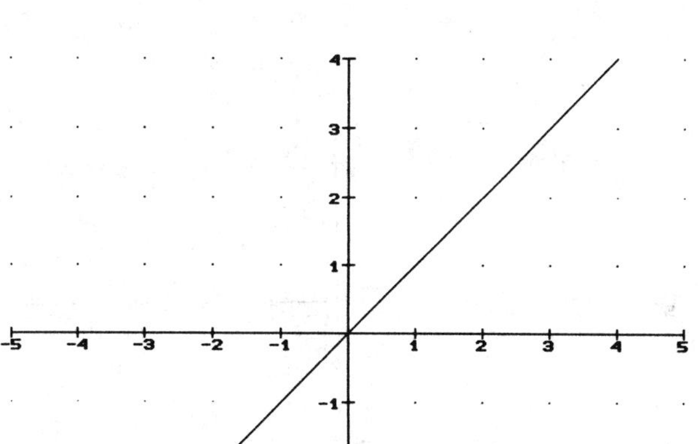

15.
$r = 3$

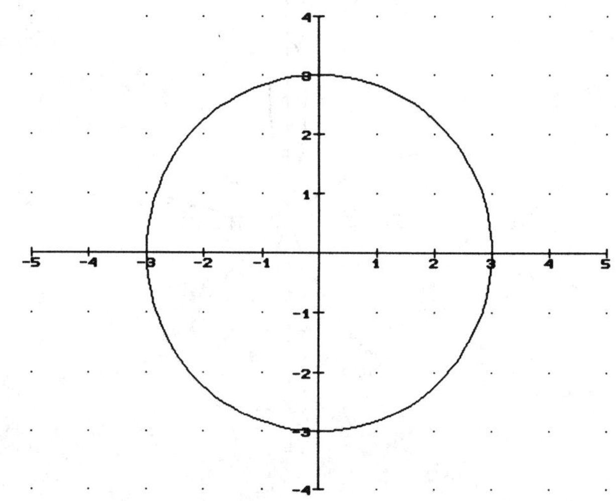

17.
$r = 2\sin\theta$

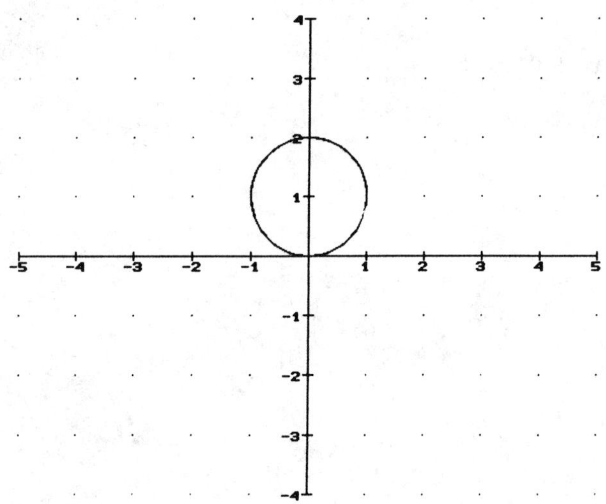

19.
$r = -3\cos\theta$

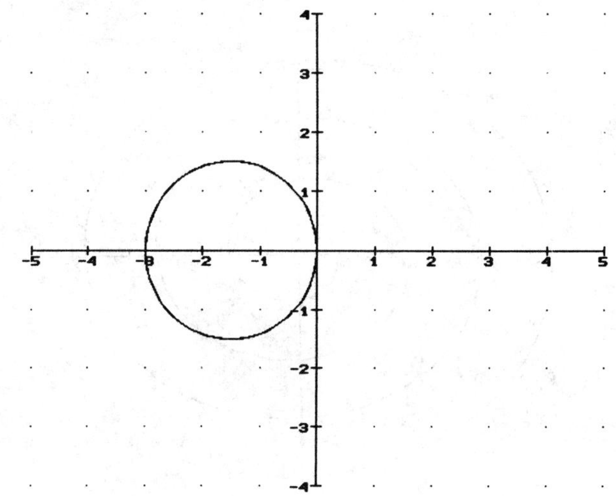

21.

$r = 3\sin 5\theta$

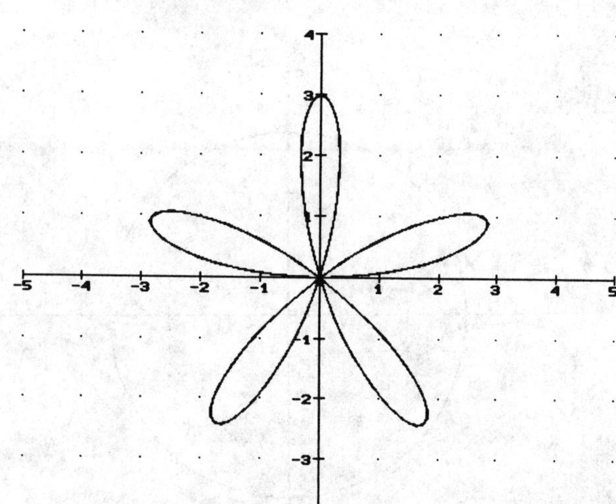

23.

$r^2 = 1 + \sin 2\theta$

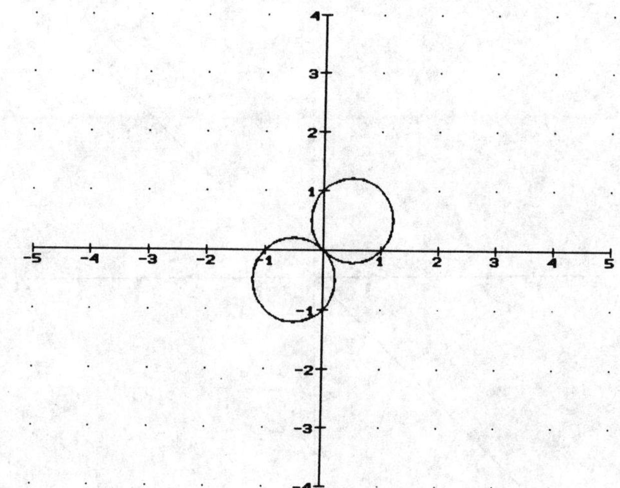

25.

$r = \theta,\ \theta \ge 0$

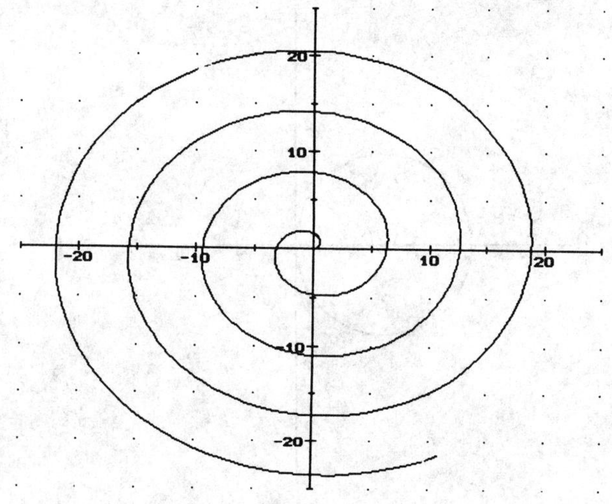

27.

$$r = 4$$
$$\sqrt{x^2 + y^2} = 4$$
$$x^2 + y^2 = 16$$

29.

$$r + 2\sin\theta = 0$$
$$r^2 + 2r\sin\theta = 0$$
$$x^2 + y^2 + 2y = 0$$

31.

$$r\cos\theta = 2$$
$$x = 2$$

33.

$$x^2 + y^2 = 25$$
$$r^2 = 25$$
$$r = 5$$

35.

$$x = -5$$
$$r\cos\theta = -5$$
$$r = -5\sec\theta$$

37.

$$y = 3x$$
$$\frac{y}{x} = 3$$
$$\tan\theta = 3$$

In Exercises 39-41, the RANGE values given show important features of the functions. Other RANGE values are also acceptable. The display on your calculator may vary from that shown.

39a).

XMIN = -1

XMAX = 1

XSCL = 1

YMIN = -1

YMAX = 1

YSCL = 1

GRAPH $r = \cos 3\theta$

$0 \le \theta \le \pi$

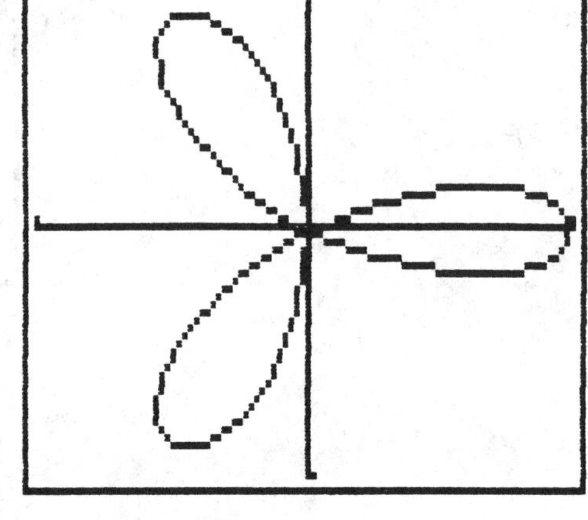

39b).

XMIN = -1

XMAX = 1

XSCL = 1

YMIN = -1

YMAX = 1

YSCL = 1

GRAPH $r = \cos 3\theta$

$0 \le \theta \le 2\pi$

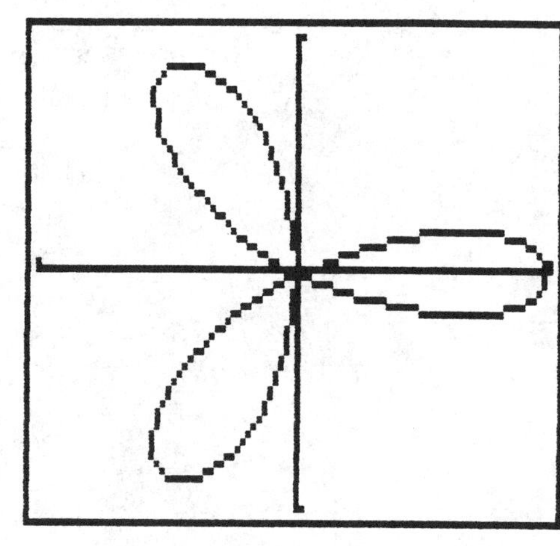

41a).

XMIN = −20

XMAX = 20

XSCL = 5

YMIN = −20

YMAX = 20

YSCL = 5

GRAPH $r = 2\theta$

$0 \le \theta \le 10$

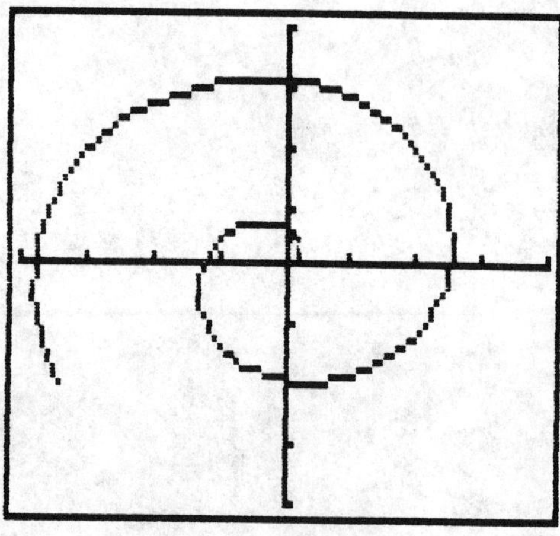

41b).

XMIN = −20

XMAX = 20

XSCL = 5

YMIN = −20

YMAX = 20

YSCL = 5

GRAPH $r = 2\theta$

$-10 \le \theta \le 10$

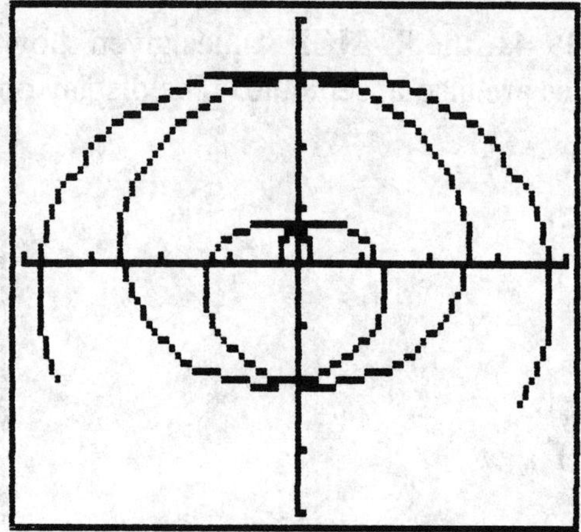

CHAPTER 9 SECTION 5

SECTION 9.5 PROGRESS CHECK

1. (Page 544)

$$\vec{\mathbf{V}}_1 + \vec{\mathbf{V}}_2 = \langle -1, 1\rangle + \langle 5, 3\rangle$$
$$= \langle -1+5, 1+3\rangle$$
$$= \langle 4, 4\rangle$$

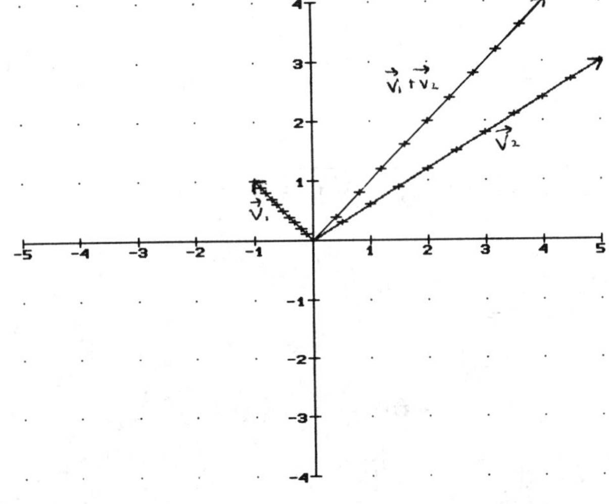

2a). (Page 545)

$$-5\vec{\mathbf{V}}_1 = -5\langle 1, 2\rangle = \langle -5(1), (-5)(2)\rangle = \langle -5, -10\rangle$$

2b).

$$3\vec{\mathbf{V}}_2 = 3\langle -3, 4\rangle = \langle 3(-3), (3)(4)\rangle = \langle -9, 12\rangle$$

2c).

$$2\vec{\mathbf{V}}_1 - 3\vec{\mathbf{V}}_2 = 2\langle 1, 2\rangle - 3\langle -3, 4\rangle$$
$$= \langle 2(1), 2(2)\rangle + \langle (-3)(-3), (-3)(4)\rangle$$
$$= \langle 2, 4\rangle + \langle 9, -12\rangle$$
$$= \langle 2+9, 4-12\rangle$$
$$= \langle 11, -8\rangle$$

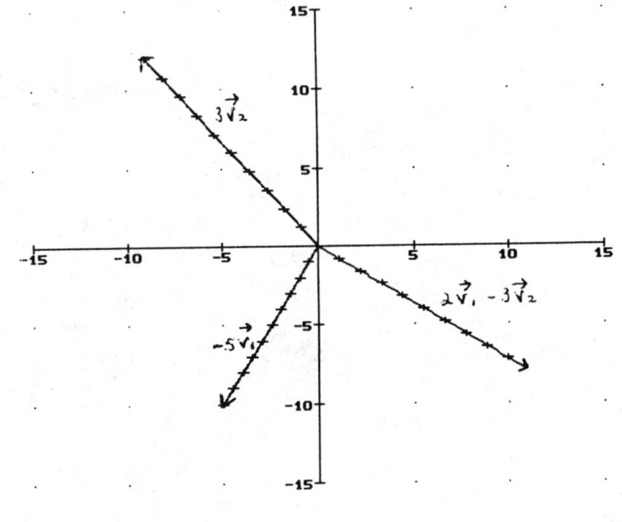

3. (Page 549)

$$\vec{\mathbf{V}} = 2\langle 3, -1\rangle - 2\langle 4, 2\rangle - 3\langle -2, 0\rangle$$
$$= \langle 6, -2\rangle + \langle -8, -4\rangle + \langle 6, 0\rangle$$
$$= \langle 4, -6\rangle$$
$$= 4\mathrm{i} - 6\mathrm{j}$$

$$\|\vec{\mathbf{V}}\| = \sqrt{4^2 + (-6)^2} = \sqrt{52} = 2\sqrt{13}$$

$$\vec{\mathbf{U}} = \frac{1}{2\sqrt{13}}(4\mathrm{i} - 6\mathrm{j})$$
$$= \frac{\sqrt{13}}{26}(4\mathrm{i} - 6\mathrm{j})$$
$$= \frac{2\sqrt{13}}{13}\mathrm{i} - \frac{3\sqrt{13}}{13}\mathrm{j}$$

4. (Page 552)

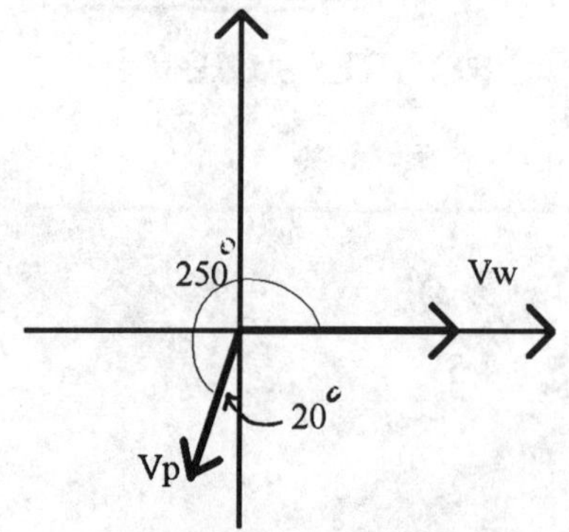

$$\vec{\mathbf{V}}_{\mathbf{p}} = \langle 400\cos 250°, 400\sin 250°\rangle$$
$$= \langle -136.81, -375.88\rangle$$
$$\vec{\mathbf{V}}_{\mathbf{w}} = \langle 20\cos 0°, 20\sin 0°\rangle$$
$$= \langle 20, 0\rangle$$

4a).

$$\vec{\mathbf{V}} = \vec{\mathbf{V}}_{\mathbf{p}} + \vec{\mathbf{V}}_{\mathbf{w}} = \langle -136.81, -375.88\rangle + \langle 20, 0\rangle$$
$$= \langle -116.81, -375.88\rangle$$

4b).

$$\text{ground speed} = \|\vec{\mathbf{V}}\|$$
$$= \sqrt{(-116.81)^2 + (-375.88)^2}$$
$$= 393.61\text{ mph}$$

4c).

$$\theta = \tan^{-1}\left(\frac{-375.88}{-116.81}\right) = 252.74°$$

$$\text{drift angle} = 252.74° - (270° - 20°) = 2.74°$$

EXERCISE SET 9.5

1.

$\langle 2, 5 \rangle$

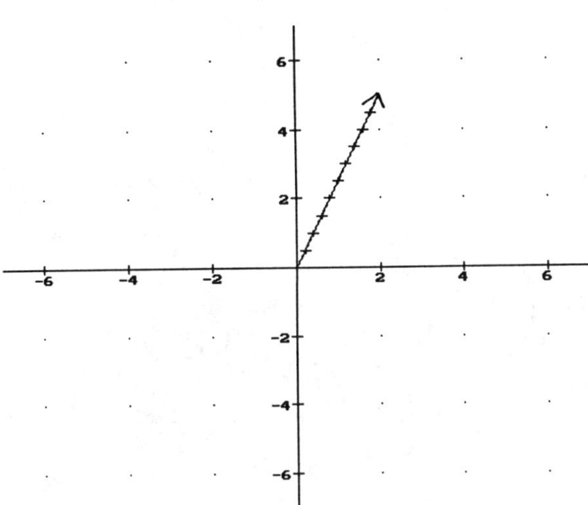

3.

$\langle -5, -1 \rangle$

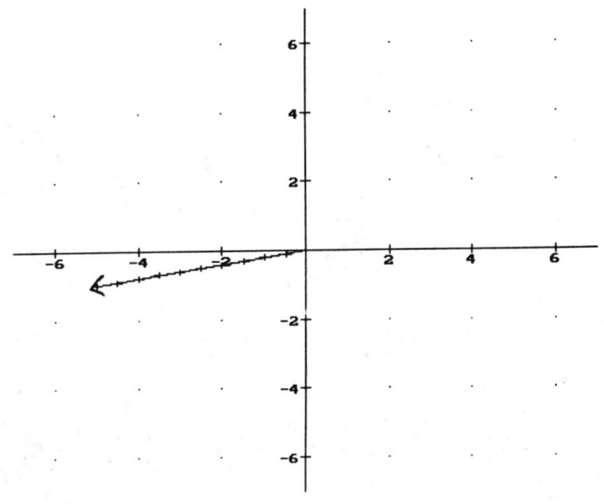

5.

$\langle -4, 1 \rangle$

7.

$P(0, 0), Q(1, 2)$

$$\overrightarrow{\mathbf{PQ}} = \langle 1-0, 2-0 \rangle = \langle 1, 2 \rangle$$

$$\left\| \overrightarrow{\mathbf{PQ}} \right\| = \sqrt{2^2 + 1^2} = \sqrt{5}$$

9.

$P(-1, 3), Q(2, -1)$

$\overrightarrow{\mathbf{PQ}} = \langle 2-(-1), -1-3\rangle = \langle 3, -4\rangle$

$\left\|\overrightarrow{\mathbf{PQ}}\right\| = \sqrt{3^2+(-4)^2} = 5$

11.

$P(4, -1), Q(-7, -3)$

$\overrightarrow{\mathbf{PQ}} = \langle -7-4, -3-(-1)\rangle = \langle -11, -2\rangle$

$\left\|\overrightarrow{\mathbf{PQ}}\right\| = \sqrt{(-11)^2+(-2)^2} = 5\sqrt{5}$

13.

$$\begin{aligned} 2\vec{\mathbf{V}}_1 + 3\vec{\mathbf{V}}_2 &= 2\langle 1, -1\rangle + 3\langle -3, 2\rangle \\ &= \langle 2, -2\rangle + \langle -9, 6\rangle \\ &= \langle -7, 4\rangle \end{aligned}$$

15.

$$\begin{aligned} -2\vec{\mathbf{V}}_1 + \vec{\mathbf{V}}_2 - 6\vec{\mathbf{V}}_3 &= -2\langle 1, -1\rangle + \langle -3, 2\rangle - 6\left\langle \frac{1}{2}, -\frac{1}{3}\right\rangle \\ &= \langle -2, 2\rangle + \langle -3, 2\rangle + \langle -3, 2\rangle \\ &= \langle -8, 6\rangle \end{aligned}$$

17.

$$\begin{aligned} 2\vec{\mathbf{V}}_4 - 3\vec{\mathbf{V}}_5 &= 2\left(7\vec{i} - 3\vec{j}\right) - 3\left(-\vec{i} + 4\vec{j}\right) \\ &= \left(14\vec{i} - 6\vec{j}\right) + \left(3\vec{i} - 12\vec{j}\right) \\ &= 17\vec{i} - 18\vec{j} \\ &= \langle 17, -18\rangle \end{aligned}$$

19.

$$\begin{aligned} \vec{\mathbf{V}}_1 + 2\vec{\mathbf{V}}_2 + 12\vec{\mathbf{V}}_3 - \vec{\mathbf{V}}_4 + \vec{\mathbf{V}}_5 &= \langle 1, -1\rangle + 2\langle -3, 2\rangle + 12\left\langle \frac{1}{2}, -\frac{1}{3}\right\rangle - \left(7\vec{i} - 3\vec{j}\right) + \left(-\vec{i} + 4\vec{j}\right) \\ &= \langle 1, -1\rangle + \langle -6, 4\rangle + \langle 6, -4\rangle + \langle -7, 3\rangle + \langle -1, 4\rangle \\ &= \langle -7, 6\rangle \end{aligned}$$

21.

$$\vec{\mathbf{V}}_1 + \vec{\mathbf{V}}_2 = \langle 1, -1\rangle + \langle -3, 2\rangle = \langle -2, 1\rangle = -2\vec{i} + \vec{j}$$

$$\vec{\mathbf{V}}_1 - \vec{\mathbf{V}}_2 = \langle 1, -1\rangle - \langle -3, 2\rangle = \langle 4, -3\rangle = 4\vec{i} - 3\vec{j}$$

$$2\vec{\mathbf{V}}_1 = 2\langle 1, -1\rangle = \langle 2, -2\rangle = 2\vec{i} - 2\vec{j}$$

$$-\vec{\mathbf{V}}_2 = -\langle -3, 2\rangle = \langle 3, -2\rangle = 3\vec{i} - 2\vec{j}$$

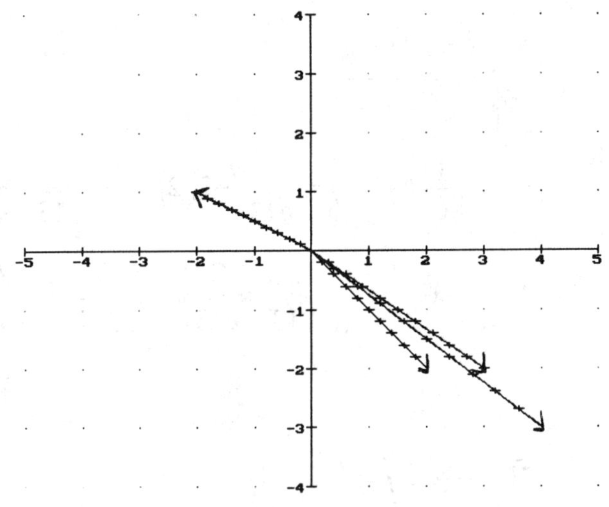

23.

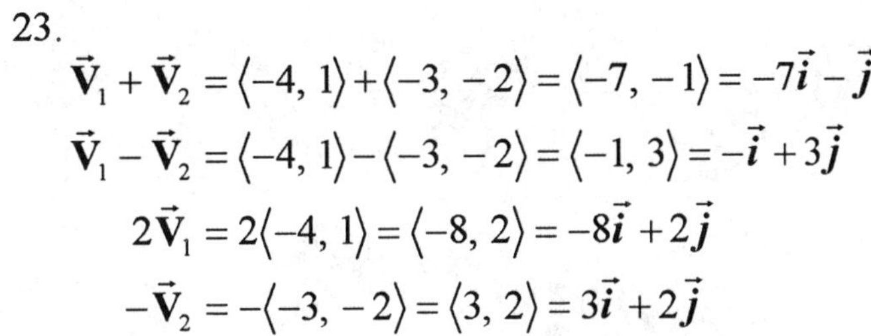

$$\vec{\mathbf{V}}_1 + \vec{\mathbf{V}}_2 = \langle -4, 1\rangle + \langle -3, -2\rangle = \langle -7, -1\rangle = -7\vec{i} - \vec{j}$$

$$\vec{\mathbf{V}}_1 - \vec{\mathbf{V}}_2 = \langle -4, 1\rangle - \langle -3, -2\rangle = \langle -1, 3\rangle = -\vec{i} + 3\vec{j}$$

$$2\vec{\mathbf{V}}_1 = 2\langle -4, 1\rangle = \langle -8, 2\rangle = -8\vec{i} + 2\vec{j}$$

$$-\vec{\mathbf{V}}_2 = -\langle -3, -2\rangle = \langle 3, 2\rangle = 3\vec{i} + 2\vec{j}$$

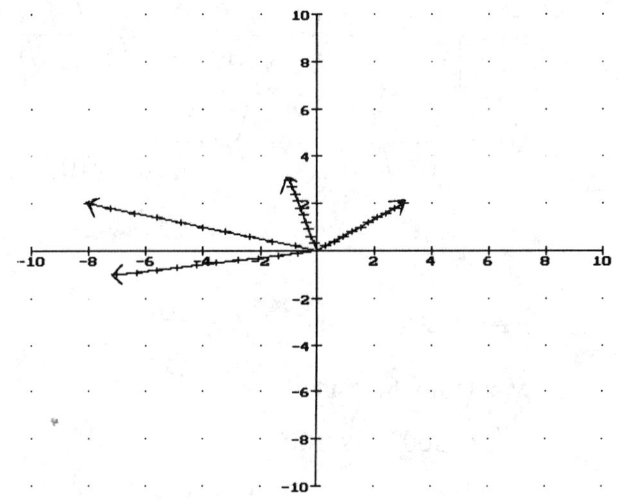

25.

$$\vec{\mathbf{V}} = \langle 2, 0\rangle$$

$$\left\|\vec{\mathbf{V}}\right\| = \sqrt{2^2 + 0^2} = 2$$

$$\vec{\mathbf{U}}_1 = \frac{1}{\left\|\vec{\mathbf{V}}\right\|}\langle 2, 0\rangle = \frac{1}{2}\langle 2, 0\rangle = \langle 1, 0\rangle$$

$$\vec{\mathbf{U}}_2 = -\vec{\mathbf{U}}_1 = -\langle 1, 0\rangle = \langle -1, 0\rangle$$

27.

$$\vec{\mathbf{V}} = \langle 5, 1\rangle$$

$$\left\|\vec{\mathbf{V}}\right\| = \sqrt{5^2 + 1^2} = \sqrt{26}$$

$$\vec{\mathbf{U}}_1 = \frac{1}{\left\|\vec{\mathbf{V}}\right\|}\langle 5, 1\rangle = \frac{1}{\sqrt{26}}\langle 5, 1\rangle = \left\langle \frac{5}{\sqrt{26}}, \frac{1}{\sqrt{26}}\right\rangle$$

$$\vec{\mathbf{U}}_2 = -\vec{\mathbf{U}}_1 = -\left\langle \frac{5}{\sqrt{26}}, \frac{1}{\sqrt{26}}\right\rangle = \left\langle -\frac{5}{\sqrt{26}}, -\frac{1}{\sqrt{26}}\right\rangle$$

29.

$$\vec{\mathbf{V}} = \langle -3, 2 \rangle$$

$$\|\vec{\mathbf{V}}\| = \sqrt{(-3)^2 + 2^2} = \sqrt{13}$$

$$\vec{\mathbf{U}}_1 = \frac{1}{\|\vec{\mathbf{V}}\|}\langle -3, 2\rangle = \frac{1}{\sqrt{13}}\langle -3, 2\rangle = \left\langle -\frac{3}{\sqrt{13}}, \frac{2}{\sqrt{13}} \right\rangle$$

$$\vec{\mathbf{U}}_2 = -\vec{\mathbf{U}}_1 = -\left\langle -\frac{3}{\sqrt{13}}, \frac{2}{\sqrt{13}} \right\rangle = \left\langle \frac{3}{\sqrt{13}}, -\frac{2}{\sqrt{13}} \right\rangle$$

31.

$$\vec{\mathbf{V}} = -3\vec{i} - \vec{j} = \langle -3, -1 \rangle$$

$$\|\vec{\mathbf{V}}\| = \sqrt{(-3)^2 + (-1)^2} = \sqrt{10}$$

$$\vec{\mathbf{U}}_1 = \frac{1}{\|\vec{\mathbf{V}}\|}\langle -3, -1\rangle = \frac{1}{\sqrt{10}}\langle -3, -1\rangle = \left\langle -\frac{3}{\sqrt{10}}, -\frac{1}{\sqrt{10}} \right\rangle$$

$$\vec{\mathbf{U}}_2 = -\left\langle -\frac{3}{\sqrt{10}}, -\frac{1}{\sqrt{10}} \right\rangle = \left\langle \frac{3}{\sqrt{10}}, \frac{1}{\sqrt{10}} \right\rangle$$

33.

$$r = \|\vec{\mathbf{V}}\| = 3$$

$$\vec{\mathbf{V}} = \langle r\cos\theta, r\sin\theta \rangle$$

$$= \langle 3\cos 60°, 3\sin 60° \rangle$$

$$= \left\langle \frac{3}{2}, \frac{3\sqrt{3}}{2} \right\rangle$$

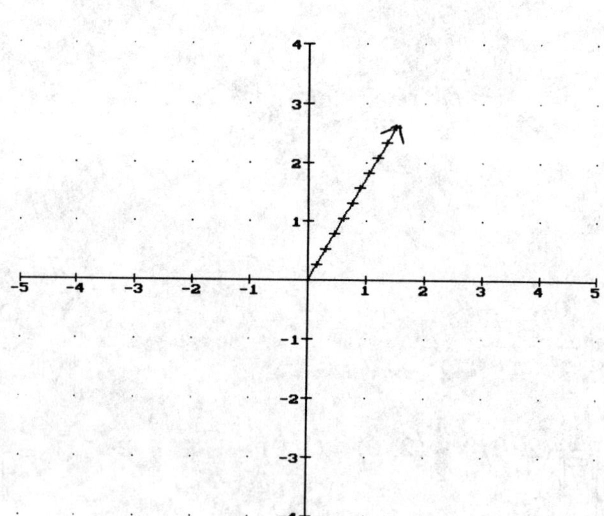

35.

$$r = \left\|\vec{\mathbf{V}}\right\| = 2$$

$$\vec{\mathbf{V}} = \langle r\cos\theta,\ r\sin\theta\rangle$$
$$= \langle 2\cos 300°,\ 2\sin 300°\rangle$$
$$= \langle 1,\ -\sqrt{3}\rangle$$

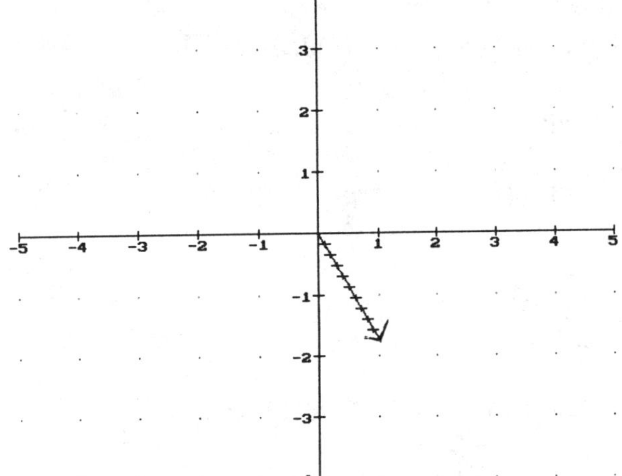

37.

$$r = \left\|\vec{\mathbf{V}}\right\| = 8$$

$$\vec{\mathbf{V}} = \langle r\cos\theta,\ r\sin\theta\rangle$$
$$= \langle 8\cos 315°,\ 8\sin 315°\rangle$$
$$= \langle 4\sqrt{2},\ -4\sqrt{2}\rangle$$

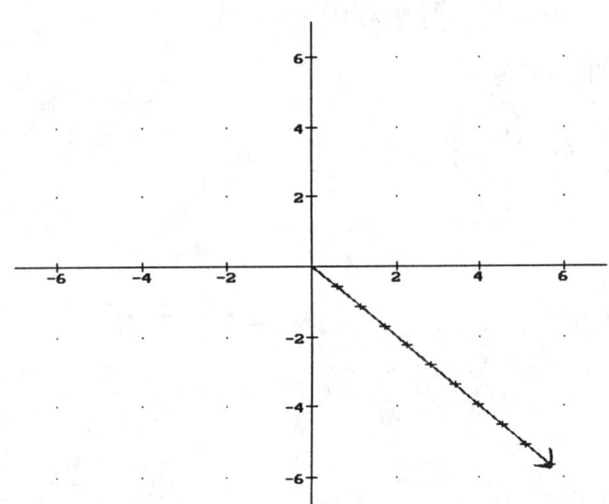

39.

$\vec{\mathbf{V}}_1 = \langle 3,\ -1\rangle$, (Quadrant IV)

$\vec{\mathbf{V}}_2 = \langle 2,\ -3\rangle$ (Quadrant IV)

$$\theta_1 = \tan^{-1}\left(-\frac{1}{3}\right) \qquad \theta_2 = \tan^{-1}\left(-\frac{3}{2}\right)$$

$$\theta_1 = 341.57° \qquad \theta_2 = 303.69°$$

$$\theta = |\theta_1 - \theta_2|$$
$$= |341.57° - 303.69°|$$
$$= 37.88°$$

41.

$\vec{\mathbf{V}}_1 = \langle -1,\ -2\rangle$, (Quadrant III)

$\vec{\mathbf{V}}_2 = \langle -2,\ -3\rangle$ (Quadrant III)

$$\theta_1 = \tan^{-1}\left(\frac{-2}{-1}\right) \qquad \theta_2 = \tan^{-1}\left(\frac{-3}{-2}\right)$$

$$\theta_1 = 243.43° \qquad \theta_2 = 236.31°$$

$$\theta = |\theta_1 - \theta_2|$$
$$= |243.43° - 236.31°|$$
$$= 7.12°$$

43.

$\vec{\mathbf{V}}_1 = \langle 2, -5 \rangle,$ (Quadrant IV)

$\theta_1 = \tan^{-1}\left(\dfrac{-5}{2}\right)$

$\theta_1 = 291.80°$

$\vec{\mathbf{V}}_2 = \langle -2, -5 \rangle$ (Quadrant III)

$\theta_2 = \tan^{-1}\left(\dfrac{-5}{-2}\right)$

$\theta_2 = 248.20°$

$$\begin{aligned}\theta &= |\theta_1 - \theta_2| \\ &= |291.80° - 248.20°| \\ &= 43.60°\end{aligned}$$

45.

$\vec{\mathbf{V}}_1 + \vec{\mathbf{V}}_2 = \langle 1, -1 \rangle + \langle 2, 0 \rangle = \langle 3, -1 \rangle$

(Quadrant IV)

$\theta = \tan^{-1}\left(\dfrac{-1}{3}\right)$

$\theta = 341.57°$

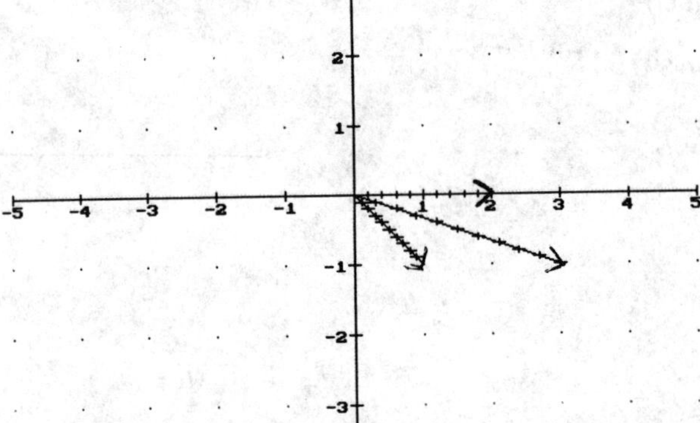

47.

$$\mathbf{V}_p = \langle 280\cos 110°,\ 280\sin 110°\rangle$$
$$= \langle -95.77,\ 263.11\rangle$$
$$\mathbf{V}_w = \langle 40\cos 270°,\ 40\sin 270°\rangle$$
$$= \langle 0,\ -40\rangle$$

47a).

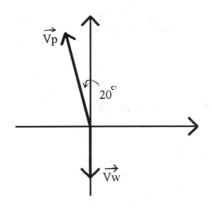

$$\vec{\mathbf{V}} = \vec{\mathbf{V}}_p + \vec{\mathbf{V}}_w = \langle -95.77,\ 263.11\rangle + \langle 0,\ -40\rangle$$
$$= \langle -95.77,\ 223.11\rangle$$

47b).

$$\text{ground speed} = \left\|\vec{\mathbf{V}}\right\| = \sqrt{(-95.77)^2 + (223.11)^2}$$
$$= 242.80 \text{ mph}$$

47c).

$$\theta = \tan^{-1}\left(\frac{223.11}{-95.77}\right) = 113.23°$$

$$\text{drift angle} = 113.23° - (90° + 20°) = 3.23°$$

49.

$$\left\|\vec{\mathbf{V}}\right\| = \sqrt{40^2 + 280^2 - 2(40)(280)\cos 20°}$$
$$= 242.80 \text{ mph}$$

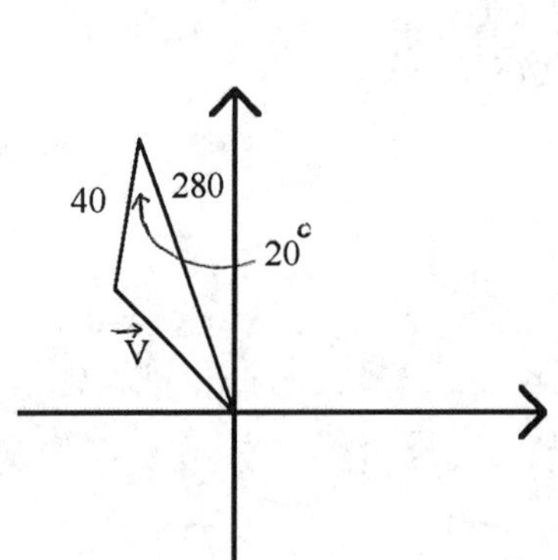

$$\frac{40}{\sin\gamma} = \frac{242.8}{\sin 20°}$$

$$\sin\gamma = \frac{40\sin 20°}{242.8}$$

$$\gamma = 3.23° \text{ (drift angle)}$$

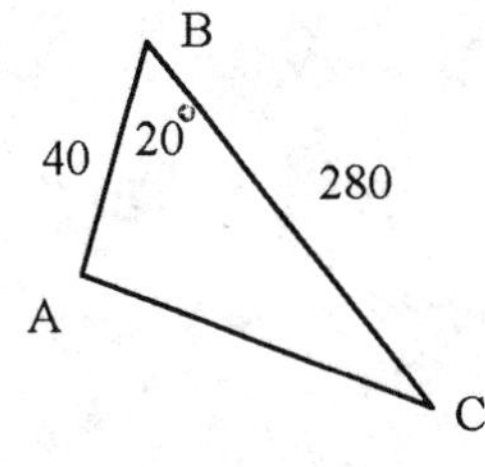

CHAPTER 9 REVIEW EXERCISES

1.

$$a^2 = b^2 + c^2 - 2bc\cos\alpha$$
$$12^2 = 7^2 + 15^2 - 2(7)(15)\cos\alpha$$
$$\cos\alpha = 0.6190$$
$$\alpha = 51.75° = 51°45'$$

2.

$$\frac{a}{\sin\alpha} = \frac{b}{\sin\beta}$$
$$\frac{20}{\sin 55°} = \frac{15}{\sin\beta}$$
$$\sin\beta = \frac{15\sin 55°}{20}$$
$$\sin\beta = 0.6144$$
$$\beta = 37.91° = 37°55'$$

3.

$$\gamma = 180° - (\alpha + \beta)$$
$$\gamma = 180° - (38° + 22°) = 120°$$
$$\frac{a}{\sin\alpha} = \frac{c}{\sin\gamma}$$
$$\frac{10}{\sin 38°} = \frac{c}{\sin 120°}$$
$$c = \frac{10\sin 120°}{\sin 38°}$$
$$c = 14.07$$

4.

$$a^2 = b^2 + c^2 - 2bc\cos\alpha$$
$$a^2 = 8^2 + 12^2 - 2(8)(12)\cos 35°$$
$$a^2 = 50.72$$
$$a = 7.12$$

5.

$$|2-i| = \sqrt{2^2 + (-1)^2} = \sqrt{5}$$

6.

$$|-3+2i| = \sqrt{(-3)^2 + 2^2} = \sqrt{13}$$

7.

$$|-4-5i| = \sqrt{(-4)^2 + (-5)^2} = \sqrt{41}$$

8.

$-3+3i$ (Quadrant II)

$a = -3 \qquad b = 3$

$$r = \sqrt{a^2+b^2} = \sqrt{(-3)^2+3^2} = 3\sqrt{2}$$

$$\tan\theta = \frac{3}{-3} = -1$$

$$\theta = \frac{3\pi}{4}$$

$$-3+3i = 3\sqrt{2}\left(\cos\frac{3\pi}{4} + i\sin\frac{3\pi}{4}\right)$$

9.

$$2(\cos 90° + i\sin 90°) = 2(0+i) = 2i$$

10.

$$\sqrt{2}(\cos 315° + i\sin 315°) = \sqrt{2}\left(\frac{\sqrt{2}}{2} - \frac{\sqrt{2}}{2}i\right) = 1-i$$

11.

-2

$a = -2 \qquad b = 0$

$$r = \sqrt{(-2)^2 + 0^2} = 2$$

$$\theta = \pi$$

$$-2 = 2(\cos\pi + i\sin\pi)$$

12.

$$4(\cos 22° + i\sin 22°)\cdot 6(\cos 15° + i\sin 15°)$$
$$= 4\cdot 6[\cos(22°+15°) + i\sin(22°+15°)]$$
$$= 24(\cos 37° + i\sin 37°)$$

13.

$$\frac{5(\cos 71° + i\sin 71°)}{3(\cos 50° + i\sin 50°)} = \frac{5}{3}[\cos(71°-50°) + i\sin(71°-50°)]$$
$$= \frac{5}{3}(\cos 21° + i\sin 21°)$$

14.

$$2(\cos 210°+i\sin 210°)\cdot(\cos 240°+i\sin 240°)$$
$$=2\cdot 1[\cos(210°+240°)+i\sin(210°-240°)]$$
$$=2(\cos 450°+i\sin 450°)$$
$$=2(\cos 90°+i\sin 90°)$$

15.

$3-3i$ (Quadrant IV)

$a=3$ $b=-3$

$$r=\sqrt{3^2+(-3)^2}=3\sqrt{2}$$
$$\theta=\tan^{-1}\left(-\frac{3}{3}\right)$$
$$\theta=\frac{7\pi}{4}$$

$$3-3i=3\sqrt{2}\left(\cos\frac{7\pi}{4}+i\sin\frac{7\pi}{4}\right)$$
$$(3-3i)^5=\left[3\sqrt{2}\left(\cos\frac{7\pi}{4}+i\sin\frac{7\pi}{4}\right)\right]^5$$
$$=(3\sqrt{2})^5\left[\cos\left(\frac{7\pi}{4}\cdot 5\right)+i\sin\left(\frac{7\pi}{4}\cdot 5\right)\right]$$
$$=972\sqrt{2}\left(\cos\frac{35\pi}{4}+i\sin\frac{35\pi}{4}\right)$$
$$=972\sqrt{2}\left(-\frac{\sqrt{2}}{2}+\frac{\sqrt{2}}{2}i\right)$$
$$=-972+972i$$

16.

$$\left[2(\cos 90°+i\sin 90°)\right]^3$$
$$=2^3\left[\cos(3\cdot 90°)+i\sin(3\cdot 90°)\right]$$
$$=8(\cos 270°+i\sin 270°)$$
$$=8(0-i)$$
$$=0-8i$$

17.

$$\sqrt{-9}=\pm 3i$$

$3i$

$a=0 \qquad b=3$

$$r=\sqrt{0^2+3^2}=3$$
$$\theta=\frac{\pi}{2}$$
$$3i=3\left(\cos\frac{\pi}{2}+i\sin\frac{\pi}{2}\right)$$
$$=3(\cos 90°+i\sin 90°)$$

$-3i$

$a=0 \qquad b=-3$

$$r=\sqrt{0^2+(-3)^2}=3$$
$$\theta=\frac{3\pi}{2}$$
$$-3i=3\left(\cos\frac{3\pi}{2}+i\sin\frac{3\pi}{2}\right)$$
$$=3(\cos 270°+i\sin 270°)$$

18.

$$z^3 - 1 = 0$$

$$z^3 = 1$$

Find all cube roots of 1.

$$1 = 1(\cos 0° + i \sin 0°)$$

$$U_0 = \sqrt[3]{1}(\cos 0° + i \sin 0°) = 1$$

$$U_1 = \sqrt[3]{1}\left(\cos\frac{0°+360°}{3} + i\sin\frac{0°+360°}{3}\right)$$

$$= 1(\cos 120° + i \sin 120°)$$

$$= -\frac{1}{2} + \frac{\sqrt{3}}{2}i$$

$$U_2 = \sqrt[3]{1}\left(\cos\frac{0°+360°\cdot 2}{3} + i\sin\frac{0°+360°\cdot 2}{3}\right)$$

$$= 1(\cos 240° + i \sin 240°)$$

$$= -\frac{1}{2} - \frac{\sqrt{3}}{2}i$$

$$1,\ -\frac{1}{2} + \frac{\sqrt{3}}{2}i,\ -\frac{1}{2} - \frac{\sqrt{3}}{2}i$$

19.

$$r = \sec\theta\tan\theta$$

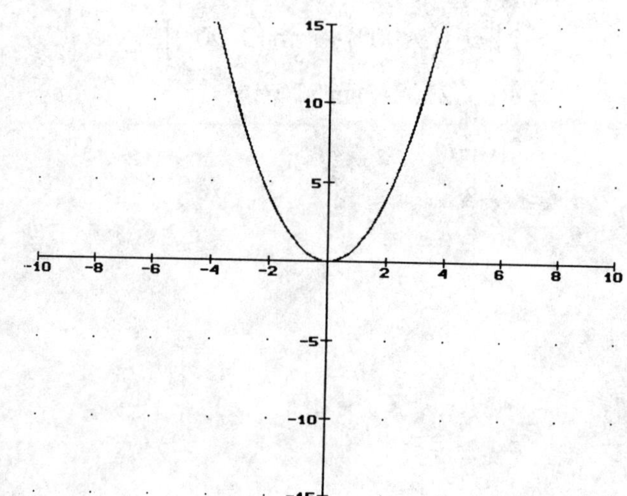

20.

$$r = |\theta|$$

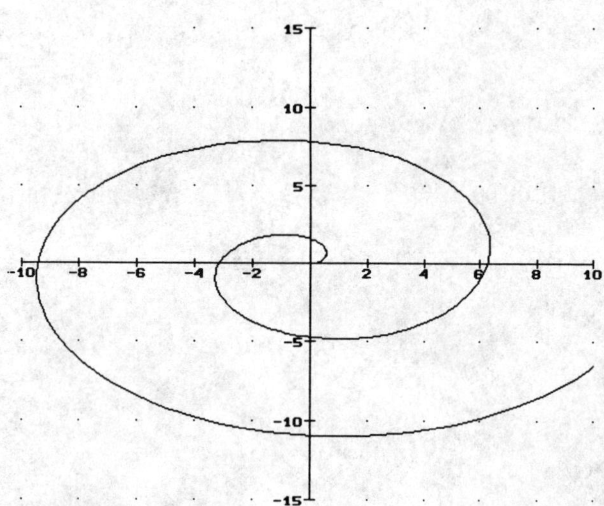

21.

$$r = 3\csc\theta$$

$$r = \frac{3}{\sin\theta}$$

$$3 = r\sin\theta$$

$$y = 3$$

22.

$$x^2+9y^2=9$$

$$(r\cos\theta)^2+9(r\sin\theta)^2=9$$

$$r^2\cos^2\theta+9r^2\sin^2\theta=9$$

$$r^2(\cos^2\theta+9\sin^2\theta)=9$$

$$r^2=\frac{9}{\cos^2\theta+9\sin^2\theta}$$

Exercise 23 graphs are not drawn in a multiple of the EQUAL viewing rectangle; hence the circles do not appear round, as they should.

23a).

XMIN $=-5$

XMAX $=5$

XSCL $=1$

YMIN $=-5$

YMAX $=5$

YSCL $=1$

GRAPH $r=3\sin\theta$

$0\le\theta\le 2\pi$

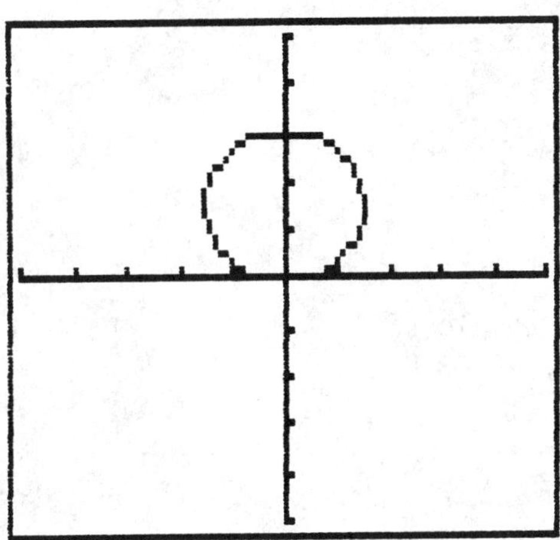

23b).

XMIN $=-5$

XMAX $=5$

XSCL $=1$

YMIN $=-5$

YMAX $=5$

YSCL $=1$

GRAPH $r=-3\cos\theta$

$0\le\theta\le 2\pi$

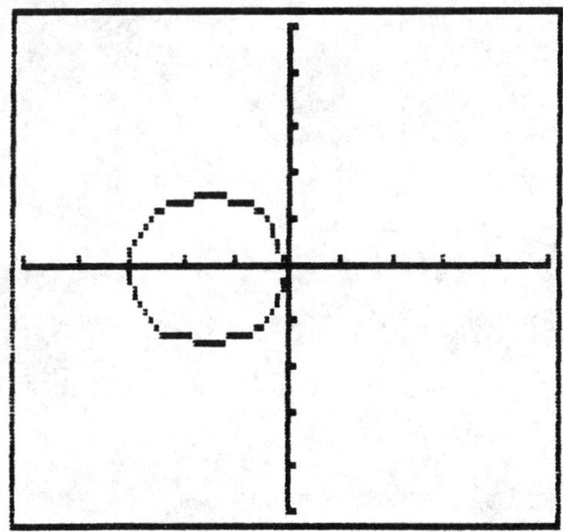

23c).

XMIN = −5

XMAX = 5

XSCL = 1

YMIN = −5

YMAX = 5

YSCL = 1

GRAPH $r = -3\sin\theta$

$0 \le \theta \le 2\pi$

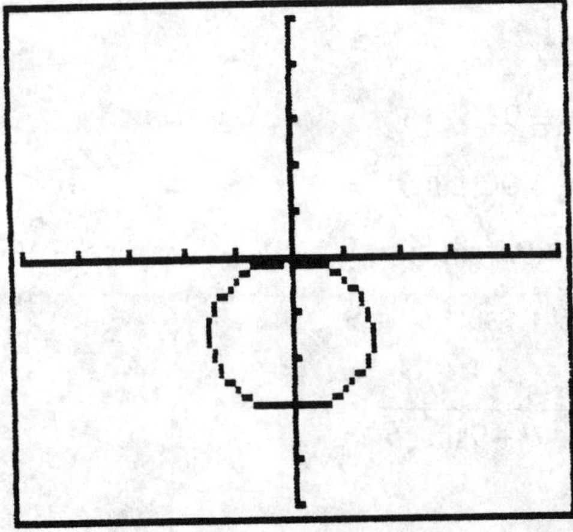

23d).

XMIN = −5

XMAX = 5

XSCL = 1

YMIN = −5

YMAX = 5

YSCL = 1

GRAPH $r = 3\cos\theta$

$0 \le \theta \le 2\pi$

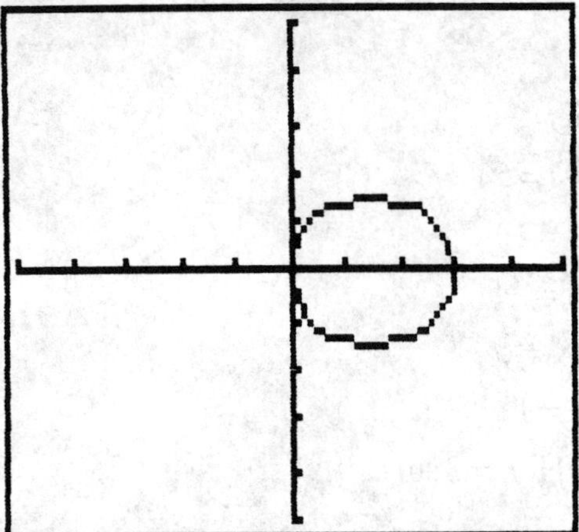

24a).

XMIN = −2

XMAX = 2

XSCL = 1

YMIN = −2

YMAX = 2

YSCL = 1

GRAPH $r = 2\sqrt{\cos(2\theta)}$

$0 \le \theta \le 2\pi$

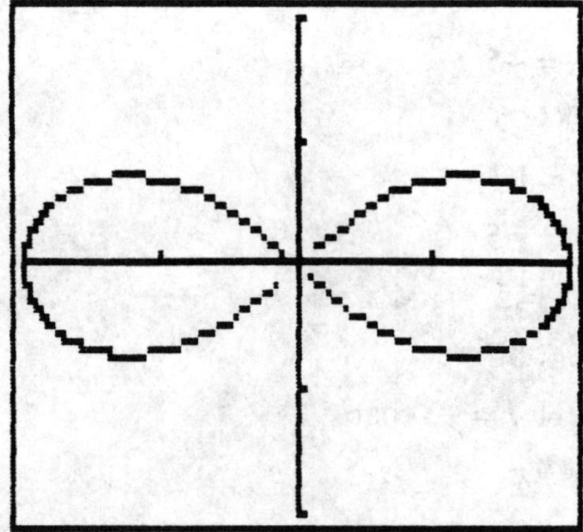

24b).

XMIN = −2

XMAX = 2

XSCL = 1

YMIN = −2

YMAX = 2

YSCL = 1

GRAPH $r = -2\sqrt{\cos(2\theta)}$

$0 \le \theta \le 2\pi$

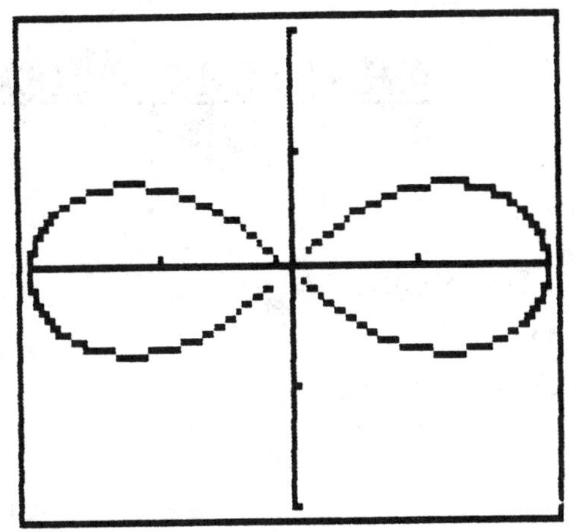

25.

$$3\vec{\mathbf{V}}_1 - 2\vec{\mathbf{V}}_2 + \vec{\mathbf{V}}_3 = 3\langle -1,\ 3\rangle - 2\langle 4,\ 2\rangle + \langle -2,\ -3\rangle$$
$$= \langle -3,\ 9\rangle + \langle -8,\ -4\rangle + \langle -2,\ -3\rangle$$
$$= \langle -13,\ 2\rangle$$
$$= -13\vec{i} + 2\vec{j}$$

26.

$\vec{\mathbf{V}}_3 = \langle -2,\ -3\rangle$ (Quadrant III)

$$\left\|\vec{\mathbf{V}}_3\right\| = \sqrt{(-2)^3 + (-3)^2} = \sqrt{13}$$
$$\tan\theta = \frac{-3}{-2}$$
$$\theta = \tan\frac{3}{2}$$
$$\theta = 236.31°$$

27.

$\vec{\mathbf{V}}_2 = \langle 4,\ 2\rangle$

(Quadrant I)

$$\tan\theta_2 = \frac{2}{4}$$
$$\theta_2 = \tan^{-1}\frac{1}{2}$$
$$\theta_2 = 26.57°$$

$\vec{\mathbf{V}}_3 = \langle -2,\ -3\rangle$

(Quadrant III)

$$\tan\theta_3 = \frac{-3}{-2}$$
$$\theta_3 = \tan^{-1}\frac{3}{2}$$
$$\theta_3 = 236.31°$$

$$\theta = 360° - (236.31° - 26.57°) \quad (0° \le \theta \le 180°)$$
$$\theta = 150.26°$$

CHAPTER 9 REVIEW TEST

1.

$$a^2 = b^2 + c^2 - 2bc\cos\alpha$$
$$2^2 = 4^2 + 5^2 - 2(4)(5)\cos\alpha$$
$$\cos\alpha = 0.925$$
$$\alpha = 22.33° = 22°20'$$

2.

$$\gamma = 180° - (\alpha + \beta)$$
$$= 180° - (15° + 28°)$$
$$= 137°$$

$$\frac{c}{\sin 137°} = \frac{10}{\sin 28°}$$
$$c = \frac{10\sin 137°}{\sin 28°}$$
$$c = 14.53$$

3.

$$a^2 = b^2 + c^2 - 2bc\cos\alpha$$
$$a^2 = 10^2 + 13^2 - 2(10)(13)\cos 54°$$
$$a^2 = 116.18$$
$$a = 10.78$$

$$\frac{10.78}{\sin 54°} = \frac{10}{\sin\beta}$$
$$\sin\beta = \frac{10\sin 54°}{10.78}$$
$$\sin\beta = 0.7505$$
$$\beta = 48.63° = 48°38'$$

4.

$$b^2 = a^2 + c^2 - 2ac\cos\beta$$
$$b^2 = 5^2 + 9^2 - 2(5)(9)\cos 36°$$
$$b^2 = 33.19$$
$$b = 5.76$$

5.

$$|3 - 4i| = \sqrt{3^2 + (-4)^2} = 5$$

6.

$$|-2 + 2i| = \sqrt{(-2)^2 + 2^2} = 2\sqrt{2}$$

7.

$$\frac{1}{2}(\cos 14°+i\sin 14°)\cdot 10(\cos 72°+i\sin 72°)$$
$$=\frac{1}{2}\cdot 10[\cos(14°+72°)+i\sin(14°+72°)]$$
$$=5(\cos 86°+i\sin 86°)$$

8.

$$\frac{3(\cos 85°+i\sin 85°)}{6(\cos 8°+i\sin 8°)}$$
$$=\frac{3}{6}[\cos(85°-8°)+i\sin(85°-8°)]$$
$$=\frac{1}{2}(\cos 77°+i\sin 77°)$$

9.

$$\left(\cos\frac{\pi}{5}+i\sin\frac{\pi}{5}\right)\cdot 5\left(\cos\frac{\pi}{8}+i\sin\frac{\pi}{8}\right)$$
$$=1\cdot 5\left(\cos\left(\frac{\pi}{5}+\frac{\pi}{8}\right)+i\sin\left(\frac{\pi}{5}+\frac{\pi}{8}\right)\right)$$
$$=5\left(\cos\frac{13\pi}{40}+i\sin\frac{13\pi}{40}\right)$$

10.

$$\frac{\frac{1}{2}\left(\cos\frac{\pi}{10}+i\sin\frac{\pi}{10}\right)}{\frac{1}{4}\left(\cos\frac{\pi}{7}+i\sin\frac{\pi}{7}\right)}$$
$$=\frac{\frac{1}{2}}{\frac{1}{4}}\left[\cos\left(\frac{\pi}{10}-\frac{\pi}{7}\right)+i\sin\left(\frac{\pi}{10}-\frac{\pi}{7}\right)\right]$$
$$=2\left[\cos\left(\frac{-3\pi}{70}\right)+i\sin\left(\frac{-3\pi}{70}\right)\right]$$
$$=2\left(\cos\frac{137\pi}{70}+i\sin\frac{137\pi}{70}\right)$$

11.

$$\left[\frac{1}{5}(\cos 120°+i\sin 120°)\right]^4$$
$$=\left(\frac{1}{5}\right)^4\left[\cos(4\cdot 120°)+i\sin(4\cdot 120°)\right]$$
$$=\frac{1}{625}(\cos 480°+i\sin 480°)$$
$$=\frac{1}{625}(\cos 120°+i\sin 120°)$$
$$=\frac{1}{625}\left(-\frac{1}{2}+\frac{\sqrt{3}}{2}i\right)$$
$$=-\frac{1}{1250}+\frac{\sqrt{3}}{1250}i$$

12.

$$(-2i)^6=\left[2\left(\cos\frac{3\pi}{2}+i\sin\frac{3\pi}{2}\right)\right]^6$$
$$=2^6\left[\cos\left(6\cdot\frac{3\pi}{2}\right)+i\sin\left(6\cdot\frac{3\pi}{2}\right)\right]$$
$$=64(\cos 9\pi+i\sin 9\pi)$$
$$=64[(-1)+0i]$$
$$=-64+0i$$

13.

$$z^3+1=0$$
$$z^3=-1$$
$$-1=1(\cos\pi+i\sin\pi)$$

$$U_0=\sqrt[3]{1}\left(\cos\frac{\pi}{3}+i\sin\frac{\pi}{3}\right)=\frac{1}{2}+\frac{\sqrt{3}}{2}i$$

$$U_1=\sqrt[3]{1}\left[\cos\left(\frac{\pi+2\pi}{3}\right)+i\sin\left(\frac{\pi+2\pi}{3}\right)\right]$$
$$=\cos\pi+i\sin\pi$$
$$=-1$$

$$U_2=\sqrt[3]{1}\left[\cos\left(\frac{\pi+2\pi\cdot 2}{3}\right)+i\sin\left(\frac{\pi+2\pi\cdot 2}{3}\right)\right]$$
$$=\cos\frac{5\pi}{3}+i\sin\frac{5\pi}{3}$$
$$=\frac{1}{2}-\frac{\sqrt{3}}{2}i$$

14.

$$z^3 - 27 = 0$$
$$z^3 = 27$$
$$27 = 27(\cos 0° + i \sin 0°)$$

$$U_0 = \sqrt[3]{27}(\cos 0° + i \sin 0°) = 3$$

$$U_1 = \sqrt[3]{27}\left[\cos\left(\frac{0°+360°}{3}\right) + i \sin\left(\frac{0°+360°}{3}\right)\right]$$
$$= 3(\cos 120° + i \sin 120°)$$
$$= 3\left(-\frac{1}{2} + \frac{\sqrt{3}}{2} i\right)$$
$$= -\frac{3}{2} + \frac{3\sqrt{3}}{2} i$$

$$U_2 = \sqrt[3]{27}\left[\cos\left(\frac{0°+2\cdot 360°}{3}\right) + i \sin\left(\frac{0°+2\cdot 360°}{3}\right)\right]$$
$$= 3(\cos 240° + i \sin 240°)$$
$$= 3\left(-\frac{1}{2} - \frac{\sqrt{3}}{2} i\right)$$
$$= -\frac{3}{2} - \frac{3\sqrt{3}}{2} i$$

15.

$$r = 4\csc\theta$$

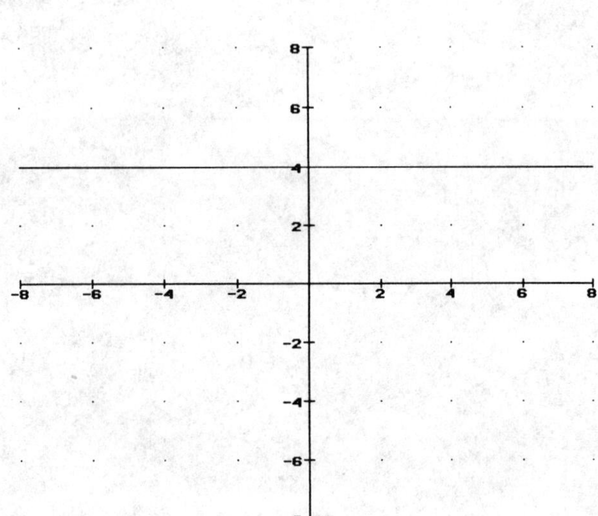

16.

$$r = \sec^2\left(\frac{\theta}{2}\right)$$

$$r = \frac{1}{\cos^2\left(\frac{\theta}{2}\right)}$$

$$1 = r\cos^2\left(\frac{\theta}{2}\right)$$

$$1 = r\left(\frac{1+\cos\theta}{2}\right)$$

$$2 = r + r\cos\theta$$

$$2 = \sqrt{x^2+y^2} + x$$

$$2 - x = \sqrt{x^2+y^2}$$

$$(2-x)^2 = x^2 + y^2$$

$$4 - 4x + x^2 = x^2 + y^2$$

$$0 = y^2 + 4x - 4$$

17.

$$x^2 + xy + y^2 = 1$$

$$\left(x^2+y^2\right) + xy = 1$$

$$r^2 + (r\cos\theta)(r\sin\theta) = 1$$

$$r^2 + r^2\cos\theta\sin\theta = 1$$

$$r^2(1+\cos\theta\sin\theta) = 1$$

$$r^2 = \frac{1}{1+\cos\theta\sin\theta}$$

$$r^2 = \frac{1}{1+\frac{1}{2}\sin 2\theta}$$

$$r^2 = \frac{2}{2+\sin 2\theta}$$

18.

$\vec{\mathbf{V}}_1 = \langle 2, -5 \rangle$ (Quadrant IV)

$\theta_1 = \tan^{-1}\left(\frac{-5}{2}\right)$

$\theta_1 = 291.80°$

$\vec{\mathbf{V}}_2 = \langle -4, 1 \rangle$ (Quadrant II)

$\theta_2 = \tan^{-1}\left(\frac{1}{-4}\right)$

$\theta_2 = 165.96°$

$$\begin{aligned}\theta &= |\theta_1 - \theta_2| \\ &= |291.80° - 165.96°| \\ &= 125.84°\end{aligned}$$

CHAPTERS 7-9 CUMULATIVE REVIEW EXERCISES

1a).

$$\sin\left(-\frac{9\pi}{5}\right) = \sin\left(-\frac{9\pi}{5} + 2\pi\right) = \sin\frac{\pi}{5}$$

1b).

$$\cos\frac{8\pi}{3} = \cos\left(\frac{8\pi}{3} - 2\pi\right) = \cos\frac{2\pi}{3}$$

$$= -\cos\left(\pi - \frac{2\pi}{3}\right) = -\cos\frac{\pi}{3}$$

1c).

$$\tan\left(-\frac{11\pi}{7}\right) = \tan\left(-\frac{11\pi}{7} + 2\pi\right) = \tan\frac{3\pi}{7}$$

2a).

$$\sec\left(-\frac{37\pi}{3}\right) = \sec\left(-\frac{37\pi}{3} + 7\cdot 2\pi\right) = \sec\frac{5\pi}{3}$$

$$= \frac{1}{\cos\frac{5\pi}{3}} = \frac{1}{\frac{1}{2}} = 2$$

2b).

$$\cot\frac{23\pi}{6} = \cot\frac{11\pi}{6} = \frac{1}{\tan\frac{11\pi}{6}} = \frac{1}{-\frac{1}{\sqrt{3}}} = -\sqrt{3}$$

2c).

$$\sin\left(-\frac{29\pi}{6}\right) = \sin\frac{7\pi}{6} = -\frac{1}{2}$$

3.

$$f(x) = 3\sin\left(2x + \frac{\pi}{2}\right)$$

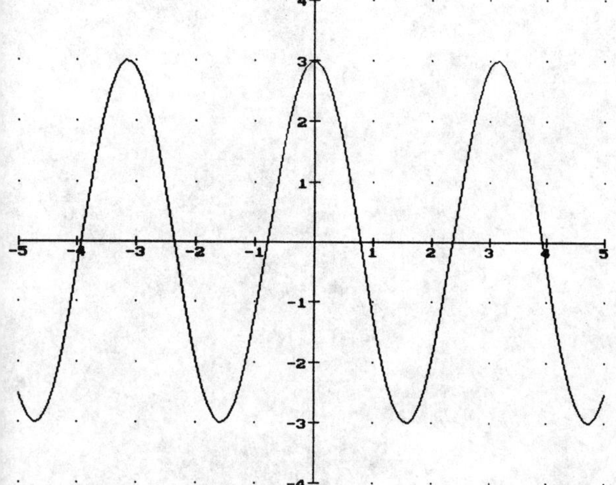

4a).

$$\cos t = -\frac{\sqrt{2}}{2}$$

reference angle: $\frac{\pi}{4}$

Quadrant II: $\pi - \frac{\pi}{4} = \frac{3\pi}{4}$

Quadrant III: $\pi + \frac{\pi}{4} = \frac{5\pi}{4}$

4b).

$\csc t = 2$

$\sin t = \dfrac{1}{\csc t} = \dfrac{1}{2}$

reference angle: $\dfrac{\pi}{6}$

Quadrant I: $\dfrac{\pi}{6}$

Quadrant II: $\pi - \dfrac{\pi}{6} = \dfrac{5\pi}{6}$

4c).

$\tan t = -\sqrt{3}$

reference angle: $\dfrac{\pi}{3}$

Quadrant II: $\pi - \dfrac{\pi}{3} = \dfrac{2\pi}{3}$

Quadrant IV: $2\pi - \dfrac{\pi}{3} = \dfrac{5\pi}{3}$

5a).

Let $\theta = \cos^{-1}\dfrac{3}{5}$ $\qquad \theta \in [0, \pi]$

$$\cos\theta = \frac{3}{5}$$

$$\begin{aligned}\sin^2 &= 1 - \cos^2\theta \\ &= 1 - \left(\frac{3}{5}\right)^2 \\ &= \frac{16}{25}\end{aligned}$$

$$\sin\theta = \frac{4}{5}$$

$$\tan\left(\cos^{-1}\frac{3}{5}\right) = \tan\theta = \frac{\sin\theta}{\cos\theta} = \frac{\frac{4}{5}}{\frac{3}{5}} = \frac{4}{3}$$

5b).

Let $\theta = \sin^{-1}\dfrac{5}{13}$ $\qquad \theta \in \left[-\dfrac{\pi}{2}, \dfrac{\pi}{2}\right]$

$$\sin\theta = \frac{5}{13}$$

$$\begin{aligned}\cos^2 &= 1 - \sin^2\theta \\ &= 1 - \left(\frac{5}{13}\right)^2 \\ &= \frac{144}{169}\end{aligned}$$

$$\cos\theta = \frac{12}{13}$$

$$\sec\theta = \frac{1}{\cos\theta} = \frac{1}{\frac{12}{13}} = \frac{13}{12}$$

5c).

$$\cot\left(-\frac{\pi}{4}\right) = -\cot\frac{\pi}{4} = -\frac{1}{\tan\frac{\pi}{4}} = -\frac{1}{1} = -1$$

$$\cos^{-1}(-1) = \pi$$

6a).

$\left.\begin{array}{l}\tan t > 0 : \text{QI and III} \\ \cos t < 0 : \text{QII and III}\end{array}\right\}$ Quadrant III

$\tan t = \sqrt{3}$

reference angle: $\dfrac{\pi}{3}$

$t = \pi + \dfrac{\pi}{3} = \dfrac{4\pi}{3}$

6b).

$$\left.\begin{array}{l}\sec t > 0 : \text{ QI and IV} \\ \tan t < 0 : \text{ QII and IV}\end{array}\right\} \quad \text{Quadrant IV}$$

$$\sec t = 2$$

$$\cos t = \frac{1}{2}$$

$$\text{reference angle: } \frac{\pi}{3}$$

$$t = 2\pi - \frac{\pi}{3} = \frac{5\pi}{3}$$

6c).

$$\left.\begin{array}{l}\cos t < 0 : \text{ QII and III} \\ \tan t > 0 : \text{ QI and III}\end{array}\right\} \quad \text{Quadrant III}$$

$$\cos t = -\sqrt{3} < -1$$

no solution

7a).

$$\sec t \cot t = \frac{1}{\cos t} \cdot \frac{\cos t}{\sin t} = \frac{1}{\sin t} = \csc t$$

7b).

$$\frac{1}{\sec t - \tan t} = \frac{1}{\dfrac{1}{\cos t} - \dfrac{\sin t}{\cos t}} = \frac{\cos t}{1 - \sin t}$$

$$= \frac{\cos t}{1 - \sin t} \cdot \frac{1 + \sin t}{1 + \sin t} = \frac{\cos t(1 + \sin t)}{1 - \sin^2 t}$$

$$= \frac{\cos t(1 + \sin t)}{\cos^2 t} = \frac{1 + \sin t}{\cos t}$$

$$= \frac{1}{\cos t} + \frac{\sin t}{\cos t} = \sec t + \tan t$$

8a).

$$f(x) = -\sin\left(\frac{x}{2} + \frac{\pi}{2}\right)$$

$$A = -1 \qquad B = \frac{1}{2} \qquad C = \frac{\pi}{2}$$

$$\text{amplitude} = |A| = |-1| = 1$$

$$\text{period} = \frac{2\pi}{|B|} = \frac{2\pi}{\left|\frac{1}{2}\right|} = 4\pi$$

$$\text{phase shift} = -\frac{C}{B} = \frac{-\frac{\pi}{2}}{\frac{1}{2}} = -\pi$$

8b).

$$f(x) = 3\cos\left(2x - \frac{\pi}{2}\right)$$

$$A = 3 \qquad B = 2 \qquad C = -\frac{\pi}{2}$$

$$\text{amplitude} = |A| = |3| = 3$$

$$\text{period} = \frac{2\pi}{|B|} = \frac{2\pi}{|2|} = \pi$$

$$\text{phase shift} = -\frac{C}{B} = \frac{-\left(-\frac{\pi}{2}\right)}{2} = \frac{\pi}{4}$$

9a).

$$3\sin^2 x = 2$$

$$\sin^2 x = \frac{2}{3}$$

$$\sin x = \pm\sqrt{\frac{2}{3}} = \frac{\pm\sqrt{6}}{3}$$

$$x = \sin^{-1}\left(\pm\frac{\sqrt{6}}{3}\right)$$

9b).

$$2\cos^2 x - \cos x - 3 = 0$$

$$(2\cos x - 3)(\cos x + 1) = 0$$

$$2\cos x - 3 = 0 \qquad \cos x = -1$$

$$\cos x = \frac{3}{2} \text{ (crossed out)} \qquad x = \cos^{-1}(-1)$$

10.

$$\tan^2 x - 4\tan x + 2 = 0$$

Let $u = \tan x$

$$u^2 - 4u + 2 = 0$$

$$u = \frac{-(-4) \pm \sqrt{(-4)^2 - 4(1)(2)}}{2(1)}$$

$$= \frac{4 \pm \sqrt{8}}{2}$$

$$= \frac{4 \pm 2\sqrt{2}}{2}$$

$$= 2 \pm \sqrt{2}$$

$$\tan x = 2 + \sqrt{2} \qquad \tan x = 2 - \sqrt{2}$$

$$x = \tan^{-1}(2 + \sqrt{2}) \qquad x = \tan^{-1}(2 - \sqrt{2})$$

$$x = 73.68° = 73°41' \qquad x = 30.36° = 30°22'$$

11a).

$$265° - 180° = 85°$$

11b).

$$-\frac{7\pi}{3} + 2\pi = \frac{-\pi}{3}$$

reference angle : $\left|-\frac{\pi}{3}\right| = \frac{\pi}{3}$

11c).

$$-120° + 360° = 240°$$

$$240° - 180° = 60°$$

12a).

$$\frac{\theta}{\pi}=\frac{5}{180}$$

$$\theta=\frac{5\pi}{180}=\frac{\pi}{36}$$

12b).

$$\frac{-\frac{\pi}{6}}{\pi}=\frac{\theta}{180}$$

$$\theta=\frac{180\left(-\frac{\pi}{6}\right)}{\pi}=-30°$$

12c).

$$\frac{2}{\pi}=\frac{\theta}{180}$$

$$\theta=\frac{2(180)}{\pi}=\left(\frac{360}{\pi}\right)^{\circ}$$

13.

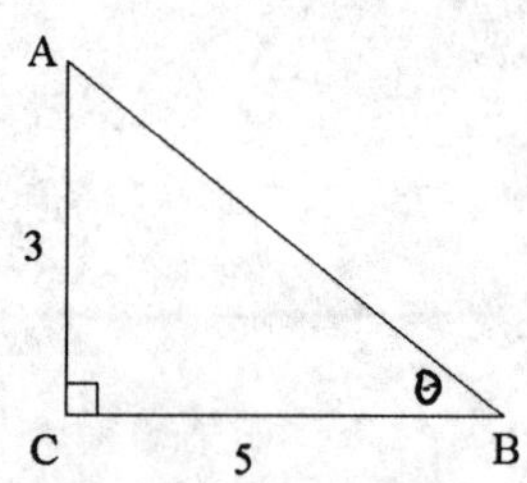

$$c^2=3^2+5^2$$

$$c^2=34$$

$$c=\sqrt{34}$$

13a).

$$\sec\theta=\frac{1}{\cos\theta}=\frac{1}{\frac{5}{\sqrt{34}}}=\frac{\sqrt{34}}{5}$$

13b).

$$\sin\left(\frac{\pi}{2}-\theta\right)=\sin\alpha=\frac{5}{\sqrt{34}}=\frac{5\sqrt{34}}{34}$$

13c).

$$\cot\theta=\frac{1}{\tan\theta}=\frac{1}{\frac{3}{5}}=\frac{5}{3}$$

14.

$$\text{radian measure of } \theta=\frac{\text{intercepted arc}}{\text{radius}}$$

$$30°=\frac{\pi}{6}\text{ radians}$$

$$\frac{\pi}{6}=\frac{20}{r}$$

$$r=\frac{120}{\pi}$$

$$\text{diameter}=2r=2\left(\frac{120}{\pi}\right)=\frac{240}{\pi}\text{ ft}$$

15.

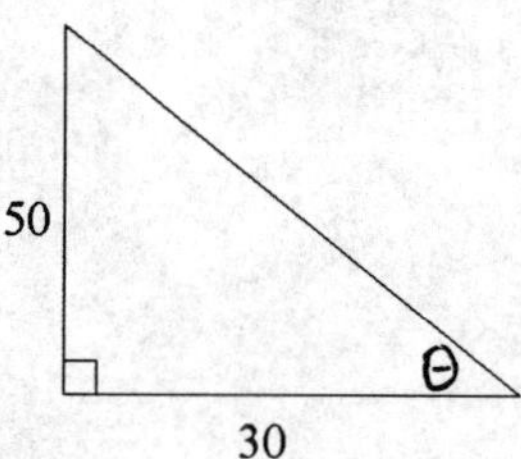

$$\tan\theta=\frac{50}{30}$$

$$\theta=\tan^{-1}\left(\frac{5}{3}\right)$$

$$\theta\approx 59°$$

16.

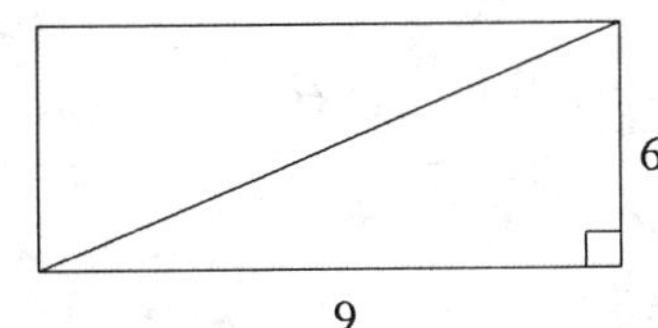

$$\tan\theta = \frac{9}{6}$$

$$\theta = \tan^{-1}\left(\frac{3}{2}\right)$$

$$\theta = 56.3°$$

17a).

$$c^2 = a^2 + b^2 - 2ab\cos\gamma$$

$$8^2 = 5^2 + 12^2 - 2(5)(12)\cos\gamma$$

$$\cos\gamma = 0.875$$

$$\gamma = 28.96° = 28°58'$$

17b).

$$\frac{a}{\sin\alpha} = \frac{b}{\sin\beta}$$

$$\frac{15}{\sin 50°} = \frac{18}{\sin\beta}$$

$$\sin\beta = \frac{18\sin 50°}{15}$$

$$\sin\beta = 0.9193$$

$\beta = 66.82° = 66°49'$ or $\beta = 113.18° = 113°11'$

$\gamma = 180° - (\alpha + \beta)$

$= 180° - (50° + 66°49')$

$= 63°11'$

$\gamma = 180° - (\alpha + \beta)$

$= 180° - (50° + 113°11')$

$= 16°49'$

18.

$$\frac{\tan^2 t-1}{\tan^2 t+1}=\frac{\tan^2 t-1}{\sec^2 t}$$
$$=\frac{\tan^2 t}{\sec^2 t}-\frac{1}{\sec^2 t}$$
$$=\frac{\frac{\sin^2 t}{\cos^2 t}}{\frac{1}{\cos^2 t}}-\cos^2 t$$
$$=\sin^2 t-\cos^2 t$$
$$=\left(1-\cos^2 t\right)-\cos^2 t$$
$$=1-2\cos^2 t$$

19a).

$$\cos\left(\frac{\pi}{6}-\theta\right)=\cos\frac{\pi}{6}\cos\theta+\sin\frac{\pi}{6}\sin\theta$$
$$=\frac{\sqrt{3}}{2}\cos\theta+\frac{1}{2}\sin\theta$$

19b).

$$\tan\left(\theta-\frac{\pi}{4}\right)=\frac{\tan\theta-\tan\frac{\pi}{4}}{1+\tan\theta\tan\frac{\pi}{4}}$$
$$=\frac{\tan\theta-1}{1+(\tan\theta)(1)}$$
$$=-\frac{1-\tan\theta}{1+\tan\theta}$$

20a).

$$\cos\left(-\frac{\pi}{8}\right)=\cos\frac{\pi}{8}$$
$$=\cos\frac{\frac{\pi}{4}}{2}$$
$$=\sqrt{\frac{1+\cos\frac{\pi}{4}}{2}} \quad \text{(Quadrant IV)}$$
$$=\sqrt{\frac{1+\frac{\sqrt{2}}{2}}{2}}$$
$$=\sqrt{\frac{2+\sqrt{2}}{4}}$$
$$=\frac{\sqrt{2+\sqrt{2}}}{2}$$

20b).

$$\tan 255°=\tan(30°+225°)$$
$$=\frac{\tan 30°+\tan 225°}{1-(\tan 30°)(\tan 225°)}$$
$$=\frac{\frac{1}{\sqrt{3}}+1}{1-\left(\frac{1}{\sqrt{3}}\right)(1)}$$
$$=\frac{1+\sqrt{3}}{\sqrt{3}-1}$$
$$=2+\sqrt{3}$$

21a).

$$\cot 62° = \tan(90° - 62°) = \tan 28°$$

21b).

$$\sin\frac{\pi}{12} = \cos\left(\frac{\pi}{2} - \frac{\pi}{12}\right) = \cos\frac{5\pi}{12}$$

22.

$$\left.\begin{array}{l}\sin t < 0 : \text{QIII and IV} \\ \cos t > 0 : \text{QI and IV}\end{array}\right\} \text{Quadrant IV}$$

22a).

$$\begin{aligned}\cos^2 t &= 1 - \sin^2 t \\ &= 1 - \left(-\frac{3}{5}\right)^2 \\ &= \frac{16}{25}\end{aligned}$$

$$\cos t = \frac{4}{5}$$

$$\begin{aligned}\sin\frac{t}{2} &= \sqrt{\frac{1-\cos t}{2}} \\ &= \sqrt{\frac{1-\frac{4}{5}}{2}} \\ &= \sqrt{\frac{1}{10}} \\ &= \frac{\sqrt{10}}{10}\end{aligned}$$

22b).

$$\tan t = \frac{\sin t}{\cos t} = \frac{-\frac{3}{5}}{\frac{4}{5}} = -\frac{3}{4}$$

$$\begin{aligned}\tan 2t &= \frac{2\tan t}{1-\tan^2 t} \\ &= \frac{2\left(-\frac{3}{4}\right)}{1-\left(-\frac{3}{4}\right)^2} \\ &= -\frac{24}{7}\end{aligned}$$

23.

$$\cos 40°+\cos 20°$$
$$= 2\cos\left(\frac{40°+20°}{2}\right)\cos\left(\frac{40°-20°}{2}\right)$$
$$= 2\cos 30°\cos 10°$$

24.

$$\sin 2t\cos 3t = \frac{1}{2}\sin(2t+3t)+\frac{1}{2}\sin(2t-3t)$$
$$= \frac{1}{2}\sin 5t+\frac{1}{2}\sin(-t)$$
$$= \frac{1}{2}\sin 5t-\frac{1}{2}\sin t$$

25.

$$\sin\theta-\cos 2\theta=0$$
$$\sin\theta-\left(1-2\sin^2\theta\right)=0$$
$$\sin\theta-1+2\sin^2\theta=0$$
$$2\sin^2\theta+\sin\theta-1=0$$
$$(2\sin\theta-1)(\sin\theta+1)=0$$
$$2\sin\theta-1=0 \qquad \sin\theta+1=0$$
$$\sin\theta=\frac{1}{2} \qquad \sin\theta=-1$$
$$\theta=\frac{\pi}{6},\frac{5\pi}{6} \qquad \theta=\frac{3\pi}{2}$$
$$\frac{\pi}{6},\frac{5\pi}{6},\frac{3\pi}{2}$$

26a).

$$|3+2i|=\sqrt{3^2+2^2}=\sqrt{13}$$

26b).

$$|6-3i|=\sqrt{6^2+(-3)^2}=\sqrt{45}=3\sqrt{5}$$

27a).

$$-3i$$
$$a=0 \qquad b=-3$$
$$r=\sqrt{0^2+(-3)^2}=3$$
$$\theta=\frac{3\pi}{2}$$
$$-3i=3\left(\cos\frac{3\pi}{2}+i\sin\frac{3\pi}{2}\right)$$

27b).

$$4-4i \qquad \text{(Quadrant IV)}$$
$$a=4 \qquad b=-4$$
$$r=\sqrt{4^2+(-4)^2}=4\sqrt{2}$$
$$\tan\theta=\frac{-4}{4}=-1$$
$$\theta=\frac{7\pi}{4}$$
$$4-4i=4\sqrt{2}\left(\cos\frac{7\pi}{4}+i\sin\frac{7\pi}{4}\right)$$

28.

$$\frac{35(\cos 210°+i\sin 210°)}{7(\cos 180°+i\sin 180°)}$$

$$=\frac{35}{7}\left[\cos(210°-180°)+i\sin(210°-180°)\right]$$

$$=5(\cos 30°+i\sin 30°)$$

$$=5\left(\frac{\sqrt{3}}{2}+\frac{1}{2}i\right)$$

$$=\frac{5\sqrt{3}}{2}+\frac{5}{2}i$$

29.

$$z^2+4=0$$

$$z^2=-4$$

$$-4=4(\cos\pi+i\sin\pi)$$

$$U_0=\sqrt[4]{4}\left(\cos\frac{\pi}{4}+i\sin\frac{\pi}{4}\right)$$

$$U_1=\sqrt[4]{4}\left[\cos\left(\frac{\pi+2\pi\cdot 1}{4}\right)+i\sin\left(\frac{\pi+2\pi\cdot 1}{4}\right)\right]$$

$$=\sqrt[4]{4}\left(\cos\frac{3\pi}{4}+i\sin\frac{3\pi}{4}\right)$$

$$U_2=\sqrt[4]{4}\left[\cos\left(\frac{\pi+2\pi\cdot 2}{4}\right)+i\sin\left(\frac{\pi+2\pi\cdot 2}{4}\right)\right]$$

$$=\sqrt[4]{4}\left(\cos\frac{5\pi}{4}+i\sin\frac{5\pi}{4}\right)$$

$$U_3=\sqrt[4]{4}\left[\cos\left(\frac{\pi+2\pi\cdot 3}{4}\right)+i\sin\left(\frac{\pi+2\pi\cdot 3}{4}\right)\right]$$

$$=\sqrt[4]{4}\left(\cos\frac{7\pi}{4}+i\sin\frac{7\pi}{4}\right)$$

30.

$$r=1-\sin\theta$$

31.

$$r^2\cos 2\theta=1$$

$$r^2(\cos^2\theta-\sin^2\theta)=1$$

$$(r\cos\theta)^2-(r\sin\theta)^2=1$$

$$x^2-y^2=1$$

hyperbola

32.

$\vec{\mathbf{V}}_1 = \langle 2, 7 \rangle$ (Quadrant I)

$\tan\theta_1 = \frac{7}{2}$

$\theta_1 = \tan^{-1}\frac{7}{2}$

$\theta_1 = 74.05°$

$\vec{\mathbf{V}}_2 = \langle -3, 5 \rangle$ (Quadrant II)

$\tan\theta_2 = \frac{5}{-3}$

$\theta_2 = \tan^{-1}\left(-\frac{5}{3}\right)$

$\theta_2 = 120.96°$

$$\begin{aligned}\theta &= |\theta_1 - \theta_2| \\ &= |74.05° - 120.96°| \\ &= 46.91°\end{aligned}$$

33.

$$\begin{aligned}\|\vec{\mathbf{U}}\| &= r \\ 1 &= r\end{aligned}$$

$$\begin{aligned}\vec{\mathbf{U}} &= \langle 1\cos 230°,\ 1\sin 230° \rangle \\ &= \langle -0.64,\ -0.77 \rangle\end{aligned}$$

CHAPTER 9 ADDITIONAL PRACTICE EXERCISES

1. Given triangle ABC with $\alpha = 28°$, $\beta = 76°$ and $a = 14.6$, find b.

2. Given triangle ABC with $\alpha = 60°$, $a = 30$ and $b = 32$, find β.

3. Given triangle ABC with $a = 12$, $b = 16$ and $c = 22$, find α.

4. Find the absolute value of $4 - 5i$.

5. Express $4 - 4i$ in trigonometric form.

6. Convert $6(\cos 225° + i \sin 225°)$ to the algebraic form $a + bi$.

7. Compute $4\left(\cos\frac{\pi}{8} + i\sin\frac{\pi}{8}\right) \cdot 3\left(\cos\frac{\pi}{2} + i\sin\frac{\pi}{2}\right)$.

8. Find the rectangular coordinates of the point with polar coordinates $\left(3, \frac{2\pi}{3}\right)$.

9. Sketch the graph of $r = 1 + \sin\theta$.

10. Given $\vec{\mathbf{V}}_1 = \langle -4, 6\rangle$ and $\vec{\mathbf{V}}_2 = \langle 3, 5\rangle$, find $2\vec{\mathbf{V}}_1 - 4\vec{\mathbf{V}}_2$.

CHAPTER 9 PRACTICE TEST

1. Given triangle ABC with $\beta = 28°$, $\gamma = 46°$ and $a = 12$, find c.

2. Given triangle ABC with $a = 4$, $b = 7$ and $c = 10$, find α.

3. Find the absolute value of $10 - 2i$.

4. Express -5 in trigonometric form.

5. Convert $\frac{3}{4}(\cos\pi + i\sin\pi)$ to the algebraic form $a + bi$.

6. Find the product of $2 - 3i$ and $5 + 6i$.

7. Find the quotient $\dfrac{9\left(\cos\frac{3\pi}{4} + i\sin\frac{3\pi}{4}\right)}{3\left(\cos\frac{\pi}{6} + i\sin\frac{\pi}{6}\right)}$.

8. Use DeMoire's Theorem to express $(3 - 3i)^5$ in the form $a + bi$.

9. Determine all the roots of $z^3 + 64 = 0$.

10. Give two other polar coordinate representations of $(5, 220°)$.

11. Give a polar coordinate representation of $\left(-2, \frac{\pi}{6}\right)$ with $r \geq 0$.

12. Find the rectangular coordinates of the point with polar coordinates $\left(-8, \frac{7\pi}{4}\right)$.

13. Find the polar coordinates of the point with rectangular coordinates $(-5, -5)$.

14. Sketch the graph of $r = 3\cos\theta$.

15.
Transform $r + 4\cos\theta = 0$ to an equation in x and y.

16.
Given points $P(-2, 3)$ and $Q(5, 11)$ find vector $\overrightarrow{\mathbf{PQ}}$ in component form $\langle \text{a}, \text{b} \rangle$.

17.
Find the magnitude of vector $\vec{\mathbf{V}} = \langle 6, 1 \rangle$.

18.
Given vectors $\vec{\mathbf{V}}_1 = \langle 18, -3 \rangle$ and $\vec{\mathbf{V}}_2 = \langle -10, 4 \rangle$, find $-3\vec{\mathbf{V}}_1 - 2\vec{\mathbf{V}}_2$.

19.
Find vector $\vec{\mathbf{V}}$ in component form $\langle a, b \rangle$ given $\|\vec{\mathbf{V}}\| = 5$ and direction angle $\theta = 30°$.

20.
Find a unit vector in the direction of $\vec{\mathbf{V}} = \langle 5, -9 \rangle$.

CHAPTER 10 SECTION 1

SECTION 10.1 PROGRESS CHECK

1a). (Page 562)

$$x^2+3y^2=12$$
$$x+3y=6$$

$$x=6-3y$$

$$(6-3y)^2+3y^2=12$$
$$36-36y+9y^2+3y^2=12$$
$$12y^2-36y+24=0$$
$$y^2-3y+2=0$$
$$(y-2)(y-1)=0$$

$y=2$ $\quad$ $y=1$

$x=6-3(2)=0$ $\quad$ $x=6-3(1)=3$

$x=0,\ y=2;$ $\quad$ $x=3,\ y=1$

1b).

$$x^2+y^2=34$$
$$x-y=2$$

$$x=y+2$$

$$(y+2)^2+y^2=34$$
$$y^2+4y+4+y^2=0$$
$$2y^2+4y-30=0$$
$$y^2+2y-15=0$$
$$(y+5)(y-3)=0$$

$y=-5$ $\quad$ $y=3$

$x=-5+2=-3$ $\quad$ $x=3+2=5$

$x=-3,\ y=-5;$ $\quad$ $x=5, y=3$

2a). (Page 563)

$$3x^2 - y^2 = 7$$
$$-9x + 3y = -2$$
$$3y = 9x - 2$$
$$y = 3x - \frac{2}{3}$$

$$3x^2 - \left(3x - \frac{2}{3}\right)^2 = 7$$
$$3x^2 - \left(9x^2 - 4x + \frac{4}{9}\right) = 7$$
$$3x^2 - 9x^2 + 4x - \frac{4}{9} = 7$$
$$-6x^2 + 4x - \frac{67}{9} = 0$$
$$-54x^2 + 36x - 67 = 0$$

$$x = \frac{-36 \pm \sqrt{36^2 - 4(-54)(-67)}}{2(-54)}$$
$$= \frac{-36 \pm \sqrt{-13176}}{-108}$$

No solution

2b).

$$-5x + 2y = -4$$
$$\frac{5}{2}x - y = 2$$
$$y = \frac{5}{2}x - 2$$

$$-5x + 2\left(\frac{5}{2}x - 2\right) = -4$$
$$-5x + 5x - 4 = -4$$
$$-4 = -4$$

All points on the line $-5x + 2y = -4$

3a). (Page 567)

$$x - y = 2$$
$$3x - 3y = -6$$
$$-3x + 3y = -6$$
$$\underline{3x - 3y = -6}$$
$$0 = -12$$

No solution

3b).

$$4x + 6y = 3$$
$$-2x - 3y = -\frac{3}{2}$$

$$4x + 6y = 3$$
$$\underline{-4x - 6y = -3}$$
$$0 = 0$$

All points on the line $4x + 6y = 3$

3c).

$$x^2-4x+y^2-4y=1$$
$$x^2-4x \qquad +y=-5$$

$$\begin{array}{r} x^2-4x+y^2-4y=1 \\ -x^2+4x \qquad -y=5 \\ \hline y^2-5y=6 \\ y^2-5y-6=0 \\ (y-6)(y+1)=0 \end{array}$$

$y=6$

$$x^2-4x+6=-5$$
$$x^2-4x+11=0$$
$$x=\frac{-(-4)\pm\sqrt{(-4)^2-4(1)(11)}}{2(1)}$$
$$=\frac{4\pm\sqrt{-28}}{2}$$

No solution

$y=-1$

$$x^2-4x-1=-5$$
$$x^2-4x+4=0$$
$$(x-2)^2=0$$
$$x=2$$

Solution: $x=2, y=-1$

EXERCISE SET 10.1

1.

3.

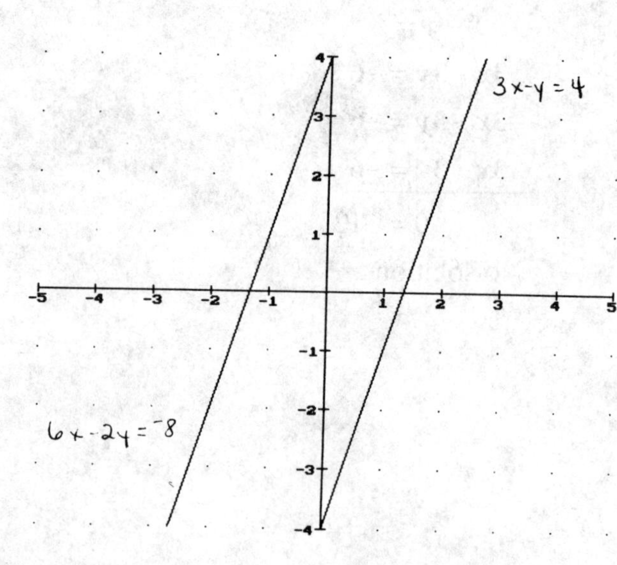

No solution

5.

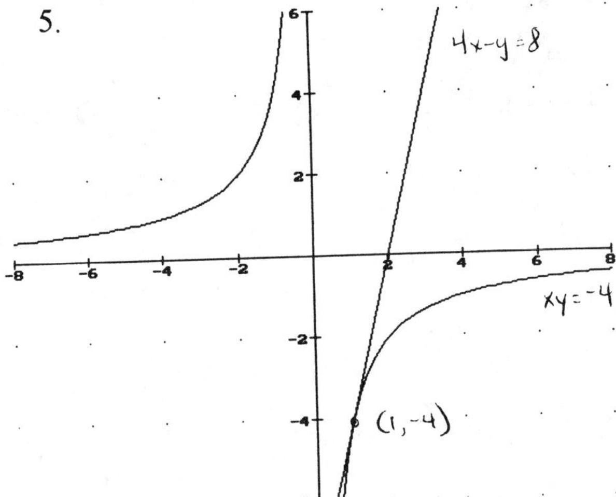

7.

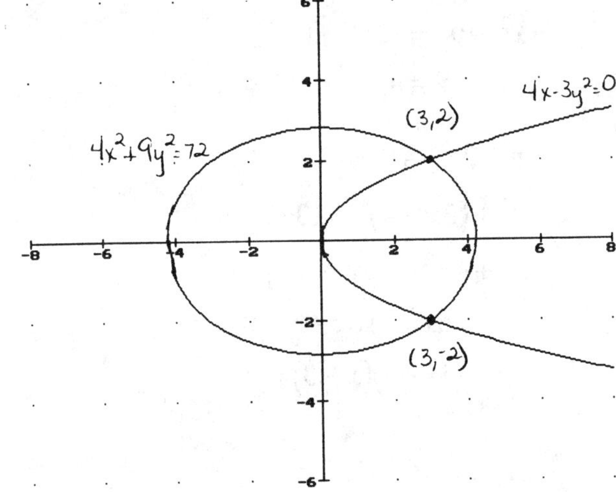

9.

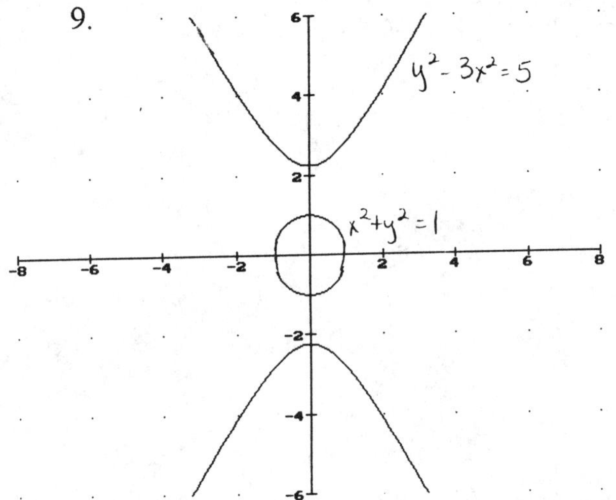

No solution

11.

$$x + y = 1$$
$$x - y = 3$$

$$x = y + 3$$
$$y + 3 + y = 1$$
$$2y = -2$$
$$y = -1$$
$$x = -1 + 3 = 2$$
$$x = 2, \quad y = -1$$

13.

$$x^2+y^2=13$$
$$2x-y=4$$

$$y=2x-4$$
$$x^2+(2x-4)^2=13$$
$$x^2+4x^2-16x+16=13$$
$$5x^2-16x+3=0$$
$$(5x-1)(x-3)=0$$
$$x=\frac{1}{5} \qquad x=3$$
$$y=2\left(\frac{1}{5}\right)-4=\frac{-18}{5} \qquad y=2(3)-4=2$$
$$x=\frac{1}{5},\ y=-\frac{18}{5}; \qquad x=3,\ y=2$$

15.

$$y^2-x=0$$
$$y-4x=-3$$

$$x=y^2$$
$$y-4y^2=-3$$
$$0=4y^2-y-3$$
$$0=(4y+3)(y-1)$$
$$y=-\frac{3}{4} \qquad y=1$$
$$x=\left(-\frac{3}{4}\right)^2=\frac{9}{16} \qquad x=(1)^2=1$$
$$x=\frac{9}{16},\ y=-\frac{3}{4}; \qquad x=1,\ y=1$$

17.

$$x^2-2x+y^2=3$$
$$2x+y=4$$

$$y=4-2x$$
$$x^2-2x+(4-2x)^2=3$$
$$x^2-2x+16-16x+4x^2=3$$
$$5x^2-18x+13=0$$
$$(5x-13)(x-1)=0$$
$$x=\frac{13}{5} \qquad x=1$$
$$y=4-2\left(\frac{13}{5}\right)=-\frac{6}{5} \qquad y=4-2(1)=2$$
$$x=\frac{13}{5},\ y=-\frac{6}{5}; \qquad x=1,\ y=2$$

19.

$$xy = 1$$
$$x - y + 1 = 0$$

$$x = y - 1$$
$$(y-1)y = 1$$
$$y^2 - y = 1$$
$$y^2 - y - 1 = 0$$
$$y = \frac{-(-1) \pm \sqrt{(-1)^2 - 4(1)(-1)}}{2(1)}$$

$$y = \frac{1 \pm \sqrt{5}}{2} \qquad y = \frac{1-\sqrt{5}}{2}$$

$$x = \frac{1+\sqrt{5}}{2} - 1 = \frac{-1+\sqrt{5}}{2} \qquad x = \frac{1-\sqrt{5}}{2} - 1 = \frac{-1-\sqrt{5}}{2}$$

$$x = \frac{-1+\sqrt{5}}{2},\ y = \frac{1+\sqrt{5}}{2};\quad x = \frac{-1-\sqrt{5}}{2},\ y = \frac{1-\sqrt{5}}{2}$$

21.

$$x + 2y = 1$$
$$5x + 2y = 13$$

$$x + 2y = 1$$
$$-5x - 2y = -13$$
$$-4x = -12$$
$$x = 3$$
$$3 + 2y = 1$$
$$2y = -2$$
$$y = -1$$
$$x = 3,\ y = -1$$

23.

$$25y^2 - 16x^2 = 400$$
$$9y^2 - 4x^2 = 36$$

$$25y^2 - 16x^2 = 400$$
$$-36y^2 + 16x^2 = -144$$
$$-9y^2 = 256$$
$$y^2 = -\frac{256}{9}$$

No solution

25.

$$2x^2+3y^2=30$$
$$x-y^2=-1$$

$$2x^2+3y^2=30$$
$$\underline{3x-3y^2=-3}$$
$$2x^2+3x=27$$
$$2x^2+3x-27=0$$
$$(2x+9)(x-3)=0$$

$$x=-\frac{9}{2} \qquad x=3$$
$$-\frac{9}{2}-y^2=-1 \qquad 3-y^2=-1$$
$$y^2=-\frac{7}{2} \qquad y^2=4$$

No solution

$$y=\pm 2$$
$$x=3,\ y=2;$$
$$x=3,\ y=-2$$

27.

$$2x+y=4$$
$$-6x-3y=-8$$

$$6x+3y=12$$
$$\underline{-6x-3y=-8}$$
$$0=4$$

No solution

29.

$$y^2-x^2=-5$$
$$\underline{3y^2+x^2=21}$$
$$4y^2=16$$
$$y^2=4$$

$$y=2 \qquad y=-2$$
$$2^2-x^2=-5 \qquad (-2)^2-x^2=-5$$
$$x^2=9 \qquad x^2=9$$
$$x=\pm 3 \qquad x=\pm 3$$

$$x=3,\ y=2; \qquad x=-3,\ y=2;$$
$$x=3,\ y=-2; \qquad x=-3,\ y=-2$$

31.

$$2x+2y=6$$
$$3x+3y=6$$

$$6x+6y=18$$
$$\underline{-6x-6y=-12}$$
$$0=6$$

Inconsistent

33.

$$y^2-8x^2=9$$
$$y^2+3x^2=-3$$

$$\begin{aligned} y^2-8x^2&=9\\ \underline{-y^2-3x^2}&\underline{=-31}\\ -11x^2&=40\\ x^2&=-\frac{40}{11}\end{aligned}$$

Inconsistent

35.

$$3x+3y=9$$
$$2x+2y=-6$$

$$\begin{aligned} -6x-6y&=-18\\ \underline{6x+6y}&\underline{=-18}\\ 0&=-36\end{aligned}$$

Inconsistent

37.

$$3x-y=18$$
$$\frac{3}{2}x-\frac{1}{2}y=9$$

$$y=3x-18$$
$$\frac{3}{2}x-\frac{1}{2}(3x-18)=9$$
$$\frac{3}{2}x-\frac{3}{2}x+9=9$$
$$9=9$$

Consistent

all points on the line $3x-y=18$

39.

$$x^2-3xy-2y^2-2=0$$
$$x-y-2=0$$

$$x=y+2$$

$$(y+2)^2-3(y+2)y-2y^2-2=0$$
$$y^2+4y+4-3y^2-6y-2y^2-2=0$$
$$-4y^2-2y+2=0$$
$$2y^2+y-1=0$$
$$(2y-1)(y+1)=0$$

$y=\frac{1}{2}$ $\qquad$ $y=-1$

$x=\frac{1}{2}+2=\frac{5}{2}$ $\qquad$ $x=-1+2=1$

$x=\frac{5}{2},\ y=\frac{1}{2};$ $\qquad$ $x=1,\ y=-1$

Consistent

41.

$$\begin{aligned} 2x - 2y &= 4 \\ x - y &= 8 \end{aligned}$$

$$\begin{aligned} 2x - 2y &= 4 \\ -2x + 2y &= -16 \\ \hline 0 &= -12 \end{aligned}$$

Inconsistent

CHAPTER 10 SECTION 2

SECTION 10.2 PROGRESS CHECK

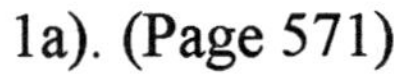

1a). (Page 571)

x liters sold at \$0.50 per liter

$R = 0.50\ x$

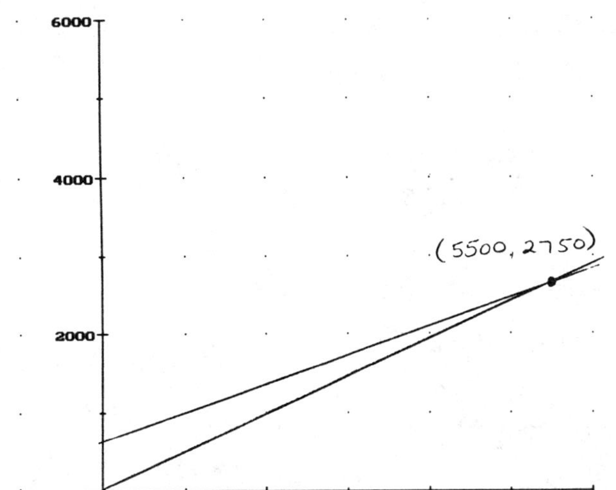

1b).

Break even : $C = R$

5500 liters

1c).

\$2750

2a). (Page 574)

$$S = 3p + 120$$
$$D = -p + 200$$

$$S = D$$
$$3p + 120 = -p + 200$$
$$4p = 80$$
$$p = 20$$

2b).

$$S = 3p + 120$$
$$D = -p + 200$$

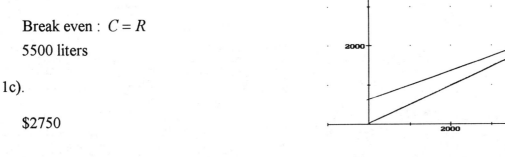

$$D = -20 + 200$$
$$= 180$$

EXERCISE SET 10.2

1.

Number of dimes : d

Number of nickels: n

$$n + d = 40$$
$$5n + 10d = 275$$

$$-5n - 5d = -200$$
$$5n + 10d = 275$$
$$5d = 75$$
$$d = 15$$

$$n + 15 = 40$$
$$n = 25$$

25 nickels, 15 dimes

3.

Cost per roll of color film : x

Cost per roll of black and white film : y

$$6x + 4y = 21$$
$$4x + 6y = 19$$

$$-12x - 8y = -42$$
$$12x + 18y = 57$$
$$10y = 15$$
$$y = 1.5$$
$$6x + 4(1.5) = 21$$
$$6x = 15$$
$$x = 2.5$$

Color film : \$2.50

Black and white film : \$1.50

5.

Amount of Bond A investment : a

Amount of Bond B investment : b

$$a+b=6000$$
$$0.08a+0.10b=520$$

$$a+b=6000$$
$$8a+10b=52000$$

$$-8a-8b=-48000$$
$$8a+10b=5200$$
$$2b=4000$$
$$b=2000$$
$$a+2000=6000$$
$$a=4000$$

\$4,000 in Bond A; \$2,000 in Bond B

7.

Number of 12-inch rolls : x

Number of 15-inch rolls : y

$$x+y=14$$
$$12x+15y=180$$

$$-12x-12y=-168$$
$$12x+15y=180$$
$$3y=12$$
$$y=4$$
$$x+4=14$$
$$x=10$$

10 12-inch rolls; 4 15-inch rolls

9.

Amount of \$1.20 / lb coffee : x

Amount of \$1.80 / lb coffee : y

$$x+y=24$$
$$1.20x+1.80y=1.60(24)$$

$$x+y=24$$
$$12x+18y=384$$

$$-12x-12y=-288$$
$$\underline{12x+18y=384}$$
$$6y=96$$
$$y=16$$
$$x+16=24$$
$$x=8$$

8 lbs of \$1.20 / lb coffee ;

16 lbs of \$1.80 / lb coffee

11.

Speed of cyclist in still air : x

Speed of wind : y

	rate	·	time	=	distance
going	$x-y$		4		$4(x-y)$
returning	$x+y$		3		$3(x+y)$

$$4(x-y)=45 \qquad 4x-4y=45$$
$$3(x+y)=45 \qquad x+y=15$$

$$4x-4y=45$$
$$\underline{4x+4y=60}$$
$$8x=105$$
$$x=\frac{105}{8}$$
$$\frac{105}{8}+y=15$$
$$y=\frac{15}{8}$$

Speed of cyclist : $\frac{105}{8}$ mph

Speed of wind : $\frac{15}{8}$ mph

13.

Tens digit : t

Units digit : u

$$t+u=7$$

digits reversed : $\underline{u}\,\underline{t}$

$$\begin{aligned}10u+t&=9+(10t+u)\\9u-9t&=9\\u-t&=1\end{aligned}$$

$$\begin{aligned}u+t&=7\\\underline{u-t}&\underline{=1}\\2u&=8\\u&=4\\4+t&=7\\t&=3\end{aligned}$$

The number is 34.

15.

Pounds of raisins : x

Pounds of nuts : y

$$\begin{aligned}x+y&=50 & x+y&=50\\2y+1.5x&=1.8(50) & 15x+20y&=900\end{aligned}$$

$$\begin{aligned}-15x-15y&=-750\\\underline{15x+20y}&\underline{=900}\\5y&=150\\y&=30\\x+30&=50\\x&=20\end{aligned}$$

20 lbs of raisins ; 30 lbs of nuts

17.

Model A investment : a

Model B investment : b

$$\begin{aligned}a+b&=18000 & a+b&=18000\\0.12a+0.18b&=0.16(18000) & 12a+18b&=288000\end{aligned}$$

$$\begin{aligned}-12a-12b&=-216000\\\underline{12a+18b}&\underline{=288000}\\6b&=72000\\b&=12000\\a+12000&=18000\\a&=6000\end{aligned}$$

$6,000 in model A ; $12,000 in model B

19.

Units of Epiline I : a

Units of Epiline II : b

$$1a + 2b = 13$$
$$2a + 3b = 22$$

$$-2a - 4b = -26$$
$$\underline{2a + 3b = 22}$$
$$-b = -4$$
$$b = 4$$
$$a + 2(4) = 13$$
$$a = 5$$

5 units of Epiline I ; 4 units of Epiline II

21.

$$C = 24000 + 55x$$

21a).

$$R = 95x$$

21b).

$$C = R$$
$$24000 + 55x = 95x$$
$$24000 = 40x$$
$$x = 600$$

600 devices

21c).

$$R(600) = 95(600)$$
$$= 57000$$

$57,000

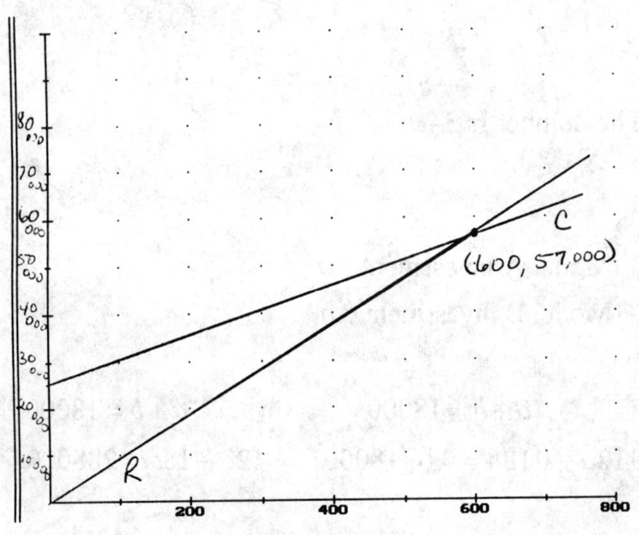

23.

$$S = 2p + 10$$
$$D = -p + 22$$

23a).

$$S = D$$
$$2p + 10 = -p + 22$$
$$3p = 12$$
$$p = 4$$

23b).

$$D = -4 + 22$$
$$= 18$$

18 mopeds

25.

First number : x

Second number : y

$$xy = 20$$
$$x + y = 9$$

$$x = 9 - y$$

$$(9 - y)y = 20$$
$$9y - y^2 = 20$$
$$0 = y^2 - 9y + 20$$
$$0 = (y - 5)(y - 4)$$
$$y = 5 \qquad y = 4$$

$$x = 9 - 5 = 4 \qquad x = 9 - 4 = 5$$

The numbers are 4 and 5.

27.

Number of nickels : n

Number of quarters : q

$$n+q=34$$
$$5n+25q=410$$

$$-5n-5q=-170$$
$$5n+25q=410$$
$$20q=240$$
$$q=12$$
$$n+12=34$$
$$n=22$$

22 nickels, 12 quarters

29.

Number of pounds of nuts : x

Number of pounds of raisins : y

$$x+y=2$$
$$2.10x+0.90y=1.62(2)$$

$$x+y=2$$
$$210x+90y=324$$

$$-90x-90y=-180$$
$$210x+90y=324$$
$$120x \quad =144$$
$$x=\frac{6}{5}$$
$$\frac{6}{5}+y=2$$
$$y=\frac{4}{5}$$

$\frac{6}{5}$ lbs of nuts ; $\frac{4}{5}$ lb of raisins.

31.

Number of type A tickets : a

Number of type B tickets : b

$$a + b = 200$$
$$10a + 1b = 1325$$

$$-a - b = -200$$
$$10a + b = 1325$$
$$9a = 1125$$
$$a = 125$$
$$125 + b = 200$$
$$b = 75$$

125 type A tickets ; 75 type B tickets

33.

Length : l

Width : w

$$l^2 + w^2 = 100$$
$$lw = 48$$

$$l = \frac{48}{w}$$

$$\left(\frac{48}{w}\right)^2 + w^2 = 100$$
$$\frac{2304}{w^2} + w^2 = 100$$
$$2304 + w^4 = 100w^2$$

$$w^4 - 100w^2 + 2304 = 0$$
$$(w^2 - 64)(w^2 - 36) = 0$$

$w^2 = 64$ $w^2 = 36$

$w = 8$ $w = 6$

$l = \frac{48}{8} = 6$ $l = \frac{48}{6} = 8$

The dimensions are 6m x 8m

CHAPTER 10 SECTION 3

SECTION 10.3 PROGRESS CHECK

1a). (Page 579)

$$\begin{aligned} 2x - 4y + 2z &= 1 \\ 3x + y + 3z &= 5 \\ x - y - 2z &= -8 \end{aligned}$$

$$\begin{array}{r} 3x + y + 3z = 5 \\ \underline{x - y - 2z = -8} \\ 4x + z = -3 \end{array} \qquad \begin{array}{r} 2x - 4y + 2z = 1 \\ \underline{12x + 4y + 12z = 20} \\ 14x + 14z = 21 \\ 2x + 2z = 3 \end{array}$$

$$\begin{array}{r} -8x - 2z = 6 \\ \underline{2x + 2z = 3} \\ -6x = 9 \\ x = -\frac{3}{2} \end{array}$$

$$2\left(-\frac{3}{2}\right) + 2z = 3$$

$$2z = 6$$

$$z = 3$$

$$-\frac{3}{2} - y - 2(3) = -8$$

$$-y = -\frac{1}{2}$$

$$y = \frac{1}{2}$$

$$x = -\frac{3}{2},\ y = \frac{1}{2},\ z = 3$$

1b).

$$\begin{aligned}-2x-3y-12z&=-11\\3x-y-15z&=11\\-x+5y+3z&=-9\end{aligned}$$

$$\begin{aligned}-2x-3y-12z&=-11\\-4x+20y+12z&=-36\\\hline-6x+17y&=-47\end{aligned}$$

$$\begin{aligned}-6x+17y&=-47\\6x-72y&=102\\\hline-55y&=55\\y&=-1\end{aligned}$$

$$\begin{aligned}3x-y-15z&=11\\-5x+25y+15z&=-45\\\hline-2x+24y&=-34\\x-12y&=17\\x-12(-1)&=17\\x&=5\\-(5)+5(-1)+3z&=-9\\3z&=1\\z&=\frac{1}{3}\end{aligned}$$

2a). (Page 581)

$$\begin{aligned}x-2y+z&=3\\2x+y-2z&=-1\\-x-8y+7z&=5\end{aligned}$$

$$\begin{aligned}x-2y+z&=3\\-x-8y+7z&=5\\\hline-10y+8z&=8\\-5y+4z&=4\end{aligned}$$

$$\begin{aligned}2x+y-2z&=-1\\-2x-16y+14z&=10\\\hline-15y+12z&=9\\-5y+4z&=3\end{aligned}$$

$$\begin{aligned}-5y+4z&=4\\5y-4z&=-3\\\hline0&=1\end{aligned}$$

Inconsistent

2b).

$$\begin{aligned}2x+y+2z&=1\\x-4y+7z&=-4\\x+y+3z&=-1\end{aligned}$$

$$\begin{aligned}2x+y+2z&=1\\x-y+3z&=-1\\\hline3x+5z&=0\end{aligned}$$

$$\begin{aligned}8x+4y+8z&=4\\x-4y+7z&=-4\\\hline9x+15z&=0\\-3x-5z&=0\end{aligned}$$

$$\begin{aligned}3x+5z&=0\\-3x-5z&=0\\\hline0&=0\end{aligned}$$

Infinite number of solutions

EXERCISE SET 10.3

1.

$$\begin{aligned} x+2y+3z&=-6 \\ 2x-3y-4z&=15 \\ 3x+4y+5z&=-8 \end{aligned}$$

$$\begin{aligned} -2x-4y-6z&=12 \\ 2x-3y-4z&=15 \\ \hline -7y-10z&=27 \end{aligned}$$

$$\begin{aligned} -3x-6y-9z&=18 \\ 3x+4y+5z&=-8 \\ \hline -2y-4z&=10 \\ y+2z&=-5 \end{aligned}$$

$$\begin{aligned} -7y-10z&=27 \\ 5y+10z&=-25 \\ \hline -2y&=2 \\ y&=-1 \\ -1+2z&=-5 \\ 2z&=-4 \\ z&=-2 \end{aligned}$$

$$\begin{aligned} x+2(-1)+3(-2)&=-6 \\ x&=2 \end{aligned}$$

$x=2,\quad y=-1,\quad z=-2$

3.

$$\begin{aligned} x+y+\ z&=1 \\ x+y-2z&=3 \\ 2x+y+z&=2 \end{aligned} \qquad \begin{aligned} x+y+\ z&=1 \\ \underline{-x-y+2z}&\underline{=-3} \\ 3z&=-2 \\ z&=-\frac{2}{3} \end{aligned} \qquad \begin{aligned} x+y+z&=1 \\ \underline{-2x-y-z}&\underline{=-2} \\ -x&=-1 \\ x&=1 \end{aligned}$$

$$\begin{aligned} 1+y-\frac{2}{3}&=1 \\ y&=\frac{2}{3} \end{aligned}$$

$$x=1,\ \ y=\frac{2}{3},\ \ z=-\frac{2}{3}$$

5.

$$\begin{aligned} x+\ y+\ z&=2 \\ x-\ y+2z&=3 \\ 3x+5y+2z&=6 \end{aligned} \qquad \begin{aligned} x+y+\ z&=2 \\ \underline{x-y+2z}&\underline{=3} \\ 2x+\qquad 3z&=5 \end{aligned} \qquad \begin{aligned} 5x-5y+10z&=15 \\ \underline{3x+5y+\ 2z}&\underline{=6} \\ 8x+\qquad 12z&=21 \end{aligned}$$

$$\begin{aligned} -8x-12z&=-20 \\ \underline{8x+12z}&\underline{=21} \\ 0&=1 \end{aligned}$$

Inconsistent

7.

$$\begin{aligned} x+2y+\ z&=7 \\ x+2y+3z&=11 \\ 2x+\ y+4z&=12 \end{aligned} \qquad \begin{aligned} x+2y+\ z&=7 \\ \underline{-x-2y-3z}&\underline{=-11} \\ -2z&=-4 \\ z&=2 \end{aligned} \qquad \begin{aligned} -2x-4y-6z&=-22 \\ \underline{2x+\ y+4z}&\underline{=12} \\ -3y-2z&=-10 \\ -3y-2(2)&=-10 \\ -3y&=-6 \\ y&=2 \\ x+2(2)+2&=7 \\ x&=1 \end{aligned}$$

$$x=1,\ \ y=2,\ \ z=2$$

9.

$$\begin{aligned} x+\ y+z&=2 \\ x+2y+z&=3 \\ x+\ y-z&=2 \end{aligned}$$

$$\begin{aligned} x+\ y+z&=2 \\ \underline{-x-2y-z}&\underline{=-3} \\ -y\qquad &=-1 \\ y&=1 \end{aligned}$$

$$\begin{aligned} x+2y+z&=3 \\ \underline{x+\ y-z}&\underline{=2} \\ 2x+3y\qquad&=5 \\ 2x+3(1)&=5 \\ 2x&=2 \\ x&=1 \end{aligned}$$

$$\begin{aligned} 1+1+z&=2 \\ z&=0 \end{aligned}$$

$x=1,\quad y=1,\quad z=0$

11.

$$\begin{aligned} 2x+y+3z&=8 \\ -x+y+\ z&=10 \\ x+y+\ z&=12 \end{aligned}$$

$$\begin{aligned} 2x+y+3z&=8 \\ \underline{x-y-\ z}&\underline{=-10} \\ 3x+\qquad 2z&=-2 \end{aligned}$$

$$\begin{aligned} -x+y+z&=10 \\ \underline{-x-y-z}&\underline{=-12} \\ -2x\qquad\quad&=-2 \\ x&=1 \\ 3(1)+2z&=-2 \\ 2z&=-5 \\ z&=-\frac{5}{2} \\ 1+y-\frac{5}{2}&=12 \\ y&=\frac{27}{2} \end{aligned}$$

$x=1,\quad y=\frac{27}{2},\quad z=-\frac{5}{2}$

13.

$$\begin{aligned} x+3y+7z&=1 \\ 3x-\ y-5z&=9 \\ 2x+\ y+\ z&=4 \end{aligned}$$

$$\begin{aligned} x+3y+\ 7z&=1 \\ \underline{9x-3y-15z}&\underline{=27} \\ 10x-\qquad 8z&=28 \\ 5x-\qquad 4z&=14 \end{aligned}$$

$$\begin{aligned} 3x-y-5z&=9 \\ \underline{2x+y+\ z}&\underline{=4} \\ 5x-\qquad 4z&=13 \end{aligned}$$

$$\begin{aligned} 5x-4z&=14 \\ \underline{-5x+4z}&\underline{=-13} \\ 0&=1 \end{aligned}$$

Inconsistent

15.

$$\begin{aligned} x-2y+3z&=-2 \\ x-5y+9z&=4 \\ 2x-y&=6 \end{aligned} \qquad \begin{aligned} -3x+6y-9z&=6 \\ x-5y+9z&=4 \\ \hline -2x+y&=10 \end{aligned}$$

$$\begin{aligned} 2x-y&=6 \\ -2x+y&=10 \\ \hline 0&=16 \end{aligned}$$

Inconsistent

17.

$$\begin{aligned} z-2y+x&=-5 \\ z+2x&=-10 \\ -z+y&=15 \end{aligned} \qquad \begin{aligned} z-2y+x&=-5 \\ -z+y&=15 \\ \hline -y+x&=10 \end{aligned} \qquad \begin{aligned} -z+2y-x&=5 \\ z+2x&=-10 \\ \hline 2y+x&=-5 \end{aligned}$$

$$\begin{aligned} y-x&=-10 \\ 2y+x&=-5 \\ \hline 3y&=-15 \\ y&=-5 \\ -5-x&=-10 \\ -x&=-5 \\ x&=5 \\ z+2(5)&=-10 \\ z&=-20 \end{aligned}$$

$x=5,\quad y=-5,\quad z=-20$

19.

Amount of dish A : a $\quad 2a+b+c=10$

Amount of dish B : b $\quad a+2b+2c=14$

Amount of dish C : c $\quad 3a+b+3c=18$

$$\begin{array}{r} -4a-2b-2c=-20 \\ a+2b+2c=14 \\ \hline -3a \qquad\quad =-6 \\ a=2 \end{array} \qquad \begin{array}{r} 2a+b+c=10 \\ -3a-b-3c=-18 \\ \hline -a-\quad 2c=-8 \\ -2-2c=-8 \\ -2c=-6 \\ c=3 \end{array} \qquad \begin{array}{r} 2(2)+b+3=10 \\ b=3 \end{array}$$

2 units of dish A,

3 units of dish B,

3 units of dish C

21.

Number of 12" televisions : x

Number of 16" televisions : y

Number of 19" televisions : z

Change all units to hours. Clear all fractions

$$\begin{array}{ll} \frac{3}{4}x+y+1.5z=17.75 & 3x+4y+6z=71 \\ \frac{1}{2}x+\frac{3}{4}y+z=12.5 & 2x+3y+4z=50 \\ \frac{1}{6}x+\frac{1}{4}y+\frac{1}{4}z=3.75 & 2x+3y+3z=45 \end{array}$$

$$\begin{array}{r} -6x-8y-12z=-142 \\ 6x+9y+12z=150 \\ \hline y=8 \end{array} \qquad \begin{array}{r} 2x+3y+4z=50 \\ -2x-3y-3z=-45 \\ \hline z=5 \end{array}$$

$$\begin{aligned} 3x+4(8)+6(5)&=71 \\ 3x&=9 \\ x&=3 \end{aligned}$$

3 12" televisions,

8 16" televisions,

5 19" televisions

CHAPTER 10 SECTION 4

EXERCISE SET 10.4

1.

$$\frac{2x-11}{(x+2)(x-3)}=\frac{A}{x+2}+\frac{B}{x-3}$$

$$2x-11=A(x-3)+B(x+2)$$

Set $x=3$: $\quad -5=A(0)+B(5)$

$$B=-1$$

Set $x=-2$: $-15=A(-5)+B(0)$

$$A=3$$

$$\frac{3}{x+2}-\frac{1}{x-3}$$

3.

$$\frac{3x-2}{6x^2-5x+1}=\frac{3x-2}{(3x-1)(2x-1)}=\frac{A}{3x-1}+\frac{B}{2x-1}$$

$$3x-2=A(2x-1)+B(3x-1)$$

Set $x=\frac{1}{2}$: $\quad \frac{3}{2}-2=A(0)+B\left(\frac{3}{2}-1\right)$

$$-\frac{1}{2}=\frac{1}{2}B$$

$$B=-1$$

Set $x=\frac{1}{3}$: $\quad -1=A\left(\frac{2}{3}-1\right)+B(0)$

$$-1=-\frac{1}{3}A$$

$$A=3$$

$$\frac{3}{3x-1}-\frac{1}{2x-1}$$

5.

$$\frac{x^2+x+2}{x^3-x}=\frac{x^2+x+2}{x(x+1)(x-1)}=\frac{A}{x}+\frac{B}{x+1}+\frac{C}{x-1}$$

$$x^2+x+2=A(x+1)(x-1)+Bx(x-1)+Cx(x+1)$$

Set $x=1$: $\quad 1^2+1+2=A(2)(0)+B(1)(0)+C(1)(2)$

$$4=2C$$

$$C=2$$

Set $x=0$: $\quad 0^2+0+2=A(1)(-1)+B(0)(-1)+C(0)(1)$

$$2=-A$$

$$A=-2$$

Set $x=-1$: $\quad (-1)^2-1+2=A(0)(-2)+B(-1)(-2)+C(-1)(0)$

$$2=2B$$

$$B=1$$

$$-\frac{2}{x}+\frac{1}{x+1}+\frac{2}{x-1}$$

7.

$$\frac{3x-2}{x^3+2x^2}=\frac{3x-2}{x(x)(x+2)}=\frac{A}{x}+\frac{B}{x^2}+\frac{C}{x+2}$$

$$3x-2=Ax(x+2)+B(x+2)+Cx^2$$

Set $x=0$: $\quad 3(0)-2=A(0)(2)+B(2)+C(0)$

$$-2=2B$$

$$B=-1$$

Set $x=-2$: $\quad 3(-2)-2=A(-2)(0)+B(0)+C(-2)^2$

$$-8=4C$$

$$C=-2$$

Set $x=-1$: $\quad 3(-1)-2=A(-1)(1)-1(1)-2(-1)^2$

$B=-1,\ C=-2$ $\quad -5=-A-1-2$

$$-2=-A$$

$$A=2$$

$$\frac{2}{x}-\frac{1}{x^2}-\frac{2}{x+2}$$

9.

$$\frac{x^2-x+2}{(x-1)(x+1)^2}=\frac{A}{x-1}+\frac{B}{x+1}+\frac{C}{(x+1)^2}$$

$$x^2-x+2=A(x+1)^2+B(x-1)(x+1)+C(x-1)$$

Set $x=1$: $\quad 1^2-1+2=A(2)^2+B(0)(2)+C(0)$

$$2=4A$$

$$A=\frac{1}{2}$$

Set $x=-1$: $\quad (-1)^2-(-1)+2=A(0)^2+B(-2)(0)+C(-2)$

$$4=-2C$$

$$C=-2$$

Set $x=0$: $\quad 0^2-0+2=\frac{1}{2}(1)^2+B(-1)(1)-2(-1)$

$A=\frac{1}{2},\ C=-2 \qquad 2=\frac{1}{2}-B+2$

$$B=\frac{1}{2}$$

$$\frac{\frac{1}{2}}{x-1}+\frac{\frac{1}{2}}{x+1}-\frac{2}{(x+1)^2}$$

11.

$$\frac{1-2x}{x^3+4x}=\frac{1-2x}{x(x^2+4)}=\frac{A}{x}+\frac{Bx+C}{x^2+4}$$

$$1-2x=A(x^2+4)+(Bx+C)x$$

Set $x=0$: $1-2(0)=A(4)+[B(0)+C](0)$

$$1=4A$$

$$A=\frac{1}{4}$$

Set $x=1$: $1-2(1)=\frac{1}{4}(1^2+4)+[B(1)+C](1)$

$A=\frac{1}{4}$ $\quad -1=\frac{5}{4}+B+C$

$$-\frac{9}{4}=B+C$$

(Continued on next page)

11. (Continued)

Set $x=-1$: $\quad 1-2(-1)=\frac{1}{4}\left[(-1)^2+4\right]+\left[B(-1)+C\right](-1)$

$A=\frac{1}{4}$

$$3=\frac{5}{4}+B-C$$

$$\frac{7}{4}=B-C$$

$$-\frac{9}{4}=B+C$$

$$\frac{7}{4}=B-C$$

$$-\frac{1}{2}=2B$$

$$B=-\frac{1}{4}$$

$$-\frac{1}{4}+C=-\frac{9}{4}$$

$$C=-2$$

$$\frac{\frac{1}{4}}{x}+\frac{-\frac{1}{4}x-2}{x^2+4}=$$

$$\frac{\frac{1}{4}}{x}-\frac{\frac{1}{4}x+2}{x^2+4}$$

13.

$$\frac{2x^3 - x^2 + x}{\left(x^2+3\right)^2} = \frac{Ax+B}{x^2+3} + \frac{Cx+D}{\left(x^2+3\right)^2}$$

$$2x^3 - x^2 + x = (Ax+B)(x^2+3) + Cx + D$$

Set $x = 0$: $2(0)^3 - 0^2 + 0 = [A(0)+B][0^2+3] + C(0) + D$

$$0 = 3B + D$$

Set $x = 1$: $2(1)^3 - 1^2 + 1 = [A(1)+B][1^2+3] + C(1) + D$

$$2 = 4(A+B) + C + D$$

$$2 = 4A + 4B + C + D$$

Set $x = -1$: $2(-1)^3 - (-1)^2 + (-1) = [A(-1)+B][(-1)^2+3] + C(-1) + D$

$$-4 = 4(-A+B) - C + D$$

$$-4 = -4A + 4B - C + D$$

Set $x = 2$: $2(2)^3 - 2^2 + 2 = [A(2)+B][2^2+3] + C(2) + D$

$$14 = 7(2A+B) + 2C + D$$

$$14 = 14A + 7B + 2C + D$$

(Continued on next page)

13. (Continued)

$$\begin{aligned} 3B + \quad D &= 0 \\ 4A + 4B + C + D &= 2 \\ -4A + 4B - C + D &= -4 \\ 14A + 7B + 2C + D &= 14 \end{aligned}$$

$$\begin{aligned} 4A + 4B + C + D &= 2 \\ \underline{-4A + 4B - C + D} &\underline{= -4} \\ 8B + \quad 2D &= -2 \\ 4B + D &= -1 \end{aligned}$$

$$\begin{aligned} -3B - D &= 0 \\ \underline{4B + D} &\underline{= -1} \\ B &= -1 \\ 4(-1) + D &= -1 \\ D &= 3 \end{aligned}$$

$$\begin{aligned} 4A - 4B + C - D &= 4 \\ \underline{14A + 7B + 2C + D} &\underline{= 14} \\ 18A + 3B + 3C \quad &= 18 \\ 18A + 3(-1) + 3C &= 18 \\ 18A + 3C &= 21 \\ 6A + C &= 7 \end{aligned}$$

$$\begin{aligned} 4A + 4(-1) + C + 3 &= 2 \\ 4A + C &= 3 \end{aligned}$$

$$\begin{aligned} -6A - C &= -7 \\ \underline{4A + C} &\underline{= 3} \\ -2A \quad &= -4 \\ A &= 2 \\ 4(2) + C &= 3 \\ C &= -5 \end{aligned}$$

$$\frac{2x-1}{x^2+3} + \frac{-5x+3}{\left(x^2+3\right)^2}$$

15.

$$\frac{-x}{x^3-2x^2-4x-1}$$

Possible rational roots: ± 1

$$\begin{array}{r|rrrr} 1 & 1 & -2 & -4 & -1 \\ & & 1 & -1 & -5 \\ \hline & 1 & -1 & -5 & |-6 \end{array}$$

Not a root

$$\begin{array}{r|rrrr} -1 & 1 & -2 & -4 & -1 \\ & & -1 & 3 & 1 \\ \hline & 1 & -3 & -1 & |\,0 \end{array}$$

$x=-1$ is a root

$$\frac{-x}{x^3-2x^2-4x-1}=\frac{-x}{(x+1)(x^2-3x-1)}=\frac{A}{x+1}+\frac{Bx+C}{x^2-3x-1}$$

(Continued on next page)

15. (Continued)

$$-x = A(x^2-3x-1)+(Bx+C)(x+1)$$

Set $x=-1$: $1 = A[(-1)^2-3(-1)-1]+[B(-1)+C](0)$

$$1 = 3A$$

$$A = \frac{1}{3}$$

Set $x=0$: $0 = \frac{1}{3}[0^2-3(0)-1]+[B(0)+C](1)$

$A = \frac{1}{3}$ $\quad 0 = -\frac{1}{3}+C$

$$C = \frac{1}{3}$$

Set $x=1$: $-1 = \frac{1}{3}[1^2-3(1)-1]+\left[B(1)+\frac{1}{3}\right](2)$

$A = \frac{1}{3},\ C = \frac{1}{3}$ $\quad -1 = -1+2B+\frac{2}{3}$

$$-\frac{2}{3} = 2B$$

$$B = -\frac{1}{3}$$

$$\frac{\frac{1}{3}}{x+1}+\frac{-\frac{1}{3}x+\frac{1}{3}}{x^2-3x-1}$$

17.

$$\frac{x^4 - x^2 - 9}{(x+1)(x^2+2)^2} = \frac{A}{x+1} + \frac{Bx+C}{x^2+2} + \frac{Dx+E}{(x^2+2)^2}$$

$$x^4 - x^2 - 9 = A(x^2+2)^2 + (Bx+C)(x+1)(x^2+2) + (Dx+E)(x+1)$$

Set $x = -1$: $(-1)^4 - (-1)^2 - 9 = A[(-1)^2 + 2]^2 + [B(-1) + C](0)(3) + [D(-1) + E](0)$

$$-9 = 9A$$

$$A = -1$$

Set $x = 0$: $0^4 - 0^2 - 9 = -1[0^2 + 2]^2 + [B(0) + C](1)(2) + [D(0) + E](1)$

$A = -1$ $\quad -9 = -4 + 2C + E$

$$-5 = 2C + E$$

Set $x = 1$: $1^4 - 1^2 - 9 = -1(1^2 + 2)^2 + [B(1) + C](2)(1^2 + 2) + [D(1) + E](2)$

$A = -1$ $\quad -9 = -9 + 6B + 6C + 2D + 2E$

$$0 = 6B + 6C + 2D + 2E$$

$$0 = 3B + 3C + D + E$$

Set $x = 2$: $2^4 - 2^2 - 9 = -1(2^2 + 2)^2 + [B(2) + C](3)(2^2 + 2) + [D(2) + E](2)$

$A = -1$ $\quad 3 = -36 + 36B + 18C + 6D + 3E$

$$1 = -12 + 12B + 6C + 2D + E$$

$$13 = 12B + 6C + 2D + E$$

(Continued on next page)

17. (Continued)

Set $x=-2$: $(-2)^4-(-2)^2-9=-1\left[(-2)^2+2\right]^2+\left[B(-2)+C\right](-1)\left[(-2)^2+2\right]+\left[D(-2)+E\right](-1)$

$A=-1$

$$3=-36+12B-6C+2D-E$$

$$39=12B-6C+2D-E$$

$$2C+E=-5$$
$$3B+3C+D+E=0$$
$$12B+6C+2D+E=13$$
$$12B-6C+2D-E=39$$

$$\begin{array}{r} 12B+6C+2D+E=13 \\ \underline{12B-6C+2D-E=39} \\ 24B+4D=52 \\ 6B+D=13 \end{array}$$

$$\begin{array}{r} -12B-6C-2D-E=-13 \\ \underline{12B-6C+2D-E=39} \\ -12C-2E=26 \\ -6C-E=13 \end{array}$$

$$\begin{array}{r} -6C-E=13 \\ \underline{2C+E=-5} \\ -4C=8 \\ C=-2 \\ 2(-2)+E=-5 \\ E=-1 \end{array}$$

$$\begin{array}{r} -6B-6C-2D-2E=0 \\ \underline{12B+6C+2D+E=13} \\ 6B-E=13 \\ 6B-(-1)=13 \\ 6B=12 \\ B=2 \\ 6(2)+D=13 \\ D=1 \end{array}$$

$$-\frac{1}{x+1}+\frac{2x-2}{x^2+2}+\frac{x-1}{\left(x^2+2\right)^2}$$

19.
$$\frac{x^4+x^3+x^2+3x-2}{(x+1)(x^2+1)}$$

$$(x+1)(x^2+1)=x^3+x^2+x+1$$

$$\begin{array}{r} x \\ x^3+x_2+x+1\overline{\big)x^4+x^3+x^2+3x-2} \\ \underline{x^4+x^3+x^2+x} \\ 2x-2 \end{array}$$

$$\frac{x^4+x^3+x^2+3x-2}{(x+1)(x^2+1)}=x+\frac{2x-2}{x^3+x^2+x+1}=x+\frac{2x-2}{(x+1)(x^2+1)}$$

$$\frac{2x-2}{(x+1)(x^2+1)}=\frac{A}{x+1}+\frac{Bx+C}{x^2+1}$$

$$2x-2=A(x^2+1)+(Bx+C)(x+1)$$

Set $x=-1$: $\quad 2(-1)-2=A\left[(-1)^2+1\right]+\left[B(-1)+C\right](0)$

$$-4=2A$$
$$A=-2$$

Set $x=0$: $\quad 2(0)-2=-2(0^2+1)+\left[B(0)+C\right](1)$

$A=-2$ $\quad -2=-2+C$

$$C=0$$

Set $x=1$: $\quad 2(1)-2=-2(1^2+1)+\left[B(1)+0\right](2)$

$A=-2, C=0$ $\quad 0=-4+2B$

$$4=2B$$
$$B=2$$

$$x-\frac{2}{x+1}+\frac{2x}{x^2+1}$$

CHAPTER 10 SECTION 5

SECTION 10.5 PROGRESS CHECK

1a). (Page 591)

$$y \le 2x + 1$$

Graph $y = 2x + 1$

Test Point : $(1, 0)$

$0 \le 2(1) + 1$?

$0 \le 3$ True

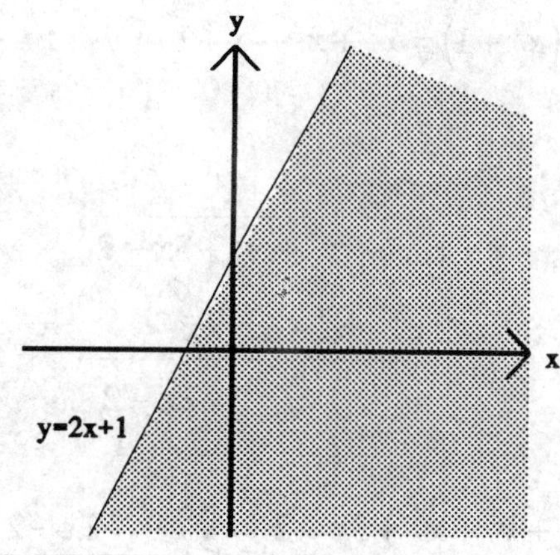

1b).

$$y + 3x > -2$$

$$y > -3x - 2$$

Graph $y = -3x - 2$

Test Point : $(0, 0)$

$0 > -3(0) - 2$?

$0 > -2$ True

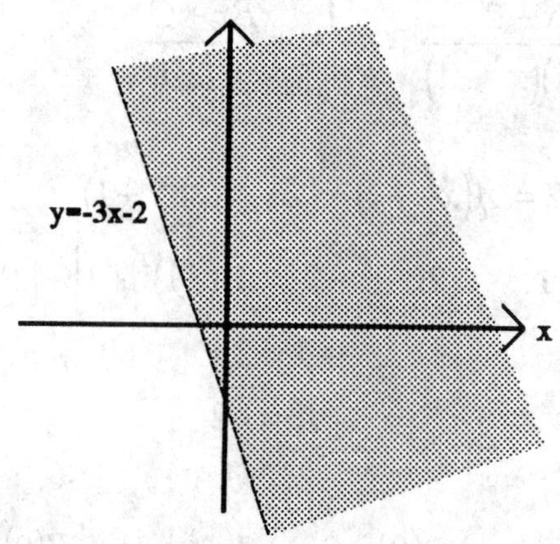

1c).

$$y \geq -x+1$$
$$\text{Graph } y = -x+1$$
$$\text{Test Point: } (0,\ 0)$$
$$0 \geq -0+1 \ ?$$
$$0 \geq 1 \quad \text{False}$$

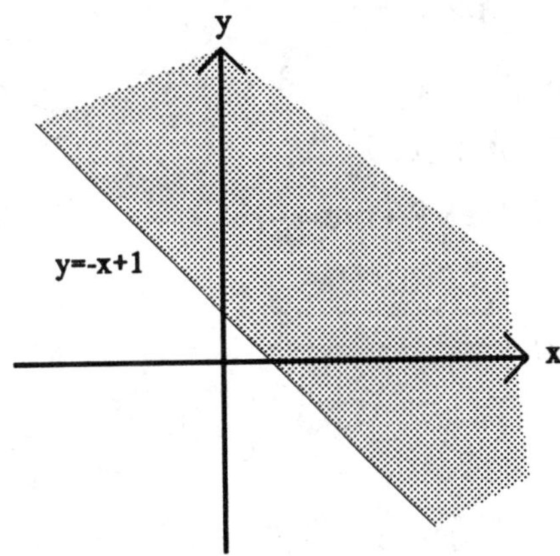

2a). (Page 593)

$$2y \geq 7$$
$$y \geq \frac{7}{2}$$

$$\text{Graph } y = \frac{7}{2}$$
$$\text{Test Point: } (0,\ 0)$$
$$0 \geq \frac{7}{2} \ ? \ \text{False}$$

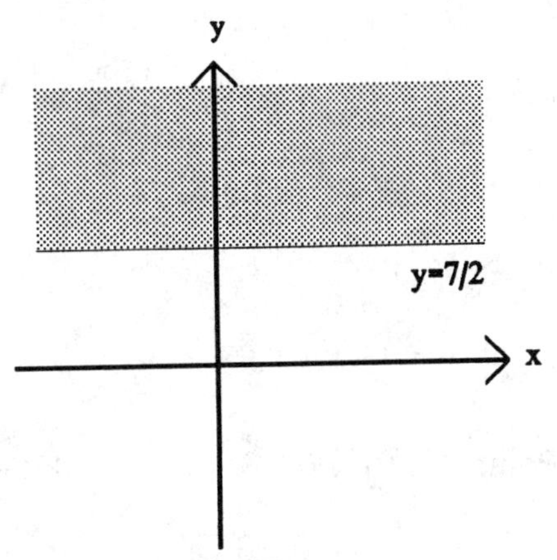

2b).

$$x < -2$$
$$\text{Graph } x = -2$$
$$\text{Test Point: } (0,\ 0)$$
$$0 < -2 \ ? \ \text{False}$$

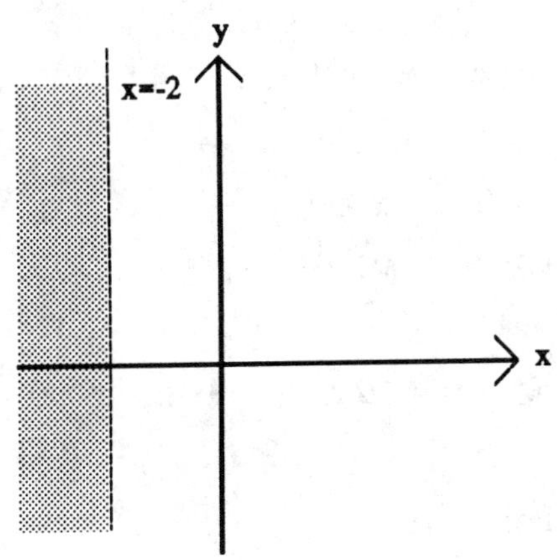

2c).

$$1 \le y < 3$$

Graph $y = 1 \quad y = 3$

Test Point : $(0, 2)$

$1 \le 2 < 3$? True

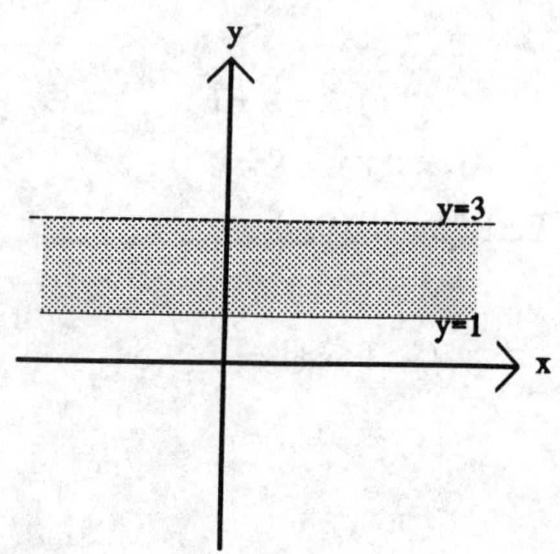

3a). (Page 595)

$x + y \ge 3 \qquad y \ge 3 - x$

$x + 2y < 8 \qquad y < \frac{1}{2}(-x + 8)$

Graph $y = 3 - x, \qquad y = -\frac{1}{2}x + 4$

Test Point : $(0, 0) \quad 0 \ge 3 - 0? \qquad 0 < -\frac{1}{2}(0) + 4$?

$0 \ge 3$ False $\qquad 0 < 4$ True

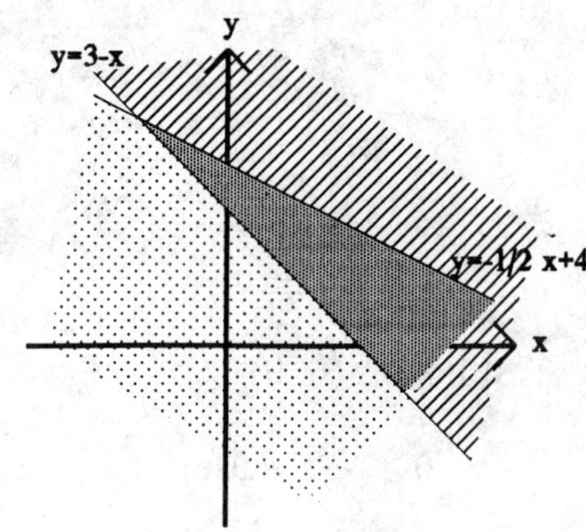

3b).

$2x + y \le 4 \qquad y \le 4 - 2x$

$x + y \le 3 \qquad y \le 3 - x$

$x \ge 0 \qquad x \ge 0$

$y \ge 0 \qquad y \ge 0$

Graph : $y = 4 - 2x, \quad y = 3 - x, \quad x = 0, \quad y = 0$

Test Point : $(0, 0)$

$0 \le 4 - 2(0)$? $\quad 0 \le 3 - 0$?

$0 \le 4$ True $\qquad 0 \le 3$ True

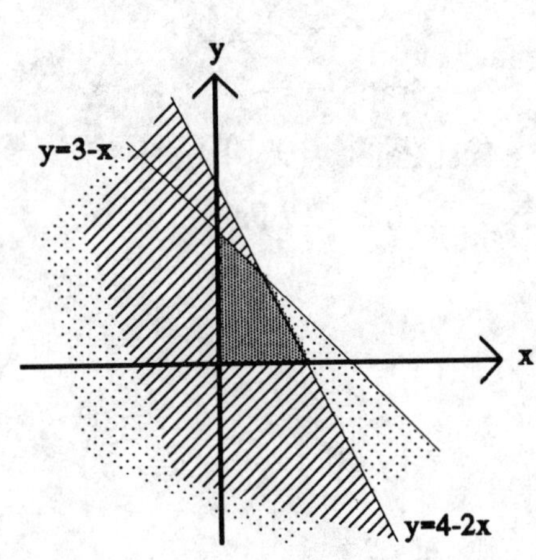

EXERCISE SET 10.5

1.

$$y \le x + 2$$

Graph $y = x + 2$

Test point : $(0, 0)$

$$0 \le 0 + 2 ?$$

$0 \le 2$ True

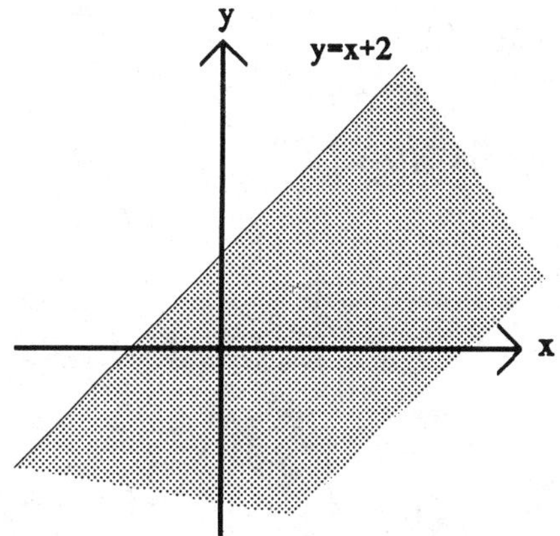

3.

$$y > x - 4$$

Graph $y = x - 4$

Test Point : $(0, 0)$

$$0 > 0 - 4 ?$$

$0 > -4$ True

y

y=x-4

x

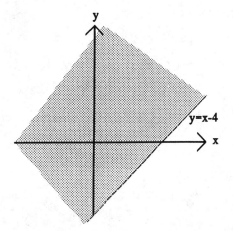

5.

$$y \le 4 - x$$

Graph $y = 4 - x$

Test Point : $(0, 0)$

$$0 \le 4 - 0\,?$$

$$0 \le 4 \quad \text{True}$$

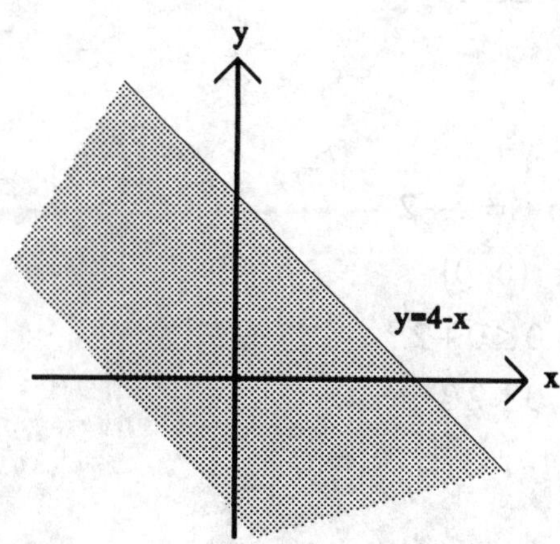

7.

$$y > x$$

Graph $y = x$

Test point : $(1, 0)$

$$0 > 1\,? \text{ False}$$

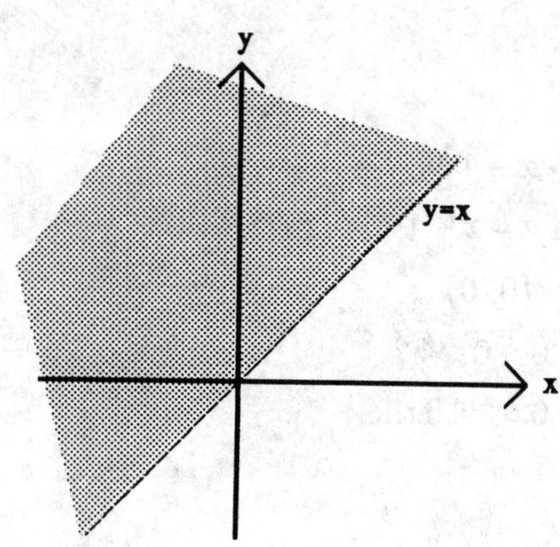

9.

$$3x - 5y > 15$$

$$3x - 15 > 5y$$

$$y < \frac{3}{5}x - 3$$

Graph $y = \frac{3}{5}x - 3$

Test point : $(0, 0)$

$$0 < \frac{3}{5}(0) - 3\,?$$

$$0 < -3 \quad \text{False}$$

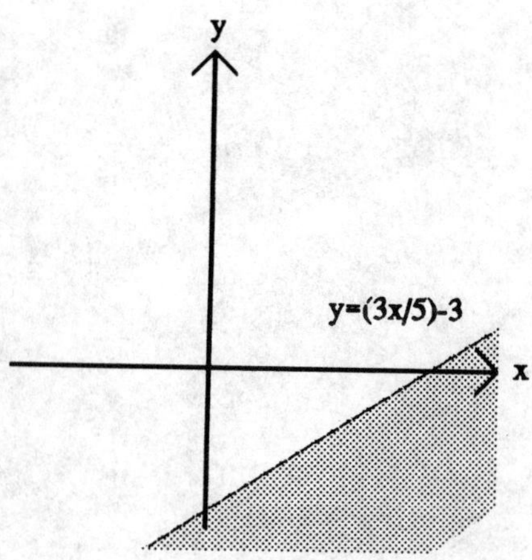

11.

$x \le 4$

Graph $x = 4$

Test point : $(0, 0)$

$0 \le 4$? True

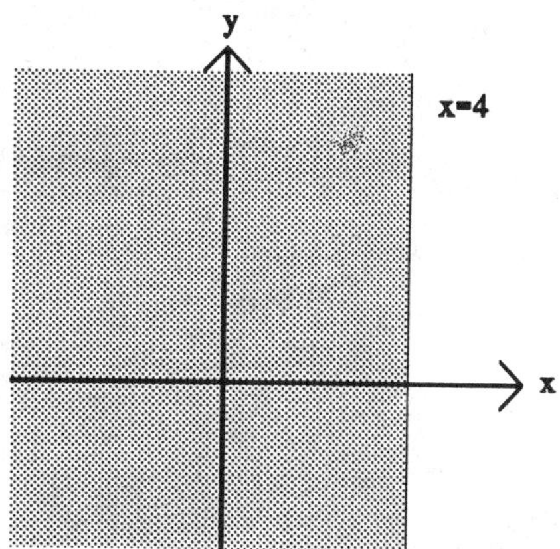

13.

$y > -3$

Graph $y = -3$

Test point : $(0, 0)$

$0 > -3$? True

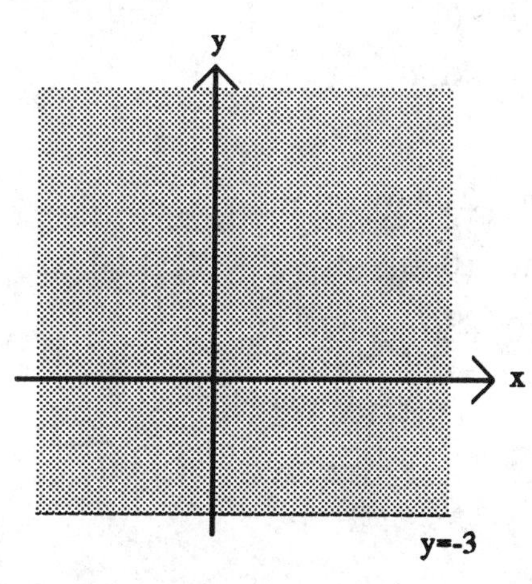

15.

$x > 0$

Graph $x = 0$

Test point : $(1, 0)$

$1 > 0$? True

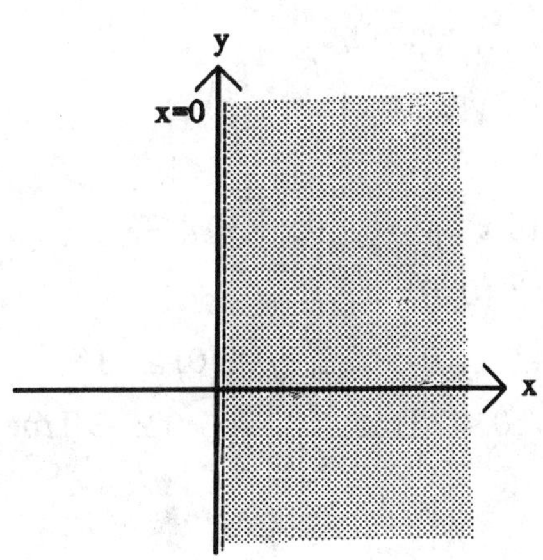

17.

$$-2 \le x \le 3$$

Graph $x = -2, \quad x = 3$

Test point : $(0, 0)$

$-2 \le 0 \le 3$? True

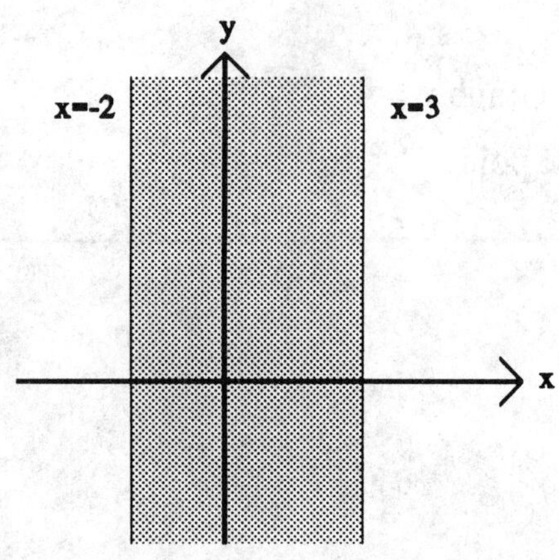

19.

$$2x + 5y \le 15$$
$$x \ge 0$$
$$y \ge 0$$

Graph $\quad 2x + 5y = 15$

Test point : $(0, 0)$

$2(0) + 5(0) \le 15$?

$0 \le 15$ True

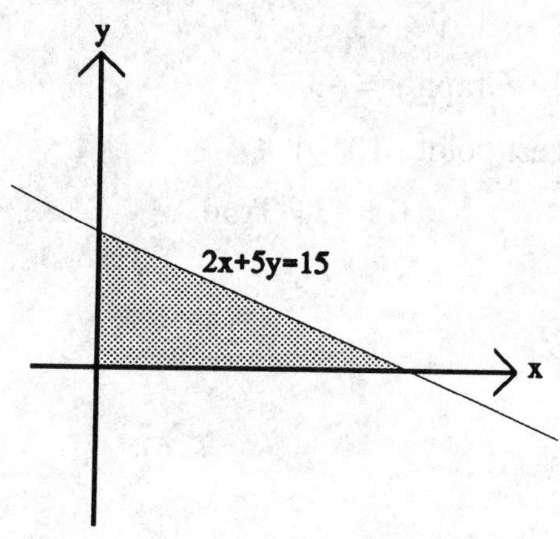

21.

$$2x - y \le 3$$
$$2x + 3y \ge -3$$

Graph $2x - y = 3, \quad 2x + 3y = -3$

Test point : $(0, 0)$

$2(0) - 0 \le 3$? $\qquad 2(0) + 3(0) \ge -3$?

$0 \le 3$ True $\qquad 0 \ge -3$ True

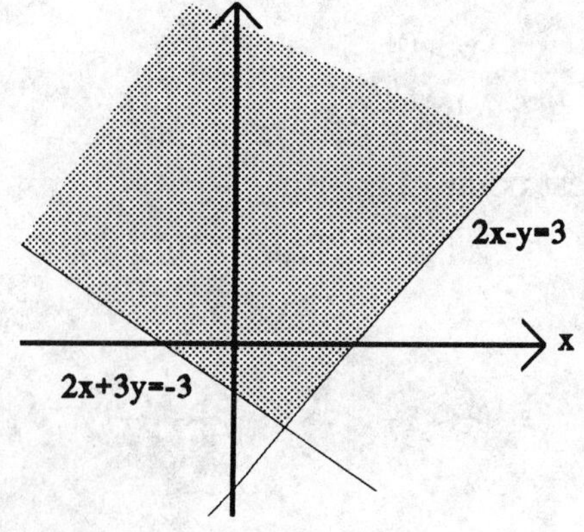

23.

$$3x - y \geq -7$$
$$3x + y \leq -2$$

Graph $3x - y = -7$, $\quad 3x + y = -2$

Test point : $(0, 0)$

$3(0) - 0 \geq -7$? $\quad 3(0) + 0 \leq -2$?

$0 \geq -7$ True $\quad 0 \leq -2$ False

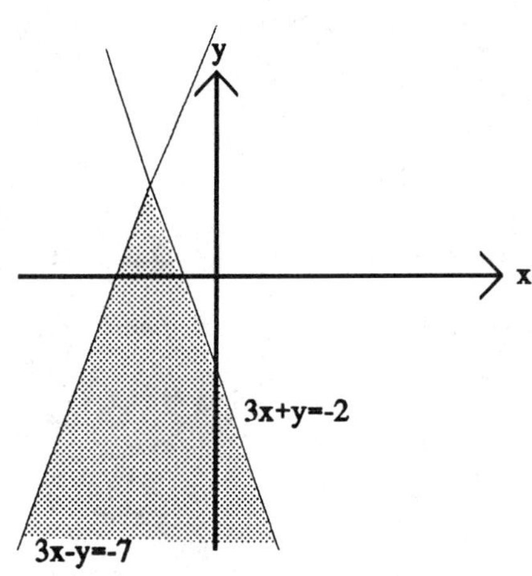

25.

$$3x - 2y \geq -5$$
$$4x - y \leq 10$$
$$y \geq 2$$

Graph $3x - 2y = -5$, $\quad 4x - y = 10$, $\quad y = 2$

Test point : $(0,0)$

$3(0) - 2(0) \geq -5$? $\quad 4(0) - 0 \leq 10$? $\quad 0 \geq 2$?

$0 \geq -5$ $\quad 0 \leq 10$ $\quad$ False

True $\quad$ True

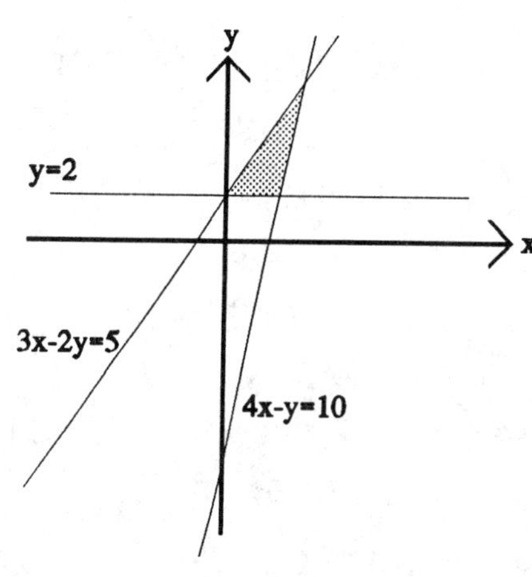

27.

$$2x - y \leq 5$$
$$x + 2y \geq 1$$
$$x \geq 0$$
$$y \geq 0$$

Graph $2x - y = 5$, $\quad x + 2y = 1$, $\quad x = 0$, $\quad y = 0$

Test point : $(0, 0)$ $\quad$ Test point : $(1, 1)$

$2(0) - 0 \leq 5$? $\quad 0 + 2(0) \geq 1$? $\quad 1 \geq 0$? $\quad 1 \geq 0$?

$0 \leq 5$ $\quad 0 \geq 1$ $\quad$ True $\quad$ True

True $\quad$ False

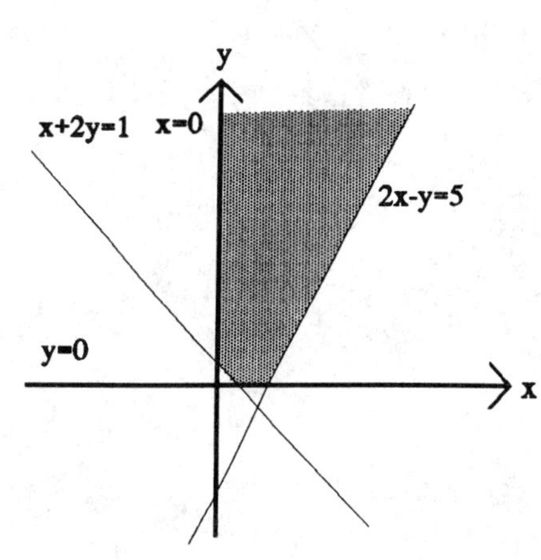

29.

$$3x + y \le 6$$
$$x - 2y \le -1$$
$$x \ge 2$$

Graph $3x + y = 6, \quad x - 2y = -1, \quad x = 2$

Test point : $(0, 0)$

$3(0) + 0 \le 6$? $\quad 0 - 2(0) \le -1$? $\quad 0 \ge 2$?

$0 \le 6 \quad\quad 0 \le -1 \quad$ False

True $\quad\quad$ False

No solution since the three regions don't all overlap.

31.

$$3x - 2y \le -6$$
$$8x + 3y \le 24$$
$$5x + 4y \ge 20$$
$$x \ge 0$$
$$y \ge 0$$

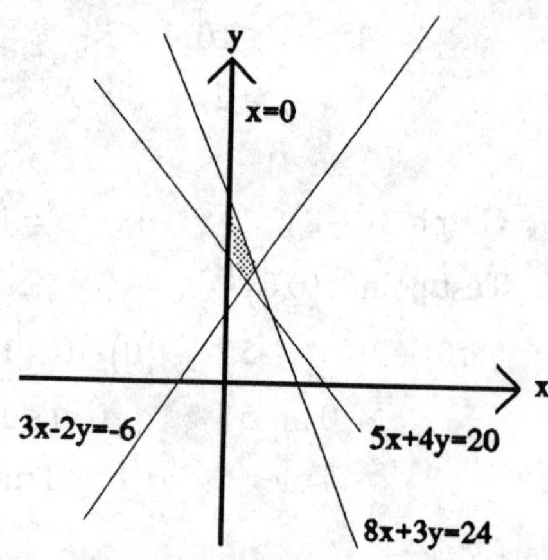

Graph $3x - 2y = -6, \; 8x + 3y = 24, \; 5x + 4y = 20$

Test point : $(0, 0)$

$3(0) - 2(0) \le -6$? $\quad 8(0) + 3(0) \le 24$? $\quad 5(0) + 4(0) \ge 20$?

$0 \le -6 \quad\quad 0 \le 24 \quad\quad 0 \ge 20$

False $\quad\quad$ True $\quad\quad$ False

Graph $x = 0, \quad y = 0$

Test point : $(1, 1)$

$1 \ge 0$? $\quad 1 \ge 0$?

True $\quad$ True

33.

Amount of ice cream : x

Amount of yogurt : y

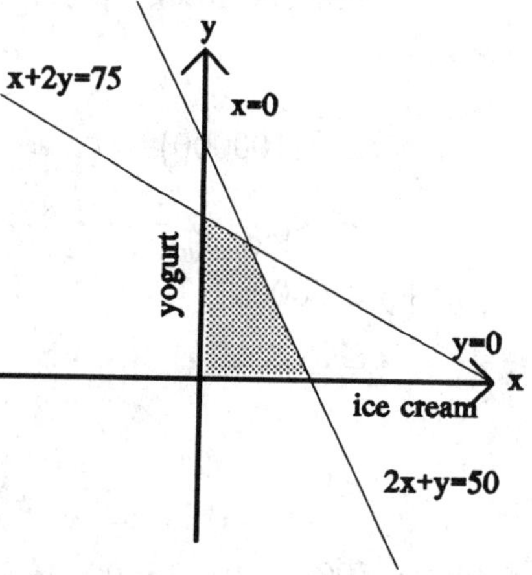

$0.4x + 0.2y \le 10$

$0.2x + 0.4y \le 15$

$x \ge 0$

$y \ge 0$

$4x + 2y = 100, \qquad 2x + 4y = 150$

Graph $2x + y = 50, \qquad x + 2y = 75$

Test point : $(0, 0)$

$0.4(0) + 0.2(0) \le 10\,? \qquad 0.2(0) + 0.4(0) \le 15\,?$

$0 \le 10 \qquad 0 \le 15$

True True

Graph $x = 0,\ y = 0$

Test point : $(1, 1)$

$1 \ge 0\,? \qquad 1 \ge 0\,?$

True True

35.

Common stock : x

Preferred stock : y

$$x \le \frac{1}{2}(100000)$$
$$y \le 35000$$
$$x + y \le 60000$$
$$x \le 2y$$
$$x \ge 0$$
$$y \ge 0$$

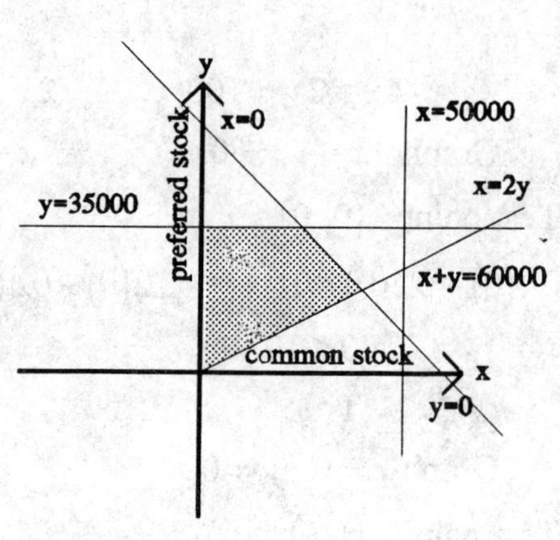

Graph $x = 50000,\ \ y = 35000,\ \ x + y = 60000$

Test point : $(0, 0)$

$0 \le 50000?$ $\quad 0 \le 35000?$ $\quad 0 + 0 \le 60000?$

True $\quad$ True $\quad$ True

Graph $x = 2y,$ $\quad x = 0,$ $\quad y = 0$

Test point : $(1, 1)$ $\quad 1 \ge 0?$ $\quad 1 \ge 0?$

$1 \le 2(1)?$ $\quad$ True $\quad$ True

$1 \le 2$

True

CHAPTER 10 SECTION 6

EXERCISE SET 10.6

1.

$x-\frac{1}{2}y$ subject to $3x-y\geq 1$

$x\geq 0$

$x\leq 5$

$y\geq 0$

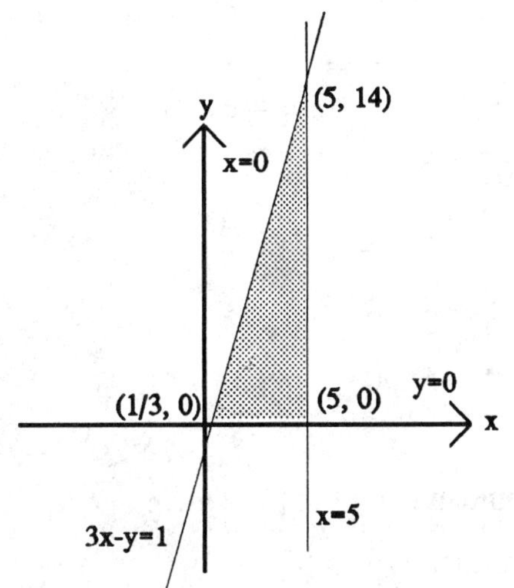

Vertex	x	y	$z=x-\frac{1}{2}y$
$\left(\frac{1}{3},0\right)$	$\frac{1}{3}$	0	$\frac{1}{3}$
$(5,0)$	5	0	5
$(5,14)$	5	14	−2

Minimum is -2 at $(5,14)$;

Maximum is 5 at $(5,0)$

3.

$\frac{1}{2}x - 2y$ subject to $x + 2y \leq 6$

$3y - 2x \leq 2$

$x \geq 0$

$y \geq 0$

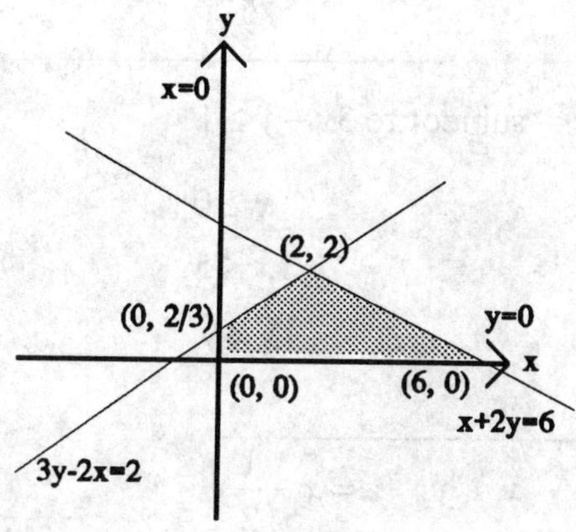

Vertex	x	y	$z = \frac{1}{2}x - 2y$
$(0, 0)$	0	0	0
$\left(0, \frac{2}{3}\right)$	0	$\frac{2}{3}$	$-\frac{4}{3}$
$(2, 2)$	2	2	-3
$(6, 0)$	6	0	3

Minimum is -3 at $(2, 2)$;

Maximum is 3 at $(6, 0)$

5.

$2x - y$ subject to $y - x \leq 0$

$4y + 3x \geq 6$

$x \leq 4$

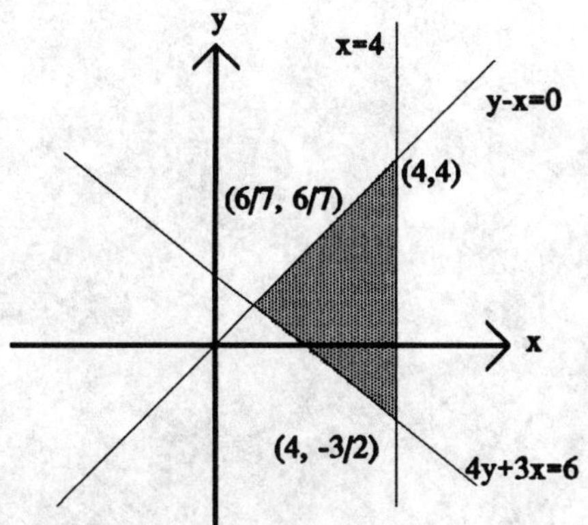

Vertex	x	y	$z = 2x - y$
$\left(\frac{6}{7}, \frac{6}{7}\right)$	$\frac{6}{7}$	$\frac{6}{7}$	$\frac{6}{7}$
$(4, 4)$	4	4	4
$\left(4, -\frac{3}{2}\right)$	4	$-\frac{3}{2}$	$\frac{19}{2}$

Minimum: $\frac{6}{7}$ at $\left(\frac{6}{7}, \frac{6}{7}\right)$;

Maximum: $\frac{19}{2}$ at $\left(4, -\frac{3}{2}\right)$

7.

$2x - y$ subject to $2y - x \le 8$

$x + 2y \ge 12$

$5x + 2y \le 44$

$x \ge 3$

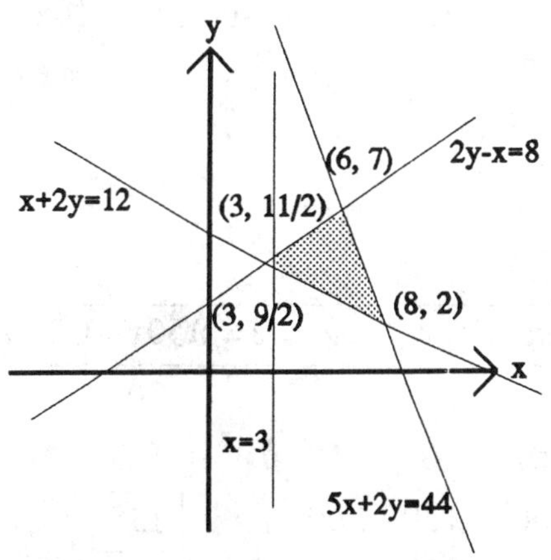

Vertex	x	y	$z = 2x - y$
$\left(3, \frac{9}{2}\right)$	3	$\frac{9}{2}$	$\frac{3}{2}$
$\left(3, \frac{11}{2}\right)$	3	$\frac{11}{2}$	$\frac{1}{2}$
$(6, 7)$	6	7	5
$(8, 2)$	8	2	14

Minimum: $\frac{1}{2}$ at $\left(3, \frac{11}{2}\right)$;

Maximum: 14 at $(8, 2)$

9.

Square feet of preferred space : x

Square feet of regular stock : y

Maximize $120x + 60y$ subject to $18x + 12y \le 1500$

$x \ge 60$

$y \ge 30$

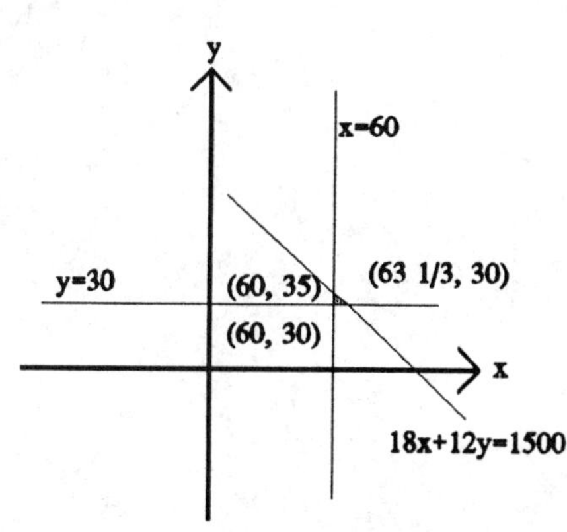

Vertex	x	y	$z = 120x + 60y$	
$(60, 30)$	60	30	9000	
$(60, 35)$	60	35	9300	
$\left(63\frac{1}{3}, 30\right)$	$63\frac{1}{3}$	30	9400	max

$63\frac{1}{3}$ square feet of preferred space

and 30 square feet of regular space

11.

Number of large containers : x

Number of small containers : y

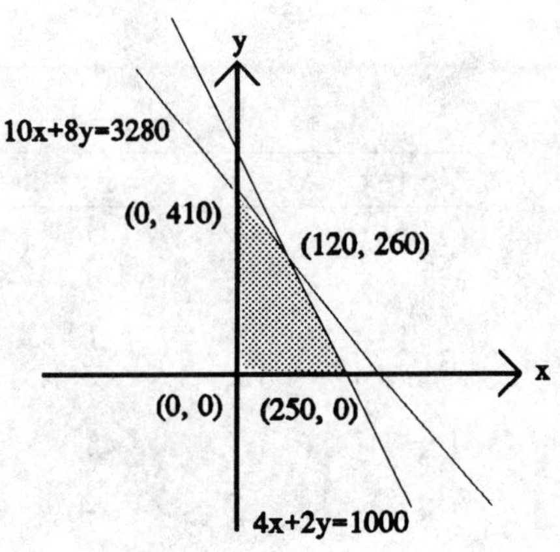

Maximize $0.50x + 0.30y$ subject to $10x + 8y \le 3280$

$$4x + 2y \le 1000$$
$$x \ge 0$$
$$y \ge 0$$

Vertex	x	y	$z = 0.50x + 0.30y$	
$(0, 0)$	0	0	0	
$(0, 410)$	0	410	123	
$(120, 260)$	120	260	$\boxed{138}$	max
$(250, 0)$	250	0	125	

120 large containers ; 260 small containers

13.

Pounds of Java : x

Pounds of Columbian : y

Revenue : $4(x+y)$

Cost : $1.50x+2y$

Profit : $4(x+y)-(1.50x+2y)=2.5x+2y$

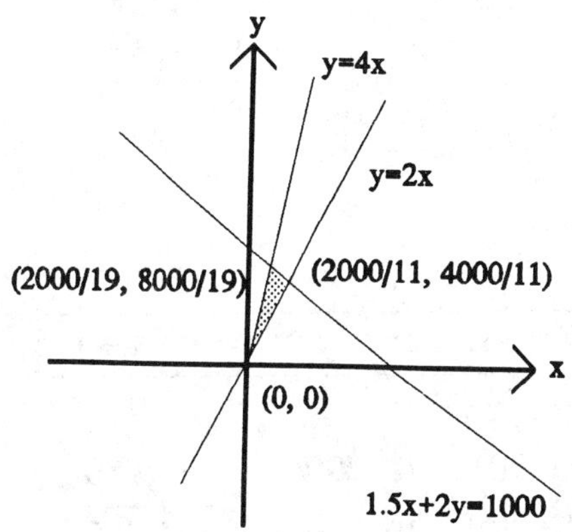

Maximize $2.5x+2y$ subject to $x \geq 0$

$$y \geq 0$$
$$1.5x+2y \leq 1000$$
$$y \geq 2x$$
$$y \leq 4x$$

Vertex	x	y	$z=2.5x+2y$	
$(0, 0)$	0	0	0	
$\left(\frac{2000}{19}, \frac{8000}{19}\right)$	$\frac{2000}{19}$	$\frac{8000}{19}$	1105.26	
$\left(\frac{2000}{11}, \frac{4000}{11}\right)$	$\frac{2000}{11}$	$\frac{4000}{11}$	$\boxed{1181.82}$	max

$\frac{2000}{11}$ lbs of Java ; $\frac{4000}{11}$ lbs of Columbian

15.

Acres of crop A : x

Acres of crop B : y

Maximize $150x + 175y$ subject to $6x + 9y \le 810$

$$20x + 15y \le 1800$$
$$x \ge 0$$
$$y \ge 0$$
$$x + y \le 100$$

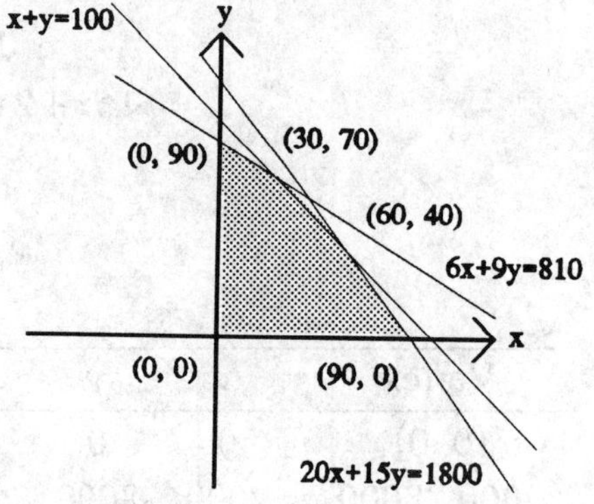

Vertex	x	y	$z = 150x + 175y$	
(0, 0)	0	0	0	
(0, 90)	0	90	15,750	
(30, 70)	30	70	16,750	max
(90, 0)	90	0	13,500	
(60, 40)	60	40	16,000	

30 acres of crop A, 70 acres of crop B

17.

Pounds of volume pack A : x

Pounds of volume pack B : y

Minimize $2.50x + 1.50y$ subject to

$$4x + 3y \geq 60$$
$$3x + 4y \geq 52$$
$$5x + y \geq 42$$
$$x \geq 0$$
$$y \geq 0$$

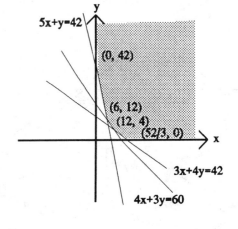

Vertex	x	y	$z = 2.50x + 1.50y$	
$(0, 42)$	0	42	63	
$(6, 12)$	6	12	$\boxed{33}$	min
$(12, 4)$	12	4	36	
$\left(\frac{52}{3}, 0\right)$	$\frac{52}{3}$	0	$43\frac{1}{3}$	

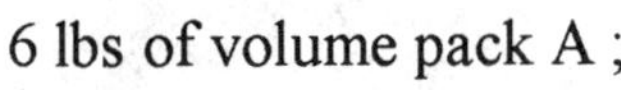

6 lbs of volume pack A ;

12 lbs of volume pack B

CHAPTER 10 REVIEW EXERCISES

1.

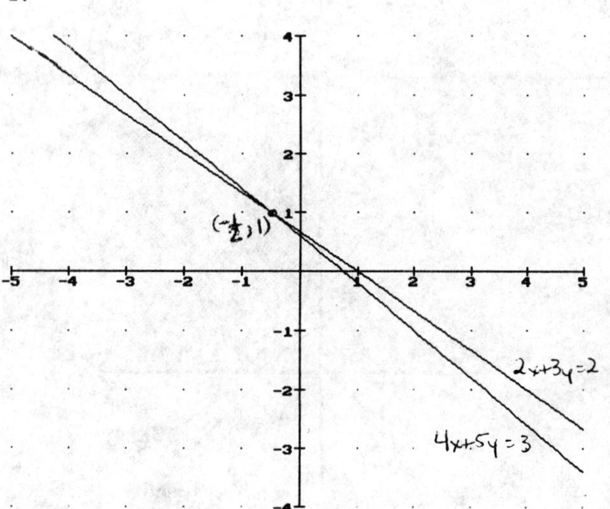

2.

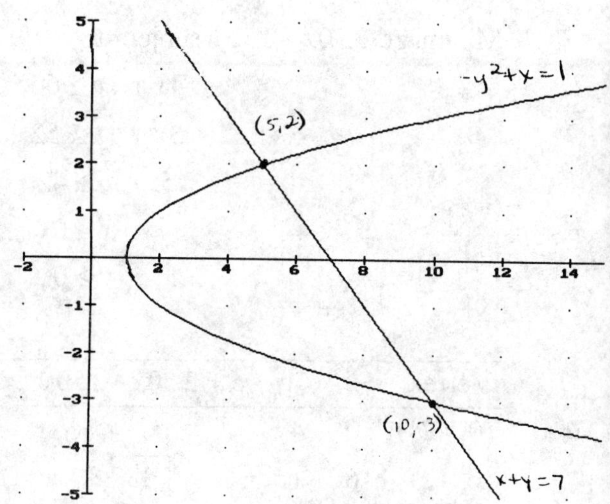

3.

$$-x+6y=-11$$
$$2x+5y=5$$

$$6y+11=x$$
$$2(6y+11)+5y=5$$
$$12y+22+5y=5$$
$$17y=-17$$
$$y=-1$$
$$x=6(-1)+11=5$$
$$x=5,\quad y=-1$$

4.

$$2x-4y=-14$$
$$-x-6y=-5$$

$$5-6y=x$$
$$2(5-6y)-4y=-14$$
$$10-12y-4y=-14$$
$$-16y=-24$$
$$y=\frac{3}{2}$$
$$x=5-6\left(\frac{3}{2}\right)=-4$$
$$x=-4,\quad y=\frac{3}{2}$$

5.

$$2x + y = 0$$
$$x - 3y = \frac{7}{4}$$

$$x = 3y + \frac{7}{4}$$
$$2\left(3y + \frac{7}{4}\right) + y = 0$$
$$6y + \frac{7}{2} + y = 0$$
$$7y = -\frac{7}{2}$$
$$y = -\frac{1}{2}$$
$$x = 3\left(-\frac{1}{2}\right) + \frac{7}{4} = \frac{1}{4}$$
$$x = \frac{1}{4}, \quad y = -\frac{1}{2}$$

6.

$$x^2 + y^2 = 25$$
$$x + 3y = 5$$

$$x = 5 - 3y$$
$$(5 - 3y)^2 + y^2 = 25$$
$$25 - 30y + 9y^2 + y^2 = 25$$
$$10y^2 - 30y = 0$$
$$10y(y - 3) = 0$$
$$y = 0 \quad y = 3$$
$$x = 5 - 3(0) = 5 \quad x = 5 - 3(3) = -4$$
$$x = 5,\ y = 0; \qquad x = -4,\ y = 3$$

7.

$$x^2 - 4y^2 = 9$$
$$y - 2x = 0$$

$$y = 2x$$
$$x^2 - 4(2x)^2 = 9$$
$$x^2 - 16x^2 = 9$$
$$-15x^2 = 9$$
$$x^2 = -\frac{9}{15}$$

No solution

8.

$$y^2 - 4x = 0$$
$$y^2 + x - 2y = 12$$

$$4x = y^2$$
$$x = \frac{y^2}{4}$$
$$y^2 + \frac{y^2}{4} - 2y = 12$$
$$4y^2 + y^2 - 8y = 48$$
$$5y^2 - 8y - 48 = 0$$
$$(5y+12)(y-4) = 0$$
$$y = \frac{-12}{5} \qquad y = 4$$
$$x = \frac{\left(-\frac{12}{5}\right)^2}{4} = \frac{36}{25} \qquad x = \frac{4^2}{4} = 4$$
$$x = \frac{36}{25},\ y = -\frac{12}{5}; \qquad x = 4,\ y = 4$$

9.

$$x + 4y = 17$$
$$2x - 3y = -21$$

$$-2x - 8y = -34$$
$$\underline{2x - 3y = -21}$$
$$-11y = -55$$
$$y = 5$$
$$x + 4(5) = 17$$
$$x = -3$$
$$x = -3,\ y = 5$$

10.

$$5x - 2y = 14$$
$$-x - 3y = 4$$

$$5x - 2y = 14$$
$$\underline{-5x - 15y = 20}$$
$$-17y = 34$$
$$y = -2$$
$$-x - 3(-2) = 4$$
$$x = 2$$
$$x = 2,\ y = -2$$

11.

$$-3x+y=-13$$
$$2x-3y=11$$

$$-9x+3y=-39$$
$$\underline{2x-3y=11}$$
$$-7x=-28$$
$$x=4$$
$$2(4)-3y=11$$
$$-3y=3$$
$$y=-1$$
$$x=4,\ y=-1$$

12.

$$7x-2y=-20$$
$$3x-y=-9$$

$$7x-2y=-20$$
$$\underline{-6x+2y=18}$$
$$x=-2$$
$$3(-2)-y=-9$$
$$-y=-3$$
$$y=3$$
$$x=-2,\ y=3$$

13.

$$y^2=2x-1$$
$$x-y=2$$

$$y^2=2x-1$$
$$-y=-x+2$$

$$y^2=2x-1$$
$$\underline{-2y=-2x+4}$$
$$y^2-2y=3$$
$$y^2-2y-3=0$$
$$(y-3)(y+1)=0$$
$$y=3 \qquad y=-1$$
$$-3=-x+2 \qquad 1=-x+2$$
$$x=5 \qquad x=1$$
$$x=5,\ y=3;\ x=1,\ y=-1$$

14.

$$x^2+y^2=9$$
$$\underline{-x^2+y=3}$$
$$y^2+y=12$$
$$y^2+y-12=0$$
$$(y+4)(y-3)=0$$
$$y=-4 \qquad y=3$$
$$-x^2-4=3 \qquad -x^2+3=3$$
$$-x^2=7 \qquad -x^2=0$$
$$\underbrace{x^2=-7} \qquad x=0$$

No solution

$$x=0,\quad y=3$$

15.

Tens digit : t

Units digit : u

Original number : $\underline{tu}$

Reverse digits : $\underline{ut}$

$$10t + u + t = 49$$
$$10u + t = 9 + 10t + u$$

$$11t + u = 49$$
$$-9t + 9u = 9$$

$$11t + u = 49$$
$$\underline{t - u = -1}$$
$$12t = 48$$
$$t = 4$$
$$4 - u = -1$$
$$u = 5$$

The number is 45.

16.

Tens digit : t

Units digit : u

$$t + u = 9$$
$$10t + u + u = 74$$

$$t + u = 9$$
$$10t + 2u = 74$$

$$t + u = 9$$
$$\underline{-5t - u = -37}$$
$$-4t = -28$$
$$t = 7$$
$$7 + u = 9$$
$$u = 2$$

The number is 72.

17.

Hamburger cost per pound : x

Steak cost per pound : y

$$5x+4y=22$$
$$3x+7y=28.15$$

$$-15x-12y=-66$$
$$15x+35y=140.75$$
$$23y=74.75$$
$$y=3.25$$
$$5x+4(3.25)=22$$
$$5x=9$$
$$x=1.80$$

Hamburger : \$1.80/ lb ; Steak : \$3.25/ lb

18.

Speed of plane in still air : x

Speed of wind : y

	r	$\cdot\ t$	$=$ d
with wind	$x+y$	5	$5(x+y)$
against wind	$x-y$	7	$7(x-y)$

$$5(x+y)=3500$$
$$7(x-y)=3500$$

$$x+y=700$$
$$x-y=500$$
$$2x=1200$$
$$x=600$$
$$600+y=700$$
$$y=100$$

The plane's speed is 600 km / hr

19.

$$S=3p+2$$
$$D=-2p+17$$

$$S=D$$
$$3p+2=-2p+17$$
$$5p=15$$
$$p=3$$

$$D=-2(3)+17$$
$$=11 \text{ faucets}$$

20.

$$C=4025+9x$$
$$R=16x$$

$$C=R$$
$$4025+9x=16x$$
$$4025=7x$$
$$x=575$$
$$R=16(575)=9200$$

575 hours; \$9,200 revenue

21.

$$\begin{aligned} -3x - y + z &= 12 \\ 2x + 5y - 2z &= -9 \\ -x + 4y + 2z &= 15 \end{aligned}$$

$$\begin{array}{r} -6x - 2y + 2z = 24 \\ 2x + 5y - 2z = -9 \\ \hline -4x + 3y = 15 \end{array} \qquad \begin{array}{r} 2x + 5y - 2z = -9 \\ -x + 4y + 2z = 15 \\ \hline x + 9y = 6 \end{array}$$

$$\begin{array}{r} -4x + 3y = 15 \\ 4x + 36y = 24 \\ \hline 39y = 39 \\ y = 1 \end{array} \qquad \begin{array}{r} x + 9(1) = 6 \\ x = -3 \\ -(-3) + 4(1) + 2z = 15 \\ 2z = 8 \\ z = 4 \end{array}$$

$x = -3,\ y = 1,\ z = 4$

22.

$$\begin{aligned} 3x + 2y - z &= -8 \\ 2x + \quad 3z &= 5 \\ x - 4y \quad &= -4 \end{aligned}$$

$$\begin{array}{r} 6x + 4y - 2z = -16 \\ x - 4y \quad = -4 \\ \hline 7x - \quad 2z = -20 \end{array} \qquad \begin{array}{r} 4x + 6z = 10 \\ 21x - 6z = -60 \\ \hline 25x \quad = -50 \\ x = -2 \\ -2 - 4y = -4 \\ -4y = -2 \\ y = \frac{1}{2} \\ 2(-2) + 3z = 5 \\ 3z = 9 \\ z = 3 \end{array}$$

$x = -2,\ y = \frac{1}{2},\ z = 3$

23.

$$\begin{aligned} 5x - \quad y + 2z &= 10 \\ -2x + 3y - \quad z &= -7 \\ 3x + \quad 2z &= 7 \end{aligned}$$

$$\begin{array}{r} 15x - 3y + 6z = 30 \\ -2x + 3y - \quad z = -7 \\ \hline 13x + \quad 5z = 23 \end{array} \qquad \begin{array}{r} 15x + 10z = 35 \\ -26x - 10z = -46 \\ \hline -11x \quad = -11 \\ x = 1 \\ 3(1) + 2z = 7 \\ 2z = 4 \\ z = 2 \\ 5(1) - y + 2(2) = 10 \\ -y = 1 \\ y = -1 \end{array}$$

$x = 1,\ y = -1,\ z = 2$

24.
$$\begin{aligned} x+4y \quad &= 4 \\ -x+ \quad 3z &= -4 \\ 2x+2y-z &= \frac{41}{6} \end{aligned}$$

$$\begin{aligned} -x+ \quad 3z &= -4 \\ 6x+6y-3z &= \frac{41}{2} \\ \hline 5x+6y \quad &= \frac{33}{2} \end{aligned}$$

$$\begin{aligned} -5x-20y &= -20 \\ 5x+\ 6y &= \frac{33}{2} \\ \hline -14y &= -\frac{7}{2} \\ y &= \frac{1}{4} \\ x+4\left(\frac{1}{4}\right) &= 4 \\ x &= 3 \\ -3+3z &= -4 \\ 3z &= -1 \\ z &= -\frac{1}{3} \end{aligned}$$

$x=3,\ y=\frac{1}{4},\ z=-\frac{1}{3}$

25.
$$\begin{aligned} 2x+3y &= 6 \\ 3x-y &= -13 \end{aligned}$$

$$\begin{aligned} y &= 3x+13 \\ 2x+3(3x+13) &= 6 \\ 2x+9x+39 &= 6 \\ 11x &= -33 \\ x &= -3 \\ y = 3(-3)+13 &= 4 \end{aligned}$$

$x=-3,\ y=4$

26.
$$\begin{aligned} x+2y &= 0 \\ -x+4y &= 5 \\ \hline 6y &= 5 \\ y &= \frac{5}{6} \\ x+2\left(\frac{5}{6}\right) &= 0 \\ x &= -\frac{5}{3} \end{aligned}$$

$x=-\frac{5}{3},\ y=\frac{5}{6}$

27.

$$\begin{aligned} 2x+3y-z&=-4\\ x-2y+2z&=-6\\ 2x-\quad\ 3z&=5 \end{aligned}$$

$$\begin{aligned} 4x+6y-2z&=-8\\ 3x-6y+6z&=-18\\ \hline 7x+4z&=-26 \end{aligned}$$

$$\begin{aligned} 8x-12z&=20\\ 21x+12z&=-78\\ \hline 29x&=-58\\ x&=-2\\ 2(-2)-3z&=5\\ -3z&=9\\ z&=-3 \end{aligned}$$

$$\begin{aligned} -2-2y+2(-3)&=-6\\ -2y&=2\\ y&=-1 \end{aligned}$$

$x=-2,\ y=-1,\ z=-3$

28.

$$\begin{aligned} 2x+2y-3z&=-4\\ 3y-\ z&=-4\\ 4x-\ y+\ z&=4 \end{aligned}$$

$$\begin{aligned} -4x-4y+6z&=8\\ 4x-\ y+\ z&=4\\ \hline -5y+7z&=12 \end{aligned}$$

$$\begin{aligned} 21y-7z&=-28\\ -5y+7z&=12\\ \hline 16y\quad\ &=-16\\ y&=-1\\ 3(-1)-z&=-4\\ -z&=-1\\ z&=1\\ 2x+2(-1)-3(1)&=-4\\ 2x&=1\\ x&=\frac{1}{2} \end{aligned}$$

$x=\frac{1}{2},\ y=-1,\ z=1$

29.

$$\frac{8-x}{2x^2+3x-2}=\frac{8-x}{(2x-1)(x+2)}=\frac{A}{2x-1}+\frac{B}{x+2}$$

$$8-x=A(x+2)+B(2x-1)$$

Set $x=-2$: $8-(-2)=A(0)+B(-5)$

$$10=-5B$$

$$B=-2$$

Set $x=\frac{1}{2}$: $8-\frac{1}{2}=A\left(\frac{5}{2}\right)+B(0)$

$$\frac{15}{2}=\frac{5}{2}A$$

$$A=3$$

$$\frac{3}{2x-1}-\frac{2}{x+2}$$

30.

$$\frac{3x^3+5x-1}{(x^2+1)^2}=\frac{Ax+B}{x^2+1}+\frac{Cx+D}{(x^2+1)^2}$$

$$3x^3+5x-1=(Ax+B)(x^2+1)+Cx+D$$

Set $x=0$: $3(0)^3+5(0)-1=[A(0)+B](1)+C(0)+D$

$$-1=B+D$$

Set $x=1$: $3(1)^3+5(1)-1=[A(1)+B](2)+C(1)+D$

$$7=2A+2B+C+D$$

Set $x=-1$: $3(-1)^3+5(-1)-1=[A(-1)+B](2)+C(-1)+D$

$$-9=-2A+2B-C+D$$

Set $x=2$: $3(2)^3+5(2)-1=[A(2)+B](5)+C(2)+D$

$$33=10A+5B+2C+D$$

30. (Continued)

$$\begin{aligned} B + \quad D &= -1 \\ 2A + 2B + C + D &= 7 \\ -2A + 2B - C + D &= -9 \\ 10A + 5B + 2C + D &= 33 \end{aligned}$$

$$\begin{aligned} 2A + 2B + C + D &= 7 \\ -2A + 2B - C + D &= -9 \\ \hline 4B + \quad 2D &= -2 \\ -2B - D &= 1 \end{aligned}$$

$$\begin{aligned} -2B - D &= 1 \\ B + D &= -1 \\ \hline -B \quad &= 0 \\ B &= 0 \\ 0 + D &= -1 \\ D &= -1 \end{aligned}$$

$$\begin{aligned} -4A + 4B - 2C + 2D &= -18 \\ 10A + 5B + 2C + \quad D &= 33 \\ \hline 6A + 9B + \quad 3D &= 15 \\ 6A + 9(0) + 3(-1) &= 15 \\ 6A &= 18 \\ A &= 3 \end{aligned}$$

$$\begin{aligned} 2(3) + 2(0) + C + (-1) &= 7 \\ C &= 2 \end{aligned}$$

$$\frac{3x}{x^2+1} + \frac{2x-1}{\left(x^2+1\right)^2}$$

31.

$$\frac{2x^3-3x^2+4x-2}{(x-1)^2}=\frac{2x^3-3x^2+4x-2}{x^2-2x+1}$$

$$\begin{array}{r} 2x+1 \\ x^2-2x+1\overline{)2x^3-3x^2+4x-2} \\ \underline{2x^3-4x^2+2x\quad} \\ x^2+2x-2 \\ \underline{x^2-2x+1} \\ 4x-3 \end{array}$$

$$2x+1+\frac{4x-3}{x^2-2x+1}$$

$$\frac{4x-3}{x^2-2x+1}=\frac{4x-3}{(x-1)^2}=\frac{A}{x-1}+\frac{B}{(x-1)^2}$$

$$4x-3=A(x-1)+B$$

Set $x=1$: $4(1)-3=A(0)+B$

$$B=1$$

Set $x=0$: $4(0)-3=A(-1)+1$

$B=1$ $\quad -3=-A+1$

$$A=4$$

$$2x+1+\frac{4}{x-1}+\frac{1}{(x-1)^2}$$

32.

$$x-2y\le 5$$

Graph $x-2y=5$

Test point : $(0,\ 0)$

$$0-2(0)\le 5\,?$$

$$0\le 5$$

True

33.

$$2x+y>4$$

Graph $2x+y=4$

Test point : $(0,\ 0)$

$$2(0)+0>4\,?$$

$$0>4$$

False

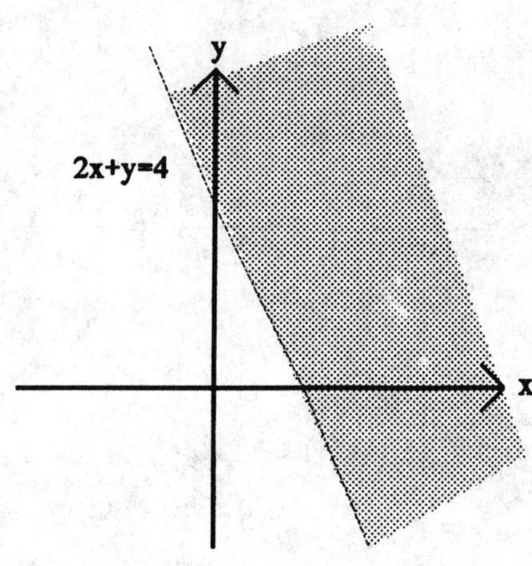

34.

$$2x+3y\le 2$$

$$x-y\ge 1$$

Graph $2x+3y=2$, $\quad x-y=1$

Test point : $(0,\ 0)$

$$2(0)+3(0)\le 2\,? \quad 0-0\ge 1\,?$$

$$0\le 2 \quad 0\ge 1$$

True False

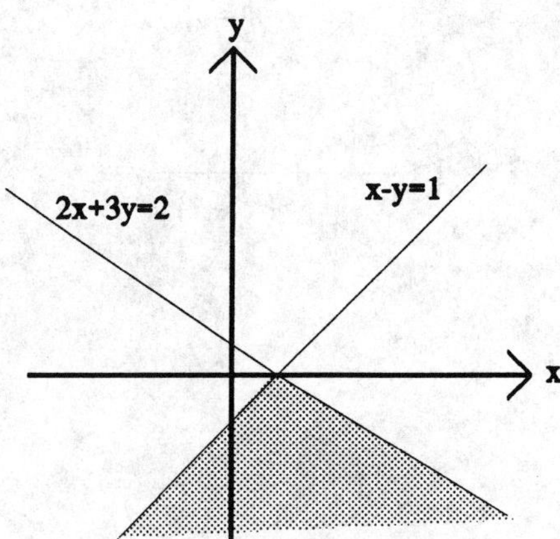

35.

$$x - 2y \geq 4$$
$$2x - y \leq 2$$

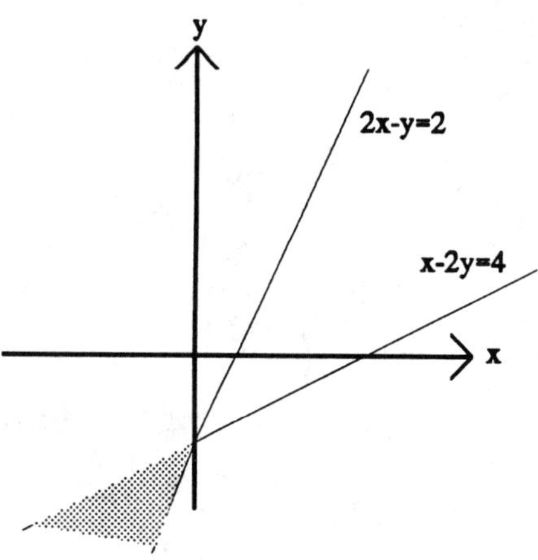

Graph $x - 2y = 4$, $\quad 2x - y = 2$

Test point : $(0, 0)$

$0 - 2(0) \geq 4$?	$2(0) - 0 \leq 2$?
$0 \geq 4$	$0 \leq 2$
False	True

36.

$$2x + 3y \leq 6$$
$$x \geq 0$$
$$y \geq 1$$

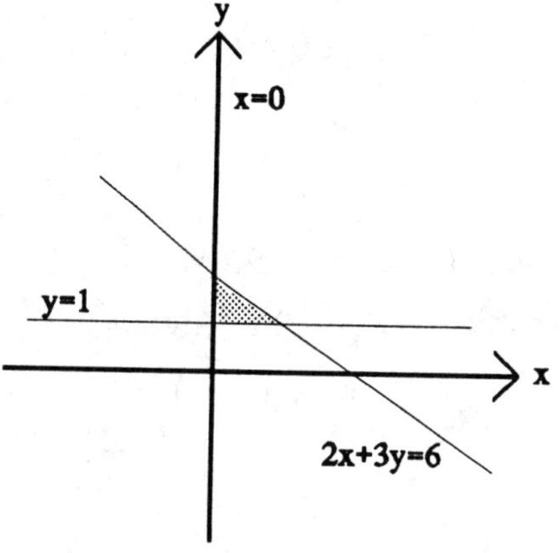

Graph $2x + 3y = 6$, $\quad x = 0$, $\quad y = 1$

Test point : $(2, 2)$

$2(2) + 3(2) \leq 6$?	$2 \geq 0$?	$2 \geq 1$?
$10 \leq 6$	True	True
False		

37.

$$2x + y \le 4$$
$$2x - y \le 3$$
$$x \ge 0$$
$$y \ge 0$$

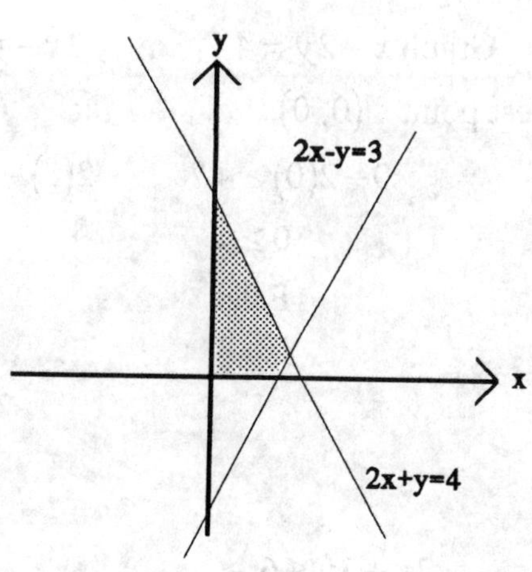

Graph $2x + y = 4$, $\quad 2x - y = 3$

Test point : $(0, 0)$

$2(0) + 0 \le 4$? $\quad 2(0) - 0 \le 3$?

$0 \le 4 \quad 0 \le 3$

True $\quad$ True

Graph $x = 0$, $\quad y = 0$

Test point : $(1, 1)$

$1 \ge 0$? $\quad 1 \ge 0$?

True $\quad$ True

38.

Maximize $z = 5y - x$

subject to $8y - 3x \le 36$

$6x + y \le 30$

$y \ge 1$

$x \ge 0$

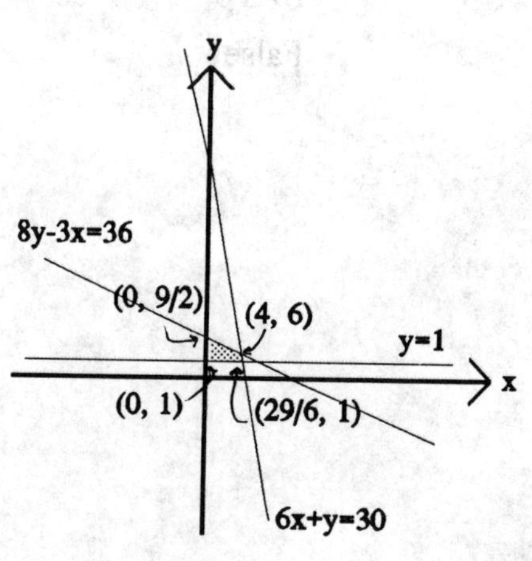

Vertex	x	y	$z = 5y - x$	
$(0, 1)$	0	1	5	
$\left(0, \frac{9}{2}\right)$	0	$\frac{9}{2}$	$\frac{45}{2}$	
$(4, 6)$	4	6	$\boxed{26}$	max
$\left(\frac{29}{6}, 1\right)$	$\frac{29}{6}$	1	$\frac{1}{6}$	

Maximum : 26 at $(4, 6)$

39.

Minimize $z = x + 4y$

subject to $4x - y \geq 8$

$4x + y \leq 24$

$5y + 4x \geq 32$

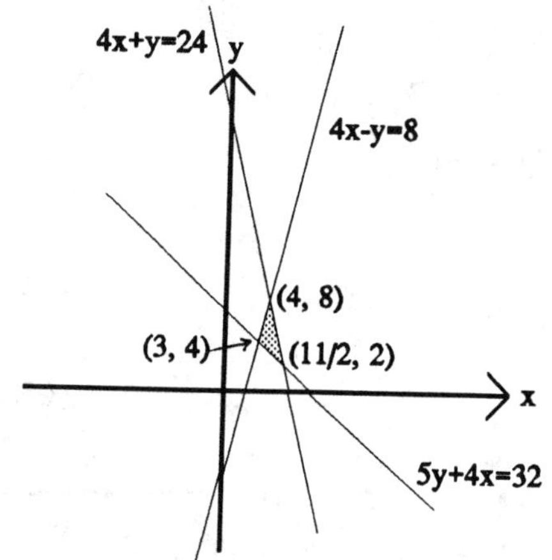

Vertex	x	y	$z = x + 4y$	
$(3, 4)$	3	4	19	
$(4, 8)$	4	8	36	
$\left(\frac{11}{2}, 2\right)$	$\frac{11}{2}$	2	$\boxed{13\frac{1}{2}}$	min

Minimum: $13\frac{1}{2}$ at $\left(\frac{11}{2}, 2\right)$

CHAPTER 10 REVIEW TEST

1.

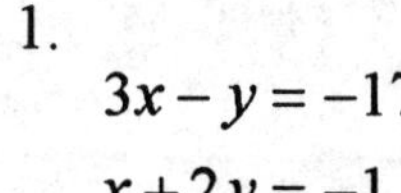

$$3x - y = -17$$
$$x + 2y = -1$$

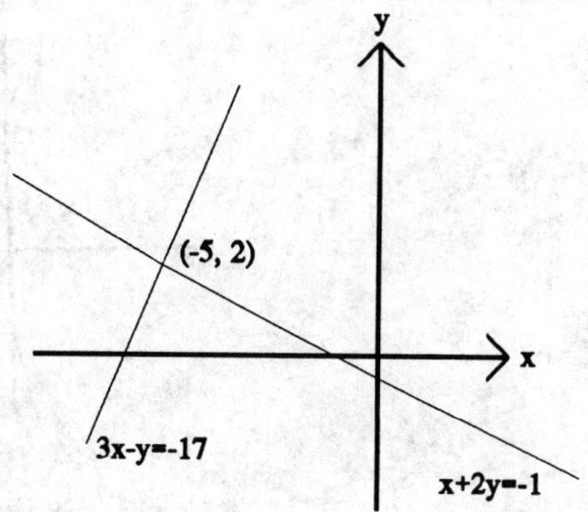

2.

$$2x + y = 4$$
$$3x - 2y = -15$$

$$y = 4 - 2x$$
$$3x - 2(4 - 2x) = -15$$
$$3x - 8 + 4x = -15$$
$$7x = -7$$
$$x = -1$$
$$y = 4 - 2(-1) = 6$$

$x = -1, \quad y = 6$

3.

$$y^2 - 5x = 0$$
$$y^2 - x^2 = 6$$

$$y^2 = 5x$$
$$5x - x^2 = 6$$
$$x^2 - 5x + 6 = 0$$
$$(x-2)(x-3) = 0$$

$x = 2$ $\qquad$ $x = 3$

$y^2 = 5(2)$ $\qquad$ $y^2 = 5(3)$

$y^2 = 10$ $\qquad$ $y^2 = 15$

$y = \pm\sqrt{10}$ $\qquad$ $y = \pm\sqrt{15}$

$x = 2,\ y = \sqrt{10};\ \ x = 2,\ y = -\sqrt{10};\ \ x = 3,\ y = \sqrt{15};\ \ x = 3,\ y = -\sqrt{15}$

4.

$$x - 2y = 7$$
$$3x + 4y = -9$$

$$2x - 4y = 14$$
$$\underline{3x + 4y = -9}$$
$$5x = 5$$
$$x = 1$$
$$1 - 2y = 7$$
$$-2y = 6$$
$$y = -3$$

$x = 1,\ \ y = -3$

5.

$$\begin{aligned} x^2+y^2&=25\\ \underline{4x^2-y^2}&\underline{=20}\\ 5x^2&=45\\ x^2&=9 \end{aligned}$$

$$x=3 \qquad\qquad x=-3$$

$$3^2+y^2=25 \qquad\qquad (-3)^2+y^2=25$$

$$y^2=16 \qquad\qquad y^2=16$$

$$y=\pm4 \qquad\qquad y=\pm4$$

$x=3,\ y=4$; $x=3,\ y=-4$; $x=-3,\ y=4$; $x=-3,\ y=-4$

6.

Tens digit : t
Units digit : u

$$t+u=11 \qquad t+u=11 \qquad -t-u=-11$$

$$(10t+u)+t=41 \qquad 11t+u=41 \qquad \underline{11t+u=41}$$

$$\begin{aligned} 10t&=30\\ t&=3\\ 3+u&=11\\ u&=8 \end{aligned}$$

The number is 38.

7.

Price of shirt : x

Price of tie : y

$$3y+7x=135$$
$$5y+3x=95$$

$$-15y-35x=-675$$
$$\underline{15y+9x=285}$$
$$-26x=-390$$
$$x=15$$
$$5y+3(15)=95$$
$$5y=50$$
$$y=10$$

Shirt's price : \$15 ;

Tie's price : \$10

8.

Number of meals : x

$$C=1375+1.25x$$
$$R=2.50x$$

$$P=R-C=0$$
$$R=C$$
$$2.50x=1375+1.25x$$
$$1.25x=1375$$
$$x=1100$$

1100 meals

9.

$$3x+2y-z=-4$$
$$x-y+3z=12$$
$$2x-y-2z=-20$$

$$3x+2y-z=-4$$
$$\underline{2x-2y+6z=24}$$
$$5x+5z=20$$

$$-x+y-3z=-12$$
$$\underline{2x-y-2z=-20}$$
$$x-5z=-32$$

$$5x+5z=20$$
$$\underline{x-5z=-32}$$
$$6x=-12$$
$$x=-2$$
$$-2-5z=-32$$
$$-5z=-30$$
$$z=6$$
$$-2-y+3(6)=12$$
$$-y=-4$$
$$y=4$$

$x=-2,\ \ y=4,\ \ z=6$

10.

$$-3x+2y=-1$$
$$6x-y=-1$$

$$-6x+4y=-2$$
$$\underline{6x-y=-1}$$
$$3y=-3$$
$$y=-1$$
$$6x-(-1)=-1$$
$$6x=-2$$
$$x=-\frac{1}{3}$$

$x=-\frac{1}{3},\ y=-1$

11.

$$3x + y - 2z = 8$$
$$3y - 4z = 14$$
$$3x + \frac{1}{2}y + z = 1$$

$$\begin{array}{r} -3x - y + 2z = -8 \\ 3x + \frac{1}{2}y + z = 1 \\ \hline -\frac{1}{2}y + 3z = -7 \end{array} \qquad \begin{array}{r} 9y - 12z = 42 \\ -2y + 12z = -28 \\ \hline 7y \qquad = 14 \end{array}$$

$$y = 2$$
$$3(2) - 4z = 14$$
$$-4z = 8$$
$$z = -2$$
$$3x + 2 - 2(-2) = 8$$
$$3x = 2$$
$$x = \frac{2}{3}$$

$x = \frac{2}{3},\ y = 2,\ z = -2$

12.

$$\frac{x-12}{x^2+x-6} = \frac{x-12}{(x+3)(x-2)} = \frac{A}{x+3} + \frac{B}{x-2}$$

$x - 12 = A(x-2) + B(x+3)$

Set $x = 2$: $2 - 12 = A(0) + B(5)$

$$-10 = 5B$$
$$B = -2$$

Set $x = -3$: $-3 - 12 = A(-5) + B(0)$

$$-15 = -5A$$
$$A = 3$$

$$\frac{3}{x+3} - \frac{2}{x-2}$$

13.

$$2x - 3y \geq 6$$
$$3x + y \leq 3$$

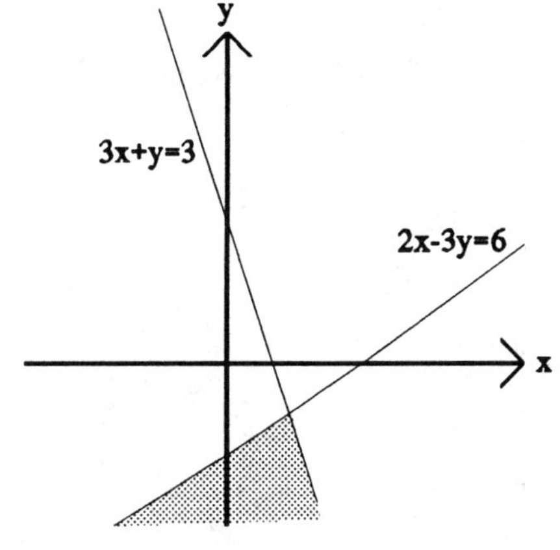

Graph $2x - 3y = 6, \qquad 3x + y = 3$

Test point : $(0, 0)$

$2(0) - 3(0) \geq 6? \qquad 3(0) + 0 \leq 3?$

$0 \geq 6 \qquad 0 \leq 3$

False $\qquad$ True

14.

$$2x + y \leq 4$$
$$2x - 5y \leq 5$$
$$y \geq 1$$

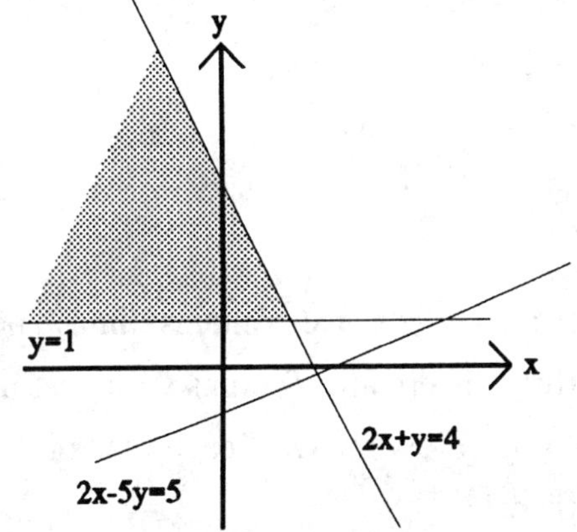

Graph $2x + y = 4, \qquad 2x - 5y = 5, \qquad y = 1$

Test point : $(0, 0)$

$2(0) + 0 \leq 4? \qquad 2(0) - 5(0) \leq 5? \qquad 0 \geq 1?$

$0 \leq 4 \qquad 0 \leq 5 \qquad$ False

True $\qquad$ True

15.

Maximize $z = x + 4y$

subject to $4x - y \geq 8$

$4x + y \leq 24$

$4x + 5y \geq 32$

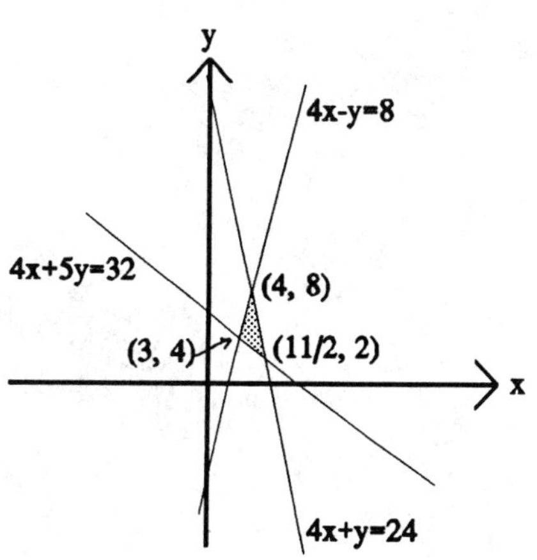

Vertex	x	y	$z = x + 4y$	
$(3, 4)$	3	4	19	
$(4, 8)$	4	8	$\boxed{36}$	max
$\left(\frac{11}{2}, 2\right)$	$\frac{11}{2}$	2	$13\frac{1}{2}$	

Maximum : 36 at $(4, 8)$

CHAPTER 10 ADDITIONAL PRACTICE EXERCISES

In Exercises 1-5, solve the system of equations.

1.
$$3x - 5y = -39$$
$$-x + 2y = 15$$

2.
$$xy = 14$$
$$3x - y = -1$$

3.
$$x^2 - y^2 = 25$$
$$x^2 + y^2 = 2$$

4.
$$3x + 4y - 5z = -10$$
$$-x + 2y + 7z = 26$$
$$2x - y + 3z = 5$$

5.
$$3x + 5z = 7$$
$$x + 2y - z = -11$$
$$9x - 10y - z = -59$$

6.
Find the partial fraction decomposition of $\dfrac{26 - 2x}{x^2 - 2x - 8}$

7.
Betty's change purse contains dimes and quarters. There are 19 coins with a value of $ 3.55. How many of each type of coin are there?

8.
Graph the solution set : $y > x + 5$

9.
Graph the solution set of the system :
$$x + y \geq 2$$
$$3x + 4y \leq 12$$

10.
Maximi ze $z = 2x + 5y$
subject to
$$x + y \geq 2$$
$$x \geq 0$$
$$y \geq 0$$
$$x \leq 5$$
$$y \leq 3$$

CHAPTER 10 PRACTICE TEST

In Exercises 1-10, solve the system of equations.

1.
$$\begin{aligned} 9x+2y&=-46 \\ -x+y&=-1 \end{aligned}$$

2.
$$\begin{aligned} x^2+y^2&=100 \\ 2x^2+y^2&=3 \end{aligned}$$

3.
$$\begin{aligned} 5x+3y&=6 \\ -\frac{5}{2}x-\frac{3}{2}y&=-3 \end{aligned}$$

4.
$$\begin{aligned} x^2+2y^2&=18 \\ xy&=4 \end{aligned}$$

5.
$$\begin{aligned} 4x+9y&=8 \\ -2x+3y&=1 \end{aligned}$$

6.
$$\begin{aligned} 2x+y+z&=2 \\ -3x-y+4z&=-28 \\ x+3y-z&=21 \end{aligned}$$

7.
$$\begin{aligned} 5x-4y+3z&=14 \\ 2x+5y-z&=10 \\ -x+y+2z&=0 \end{aligned}$$

8.
$$\begin{aligned} x+y-z&=6 \\ 2x+y-3z&=5 \\ x-2z&=1 \end{aligned}$$

9.
$$\begin{aligned} x-3z&=7 \\ 4x+5y&=6 \\ 6y-7z&=-5 \end{aligned}$$

10.
$$\begin{aligned} 7x-y+z&=3 \\ 2x+3y-z&=4 \\ 5x-4y+2z&=-1 \end{aligned}$$

11. The sum of the digits of a two-digit number is 11. If the digits are reversed, the resulting number exceeds the given number by 45. Find the number.

12. Find the dimensions of a rectangle with an area of 143 square centimeters and a perimeter of 48 centimeters.

13.
Joe travels 15 miles per hour faster than Sam. If Joe travels 360 miles in the same time that Sam travels 270 miles, what is the speed of each car ?

In Exercises 14-15, find the partial fraction decomposition of the given rational function.

14.
$$\frac{-x+7}{6x^2+11x+4}$$

15.
$$\frac{5x^2+10x+7}{x^3+3x^2+2x+6}$$

16.
Graph the solution set of $2x+5y>10$

17.
Graph the solution set of $y \le 3x+1$

In Exercises 18-19, graph the solution set of the system of linear inequalities.

18.
$$x+2y \ge 6$$
$$x-y \le 2$$

19.
$$4x+y \le 4$$
$$x+y \ge 1$$

20.
Maximize $z = x+2y$ subject to $x \ge 0$
$$y \ge 0$$
$$x+y \ge 2$$
$$2x+3y \le 6$$
$$3x+2y \le 6$$

CHAPTER 11 SECTION 1

SECTION 11.1 PROGRESS CHECK

1a). (Page 612)

b_{11} : 1st row, 1st column: 4

1b).

b_{23} : 2nd row, 3rd column: 3

1c).

b_{31} : 3rd row, 1st column: -8

1d).

b_{42} : 4th row, 2nd column: 1

2. (Page 615)

$$2x+4y-z=0$$
$$x-2y-2z=2$$
$$-5x-8y+3z=-2$$

$$R_1 \leftrightarrow R_2 \left[\begin{array}{ccc|c} 2 & 4 & -1 & 0 \\ 1 & -2 & -2 & 2 \\ -5 & -8 & 3 & -2 \end{array}\right]$$

$$\begin{array}{l} -2R_1+R_2 \to R_2 \\ 5R_1+R_3 \to R_3 \end{array} \left[\begin{array}{ccc|c} 1 & -2 & -2 & 2 \\ 2 & 4 & -1 & 0 \\ -5 & -8 & 3 & -2 \end{array}\right]$$

$$9R_2+4R_3 \to R_3 \left[\begin{array}{ccc|c} 1 & -2 & -2 & 2 \\ 0 & 8 & 3 & -4 \\ 0 & -18 & -7 & 8 \end{array}\right]$$

$$\left[\begin{array}{ccc|c} 1 & -2 & -2 & 2 \\ 0 & 8 & 3 & -4 \\ 0 & 0 & -1 & -4 \end{array}\right]$$

$$-z=-4$$
$$z=4$$
$$8y+3(4)=-4$$
$$8y=-16$$
$$y=-2$$
$$x-2(-2)-2(4)=2$$
$$x=6$$
$$x=6, \quad y=-2, \quad z=4$$

EXERCISE SET 11.1

1.
2 rows, 2 columns : 2×2

3.
4 rows, 3 columns : 4×3

5.
3 rows, 3 columns : 3×3

7a).
1st row, 2nd column : -4

7b).
2nd row, 2nd column : 7

7c).
2nd row, 3rd column : 6

7d).
3rd row, 4th column : -3

9.
$$\begin{bmatrix} 3 & -2 \\ 5 & 1 \end{bmatrix};\ \left[\begin{array}{cc|c} 3 & -2 & 12 \\ 5 & 1 & -8 \end{array}\right]$$

11.
$$\begin{bmatrix} \frac{1}{2} & 1 & 1 \\ 2 & -1 & -4 \\ 4 & 2 & -3 \end{bmatrix};\ \left[\begin{array}{ccc|c} \frac{1}{2} & 1 & 1 & 4 \\ 2 & -1 & -4 & 6 \\ 4 & 2 & -3 & 8 \end{array}\right]$$

13.
$$\frac{3}{2}x+6y=-1$$
$$4x+5y=3$$

15.
$$x+y+3z=-4$$
$$-3x+4y=8$$
$$2x+7z=6$$

17.
$$z=2$$
$$y-2(2)=4$$
$$y=8$$
$$x+2(8)=3$$
$$x=-13$$
$$x=-13,\ y=8,\ z=2$$

19.
$$z=-4$$
$$y+3(-4)=2$$
$$y=14$$
$$x-2(14)-4=3$$
$$x=35$$
$$x=35,\ y=14,\ z=-4$$

21.

$$x-2y=-4$$
$$2x+3y=13$$

$$-2R_1+R_2 \to R_2\begin{bmatrix}1 & -2 & | & -4\\ 2 & 3 & | & 13\end{bmatrix}$$

$$\begin{bmatrix}1 & -2 & | & -4\\ 0 & 7 & | & 21\end{bmatrix}$$

$$7y=21$$
$$y=3$$
$$x-2(3)=-4$$
$$x=2$$
$$x=2,\quad y=3$$

23.

$$x+y+z=4$$
$$2x-y+2z=11$$
$$x+2y+2z=6$$

$$\begin{matrix}-2R_1+R_2 \to R_2\\ -R_1+R_3 \to R_3\end{matrix}\begin{bmatrix}1 & 1 & 1 & | & 4\\ 2 & -1 & 2 & | & 11\\ 1 & 2 & 2 & | & 6\end{bmatrix}$$

$$3R_3+R_2 \to R_3\begin{bmatrix}1 & 1 & 1 & | & 4\\ 0 & -3 & 0 & | & 3\\ 0 & 1 & 1 & | & 2\end{bmatrix}$$

$$\begin{bmatrix}1 & 1 & 1 & | & 4\\ 0 & -3 & 0 & | & 3\\ 0 & 0 & 3 & | & 9\end{bmatrix}$$

$$3z=9$$
$$z=3$$
$$-3y=3$$
$$y=-1$$
$$x+(-1)+3=4$$
$$x=2$$

$$x=2,\quad y=-1,\quad z=3$$

25.

$$\begin{aligned} 2x + y - z &= 9 \\ x - 2y + 2z &= -3 \\ 3x + 3y + 4z &= 11 \end{aligned}$$

$$R_1 \leftrightarrow R_2 \left[\begin{array}{ccc|c} 2 & 1 & -1 & 9 \\ 1 & -2 & 2 & -3 \\ 3 & 3 & 4 & 11 \end{array}\right]$$

$$\begin{array}{l} -2R_1 + R_2 \to R_2 \\ -3R_1 + R_3 \to R_3 \end{array} \left[\begin{array}{ccc|c} 1 & -2 & 2 & -3 \\ 2 & 1 & -1 & 9 \\ 3 & 3 & 4 & 11 \end{array}\right]$$

$$\frac{1}{5}R_2 \to R_2 \left[\begin{array}{ccc|c} 1 & -2 & 2 & -3 \\ 0 & 5 & -5 & 15 \\ 0 & 9 & -2 & 20 \end{array}\right]$$

25. (Continued)

$$-9R_2 + R_3 \to R_3 \left[\begin{array}{ccc|c} 1 & -2 & 2 & -3 \\ 0 & 1 & -1 & 3 \\ 0 & 9 & -2 & 20 \end{array}\right]$$

$$\left[\begin{array}{ccc|c} 1 & -2 & 2 & -3 \\ 0 & 1 & -1 & 3 \\ 0 & 0 & 7 & -7 \end{array}\right]$$

$$\begin{aligned} 7z &= -7 \\ z &= -1 \\ y - (-1) &= 3 \\ y &= 2 \\ x - 2(2) + 2(-1) &= -3 \\ x &= 3 \end{aligned}$$

$x = 3, \quad y = 2, \quad z = -1$

27.

$$\begin{aligned} -x - y + 2z &= 9 \\ x + 2y - 2z &= -7 \\ 2x - y + z &= -9 \end{aligned}$$

$$\begin{array}{r} \\ R_1 + R_2 \to R_2 \\ 2R_1 + R_2 \to R_3 \end{array} \left[\begin{array}{ccc|c} -1 & -1 & 2 & 9 \\ 1 & 2 & -2 & -7 \\ 2 & -1 & 1 & -9 \end{array}\right]$$

$$3R_2 + R_3 \to R_3 \left[\begin{array}{ccc|c} -1 & -1 & 2 & 9 \\ 0 & 1 & 0 & 2 \\ 0 & -3 & 5 & 9 \end{array}\right]$$

$$\left[\begin{array}{ccc|c} -1 & -1 & 2 & 9 \\ 0 & 1 & 0 & 2 \\ 0 & 0 & 5 & 15 \end{array}\right]$$

$$\begin{aligned} 5z &= 15 \\ z &= 3 \\ y &= 2 \\ -x - 2 + 2(3) &= 9 \\ -x &= 5 \\ x &= -5 \end{aligned}$$

$$x = -5, \quad y = 2, \quad z = 3$$

29.

$$\begin{aligned} x + y - z + 2w &= 0 \\ 2x + y \quad - w &= -2 \\ 3x \quad + 2z \quad &= -3 \\ -x + 2y \quad + 3w &= 1 \end{aligned}$$

$$\begin{matrix} \\ -2R_1 + R_2 \to R_2 \\ 3R_4 + R_3 \to R_3 \\ R_1 + R_4 \to R_4 \end{matrix} \left[\begin{array}{cccc|c} 1 & 1 & -1 & 2 & 0 \\ 2 & 1 & 0 & -1 & -2 \\ 3 & 0 & 2 & 0 & -3 \\ -1 & 2 & 0 & 3 & 1 \end{array}\right]$$

$$\begin{matrix} \\ 6R_2 + R_3 \to R_3 \\ 3R_2 + R_4 \to R_4 \\ \\ \end{matrix} \left[\begin{array}{cccc|c} 1 & 1 & -1 & 2 & 0 \\ 0 & -1 & 2 & -5 & -2 \\ 0 & 6 & 2 & 9 & 0 \\ 0 & 3 & -1 & 5 & 1 \end{array}\right]$$

$$R_3 \leftrightarrow \frac{1}{5}R_4 \left[\begin{array}{cccc|c} 1 & 1 & -1 & 2 & 0 \\ 0 & -1 & 2 & -5 & -2 \\ 0 & 0 & 14 & -21 & -12 \\ 0 & 0 & 5 & -10 & -5 \end{array}\right]$$

29. (Continued)

$$-14R_3 + R_4 \to R_4 \left[\begin{array}{cccc|c} 1 & 1 & -1 & 2 & 0 \\ 0 & -1 & 2 & -5 & -2 \\ 0 & 0 & 1 & -2 & -1 \\ 0 & 0 & 14 & -21 & -12 \end{array}\right]$$

$$\left[\begin{array}{cccc|c} 1 & 1 & -1 & 2 & 0 \\ 0 & -1 & 2 & -5 & -2 \\ 0 & 0 & 1 & -2 & -1 \\ 0 & 0 & 0 & 7 & 2 \end{array}\right]$$

$$7w = 2$$

$$w = \frac{2}{7}$$

$$z - 2\left(\frac{2}{7}\right) = -1$$

$$z = -\frac{3}{7}$$

$$-y + 2\left(-\frac{3}{7}\right) - 5\left(\frac{2}{7}\right) = -2$$

$$-y = \frac{2}{7}$$

$$y = -\frac{2}{7}$$

$$x - \frac{2}{7} + \frac{3}{7} + 2\left(\frac{2}{7}\right) = 0$$

$$x = -\frac{5}{7}$$

$$x = -\frac{5}{7},\ y = -\frac{2}{7},\ z = -\frac{3}{7},\ w = \frac{2}{7}$$

In Exercises 31-39, the solutions will start where Exercises 21-29 stopped with matrix manipulations.

31.
See Exercise 21

$$\frac{1}{7}R_2 \to R_2 \left[\begin{array}{cc|c} 1 & -2 & -4 \\ 0 & 7 & 21 \end{array}\right]$$

$$2R_2 + R_1 \to R_1 \left[\begin{array}{cc|c} 1 & -2 & -4 \\ 0 & 1 & 3 \end{array}\right]$$

$$\left[\begin{array}{cc|c} 1 & 0 & 2 \\ 0 & 1 & 3 \end{array}\right]$$

$x = 2,\ y = 3$

33.
See Exercise 23

$$\begin{array}{l} -\frac{1}{3}R_2 \to R_2 \\ \\ \frac{1}{3}R_3 \to R_3 \end{array} \left[\begin{array}{ccc|c} 1 & 1 & 1 & 4 \\ 0 & -3 & 0 & 3 \\ 0 & 0 & 3 & 9 \end{array}\right]$$

$$R_1 + R_2 \to R_1 \left[\begin{array}{ccc|c} 1 & 1 & 1 & 4 \\ 0 & 1 & 0 & -1 \\ 0 & 0 & 1 & 3 \end{array}\right]$$

$$R_1 - R_3 \to R_1 \left[\begin{array}{ccc|c} 1 & 0 & 1 & 5 \\ 0 & 1 & 0 & -1 \\ 0 & 0 & 1 & 3 \end{array}\right]$$

$$\left[\begin{array}{ccc|c} 1 & 0 & 0 & 2 \\ 0 & 1 & 0 & -1 \\ 0 & 0 & 1 & 3 \end{array}\right]$$

$x = 2,\ y = -1,\ z = 3$

35.

See Exercise 25

$$\begin{array}{l} R_1+2R_2 \to R_1 \\ \\ \frac{1}{7}R_3 \to R_3 \end{array} \left[\begin{array}{ccc|c} 1 & -2 & 2 & -3 \\ 0 & 1 & -1 & 3 \\ 0 & 0 & 7 & -7 \end{array}\right]$$

$$R_2+R_3 \to R_2 \left[\begin{array}{ccc|c} 1 & 0 & 0 & 3 \\ 0 & 1 & -1 & 3 \\ 0 & 0 & 1 & -1 \end{array}\right]$$

$$\left[\begin{array}{ccc|c} 1 & 0 & 0 & 3 \\ 0 & 1 & 0 & 2 \\ 0 & 0 & 1 & -1 \end{array}\right]$$

$x=3, \quad y=2, \quad z=-1$

37.

See Exercise 27

$$\begin{array}{l} -R_1 \to R_1 \\ \\ \frac{1}{5}R_3 \to R_3 \end{array} \left[\begin{array}{ccc|c} -1 & -1 & 2 & 9 \\ 0 & 1 & 0 & 2 \\ 0 & 0 & 5 & 15 \end{array}\right]$$

$$R_1-R_2 \to R_1 \left[\begin{array}{ccc|c} 1 & 1 & -2 & -9 \\ 0 & 1 & 0 & 2 \\ 0 & 0 & 1 & 3 \end{array}\right]$$

$$2R_3+R_1 \to R_1 \left[\begin{array}{ccc|c} 1 & 0 & -2 & -11 \\ 0 & 1 & 0 & 2 \\ 0 & 0 & 1 & 3 \end{array}\right]$$

$$\left[\begin{array}{ccc|c} 1 & 0 & 0 & -5 \\ 0 & 1 & 0 & 2 \\ 0 & 0 & 1 & 3 \end{array}\right]$$

$x=-5, \quad y=2, \quad z=3$

39.
See Exercise 29

$$\begin{array}{l} R_1+R_2 \to R_1 \\ -R_2 \to R_2 \\ \\ \frac{1}{7}R_4 \to R_4 \end{array} \left[\begin{array}{cccc|c} 1 & 1 & -1 & 2 & 0 \\ 0 & -1 & 2 & -5 & -2 \\ 0 & 0 & 1 & -2 & -1 \\ 0 & 0 & 0 & 7 & 2 \end{array}\right]$$

$$\begin{array}{l} R_1-R_3 \to R_1 \\ \\ 2R_3+R_2 \to R_2 \end{array} \left[\begin{array}{cccc|c} 1 & 0 & 1 & -3 & -2 \\ 0 & 1 & -2 & 5 & 2 \\ 0 & 0 & 1 & -2 & -1 \\ 0 & 0 & 0 & 1 & \frac{2}{7} \end{array}\right]$$

$$\begin{array}{l} R_1+R_4 \to R_1 \\ R_2-R_4 \to R_2 \\ R_3+2R_4 \to R_3 \end{array} \left[\begin{array}{cccc|c} 1 & 0 & 0 & -1 & -1 \\ 0 & 1 & 0 & 1 & 0 \\ 0 & 0 & 1 & -2 & -1 \\ 0 & 0 & 0 & 1 & \frac{2}{7} \end{array}\right]$$

$$\left[\begin{array}{cccc|c} 1 & 0 & 0 & 0 & -\frac{5}{7} \\ 0 & 1 & 0 & 0 & -\frac{2}{7} \\ 0 & 0 & 1 & 0 & -\frac{3}{7} \\ 0 & 0 & 0 & 1 & \frac{2}{7} \end{array}\right]$$

$x=-\frac{5}{7},\ y=-\frac{2}{7},\ z=-\frac{3}{7},\ w=\frac{2}{7}$

CHAPTER 11 SECTION 2

SECTION 11.2 PROGRESS CHECK

1. (Page 624)

$$\begin{bmatrix} -2 & -1 & 2 \\ 4 & 3 & 1 \end{bmatrix}\begin{bmatrix} 5 & -4 \\ 3 & 1 \\ -1 & 0 \end{bmatrix}$$

$$=\begin{bmatrix} (-2)(5)+(-1)(3)+2(-1) & (-2)(-4)+(-1)(1)+2(0) \\ 4(5)+3(3)+1(-1) & 4(-4)+3(1)+1(0) \end{bmatrix}$$

$$=\begin{bmatrix} -15 & 7 \\ 28 & -13 \end{bmatrix}$$

2. (Page 624)

$$\begin{array}{ccc} C & \cdot\ D & = CD \\ 2\times 3 & 3\times 1 & 2\times 1 \end{array}$$

$$\begin{array}{ccc} C & \cdot\ B & CB \text{ not defined} \\ 2\times 3 & 2\times 2 & \end{array}$$

$\neq$

3. (Page 625)

$$A(BC)=\begin{bmatrix} 1 & -1 \\ 2 & 3 \end{bmatrix}\left(\begin{bmatrix} 5 & -3 \\ -2 & 2 \end{bmatrix}\begin{bmatrix} 3 & -1 & -2 \\ 1 & 0 & 4 \end{bmatrix}\right)$$

$$=\begin{bmatrix} 1 & -1 \\ 2 & 3 \end{bmatrix}\begin{bmatrix} 12 & -5 & -22 \\ -4 & 2 & 12 \end{bmatrix}$$

$$=\begin{bmatrix} 16 & -7 & -34 \\ 12 & -4 & -8 \end{bmatrix}$$

$$(AB)C=\left(\begin{bmatrix} 1 & -1 \\ 2 & 3 \end{bmatrix}\begin{bmatrix} 5 & -3 \\ -2 & 2 \end{bmatrix}\right)\begin{bmatrix} 3 & -1 & -2 \\ 1 & 0 & 4 \end{bmatrix}$$

$$=\begin{bmatrix} 7 & -5 \\ 4 & 0 \end{bmatrix}\begin{bmatrix} 3 & -1 & -2 \\ 1 & 0 & 4 \end{bmatrix}$$

$$=\begin{bmatrix} 16 & -7 & -34 \\ 12 & -4 & -8 \end{bmatrix}$$

$$A(BC)=(AB)C$$

EXERCISE SET 11.2

1.

$$A=\begin{bmatrix} a & b \\ 6 & -2 \end{bmatrix} \qquad B=\begin{bmatrix} 3 & -4 \\ c & d \end{bmatrix}$$

$A=B$ if and only if $a=3$

$b=-4$

$c=6$

$d=-2$

3.

$$C+E=\begin{bmatrix} 1 & 2 & 3 \\ 4 & -1 & 2 \\ 3 & 2 & 5 \end{bmatrix}+\begin{bmatrix} 1 & -3 & 2 \\ 3 & 2 & 4 \\ 1 & 1 & 2 \end{bmatrix}$$

$$=\begin{bmatrix} 1+1 & 2+(-3) & 3+2 \\ 4+3 & -1+2 & 2+4 \\ 3+1 & 2+1 & 5+2 \end{bmatrix}$$

$$=\begin{bmatrix} 2 & -1 & 5 \\ 7 & 1 & 6 \\ 4 & 3 & 7 \end{bmatrix}$$

5.

$$2A+3G=2\begin{bmatrix} 2 & 3 & 1 \\ -3 & 4 & 1 \end{bmatrix}+3\begin{bmatrix} -2 & 4 & 2 \\ 1 & 0 & 3 \end{bmatrix}$$

$$=\begin{bmatrix} 4 & 6 & 2 \\ -6 & 8 & 2 \end{bmatrix}+\begin{bmatrix} -6 & 12 & 6 \\ 3 & 0 & 9 \end{bmatrix}$$

$$=\begin{bmatrix} -2 & 18 & 8 \\ -3 & 8 & 11 \end{bmatrix}$$

7.

$A+F$ $\quad A: 2\times 3$

$F: 2\times 2$

$A+F$: not possible

9.

$$\text{AB}=\begin{bmatrix} 2 & 3 & 1 \\ -3 & 4 & 1 \end{bmatrix}\begin{bmatrix} 2 & -1 \\ 3 & 2 \\ 4 & 1 \end{bmatrix}$$

$$=\begin{bmatrix} 2(2)+3(3)+1(4) & 2(-1)+3(2)+1(1) \\ -3(2)+4(3)+1(4) & -3(-1)+4(2)+1(1) \end{bmatrix}$$

$$=\begin{bmatrix} 17 & 5 \\ 10 & 12 \end{bmatrix}$$

11.

$CB+D$ $\quad C: 3\times 3$

$B: 3\times 2$

$CB: 3\times 2$

$D: 2\times 2$

$CB+D$: not possible

13.

$$DF + AB = \begin{bmatrix} -3 & 2 \\ 4 & 1 \end{bmatrix}\begin{bmatrix} 1 & 3 \\ -2 & 4 \end{bmatrix} + \begin{bmatrix} 2 & 3 & 1 \\ -3 & 4 & 1 \end{bmatrix}\begin{bmatrix} 2 & -1 \\ 3 & 2 \\ 4 & 1 \end{bmatrix}$$

$$= \begin{bmatrix} -7 & -1 \\ 2 & 16 \end{bmatrix} + \begin{bmatrix} 17 & 5 \\ 10 & 12 \end{bmatrix} = \begin{bmatrix} 10 & 4 \\ 12 & 28 \end{bmatrix}$$

15.

$DA + EB$ $D: 2\times 2$

$A: 2\times 3$

$DA: 2\times 3$

$E: 3\times 3$ $\neq$

$B: 3\times 2$

$EB: 3\times 2$

$DA + EB$: not possible

17.

$$2GE - 3A = 2\begin{bmatrix} -2 & 4 & 2 \\ 1 & 0 & 3 \end{bmatrix}\begin{bmatrix} 1 & -3 & 2 \\ 3 & 2 & 4 \\ 1 & 1 & 2 \end{bmatrix} - 3\begin{bmatrix} 2 & 3 & 1 \\ -3 & 4 & 1 \end{bmatrix}$$

$$= 2\begin{bmatrix} 12 & 16 & 16 \\ 4 & 0 & 8 \end{bmatrix} - \begin{bmatrix} 6 & 9 & 3 \\ -9 & 12 & 3 \end{bmatrix}$$

$$= \begin{bmatrix} 24 & 32 & 32 \\ 8 & 0 & 16 \end{bmatrix} - \begin{bmatrix} 6 & 9 & 3 \\ -9 & 12 & 3 \end{bmatrix}$$

$$= \begin{bmatrix} 18 & 23 & 29 \\ 17 & -12 & 13 \end{bmatrix}$$

19.

$$AB = \begin{bmatrix} -2 & 3 \\ 2 & -3 \end{bmatrix}\begin{bmatrix} -1 & 3 \\ 2 & 0 \end{bmatrix} = \begin{bmatrix} 8 & -6 \\ -8 & 6 \end{bmatrix}$$

$$AC = \begin{bmatrix} -2 & 3 \\ 2 & -3 \end{bmatrix}\begin{bmatrix} -4 & -3 \\ 0 & -4 \end{bmatrix} = \begin{bmatrix} 8 & -6 \\ -8 & 6 \end{bmatrix}$$

$$AB = AC$$

21.

$$AB = \begin{bmatrix} -2 & 3 \\ 2 & -3 \end{bmatrix}\begin{bmatrix} 3 & 6 \\ 2 & 4 \end{bmatrix}$$

$$= \begin{bmatrix} (-2)(3)+(3)(2) & (-2)(6)+(3)(4) \\ (2)(3)+(-3)(2) & (2)(6)+(-3)(4) \end{bmatrix}$$

$$= \begin{bmatrix} 0 & 0 \\ 0 & 0 \end{bmatrix}$$

23.

$$AI = \begin{bmatrix} a_{11} & a_{12} & a_{13} \\ a_{21} & a_{22} & a_{23} \\ a_{31} & a_{32} & a_{33} \end{bmatrix}\begin{bmatrix} 1 & 0 & 0 \\ 0 & 1 & 0 \\ 0 & 0 & 1 \end{bmatrix}$$

$$= \begin{bmatrix} a_{11}(1)+a_{12}(0)+a_{13}(0) & a_{11}(0)+a_{12}(1)+a_{13}(0) & a_{11}(0)+a_{12}(0)+a_{13}(1) \\ a_{21}(1)+a_{22}(0)+a_{23}(0) & a_{21}(0)+a_{22}(1)+a_{23}(0) & a_{21}(0)+a_{22}(0)+a_{23}(1) \\ a_{31}(1)+a_{32}(0)+a_{33}(0) & a_{31}(0)+a_{32}(1)+a_{33}(0) & a_{31}(0)+a_{32}(0)+a_{33}(1) \end{bmatrix}$$

$$= \begin{bmatrix} a_{11} & a_{12} & a_{13} \\ a_{21} & a_{22} & a_{23} \\ a_{31} & a_{32} & a_{33} \end{bmatrix} = A$$

$$IA = \begin{bmatrix} 1 & 0 & 0 \\ 0 & 1 & 0 \\ 0 & 0 & 1 \end{bmatrix}\begin{bmatrix} a_{11} & a_{12} & a_{13} \\ a_{21} & a_{22} & a_{23} \\ a_{31} & a_{32} & a_{33} \end{bmatrix}$$

$$= \begin{bmatrix} 1(a_{11})+0(a_{21})+0(a_{31}) & 1(a_{12})+0(a_{22})+0(a_{32}) & 1(a_{13})+0(a_{23})+0(a_{33}) \\ 0(a_{11})+1(a_{21})+0(a_{31}) & 0(a_{12})+1(a_{22})+0(a_{32}) & 0(a_{13})+1(a_{23})+0(a_{33}) \\ 0(a_{11})+0(a_{21})+1(a_{31}) & 0(a_{12})+0(a_{22})+1(a_{32}) & 0(a_{13})+0(a_{23})+1(a_{33}) \end{bmatrix}$$

$$= \begin{bmatrix} a_{11} & a_{12} & a_{13} \\ a_{21} & a_{22} & a_{23} \\ a_{31} & a_{32} & a_{33} \end{bmatrix} = A$$

25.

The amount of pesticide 2 eaten by herbivore 3.

27.

$$A=\begin{bmatrix}3 & 4\\ 3 & -1\end{bmatrix} \quad X=\begin{bmatrix}x\\ y\end{bmatrix} \quad B=\begin{bmatrix}-3\\ 5\end{bmatrix}$$

29.

$$A=\begin{bmatrix}3 & -1 & 4\\ 2 & 2 & \frac{3}{4}\\ 1 & -\frac{1}{4} & 1\end{bmatrix} \quad X=\begin{bmatrix}x\\ y\\ z\end{bmatrix} \quad B=\begin{bmatrix}5\\ -1\\ \frac{1}{2}\end{bmatrix}$$

31.

$$\begin{bmatrix}1 & -5\\ 4 & 3\end{bmatrix}\begin{bmatrix}x_1\\ x_2\end{bmatrix}=\begin{bmatrix}0\\ 2\end{bmatrix}$$

$$\begin{bmatrix}x_1-5x_2\\ 4x_1+3x_2\end{bmatrix}=\begin{bmatrix}0\\ 2\end{bmatrix}$$

$$x_1-5x_2=0$$
$$4x_1+3x_2=2$$

33.

$$\begin{bmatrix}4 & 5 & -2\\ 0 & 3 & -1\\ 0 & 0 & 2\end{bmatrix}\begin{bmatrix}x_1\\ x_2\\ x_3\end{bmatrix}=\begin{bmatrix}2\\ -5\\ 4\end{bmatrix}$$

$$\begin{bmatrix}4x_1+5x_2-2x_3\\ 3x_2-x_3\\ 2x_3\end{bmatrix}=\begin{bmatrix}2\\ -5\\ 4\end{bmatrix}$$

$$4x_1+5x_2-2x_3=2$$
$$3x_2-x_3=-5$$
$$2x_3=4$$

35.

$$AI_n=\begin{bmatrix}a_{11} & a_{12} & \cdots & a_{1n}\\ a_{21} & a_{22} & \cdots & a_{2n}\\ \vdots & & & \vdots\\ a_{n1} & a_{n2} & \cdots & a_{nn}\end{bmatrix}\begin{bmatrix}1 & 0 & \cdots & 0\\ 0 & 1 & 0\cdots & 0\\ \vdots & & \ddots & \vdots\\ 0 & 0 & \cdots 0 & \ddots 1\end{bmatrix}$$

$$=\begin{bmatrix}a_{11} & a_{12} & \cdots & a_{1n}\\ a_{21} & a_{22} & \cdots & a_{2n}\\ \vdots & & & \vdots\\ a_{n1} & a_{n2} & \cdots & a_{nn}\end{bmatrix}=A$$

$$I_nA=\begin{bmatrix}1 & 0 & \cdots & 0\\ 0 & 1 & 0\cdots & 0\\ \vdots & & \ddots & \vdots\\ 0 & 0 & \cdots 0 & 1\end{bmatrix}\begin{bmatrix}a_{11} & a_{12} & \cdots & a_{1n}\\ a_{21} & a_{22} & \cdots & a_{2n}\\ \vdots & & & \vdots\\ a_{n1} & a_{n2} & \cdots & a_{nn}\end{bmatrix}$$

$$=\begin{bmatrix}a_{11} & a_{12} & \cdots & a_{1n}\\ a_{21} & a_{22} & \cdots & a_{2n}\\ \vdots & & & \vdots\\ a_{n1} & a_{n2} & \cdots & a_{nn}\end{bmatrix}=A$$

CHAPTER 11 SECTION 3

SECTION 11.3 PROGRESS CHECK

1. (Page 629)

$$AB = \begin{bmatrix} 4 & 5 \\ 2 & 2 \end{bmatrix} \begin{bmatrix} -1 & \frac{5}{2} \\ 1 & -2 \end{bmatrix} = \begin{bmatrix} 1 & 0 \\ 0 & 1 \end{bmatrix}$$

$$BA = \begin{bmatrix} -1 & \frac{5}{2} \\ 1 & -2 \end{bmatrix} \begin{bmatrix} 4 & 5 \\ 2 & 2 \end{bmatrix} = \begin{bmatrix} 1 & 0 \\ 0 & 1 \end{bmatrix}$$

$AB = I_2 = BA$, thus A and B are inverses.

2. (Page 632)

$$\begin{array}{r} \\ -3R_1 + R_2 \to R_2 \\ -5R_1 + R_3 \to R_3 \end{array} \left[\begin{array}{ccc|ccc} 1 & 2 & -3 & 1 & 0 & 0 \\ 3 & 2 & 1 & 0 & 1 & 0 \\ 5 & 6 & -5 & 0 & 0 & 1 \end{array}\right]$$

$$R_2 - R_3 \to R_3 \left[\begin{array}{ccc|ccc} 1 & 2 & -3 & 1 & 0 & 0 \\ 0 & -4 & 10 & -3 & 1 & 0 \\ 0 & -4 & 10 & -5 & 0 & 1 \end{array}\right]$$

$$\left[\begin{array}{ccc|ccc} 1 & 2 & -3 & 1 & 0 & 0 \\ 0 & -4 & 10 & -3 & 1 & 0 \\ 0 & 0 & 0 & 2 & 1 & -1 \end{array}\right]$$

Since row 3 consists of all zeros, A is not invertible.

3. (Page 634)

$$\begin{matrix} \\ R_1 - R_2 \to R_2 \\ R_1 + R_3 \to R_3 \end{matrix} \left[\begin{array}{ccc|ccc} 1 & -2 & 1 & 1 & 0 & 0 \\ 1 & 3 & 2 & 0 & 1 & 0 \\ -1 & 0 & 1 & 0 & 0 & 1 \end{array}\right]$$

$$R_2 \leftrightarrow -\frac{1}{2}R_3 \left[\begin{array}{ccc|ccc} 1 & -2 & 1 & 1 & 0 & 0 \\ 0 & -5 & -1 & 1 & -1 & 0 \\ 0 & -2 & 2 & 1 & 0 & 1 \end{array}\right]$$

$$\begin{matrix} \\ 2R_2 + R_1 \to R_1 \\ 5R_2 + R_3 \to R_3 \end{matrix} \left[\begin{array}{ccc|ccc} 1 & -2 & 1 & 1 & 0 & 0 \\ 0 & 1 & -1 & -\frac{1}{2} & 0 & -\frac{1}{2} \\ 0 & -5 & -1 & 1 & -1 & 0 \end{array}\right]$$

$$-\frac{1}{6}R_3 \to R_3 \left[\begin{array}{ccc|ccc} 1 & 0 & -1 & 0 & 0 & -1 \\ 0 & 1 & -1 & -\frac{1}{2} & 0 & -\frac{1}{2} \\ 0 & 0 & -6 & -\frac{3}{2} & -1 & -\frac{5}{2} \end{array}\right]$$

3. (Continued)

$$\begin{matrix} R_1 + R_3 \to R_1 \\ R_2 + R_3 \to R_2 \end{matrix} \left[\begin{array}{ccc|ccc} 1 & 0 & -1 & 0 & 0 & -1 \\ 0 & 1 & -1 & -\frac{1}{2} & 0 & -\frac{1}{2} \\ 0 & 0 & 1 & \frac{1}{4} & \frac{1}{6} & \frac{5}{12} \end{array}\right]$$

$$\left[\begin{array}{ccc|ccc} 1 & 0 & 0 & \frac{1}{4} & \frac{1}{6} & -\frac{7}{12} \\ 0 & 1 & 0 & -\frac{1}{4} & \frac{1}{6} & -\frac{1}{12} \\ 0 & 0 & 1 & \frac{1}{4} & \frac{1}{6} & \frac{5}{12} \end{array}\right]$$

$$\begin{bmatrix} x \\ y \\ z \end{bmatrix} = X = A^{-1}B = \begin{bmatrix} \frac{1}{4} & \frac{1}{6} & -\frac{7}{12} \\ -\frac{1}{4} & \frac{1}{6} & -\frac{1}{12} \\ \frac{1}{4} & \frac{1}{6} & \frac{5}{12} \end{bmatrix} \begin{bmatrix} 1 \\ 2 \\ -11 \end{bmatrix}$$

$$= \begin{bmatrix} 7 \\ 1 \\ -4 \end{bmatrix}$$

$x = 7, \quad y = 1, \quad z = -4$

EXERCISE SET 11.3

1.

$$AB = \begin{bmatrix} 2 & \frac{1}{2} \\ -1 & 3 \end{bmatrix} \begin{bmatrix} 1 & -1 \\ -2 & 4 \end{bmatrix} = \begin{bmatrix} 1 & 0 \\ -7 & 13 \end{bmatrix} \neq I_2$$

Not inverses

3.

$$AB = \begin{bmatrix} 1 & 2 & 2 \\ -1 & 3 & 0 \\ 0 & 2 & 1 \end{bmatrix} \begin{bmatrix} 3 & 2 & -6 \\ 1 & 1 & -2 \\ -2 & -2 & 5 \end{bmatrix} = \begin{bmatrix} 1 & 0 & 0 \\ 0 & 1 & 0 \\ 0 & 0 & 1 \end{bmatrix}$$

$$BA = \begin{bmatrix} 3 & 2 & -6 \\ 1 & 1 & -2 \\ -2 & -2 & 5 \end{bmatrix} \begin{bmatrix} 1 & 2 & 2 \\ -1 & 3 & 0 \\ 0 & 2 & 1 \end{bmatrix} = \begin{bmatrix} 1 & 0 & 0 \\ 0 & 1 & 0 \\ 0 & 0 & 1 \end{bmatrix}$$

$$AB = I_3 = BA$$

Inverses

5.

$$\begin{array}{r} -R_1 \to R_1 \\ 2R_1 + R_2 \to R_2 \end{array} \left[\begin{array}{cc|cc} -1 & 5 & 1 & 0 \\ 2 & -4 & 0 & 1 \end{array}\right]$$

$$\frac{1}{6}R_2 \to R_2 \left[\begin{array}{cc|cc} 1 & -5 & -1 & 0 \\ 0 & 6 & 2 & 1 \end{array}\right]$$

$$5R_2 + R_1 \to R_1 \left[\begin{array}{cc|cc} 1 & -5 & -1 & 0 \\ 0 & 1 & \frac{1}{3} & \frac{1}{6} \end{array}\right]$$

$$\left[\begin{array}{cc|cc} 1 & 0 & \frac{2}{3} & \frac{5}{6} \\ 0 & 1 & \frac{1}{3} & \frac{1}{6} \end{array}\right]$$

$$\begin{bmatrix} \frac{2}{3} & \frac{5}{6} \\ \frac{1}{3} & \frac{1}{6} \end{bmatrix}$$

7.

$$\begin{array}{r} -R_1 \to R_1 \\ -2R_1 + R_2 \to R_2 \end{array} \left[\begin{array}{cc|cc} -1 & 1 & 1 & 0 \\ -2 & 1 & 0 & 1 \end{array}\right]$$

$$\begin{array}{r} R_1 - R_2 \to R_1 \\ -R_2 \to R_2 \end{array} \left[\begin{array}{cc|cc} 1 & -1 & -1 & 0 \\ 0 & -1 & -2 & 1 \end{array}\right]$$

$$\left[\begin{array}{cc|cc} 1 & 0 & 1 & -1 \\ 0 & 1 & 2 & -1 \end{array}\right]$$

$$\begin{bmatrix} 1 & -1 \\ 2 & -1 \end{bmatrix}$$

9.

$$R_1 + R_2 \to R_2 \left[\begin{array}{ccc|ccc} 1 & -2 & 3 & 1 & 0 & 0 \\ -1 & 3 & -4 & 0 & 1 & 0 \\ 0 & 5 & -4 & 0 & 0 & 1 \end{array}\right]$$

$$\begin{array}{l} 2R_2 + R_1 \to R_1 \\ -5R_2 + R_3 \to R_3 \end{array} \left[\begin{array}{ccc|ccc} 1 & -2 & 3 & 1 & 0 & 0 \\ 0 & 1 & -1 & 1 & 1 & 0 \\ 0 & 5 & -4 & 0 & 0 & 1 \end{array}\right]$$

$$\begin{array}{l} R_1 - R_3 \to R_1 \\ R_2 + R_3 \to R_2 \end{array} \left[\begin{array}{ccc|ccc} 1 & 0 & 1 & 3 & 2 & 0 \\ 0 & 1 & -1 & 1 & 1 & 0 \\ 0 & 0 & 1 & -5 & -5 & 1 \end{array}\right]$$

$$\left[\begin{array}{ccc|ccc} 1 & 0 & 0 & 8 & 7 & -1 \\ 0 & 1 & 0 & -4 & -4 & 1 \\ 0 & 0 & 1 & -5 & -5 & 1 \end{array}\right]$$

$$\begin{bmatrix} 8 & 7 & -1 \\ -4 & -4 & 1 \\ -5 & -5 & 1 \end{bmatrix}$$

11.

$$R_1 + R_2 \to R_2 \left[\begin{array}{cc|cc} 1 & 3 & 1 & 0 \\ -1 & 4 & 0 & 1 \end{array}\right]$$

$$\frac{1}{7} R_2 \to R_2 \left[\begin{array}{cc|cc} 1 & 3 & 1 & 0 \\ 0 & 7 & 1 & 1 \end{array}\right]$$

$$-3R_2 + R_1 \to R_1 \left[\begin{array}{cc|cc} 1 & 3 & 1 & 0 \\ 0 & 1 & \frac{1}{7} & \frac{1}{7} \end{array}\right]$$

$$\left[\begin{array}{cc|cc} 1 & 0 & \frac{4}{7} & -\frac{3}{7} \\ 0 & 1 & \frac{1}{7} & \frac{1}{7} \end{array}\right]$$

$$\begin{bmatrix} \frac{4}{7} & -\frac{3}{7} \\ \frac{1}{7} & \frac{1}{7} \end{bmatrix}$$

13.

$$\begin{array}{l} -2R_1 + R_2 \to R_2 \\ R_1 + R_3 \to R_3 \end{array} \left[\begin{array}{ccc|ccc} 1 & 1 & 3 & 1 & 0 & 0 \\ 2 & -8 & -4 & 0 & 1 & 0 \\ -1 & 2 & 0 & 0 & 0 & 1 \end{array}\right]$$

$$-\frac{1}{10}R_2 \to R_2 \left[\begin{array}{ccc|ccc} 1 & 1 & 3 & 1 & 0 & 0 \\ 0 & -10 & -10 & -2 & 1 & 0 \\ 0 & 3 & 3 & 1 & 0 & 1 \end{array}\right]$$

$$-3R_2 + R_3 \to R_3 \left[\begin{array}{ccc|ccc} 1 & 1 & 3 & 1 & 0 & 0 \\ 0 & 1 & 1 & \frac{1}{5} & -\frac{1}{10} & 0 \\ 0 & 3 & 3 & 1 & 0 & 1 \end{array}\right]$$

$$\left[\begin{array}{ccc|ccc} 1 & 1 & 3 & 1 & 0 & 0 \\ 0 & 1 & 1 & \frac{1}{5} & -\frac{1}{10} & 0 \\ 0 & 0 & 0 & \frac{2}{5} & \frac{3}{10} & 1 \end{array}\right]$$

Not invertible

15.

$$\begin{array}{l} \frac{1}{2}R_1 \to R_1 \\ -\frac{1}{3}R_2 \to R_2 \end{array} \left[\begin{array}{cc|cc} 2 & 0 & 1 & 0 \\ 0 & -3 & 0 & 1 \end{array}\right]$$

$$\left[\begin{array}{cc|cc} 1 & 0 & \frac{1}{2} & 0 \\ 0 & 1 & 0 & -\frac{1}{3} \end{array}\right]$$

$$\begin{bmatrix} \frac{1}{2} & 0 \\ 0 & -\frac{1}{3} \end{bmatrix}$$

17.

$$-2R_1+R_2\to R_2\left[\begin{array}{ccc|ccc}1&0&-1&1&0&0\\2&1&0&0&1&0\\0&1&1&0&0&1\end{array}\right]$$

$$R_2-R_3\to R_3\left[\begin{array}{ccc|ccc}1&0&-1&1&0&0\\0&1&2&-2&1&0\\0&1&1&0&0&1\end{array}\right]$$

$$\begin{array}{l}R_1+R_3\to R_1\\-2R_3+R_2\to R_2\end{array}\left[\begin{array}{ccc|ccc}1&0&-1&1&0&0\\0&1&2&-2&1&0\\0&0&1&-2&1&-1\end{array}\right]$$

$$\left[\begin{array}{ccc|ccc}1&0&0&-1&1&-1\\0&1&0&2&-1&2\\0&0&1&-2&1&-1\end{array}\right]$$

$$\begin{bmatrix}-1&1&-1\\2&-1&2\\-2&1&-1\end{bmatrix}$$

19.

$$R_1\leftrightarrow R_2\left[\begin{array}{cc|cc}2&1&1&0\\1&-3&0&1\end{array}\right]$$

$$-2R_1+R_2\to R_2\left[\begin{array}{cc|cc}1&-3&0&1\\2&1&1&0\end{array}\right]$$

$$\frac{1}{7}R_2\to R_2\left[\begin{array}{cc|cc}1&-3&0&1\\0&7&1&-2\end{array}\right]$$

$$3R_2+R_1\to R_1\left[\begin{array}{cc|cc}1&-3&0&1\\0&1&\frac{1}{7}&-\frac{2}{7}\end{array}\right]$$

$$\left[\begin{array}{cc|cc}1&0&\frac{3}{7}&\frac{1}{7}\\0&1&\frac{1}{7}&-\frac{2}{7}\end{array}\right]$$

$$X=A^{-1}B$$

$$\begin{bmatrix}x\\y\end{bmatrix}=\begin{bmatrix}\frac{3}{7}&\frac{1}{7}\\\frac{1}{7}&-\frac{2}{7}\end{bmatrix}\begin{bmatrix}5\\6\end{bmatrix}=\begin{bmatrix}3\\-1\end{bmatrix}$$

$x=3,\quad y=-1$

21.

$$R_2 \leftrightarrow R_1 \left[\begin{array}{ccc|ccc} 3 & 1 & -1 & 1 & 0 & 0 \\ 1 & -2 & 0 & 0 & 1 & 0 \\ 0 & 3 & 1 & 0 & 0 & 1 \end{array}\right]$$

$$-3R_1 + R_2 \to R_2 \left[\begin{array}{ccc|ccc} 1 & -2 & 0 & 0 & 1 & 0 \\ 3 & 1 & -1 & 1 & 0 & 0 \\ 0 & 3 & 1 & 0 & 0 & 1 \end{array}\right]$$

$$\frac{1}{7}R_2 \to R_2 \left[\begin{array}{ccc|ccc} 1 & -2 & 0 & 0 & 1 & 0 \\ 0 & 7 & -1 & 1 & -3 & 0 \\ 0 & 3 & 1 & 0 & 0 & 1 \end{array}\right]$$

$$\begin{array}{l} 2R_2 + R_1 \to R_1 \\ -3R_2 + R_3 \to R_3 \end{array} \left[\begin{array}{ccc|ccc} 1 & -2 & 0 & 0 & 1 & 0 \\ 0 & 1 & -\frac{1}{7} & \frac{1}{7} & -\frac{3}{7} & 0 \\ 0 & 3 & 1 & 0 & 0 & 1 \end{array}\right]$$

$$\frac{7}{10}R_3 \to R_3 \left[\begin{array}{ccc|ccc} 1 & 0 & -\frac{2}{7} & \frac{2}{7} & \frac{1}{7} & 0 \\ 0 & 1 & -\frac{1}{7} & \frac{1}{7} & -\frac{3}{7} & 0 \\ 0 & 0 & \frac{10}{7} & -\frac{3}{7} & \frac{9}{7} & 1 \end{array}\right]$$

21. (Continued).

$$\begin{array}{l} \frac{2}{7}R_3 + R_1 \to R_1 \\ \frac{1}{7}R_3 + R_2 \to R_2 \end{array} \left[\begin{array}{ccc|ccc} 1 & 0 & -\frac{2}{7} & \frac{2}{7} & \frac{1}{7} & 0 \\ 0 & 1 & -\frac{1}{7} & \frac{1}{7} & -\frac{3}{7} & 0 \\ 0 & 0 & 1 & -\frac{3}{10} & \frac{9}{10} & \frac{7}{10} \end{array}\right]$$

$$\left[\begin{array}{ccc|ccc} 1 & 0 & 0 & \frac{1}{5} & \frac{2}{5} & \frac{1}{5} \\ 0 & 1 & 0 & \frac{1}{10} & -\frac{3}{10} & \frac{1}{10} \\ 0 & 0 & 1 & -\frac{3}{10} & \frac{9}{10} & \frac{7}{10} \end{array}\right]$$

$X = A^{-1}B$

$$\begin{bmatrix} x \\ y \\ z \end{bmatrix} = \begin{bmatrix} \frac{1}{5} & \frac{2}{5} & \frac{1}{5} \\ \frac{1}{10} & -\frac{3}{10} & \frac{1}{10} \\ -\frac{3}{10} & \frac{9}{10} & \frac{7}{10} \end{bmatrix} \begin{bmatrix} 2 \\ 8 \\ -8 \end{bmatrix} = \begin{bmatrix} 2 \\ -3 \\ 1 \end{bmatrix}$$

$x = 2, \quad y = -3, \quad z = 1$

23.

$$R_1 \leftrightarrow R_3 \left[\begin{array}{ccc|ccc} 2 & -1 & 3 & 1 & 0 & 0 \\ 3 & -1 & 1 & 0 & 1 & 0 \\ 1 & 1 & 1 & 0 & 0 & 1 \end{array}\right]$$

$$\begin{array}{l} -3R_1 + R_2 \to R_2 \\ -2R_1 + R_3 \to R_3 \end{array} \left[\begin{array}{ccc|ccc} 1 & 1 & 1 & 0 & 0 & 1 \\ 3 & -1 & 1 & 0 & 1 & 0 \\ 2 & -1 & 3 & 1 & 0 & 0 \end{array}\right]$$

$$-\frac{1}{4}R_2 \to R_2 \left[\begin{array}{ccc|ccc} 1 & 1 & 1 & 0 & 0 & 1 \\ 0 & -4 & -2 & 0 & 1 & -3 \\ 0 & -3 & 1 & 1 & 0 & -2 \end{array}\right]$$

$$\begin{array}{l} R_1 - R_2 \to R_1 \\ 3R_2 + R_3 \to R_3 \end{array} \left[\begin{array}{ccc|ccc} 1 & 1 & 1 & 0 & 0 & 1 \\ 0 & 1 & \frac{1}{2} & 0 & -\frac{1}{4} & \frac{3}{4} \\ 0 & -3 & 1 & 1 & 0 & -2 \end{array}\right]$$

$$\frac{2}{5}R_3 \to R_3 \left[\begin{array}{ccc|ccc} 1 & 0 & \frac{1}{2} & 0 & \frac{1}{4} & \frac{1}{4} \\ 0 & 1 & \frac{1}{2} & 0 & -\frac{1}{4} & \frac{3}{4} \\ 0 & 0 & \frac{5}{2} & 1 & -\frac{3}{4} & \frac{1}{4} \end{array}\right]$$

23. (Continued)

$$\begin{array}{l} -\frac{1}{2}R_3 + R_1 \to R_1 \\ -\frac{1}{2}R_3 + R_2 \to R_2 \end{array} \left[\begin{array}{ccc|ccc} 1 & 0 & \frac{1}{2} & 0 & \frac{1}{4} & \frac{1}{4} \\ 0 & 1 & \frac{1}{2} & 0 & -\frac{1}{4} & \frac{3}{4} \\ 0 & 0 & 1 & \frac{2}{5} & -\frac{3}{10} & \frac{1}{10} \end{array}\right]$$

$$\left[\begin{array}{ccc|ccc} 1 & 0 & 0 & -\frac{1}{5} & \frac{2}{5} & \frac{1}{5} \\ 0 & 1 & 0 & -\frac{1}{5} & -\frac{1}{10} & \frac{7}{10} \\ 0 & 0 & 1 & \frac{2}{5} & -\frac{3}{10} & \frac{1}{10} \end{array}\right]$$

$X = A^{-1}B$

$$\begin{bmatrix} x \\ y \\ z \end{bmatrix} = \begin{bmatrix} -\frac{1}{5} & \frac{2}{5} & \frac{1}{5} \\ -\frac{1}{5} & -\frac{1}{10} & \frac{7}{10} \\ \frac{2}{5} & -\frac{3}{10} & \frac{1}{10} \end{bmatrix} \begin{bmatrix} -11 \\ -5 \\ -1 \end{bmatrix} = \begin{bmatrix} 0 \\ 2 \\ -3 \end{bmatrix}$$

$x = 0, \quad y = 2, \quad z = -3$

25.

$$x-2y=-4$$
$$2x+3y=13$$

$$-2R_1+R_2\to R_2\left[\begin{array}{cc|cc}1 & -2 & 1 & 0\\ 2 & 3 & 0 & 1\end{array}\right]$$

$$\frac{1}{7}R_2\to R_2\left[\begin{array}{cc|cc}1 & -2 & 1 & 0\\ 0 & 7 & -2 & 1\end{array}\right]$$

$$2R_2+R_1\to R_1\left[\begin{array}{cc|cc}1 & -2 & 1 & 0\\ 0 & 1 & -\frac{2}{7} & \frac{1}{7}\end{array}\right]$$

$$\left[\begin{array}{cc|cc}1 & 0 & \frac{3}{7} & \frac{2}{7}\\ 0 & 1 & -\frac{2}{7} & \frac{1}{7}\end{array}\right]$$

$$X=A^{-1}B$$

$$\begin{bmatrix}x\\ y\end{bmatrix}=\begin{bmatrix}\frac{3}{7} & \frac{2}{7}\\ -\frac{2}{7} & \frac{1}{7}\end{bmatrix}\begin{bmatrix}-4\\ 13\end{bmatrix}=\begin{bmatrix}2\\ 3\end{bmatrix}$$

$$x=2,\quad y=3$$

27.

$$x + y + z = 4$$
$$2x - y + 2z = 11$$
$$x + 2y + 2z = 6$$

$$\begin{array}{r} \\ -2R_1 + R_2 \to R_2 \\ R_1 - R_3 \to R_3 \end{array} \left[\begin{array}{ccc|ccc} 1 & 1 & 1 & 1 & 0 & 0 \\ 2 & -1 & 2 & 0 & 1 & 0 \\ 1 & 2 & 2 & 0 & 0 & 1 \end{array}\right]$$

$$\begin{array}{r} R_1 + R_3 \to R_1 \\ -R_3 \leftrightarrow R_2 \\ \\ \end{array} \left[\begin{array}{ccc|ccc} 1 & 1 & 1 & 1 & 0 & 0 \\ 0 & -3 & 0 & -2 & 1 & 0 \\ 0 & -1 & -1 & 1 & 0 & -1 \end{array}\right]$$

$$3R_2 + R_3 \to R_3 \left[\begin{array}{ccc|ccc} 1 & 0 & 0 & 2 & 0 & -1 \\ 0 & 1 & 1 & -1 & 0 & 1 \\ 0 & -3 & 0 & -2 & 1 & 0 \end{array}\right]$$

$$\frac{1}{3}R_3 \to R_3 \left[\begin{array}{ccc|ccc} 1 & 0 & 0 & 2 & 0 & -1 \\ 0 & 1 & 1 & -1 & 0 & 1 \\ 0 & 0 & 3 & -5 & 1 & 3 \end{array}\right]$$

27. (Continued)

$$R_2 - R_3 \to R_2 \left[\begin{array}{ccc|ccc} 1 & 0 & 0 & 2 & 0 & -1 \\ 0 & 1 & 1 & -1 & 0 & 1 \\ 0 & 0 & 1 & -\frac{5}{3} & \frac{1}{3} & 1 \end{array}\right]$$

$$\left[\begin{array}{ccc|ccc} 1 & 0 & 0 & 2 & 0 & -1 \\ 0 & 1 & 0 & \frac{2}{3} & -\frac{1}{3} & 0 \\ 0 & 0 & 1 & -\frac{5}{3} & \frac{1}{3} & 1 \end{array}\right]$$

$$X = A^{-1}B$$

$$\begin{bmatrix} x \\ y \\ z \end{bmatrix} = \begin{bmatrix} 2 & 0 & -1 \\ \frac{2}{3} & -\frac{1}{3} & 0 \\ -\frac{5}{3} & \frac{1}{3} & 1 \end{bmatrix} \begin{bmatrix} 4 \\ 11 \\ 6 \end{bmatrix} = \begin{bmatrix} 2 \\ -1 \\ 3 \end{bmatrix}$$

$x = 2, \quad y = -1, \quad z = 3$

29.

$$\begin{aligned} 2x + y - z &= 9 \\ x - 2y + 2z &= -3 \\ 3x + 3y + 4z &= 11 \end{aligned}$$

$$R_3 - R_1 \to R_1 \left[\begin{array}{ccc|ccc} 2 & 1 & -1 & 1 & 0 & 0 \\ 1 & -2 & 2 & 0 & 1 & 0 \\ 3 & 3 & 4 & 0 & 0 & 1 \end{array}\right]$$

$$\begin{array}{l} R_1 - R_2 \to R_2 \\ -3R_1 + R_3 \to R_3 \end{array} \left[\begin{array}{ccc|ccc} 1 & 2 & 5 & -1 & 0 & 1 \\ 1 & -2 & 2 & 0 & 1 & 0 \\ 3 & 3 & 4 & 0 & 0 & 1 \end{array}\right]$$

$$R_2 + R_3 \to R_2 \left[\begin{array}{ccc|ccc} 1 & 2 & 5 & -1 & 0 & 1 \\ 0 & 4 & 3 & -1 & -1 & 1 \\ 0 & -3 & -11 & 3 & 0 & -2 \end{array}\right]$$

$$\begin{array}{l} -2R_2 + R_1 \to R_1 \\ 3R_2 + R_3 \to R_3 \end{array} \left[\begin{array}{ccc|ccc} 1 & 2 & 5 & -1 & 0 & 1 \\ 0 & 1 & -8 & 2 & -1 & -1 \\ 0 & -3 & -11 & 3 & 0 & -2 \end{array}\right]$$

$$-\frac{1}{35}R_3 \to R_3 \left[\begin{array}{ccc|ccc} 1 & 0 & 21 & -5 & 2 & 3 \\ 0 & 1 & -8 & 2 & -1 & -1 \\ 0 & 0 & -35 & 9 & -3 & -5 \end{array}\right]$$

29. (Continued)

$$\begin{array}{l} -21R_3 + R_1 \to R_1 \\ 8R_3 + R_2 \to R_2 \end{array} \left[\begin{array}{ccc|ccc} 1 & 0 & 21 & -5 & 2 & 3 \\ 0 & 1 & -8 & 2 & -1 & -1 \\ 0 & 0 & 1 & -\frac{9}{35} & \frac{3}{35} & \frac{1}{7} \end{array}\right]$$

$$\left[\begin{array}{ccc|ccc} 1 & 0 & 0 & \frac{2}{5} & \frac{1}{5} & 0 \\ 0 & 1 & 0 & -\frac{2}{35} & -\frac{11}{35} & \frac{1}{7} \\ 0 & 0 & 1 & -\frac{9}{35} & \frac{3}{35} & \frac{1}{7} \end{array}\right]$$

$$X = A^{-1}B$$

$$\begin{bmatrix} x \\ y \\ z \end{bmatrix} = \begin{bmatrix} \frac{2}{5} & \frac{1}{5} & 0 \\ -\frac{2}{35} & -\frac{11}{35} & \frac{1}{7} \\ -\frac{9}{35} & \frac{3}{35} & \frac{1}{7} \end{bmatrix} \begin{bmatrix} 9 \\ -3 \\ 11 \end{bmatrix} = \begin{bmatrix} 3 \\ 2 \\ -1 \end{bmatrix}$$

$x = 3, \quad y = 2, \quad z = -1$

31.

$$\begin{aligned} -x - y + 2z &= 9 \\ x + 2y - 2z &= -7 \\ 2x - y + z &= -9 \end{aligned}$$

$$\begin{array}{l} -R_1 \to R_1 \\ R_1 + R_2 \to R_2 \\ 2R_1 + R_3 \to R_3 \end{array} \left[\begin{array}{ccc|ccc} -1 & -1 & 2 & 1 & 0 & 0 \\ 1 & 2 & -2 & 0 & 1 & 0 \\ 2 & -1 & 1 & 0 & 0 & 1 \end{array}\right]$$

$$\begin{array}{l} R_1 - R_2 \to R_1 \\ 3R_2 + R_3 \to R_3 \\ \end{array} \left[\begin{array}{ccc|ccc} 1 & 1 & -2 & -1 & 0 & 0 \\ 0 & 1 & 0 & 1 & 1 & 0 \\ 0 & -3 & 5 & 2 & 0 & 1 \end{array}\right]$$

$$\frac{1}{5}R_3 \to R_3 \left[\begin{array}{ccc|ccc} 1 & 0 & -2 & -2 & -1 & 0 \\ 0 & 1 & 0 & 1 & 1 & 0 \\ 0 & 0 & 5 & 5 & 3 & 1 \end{array}\right]$$

31. (Continued)

$$2R_3 + R_1 \to R_1 \left[\begin{array}{ccc|ccc} 1 & 0 & -2 & -2 & -1 & 0 \\ 0 & 1 & 0 & 1 & 1 & 0 \\ 0 & 0 & 1 & 1 & \frac{3}{5} & \frac{1}{5} \end{array}\right]$$

$$\left[\begin{array}{ccc|ccc} 1 & 0 & 0 & 0 & \frac{1}{5} & \frac{2}{5} \\ 0 & 1 & 0 & 1 & 1 & 0 \\ 0 & 0 & 1 & 1 & \frac{3}{5} & \frac{1}{5} \end{array}\right]$$

$X = A^{-1}B$

$$\begin{bmatrix} x \\ y \\ z \end{bmatrix} = \begin{bmatrix} 0 & \frac{1}{5} & \frac{2}{5} \\ 1 & 1 & 0 \\ 1 & \frac{3}{5} & \frac{1}{5} \end{bmatrix} \begin{bmatrix} 9 \\ -7 \\ -9 \end{bmatrix} = \begin{bmatrix} -5 \\ 2 \\ 3 \end{bmatrix}$$

$x = -5, \quad y = 2, \quad z = 3$

33.

$$\begin{aligned} x+y \quad -z+2w &= 0 \\ 2x+y \qquad -w &= -2 \\ 3x \quad +2z \qquad &= -3 \\ -x+2y \quad +3w &= 1 \end{aligned}$$

$$\begin{array}{r} \\ -2R_1+R_2\to R_2 \\ -3R_1+R_3\to R_3 \\ R_1+R_4\to R_4 \end{array}\left[\begin{array}{cccc|cccc} 1 & 1 & -1 & 2 & 1 & 0 & 0 & 0 \\ 2 & 1 & 0 & -1 & 0 & 1 & 0 & 0 \\ 3 & 0 & 2 & 0 & 0 & 0 & 1 & 0 \\ -1 & 2 & 0 & 3 & 0 & 0 & 0 & 1 \end{array}\right]$$

$$\begin{array}{r} R_1+R_2\to R_1 \\ -R_2\to R_2 \\ R_3+R_4\to R_3 \\ 3R_2+R_4\to R_4 \end{array}\left[\begin{array}{cccc|cccc} 1 & 1 & -1 & 2 & 1 & 0 & 0 & 0 \\ 0 & -1 & 2 & -5 & -2 & 1 & 0 & 0 \\ 0 & -3 & 5 & -6 & -3 & 0 & 1 & 0 \\ 0 & 3 & -1 & 5 & 1 & 0 & 0 & 1 \end{array}\right]$$

$$\begin{array}{r} \\ R_4-R_3\to R_3 \\ \\ \\ \end{array}\left[\begin{array}{cccc|cccc} 1 & 0 & 1 & -3 & -1 & 1 & 0 & 0 \\ 0 & 1 & -2 & 5 & 2 & -1 & 0 & 0 \\ 0 & 0 & 4 & -1 & -2 & 0 & 1 & 1 \\ 0 & 0 & 5 & -10 & -5 & 3 & 0 & 1 \end{array}\right]$$

$$\begin{array}{r} R_1-R_3\to R_1 \\ 2R_3+R_2\to R_2 \\ -5R_3+R_4\to R_4 \\ \\ \end{array}\left[\begin{array}{cccc|cccc} 1 & 0 & 1 & -3 & -1 & 1 & 0 & 0 \\ 0 & 1 & -2 & 5 & 2 & -1 & 0 & 0 \\ 0 & 0 & 1 & -9 & -3 & 3 & -1 & 0 \\ 0 & 0 & 5 & -10 & -5 & 3 & 0 & 1 \end{array}\right]$$

$$\begin{array}{r} \\ \\ \frac{1}{35}R_4\to R_4 \\ \\ \end{array}\left[\begin{array}{cccc|cccc} 1 & 0 & 0 & 6 & 2 & -2 & 1 & 0 \\ 0 & 1 & 0 & -13 & -4 & 5 & -2 & 0 \\ 0 & 0 & 1 & -9 & -3 & 3 & -1 & 0 \\ 0 & 0 & 0 & 35 & 10 & -12 & 5 & 1 \end{array}\right]$$

33. (Continued)

$$\begin{matrix} \\ -6R_4+R_1\to R_1 \\ 13R_4+R_2\to R_2 \\ 9R_4+R_3\to R_3 \end{matrix} \left[\begin{array}{cccc|cccc} 1 & 0 & 0 & 6 & 2 & -2 & 1 & 0 \\ 0 & 1 & 0 & -13 & -4 & 5 & -2 & 0 \\ 0 & 0 & 1 & -9 & -3 & 3 & -1 & 0 \\ 0 & 0 & 0 & 1 & \frac{2}{7} & -\frac{12}{35} & \frac{1}{7} & \frac{1}{35} \end{array}\right]$$

$$\left[\begin{array}{cccc|cccc} 1 & 0 & 0 & 0 & \frac{2}{7} & \frac{2}{35} & \frac{1}{7} & -\frac{6}{35} \\ 0 & 1 & 0 & 0 & -\frac{2}{7} & \frac{19}{35} & -\frac{1}{7} & \frac{13}{35} \\ 0 & 0 & 1 & 0 & -\frac{3}{7} & -\frac{3}{35} & \frac{2}{7} & \frac{9}{35} \\ 0 & 0 & 0 & 1 & \frac{2}{7} & -\frac{12}{35} & \frac{1}{7} & \frac{1}{35} \end{array}\right]$$

$X = A^{-1}B$

$$\begin{bmatrix} x \\ y \\ z \\ w \end{bmatrix} = \begin{bmatrix} \frac{2}{7} & \frac{2}{35} & \frac{1}{7} & -\frac{6}{35} \\ -\frac{2}{7} & \frac{19}{35} & -\frac{1}{7} & \frac{13}{35} \\ -\frac{3}{7} & -\frac{3}{35} & \frac{2}{7} & \frac{9}{35} \\ \frac{2}{7} & -\frac{12}{35} & \frac{1}{7} & \frac{1}{35} \end{bmatrix} \begin{bmatrix} 0 \\ -2 \\ -3 \\ 1 \end{bmatrix} = \begin{bmatrix} -\frac{5}{7} \\ -\frac{2}{7} \\ -\frac{3}{7} \\ \frac{2}{7} \end{bmatrix}$$

$x = -\frac{5}{7},\ \ y = -\frac{2}{7},\ \ z = -\frac{3}{7},\ \ w = \frac{2}{7}$

35.

$$AX = B_1$$
$$X = A^{-1}B_1$$

$$X = \begin{bmatrix} 3 & -2 & 4 \\ 2 & -1 & 0 \\ 0 & 4 & 1 \end{bmatrix} \begin{bmatrix} 1 \\ -1 \\ 5 \end{bmatrix} = \begin{bmatrix} 25 \\ 3 \\ 1 \end{bmatrix}$$

$$AX = B_2$$
$$X = A^{-1}B_2$$
$$X = \begin{bmatrix} 3 & -2 & 4 \\ 2 & -1 & 0 \\ 0 & 4 & 1 \end{bmatrix} \begin{bmatrix} 4 \\ 3 \\ -2 \end{bmatrix} = \begin{bmatrix} -2 \\ 5 \\ 10 \end{bmatrix}$$

37.

We cannot perform any elementary row operations upon rows 1 and 3 that will produce a nonzero pivot element for a_{22}.

CHAPTER 11 SECTION 4

SECTION 11.4 PROGRESS CHECK

1a). (Page 638)

$$\begin{vmatrix} -6 & 2 \\ -1 & -2 \end{vmatrix} = (-6)(-2)-(-1)(2)=14$$

1b).

$$\begin{vmatrix} \frac{1}{2} & \frac{1}{4} \\ -4 & -2 \end{vmatrix} = \left(\frac{1}{2}\right)(-2)-(-4)\left(\frac{1}{4}\right)=0$$

2. (Page 640)

Cofactor of $-9: (-1)^{1+2}\begin{vmatrix} -5 & 0 \\ -3 & -1 \end{vmatrix} = (-1)(5) = -5$

Cofactor of 2: $(-1)^{2+2}\begin{vmatrix} 16 & 3 \\ -3 & -1 \end{vmatrix} = -16-(-9) = -7$

Cofactor of 4: $(-1)^{3+2}\begin{vmatrix} 16 & 3 \\ -5 & 0 \end{vmatrix} = -1[0-(-15)] = -15$

3. (Page 641)

$$6(-1)^{2+1}\begin{vmatrix} 7 & 2 \\ 10 & -3 \end{vmatrix} - 6(-1)^{2+2}\begin{vmatrix} -2 & 2 \\ 4 & -3 \end{vmatrix} + 0(-1)^{2+3}\begin{vmatrix} -2 & 7 \\ 4 & 10 \end{vmatrix}$$
$$= -6(-21-20)-6(6-8)+0$$
$$= 258$$

4. (Page 642)

Expand about row 1

$$a(-1)^{1+1}\begin{vmatrix} b & c \\ e & f \end{vmatrix} + b(-1)^{1+2}\begin{vmatrix} a & c \\ d & f \end{vmatrix} + c(-1)^{1+3}\begin{vmatrix} a & b \\ d & e \end{vmatrix}$$
$$= a(bf-ec)-b(af-cd)+c(ae-bd)$$
$$= abf-aec-baf+bcd+aec-bdc$$
$$= 0$$

5. (Page 643)

Expand about fourth row :

$$1(-1)^{4+1}\begin{vmatrix}-1 & 0 & 2\\ 0 & 4 & 0\\ 5 & 0 & -3\end{vmatrix} + 0 + 1(-1)^{4+3}\begin{vmatrix}0 & -1 & 2\\ 3 & 0 & 0\\ 0 & 5 & -3\end{vmatrix} + 0$$

Expand about 2nd row   Expand about 2nd row

$$= (-1)\left[4(-1)^{2+2}\begin{vmatrix}-1 & 2\\ 5 & -3\end{vmatrix}\right] - 1\left[3(-1)^{2+1}\begin{vmatrix}-1 & 2\\ 5 & -3\end{vmatrix}\right]$$

$$= (-1)[4(3-10)] - 1[-3(3-10)] = 7$$

EXERCISE SET 11.4

1.

$$\begin{vmatrix}2 & -3\\ 4 & 5\end{vmatrix} = (2)(5) - (4)(-3) = 22$$

3.

$$\begin{vmatrix}-4 & 1\\ 0 & 2\end{vmatrix} = (-4)(2) - (0)(1) = -8$$

5.

$$\begin{vmatrix} 0 & 0 \\ 1 & 3 \end{vmatrix} = (0)(3) - (1)(0) = 0$$

7a).

$$\text{Minor of } a_{11} : \begin{vmatrix} 3 & -1 & 2 \\ 4 & 1 & -3 \\ 5 & -2 & 0 \end{vmatrix} = \begin{vmatrix} 1 & -3 \\ -2 & 0 \end{vmatrix}$$

$$= 0 - 6 = -6$$

7b).

$$\text{Minor of } a_{23} : \begin{vmatrix} 3 & -1 & 2 \\ 4 & 1 & -3 \\ 5 & -2 & 0 \end{vmatrix} = \begin{vmatrix} 3 & -1 \\ 5 & -2 \end{vmatrix}$$

$$= -6 - (-5) = -1$$

7c).

$$\text{Minor of } a_{31} : \begin{vmatrix} 3 & -1 & 2 \\ 4 & 1 & -3 \\ 5 & -2 & 0 \end{vmatrix} = \begin{vmatrix} -1 & 2 \\ 1 & -3 \end{vmatrix} = 3 - 2 = 1$$

7d).

$$\text{Minor of } a_{33} : \begin{vmatrix} 3 & -1 & 2 \\ 4 & 1 & -3 \\ 5 & -2 & 0 \end{vmatrix} = \begin{vmatrix} 3 & -1 \\ 4 & 1 \end{vmatrix}$$

$$= 3 - (-4) = 7$$

9a).

$$\text{Cofactor of } a_{11} : (-1)^{1+1} \cdot \text{Minor of } a_{11}$$
$$= (1)(-6) = -6$$

9b).

$$\text{Cofactor of } a_{23} : (-1)^{2+3} \cdot \text{Minor of } a_{23}$$
$$= (-1)(-1) = 1$$

9c).

$$\text{Cofactor of } a_{31} : (-1)^{3+1} \cdot \text{Minor of } a_{31}$$
$$= (1)(1) = 1$$

9d).

$$\text{Cofactor of } a_{33} : (-1)^{3+3} \cdot \text{Minor of } a_{33}$$
$$= (1)(7) = 7$$

11.

Expand about third row :

$$2(-1)^{3+1}\begin{vmatrix} -2 & 5 \\ 2 & 0 \end{vmatrix} + 0 + 4(-1)^{3+3}\begin{vmatrix} 4 & -2 \\ 5 & 2 \end{vmatrix}$$

$$= 2(0 - 10) + 4(8 + 10) = 52$$

13.

Expand about first row :

$$(-1)(-1)^{1+1}\begin{vmatrix}4 & 1\\ 5 & 2\end{vmatrix}+2(-1)^{1+2}\begin{vmatrix}3 & 1\\ 6 & 2\end{vmatrix}+0$$

$$=(-1)(8-5)-2(6-6)=-3$$

15.

Expand about first row :

$$0+(-1)(-1)^{1+2}\begin{vmatrix}0 & 2 & 1\\ 2 & 2 & 3\\ 3 & 1 & 0\end{vmatrix}+0+3(-1)^{1+4}\begin{vmatrix}0 & 1 & 2\\ 2 & -2 & 2\\ 3 & 3 & 1\end{vmatrix}$$

Expand about first row | Expand about first row

$$=2(-1)^{1+2}\begin{vmatrix}2 & 3\\ 3 & 0\end{vmatrix}+1(-1)^{1+3}\begin{vmatrix}2 & 2\\ 3 & 1\end{vmatrix}-3\left[1(-1)^{1+2}\begin{vmatrix}2 & 2\\ 3 & 1\end{vmatrix}+2(-1)^{1+3}\begin{vmatrix}2 & -2\\ 3 & 3\end{vmatrix}\right]$$

$$=-2(0-9)+1(2-6)-3[(-1)(2-6)+2(6+6)]$$
$$=-70$$

17.

Expand about second column :

$$1(-1)^{1+2}\underbrace{\begin{vmatrix}2&3&-5\\1&2&2\\0&1&3\end{vmatrix}}_{\text{Expand about third row}}-1(-1)^{3+2}\underbrace{\begin{vmatrix}2&-3&1\\2&3&-5\\0&1&3\end{vmatrix}}_{\text{Expand about third row}}-1(-1)^{4+2}\underbrace{\begin{vmatrix}2&-3&1\\2&3&-5\\1&2&2\end{vmatrix}}_{\text{Expand about first row}}$$

$$=(-1)\left[1(-1)^{3+2}\begin{vmatrix}2&-5\\1&2\end{vmatrix}+3(-1)^{3+3}\begin{vmatrix}2&3\\1&2\end{vmatrix}\right]+\left[(1)(-1)^{3+2}\begin{vmatrix}2&1\\2&-5\end{vmatrix}+3(-1)^{3+3}\begin{vmatrix}2&-3\\2&3\end{vmatrix}\right]$$

$$(-1)\left[2(-1)^{1+1}\begin{vmatrix}3&-5\\2&2\end{vmatrix}-3(-1)^{1+2}\begin{vmatrix}2&-5\\1&2\end{vmatrix}+1(-1)^{1+3}\begin{vmatrix}2&3\\1&2\end{vmatrix}\right]$$

$$=(-1)[(-1)(4+5)+3(4-3)]+[(-1)(-10-2)+3(6+6)]-[2(6+10)+3(4+5)+1(4-3)]$$

$$=-6$$

19.

Expand about first row :

$$2(-1)^{1+3}\underbrace{\begin{vmatrix}0&1&1\\5&1&3\\3&3&0\end{vmatrix}}_{\text{Expand about first row}}+4(-1)^{1+4}\underbrace{\begin{vmatrix}0&1&2\\5&1&3\\3&3&1\end{vmatrix}}_{\text{Expand about first row}}$$

$$2\left[1(-1)^{1+2}\begin{vmatrix}5&3\\3&0\end{vmatrix}+1(-1)^{1+3}\begin{vmatrix}5&1\\3&3\end{vmatrix}\right]-4\left[1(-1)^{1+2}\begin{vmatrix}5&3\\3&1\end{vmatrix}+2(-1)^{1+3}\begin{vmatrix}5&1\\3&3\end{vmatrix}\right]$$

$$=2[(-1)(0-9)+(15-3)]-4[(-1)(5-9)+2(15-3)]$$

$$=-70$$

CHAPTER 11 SECTION 5

SECTION 11.5 PROGRESS CHECK

1. (Page 647)

$$\begin{array}{r} \\ 4R_4+R_2\to R_2 \\ R_3+R_4\to R_3 \\ \\ \end{array}\begin{vmatrix} 4 & 0 & 0 & 3 \\ 2 & 4 & 5 & 8 \\ -2 & 1 & 0 & 2 \\ -4 & -1 & -2 & -3 \end{vmatrix}$$

$$= \begin{vmatrix} 4 & 0 & 0 & 3 \\ -14 & 0 & -3 & -4 \\ -6 & 0 & -2 & -1 \\ -4 & -1 & -2 & -3 \end{vmatrix}$$

Expand about second column :

$$(-1)(-1)^{4+2}\begin{vmatrix} 4 & 0 & 3 \\ -14 & -3 & -4 \\ -6 & -2 & -1 \end{vmatrix} \begin{array}{l} 3R_3+R_1\to R_1 \\ -4R_3+R_2\to R_2 \\ \\ \end{array}$$

$$= -1\begin{vmatrix} -14 & -6 & 0 \\ 10 & 5 & 0 \\ -6 & -2 & -1 \end{vmatrix}$$

Expand about third column :

$$=-1\left[(-1)(-1)^{1+3}\begin{vmatrix} -14 & -6 \\ 10 & 5 \end{vmatrix}\right]$$

$$=-1[(-1)(-70+60)]=-10$$

EXERCISE SET 11.5

1.

$$\begin{matrix} -2R_3+R_1\to R_1 \\ -3R_3+R_2\to R_2 \\ \end{matrix}\begin{vmatrix} 2 & 2 & 4 \\ 3 & 8 & 1 \\ 1 & 1 & 2 \end{vmatrix} = \begin{vmatrix} 0 & 0 & 0 \\ 0 & 5 & -5 \\ 1 & 1 & 2 \end{vmatrix} = 0$$

3.

$$-\frac{3}{2}R_3+R_4\to R_4 \begin{vmatrix} 3 & 2 & 1 & 0 \\ -1 & -3 & -1 & 0 \\ 0 & 0 & 2 & 2 \\ 4 & 1 & 3 & 3 \end{vmatrix}$$

$$= \begin{vmatrix} 3 & 2 & 1 & 0 \\ -1 & -3 & -1 & 0 \\ 0 & 0 & 2 & 2 \\ 4 & 1 & 0 & 0 \end{vmatrix}$$

$$= 2(-1)^{3+4} \begin{vmatrix} 3 & 2 & 1 \\ -1 & -3 & -1 \\ 4 & 1 & 0 \end{vmatrix} \quad R^1+R^2\to R_2$$

$$= -2\begin{vmatrix} 3 & 2 & 1 \\ 2 & -1 & 0 \\ 4 & 1 & 0 \end{vmatrix}$$

$$= -2(1)(-1)^{1+3}\begin{vmatrix} 2 & -1 \\ 4 & 1 \end{vmatrix}$$

$$= -2(2+4) = -12$$

5.

$$\begin{vmatrix} 2 & -3 & 2 & -4 \\ 0 & 4 & -1 & 9 \\ 0 & 1 & 2 & 0 \\ 0 & 1 & 3 & -1 \end{vmatrix} \underset{9R_3 + R_1 \to R_1}{=} 2(-1)^{1+1} \begin{vmatrix} 4 & -1 & 9 \\ 1 & 2 & 0 \\ 1 & 3 & -1 \end{vmatrix}$$

$$= 2\begin{vmatrix} 13 & 26 & 0 \\ 1 & 2 & 0 \\ 1 & 3 & -1 \end{vmatrix}$$

$$= 2(-1)(-1)^{3+3}\begin{vmatrix} 13 & 26 \\ 1 & 2 \end{vmatrix}$$

$$= -2(26 - 26) = 0$$

7.

$$\begin{vmatrix} a_1 + b_1 & a_2 + b_2 \\ c & d \end{vmatrix} = d(a_1 + b_1) - c(a_2 + b_2)$$

$$= a_1 d + b_1 d - a_2 c - b_2 c$$

$$\begin{vmatrix} a_1 & a_2 \\ c & d \end{vmatrix} = a_1 d - a_2 c \qquad \begin{vmatrix} b_1 & b_2 \\ c & d \end{vmatrix} = b_1 d - b_2 c$$

$$\begin{vmatrix} a_1 & a_2 \\ c & d \end{vmatrix} + \begin{vmatrix} b_1 & b_2 \\ c & d \end{vmatrix} = a_1 d - a_2 c + b_1 d - b_2 c$$

Hence $\begin{vmatrix} a_1 + b_1 & a_2 + b_2 \\ c & d \end{vmatrix} = \begin{vmatrix} a_1 & a_2 \\ c & d \end{vmatrix} + \begin{vmatrix} b_1 & b_2 \\ c & d \end{vmatrix}$

9.

$$A = \begin{bmatrix} a_{11} & a_{12} & \cdots & a_{1n} \\ a_{21} & a_{22} & \cdots & a_{2n} \\ \vdots & & & \vdots \\ a_{i1} & a_{i2} & \cdots & a_{in} \\ \vdots & & & \vdots \\ a_{n1} & a_{n2} & \cdots & a_{nn} \end{bmatrix}$$

$$B = \begin{bmatrix} a_{11} & a_{12} & \cdots & a_{1n} \\ a_{21} & a_{22} & \cdots & a_{2n} \\ \vdots & & & \vdots \\ ka_{i1} & ka_{i2} & \cdots & ka_{in} \\ \vdots & & & \vdots \\ a_{n1} & a_{n2} & \cdots & a_{nn} \end{bmatrix}$$

$$|B| = ka_{i1} \cdot \text{ cofactor } b_{i1} + ka_{i2} \cdot \text{ cofactor } b_{i2} + \cdots + ka_{in} \cdot \text{ cofactor } b_{in}$$

$$= k[a_{i1} \cdot \text{ cofactor } b_{i1} + a_{i2} \cdot \text{ cofactor } b_{i2} + \cdots + a_{in} \cdot \text{ cofactor } b_{in}] = k|A|$$

Note : Cofactor a_{ij} = cofactor b_{ij} since when you cross out the $i^{\underline{th}}$ row and $j^{\underline{th}}$ column to find the corresponding minor, matrix A is identical to matrix B.

11.

From Exercise 9, if one row of A is multiplied by k to obtain matrix B then $|B| = k|A|$. If each of the n rows of A is multiplied by k to obtain matrix B then $|B| = \underbrace{k \cdot k \cdots k}_{n \text{ times}}|A| = k^n|A|$.

CHAPTER 11 SECTION 6

SECTION 11.6 PROGRESS CHECK

1. (Page 650)

$$2x+3y=-4$$
$$3x+4y=-7$$

$$|A|=\begin{vmatrix}2 & 3\\ 3 & 4\end{vmatrix}=8-9=-1$$

$$|A_1|=\begin{vmatrix}-4 & 3\\ -7 & 4\end{vmatrix}=-16+21=5$$

$$|A_2|=\begin{vmatrix}2 & -4\\ 3 & -7\end{vmatrix}=-14+12=-2$$

$$x=\frac{|A_1|}{|A|}=\frac{5}{-1}=-5 \qquad y=\frac{|A_2|}{|A|}=\frac{-2}{-1}=2$$

2. (Page 651)

$$\begin{aligned}3x \qquad\quad -z&=1\\ -6x+2y \qquad &=-5\\ -4y+3z&=5\end{aligned}$$

$$|A|=\begin{vmatrix}3 & 0 & -1\\ -6 & 2 & 0\\ 0 & -4 & 3\end{vmatrix}=-6$$

$$|A_1|=\begin{vmatrix}1 & 0 & -1\\ -5 & 2 & 0\\ 5 & -4 & 3\end{vmatrix}=-4$$

$$|A_2|=\begin{vmatrix}3 & 1 & -1\\ -6 & -5 & 0\\ 0 & 5 & 3\end{vmatrix}=3$$

$$|A_3|=\begin{vmatrix}3 & 0 & 1\\ -6 & 2 & -5\\ 0 & -4 & 5\end{vmatrix}=-6$$

$$x=\frac{|A_1|}{|A|}=\frac{-4}{-6}=\frac{2}{3}$$

$$y=\frac{|A_2|}{|A|}=\frac{3}{-6}=-\frac{1}{2}$$

$$z=\frac{|A_3|}{|A|}=\frac{-6}{-6}=1$$

EXERCISE SET 11.6

1.

$$\begin{aligned} 2x + y + z &= -1 \\ 2x - y + 2z &= 2 \\ x + 2y + z &= -4 \end{aligned}$$

$$|A| = \begin{vmatrix} 2 & 1 & 1 \\ 2 & -1 & 2 \\ 1 & 2 & 1 \end{vmatrix} = -5 \qquad |A_2| = \begin{vmatrix} 2 & -1 & 1 \\ 2 & 2 & 2 \\ 1 & -4 & 1 \end{vmatrix} = 10$$

$$|A_1| = \begin{vmatrix} -1 & 1 & 1 \\ 2 & -1 & 2 \\ -4 & 2 & 1 \end{vmatrix} = -5 \qquad |A_3| = \begin{vmatrix} 2 & 1 & -1 \\ 2 & -1 & 2 \\ 1 & 2 & -4 \end{vmatrix} = 5$$

$$x = \frac{|A_1|}{|A|} = \frac{-5}{-5} = 1; \quad y = \frac{|A_2|}{|A|} = \frac{10}{-5} = -2; \quad z = \frac{|A_3|}{|A|} = \frac{5}{-5} = -1$$

3.

$$\begin{aligned} 2x + y - z &= 9 \\ x - 2y + 2z &= -3 \\ 3x + 3y + 4z &= 11 \end{aligned}$$

$$|A| = \begin{vmatrix} 2 & 1 & -1 \\ 1 & -2 & 2 \\ 3 & 3 & 4 \end{vmatrix} = -35 \qquad |A_2| = \begin{vmatrix} 2 & 9 & -1 \\ 1 & -3 & 2 \\ 3 & 11 & 4 \end{vmatrix} = -70$$

$$|A_1| = \begin{vmatrix} 9 & 1 & -1 \\ -3 & -2 & 2 \\ 11 & 3 & 4 \end{vmatrix} = -105 \qquad |A_3| = \begin{vmatrix} 2 & 1 & 9 \\ 1 & -2 & -3 \\ 3 & 3 & 11 \end{vmatrix} = 35$$

$$x\frac{|A_1|}{|A|} = \frac{-105}{-35} = 3; \quad y = \frac{|A_2|}{|A|} = \frac{-70}{-35} = 2; \quad z = \frac{|A_3|}{|A|} = \frac{35}{-35} = -1$$

5.

$$\begin{aligned} -x \ - y + 2z &= 7 \\ x + 2y - 2z &= -7 \\ 2x \ - y \ + z &= -4 \end{aligned}$$

$$|A| = \begin{vmatrix} -1 & -1 & 2 \\ 1 & 2 & -2 \\ 2 & -1 & 1 \end{vmatrix} = -5 \qquad |A_2| = \begin{vmatrix} -1 & 7 & 2 \\ 1 & -7 & -2 \\ 2 & -4 & 1 \end{vmatrix} = 0$$

$$|A_1| = \begin{vmatrix} 7 & -1 & 2 \\ -7 & 2 & -2 \\ -4 & -1 & 1 \end{vmatrix} = 15 \qquad |A_3| = \begin{vmatrix} -1 & -1 & 7 \\ 1 & 2 & -7 \\ 2 & -1 & -4 \end{vmatrix} = -10$$

$$x = \frac{|A_1|}{|A|} = \frac{15}{-5} = -3; \quad y = \frac{|A_2|}{|A|} = \frac{0}{-5} = 0; \quad z = \frac{|A_3|}{|A|} = \frac{-10}{-5} = 2$$

7.

$$\begin{aligned} x+y\quad -z+2w &= 0 \\ 2x\ +y\qquad -w &= -2 \\ 3x\qquad +2z\qquad &= -3 \\ -x+2y\qquad +3w &= 1 \end{aligned}$$

$$|A| = \begin{vmatrix} 1 & 1 & -1 & 2 \\ 2 & 1 & 0 & -1 \\ 3 & 0 & 2 & 0 \\ -1 & 2 & 0 & 3 \end{vmatrix} = 35$$

$$|A_1| = \begin{vmatrix} 0 & 1 & -1 & 2 \\ -2 & 1 & 0 & -1 \\ -3 & 0 & 2 & 0 \\ 1 & 2 & 0 & 3 \end{vmatrix} = -25$$

$$|A_2| = \begin{vmatrix} 1 & 0 & -1 & 2 \\ 2 & -2 & 0 & -1 \\ 3 & -3 & 2 & 0 \\ -1 & 1 & 0 & 3 \end{vmatrix} = -10$$

$$|A_3| = \begin{vmatrix} 1 & 1 & 0 & 2 \\ 2 & 1 & -2 & -1 \\ 3 & 0 & -3 & 0 \\ -1 & 2 & 1 & 3 \end{vmatrix} = -15$$

$$|A_4| = \begin{vmatrix} 1 & 1 & -1 & 0 \\ 2 & 1 & 0 & -2 \\ 3 & 0 & 2 & -3 \\ -1 & 2 & 0 & 1 \end{vmatrix} = 10$$

$$x = \frac{|A_1|}{|A|} = \frac{-25}{35} = -\frac{5}{7};\ y = \frac{|A_2|}{|A|} = \frac{-10}{35} = -\frac{2}{7};$$

$$z = \frac{|A_3|}{|A|} = \frac{-15}{35} = -\frac{3}{7};\ w = \frac{|A_4|}{|A|} = \frac{10}{35} = \frac{2}{7}$$

CHAPTER 11 REVIEW EXERCISES

1.

3 rows, 5 columns : 3×5

2.

a_{24}, element in second row, fourth column : -1

3.

a_{31}, element in third row, first column : 4

4.

a_{15}, element in first row, fifth column : 8

5.

$$\begin{bmatrix} 3 & -7 \\ 1 & 4 \end{bmatrix}$$

6.

$$\left[\begin{array}{cc|c} 3 & -7 & 14 \\ 1 & 4 & 6 \end{array}\right]$$

7.

$$\begin{aligned} 4x - y &= 3 \\ 2x + 5y &= 0 \end{aligned}$$

8.

$$\begin{aligned} -2x + 4y + 5z &= 0 \\ 6x - 9y + 4z &= 0 \\ 3x + 2y - z &= 0 \end{aligned}$$

9.

$$\left[\begin{array}{cc|c} 1 & -2 & 7 \\ 0 & 1 & -4 \end{array}\right]$$

$$\begin{aligned} y &= -4 \\ x - 2(-4) &= 7 \\ x &= -1 \end{aligned}$$

$x = -1,\ y = -4$

10.

$$\left[\begin{array}{cc|c} 1 & 2 & \frac{21}{2} \\ 0 & 1 & 5 \end{array}\right]$$

$$\begin{aligned} y &= 5 \\ x + 2(5) &= \frac{21}{2} \\ x &= \frac{1}{2} \end{aligned}$$

$x = \frac{1}{2},\ y = 5$

11.

$$\left[\begin{array}{ccc|c} 1 & -4 & 2 & -18 \\ 0 & 1 & -2 & 5 \\ 0 & 0 & 1 & -1 \end{array}\right]$$

$$\begin{aligned} z &= -1 \\ y-2(-1) &= 5 \\ y &= 3 \\ x-4(3)+2(-1) &= -18 \\ x &= -4 \end{aligned}$$

$x=-4, \quad y=3, \quad z=-1$

12.

$$\left[\begin{array}{ccc|c} 1 & -2 & 2 & -9 \\ 0 & 1 & 3 & -8 \\ 0 & 0 & 1 & -3 \end{array}\right]$$

$$\begin{aligned} z &= -3 \\ y+3(-3) &= -8 \\ y &= 1 \\ x-2(1)+2(-3) &= -9 \\ x &= -1 \end{aligned}$$

$x=-1, \quad y=1, \quad z=-3$

13.

$$-2R_1+R_2 \to R_2 \left[\begin{array}{cc|c} 1 & 1 & 2 \\ 2 & -4 & -5 \end{array}\right]$$

$$-\frac{1}{6}R_2 \to R_2 \left[\begin{array}{cc|c} 1 & 1 & 2 \\ 0 & -6 & -9 \end{array}\right]$$

$$R_1-R_2 \to R_1 \left[\begin{array}{cc|c} 1 & 1 & 2 \\ 0 & 1 & \frac{3}{2} \end{array}\right]$$

$$\left[\begin{array}{cc|c} 1 & 0 & \frac{1}{2} \\ 0 & 1 & \frac{3}{2} \end{array}\right]$$

$$x=\frac{1}{2},\ y=\frac{3}{2}$$

14.

$$R_1-R_2 \to R_1 \left[\begin{array}{cc|c} 3 & -1 & -17 \\ 2 & 3 & -4 \end{array}\right]$$

$$-2R_1+R_2 \to R_2 \left[\begin{array}{cc|c} 1 & -4 & -13 \\ 2 & 3 & -4 \end{array}\right]$$

$$\frac{1}{11}R_2 \to R_2 \left[\begin{array}{cc|c} 1 & -4 & -13 \\ 0 & 11 & 22 \end{array}\right]$$

$$4R_2+R_1 \to R_1 \left[\begin{array}{cc|c} 1 & -4 & -13 \\ 0 & 1 & 2 \end{array}\right]$$

$$\left[\begin{array}{cc|c} 1 & 0 & -5 \\ 0 & 1 & 2 \end{array}\right]$$

$x=-5, \quad y=2$

15.

$$\begin{array}{l} 2R_1+R_2\to R_2 \\ R_2+R_3\to R_3 \end{array}\left[\begin{array}{ccc|c} 1 & 3 & 2 & 0 \\ -2 & 0 & 3 & -12 \\ 2 & -6 & -1 & 6 \end{array}\right]$$

$$\begin{array}{l} \frac{1}{6}R_2\to R_2 \\ R_2+R_3\to R_3 \end{array}\left[\begin{array}{ccc|c} 1 & 3 & 2 & 0 \\ 0 & 6 & 7 & -12 \\ 0 & -6 & 2 & -6 \end{array}\right]$$

$$\begin{array}{l} -3R_2+R_1\to R_1 \\ \frac{1}{9}R_3\to R_3 \end{array}\left[\begin{array}{ccc|c} 1 & 3 & 2 & 0 \\ 0 & 1 & \frac{7}{6} & -2 \\ 0 & 0 & 9 & -18 \end{array}\right]$$

15. (Continued)

$$\begin{array}{l} \frac{3}{2}R_3+R_1\to R_1 \\ -\frac{7}{6}R_3+R_2\to R_2 \end{array}\left[\begin{array}{ccc|c} 1 & 0 & -\frac{3}{2} & 6 \\ 0 & 1 & \frac{7}{6} & -2 \\ 0 & 0 & 1 & -2 \end{array}\right]$$

$$\left[\begin{array}{ccc|c} 1 & 0 & 0 & 3 \\ 0 & 1 & 0 & \frac{1}{3} \\ 0 & 0 & 1 & -2 \end{array}\right]$$

$x=3,\quad y=\frac{1}{3},\quad z=-2$

16.

$$\begin{array}{l}\frac{1}{2}R_1 \to R_1 \\ R_1 + R_2 \to R_2\end{array}\left[\begin{array}{ccc|c} 2 & -1 & -2 & 3 \\ -2 & 3 & 1 & 3 \\ 0 & 2 & -1 & 6 \end{array}\right]$$

$$R_2 - R_3 \to R_3 \left[\begin{array}{ccc|c} 1 & -\frac{1}{2} & -1 & \frac{3}{2} \\ 0 & 2 & -1 & 6 \\ 0 & 2 & -1 & 6 \end{array}\right]$$

$$\frac{1}{2}R_2 \to R_2 \left[\begin{array}{ccc|c} 1 & -\frac{1}{2} & -1 & \frac{3}{2} \\ 0 & 2 & -1 & 6 \\ 0 & 0 & 0 & 0 \end{array}\right]$$

$$\frac{1}{2}R_2 + R_1 \to R_1 \left[\begin{array}{ccc|c} 1 & -\frac{1}{2} & -1 & \frac{3}{2} \\ 0 & 1 & -\frac{1}{2} & 3 \\ 0 & 0 & 0 & 0 \end{array}\right]$$

16. (Continued)

$$\left[\begin{array}{ccc|c} 1 & 0 & -\frac{5}{4} & 3 \\ 0 & 1 & -\frac{1}{2} & 3 \\ 0 & 0 & 0 & 0 \end{array}\right]$$

$$y - \frac{1}{2}z = 3$$
$$y = \frac{1}{2}z + 3$$
$$x - \frac{5}{4}z = 3$$
$$x = \frac{5}{4}z + 3$$

$x = \frac{5}{4}t + 3,\quad y = \frac{1}{2}t + 3,\quad z = t,$

t is any real number

17.

$$\begin{bmatrix} 5 & -1 \\ 3 & 2x \end{bmatrix} = \begin{bmatrix} 5 & -1 \\ 3 & -6 \end{bmatrix}$$

$$2x = -6$$
$$x = -3$$

18.

$$\begin{bmatrix} 6 & x^2 \\ 4x & -2 \end{bmatrix} = \begin{bmatrix} 6 & 9 \\ -12 & -2 \end{bmatrix}$$

$x^2 = 9 \qquad 4x = -12$

$x = \pm 3 \qquad x = -3$

Use the value of x that both equations have in common : $x = -3$

19.

$$A+B=\begin{bmatrix}2 & -1\\3 & 2\end{bmatrix}+\begin{bmatrix}-1 & 5\\4 & -3\end{bmatrix}=\begin{bmatrix}1 & 4\\7 & -1\end{bmatrix}$$

20.

$$B-A=\begin{bmatrix}-1 & 5\\4 & -3\end{bmatrix}-\begin{bmatrix}2 & -1\\3 & 2\end{bmatrix}=\begin{bmatrix}-3 & 6\\1 & -5\end{bmatrix}$$

21.

$A+C$ $\quad A:2\times 2$, $C:3\times 2$ $\neq$

$A+C$: not possible

22.

$$5D=5\begin{bmatrix}1 & 3 & 4\\-1 & 0 & -6\end{bmatrix}=\begin{bmatrix}5 & 15 & 20\\-5 & 0 & -30\end{bmatrix}$$

23.

$$CD=\begin{bmatrix}-1 & 0\\0 & 4\\2 & -2\end{bmatrix}\begin{bmatrix}1 & 3 & 4\\-1 & 0 & -6\end{bmatrix}$$

$$=\begin{bmatrix}(-1)(1)+(0)(-1) & (-1)(3)+(0)(0) & (-1)(4)+0(-6)\\(0)(1)+(4)(-1) & (0)(3)+(4)(0) & (0)(4)+(4)(-6)\\(2)(1)+(-2)(-1) & (2)(3)+(-2)(0) & (2)(4)+(-2)(-6)\end{bmatrix}$$

$$=\begin{bmatrix}-1 & -3 & -4\\-4 & 0 & -24\\4 & 6 & 20\end{bmatrix}$$

24.

$$DC=\begin{bmatrix}1 & 3 & 4\\-1 & 0 & -6\end{bmatrix}\begin{bmatrix}-1 & 0\\0 & 4\\2 & -2\end{bmatrix}$$

$$=\begin{bmatrix}(1)(-1)+3(0)+4(2) & (1)(0)+(3)(4)+(4)(-2)\\(-1)(-1)+(0)(0)+(-6)(2) & (-1)(0)+(0)(4)+(-6)(-2)\end{bmatrix}$$

$$=\begin{bmatrix}7 & 4\\-11 & 12\end{bmatrix}$$

25.

$BC \quad B:2\times 2$

$C:3\times 2$

BC: not possible

26.

$$CB=\begin{bmatrix}-1 & 0\\ 0 & 4\\ 2 & -2\end{bmatrix}\begin{bmatrix}-1 & 5\\ 4 & -3\end{bmatrix}$$

$$=\begin{bmatrix}(-1)(-1)+(0)(4) & (-1)(5)+(0)(-3)\\ (0)(-1)+(4)(4) & (0)(5)+(4)(-3)\\ (2)(-1)+(-2)(4) & (2)(5)+(-2)(-3)\end{bmatrix}$$

$$=\begin{bmatrix}1 & -5\\ 16 & -12\\ -10 & 16\end{bmatrix}$$

27.

$$A+2B=\begin{bmatrix}2 & -1\\ 3 & 2\end{bmatrix}+2\begin{bmatrix}-1 & 5\\ 4 & -3\end{bmatrix}$$

$$=\begin{bmatrix}2 & -1\\ 3 & 2\end{bmatrix}+\begin{bmatrix}-2 & 10\\ 8 & -6\end{bmatrix}$$

$$=\begin{bmatrix}0 & 9\\ 11 & -4\end{bmatrix}$$

28.

$$-AB=-\begin{bmatrix}2 & -1\\ 3 & 2\end{bmatrix}\begin{bmatrix}-1 & 5\\ 4 & -3\end{bmatrix}$$

$$=\begin{bmatrix}-2 & 1\\ -3 & -2\end{bmatrix}\begin{bmatrix}-1 & 5\\ 4 & -3\end{bmatrix}$$

$$=\begin{bmatrix}(-2)(-1)+(1)(4) & (-2)(5)+(1)(-3)\\ (-3)(-1)+(-2)(4) & (-3)(5)+(-2)(-3)\end{bmatrix}$$

$$=\begin{bmatrix}6 & -13\\ -5 & -9\end{bmatrix}$$

29.

$$R_1 \leftrightarrow R_2 \left[\begin{array}{cc|cc} -2 & 3 & 1 & 0 \\ 1 & 4 & 0 & 1 \end{array}\right]$$

$$2R_1 + R_2 \rightarrow R_2 \left[\begin{array}{cc|cc} 1 & 4 & 0 & 1 \\ -2 & 3 & 1 & 0 \end{array}\right]$$

$$\frac{1}{11}R_2 \rightarrow R_2 \left[\begin{array}{cc|cc} 1 & 4 & 0 & 1 \\ 0 & 11 & 1 & 2 \end{array}\right]$$

$$-4R_2 + R_1 \rightarrow R_1 \left[\begin{array}{cc|cc} 1 & 4 & 0 & 1 \\ 0 & 1 & \frac{1}{11} & \frac{2}{11} \end{array}\right]$$

$$\left[\begin{array}{cc|cc} 1 & 0 & -\frac{4}{11} & \frac{3}{11} \\ 0 & 1 & \frac{1}{11} & \frac{2}{11} \end{array}\right]$$

$$\begin{bmatrix} -\frac{4}{11} & \frac{3}{11} \\ \frac{1}{11} & \frac{2}{11} \end{bmatrix}$$

30.

$$\begin{array}{r} \\ 5R_1+R_2\to R_2 \\ -4R_1+R_3\to R_3 \end{array}\left[\begin{array}{ccc|ccc} 1 & 1 & -4 & 1 & 0 & 0 \\ -5 & -2 & 0 & 0 & 1 & 0 \\ 4 & 2 & -1 & 0 & 0 & 1 \end{array}\right]$$

$$R_2+R_3\to R_2\left[\begin{array}{ccc|ccc} 1 & 1 & -4 & 1 & 0 & 0 \\ 0 & 3 & -20 & 5 & 1 & 0 \\ 0 & -2 & 15 & -4 & 0 & 1 \end{array}\right]$$

$$\begin{array}{r} R_1-R_2\to R_1 \\ 2R_2+R_3\to R_3 \\ \\ \end{array}\left[\begin{array}{ccc|ccc} 1 & 1 & -4 & 1 & 0 & 0 \\ 0 & 1 & -5 & 1 & 1 & 1 \\ 0 & -2 & 15 & -4 & 0 & 1 \end{array}\right]$$

$$\begin{array}{r} \\ R_2+R_3\to R_2 \\ \frac{1}{5}R_3\to R_3 \end{array}\left[\begin{array}{ccc|ccc} 1 & 0 & 1 & 0 & -1 & -1 \\ 0 & 1 & -5 & 1 & 1 & 1 \\ 0 & 0 & 5 & -2 & 2 & 3 \end{array}\right]$$

30. (Continued)

$$R_1-R_3\to R_1\left[\begin{array}{ccc|ccc} 1 & 0 & 1 & 0 & -1 & -1 \\ 0 & 1 & 0 & -1 & 3 & 4 \\ 0 & 0 & 1 & -\frac{2}{5} & \frac{2}{5} & \frac{3}{5} \end{array}\right]$$

$$\left[\begin{array}{ccc|ccc} 1 & 0 & 0 & \frac{2}{5} & -\frac{7}{5} & -\frac{8}{5} \\ 0 & 1 & 0 & -1 & 3 & 4 \\ 0 & 0 & 1 & -\frac{2}{5} & \frac{2}{5} & \frac{3}{5} \end{array}\right]$$

$$\begin{bmatrix} \frac{2}{5} & -\frac{7}{5} & -\frac{8}{5} \\ -1 & 3 & 4 \\ -\frac{2}{5} & \frac{2}{5} & \frac{3}{5} \end{bmatrix}$$

31.

$$2x - y = 1$$
$$x + y = 5$$

$$R_1 - R_2 \to R_1 \left[\begin{array}{cc|cc} 2 & -1 & 1 & 0 \\ 1 & 1 & 0 & 1 \end{array}\right]$$

$$R_1 - R_2 \to R_2 \left[\begin{array}{cc|cc} 1 & -2 & 1 & -1 \\ 1 & 1 & 0 & 1 \end{array}\right]$$

$$-\frac{1}{3}R_2 \to R_2 \left[\begin{array}{cc|cc} 1 & -2 & 1 & -1 \\ 0 & -3 & 1 & -2 \end{array}\right]$$

31. (Continued)

$$2R_2 + R_1 \to R_1 \left[\begin{array}{cc|cc} 1 & -2 & 1 & -1 \\ 0 & 1 & -\frac{1}{3} & \frac{2}{3} \end{array}\right]$$

$$\left[\begin{array}{cc|cc} 1 & 0 & \frac{1}{3} & \frac{1}{3} \\ 0 & 1 & -\frac{1}{3} & \frac{2}{3} \end{array}\right]$$

$$X = A^{-1}B = \begin{bmatrix} \frac{1}{3} & \frac{1}{3} \\ -\frac{1}{3} & \frac{2}{3} \end{bmatrix} \begin{bmatrix} 1 \\ 5 \end{bmatrix} = \begin{bmatrix} 2 \\ 3 \end{bmatrix}$$

$x = 2, \ y = 3$

32.

$$\begin{aligned} x+2y-2z&=-4\\ 3x-y\quad&=-2\\ y+4z&=1 \end{aligned}$$

$$-3R_1+R_2\to R_2\left[\begin{array}{ccc|ccc}1&2&-2&1&0&0\\3&-1&0&0&1&0\\0&1&4&0&0&1\end{array}\right]$$

$$\begin{array}{r}-2R_3+R_1\to R_1\\ R_2\leftrightarrow R_3\end{array}\left[\begin{array}{ccc|ccc}1&2&-2&1&0&0\\0&-7&6&-3&1&0\\0&1&4&0&0&1\end{array}\right]$$

$$7R_2+R_3\to R_3\left[\begin{array}{ccc|ccc}1&0&-10&1&0&-2\\0&1&4&0&0&1\\0&-7&6&-3&1&0\end{array}\right]$$

$$\frac{1}{34}R_3\to R_3\left[\begin{array}{ccc|ccc}1&0&-10&1&0&-2\\0&1&4&0&0&1\\0&0&34&-3&1&7\end{array}\right]$$

32. (Continued)

$$\begin{array}{r}10R_3+R_1\to R_1\\ -4R_3+R_2\to R_2\end{array}\left[\begin{array}{ccc|ccc}1&0&-10&1&0&-2\\0&1&4&0&0&1\\0&0&1&-\frac{3}{34}&\frac{1}{34}&\frac{7}{34}\end{array}\right]$$

$$\left[\begin{array}{ccc|ccc}1&0&0&\frac{2}{17}&\frac{5}{17}&\frac{1}{17}\\0&1&0&\frac{6}{17}&-\frac{2}{17}&\frac{3}{17}\\0&0&1&-\frac{3}{34}&\frac{1}{34}&\frac{7}{34}\end{array}\right]$$

$$X=A^{-1}B=\begin{bmatrix}\frac{2}{17}&\frac{5}{17}&\frac{1}{17}\\\frac{6}{17}&-\frac{2}{17}&\frac{3}{17}\\-\frac{3}{34}&\frac{1}{34}&\frac{7}{34}\end{bmatrix}\begin{bmatrix}-4\\-2\\1\end{bmatrix}=\begin{bmatrix}-1\\-1\\\frac{1}{2}\end{bmatrix}$$

$$x=-1,\quad y=-1,\quad z=\frac{1}{2}$$

33.

$$\begin{vmatrix}3&1\\-4&2\end{vmatrix}=(3)(2)-(-4)(1)=10$$

34.

$$\begin{vmatrix}-1&2\\0&6\end{vmatrix}=(-1)(6)-(0)(2)=-6$$

35.

$$\begin{vmatrix}2&-1\\6&-3\end{vmatrix}=(2)(-3)-(6)(-1)=0$$

36.

$$\begin{vmatrix}1&0&-1\\2&3&-5\\0&4&0\end{vmatrix}=4(-1)^{3+2}\begin{vmatrix}1&-1\\2&-5\end{vmatrix}=-4(-5+2)=12$$

37.

$$-2R_1+R_3\to R_3\begin{vmatrix}1&-1&2\\0&5&4\\2&3&8\end{vmatrix}=\begin{vmatrix}1&-1&2\\0&5&4\\0&5&4\end{vmatrix}$$

$$=1(-1)^{1+1}\begin{vmatrix}5&4\\5&4\end{vmatrix}$$

$$=20-20=0$$

38.

$$\begin{vmatrix}1&2&-1\\0&3&4\\0&0&-1\end{vmatrix}=1(-1)^{1+1}\begin{vmatrix}3&4\\0&-1\end{vmatrix}=-3-0=-3$$

39.

$$2x-y=-3$$
$$-2x+3y=11$$

$$|A|=\begin{vmatrix}2&-1\\-2&3\end{vmatrix}=4\qquad |A_2|=\begin{vmatrix}2&-3\\-2&11\end{vmatrix}=16$$

$$|A_1|=\begin{vmatrix}-3&-1\\11&3\end{vmatrix}=2$$

$$x=\frac{|A_1|}{|A|}=\frac{2}{4}=\frac{1}{2}\qquad y=\frac{|A_2|}{|A|}=\frac{16}{4}=4$$

40.

$$3x-y=7$$
$$2x+5y=-18$$

$$|A|=\begin{vmatrix}3&-1\\2&5\end{vmatrix}=17\qquad |A_2|=\begin{vmatrix}3&7\\2&-18\end{vmatrix}=-68$$

$$|A_1|=\begin{vmatrix}7&-1\\-18&5\end{vmatrix}=17$$

$$x=\frac{|A_1|}{|A|}=\frac{17}{17}=1;\qquad y=\frac{|A_2|}{|A|}=\frac{-68}{17}=-4$$

41.

$$x+2y=2$$
$$2x-7y=48$$

$$|A|=\begin{vmatrix}1&2\\2&-7\end{vmatrix}=-11\qquad |A_2|=\begin{vmatrix}1&2\\2&48\end{vmatrix}=44$$

$$|A_1|=\begin{vmatrix}2&2\\48&-7\end{vmatrix}=-110$$

$$x=\frac{|A_1|}{|A|}=\frac{-110}{-11}=10;\quad y=\frac{|A_2|}{|A|}=\frac{44}{-11}=-4$$

42.

$$\begin{aligned} 2x+3y\ -z&=-3\\ -3x\qquad +4z&=16\\ 2y+5z&=9 \end{aligned}$$

$$|A|=\begin{vmatrix} 2 & 3 & -1\\ -3 & 0 & 4\\ 0 & 2 & 5 \end{vmatrix}=35 \qquad |A_2|=\begin{vmatrix} 2 & -3 & -1\\ -3 & 16 & 4\\ 0 & 9 & 5 \end{vmatrix}=70$$

$$|A_1|=\begin{vmatrix} -3 & 3 & -1\\ 16 & 0 & 4\\ 9 & 2 & 5 \end{vmatrix}=-140 \qquad |A_3|=\begin{vmatrix} 2 & 3 & -3\\ -3 & 0 & 16\\ 0 & 2 & 9 \end{vmatrix}=35$$

$$x=\frac{|A_1|}{|A|}=\frac{-140}{35}=-4;\quad y=\frac{|A_2|}{|A|}=\frac{70}{35}=2;\quad z=\frac{|A_3|}{|A|}=\frac{35}{35}=1$$

43.

$$\begin{aligned} 3x\qquad +z&=0\\ x\ +y\ +z&=0\\ -3y+2z&=-4 \end{aligned}$$

$$|A|=\begin{vmatrix} 3 & 0 & 1\\ 1 & 1 & 1\\ 0 & -3 & 2 \end{vmatrix}=12 \qquad |A_2|=\begin{vmatrix} 3 & 0 & 1\\ 1 & 0 & 1\\ 0 & -4 & 2 \end{vmatrix}=8$$

$$|A_1|=\begin{vmatrix} 0 & 0 & 1\\ 0 & 1 & 1\\ -4 & -3 & 2 \end{vmatrix}=4 \qquad |A_3|=\begin{vmatrix} 3 & 0 & 0\\ 1 & 1 & 0\\ 0 & -3 & -4 \end{vmatrix}=-12$$

$$x=\frac{|A_1|}{|A|}=\frac{4}{12}=\frac{1}{3};\quad y=\frac{|A_2|}{|A|}=\frac{8}{12}=\frac{2}{3};\quad z=\frac{|A_3|}{|A|}=\frac{-12}{12}=-1$$

44.

$$\begin{aligned} 2x+3y\quad +z &= -5 \\ 2y+2z &= -3 \\ 4x\quad +y-2z &= -2 \end{aligned}$$

$$|A| = \begin{vmatrix} 2 & 3 & 1 \\ 0 & 2 & 2 \\ 4 & 1 & -2 \end{vmatrix} = 4 \qquad |A_2| = \begin{vmatrix} 2 & -5 & 1 \\ 0 & -3 & 2 \\ 4 & -2 & -2 \end{vmatrix} = -8$$

$$|A_1| = \begin{vmatrix} -5 & 3 & 1 \\ -3 & 2 & 2 \\ -2 & 1 & -2 \end{vmatrix} = 1 \qquad |A_3| = \begin{vmatrix} 2 & 3 & -5 \\ 0 & 2 & -3 \\ 4 & 1 & -2 \end{vmatrix} = 2$$

$$x = \frac{|A_1|}{|A|} = \frac{1}{4}; \quad y = \frac{|A_2|}{|A|} = \frac{-8}{4} = -2; \quad z = \frac{|A_3|}{|A|} = \frac{2}{4} = \frac{1}{2}$$

CHAPTER 11 REVIEW TEST

1.
3 rows, 2 columns : 3×2

2.
a_{31}, third row and first column : 0

3.
$$\left[\begin{array}{ccc|c} -7 & 0 & 6 & 3 \\ 0 & 2 & -1 & 10 \\ 1 & -1 & 1 & 5 \end{array}\right]$$

4.
$$-5x + 2y = 4$$
$$3x - 4y = 4$$

5.
$$\left[\begin{array}{cc|c} 1 & 1 & 0 \\ 0 & 1 & \frac{1}{2} \end{array}\right]$$

$$y = \frac{1}{2}$$
$$x + \frac{1}{2} = 0$$
$$x = -\frac{1}{2}$$
$$x = -\frac{1}{2},\ y = \frac{1}{2}$$

6.
$$-x + 2y = 2$$
$$\frac{1}{2}x + 2y = -7$$

$$\begin{array}{l} -R_1 \to R_1 \\ \frac{1}{2}R_1 + R_2 \to R_2 \end{array} \left[\begin{array}{cc|c} -1 & 2 & 2 \\ \frac{1}{2} & 2 & -7 \end{array}\right]$$

$$\frac{1}{3}R_2 \to R_2 \left[\begin{array}{cc|c} 1 & -2 & -2 \\ 0 & 3 & -6 \end{array}\right]$$

$$3y = -6$$
$$y = -2$$
$$x - 2(-2) = -2$$
$$x = -6$$
$$x = -6,\ y = -2$$

7.

$$2x-y+3z=2$$
$$x+2y-z=1$$
$$-x+y+4z=2$$

$$\begin{array}{l} R_1-R_2 \to R_1 \\ R_2+R_3 \to R_2 \\ R_1+2R_3 \to R_3 \end{array} \left[\begin{array}{ccc|c} 2 & -1 & 3 & 2 \\ 1 & 2 & -1 & 1 \\ -1 & 1 & 4 & 2 \end{array}\right]$$

$$\begin{array}{l} R_1+R_2 \to R_1 \\ \frac{1}{3}R_2 \to R_2 \end{array} \left[\begin{array}{ccc|c} 1 & -3 & 4 & 1 \\ 0 & 3 & 3 & 3 \\ 0 & 1 & 11 & 6 \end{array}\right]$$

$$R_2-R_3 \to R_3 \left[\begin{array}{ccc|c} 1 & 0 & 7 & 4 \\ 0 & 1 & 1 & 1 \\ 0 & 1 & 11 & 6 \end{array}\right]$$

7. (Continued)

$$-\frac{1}{10}R_3 \to R_3 \left[\begin{array}{ccc|c} 1 & 0 & 7 & 4 \\ 0 & 1 & 1 & 1 \\ 0 & 0 & -10 & -5 \end{array}\right]$$

$$\begin{array}{l} -7R_3+R_1 \to R_1 \\ R_2-R_3 \to R_2 \end{array} \left[\begin{array}{ccc|c} 1 & 0 & 7 & 4 \\ 0 & 1 & 1 & 1 \\ 0 & 0 & 1 & \frac{1}{2} \end{array}\right]$$

$$\left[\begin{array}{ccc|c} 1 & 0 & 0 & \frac{1}{2} \\ 0 & 1 & 0 & \frac{1}{2} \\ 0 & 0 & 1 & \frac{1}{2} \end{array}\right]$$

$$x=\frac{1}{2}, \quad y=\frac{1}{2}, \quad z=\frac{1}{2}$$

8.

$$\begin{bmatrix} 2x-1 & 0 \\ 1 & -3 \end{bmatrix} = \begin{bmatrix} 5 & 0 \\ 1 & -3 \end{bmatrix}$$

$$2x-1=5$$
$$2x=6$$
$$x=3$$

9.

$$C-2D=\begin{bmatrix} 4 & 2 \\ -2 & 0 \\ 3 & -1 \end{bmatrix} - 2\begin{bmatrix} 1 & -6 \\ 0 & 2 \\ 4 & -1 \end{bmatrix}$$

$$=\begin{bmatrix} 4 & 2 \\ -2 & 0 \\ 3 & -1 \end{bmatrix} + \begin{bmatrix} -2 & 12 \\ 0 & -4 \\ -8 & 2 \end{bmatrix}$$

$$=\begin{bmatrix} 2 & 14 \\ -2 & -4 \\ -5 & 1 \end{bmatrix}$$

10.

$$AC = \begin{bmatrix} -4 & 0 & 3 \\ 6 & 2 & -3 \end{bmatrix} \begin{bmatrix} 4 & 2 \\ -2 & 0 \\ 3 & -1 \end{bmatrix}$$

$$= \begin{bmatrix} (-4)(4)+(0)(-2)+(3)(3) & (-4)(2)+(0)(0)+(3)(-1) \\ (6)(4)+(2)(-2)+(-3)(3) & (6)(2)+(2)(0)+(-3)(-1) \end{bmatrix}$$

$$= \begin{bmatrix} -7 & -11 \\ 11 & 15 \end{bmatrix}$$

11.

$$CB = \begin{bmatrix} 4 & 2 \\ -2 & 0 \\ 3 & -1 \end{bmatrix} \begin{bmatrix} -1 \\ -3 \end{bmatrix} = \begin{bmatrix} (4)(-1)+(2)(-3) \\ (-2)(-1)+(0)(-3) \\ (3)(-1)+(-1)(-3) \end{bmatrix}$$

$$= \begin{bmatrix} -10 \\ 2 \\ 0 \end{bmatrix}$$

12.

BA $\quad B: 2 \times 1$

$A: 2 \times 3$

BA: not possible

13.

$$\begin{array}{r} \\ 2R_1+R_2\to R_2 \\ R_1+R_3\to R_3 \end{array}\left[\begin{array}{ccc|ccc} -1 & 0 & 4 & 1 & 0 & 0 \\ 2 & 1 & -1 & 0 & 1 & 0 \\ 1 & -3 & 2 & 0 & 0 & 1 \end{array}\right]$$

$$3R_2+R_3\to R_3\left[\begin{array}{ccc|ccc} 1 & 0 & -4 & -1 & 0 & 0 \\ 0 & 1 & 7 & 2 & 1 & 0 \\ 0 & -3 & 6 & 1 & 0 & 1 \end{array}\right]$$

$$\frac{1}{27}R_3\to R_3\left[\begin{array}{ccc|ccc} 1 & 0 & -4 & -1 & 0 & 0 \\ 0 & 1 & 7 & 2 & 1 & 0 \\ 0 & 0 & 27 & 7 & 3 & 1 \end{array}\right]$$

$$\begin{array}{r} 4R_3+R_1\to R_1 \\ -7R_3+R_2\to R_2 \end{array}\left[\begin{array}{ccc|ccc} 1 & 0 & -4 & -1 & 0 & 0 \\ 0 & 1 & 7 & 2 & 1 & 0 \\ 0 & 0 & 1 & \frac{7}{27} & \frac{1}{9} & \frac{1}{27} \end{array}\right]$$

$$\left[\begin{array}{ccc|ccc} 1 & 0 & 0 & \frac{1}{27} & \frac{4}{9} & \frac{4}{27} \\ 0 & 1 & 0 & \frac{5}{27} & \frac{2}{9} & -\frac{7}{27} \\ 0 & 0 & 1 & \frac{7}{27} & \frac{1}{9} & \frac{1}{27} \end{array}\right]$$

$$\begin{bmatrix} \frac{1}{27} & \frac{4}{9} & \frac{4}{27} \\ \frac{5}{27} & \frac{2}{9} & -\frac{7}{27} \\ \frac{7}{27} & \frac{1}{9} & \frac{1}{27} \end{bmatrix}$$

14.

$$3x-2y=-8$$
$$2x+3y=-1$$

$$R_1-R_2\to R_1\left[\begin{array}{cc|cc} 3 & -2 & 1 & 0 \\ 2 & 3 & 0 & 1 \end{array}\right]$$

$$-2R_1+R_2\to R_2\left[\begin{array}{cc|cc} 1 & -5 & 1 & -1 \\ 2 & 3 & 0 & 1 \end{array}\right]$$

$$\frac{1}{13}R_2\to R_2\left[\begin{array}{cc|cc} 1 & -5 & 1 & -1 \\ 0 & 13 & -2 & 3 \end{array}\right]$$

$$5R_2+R_1\to R_1\left[\begin{array}{cc|cc} 1 & -5 & 1 & -1 \\ 0 & 1 & -\frac{2}{13} & \frac{3}{13} \end{array}\right]$$

$$\left[\begin{array}{cc|cc} 1 & 0 & \frac{3}{13} & \frac{2}{13} \\ 0 & 1 & -\frac{2}{13} & \frac{3}{13} \end{array}\right]$$

$$X=A^{-1}B=\begin{bmatrix} \frac{3}{13} & \frac{2}{13} \\ -\frac{2}{13} & \frac{3}{13} \end{bmatrix}\begin{bmatrix} -8 \\ -1 \end{bmatrix}=\begin{bmatrix} -2 \\ 1 \end{bmatrix}$$

$x=-2,\quad y=1$

15.

$$\begin{vmatrix} -6 & -2 \\ 2 & 1 \end{vmatrix} = (-6)(1) - (2)(-2) = -2$$

16.

$$-2R_3 + R_2 \rightarrow R_2 \begin{vmatrix} 0 & -1 & 2 \\ 2 & -2 & 3 \\ 1 & 4 & 5 \end{vmatrix} = \begin{vmatrix} 0 & -1 & 2 \\ 0 & -10 & -7 \\ 1 & 4 & 5 \end{vmatrix}$$

$$= 1(-1)^{3+1}\begin{vmatrix} -1 & 2 \\ -10 & -7 \end{vmatrix} = 7 + 20 = 27$$

17.

$$x + 2y = -2$$
$$-2x - 3y = 1$$

$$|A| = \begin{vmatrix} 1 & 2 \\ -2 & -3 \end{vmatrix} = 1 \qquad |A_2| = \begin{vmatrix} 1 & -2 \\ -2 & 1 \end{vmatrix} = -3$$

$$|A_1| = \begin{vmatrix} -2 & 2 \\ 1 & -3 \end{vmatrix} = 4$$

$$x = \frac{|A_1|}{|A|} = \frac{4}{1} = 4 \qquad y = \frac{|A_2|}{|A|} = \frac{-3}{1} = -3$$

CHAPTER 11 ADDITIONAL PRACTICE EXERCISES

1.
Find the dimension of matrix A given

$$A=\begin{bmatrix}1 & 7 & -2 & 4\\ 3 & 0 & 6 & -5\end{bmatrix}$$

2.
Write the augmented matrix of the linear system

$$\begin{aligned}4x-2y+3z&=2\\ -y+z&=8\\ 2x+9z&=4\end{aligned}$$

3.
Use back substitution to solve the linear system corresponding to the augmented matrix

$$\left[\begin{array}{cc|c}1 & 2 & 6\\ 0 & 1 & 3\end{array}\right].$$

4.
Solve for x.

$$\begin{bmatrix}2 & 3x+1\\ -1 & 4\end{bmatrix}=\begin{bmatrix}2 & 7\\ -1 & 4\end{bmatrix}$$

5.
Solve the linear system

$$\begin{aligned}x-y+3z&=-4\\ 2x+y+2z&=5\\ -x-y+z&=-6\end{aligned}$$

by applying Gauss - Jordan Elimination to the augmented matrix.

6.
Find the inverse of $\begin{bmatrix}1 & 0 & 2\\ -1 & 2 & 3\\ 1 & -1 & 0\end{bmatrix}$.

7.
Solve the given linear system by finding the inverse of the coefficient matrix.

$$\begin{aligned}4x+3y&=18\\ -5x+y&=-32\end{aligned}$$

8.
Evaluate the determinant, $\begin{vmatrix}5 & -3\\ 2 & 1\end{vmatrix}$.

9.
Use Cramer's Rule to solve the linear system

$$\begin{aligned}3x+y&=31\\ -2x+7y&=33\end{aligned}$$

10.
Given $A=\begin{bmatrix}2 & -1 & 6\\ 5 & -3 & 8\end{bmatrix}$ and $B=\begin{bmatrix}0 & 3 & 9\\ -5 & 4 & 10\end{bmatrix}$

find $3A-B$.

CHAPTER 11 PRACTICE TEST

Exercises 1 and 2 refer to the matrix $A = \begin{bmatrix} 3 & -4 & 9 & 7 \\ 1 & 0 & 2 & -1 \\ 6 & 8 & 11 & 13 \end{bmatrix}$

1. Find the dimension of matrix A.

2. Find a_{34}.

3. Write the linear system whose augmented matrix is

$$\left[\begin{array}{cc|c} 6 & 1 & -8 \\ 3 & -2 & 7 \end{array}\right].$$

4. Solve the linear system

$$\begin{aligned} x+y+2z &= 5 \\ 3x-y+z &= 3 \\ -x+y-z &= -5 \end{aligned}$$

by applying Gaussian Elimination to the augmented matrix.

5. For what values of a, b, c and d are the matrices A and B equal ?

$$A = \begin{bmatrix} a-b & -3c \\ 2a & c+d \end{bmatrix} \quad B = \begin{bmatrix} 2 & -9 \\ 8 & 7 \end{bmatrix}$$

In Exercises 6-8, the following matrices are given.

$$A = \begin{bmatrix} -1 & 4 \\ 6 & -9 \end{bmatrix} \quad B = \begin{bmatrix} 1 & 0 & 4 \\ -3 & 2 & -1 \end{bmatrix} \quad C = \begin{bmatrix} 5 & 10 \\ 1 & 0 \end{bmatrix} \quad D = \begin{bmatrix} 8 \\ 2 \\ 3 \end{bmatrix}$$

In Exercises 6-8, if possible, compute the indicated matrix.

6. $3A+2C$

7. BD

8. BC

In Exercises 9 and 10, find the inverse of the given matrix.

9.
$$\begin{bmatrix} -1 & 2 \\ 3 & -4 \end{bmatrix}$$

10.
$$\begin{bmatrix} 1 & 0 & 2 \\ -1 & 2 & 3 \\ 1 & -1 & 0 \end{bmatrix}$$

11.
Solve the linear system

$$\begin{aligned} 3x + 2y &= 5 \\ -x + 7y &= -40 \end{aligned}$$

by finding the inverse of the coefficient matrix.

In Exercises 12-16, evaluate the determinant of the given matrix.

12.
$$\begin{vmatrix} 5 & -10 \\ 2 & 1 \end{vmatrix}$$

13.
$$\begin{vmatrix} 3 & 2 & -1 \\ 6 & 1 & -2 \\ 5 & 4 & 0 \end{vmatrix}$$

14.
$$\begin{vmatrix} 2 & -1 & 6 & 5 \\ 0 & 5 & 1 & 3 \\ 0 & 0 & -4 & 2 \\ 0 & 0 & 0 & 3 \end{vmatrix}$$

15.
$$\begin{vmatrix} 7 & 0 & 1 \\ 0 & 5 & -1 \\ 2 & 3 & 0 \end{vmatrix}$$

16.
$$\begin{vmatrix} 2 & 3 & 9 \\ -1 & 8 & 6 \\ 5 & -2 & 12 \end{vmatrix}$$

In Exercises 17 - 20, solve the linear system by using Cramer's Rule.

17.
$$\begin{aligned} 3x+5y&=2 \\ -2x+7y&=-22 \end{aligned}$$

18.
$$\begin{aligned} 4x+7y&=33 \\ 2x-3y&=23 \end{aligned}$$

19.
$$\begin{aligned} 5x+y+z&=0 \\ 6y-7z&=26 \\ 4x+3z&=-6 \end{aligned}$$

20.
$$\begin{aligned} x+y+z&=3 \\ x-y+z&=5 \\ -x-y+z&=3 \end{aligned}$$

CHAPTER 12 SECTION 1

SECTION 12.1 PROGRESS CHECK

1a). (Page 659)

$$a_n = 3(1-n)$$
$$a_1 = 3(1-1) = 0$$
$$a_2 = 3(1-2) = -3$$
$$a_3 = 3(1-3) = -6$$
$$a_{12} = 3(1-12) = -33$$

1b).

$$a_n = n^2 + n + 1$$
$$a_1 = 1^2 + 1 + 1 = 3$$
$$a_2 = 2^2 + 2 + 1 = 7$$
$$a_3 = 3^2 + 3 + 1 = 13$$
$$a_{12} = 12^2 + 12 + 1 = 157$$

1c).

$$a_n = 5$$
$$a_1 = a_2 = a_3 = a_{12} = 5$$

2. (Page 659)

$$a_n = 2a_{n-1} - 1$$
$$a_1 = -1$$
$$a_2 = 2(-1) - 1 = -3$$
$$a_3 = 2(-3) - 1 = -7$$
$$a_4 = 2(-7) - 1 = -15$$

3a). (Page 662)

$$x_1^{\ 2} + x_2^{\ 2} + x_3^{\ 2} + \cdots + x_{20}^{\ 2} = \sum_{k=1}^{20} x_k^2$$

3b).

$$2^3 + 3^4 + 4^5 + 5^6 = \sum_{k=2}^{5} k^{k+1}$$

EXERCISE SET 12.1

1.

$$a_n = 2n$$
$$a_1 = 2(1) = 2$$
$$a_2 = 2(2) = 4$$
$$a_3 = 2(3) = 6$$
$$a_4 = 2(4) = 8$$
$$a_{20} = 2(20) = 40$$

3.

$$a_n = 4n - 3$$
$$a_1 = 4(1) - 3 = 1$$
$$a_2 = 4(2) - 3 = 5$$
$$a_3 = 4(3) - 3 = 9$$
$$a_4 = 4(4) - 3 = 13$$
$$a_{20} = 4(20) - 3 = 77$$

5.

$$a_n = 5$$
$$a_1 = a_2 = a_3 = a_4 = a_{20} = 5$$

7.

$$a_n = \frac{n}{n+1}$$
$$a_1 = \frac{1}{1+1} = \frac{1}{2}$$
$$a_2 = \frac{2}{2+1} = \frac{2}{3}$$
$$a_3 = \frac{3}{3+1} = \frac{3}{4}$$
$$a_4 = \frac{4}{4+1} = \frac{4}{5}$$
$$a_{20} = \frac{20}{20+1} = \frac{20}{21}$$

9.

$$a_n = 2 + (0.1)^n$$
$$a_1 = 2 + (0.1)^1 = 2.1$$
$$a_2 = 2 + (0.1)^2 = 2.01$$
$$a_3 = 2 + (0.1)^3 = 2.001$$
$$a_4 = 2 + (0.1)^4 = 2.0001$$
$$a_{20} = 2 + (0.1)^{20} = 2 + (0.1)^{20}$$

11.

$$a_n = \frac{n^2}{2n+1}$$
$$a_1 = \frac{1^2}{2(1)+1} = \frac{1}{3}$$
$$a_2 = \frac{2^2}{2(2)+1} = \frac{4}{5}$$
$$a_3 = \frac{3^2}{2(3)+1} = \frac{9}{7}$$
$$a_4 = \frac{4^2}{2(4)+1} = \frac{16}{9}$$
$$a_{20} = \frac{20^2}{2(20)+1} = \frac{400}{41}$$

13.

$$a_n = 2a_{n-1} - 1$$
$$a_1 = 2$$
$$a_2 = 2(2) - 1 = 3$$
$$a_3 = 2(3) - 1 = 5$$
$$a_4 = 2(5) - 1 = 9$$

15.

$$a_n = \frac{1}{a_{n-1} + 1}$$
$$a_3 = 2$$
$$a_4 = \frac{1}{2+1} = \frac{1}{3}$$
$$a_5 = \frac{1}{\frac{1}{3}+1} = \frac{3}{4}$$
$$a_6 = \frac{1}{\frac{3}{4}+1} = \frac{4}{7}$$

17.

$$a_n = (a_{n-1})^2$$
$$a_1 = 2$$
$$a_2 = 2^2 = 4$$
$$a_3 = 4^2 = 16$$
$$a_4 = 16^2 = 256$$

19.

$$\sum_{k=1}^{5} (3k - 1)$$
$$= [3(1) - 1] + [3(2) - 1] + [3(3) - 1] + [3(4) - 1] + [3(5) - 1]$$
$$= 40$$

21.

$$\sum_{k=1}^{6} (k^2 + 1)$$
$$= (1^2 + 1) + (2^2 + 1) + (3^2 + 1) + (4^2 + 1) + (5^2 + 1) + (6^2 + 1)$$
$$= 97$$

23.

$$\sum_{k=3}^{5} \frac{k}{k-1}$$
$$= \frac{3}{3-1} + \frac{4}{4-1} + \frac{5}{5-1}$$
$$= \frac{49}{12}$$

25.

$$\sum_{i=1}^{4} 20 = (4)(20) = 80$$

27.

$$1+3+5+7+9$$
$$= [2(1)-1]+[2(2)-1]+[2(3)-1]+[2(4)-1]+[2(5)-1]$$
$$= \sum_{k=1}^{5} (2k-1)$$

29.

$$1+4+9+16+25$$
$$= 1^2 + 2^2 + 3^2 + 4^2 + 5^2$$
$$= \sum_{k=1}^{5} k^2$$

31.

$$-1 + \frac{1}{\sqrt{2}} - \frac{1}{\sqrt{3}} + \frac{1}{\sqrt{4}}$$
$$= (-1)^1\left(\frac{1}{\sqrt{1}}\right) + (-1)^2\left(\frac{1}{\sqrt{2}}\right) + (-1)^3\left(\frac{1}{\sqrt{3}}\right) + (-1)^4\left(\frac{1}{\sqrt{4}}\right)$$
$$= \sum_{k=1}^{4} (-1)^k\left(\frac{1}{\sqrt{k}}\right) = \sum_{k=1}^{4} \frac{(-1)^k}{\sqrt{k}}$$

33.

$$\frac{1}{1^2+1} - \frac{2}{2^2+1} + \frac{3}{3^2+1} - \frac{4}{4^2+1} = \sum_{k=1}^{4} \frac{(-1)^{k+1} k}{k^2+1}$$

35.

$$1 + \frac{1}{x} + \frac{1}{x^2} + \frac{1}{x^3} + \cdots + \frac{1}{x^n} = \sum_{k=0}^{n} \frac{1}{x^k}$$

CHAPTER 12 SECTION 2

SECTION 12.2 PROGRESS CHECK

1. (Page 668)

$$a_1 = 4$$
$$a_2 = 4 - \frac{1}{3} = \frac{11}{3}$$
$$a_3 = \frac{11}{3} - \frac{1}{3} = \frac{10}{3}$$
$$a_4 = \frac{10}{3} - \frac{1}{3} = 3$$

2. (Page 669)

$$a_n = -5 + (n-1)\left(\frac{1}{2}\right)$$
$$a_{16} = -5 + (16-1)\left(\frac{1}{2}\right) = \frac{5}{2}$$

3. (Page 670)

$$a_n = 3 + (n-1)d$$
$$a_{10} = 3 + (10-1)d = -\frac{3}{2}$$
$$9d = -\frac{9}{2}$$
$$d = -\frac{1}{2}$$
$$a_n = 3 + (n-1)\left(-\frac{1}{2}\right)$$
$$a_{60} = 3 + (60-1)\left(-\frac{1}{2}\right) = -\frac{53}{2}$$

4. (Page 671)

$$a_n = 2 + (n-1)\left(-\frac{1}{2}\right)$$
$$a_{10} = 2 + (10-1)\left(-\frac{1}{2}\right) = -\frac{5}{2}$$
$$S_n = \frac{n}{2}[a_1 + a_n]$$
$$S_{10} = \frac{10}{2}\left[2 - \frac{5}{2}\right] = -\frac{5}{2}$$

5. (Page 672)

$$S_n = \frac{n}{2}[a_1 + a_n]$$
$$\frac{77}{2} = \frac{n}{2}(6+1)$$
$$77 = 7n$$
$$n = 11$$
$$a_n = a_1 + (n-1)d$$
$$a_{11} = 1 = 6 + (11-1)d$$
$$-5 = 10d$$
$$d = -\frac{1}{2}$$

EXERCISE SET 12.2

1.

3, 6, 9, 12, ...

$d = 6 - 3 = 3$

$a_5 = 12 + 3 = 15$

$a_6 = 15 + 3 = 18$

15, 18

3.

$0, \frac{1}{4}, \frac{1}{2}, \frac{3}{4}, \ldots$

$d = \frac{1}{4} - 0 = \frac{1}{4}$

$a_5 = \frac{3}{4} + \frac{1}{4} = 1$

$a_6 = 1 + \frac{1}{4} = \frac{5}{4}$

$1, \frac{5}{4}$

5.

0, log 10, log 100, log 1000,...

$d = \log 10 - 0$

$= \log 10 = 1$

$a_5 = \log 1000 + \log 10$

$= \log(1000 \cdot 10) = \log 10{,}000 = 4$

$a_6 = \log 10{,}000 + \log 10$

$= \log(10{,}000 \cdot 10)$

$= \log 100{,}000 = 5$

4, 5

7.

$\sqrt{5} - 2, \sqrt{5}, \sqrt{5} + 2, \sqrt{5} + 4, \ldots$

$d = \sqrt{5} - (\sqrt{5} - 2) = 2$

$a_5 = \sqrt{5} + 4 + 2 = \sqrt{5} + 6$

$a_6 = \sqrt{5} + 6 + 2 = \sqrt{5} + 8$

$\sqrt{5} + 6, \sqrt{5} + 8$

9.

$a_1 = 2$

$a_2 = 2 + 4 = 6$

$a_3 = 6 + 4 = 10$

$a_4 = 10 + 4 = 14$

2, 6, 10, 14

11.

$a_1 = 3$

$a_2 = 3 - \frac{1}{2} = \frac{5}{2}$

$a_3 = \frac{5}{2} - \frac{1}{2} = 2$

$a_4 = 2 - \frac{1}{2} = \frac{3}{2}$

$3, \frac{5}{2}, 2, \frac{3}{2}$

13.

$$a_1 = \frac{1}{3}$$
$$a_2 = \frac{1}{3} - \frac{1}{3} = 0$$
$$a_3 = 0 - \frac{1}{3} = -\frac{1}{3}$$
$$a_4 = -\frac{1}{3} - \frac{1}{3} = -\frac{2}{3}$$
$$\frac{1}{3},\ 0,\ -\frac{1}{3},\ -\frac{2}{3}$$

15.

$$a_n = a_1 + (n-1)d$$
$$a_n = 4 + (n-1)(3)$$
$$a_8 = 4 + (8-1)(3) = 25$$

17.

$$a_n = a_1 + (n-1)d$$
$$a_n = 14 + (n-1)(-2)$$
$$a_{12} = 14 + (12-1)(-2) = -8$$

19.

$$a_n = a_1 + (n-1)d$$
$$a_n = -2 + (n-1)d$$
$$-2 = -2 + (20-1)d$$
$$0 = 19d$$
$$d = 0$$
$$a_{24} = -2$$

21.

$$a_n = a_1 + (n-1)d$$
$$a_n = 0 + (n-1)d$$
$$20 = (61-1)d$$
$$20 = 60d$$
$$d = \frac{1}{3}$$
$$a_n = (n-1)\left(\frac{1}{3}\right)$$
$$a_{20} = (20-1)\left(\frac{1}{3}\right) = \frac{19}{3}$$

23.

$$a_n = a_1 + (n-1)d$$
$$a_n = -\frac{1}{4} + (n-1)d$$
$$10 = -\frac{1}{4} + (41-1)d$$
$$\frac{41}{4} = 40d$$
$$d = \frac{41}{160}$$
$$a_n = -\frac{1}{4} + (n-1)\left(\frac{41}{160}\right)$$
$$a_{22} = -\frac{1}{4} + (22-1)\left(\frac{41}{160}\right) = \frac{821}{160}$$

25.

$$S_n = \frac{n}{2}\left[2a_1 + (n-1)d\right]$$

$$S_{20} = \frac{20}{2}\left[2(3) + (20-1)2\right] = 440$$

27.

$$S_n = \frac{n}{2}\left[2a_1 + (n-1)d\right]$$

$$S_{12} = \frac{12}{2}\left[2\left(\frac{1}{2}\right) + (12-1)(-2)\right]$$

$$= -126$$

29.

$$S_n = \frac{n}{2}\left[2a_1 + (n-1)d\right]$$

$$S_{40} = \frac{40}{2}\left[2(82) + (40-1)(-2)\right]$$

$$= 1720$$

31.

$$d = 4 - 2 = 2$$

$$S_n = \frac{n}{2}\left[2a_1 + (n-1)d\right]$$

$$930 = \frac{n}{2}\left[2(2) + (n-1)2\right]$$

$$930 = n(2 + n - 1)$$

$$930 = n + n^2$$

$$n^2 + n - 930 = 0$$

$$(n-30)(n+31) = 0$$

$$n = 30 \quad (n > 0)$$

33.

$$S_n = \frac{n}{2}(a_1 + a_n)$$

$$1395 = \frac{n}{2}(3 + 90)$$

$$2790 = 93n$$

$$n = 30$$

$$a_n = a_1 + (n-1)d$$

$$90 = 3 + (30-1)d$$

$$87 = 29d$$

$$d = 3$$

35.

$$S_n = \frac{n}{2}(a_1 + a_n)$$

$$\frac{27}{4} = \frac{n}{2}\left(\frac{1}{2} + \frac{7}{4}\right)$$

$$\frac{27}{4} = \frac{9n}{8}$$

$$54 = 9n$$

$$n = 6$$

$$a_n = a_1 + (n-1)d$$

$$\frac{7}{4} = \frac{1}{2} + (6-1)d$$

$$\frac{5}{4} = 5d$$

$$d = \frac{1}{4}$$

37.

$$a_n = a_1 + (n-1)d$$

$$-\frac{5}{4} = a_1 + (4-1)d \qquad \frac{1}{4} = a_1 + (10-1)d$$

$$-\frac{5}{4} = a_1 + 3d \qquad \frac{1}{4} = a_1 + 9d$$

$$\frac{5}{4} = -a_1 - 3d$$

$$\frac{1}{4} = a_1 + 9d$$

$$\frac{3}{2} = 6d$$

$$d = \frac{1}{4}$$

$$\frac{1}{4} = a_1 + 9\left(\frac{1}{4}\right)$$

$$a_1 = -2$$

$$S_n = \frac{n}{2}\left[2a_1 + (n-1)d\right]$$

$$S_{16} = \frac{16}{2}\left[2(-2) + (16-1)\left(\frac{1}{4}\right)\right]$$

$$= -2$$

39.

$$a_1 = 1 \qquad a_n = n$$

$$S_n = \frac{n}{2}(a_1 + a_n)$$

$$S_n = \frac{n}{2}(1+n) = \frac{n(n+1)}{2}$$

CHAPTER 12 SECTION 3

SECTION 12.3 PROGRESS CHECK

1. (Page 674)

$$r = -\frac{6}{2} = -3$$
$$a_n = a_1 r^{n-1}$$
$$a_6 = 2(-3)^{6-1} = -486$$

2. (Page 675)

$$5, a_2, a_3, \frac{8}{25}$$

$$a_n = a_1 r^{n-1}$$
$$\frac{8}{25} = 5r^{4-1}$$
$$\frac{8}{125} = r^3$$
$$r = \frac{2}{5}$$
$$5, 2, \frac{4}{5}, \frac{8}{25}$$

3. (Page 676)

$$r = \frac{-\frac{4}{3}}{2} = -\frac{2}{3}$$
$$S_n = \frac{a_1(1-r^n)}{1-r}$$
$$S_5 = \frac{2\left[1-\left(-\frac{2}{3}\right)^5\right]}{1-\left(-\frac{2}{3}\right)} = \frac{110}{81}$$

4. (Page 677)

$$a_1 = 64, \qquad r = \frac{1}{2}$$

Top of fifth bounce : a_6

$$a_n = a_1 r^{n-1}$$
$$a_6 = 64\left(\frac{1}{2}\right)^{6-1} = 2$$

2 feet

$$S_n = \frac{a_1(1-r^n)}{1-r}$$
$$S_6 = \frac{64\left[1-\left(\frac{1}{2}\right)^6\right]}{1-\frac{1}{2}} = 126$$

126 feet

5. (Page 679)

$$4-1+\frac{1}{4}-\frac{1}{16}+\cdots$$

$$r=-\frac{1}{4}$$

$$S=\frac{a_1}{1-r}$$

$$S=\frac{4}{1-\left(-\frac{1}{4}\right)}=\frac{16}{5}$$

6. (Page 679)

$$2.545\overline{454}=2+0.54+0.0054+\cdots$$

Consider $0.54+0.0054+\cdots$

$$a_1=0.54, r=0.01$$

$$S=\frac{a_1}{1-r}=\frac{0.54}{1-0.01}=\frac{0.54}{0.99}=\frac{54}{99}=\frac{6}{11}$$

Thus $2.545\overline{454}=2+\frac{6}{11}=2\frac{6}{11}=\frac{28}{11}$

EXERCISE SET 12.3

1.

3, 6, 12, 24, ...

$$r=\frac{6}{3}=2$$

$$a_5=(24)(2)=48$$

3.

$$-4,\ 3,\ -\frac{9}{4},\ \frac{27}{16},\ \ldots$$

$$r=\frac{3}{-4}$$

$$a_5=\left(\frac{27}{16}\right)\left(-\frac{3}{4}\right)=-\frac{81}{64}$$

5.

1.2, 0.24, 0.048, ...

$$r=\frac{0.24}{1.2}=0.2$$

$$a_4=(0.048)(0.2)=0.0096$$

7.

$$a_1=3$$

$$a_2=3(3)=9$$

$$a_3=9(3)=27$$

$$a_4=27(3)=81$$

3, 9, 27, 81

9.

$$a_1 = 4$$
$$a_2 = 4\left(\frac{1}{2}\right) = 2$$
$$a_3 = 2\left(\frac{1}{2}\right) = 1$$
$$a_4 = 1\left(\frac{1}{2}\right) = \frac{1}{2}$$

$4, 2, 1, \frac{1}{2}$

11.

$$a_1 = -3$$
$$a_2 = (-3)(2) = -6$$
$$a_3 = (-6)(2) = -12$$
$$a_4 = (-12)(2) = -24$$

$-3, -6, -12, -24$

13.

$$a_n = a_1 r^{n-1}$$
$$a_n = 3(-2)^{n-1}$$
$$a_8 = 3(-2)^{8-1} = -384$$

15.

$$r = \frac{8}{16} = \frac{1}{2}$$
$$a_n = a_1 r^{n-1}$$
$$a_n = 16\left(\frac{1}{2}\right)^{n-1}$$
$$a_7 = 16\left(\frac{1}{2}\right)^{7-1} = \frac{1}{4}$$

17.

$$a_n = a_1 r^{n-1}$$
$$a_n = 3r^{n-1}$$
$$\frac{1}{27} = 3r^{5-1}$$
$$\frac{1}{81} = r^4$$
$$r = \frac{1}{3}$$
$$a_n = 3\left(\frac{1}{3}\right)^{n-1}$$
$$a_7 = 3\left(\frac{1}{3}\right)^{7-1} = \frac{1}{243}$$

19.

$$a_n = a_1 r^{n-1}$$
$$a_n = \frac{16}{81} r^{n-1}$$
$$\frac{3}{2} = \frac{16}{81} r^{6-1}$$
$$\frac{243}{32} = r^5$$
$$r = \frac{3}{2}$$
$$a_n = \frac{16}{81}\left(\frac{3}{2}\right)^{n-1}$$
$$a_8 = \frac{16}{81}\left(\frac{3}{2}\right)^{8-1} = \frac{27}{8}$$

21.

$$a_n = a_1 r^{n-1}$$

$$4 = a_1 r^{2-1} \qquad 256 = a_1 r^{8-1}$$

$$4 = a_1 r \qquad 256 = a_1 r^7$$

$$a_1 = \frac{4}{r} \qquad 256 = \left(\frac{4}{r}\right) r^7$$

$$256 = 4r^6$$

$$64 = r^6$$

$$r = \pm 2$$

23.

$$a_n = a_1 r^{n-1}$$

$$32 = \frac{1}{2}(2)^{n-1}$$

$$64 = 2^{n-1}$$

$$2^6 = 2^{n-1}$$

$$6 = n-1$$

$$n = 7$$

25.

$$\frac{1}{3},\ a_2,\ a_3,\ 9$$

$$a_n = a_1 r^{n-1}$$

$$9 = \frac{1}{3} r^{4-1}$$

$$27 = r^3$$

$$r = 3$$

$$a_2 = \frac{1}{3}(3) = 1$$

$$a_3 = 1(3) = 3$$

1, 3

27.

$$1,\ a_2,\ a_3,\ \frac{1}{64}$$

$$a_n = a_1 r^{n-1}$$

$$\frac{1}{64} = 1r^{4-1}$$

$$\frac{1}{64} = r^3$$

$$r = \frac{1}{4}$$

$$a_2 = 1\left(\frac{1}{4}\right) = \frac{1}{4}$$

$$a_3 = \left(\frac{1}{4}\right)\left(\frac{1}{4}\right) = \frac{1}{16}$$

$$\frac{1}{4},\ \frac{1}{16}$$

29.

$$r = \frac{1}{3}$$

$$S_n = \frac{a_1(1-r^n)}{1-r}$$

$$S_7 = \frac{3\left[1-\left(\frac{1}{3}\right)^7\right]}{1-\frac{1}{3}} = \frac{1093}{243}$$

31.

$$r = \frac{\frac{6}{5}}{-3} = -\frac{2}{5}$$

$$S_n = \frac{a_1(1-r^n)}{1-r}$$

$$S_5 = \frac{(-3)\left[1-\left(-\frac{2}{5}\right)^5\right]}{1-\left(-\frac{2}{5}\right)} = -\frac{1353}{625}$$

33.

$$S_n = \frac{a_1(1-r^n)}{1-r}$$

$$S_8 = \frac{4(1-2^8)}{1-2} = 1020$$

35.

$$a_n = a_1 r^{n-1}$$

$$-\frac{54}{8} = 2r^{4-1}$$

$$-\frac{27}{8} = r^3$$

$$r = -\frac{3}{2}$$

$$S_n = \frac{a_1(1-r^n)}{1-r}$$

$$S_5 = \frac{2\left[1-\left(-\frac{3}{2}\right)^5\right]}{1-\left(-\frac{3}{2}\right)} = \frac{55}{8}$$

37.

$a_1 = 5, \quad r = 2$

$$S_{11} = \frac{5(1-2^{11})}{1-2} = 10235$$

\$10,235

39.

1980: $a_1 = 30000$

$r = 1.25$

find a_4 (2010)

$$a_n = a_1 r^{n-1}$$

$$a_4 = 30000(1.25)^{4-1} = 58594$$

58,594 people

41.

$$1+\frac{1}{2}+\frac{1}{4}+\frac{1}{8}+\cdots$$

$$r=\frac{\frac{1}{2}}{1}=\frac{1}{2}$$

$$S=\frac{a_1}{1-r}=\frac{1}{1-\frac{1}{2}}=2$$

43.

$$1-\frac{1}{3}+\frac{1}{9}+\frac{1}{27}+\cdots$$

$$r=\frac{-\frac{1}{3}}{1}=-\frac{1}{3}$$

$$S=\frac{a_1}{1-r}=\frac{1}{1-\left(-\frac{1}{3}\right)}=\frac{3}{4}$$

45.

$$2+\frac{1}{2}+\frac{1}{8}+\frac{1}{32}+\cdots$$

$$r=\frac{\frac{1}{2}}{2}=\frac{1}{4}$$

$$S=\frac{a_1}{1-r}=\frac{2}{1-\frac{1}{4}}=\frac{8}{3}$$

47.

$$0.5+(0.5)^2+(0.5)^3+(0.5)^4+\cdots$$

$$r=\frac{(0.5)^2}{0.5}=0.5$$

$$S=\frac{a_1}{1-r}=\frac{0.5}{1-0.5}=1$$

49.

$$\frac{1}{3}-\frac{2}{9}+\frac{4}{27}-\frac{8}{81}+\cdots$$

$$r=\frac{-\frac{2}{9}}{\frac{1}{3}}=-\frac{2}{3}$$

$$S=\frac{a_1}{1-r}=\frac{\frac{1}{3}}{1-\left(-\frac{2}{3}\right)}=\frac{1}{5}$$

51.

$$0.3676\overline{767}=0.3+0.067+0.00067+\cdots$$

Consider $0.067+0.00067+\cdots$

$$r=0.01$$

$$S=\frac{a_1}{1-r}=\frac{0.067}{1-0.01}=\frac{0.067}{0.99}=\frac{67}{990}$$

$$0.3676\overline{767}=\frac{3}{10}+\frac{67}{990}=\frac{364}{990}=\frac{182}{495}$$

53.

$0.325\overline{325} = 0.325 + 0.000325 + \cdots$

$$r = 0.001$$

$$S = \frac{a_1}{1-r} = \frac{0.325}{1-0.001} = \frac{0.325}{0.999} = \frac{325}{999}$$

$$0.325\overline{325} = \frac{325}{999}$$

CHAPTER 12 SECTION 4

EXERCISE SET 12.4

1.

$$2+4+6+\cdots+2n=n(n+1)$$

$$n=1:\ 2=1(1+1)\ ?$$
$$2=2$$

$$n=k:\ 2+4+6+\cdots+2k=k(k+1)$$
$$2+4+6+\cdots+2k+(2k+2)$$
$$=k(k+1)+(2k+2)$$
$$=k(k+1)+2(k+1)$$
$$=(k+1)(k+2)$$

Thus the formula holds for $n=k+1$.

3.

$$2+5+8+\cdots+(3n-1)=\frac{n(3n+1)}{2}$$

$$n=1:\ 2=\frac{1[3(1)+1]}{2}\ ?$$
$$2=2$$

$$n=k:\ 2+5+8+\cdots+(3k-1)=\frac{k(3k+1)}{2}$$
$$2+5+8+\cdots+(3k-1)+(3k+2)$$
$$=\frac{k(3k+1)}{2}+3k+2$$
$$=\frac{k(3k+1)+2(3k+2)}{2}$$
$$=\frac{3k^2+k+6k+4}{2}$$
$$=\frac{3k^2+7k+4}{2}$$
$$=\frac{(k+1)(3k+4)}{2}$$
$$=\frac{(k+1)[3(k+1)+1]}{2}$$

Thus the formula holds for $n=k+1$.

5.

$$5+10+15+\cdots+5n=\frac{5n(n+1)}{2}$$

$$n=1:\ 5=\frac{5(1)(1+1)}{2}\ ?$$
$$5=5$$

$$n=k:\ 5+10+15+\cdots+5k=\frac{5k(k+1)}{2}$$
$$5+10+15+\cdots+5k+(5k+5)$$
$$=\frac{5k(k+1)}{2}+(5k+5)$$
$$=\frac{5k(k+1)+2(5k+5)}{2}$$
$$=\frac{5(k+1)(k+2)}{2}$$

Thus the formula holds for $n=k+1$.

7.

$$1\cdot2+2\cdot3+3\cdot4+\cdots+n(n+1)=\frac{n(n+1)(n+2)}{3}$$

$$n=1:\ 1\cdot2=\frac{1(1+1)(1+2)}{3}\ ?$$
$$2=2$$

$$n=k:\ 1\cdot2+2\cdot3+3\cdot4+\cdots+k(k+1)$$
$$=\frac{k(k+1)(k+2)}{3}$$
$$1\cdot2+2\cdot3+3\cdot4+\cdots+k(k+1)+(k+1)(k+2)$$
$$=\frac{k(k+1)(k+2)}{3}+(k+1)(k+2)$$
$$=\frac{k(k+1)(k+2)+3(k+1)(k+2)}{3}$$
$$=\frac{(k+1)(k+2)(k+3)}{3}$$

Thus the formula holds for $n=k+1$.

9.

$$1+5+9+\cdots+(4n-3)=n(2n-1)$$

$$n=1:\ 1=1[2(1)-1]\ ?$$
$$1=1$$

$$n=k:\ 1+5+9+\cdots+(4k-3)=k(2k-1)$$
$$1+5+9+\cdots+(4k-3)+(4k+1)$$
$$=k(2k-1)+(4k+1)$$
$$=2k^2-k+4k+1$$
$$=2k^2+3k+1$$
$$=(k+1)(2k+1)$$
$$=(k+1)[2(k+1)-1]$$

Thus the formula holds for $n=k+1$.

11.

$$a_n=a_1+(n-1)d$$

$$n=1:\ a_1=a_1$$

$$n=k:\ a_k=a_1+(k-1)d$$
$$a_{k+1}=a_k+d=a_1+(k-1)d+d$$
$$=a_1+d(k-1+1)$$
$$=a_1+kd$$

Thus the statement holds true for $n=k+1$.

13.

$$2+2^2+2^3+\cdots+2^n=2^{n+1}-2$$

$n=1$: $\quad 2=2^{1+1}-2\ ?$

$\quad 2=2$

$n=k$: $\quad 2+2^2+\cdots+2^k=2^{k+1}-2$

$$\begin{aligned}2+2^2+\cdots+2^k+2^{k+1}&=2^{k+1}-2+2^{k+1}\\&=2\cdot 2^{k+1}-2\\&=2^{k+2}-2\end{aligned}$$

Thus the formula holds true for $n=k+1$.

15.

$$x^n-1=(x-1)Q(x)$$

$n=1$: $\quad x^1-1=(x-1)(1)$

$n=k$: $\quad x^k-1=(x-1)Q(x)$

$$Q(x)=\frac{x^k-1}{x-1}$$

$$\begin{aligned}x^{k+1}-1&=(x-1)\left(x^k+\frac{x^k-1}{x-1}\right)\\&=(x-1)\left(x^k+Q(x)\right)\\&=(x-1)P(x)\end{aligned}$$

Thus the formula holds true for $n=k+1$.

17.

$$(1+\mathrm{a})^n\geq 1+n\mathrm{a}$$

$n=1$: $(1+\mathrm{a})^1\geq 1+1\mathrm{a}\ ?$

$\quad 1+\mathrm{a}\ \geq\ 1+\mathrm{a}$

$n=k$: $(1+\mathrm{a})^k\geq 1+k\mathrm{a}$

$$\mathrm{a}>-1$$

$$\mathrm{a}+1>0$$

$$(1+\mathrm{a})^k(1+\mathrm{a})\geq(1+k\mathrm{a})(1+\mathrm{a})$$

$$\begin{aligned}(1+\mathrm{a})^{k+1}&\geq 1+\mathrm{a}+k\mathrm{a}+k\mathrm{a}^2\geq 1+\mathrm{a}+k\mathrm{a}\\&=1+(k+1)\mathrm{a}\\(1+\mathrm{a})^{k+1}&\geq 1+(k+1)\mathrm{a}\end{aligned}$$

Thus the inequality holds true for $n=k+1$.

CHAPTER 12 SECTION 5

SECTION 12.5 PROGRESS CHECK

1. (Page 686)

$$\begin{array}{cr} 1\ \ 5\ \ 10\ \ 10\ \ 5\ \ 1 & n=5 \\ 1\ \ 6\ \ 15\ \ 20\ \ 15\ \ 6\ \ 1 & n=6 \\ 1\ \ 7\ \ 21\ \ 35\ \ 35\ \ 21\ \ 7\ \ 1 & n=7 \\ 1\ \ 8\ \ 28\ \ 56\ \ 70\ \ 56\ \ 28\ \ 8\ \ 1 & n=8 \\ 1\ \ 9\ \ 36\ \ 84\ \ 126\ \ 126\ \ 84\ \ 36\ \ 9\ \ 1 & n=9 \\ 1\ \ 10\ \ 45\ \ 120\ \ 210\ \ 252\ \ 210\ \ 120\ \ 45\ \ 10\ \ 1 & n=10 \end{array}$$

$$(a+b)^{10} = a^{10} + 10a^9b + 45a^8b^2 + 120a^7b^3 + 210a^6b^4 + \cdots$$

2. (Page 686)

$$\begin{aligned}(x^2-2)^4 &= (x^2)^4 + \frac{4}{1}(x^2)^{4-1}(-2) \\ &+ \frac{4\cdot 3}{1\cdot 2}(x^2)^{4-2}(-2)^2 \\ &+ \frac{4\cdot 3\cdot 2}{1\cdot 2\cdot 3}(x^2)^{4-3}(-2)^3 \\ &+ \frac{4\cdot 3\cdot 2\cdot 1}{1\cdot 2\cdot 3\cdot 4}(x^2)^{4-4}(-2)^4\end{aligned}$$

$$= x^8 - 8x^6 + 24x^4 - 32x^2 + 16$$

3a). (Page 688)

$$\frac{12!}{10!} = \frac{12\cdot 11\cdot 10!}{10!} = 12\cdot 11 = 132$$

3b).

$$\frac{6!}{4!2!} = \frac{6\cdot 5\cdot 4!}{4!\cdot 2\cdot 1} = \frac{6\cdot 5}{2} = 15$$

3c).

$$\frac{10!8!}{9!7!} = \frac{10\cdot 9!\cdot 8\cdot 7!}{9!7!} = 10\cdot 8 = 80$$

3d).

$$\frac{n!(n-1)!}{(n+1)!(n-2)!} = \frac{n!(n-1)(n-2)!}{(n+1)n!(n-2)!} = \frac{n-1}{n+1}$$

3e).

$$\frac{8!}{6!(3-3)!} = \frac{8\cdot 7\cdot 6!}{6!0!} = \frac{8\cdot 7}{1} = 56$$

4. (Page 690)

$$\left(\frac{x}{2}-1\right)^8$$

third term : $\binom{8}{2}\left(\frac{x}{2}\right)^6(-1)^2$

$$=\frac{8!}{6!2!}\cdot\frac{x^6}{2^6}$$

$$=\frac{8\cdot 7\cdot 6\cdot !}{6!\cdot 2\cdot 1}\cdot\frac{x^6}{2^6}$$

$$=\frac{7}{16}x^6$$

5. (Page 690)

$x^6=\left(x^3\right)^2$ occurs in 4th term

$$\binom{5}{3}\left(x^3\right)^2\left(-\sqrt{2}\right)^3$$

$$=\frac{5!}{2!3!}x^6\left(-2\sqrt{2}\right)$$

$$=\frac{5\cdot 4\cdot 3!}{2\cdot 1\cdot 3!}x^6\left(-2\sqrt{2}\right)$$

$$=-20\sqrt{2}x^6$$

EXERCISE SET 12.5

1.

$$(3x+2y)^5=(3x)^5+5(3x)^4(2y)+10(3x)^3(2y)^2+10(3x)^2(2y)^3+5(3x)(2y)^4+(2y)^5$$
$$=243x^5+810x^4y+1080x^3y^2+720x^2y^3+240xy^4+32y^5$$

3.

$$(4x-y)^4=(4x)^4+4(4x)^3(-y)+6(4x)^2(-y)^2+4(4x)(-y)^3+(-y)^4$$
$$=256x^4-256x^3y+96x^2y^2-16xy^3+y^4$$

5.

$$(2-xy)^5=(2)^5+5(2)^4(-xy)+10(2)^3(-xy)^2+10(2)^2(-xy)^3+5(2)(-xy)^4+(-xy)^5$$
$$=32-80xy+80x^2y^2-40x^3y^3+10x^4y^4-x^5y^5$$

7.

$$\left(a^2b+3\right)^4=\left(a^2b\right)^4+4\left(a^2b\right)^3(3)+6\left(a^2b\right)^2(3)^2+4\left(a^2b\right)(3)^3+3^4$$
$$=a^8b^4+12a^6b^3+54a^4b^2+108a^2b+81$$

9.

$$\begin{aligned}(a-2b)^8 &= a^8+8a^7(-2b)+28a^6(-2b)^2+56a^5(-2b)^3+70a^4(-2b)^4+56a^3(-2b)^5+28a^2(-2b)^6\\ &\quad +8a(-2b)^7+(-2b)^8\\ &= a^8-16a^7b+112a^6b^2-448a^5b^3+1120a^4b^4-1792a^3b^5+1792a^2b^6-1024ab^7+256b^8\end{aligned}$$

11.

$$\begin{aligned}\left(\frac{1}{3}x+2\right)^3 &= \left(\frac{1}{3}x\right)^3+3\left(\frac{1}{3}x\right)^2(2)+3\left(\frac{1}{3}x\right)(2)^2+(2)^3\\ &= \frac{x^3}{27}+\frac{2x^2}{3}+4x+8\end{aligned}$$

13.

$$\begin{aligned}(2+x)^{10} &= 2^{10}+10(2)^9x+45(2)^8x^2+120(2)^7x^3+\cdots\\ &= 1024+5120x+11520x^2+15360x^3+\cdots\end{aligned}$$

15.

$$\begin{aligned}(3-2a)^9 &= 3^9+9(3)^8(-2a)+\frac{9\cdot 8}{1\cdot 2}(3)^7(-2a)^2+\frac{9\cdot 8\cdot 7}{1\cdot 2\cdot 3}(3)^6(-2a)^3+\cdots\\ &= 19683-118098a+314928a^2-489888a^3+\cdots\end{aligned}$$

17.

$$\begin{aligned}(2x-3y)^{14} &= (2x)^{14}+14(2x)^{13}(-3y)+\frac{14\cdot 13}{1\cdot 2}(2x)^{12}(-3y)^2+\frac{14\cdot 13\cdot 12}{1\cdot 2\cdot 3}(2x)^{11}(-3y)^3+\cdots\\ &= 16384x^{14}-344064x^{13}y+3354624x^{12}y^2-20127744x^{11}y^3+\cdots\end{aligned}$$

19.

$$\begin{aligned}(2x-yz)^{13} &= (2x)^{13}+13(2x)^{12}(-yz)+\frac{13\cdot 12}{1\cdot 2}(2x)^{11}(-yz)^2+\frac{13\cdot 12\cdot 11}{1\cdot 2\cdot 3}(2x)^{10}(-yz)^3+\cdots\\ &= 8192x^{13}-53248x^{12}yz+159744x^{11}y^2z^2-292864x^{10}y^3z^3+\cdots\end{aligned}$$

21.

$$5! = 5\cdot 4\cdot 3\cdot 2\cdot 1 = 120$$

23.

$$\frac{12!}{11!} = \frac{12\cdot 11!}{11!} = 12$$

25.

$$\frac{11!}{8!}=\frac{11\cdot 10\cdot 9\cdot 8!}{8!}=990$$

27.

$$\frac{10!}{6!}=\frac{10\cdot 9\cdot 8\cdot 7\cdot 6!}{6!}=5040$$

29.

$$\frac{6!}{3!}=\frac{6\cdot 5\cdot 4\cdot 3!}{3!}=120$$

31.

$$\binom{10}{6}=\frac{10!}{(10-6)!6!}=\frac{10!}{4!6!}=\frac{10\cdot 9\cdot 8\cdot 7\cdot 6!}{4\cdot 3\cdot 2\cdot 1\cdot 6!}=210$$

33.

$(2x-4)^7$

$4^{\underline{\text{th}}}$ term : $\frac{7\cdot 6\cdot 5}{1\cdot 2\cdot 3}(2x)^{7-3}(-4)^3=-35840x^4$

35.

$\left(\frac{1}{2}x-y\right)^{12}$

$5^{\underline{\text{th}}}$ term : $\frac{12\cdot 11\cdot 10\cdot 9}{1\cdot 2\cdot 3\cdot 4}\left(\frac{1}{2}x\right)^{12-4}(-y)^4=\frac{495}{256}x^8y^4$

37.

$\left(\frac{1}{x}-2\right)^9$

$5^{\underline{\text{th}}}$ term : $\frac{9\cdot 8\cdot 7\cdot 6}{1\cdot 2\cdot 3\cdot 4}\left(\frac{1}{x}\right)^{9-4}(-2)^4=\frac{2016}{x^5}$

39.

$(x-3y)^6$

middle term : $20x^3(-3y)^3=-540x^3y^3$

41.

$(3x+4y)^7$

x^4 occurs in the $4^{\underline{\text{th}}}$ term : $\binom{7}{3}(3x)^4(4y)^3$

$=\frac{7!}{(7-3)!3!}(81x^4)(64y^3)=181440x^4y^3$

43.

$(2x^3-1)^9$

$x^6=(x^3)^2$ occurs in the $8^{\underline{\text{th}}}$ term

$$\binom{9}{7}(2x^3)^{9-7}(-1)^7=\frac{9!}{(9-7)!7!}(4x^6)(-1)$$

$$=-144x^6$$

45.

$$\left(x^3+\frac{1}{2}\right)^7$$

$x^{12}=\left(x^3\right)^4$ occurs in the 4th term

$$\binom{7}{3}\left(x^3\right)^{7-3}\left(\frac{1}{2}\right)^3=\frac{7!}{(7-3)!3!}x^{12}\left(\frac{1}{8}\right)$$

$$=\frac{35x^{12}}{8}$$

47.

$$(1.3)^6=(1+0.3)^6$$

$$=1^6+6(1)^5(0.3)+15(1)^4(0.3)^2+20(1)^3(0.3)^3+15(1)^3(0.3)^4+6(1)^2(0.3)^5+(0.3)^6$$

$$=1+1.8+1.35+0.54+0.1215+0.0146+0.0007$$

$$=4.8268$$

CHAPTER 12 SECTION 6

SECTION 12.6 PROGRESS CHECK

1. (Page 693)

$\underline{4}\cdot\underline{3}\cdot\underline{2}\cdot\underline{1}=24$

2. (Page 693)

$\underline{15}\cdot\underline{14}\cdot\underline{13}=2730$

3a). (Page 695)

$$P(4,\ 4)=\frac{4!}{(4-4)!}=\frac{4\cdot3\cdot2\cdot1}{1}=24$$

3b).

$$P(6,\ 3)=\frac{6!}{(6-3)!}=\frac{6!}{3!}=\frac{6\cdot5\cdot4\cdot3!}{3!}=120$$

3c).

$$\frac{2P(6,\ 4)}{2!}=\frac{2P(6,\ 4)}{2\cdot1}=P(6,\ 4)=\frac{6!}{(6-4)!}$$

$$=\frac{6!}{2!}=\frac{6\cdot5\cdot4\cdot3\cdot2!}{2!}=360$$

4. (Page 695)

$$P(6,\ 4)=\frac{6!}{(6-4)!}=\frac{6!}{2!}=\frac{6\cdot5\cdot4\cdot3\cdot2!}{2!}=360$$

5. (Page 696)

2 groupings : $P(2,\ 2)$
3 out of 7 math books : $P(7,\ 3)$
2 out of 4 biology books : $P(4,\ 2)$

$$P(2,\ 2)\cdot P(7,\ 3)\cdot P(4,\ 2)$$

$$=\frac{2!}{(2-2)!}\cdot\frac{7!}{(7-3)!}\cdot\frac{4!}{(4-2)!}$$

$$=\frac{2!}{0!}\cdot\frac{7!}{4!}\cdot\frac{4!}{2!}$$

$$=\frac{7!}{1}=5040$$

6. (Page 696)

Alaska
$3a's$: 3! permutations

$$\frac{6!}{3!}=\frac{6\cdot5\cdot4\cdot3!}{3!}=120$$

7. (Page 697)

$a,\ b,\ c,\ d$

$ab,\ ac,\ ad,\ bc,\ bd,\ cd$

8a). (Page 698)

$$C(6,\ 2)=\frac{6!}{2!(6-2)!}=\frac{6!}{2!4!}=\frac{6\cdot5\cdot4!}{2\cdot1\cdot4!}=15$$

8b).

$$C(10,\,10)=\frac{10!}{10!(10-10)!}=\frac{10!}{10!0!}=1$$

8c).

$$P(3,\,2)=\frac{3!}{(3-2)!}=\frac{3!}{1!}=6$$

$$C(5,\,4)=\frac{5!}{4!(5-4)!}=\frac{5!}{4!\,1!}=\frac{5\cdot 4!}{4!}=5$$

$$\frac{P(3,\,2)}{3!C(5,\,4)}=\frac{6}{3!\cdot 5}=\frac{1}{5}$$

9. (Page 699)

$$C(52,\,5)=\frac{52!}{5!(52-5)!}=\frac{52!}{5!47!}$$

$$=\frac{52\cdot 51\cdot 50\cdot 49\cdot 48\cdot 47!}{5!47!}=2{,}598{,}960$$

10. (Page 700)

$$C(5,\,1)\cdot C(8,\,3)=\frac{5!}{1!(5-1)!}\cdot\frac{8!}{3!(8-3)!}$$

$$=\frac{5!}{4!}\cdot\frac{8!}{3!5!}=\frac{8\cdot 7\cdot 6\cdot 5\cdot 4!}{4!\cdot 3\cdot 2\cdot 1}=280$$

11. (Page 701)

$$C(6,\,3)\cdot C(4,\,2)\cdot P(5,\,5)$$

$$=\frac{6!}{3!(6-3)!}\cdot\frac{4!}{2!(4-2)!}\cdot\frac{5!}{(5-5)!}$$

$$=\frac{6!}{3!3!}\cdot\frac{4!}{2!2!}\cdot\frac{5!}{0!}$$

$$=\frac{6\cdot 5\cdot 4\cdot 3!}{3!\cdot 3\cdot 2\cdot 1}\cdot\frac{4\cdot 3\cdot 2!}{2!\cdot 2\cdot 1}\cdot\frac{5\cdot 4\cdot 3!}{1}$$

$$=14{,}400$$

EXERCISE SET 12.6

1.
$\underline{5}\cdot\underline{4}\cdot\underline{3}\cdot\underline{2}\cdot\underline{1}=120$

3.
$\underline{26}\cdot\underline{26}\cdot\underline{6}\cdot\underline{6}\cdot\underline{6}=146{,}016$

5.
$\underline{2}\cdot\underline{2}\cdot\underline{2}\cdot\underline{2}\cdot\underline{2}\cdot\underline{2}\cdot\underline{2}\cdot\underline{2}=2^8=256$

7.
$\underline{7}\cdot\underline{6}\cdot\underline{5}\cdot\underline{4}\cdot\underline{3}\cdot\underline{2}\cdot\underline{1}=5{,}040$

9.
$$P(6,5)=\frac{6!}{(6-5)!}=\frac{6!}{1!}=720$$

11.
$$P(8,3)=\frac{8!}{(8-3)!}=\frac{8!}{5!}=\frac{8\cdot7\cdot6\cdot5!}{5!}=336$$

13.
$$P(10,2)=\frac{10!}{(10-2)!}=\frac{10\cdot9\cdot8!}{8!}=90$$

15.
$$\frac{P(9,3)}{3!}=\frac{\frac{9!}{(9-3)!}}{3!}=\frac{\frac{9!}{6!}}{3!}=\frac{9!}{6!3!}$$
$$=\frac{9\cdot8\cdot7\cdot6!}{6!\cdot3\cdot2\cdot1}=84$$

17.
$$P(3,1)=\frac{3!}{(3-1)!}=\frac{3!}{2!}=\frac{3\cdot2!}{2!}=3$$

19.
$$\frac{P(10,4)}{4!}=\frac{\frac{10!}{(10-4)!}}{4!}=\frac{\frac{10!}{6!}}{4!}=\frac{10!}{6!4!}$$
$$=\frac{10\cdot9\cdot8\cdot7\cdot6!}{6!\cdot4\cdot3\cdot2\cdot1}=210$$

21.
money
$$P(5,5)=\frac{5!}{(5-5)!}=\frac{5!}{0!}=120$$

23.
needed $\dfrac{6!}{3!2!}=\dfrac{6\cdot5\cdot4\cdot3!}{3!\cdot2\cdot1}=60$

25.
$$P(8,3)=\frac{8!}{(8-3)!}=\frac{8!}{5!}=\frac{8\cdot7\cdot6\cdot5!}{5!}=336$$

27.
$$P(5,5)=\frac{5!}{(5-5)!}=\frac{5!}{0!}=120$$

29.

$$C(9,\,3)=\frac{9!}{3!(9-3)!}=\frac{9!}{3!6!}=\frac{9\cdot 8\cdot 7\cdot 6!}{3\cdot 2\cdot 1\cdot 6!}=84$$

31.

$$C(10,\,2)=\frac{10!}{2!(10-2)!}=\frac{10!}{2!8!}=\frac{10\cdot 9\cdot 8!}{2\cdot 1\cdot 8!}=45$$

33.

$$C(7,\,7)=\frac{7!}{7!(7-7)!}=\frac{7!}{7!\,0!}=1$$

35.

$$C(n,\,n-1)=\frac{n!}{(n-1)![n-(n-1)]!}$$
$$=\frac{n!}{(n-1)!1!}$$
$$=\frac{n(n-1)!}{(n-1)!\cdot 1}=n$$

37.

$$C(n+1,\,n-1)=\frac{(n+1)!}{(n-1)![(n+1)-(n-1)]!}$$
$$=\frac{(n+1)!}{(n-1)!2!}=\frac{(n+1)n(n-1)!}{(n-1)!\cdot 2\cdot 1}$$
$$=\frac{(n+1)n}{2}=\frac{n^2+n}{2}$$

39.

$$C(15,\,5)=\frac{15!}{5!(15-5)!}=\frac{15!}{5!10!}=3003$$

41a).

$$P(26,\,3)=\frac{26!}{(26-3)!}=\frac{26!}{23!}=15{,}600$$

41b).

$$\underline{26}\cdot\underline{26}\cdot\underline{26}=17{,}576$$

43.

$$C(4,\,4)\cdot C(48,\,6)$$
$$=\frac{4!}{4!(4-4)!}\cdot\frac{48!}{6!(48-6)!}$$
$$=\frac{4!}{4!0!}\cdot\frac{48!}{6!42!}=12{,}271{,}512$$

45.

$N\,(P\,D)\,Q\,H\,S$

$$P(5,\,5)\cdot P(2,\,2)=\frac{5!}{(5-5)!}\cdot\frac{2!}{(2-2)!}=\frac{5!}{0!}\cdot\frac{2!}{0!}=240$$

47.

$$C(12, 4)\cdot C(10, 3)=\frac{12!}{4!(12-4)!}\cdot\frac{10!}{3!(10-3)!}=\frac{12!}{4!8!}\cdot\frac{10!}{3!7!}=59{,}400$$

49.

$$C(26, 6)\cdot C(26, 4)=\frac{26!}{6!(26-6)!}\cdot\frac{26!}{4!(26-4)!}=\frac{26!}{6!20!}\cdot\frac{26!}{4!22!}=3.4419\times 10^{9}$$

CHAPTER 12 SECTION 7

SECTION 12.7 PROGRESS CHECK

1. (Page 704)

Number of aces : 4

$$\frac{\text{Number of successful outcomes}}{\text{Total number of outcomes}} = \frac{4}{52} = \frac{1}{13}$$

2. (Page 704)

$$\frac{\text{Number of successful outcomes}}{\text{Total number of outcomes}} = \frac{4+5}{4+5+10} = \frac{9}{19}$$

3. (Page 705)

Probability of rolling a 4 : $\frac{1}{6}$

Probability that it's not a 4 : $1 - \frac{1}{6} = \frac{5}{6}$

4. (Page 706)

$$\frac{C(4,\,3)}{C(52,\,3)} = \frac{\frac{4!}{3!1!}}{\frac{52!}{3!49!}} = \frac{4}{22100} = \frac{1}{5525}$$

5. (Page 707)

The favorable outcomes are 2, 3, 4, 5.
From Table 3 this can occur in
$1+2+3+4=10$ ways

Probability $= \frac{10}{36} = \frac{5}{18}$

6. (Page 708)

$$\frac{2}{36} \cdot \frac{2}{36} = \frac{1}{324}$$

7. (Page 708)

Probability of throwing a 7 or an 11 on
1st throw : $\frac{6+2}{36} = \frac{8}{36} = \frac{2}{9}$

Probability of not throwing a 7 or an 11 on
2nd throw : $1 - \frac{2}{9} = \frac{7}{9}$

Probability of desired result : $\frac{2}{9} \cdot \frac{7}{9} = \frac{14}{81}$

8. (Page 709)

Rainy Monday : $\frac{1}{4}$

Dry Tuesday : $1 - \frac{1}{4} = \frac{3}{4}$

Rainy Wednesday : $\frac{1}{4}$

Probability of desired result : $\frac{1}{4} \cdot \frac{3}{4} \cdot \frac{1}{4} = \frac{3}{64}$

EXERCISE SET 12.7

1.

$$\frac{\text{Number of successful outcomes}}{\text{Total number of outcomes}} = \frac{3}{6} = \frac{1}{2}$$

3a).

$$\frac{26}{52} = \frac{1}{2}$$

3b).

$$\frac{13}{52} = \frac{1}{4}$$

3c).

$$\frac{4}{52} = \frac{1}{13}$$

5a).

4 on 1st die and not a 4 on 2nd : 5

4 on 2nd die and not a 4 on 1st : 5

4 on both : 1

Total 11

$$\text{Probability} = \frac{11}{36}$$

5b).

(2, 6), (3, 5), (4, 4), (5, 3), (6, 2)

$$\text{Probability} = \frac{5}{36}$$

5c).

(1, 1), (1, 2), (1, 5), (1, 6)

(2, 1), (2, 2), (2, 5), (2, 6)

(5, 1), (5, 2), (5, 5), (5, 6)

(6, 1), (6, 2), (6, 5), (6, 6)

$$\text{Probability } \frac{16}{36} = \frac{4}{9}$$

7a).

1% defective impli es 99% or $\frac{99}{100}$ are good.

$$\text{Probability} = \frac{99}{100}$$

7b).

1% implies $\frac{1}{100}$

9a).

Ways to choose 3 defective bulbs :

$$C(5, 3) = \frac{5!}{3!2!} = 10$$

Ways to choose 3 bulbs : $C(102, 3) = \frac{102!}{3!99!}$

$$= 171700$$

Probability all 3 are defective : $\frac{10}{171700}$

$$= \frac{1}{17170}$$

9b).

Ways to choose exactly 1 defective bulb :

$$C(5, 1) = \frac{5!}{1!4!} = 5$$

Ways to choose 2 good bulbs :

$$C(97, 2) = \frac{97!}{2!95!} = 4656$$

Probability of exactly 1 defective bulb :

$$\frac{C(5, 1)C(97, 2)}{C(102, 3)}$$

$$= \frac{5 \cdot 4656}{171700} = \frac{1164}{8585}$$

9c).

None defective implies all 3 are good :

$$C(97, 3) = \frac{97!}{3!94!}$$

$$= 147440$$

Probability of none defective : $\frac{147440}{171700}$

$$= \frac{7372}{8585}$$

11.

Ways to choose 4 hearts : $13 \cdot 12 \cdot 11 \cdot 10$

Ways to choose 4 cards : $52 \cdot 51 \cdot 50 \cdot 49$

Probability of 4 hearts : $\frac{13 \cdot 12 \cdot 11 \cdot 10}{52 \cdot 51 \cdot 50 \cdot 49} = \frac{11}{4165}$

13.

Number of ways to choose 3 women :

$$C(14, 3)=\frac{14!}{3!11!}=364$$

Number of ways to choose 3 men :

$$C(12, 3)=\frac{12!}{3!9!}=220$$

Number of ways to choose 6 people :

$$C(26, 6)=\frac{26!}{6!20!}=230230$$

$$\text{Probability } =\frac{364\cdot 220}{230230}=\frac{8008}{23023}=\frac{8}{23}$$

15.

$$\frac{C(6, 2)\cdot C(5, 2)\cdot C(7, 1)}{C(18, 5)}=\frac{\frac{6!}{2!4!}\cdot\frac{5!}{2!3!}\cdot\frac{7!}{1!6!}}{\frac{18!}{5!13!}}$$

$$=\frac{15\cdot 10\cdot 7}{8568}=\frac{25}{204}$$

17.

2% defective impli es 98% good

17a).

4 good cameras : $(0.98)^4 = 0.922$

17b).

4 defective cameras: $(0.02)^4 = 1.6\times 10^{-7}$

CHAPTER 12 REVIEW EXERCISES

1.

$$a_n = n^2 + n + 1$$
$$a_1 = 1^2 + 1 + 1 = 3$$
$$a_2 = 2^2 + 2 + 1 = 7$$
$$a_3 = 3^2 + 3 + 1 = 13$$
$$a_{10} = 10^2 + 10 + 1 = 111$$

2.

$$a_n = \frac{n^3 - 1}{n + 1}$$
$$a_1 = \frac{1^3 - 1}{1 + 1} = 0$$
$$a_2 = \frac{2^3 - 1}{2 + 1} = \frac{7}{3}$$
$$a_3 = \frac{3^3 - 1}{3 + 1} = \frac{13}{2}$$
$$a_{10} = \frac{10^3 - 1}{10 + 1} = \frac{999}{11}$$

3.

$$a_n = n - a_{n-1}$$
$$a_1 = 0$$
$$a_2 = 2 - 0 = 2$$
$$a_3 = 3 - 2 = 1$$
$$a_4 = 4 - 1 = 3$$
$$a_5 = 5 - 3 = 2$$

4.

$$a_n = na_{n-1}$$
$$a_1 = 1$$
$$a_2 = 2(1) = 2$$
$$a_3 = 3(2) = 6$$
$$a_4 = 4(6) = 24$$
$$a_5 = 5(24) = 120$$

5.

$$\sum_{k-1}^{4}(1-2k) = [1-2(1)] + [1-2(2)] + [1-2(3)] + [1-2(4)]$$
$$= -1 - 3 - 5 - 7 = -16$$

6.

$$\sum_{k=3}^{5}[3(3+1)] + [4(4+1)] + [5(5+1)] = 12 + 20 + 30 = 62$$

7.

$$\sum_{i=1}^{5} 10 = 5(10) = 50$$

8.

$$\frac{1}{3} + \frac{2}{4} + \frac{3}{5} + \frac{4}{6} = \sum_{k=1}^{4} \frac{k}{k+2}$$

9.

$$1-x+x^2-x^3+x^4=\sum_{k=0}^{4}(-x)^k$$

10.

$$\log x+\log 2x+\log 3x+\cdots+\log nx=\sum_{k=1}^{n}\log(kx)$$

11.

$$a_n=a_1+(n-1)d$$
$$a_{21}=-2+(21-1)(2)=38$$

12.

$$a_n=a_1+(n-1)d$$
$$a_{16}=6+(16-1)(-1)=-9$$

13.

$$a_n=a_1+(n-1)d$$
$$9=4+(16-1)d$$
$$5=15d$$
$$d=\frac{1}{3}$$
$$a_n=4+(n-1)\left(\frac{1}{3}\right)$$
$$a_{13}=4+(13-1)\left(\frac{1}{3}\right)=8$$

14.

$$a_n=a_1+(n-1)d$$
$$-15=-4+(23-1)d$$
$$-11=22d$$
$$d=-\frac{1}{2}$$
$$a_n=-4+(n-1)\left(-\frac{1}{2}\right)$$
$$a_{26}=-4+(26-1)\left(-\frac{1}{2}\right)=-\frac{33}{2}$$

15.

$$S_n=\frac{n}{2}[2a_1+(n-1)d]$$
$$S_{25}=\frac{25}{2}\left[2\left(-\frac{1}{3}\right)+(25-1)\left(\frac{1}{3}\right)\right]$$
$$=\frac{275}{3}$$

16.

$$S_n=\frac{n}{2}[2a_1+(n-1)d]$$
$$S_{25}=\frac{25}{2}[2(6)+(25-1)(-2)]$$
$$=-450$$

17.

$$r=-\frac{6}{2}=-3$$

18.

$$r=\frac{\frac{3}{4}}{-\frac{1}{2}}=-\frac{3}{2}$$

19.

$$a_n = r_{n-1}$$
$$a_1 = 5$$
$$a_2 = \left(\frac{1}{5}\right)(5) = 1$$
$$a_3 = \left(\frac{1}{5}\right)(1) = \frac{1}{5}$$
$$a_4 = \left(\frac{1}{5}\right)\left(\frac{1}{5}\right) = \frac{1}{25}$$

20.

$$a_n = r_{n-1}$$
$$a_1 = -2$$
$$a_2 = (-1)(-2) = 2$$
$$a_3 = (-1)(2) = -2$$
$$a_4 = (-1)(-2) = 2$$

21.

$$r = -\frac{6}{4} = -\frac{3}{2}$$
$$a_n = a_1 r^{n-1}$$
$$a_6 = -4\left(-\frac{3}{2}\right)^{6-1} = \frac{243}{8}$$

22.

$$a_n = a_1 r^{n-1}$$
$$-32 = -2r^{5-1}$$
$$16 = r^4$$
$$r = \pm 2$$
$$a_n = -2(\pm 2)^{n-1}$$
$$a_8 = -2(\pm 2)^{8-1} = \pm 256$$

23.

$$3, a_2, a_3, \frac{1}{72}$$
$$a_n = a_1 r^{n-1}$$
$$\frac{1}{72} = 3r^{4-1}$$
$$\frac{1}{216} = r^3$$
$$r = \frac{1}{6}$$
$$3, \frac{1}{2}, \frac{1}{12}, \frac{1}{72}$$

24.

$$r = \frac{\frac{1}{6}}{\frac{1}{3}} = \frac{1}{2}$$
$$S_n = \frac{a_1(1-r^n)}{1-r}$$
$$S_6 = \frac{\frac{1}{3}\left[1-\left(\frac{1}{2}\right)^6\right]}{1-\frac{1}{2}} = \frac{21}{32}$$

25.

$$S_n = \frac{a_1(1-r^n)}{1-r}$$

$$S_6 = \frac{-2[1-(3)^6]}{1-3} = -728$$

26.

$$5 + \frac{5}{2} + \frac{5}{4} + \cdots$$

$$r = \frac{\frac{5}{2}}{5} = \frac{1}{2}$$

$$s_n = \frac{a_1}{1-r} = \frac{5}{1-\frac{1}{2}} = 10$$

27.

$$3 - 2 + \frac{4}{3} - \cdots$$

$$r = -\frac{2}{3}$$

$$S_n = \frac{a_1}{1-r} = \frac{3}{1-\left(-\frac{2}{3}\right)} = \frac{9}{5}$$

28.

$$3 + 6 + 9 + \cdots + 3n = \frac{3n(n+1)}{2}$$

$$n = 1: \quad 3 = \frac{3(1)(1+1)}{2} \quad ?$$

$$3 = 3$$

$$n = k: \quad 3 + 6 + 9 + \cdots + 3k = \frac{3k(k+1)}{2}$$

$$3 + 6 + 9 + \cdots + 3k + (3k+3)$$

$$= \frac{3k(k+1)}{2} + (3k+3)$$

$$= \frac{3k(k+1)}{2} + \frac{2(3)(k+1)}{2}$$

$$= \frac{3(k+1)(k+2)}{2}$$

Thus the formula holds for $n = k + 1$.

29.

$$(2x-y)^4 = (2x)^4 + 4(2x)^3(-y) + \frac{4 \cdot 3}{1 \cdot 2}(2x)^2(-y)^2 + \frac{4 \cdot 3 \cdot 2}{1 \cdot 2 \cdot 3}(2x)(-y)^3 + (-y)^4$$

$$= 16x^4 - 32x^3y + 24x^2y^2 - 8xy^3 + y^4$$

30.

$$\left(\frac{x}{2} - 2\right)^4 = \left(\frac{x}{2}\right)^4 + 4\left(\frac{x}{2}\right)^3(-2) + \frac{4 \cdot 3}{1 \cdot 2}\left(\frac{x}{2}\right)^2(-2)^2 + \frac{4 \cdot 3 \cdot 2}{1 \cdot 2 \cdot 3}\left(\frac{x}{2}\right)(-2)^3 + (-2)^4$$

$$= \frac{x^4}{16} - x^3 + 6x^2 - 16x + 16$$

31.
$$(x^2+1)^3 = (x^2)^3 + 3(x^2)^2(1) + \frac{3\cdot 2}{1\cdot 2}(x^2)(1)^2 + (1)^3$$
$$= x^6 + 3x^4 + 3x^2 + 1$$

32.
$$6! = 6\cdot 5\cdot 4\cdot 3\cdot 2\cdot 1 = 720$$

33.
$$\frac{13!}{11!2!} = \frac{13\cdot 12\cdot 11!}{11!\cdot 2\cdot 1} = 78$$

34.
$$\frac{(n-1)!(n+1)!}{n!n!} = \frac{(n-1)!(n+1)n!}{n(n-1)!n!} = \frac{n+1}{n}$$

35.
$$\binom{6}{4} = \frac{6!}{4!(6-4)!} = \frac{6!}{4!2!} = \frac{6\cdot 5\cdot 4!}{4!\cdot 2\cdot 1} = 15$$

36.
$$\binom{3}{0} = \frac{3!}{0!(3-0)!} = \frac{3!}{1\cdot 3!} = 1$$

37.
$$\binom{10}{8} = \frac{10!}{8!(10-8)!} = \frac{10!}{8!2!} = \frac{10\cdot 9\cdot 8!}{8!\cdot 2\cdot 1} = 45$$

38.
$$P(4,4) = \frac{4!}{(4-4)!} = \frac{4!}{0!} = \frac{24}{1} = 24$$

39.
$$\frac{P(6,\ 6)}{P(2,\ 2)} = \frac{6!}{2!} = 360$$

40.
$$C(10,\ 6) = \frac{10!}{6!(10-6)!} = \frac{10!}{6!4!} = 210$$

41.
$$C(3,\ 1)C(3,\ 1) = \left(\frac{3!}{1!(3-1)!}\right)^2 = \left(\frac{3!}{1!2!}\right)^2 = 9$$

42.
(1, 6), (2, 5), (3, 4), (4, 3), (5, 2), (6, 1)
(5, 6), (6, 5)
8 favorable outcomes
$6\cdot 6 = 36$ possible outcomes

Probability $= \frac{8}{36} = \frac{2}{9}$

43.

Favorable outcomes : $C(4, 2) = \frac{4!}{2!(4-2)!}$

$= \frac{4!}{2!2!} = 6$

Possible outcomes : $C(7, 2) = \frac{7!}{2!(7-2)!}$

$= \frac{7!}{2!5!} = 21$

Probability : $\frac{6}{21} = \frac{2}{7}$

44.

$$\frac{C(4, 1)C(4, 1)}{C(52, 2)} = \frac{\frac{4!}{1!3!} \cdot \frac{4!}{1!3!}}{\frac{52!}{2!50!}} = \frac{4 \cdot 4}{26 \cdot 51} = \frac{8}{663}$$

45.

10% diseased impli es 90% are free of disease.

$(0.90)(0.90) = 0.81$

46.

$$\frac{C(6, 2)}{C(12, 4)} = \frac{\frac{6!}{2!4!}}{\frac{12!}{4!8!}} = \frac{15}{495} = \frac{1}{33}$$

CHAPTER 12 REVIEW TEST

1.

$$a_n = \frac{n}{(n+1)^2}$$

$$a_1 = \frac{1}{(1+1)^2} = \frac{1}{4}$$

$$a_2 = \frac{2}{(2+1)^2} = \frac{2}{9}$$

$$a_3 = \frac{3}{(3+1)^2} = \frac{3}{16}$$

$$a_4 = \frac{4}{(4+1)^2} = \frac{4}{25}$$

2.

$$\sum_{j=2}^{4} \frac{j}{j-1} = \frac{2}{2-1} + \frac{3}{3-1} + \frac{4}{4-1} = 2 + \frac{3}{2} + \frac{4}{3} = \frac{29}{6}$$

3.

$$a_n = a_{n-1} + d$$

$$a_1 = -1$$

$$a_2 = -1 + \frac{3}{2} = \frac{1}{2}$$

$$a_3 = \frac{1}{2} + \frac{3}{2} = 2$$

$$a_4 = 2 + \frac{3}{2} = \frac{7}{2}$$

4.

$$a_n = a_1 + (n-1)d$$

$$a_{25} = -4 + (25-1)\left(\frac{1}{2}\right) = 8$$

5.

$$a_n = a_1 + (n-1)d$$
$$26 = -1 + (10-1)d$$
$$27 = 9d$$
$$d = 3$$
$$a_n = -1 + (n-1)(3)$$
$$a_{15} = -1 + (15-1)(3) = 41$$

6.

$$a_n = a_1 + (n-1)d$$
$$8 = -4 + (9-1)d$$
$$12 = 8d$$
$$d = \frac{3}{2}$$
$$a_n = -4 + (n-1)\left(\frac{3}{2}\right)$$
$$S_n = \frac{n}{2}[2a_1 + (n-1)d]$$
$$S_{10} = \frac{10}{2}\left[2(-4) + (10-1)\left(\frac{3}{2}\right)\right]$$
$$= \frac{55}{2}$$

7.

$$r = \frac{4}{12} = \frac{1}{3}$$

8.

$$a_n = r_{n-1}$$
$$a_1 = -\frac{2}{3}$$
$$a_2 = 2\left(-\frac{2}{3}\right) = -\frac{4}{3}$$
$$a_3 = 2\left(-\frac{4}{3}\right) = -\frac{8}{3}$$
$$a_4 = 2\left(-\frac{8}{3}\right) = -\frac{16}{3}$$

9.

Geometric with $r = \frac{-2}{2} = -1$

$a_n = a_1 r^{n-1}$

$a_{10} = 2(-1)^{10-1} = -2$

10.

$-4, a_2, a_3, 32$

$a_n = a_1 r^{n-1}$

$32 = -4r^{4-1}$

$-8 = r^3$

$r = -2$

$a_2 = (-4)(-2) = 8$

$a_3 = (8)(-2) = -16$

8, −16

11.

$r = \frac{4}{-8} = -\frac{1}{2}$

$S_n = \frac{a_1(1-r^n)}{1-r}$

$S_7 = \frac{(-8)\left[1-\left(-\frac{1}{2}\right)^7\right]}{1-\left(-\frac{1}{2}\right)}$

$= -\frac{43}{8}$

12.

$r = \frac{-\frac{4}{3}}{-4} = \frac{1}{3}$

$S = \frac{a_1}{1-r}$

$S = \frac{-4}{1-\frac{1}{3}} = -6$

13.

$2+6+10+\cdots+(4n-2) = 2n^2$

$n=1$: $2 = 2(1)^2$?

$2 = 2$

$n=k$: $2+6+10+\cdots+(4k-2) = 2k^2$

$2+6+10+\cdots(4k-2)+(4k+2)$

$= 2k^2 + (4k+2) = 2(k^2+2k+1)$

$= 2(k+1)^2$

Thus the formula holds for $n = k+1$.

14.

$$\left(a+\frac{1}{b}\right)^{10}=a^{10}+10a^{9}\left(\frac{1}{b}\right)+\frac{10\cdot 9}{1\cdot 2}a^{8}\left(\frac{1}{b}\right)^{2}+\frac{10\cdot 9\cdot 8}{1\cdot 2\cdot 3}a^{7}\left(\frac{1}{b}\right)^{3}+\cdots$$

$$=a^{10}+\frac{10a^{9}}{b}+\frac{45a^{8}}{b^{2}}+\frac{120a^{7}}{b^{3}}+\cdots$$

15.

$$\frac{12!}{10!3!}=\frac{12\cdot 11\cdot 10!}{10!3\cdot 2\cdot 1}=22$$

16.

$$P(6,\,4)=\frac{6!}{(6-4)!}=\frac{6!}{2!}=\frac{6\cdot 5\cdot 4\cdot 3\cdot 2!}{2!}=360$$

17.

$$C(n+1,\,n)=\frac{(n+1)!}{n![(n+1)-n]!}=\frac{(n+1)!}{n!1!}$$

$$=\frac{(n+1)n!}{n!}=n+1$$

18.

$$P(4,\,3)=\frac{4!}{(4-3)!}=\frac{4!}{1!}=24$$

19.

$$C(40,\,33)\cdot\frac{33!}{15!8!10!}=\frac{40!}{33!7!}\cdot\frac{33!}{15!8!10!}$$

$$=8.46\times 10^{20}$$

20.

Favorable outcomes : $(w,\,w),\,(b,\,b),\,(g,\,g)$

Possible outcomes : $\underline{3}\cdot\underline{3}=9$

Probability : $\frac{3}{9}=\frac{1}{3}$

21.

Favorable outcomes : $C(4,\,2)\cdot C(3,\,2)$

$$=\frac{4!}{2!2!}\cdot\frac{3!}{2!1!}=18$$

Possible outcomes : $C(10,\,4)=\frac{10!}{4!6!}=210$

Probability : $\frac{18}{210}=\frac{3}{35}$

CUMULATIVE REVIEW EXERCISES: CHAPTERS 10-12

1.

$$x^2 - y^2 = 9$$
$$+\ x^2 + y^2 = 41$$
$$2x^2 = 50$$
$$x^2 = 25$$
$$x = 5 \qquad x = -5$$
$$5^2 - y^2 = 9 \qquad (-5)^2 + y^2 = 41$$
$$y^2 = 16 \qquad y^2 = 16$$
$$y = \pm 4 \qquad y = \pm 4$$

$x = 5,\ y = 4$; $x = 5,\ y = -4$; $x = -5,\ y = 4$;
$x = -5,\ y = -4$

2.

$$2x - 3y = 11 \qquad 10x - 15y = 55$$
$$3x + 5y = -12 \qquad +\ 9x + 15y = -36$$
$$19x = 19$$
$$x = 1$$
$$2(1) - 3y = 11$$
$$-3y = 9$$
$$y = -3$$
$$x = 1,\ \ y = -3$$

3.

$$x^2 + 3y^2 = 12$$
$$x + 3y = 6$$

$$x = 6 - 3y$$
$$(6 - 3y)^2 + 3y^2 = 12$$
$$36 - 36y + 9y^2 + 3y^2 = 12$$
$$12y^2 - 36y + 24 = 0$$
$$y^2 - 3y + 2 = 0$$
$$(y - 2)(y - 1) = 0$$
$$y = 2 \qquad y = 1$$
$$x = 6 - 3(2) = 0 \qquad x = 6 - 3(1) = 3$$
$$x = 0,\ y = 2; \qquad x = 3,\ y = 1$$

4.

Adult tickets : x
Children's tickets : y

$$x + y = 600 \qquad x + y = 600$$
$$6x + 3y = 2700 \qquad -2x - y = -900$$
$$-x = -300$$
$$x = 300$$
$$300 + y = 600$$
$$y = 300$$

300 adult tickets ; 300 children's tickets

5.

$$\frac{x+5}{x^2-x-2}=\frac{x+5}{(x-2)(x+1)}=\frac{A}{x-2}+\frac{B}{x+1}$$

$$x+5=A(x+1)+B(x-2)$$

$$x=-1:\ \ -1+5=A(0)+B(-3)$$

$$4=-3B$$

$$B=-4/3$$

$$x=2:\ \ 2+5=A(3)+B(0)$$

$$7=3A$$

$$A=\frac{7}{3}$$

$$\frac{\frac{7}{3}}{x-2}-\frac{\frac{4}{3}}{x+1}$$

6.

$$z=-3$$

$$y-2(-3)=5$$

$$y=-1$$

$$x+3(-1)-(-3)=0$$

$$x=0$$

$$x=0,\ \ y=-1,\ \ z=-3$$

7.

$$4x+y-z=0$$

$$-x+3y+2z=3$$

$$3x-2y-2z=-3$$

$$\begin{array}{r}8x+2y-2z=0\\ -x+3y+2z=3\\ \hline 7x+5y=3\end{array}\qquad\begin{array}{r}-x+3y+2z=3\\ 3x-2y-2z=-3\\ \hline 2x+y=0\end{array}$$

$$\begin{array}{r}7x+5y=3\\ -10x-5y=0\\ \hline -3x=3\end{array}$$

$$x=-1$$

$$2(-1)+y=0$$

$$y=2$$

$$4(-1)+2-z=0$$

$$z=-2$$

$$x=-1,\ \ y=2,\ \ z=-2$$

8.

$$x-y=4$$

$$3x+2ky=3$$

8a).

$$\begin{array}{r}-3x+3y=-12\\ 3x+2ky=3\\ \hline 3y+2ky=-9\end{array}$$

$$y(3+2k)=-9$$

If $3+2k=0$, there is no solution.

$$k=-3/2$$

8b).

For an infinite number of solutions, $y(3+2k)$ would equal -9 for all values of y and this cannot happen.

9.

$$x+y\le 2$$
$$2x+y\ge 0$$
$$y\ge 0$$

Graph $x+y=2,\quad 2x+y=0,\quad y=0$

Test point : $(1,\ 2)$

$$1+2\le 2\,? \qquad 2(1)+2\ge 0\,? \qquad 1\ge 0\,?$$
$$3\le 2 \qquad\qquad 4\ge 0 \qquad\qquad \text{True}$$
$$\text{False} \qquad\qquad \text{True}$$

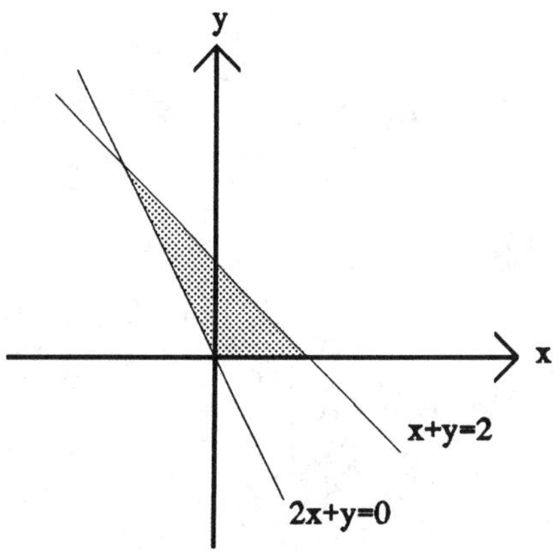

10a).

$$AB=\begin{bmatrix}1 & -1 & 2\\ 0 & 2 & 3\\ -2 & 1 & -2\end{bmatrix}\begin{bmatrix}2 & -1\\ 0 & 2\\ -1 & 4\end{bmatrix}$$

$$=\begin{bmatrix}(1)(2)+(-1)(0)+(2)(-1) & (1)(-1)+(-1)(2)+(2)(4)\\ (0)(2)+(2)(0)+(3)(-1) & (0)(-1)+(2)(2)+(3)(4)\\ (-2)(2)+(1)(0)+(-2)(-1) & (-2)(-1)+(1)(2)+(-2)(4)\end{bmatrix}$$

$$=\begin{bmatrix}0 & 5\\ -3 & 16\\ -2 & -4\end{bmatrix}$$

10b).

$$|A|=\begin{vmatrix}1 & -1 & 2\\ 0 & 2 & 3\\ -2 & 1 & -2\end{vmatrix}=\begin{vmatrix}1 & -1 & 2\\ 0 & 2 & 3\\ 0 & -1 & 2\end{vmatrix}$$
$$2R_1+R_3\to R_3$$

$$=(1)(-1)^{1+1}\begin{vmatrix}2 & 3\\ -1 & 2\end{vmatrix}$$

$$=4-(-3)=7$$

10c).

$$AX = C$$

$$X = A^{-1}C$$

$$2R_1 + R_3 \to R_3 \left[\begin{array}{ccc|ccc} 1 & -1 & 2 & 1 & 0 & 0 \\ 0 & 2 & 3 & 0 & 1 & 0 \\ -2 & 1 & -2 & 0 & 0 & 1 \end{array}\right]$$

$$\begin{array}{l} R_2 + R_3 \to R_2 \\ R_1 - R_3 \to R_1 \end{array} \left[\begin{array}{ccc|ccc} 1 & -1 & 2 & 1 & 0 & 0 \\ 0 & 2 & 3 & 0 & 1 & 0 \\ 0 & -1 & 2 & 2 & 0 & 1 \end{array}\right]$$

$$R_2 + R_3 \to R_3 \left[\begin{array}{ccc|ccc} 1 & 0 & 0 & -1 & 0 & -1 \\ 0 & 1 & 5 & 2 & 1 & 1 \\ 0 & -1 & 2 & 2 & 0 & 1 \end{array}\right]$$

$$\frac{1}{7}R_3 \to R_3 \left[\begin{array}{ccc|ccc} 1 & 0 & 0 & -1 & 0 & -1 \\ 0 & 1 & 5 & 2 & 1 & 1 \\ 0 & 0 & 7 & 4 & 1 & 2 \end{array}\right]$$

10c) (Continued).

$$-5R_3 + R_2 \to R_2 \left[\begin{array}{ccc|ccc} 1 & 0 & 0 & -1 & 0 & -1 \\ 0 & 1 & 5 & 2 & 1 & 1 \\ 0 & 0 & 1 & \frac{4}{7} & \frac{1}{7} & \frac{2}{7} \end{array}\right]$$

$$\left[\begin{array}{ccc|ccc} 1 & 0 & 0 & -1 & 0 & -1 \\ 0 & 1 & 0 & -\frac{6}{7} & \frac{2}{7} & -\frac{3}{7} \\ 0 & 0 & 1 & \frac{4}{7} & \frac{1}{7} & \frac{2}{7} \end{array}\right]$$

$$A^{-1} = \begin{bmatrix} -1 & 0 & -1 \\ -\frac{6}{7} & \frac{2}{7} & -\frac{3}{7} \\ \frac{4}{7} & \frac{1}{7} & \frac{2}{7} \end{bmatrix}$$

$$X = A^{-1}C = \begin{bmatrix} -1 & 0 & -1 \\ -\frac{6}{7} & \frac{2}{7} & -\frac{3}{7} \\ \frac{4}{7} & \frac{1}{7} & \frac{2}{7} \end{bmatrix} \begin{bmatrix} -1 \\ -3 \\ 0 \end{bmatrix} = \begin{bmatrix} 1 \\ 0 \\ -1 \end{bmatrix}$$

$x = 1, \quad y = 0, \quad z = -1$

11.

Tens digit : t

Ones digit : u

$t+u=10$

$t^2=1+16u$

$u=10-t$

$t^2=1+16(10-t)$

$t^2=1+160-16t$

$t^2+16t-161=0$

$(t+23)(t-7)=0$

~~$t=-23$~~ Not a single digit

$t=7$

$u=10-7=3$

The number is 73.

12.

Tens digit : t

Ones digit : u

Digits reversed : $\underline{ut}$

Value of new number : $10u+t$

Value of original number : $10t+u$

$3t=2+5u$

$10u+t=10t+u-18$

$3t-5u=2$

$-9t+9u=-18$

$9t-15u=6$

$\underline{-9t+9u=-18}$

$-6u=-12$

$u=2$

$3t-5(2)=2$

$t=4$

The new number is 24.

13a).

$$\sum_{k=1}^{3}(2k-1)^2=[2(1)-1]^2+[2(2)-1]^2+[2(3)-1]^2=1+9+25=35$$

13b).

$$\sum_{k=0}^{3}k(k-1)=[0(0-1)]+[1(1-1)]+[2(2-1)]+[3(3-1)]=0+0+2+6=8$$

14.

$$a_n = a_1 + (n-1)d$$

$$4 = a_1 + (4-1)d \qquad 3 = a_1 + (7-1)d$$

$$4 = a_1 + 3d \qquad 3 = a_1 + 6d$$

$$-4 = -a_1 - 3d$$

$$3 = a_1 + 6d$$

$$-1 = 3d$$

$$d = -\frac{1}{3}$$

$$4 = a_1 + 3\left(-\frac{1}{3}\right)$$

$$a_1 = 5$$

15.

$$a_2 - a_1 = a_3 - a_2$$

$$(2k-1)-(3+2k) = (-10k+1)-(2k-1)$$

$$-4 = -12k + 2$$

$$-6 = -12k$$

$$k = \frac{1}{2}$$

16.

$$a_1 = 5,\ a_3 = 2$$

$$a_n = a_1 r^{n-1}$$

$$2 = 5r^{3-1}$$

$$\frac{2}{5} = r^2$$

$$r = \pm\sqrt{\frac{2}{5}} = \pm\frac{\sqrt{10}}{5}$$

$$a_2 = 5\left(\pm\frac{\sqrt{10}}{5}\right) = \pm\sqrt{10}$$

17.

$$a_1 = 3,\ a_2 = 1$$

$$r = \frac{1}{3}$$

$$S_n = \frac{a_1(1-r^n)}{1-r}$$

$$S_4 = \frac{3\left[1-\left(\frac{1}{3}\right)^4\right]}{1-\frac{1}{3}} = \frac{40}{9}$$

18.

$$S = \frac{a_1}{1-r}$$

$$-6 = \frac{-5}{1-r}$$

$$-6 + 6r = -5$$

$$6r = 1$$

$$r = \frac{1}{6}$$

19.

$$\left(1-\frac{1}{x}\right)^5$$

Third term : $\frac{5\cdot 4}{1\cdot 2}(1)^3\left(-\frac{1}{x}\right)^2 = \frac{10}{x^2}$

20.

$$n < 2^n \quad \text{for } n \geq 0$$

$n = 0$: $\quad 0 < 2^0$?

$0 < 1$

$n = 1$: $\quad 1 < 2^1$?

$1 < 2$

$n = k$: $\quad k < 2^k$

$1 < 2^k$

Adding: $k + 1 < 2^k + 2^k$

$k + 1 < 2 \cdot 2^k$

$k + 1 < 2^{k+1}$

Thus the formula holds for $n = k + 1$.

CHAPTER 12 ADDITIONAL PRACTICE EXERCISES

1. Write the first three terms of the sequence whose nth term is $a_n = n^2 - 2n + 5$.

2. Evaluate $\sum_{k=0}^{2}(6-5k)$.

3. Find the tenth term of the arithmetic sequence whose first term is -8 and whose common difference is 3.

4. In an arithmetic sequence given $a_1 = 2$ and $a_{12} = 24$, find a_9.

5. Given the geometric sequence $2, \frac{2}{5}, \frac{2}{25}, \frac{2}{125}, \ldots$, find the common ratio.

6. Find the eighth term of the geometric sequence $-2, 1, -\frac{1}{2}, \frac{1}{4}, \ldots$

7. Find the sum of the infinite geometric series $6 + 2 + \frac{2}{3} + \cdots$

8. Expand $(x-3y)^3$.

9. Evaluate 8!.

10. Evaluate $\binom{9}{5}$.

CHAPTER 12 PRACTICE TEST

1. Write the first three terms of the sequence $a_n = \dfrac{n+2}{n+3}$.

2. Evaluate $\sum_{k=1}^{4} k(k+4)$.

3. Express the sum $\frac{1}{3}+\frac{1}{4}+\frac{1}{5}+\frac{1}{6}$ in sigma notation.

4. Find a_7 in the arithmetic sequence whose first term is -5 and whose common difference is -8.

5. Find a_9 in the geometric sequence whose first term is -4 and whose common ratio is 6.

6. Find the sum of the first fourteen terms of the arithmetic sequence whose first term is 3 and whose common difference is 7.

7. Determine the common ratio of the geometric sequence $3, -27, 243, -2187, \ldots$

8. Find the seventh term of the geometric sequence $3, -3, 3, -3, \ldots$

9. Insert two geometric means between 2 and $\frac{16}{27}$.

10. Find the sum of the infinite geometric series $4+\frac{4}{7}+\frac{4}{49}+\cdots$

11. Expand $(2x+3y)^3$.

12. Evaluate $5!$.

13. Evaluate $\dfrac{8!}{5!2!}$.

14. Evaluate $\dfrac{(2n)!}{(2n-1)!}$.

15.

Evaluate $\binom{7}{4}$.

16.

Evaluate $P(6, 3)$.

17.

Evaluate $C(9, 9)$.

18.

How many different ways are there to arrange the letters in the word MATH ?

19.

In how many ways can a debate team of 5 members be selected from among 12 candidates ?

20.

If the probability of getting an A on an exam is 0.25, what is the probability of not getting an A ?

CHAPTER 1 ADDITIONAL PRACTICE EXERCISES

1. $a^3b^2 - 11a^2b^3 + 7a + 5b$ **2.** $45x^3 - 30x^2 + 5x$ **3.** $\frac{4}{5(a+2)}$ **4.** $\frac{y^2+7y+9}{2y+3}$ **5.** $(4a+11b)(4a-11b)$

6. $(3x+4y)(x-2)$ **7.** $-2(x+5)(x^2-5x+25)$ **8.** $\frac{y^6}{125x^9}$ **9.** $2ab^2\sqrt[3]{7abc}$ **10.** $\frac{a(\sqrt{a}+b)}{a-b^2}$

CHAPTER 1 PRACTICE TEST

1. 35 **2.** 3 **3.** $-\frac{13}{10}$ **4.** $(a-5)(a+1)(a+3)^2$ **5.** $-24xy-3x-3y$

6. $c^3+3c^2-20c+6$ **7.** $\frac{5}{12y^6}$ **8.** $\frac{x-5}{(x-1)(x-2)}$ **9.** $\frac{8x-2}{(x+1)(x-1)}$ **10.** $\frac{a^2+6}{(a-2)(a+2)(a-3)}$

11. $\frac{x^2-16}{x^2-9}$ **12.** $\frac{5x-3}{(x-3)^2(x-1)}$ **13.** $\frac{13x+8}{3x+2}$ **14.** $2(2x-3)(4x^2+6x+9)$ **15.** $(a+b)(2c-3d)$

16. $(8x-3y)(x+5y)$ **17.** $3(y+3)(3x-8)$ **18.** $\frac{x^2}{16y^4}$ **19.** $-\sqrt{15}+2\sqrt{5}+3\sqrt{2}-2\sqrt{6}$ **20.** $\frac{11}{50}+\frac{23}{50}i$

CHAPTER 2 ADDITIONAL PRACTICE EXERCISES

1. -2 **2.** -1 **3.** $\frac{3\pm\sqrt{73}}{4}$ **4.** 2 **5.** 3, 11 **6.** $-\frac{1}{3}, -\frac{2}{3}$ **7.** $\left(\frac{5}{18}, \infty\right)$ **8.** $[-9, 6]$

9. $\left(-\infty, -\frac{5}{4}\right], \left[\frac{1}{4}, \infty\right)$ **10.** 6, 23

CHAPTER 2 PRACTICE TEST

1. 11 **2.** $\pm\frac{\sqrt{5}}{5}i$ **3.** $\frac{1}{5}, -\frac{7}{4}$ **4.** $\frac{3\pm 3\sqrt{3}i}{2}$ **5.** 1 **6.** 0, -12 **7.** No solution **8.** $-\frac{3}{4}, \frac{1}{2}$

9. $h = \frac{V}{\pi r^2}$ **10.** Two irrational, real roots.

11. $-2 \le x \le \frac{8}{3}$

8/3
-2 0 2 3

12. $(-\infty, -14), (10, \infty)$ **13.** $x \le -2$ or $x \ge \frac{1}{3}$ **14.** $[-17, \infty)$ **15.** $(-\infty, 4)$

16. $\left(-\infty, -\frac{1}{2}\right), (3, 5)$ **17.** width : 8 cm ; length 21 cm. **18.** 22, 24, 26

19. 6 hours **20.** 12, 13

CHAPTER 3 ADDITIONAL PRACTICE EXERCISES

1. $5\sqrt{5}$ **2.** 13 **3.** $\left\{x \middle| x \geq -\frac{3}{2}\right\}$ **4.** $-\frac{11}{8}$ **5.** $y=\frac{5}{2}x+6$ **6.** $f^{-1}(x)=\frac{x-2}{3}$

7. $2x^2+1$ **8.** 31 **9.** $3x^2-3x-9$ **10.** (a), (b)

CHAPTER 3 PRACTICE TEST

1. $\left(\frac{11}{2}, -1\right)$

2. x-intercept $=\frac{4}{3}$

y-intercept $= 4$

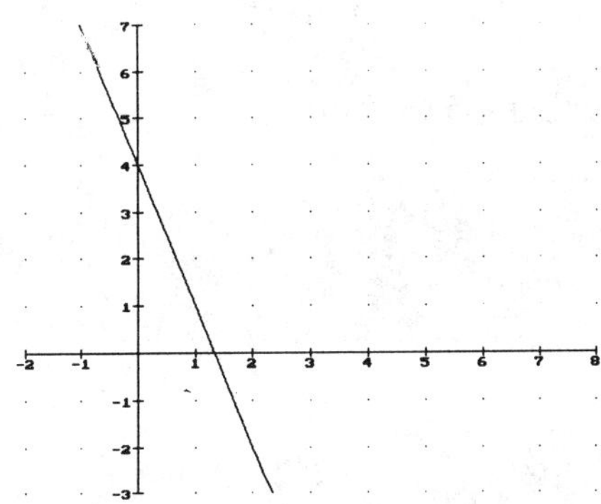

3. -4 **4.** $\left\{x \middle| x \neq \frac{6}{7}\right\}$ **5.** a^2-a-1

6.

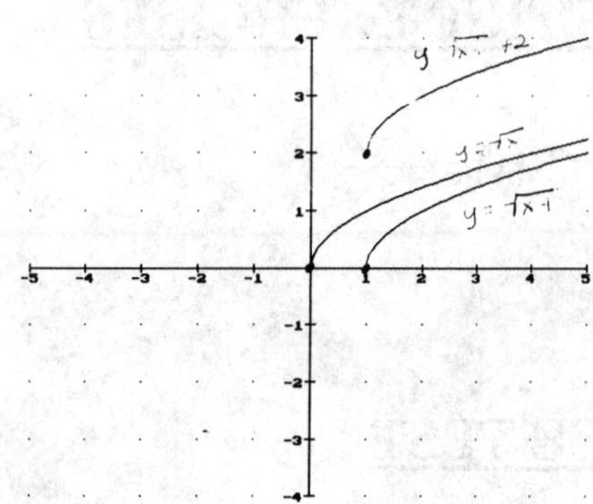

7.

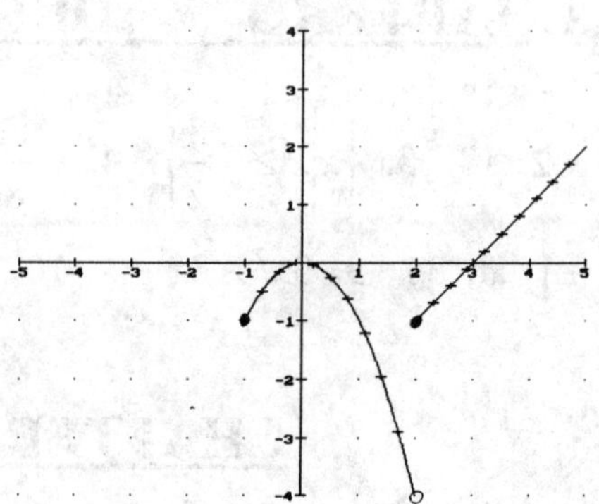

8. $-\dfrac{6}{5}$ **9.** $y=-3x+8$ **10.** $y=\dfrac{5}{3}x-\dfrac{31}{3}$ **11.** -12 **12.** 2 **13.** $\sqrt{2x+3}$

14. $f^{-1}(x)=-\dfrac{1}{5}x+\dfrac{6}{5}$ **15.** 360 **16.** $\dfrac{5}{4}$ **17.** $(b),(c)$ **18.** 4

19. $f[g(x)]=f\left(\dfrac{x+4}{5}\right)=5\left(\dfrac{x+4}{5}\right)-4=x+4-4=x$ **20.** $\sqrt{82}+\sqrt{58}+2\sqrt{2}$

$$g[f(x)]=g(5x-4)=\frac{5x-4+4}{5}=\frac{5x}{5}=x$$

f and g are inverses.

CHAPTER 4 ADDITIONAL PRACTICE EXERCISES

1. $\frac{3}{13}+\frac{2}{13}i$ **2.** Vertex: $(-1,\,-16)$

x-intercepts: $-5,\,3$

y-intercept: -15

3. $Q(x)=8x^3+19x^2+38x+74,\ R=154$ **4.** $P(-2)=-91$ **5.** $P(3)=0$, hence $x-3$ is a factor.

6. $\frac{1}{3}+\frac{7}{3}i$ **7.** $P(x)=x^3-3x^2-18x+40$ **8.** Maximum number of positive real roots: 3

Maximum number of negative real roots: 1

9. $-2,\,-2,\,1,\,-1,\,2$ **10.** $\{\{[(4x-3)x+0]x+2\}x-1\}x+7$

CHAPTER 4 PRACTICE TEST

1. Vertex: $\left(\frac{5}{4},\,-\frac{121}{8}\right)$ **2.** 30 and 30 **3a).** maximum value

x-intercepts: $-\frac{3}{2},\,4$ **3b).** $x=4$

y-intercept: -12 **3c).** 2

4. $P(0)<0;\ P(1)>0$ **5.** The graph extends infinitely downward.

6. $Q(x)=5x^2-3x+9;\ R(x)=-3x+3$ **7.** $Q(x)=x^4-8x^3+20x^2-57x+56;\ R=-169$

8. $Q(x)=x^5+x^4+x^3+x^2+x+1;\ R=2$ **9.** $P(-3)=161$ **10.** No, since $R=35$.

11. $3,\,5,\,\frac{2}{3},\,\frac{2}{3},\,\frac{2}{3}$ **12.** $k=-4$ **13.** $\frac{7}{61}-\frac{16}{61}i$ **14.** $\frac{3}{10}-\frac{9}{10}i$ **15.** $P(x)=x^4-7x^2+12$

16. $-1,\,3,\,-3$ **17.** 2 **18.** $-3,\,-1,\,1,\,\pm 3i$ **19.** $\{[(9x-8)x+0]x+6\}x-7$ **20.** 0.65

CHAPTER 5 ADDITIONAL PRACTICE EXERCISES

1. Domain: $\{x \mid x \neq \pm 5\}$

x-intercept: $x = \frac{3}{2}$

y-intercept: $y = \frac{3}{25}$

2. $(x+6)^2 + (y-4)^2 = \frac{1}{16}$ **3.** Center: $(1, -3)$ radius: $\sqrt{19}$ **4.** $(x+1)^2 + (y+5)^2 = 2$

5. $y^2 = -8x$

6. vertex: $\left(\frac{9}{2}, -2\right)$ axis: $y = -2$

7. x-intercepts: $x = \pm 2$

y-intercepts: $y = \pm 4$

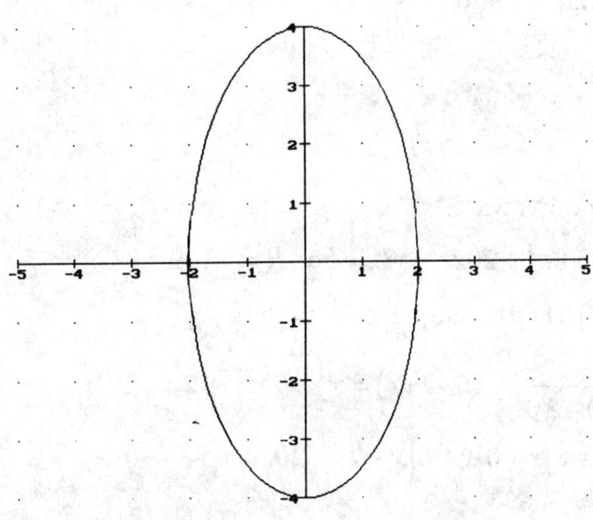

8. x-intercepts: $x = \pm 2$

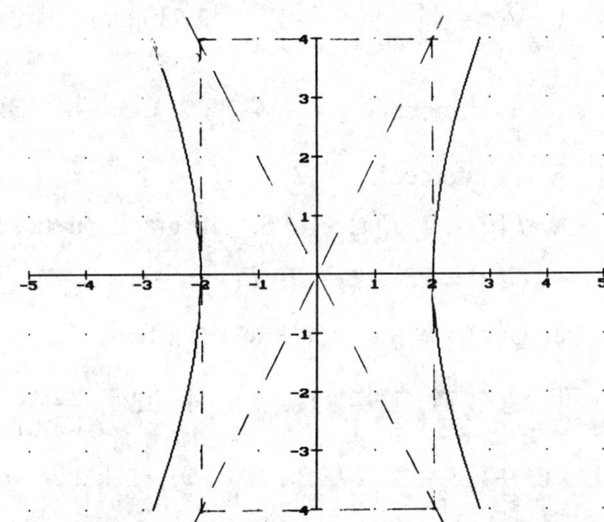

9. Hyperbola **10.** Parabola

CHAPTER 5 PRACTICE TEST

1. Domain: $\{x \mid x \neq 5, x \neq -6\}$

x-intercept: $x = -7$

y-intercept: $y = -\dfrac{7}{30}$

2. Focus: $\left(0, \dfrac{3}{2}\right)$ directrix: $y = -\dfrac{3}{2}$ **3.** $y^2 = -2x$ **4.** $y^2 = 12x$ **5.** opens up

6. $(x+2)^2 + (y-9)^2 = 9$ **7.** $(x+2)^2 + (y+5)^2 = 73$ **8.** $(x-6)^2 + \left(y - \dfrac{1}{2}\right)^2 = \dfrac{25}{4}$

9. center: $(5, 0)$ radius: $\sqrt{33}$ **10.** $2\sqrt{3}\pi$ **11.** Circle **12.** Parabola **13.** Hyperbola

14. ellipse **15.** ellipse

16.

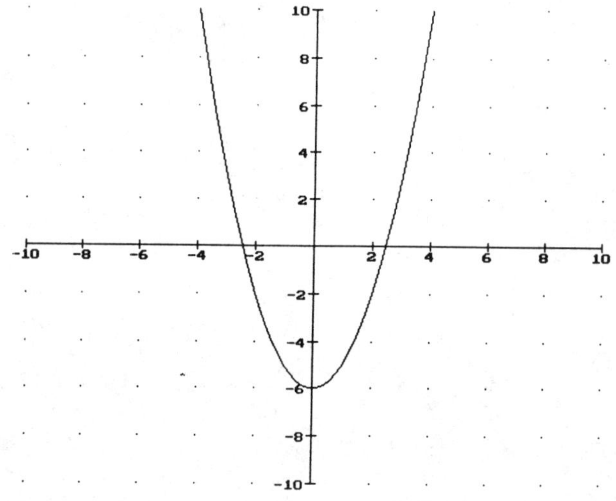

17.

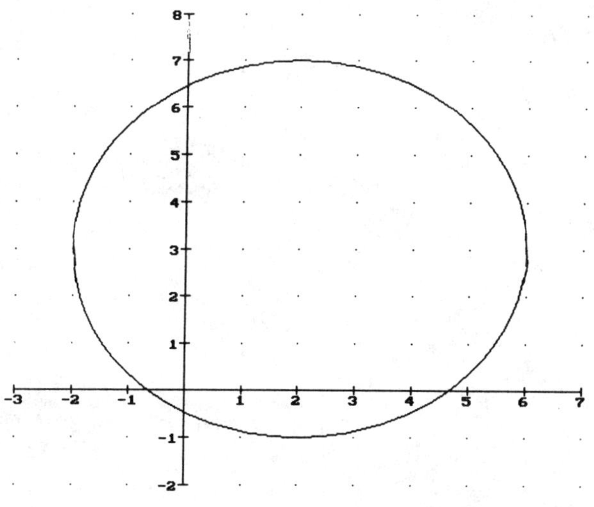

18.

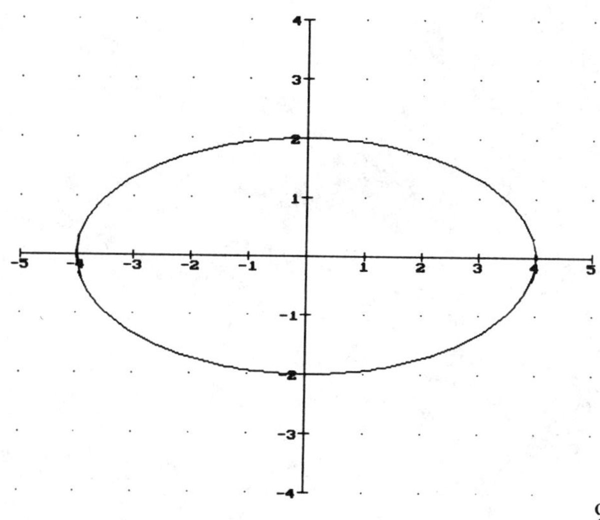

19.

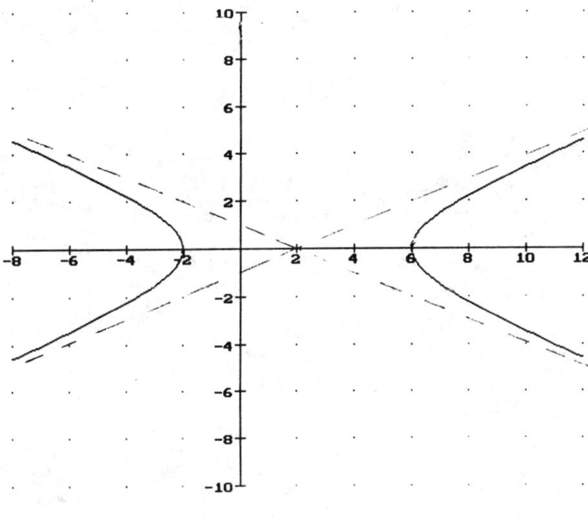

20.

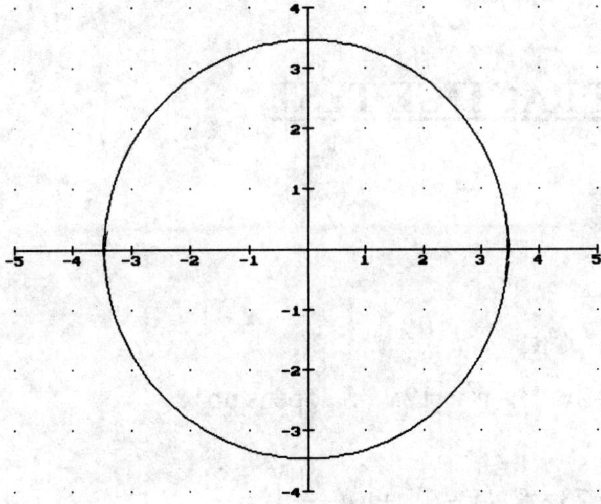

CHAPTER 6 ADDITIONAL PRACTICE EXERCISES

1. $f^{-1}(x)=\frac{4}{3}(x+1)$ **2.** 0.638 **3.** \$6,615.65 **4.** $10^{-3}=0.001$ **5.** $\log_6 216=3$

6. $3x+2$ **7.** $\frac{1}{729}$ **8.** 4 **9.** $\ln x+3\ln y+\frac{1}{2}\ln z$ **10.** $\log\frac{x^3}{y^{\frac{1}{4}}}$

CHAPTER 6 PRACTICE TEST

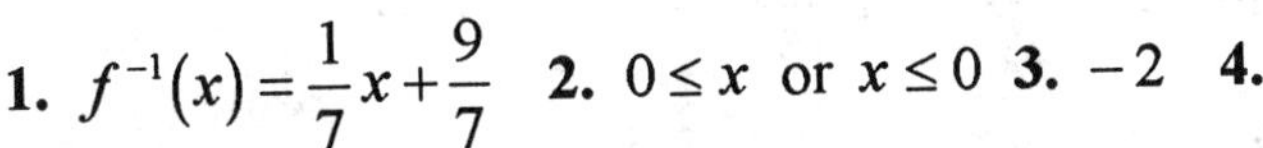

1. $f^{-1}(x)=\frac{1}{7}x+\frac{9}{7}$ **2.** $0\le x$ or $x\le 0$ **3.** -2 **4.**

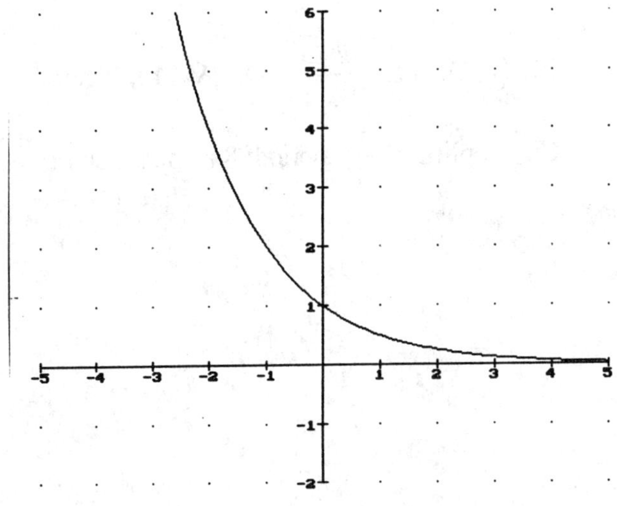

5. 1.8134 **6.** \$841.01 **7.** \$2,587.07 **8.** $5^4=625$ **9.** $\log_8 512=3$ **10.** $\frac{1}{3}$ **11.** $\frac{1}{3}$

12. 1.4037 **13.** 3 **14.** $e^5\approx 148.41$ **15.** 4 **16.** x^5 **17.** $\log\frac{x^4\sqrt{z}}{\sqrt[3]{y}}$

18. $3\log_a x+\frac{1}{2}\log_a y+\frac{5}{2}\log_a z$ **19.** $-\frac{2}{3}$ **20.** $\frac{5}{2}$

CHAPTER 7 ADDITIONAL PRACTICE EXERCISES

1. IV **2.** $\frac{13\pi}{36}$ **3.** Coterminal **4.** $\frac{3\pi}{8}$ **5.** 26.9 **6.** $(0, 1)$ **7.** II **8.** $\frac{3}{4}$
9. amplitude: 5; period: $\frac{\pi}{3}$ **10.** $-\frac{\pi}{2}$

CHAPTER 7 PRACTICE TEST

1. III **2.** 40° **3.** 88° **4.** Not coterminal **5.** 2π cm **6.** $\frac{5\sqrt{29}}{29}$ **7.** $\frac{8\pi}{5}$ **8.** $-\frac{4}{3}$ **9.** $-\sqrt{3}$
10. $(0, 1)$ **11.** $\frac{-2\sqrt{3}}{3}$ **12.** IV **13.** 1.269 **14.** $-\frac{3}{4}$ **15.** $-\frac{\sqrt{5}}{2}$ **16.** -2.7262
17. amplitude: $\frac{1}{3}$; period: 8π; phase shift: $\frac{4\pi}{3}$ **18.** $\frac{\pi}{6}$ **19.** 52° **20.** 18°26′6″

CHAPTER 8 ADDITIONAL PRACTICE EXERCISES

1. $\tan x \csc x = \dfrac{\sin x}{\cos x} \cdot \dfrac{1}{\sin x} = \dfrac{1}{\cos x} = \sec x$ **2.** $\dfrac{\sin x + \cos x}{\sin x} = \dfrac{\sin x}{\sin x} + \dfrac{\cos x}{\sin x} = 1 + \cot x$

3. $\cot 51^\circ$ **4.** $-\dfrac{1}{2}$ **5.** $\dfrac{3}{5}$ **6.** $-\dfrac{7}{25}$ **7.** $-\dfrac{\sqrt{2-\sqrt{2}}}{2}$ **8.** $\dfrac{3}{2}(\sin 9\alpha - \sin \alpha)$

9. $2\cos 4\theta \cos 2\theta$ **10.** $\dfrac{\pi}{4} + \pi n,\ \dfrac{3\pi}{4} + \pi n$ or $\dfrac{\pi}{4} + \dfrac{\pi n}{2}$

CHAPTER 8 PRACTICE TEST

1. $\dfrac{\tan^2\theta}{\sec\theta - 1} = \dfrac{\sec^2\theta - 1}{\sec\theta - 1} = \dfrac{(\sec\theta + 1)(\sec\theta - 1)}{\sec\theta - 1} = \sec\theta + 1$

2. $\sin(-\theta)\csc\theta = -\sin\theta\left(\dfrac{1}{\sin\theta}\right) = -1$

3.
$$\frac{\tan x - 1}{1 - \cot x} = \frac{\dfrac{\sin x}{\cos x} - 1}{1 - \dfrac{\cos x}{\sin x}} = \frac{\dfrac{\sin x}{\cos x} - 1}{1 - \dfrac{\cos x}{\sin x}} \cdot \frac{\sin x \cos x}{\sin x \cos x}$$
$$= \frac{\sin^2 x - \sin x \cos x}{\sin x \cos x - \cos^2 x} = \frac{\sin x(\sin x - \cos x)}{\cos x(\sin x - \cos x)} = \sin x\left(\frac{1}{\cos x}\right)$$
$$= \sin x \sec x$$

4. $\dfrac{\sqrt{6}+\sqrt{2}}{4}$ **5.** $2+\sqrt{3}$ **6.** $\csc 22^\circ$ **7.** $\cos\dfrac{7\pi}{18}$ **8.** $\dfrac{\sqrt{6}}{3}$ **9.** $\dfrac{2\sqrt{5}}{5}$

10. $4\sin^2 2x + 2\cos 4x = 4\sin^2 2x + 2(1 - 2\sin^2 2x) = 2$ **11.** $\dfrac{\sqrt{2+\sqrt{2}}}{2}$ **12.** $-\dfrac{\sqrt{2+\sqrt{3}}}{\sqrt{2-\sqrt{3}}}$

13. $2(\sin 5\theta + \sin\theta)$ **14.** $\dfrac{\cos 9x + \cos x}{2}$ **15.** $-2\cos 2x \sin\dfrac{x}{2}$ **16.** $2\cos\dfrac{11\alpha}{2}\cos\dfrac{7\alpha}{2}$

17. $0, \pi$ **18.** $\dfrac{\pi}{3}, \pi, \dfrac{5\pi}{3}$ **19.** $\dfrac{\pi}{4} + \dfrac{\pi n}{2}$

20. $30^\circ + 360^\circ n,\ 150^\circ + 360^\circ n,\ 90^\circ + 360^\circ n$

CHAPTER 9 ADDITIONAL PRACTICE EXERCISES

1. 30.18 **2.** 67.5° or 112.5° **3.** 32.2° **4.** $\sqrt{41}$ **5.** $4\sqrt{2}\left(\cos\frac{7\pi}{4}+i\sin\frac{7\pi}{4}\right)$

6. $-3\sqrt{2}-3\sqrt{2}i$ **7.** $12\left(\cos\frac{5\pi}{8}+i\sin\frac{5\pi}{8}\right)$ **8.** $\left(-\frac{3}{2},\frac{3\sqrt{3}}{2}\right)$

9.

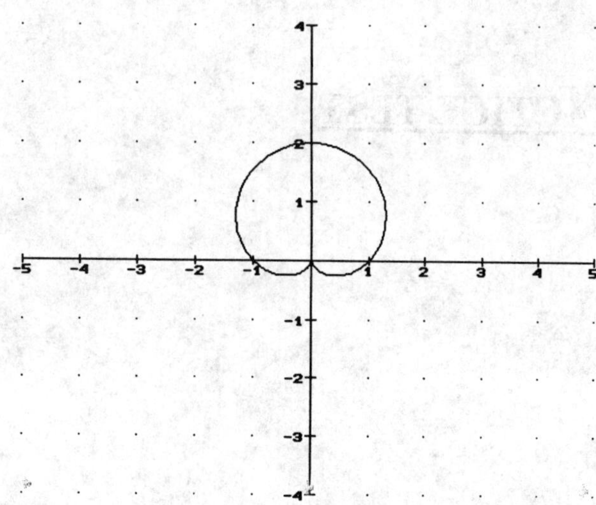

10. $\langle -20,-8\rangle$

CHAPTER 9 PRACTICE TEST

1. 8.98 **2.** 18.19° **3.** $2\sqrt{26}$ **4.** $5(\cos\pi+i\sin\pi)$ **5.** $-\frac{3}{4}$ **6.** $28-3i$

7. $3\left(\cos\frac{7\pi}{12}+i\sin\frac{7\pi}{12}\right)$ **8.** $-972+972i$ **9.** $-4,\ 2\pm2\sqrt{3}i$ **10.** $(5,-140°), (-5, 40°)$

11. $\left(2,\frac{7\pi}{6}\right)$ **12.** $\left(-4\sqrt{2},4\sqrt{2}\right)$ **13.** $\left(5\sqrt{2},\frac{5\pi}{4}\right)$

14.

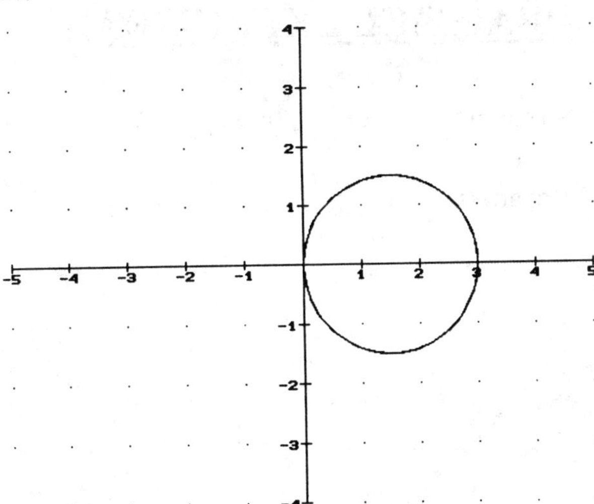

15. $x^2 + y^2 + 4x = 0$

16. $\langle 7, 8 \rangle$ **17.** $\sqrt{37}$ **18.** $\langle -34, 1 \rangle$ **19.** $\left(\frac{5\sqrt{3}}{2}, \frac{5}{2} \right)$ **20.** $\left\langle \frac{5}{\sqrt{106}}, \frac{-9}{\sqrt{106}} \right\rangle$

CHAPTER 10 ADDITIONAL PRACTICE EXERCISES

1. $x=-3,\ y=6$ **2.** $x=2,\ y=7;\ x=-\frac{7}{3},\ y=-6$ **3.** No solution **4.** $x=-1,\ y=2,\ z=3$

5. $x=-6,\ y=0,\ z=5$ **6.** $\frac{3}{x-4}-\frac{5}{x+2}$ **7.** 8 dimes ; 11 quarters

8.

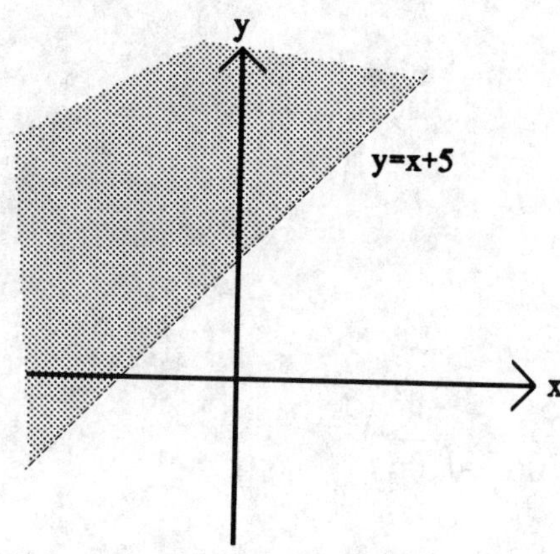

9.

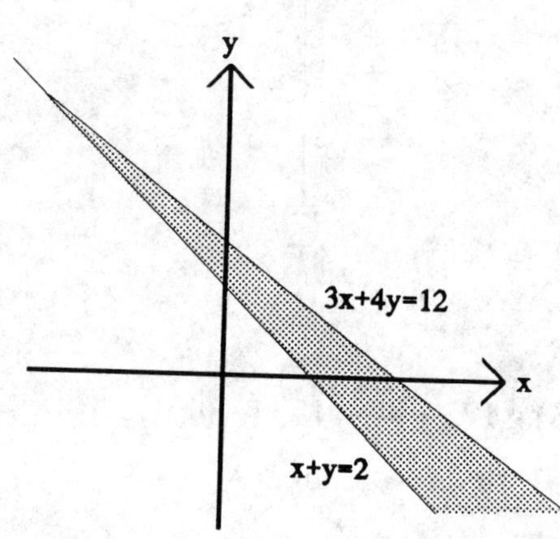

10. 25 at $(5, 3)$

CHAPTER 10 PRACTICE TEST

1. $x=-4,\ y=-5$ **2.** No solution **3.** All points on the line $5x+3y=6$

4. $x=4,\ y=1;\ y=-4,\ y=-1;\ x=\sqrt{2},\ y=2\sqrt{2};\ x=-\sqrt{2},\ y=-2\sqrt{2}$ **5.** $x=\frac{1}{2},\ y=\frac{2}{3}$

6. $x=1,\ y=5,\ z=-5$ **7.** $x=3,\ y=1,\ z=1$ **8.** No solution **9.** $x=4,\ y=-2,\ z=-1$

10. Infinite number of solutions **11.** 38 **12.** 11cm by 13cm **13.** Joe : 60 mph , Sam : 45 mph

14. $\frac{3}{2x+1}-\frac{5}{3x+4}$ **15.** $\frac{2}{x+3}+\frac{3x+1}{x^2+2}$

16.

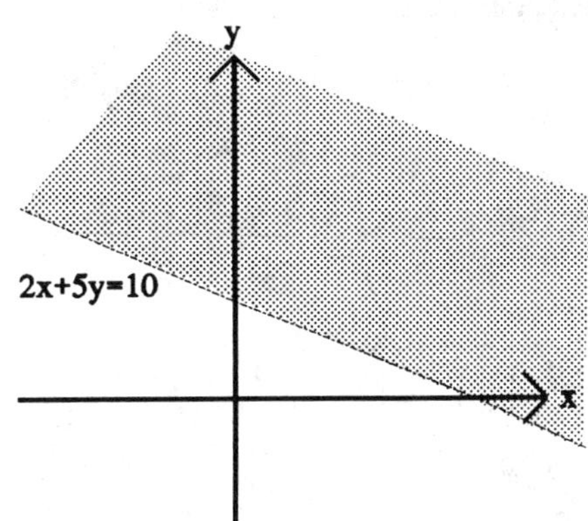

17.

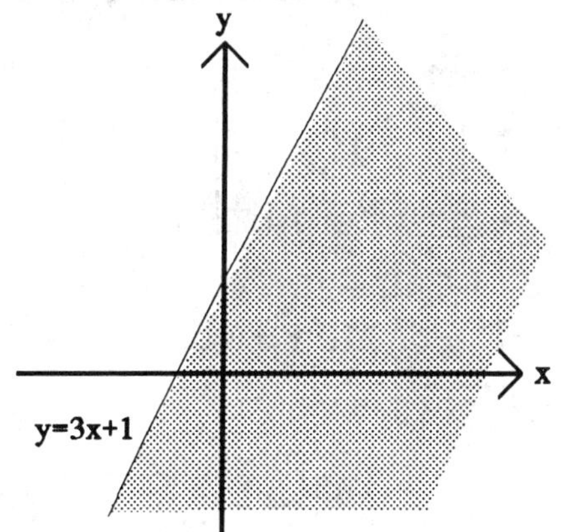

18.

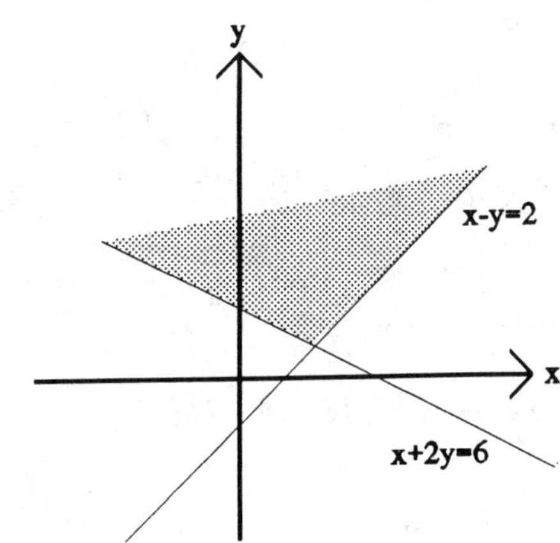

19.

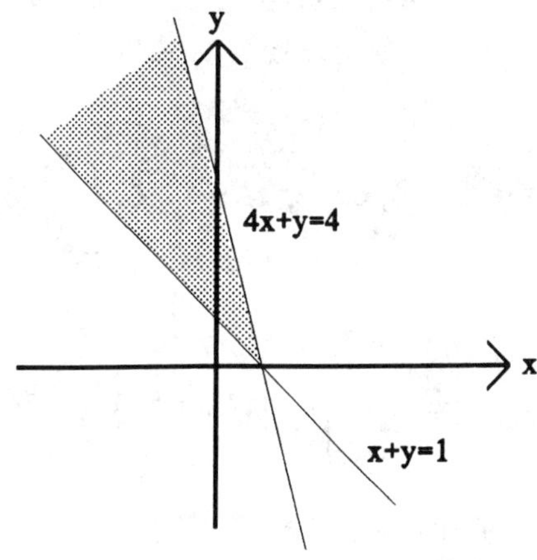

20. 4 at $(0, 2)$

CHAPTER 11 ADDITIONAL PRACTICE EXERCISES

1. 2×4 **2.** $\left[\begin{array}{ccc|c} 4 & -2 & 3 & 2 \\ 0 & -1 & 1 & 8 \\ 2 & 0 & 9 & 4 \end{array}\right]$ **3.** $x = 0, y = 3$ **4.** 2 **5.** $x = 2, y = 3, z = -1$

6. $\begin{bmatrix} 3 & -2 & -4 \\ 3 & -2 & -5 \\ -1 & 1 & 2 \end{bmatrix}$ **7.** $x = 6, y = -2$ **8.** 11 **9.** $x = 8, y = 7$ **10.** $\begin{bmatrix} 6 & -6 & 9 \\ 20 & -13 & 14 \end{bmatrix}$

CHAPTER 11 PRACTICE TEST

1. 3×4 **2.** 13 **3.** $\begin{array}{l} 6x + y = -8 \\ 3x - 2y = 7 \end{array}$ **4.** $x = -1, y = -2, z = 4$

5. $a = 4, b = 2, c = 3, d = 4$ **6.** $\begin{bmatrix} 7 & 32 \\ 20 & -27 \end{bmatrix}$ **7.** $\begin{bmatrix} 20 \\ -23 \end{bmatrix}$ **8.** Not possible **9.** $\begin{bmatrix} 2 & 1 \\ \frac{3}{2} & \frac{1}{2} \end{bmatrix}$

10. $\begin{bmatrix} 3 & -2 & -4 \\ 3 & -2 & -5 \\ -1 & 1 & 2 \end{bmatrix}$ **11.** $x = 5, y = -5$ **12.** 25 **13.** -15 **14.** -120 **15.** 11 **16.** 0

17. $x = 4, y = -2$ **18.** $x = 10, y = -1$ **19.** $x = 0, y = 2, z = -2$ **20.** $x = 1, y = -1, z = 3$

CHAPTER 12 ADDITIONAL PRACTICE EXERCISE

1. 4, 5, 8 **2.** 3 **3.** 19 **4.** 18 **5.** $\frac{1}{5}$ **6.** $\frac{1}{64}$ 7. 9 **8.** $x^3 - 9x^2y + 27xy^2 - 27y^3$

9. 40,320 **10.** 126

CHAPTER 12 PRACTICE TEST

1. $\frac{3}{4}, \frac{4}{5}, \frac{5}{6}$ **2.** 70 **3.** $\sum_{k=3}^{6} \frac{1}{k}$ **4.** -53 **5.** $-6{,}718{,}464$ **6.** 679 **7.** -9 **8.** 3 **9.** $\frac{4}{3}, \frac{8}{9}$

10. $\frac{14}{3}$ **11.** $8x^3 + 36x^2y + 54xy^2 + 27y^3$ **12.** 120 **13.** 168 **14.** $2n$ **15.** 35

16. 120 **17.** 1 **18.** 24 **19.** 792 **20.** 0.75

WRITING EXERCISES

Chapter 1 The Foundations of Algebra

1. Evaluate (8)(1.4142) and (8)($\sqrt{2}$). Are these results close to one another? Why?

2. Discuss the need for the complex number system.

3. Compare and contrast the properties of the complex numbers with those of the real numbers.

4. Discuss why division by zero is not permitted.

Chapter 2 Equations and Inequalities

In Exercises 1-3, write in complete sentences the procedure that you follow in solving the following problems.

1. An automatic teller machine gives you \$150 in five and ten dollar bills. There are 2 more than twice as many five dollar bills as there are ten dollar bills. How many of each denomination are there?

2. A pound of raisins costs \$2.50 per pound whereas a pound of chocolate bits costs \$4.00 per pound. If we want to make a one-pound mixture of these items to sell for \$3.25, how much of each item must be used?

3. Two students work in a library shelving books. Eric can shelve 50 books per hour and Steve can shelve 40 books per hour. If Steve starts work in the morning and then is relieved by Eric later in the day, how long did each student work if 355 books were shelved in an 8-hour day?

4. Why is it good practice to check your answers? Give an example to show how not following this practice can lead to a wrong conclusion.

Chapter 3 Functions

1. Given the equation $y-y_1=m(x-x_1)$, discuss how you would find the y-intercept.

2. Discuss how to use symmetry in sketching the graph of a function.

3. If we know the symmetries of the graph of a function that has an inverse, discuss the symmetries of the graph of the inverse function.

4. Discuss the relationship between the domain of $\frac{f}{g}$ with the domains of f and g.

5. Discuss in words the meaning of the following: The graph of a function is symmetric with respect to the x-axis, the y-axis or the origin.

Chapter 4 Polynomial Functions

1. Make up a word problem which uses a quadratic function and requires finding a maximum value.

2. Discuss under what conditions a polynomial function has a maximum (minimum).

3. When using the method of bisection for approximating the root of a polynomial, one has to decide when to stop the procedure. Describe criteria which can be used to make this decision.

4. Given a polynomial, describe a procedure for determining the number and type of roots (real, complex, rational, irrational). Justify each step in your explanation.

Chapter 5 Rational Functions and Conic Sections

1. We define the eccentricity of the ellipse $\frac{x^2}{a^2}+\frac{y^2}{b^2}=1$ as $\frac{\sqrt{a^2-b^2}}{a}$. Describe what happens to the shape of this ellipse when its eccentricity varies from 0 to 1

2. We define the eccentricity of the hyperbola $\frac{x^2}{a^2}-\frac{y^2}{b^2}=1$ as $\frac{\sqrt{a^2+b^2}}{a}$. Describe what happens to the shape of this hyperbola when its eccentricity increases.

3. Discuss how circles, parabolas and ellipses are used in everyday life.

4. Suppose you know that the rational function $f(x)$ has the line $y = c$ as a horizontal asymptote. Discuss what can be deduced from this information.

5. Suppose you know that the rational function $f(x)$ has the line $x = c$ as a vertical asymptote. Discuss what can be deduced from this information.

6. The acronym for long-distance radio navigation is LORAN. Research how this system uses hyperbolas to determine the location of ships.

Chapter 6 Exponential and Logarithmic Functions

1. Write a brief report illustrating how exponential and logarithmic functions occur in everyday life.

2. Discuss how the invention of logarithms enabled seventeenth century astronomers in Europe to simplify their mathematical computations.

3. Discuss the environmental impact of the exponential growth of the world's population.

4. A newscaster stated that the 1989 earthquake in Southern California, with a reading on the Richter scale of 7.1 was twice as powerful as a subsequent earthquake whose reading on the Richter scale was 3.05. Comment on the newscaster's understanding of the Richter scale.

5. Write a report on Napier's "bones".

6. Suppose you borrow \$6000 at simple interest of 5% for one year. The bank gives you \$5700. You must repay the loan in one payment when the loan is due. Explain in words how you would find the true rate of interest.

7. Explain how a slide rule works.

Chapter 7 The Trigonometric Functions

1. Compare and contrast the measurement of angles by radians and degrees.

2. Explain how to solve a right triangle.

3. Give four examples of periodic behavior in everyday life.

4. Discuss the difficulties encountered in defining inverses of trigonometric functions and how these difficulties are overcome.

5. Discuss the use of the reference angle.

6. Compare the graphs of the sine and tangent functions with respect to asymptotes, amplitude, intercepts, etc.

7. In sketching the graph of $y = A\tan(Bx + C)$, discuss how one can use the graph of $y = \tan x$.

Chapter 8 Analytic Trigonometry

1. Discuss several approaches to verifying trigonometric identities.

2. Discuss the difference between solutions to algebraic equations and trigonometric equations.

Chapter 9 Applications of Trigonometry

1. Explain how to solve a triangle.

2. State De Moivre's Theorem and interpret it geometrically. Furthermore, give a geometric interpretation of how it is used in computing with complex numbers.

3. Discuss three occurrences of vectors in everyday life.

4. Explain the geometric interpretation of the parallelogram law:

$$\|\vec{v_1}+\vec{v_2}\|^2+\|\vec{v_1}-\vec{v_2}\|^2=2(\|\vec{v_1}\|^2+\|\vec{v_2}\|^2)$$

Chapter 10 Systems of Equations and Inequalities

1. Write in complete sentences the procedure that you follow in solving the following problem using two unknowns: A stamp machine dispenses 29¢ stamps and 23¢ stamps. When you deposit a five dollar bill, you receive a packet containing 50% more 29¢ stamps than 23¢ stamps with 30¢ change.

2. Write a report on the origins of linear programming and its connection to World War II.

3. Describe an example of linear programming in everyday life.

4. Construct an example of a linear programming problem in two variables. Describe in words the geometric method used to solve it.

5. What are the limitations of solving linear programming problems geometrically?

6. Describe the method of elimination for solving a system of three linear equations in three unknowns.

Chapter 11 Matrices, Linear Systems and Determinants

1. Discuss how to solve a linear system in three unknowns if Cramer's Rule fails to hold.

2. Compare and contrast the additive properties of matrices with the additive properties of the real numbers.

3. Compare and contrast the multiplicative properties of square matrices with the multiplicative properties of the real numbers.

4. Compare and contrast Gauss-Jordan Elimination and Gaussian Elimination.

Chapter 12 Topics in Algebra

1. Sometimes on the radio, the person announcing the weather says, "There is no probability of precipitation tomorrow." Comment on the announcer's understanding of probability.

2. What does it mean when the person announcing the weather says that there is a 50% chance of rain?

3. Explain the difference between a permutation and a combination. Give illustrations of each.

4. Describe an experimental procedure to determine if a coin toss is fair.

5. Write a report on the use of probability in the insurance industry.